STUDENT'S SOLUTION MANUAL

BEVERLY FUSFIELD

COLLEGE ALGEBRA
FOURTH EDITION

J. S. Ratti
University of South Florida

Marcus McWaters
University of South Florida

Leslaw A. Skrzypek
University of South Florida

The author and publisher of this book have used their best efforts in preparing this book. These efforts include the development, research, and testing of the theories and programs to determine their effectiveness. The author and publisher make no warranty of any kind, expressed or implied, with regard to these programs or the documentation contained in this book. The author and publisher shall not be liable in any event for incidental or consequential damages in connection with, or arising out of, the furnishing, performance, or use of these programs.

Reproduced by Pearson from electronic files supplied by the author.

1 18

ISBN-13: 978-0-13-469847-2
ISBN-10: 0-13-469847-9

CONTENTS

Chapter P Basic Concepts of Algebra

P.1 The Real Numbers and Their Properties

P.1 Practice Problems

1. Let $x = 2.132132132....$
   Then, $1000x = 2132.132132...$

   $$\begin{aligned}
   1000x &= 2132.132132... \\
   x &= 2.132132... \\
   \hline
   999x &= 2130 \\
   x &= \frac{2130}{999} = \frac{710}{333}
   \end{aligned}$$

2. a. Natural numbers: 2, 7

   b. Whole numbers: 0, 2, 7

   c. Integers: $-6, -\frac{21}{7} = -3, 0, 2, 7$

   d. Rational numbers:
      $-6, -\frac{21}{7} = -3, -\frac{1}{2}, 0, \frac{4}{3}, 2, 7$

   e. Irrational numbers: $\sqrt{3}, \sqrt{17}$

   f. Real numbers: the set B

3. a. $2^3 = 8$

   b. $(3a)^2 = (3a)(3a) = 9a^2$

   c. $\left(\frac{1}{2}\right)^4 = \frac{1}{2} \cdot \frac{1}{2} \cdot \frac{1}{2} \cdot \frac{1}{2} = \frac{1}{16}$

4. a. True   b. True

   c. False

5. $A = \{-3, -1, 0, 1, 3\}$, $B = \{-4, -2, 0, 2, 4\}$
   $A \cap B = \{0\}$,
   $A \cup B = \{-4, -3, -2, -1, 0, 1, 2, 3, 4\}$

6. a. $I_1 \cup I_2 = (-\infty, \infty)$

   b. $I_1 \cap I_2 = [-2, 5)$

7. a. $|-10| = 10$

   b. $|3 - 4| = |-1| = 1$

   c. $|2(-3) + 7| = |1| = 1$

8.

$$d(-7, 2) = |-7 - 2| = |-9| = 9$$

9. a. $(-3) \cdot 5 + 20 = -15 + 20 = 5$

   b. $5 - 12 \div 6 \cdot 2 = 5 - 2 \cdot 2 = 5 - 4 = 1$

   c. $\frac{9-1}{4} - 5 \cdot 7 = \frac{8}{4} - 5 \cdot 7 = 2 - 35 = -33$

   d. $-3 + (x - 4)^2$ for $x = 6$
      $-3 + (6 - 4)^2 = -3 + 2^2 = -3 + 4 = 1$

10. a. $\frac{7}{4} + \frac{3}{8} = \frac{14}{8} + \frac{3}{8} = \frac{17}{8}$

    b. $\frac{8}{3} - \frac{2}{5} = \frac{40}{15} - \frac{6}{15} = \frac{34}{15}$

    c. $\frac{9}{14} \cdot \frac{7}{3} = \frac{\overset{3}{\cancel{9}}}{\underset{2}{\cancel{14}}} \cdot \frac{\overset{1}{\cancel{7}}}{\underset{1}{\cancel{3}}} = \frac{3}{2}$

    d. $\frac{\frac{5}{8}}{\frac{15}{16}} = \frac{5}{8} \div \frac{15}{16} = \frac{5}{8} \cdot \frac{16}{15} = \frac{\overset{1}{\cancel{5}}}{\underset{1}{\cancel{8}}} \cdot \frac{\overset{2}{\cancel{16}}}{\underset{3}{\cancel{15}}} = \frac{2}{3}$

11. a. $(x - 2) \div 3 + x = (3 - 2) \div 3 + 3$
       $= 1 \div 3 + 3 = \frac{1}{3} + 3 = \frac{10}{3}$

    b. $7 - \frac{x}{|y|} = 7 - \frac{-1}{|3|} = 7 + \frac{1}{3} = \frac{22}{3}$

12. $°C = \frac{39}{3} + 4 = 13 + 4 = 17°C$

P.1 Concepts and Vocabulary

1. Whole numbers are formed by adding the number <u>zero</u> to the set of natural numbers.

3. If $a < b$, then a is to the <u>left</u> of b on the number line.

5. True

7. True

P.1 Building Skills

9. $0.\overline{3}$, repeating

11. -0.8, terminating

13. $0.\overline{27}$, repeating

15. $3.1\overline{6}$, repeating

17. $3.75 = \dfrac{375}{100} = \dfrac{15}{4}$

19. $10x = 53.\overline{3}$
    $\underline{x = 5.\overline{3}}$
    $9x = 48$
    $x = \dfrac{48}{9} = \dfrac{16}{3}$
    $-5.\overline{3} = -\dfrac{16}{3}$

21. $100x = 213.\overline{13}$
    $\underline{x = 2.\overline{13}}$
    $99x = 211$
    $x = \dfrac{211}{99}$

23. $100x = 452.3\overline{23}$
    $\underline{x = 4.5\overline{23}}$
    $99x = 447.8$
    $x = \dfrac{447.8}{99} = \dfrac{4478}{990} = \dfrac{2239}{495}$

25. Rational   27. Rational

29. Rational   31. Irrational

Exercises 33-38 refer to the set
$$A = \left\{-19, -\dfrac{12}{3}, -\sqrt{3}, 0, 2, \sqrt{10}, \dfrac{17}{4}, 11\right\}$$

33. Natural numbers: 2, 11

35. Integers: $-19, -\dfrac{12}{3} = -4, \ 0, \ 2, \ 11$

37. Irrational numbers: $-\sqrt{3}, \sqrt{10}$

39. 10^3
    base: 10; exponent: 3
    $10^3 = 10 \cdot 10 \cdot 10 = 1000$

41. $\left(\dfrac{2}{3}\right)^3$
    base: $\dfrac{2}{3}$; exponent: 3
    $\left(\dfrac{2}{3}\right)^3 = \dfrac{2}{3} \cdot \dfrac{2}{3} \cdot \dfrac{2}{3} = \dfrac{8}{27}$

43. $(-2)^3$
    base: -2; exponent: 3
    $(-2)^3 = (-2) \cdot (-2) \cdot (-2) = -8$

45. $2 \cdot 3^4$
    base: 3; exponent: 4
    $2 \cdot 3^4 = 2 \cdot 3 \cdot 3 \cdot 3 \cdot 3 = 2 \cdot 81 = 162$

47. $-2 \cdot 3^4$
    base: 3; exponent: 4
    $-2 \cdot 3^4 = -2 \cdot 3 \cdot 3 \cdot 3 \cdot 3 = -2 \cdot 81 = -162$

49. $-3 \cdot (-2)^5$
    base: -2; exponent: 5
    $-3 \cdot (-2)^5 = -3(-2)(-2)(-2)(-2)(-2)$
    $ = -3 \cdot (-32) = 96$

51. $3 > -2$ 53. $\dfrac{1}{2} \geq \dfrac{1}{2}$

55. $5 \leq 2x$ 57. $-x > 0$

59. $2x + 7 \leq 14$ 61. $4 = \dfrac{24}{6}$

63. $-4 < 0$

65. $A \cup B = \{-4, -3, -2, 0, 1, 2, 3, 4\}$

67. $A \cap C = \{-4, -2, \ 0, 2\}$

69. $(B \cap C) \cup A = \{-3, 0, 2\} \cup A$
    $ = \{-4, -3, -2, 0, 2, 4\}$

71. $(A \cup B) \cap C = \{-4, -3, -2, 0, 1, 2, 3, 4\} \cap C$
    $ = \{-4, -3, -2, 0, 2\}$

73. $I_1 \cup I_2 = (-2, 5); \ I_1 \cap I_2 = [1, 3]$

75. $I_1 \cup I_2 = (-6, 10); \ I_1 \cap I_2 = \varnothing$

77. $I_1 \cup I_2 = (-\infty, \infty); \ I_1 \cap I_2 = [2, 5)$

79. $I_1 \cup I_2 = (-\infty, \infty); \ I_1 \cap I_2 = [-1, 3) \cup [5, 7]$

81. $|-4| = 4$

83. $\left|\dfrac{5}{-7}\right| = \dfrac{5}{7}$

85. $|5 - \sqrt{2}| = 5 - \sqrt{2}$

87. $|\sqrt{3} - 2| = 2 - \sqrt{3}$

89. $\dfrac{8}{|-8|} = \dfrac{8}{8} = 1$

91. $|5 - |-7|| = |5 - 7| = |-2| = 2$

93.

$$d(3, 8) = |3 - 8| = |-5| = 5$$

95.

$$d(-6, 9) = |-6 - 9| = |-15| = 15$$

97.

$$d(-20, -6) = |-20 - (-6)| = |-14| = 14$$

99.

$$d\left(\dfrac{22}{7}, -\dfrac{4}{7}\right) = \left|\dfrac{22}{7} - \left(-\dfrac{4}{7}\right)\right| = \left|\dfrac{26}{7}\right| = \dfrac{26}{7}$$

101.

$$-3 < x \le 1$$

103.

$$x \ge -3$$

105.

$$x \le 5$$

107.

$$-\dfrac{3}{4} < x < \dfrac{9}{4}$$

109. $4(x + 1) = 4x + 4$

111. $5(x - y + 1) = 5x - 5y + 5$

113. Additive inverse: -5; reciprocal: $\dfrac{1}{5}$

115. Additive inverse: 0; no reciprocal

117. Additive inverse

119. Multiplicative identity

121. Associative property of multiplication

123. Multiplicative inverse

125. Additive identity

127. Associative property of addition

129. $\dfrac{3}{5} + \dfrac{4}{3} = \dfrac{9}{15} + \dfrac{20}{15} = \dfrac{29}{15}$

131. $\dfrac{6}{5} + \dfrac{5}{7} = \dfrac{42}{35} + \dfrac{25}{35} = \dfrac{67}{35}$

133. $\dfrac{5}{6} + \dfrac{3}{10} = \dfrac{25}{30} + \dfrac{9}{30} = \dfrac{34}{30} = \dfrac{17}{15}$

135. $\dfrac{5}{8} - \dfrac{9}{10} = \dfrac{25}{40} - \dfrac{36}{40} = -\dfrac{11}{40}$

137. $\dfrac{5}{9} - \dfrac{7}{11} = \dfrac{55}{99} - \dfrac{63}{99} = -\dfrac{8}{99}$

139. $\dfrac{2}{5} - \dfrac{1}{2} = \dfrac{4}{10} - \dfrac{5}{10} = -\dfrac{1}{10}$

141. $\dfrac{3}{4} \cdot \dfrac{8}{27} = \dfrac{2}{9}$

143. $\dfrac{\frac{8}{5}}{\frac{16}{15}} = \dfrac{8}{5} \div \dfrac{16}{15} = \dfrac{8}{5} \cdot \dfrac{15}{16} = \dfrac{3}{2}$

145. $\dfrac{\frac{7}{8}}{\frac{21}{16}} = \dfrac{7}{8} \div \dfrac{21}{16} = \dfrac{7}{8} \cdot \dfrac{16}{21} = \dfrac{2}{3}$

147. $5 \cdot \dfrac{3}{10} - \dfrac{1}{2} = \dfrac{3}{2} - \dfrac{1}{2} = \dfrac{2}{2} = 1$

149. $3 \cdot \dfrac{2}{15} - \dfrac{1}{3} = \dfrac{2}{5} - \dfrac{1}{3} = \dfrac{6}{15} - \dfrac{5}{15} = \dfrac{1}{15}$

151. $2(x + y) - 3y = 2(3 + (-5)) - 3(-5)$
$ = 2(-2) - (-15) = -4 + 15 = 11$

153. $3|x| - 2|y| = 3|3| - 2|-5| = 3(3) - 2(5)$
$ = 9 - 10 = -1$

155. $\dfrac{x-3y}{2}+xy=\dfrac{3-3(-5)}{2}+3(-5)$

$=\dfrac{3-(-15)}{2}+(-15)$

$=\dfrac{18}{2}+(-15)=9+(-15)=-6$

157. $\dfrac{2(1-2x)}{y}-(-x)y=\dfrac{2(1-2(3))}{-5}-(-3)(-5)$

$=\dfrac{2(-5)}{-5}-15=2-15=-13$

159. $\dfrac{\frac{14}{x}+\frac{1}{2}}{\frac{-y}{4}}=\dfrac{\frac{14}{3}+\frac{1}{2}}{\frac{-(-5)}{4}}=\dfrac{31}{6}\cdot\dfrac{4}{5}=\dfrac{62}{15}$

161. $\dfrac{x}{y}+\dfrac{x}{3}=\dfrac{3x+xy}{3y}$

163. $5(x+3)=5x+5(3)=5x+15$.

165. $x-(3y+2)=x-3y-2$

167. $\dfrac{x+y}{x}=\dfrac{x}{x}+\dfrac{y}{x}=1+\dfrac{y}{x}$

169. $(x+1)(y+1)=xy+x+y+1$

171. $d(P,M)=\left|\dfrac{a+b}{2}-a\right|=\left|\dfrac{a+b-2a}{2}\right|=\left|\dfrac{-a+b}{2}\right|$

$d(Q,M)=\left|b-\dfrac{a+b}{2}\right|=\left|\dfrac{2b-a+b}{2}\right|=\left|\dfrac{b-a}{2}\right|$

Since

$d(P,M)=\left|\dfrac{-a+b}{2}\right|=\left|\dfrac{b-a}{2}\right|=d(Q,M)$, M is

the midpoint of the line segment PQ.

P.1 Applying the Concepts

173. a. people who own either MP3 players or people who own DVD players.

b. people who own both MP3 players and DVD players.

175. $119.5\le x\le 134.5$

176. $30\le x\le 107$

177. a. $|124-120|=4$

b. $|137-120|=17$

c. $|114-120|=|-6|=6$

179. Let $x=$ the number of calories from broccoli. Then we have

$522.5-55x=0\Rightarrow 522.5=55x\Rightarrow 9.5=x$

The number of grams of broccoli is $9.5\times 100=950$ grams.

P.1 Beyond the Basics

181. True

183. a. False. For example, $0\cdot\sqrt{2}=0$.

b. The products of a nonzero rational number and an irrational number is an irrational number.

185. False. For example, $(2+\sqrt{3})-(6+\sqrt{3})=-4$.

187. True

For exercises 189–190, use the following definition: An integer P is *even* if $p=2n$ for some integer n. An integer q is odd if $q=2k+1$ for some integer k.

189. a. If a is odd, then $a=2m+1$ for some integer m.

$a^2=(2m+1)^2=(2m+1)(2m+1)$

$=4m^2+4m+1=4m(m+1)+1$

$=2(2m(m+1))+1,$

which is of the format $2k+1$. Therefore, a^2 is an odd integer.

b. Assume that b is odd. Then, from part (a), b^2 is also odd. However, b^2 is an even integer, so, by contradiction, b must be an even integer.

P.1 Getting Ready for the Next Section

191. a. $a^2\cdot a^3=a^{\underline{5}}$

b. $a^4\cdot a^7=a^{\underline{11}}$

c. $a^m\cdot a^n=a^{\underline{m+n}}$

193. a. $(a^2)^3=a^{\underline{6}}$

b. $(a^4)^2=a^{\underline{8}}$

c. $(a^m)^n=a^{\underline{mn}}$

P.2 Integer Exponents and Scientific Notation

P.2 Practice Exercises

1. a. $2^{-1} = \dfrac{1}{2}$

   b. $\left(\dfrac{4}{5}\right)^0 = 1$

   c. $\left(\dfrac{3}{2}\right)^{-2} = \left(\dfrac{2}{3}\right)^2 = \dfrac{4}{9}$

2. a. $x^2 \cdot 3x^7 = 3x^9$

   b. $\left(2^2 x^3\right)\left(4x^{-3}\right) = 4x^3 \cdot 4x^{-3} = 16x^{3+(-3)}$
   $$= 16x^0 = 16 \cdot 1 = 16$$

3. a. $\dfrac{3^4}{3^0} = 3^{4-0} = 3^4 = 81$

   b. $\dfrac{5}{5^{-2}} = 5^{1-(-2)} = 5^3 = 125$

   c. $\dfrac{2x^3}{3x^{-4}} = \dfrac{2x^{3-(-4)}}{3} = \dfrac{2x^7}{3}$

4. a. $\left(7^{-5}\right)^0 = 7^{(-5)(0)} = 7^0 = 1$

   b. $\left(7^0\right)^{-5} = 7^{(0)(-5)} = 7^0 = 1$

   c. $\left(x^{-1}\right)^8 = x^{(-1)(8)} = x^{-8} = \dfrac{1}{x^8}$

   d. $\left(x^{-2}\right)^{-5} = x^{(-2)(-5)} = x^{10}$

5. a. $\left(\dfrac{1}{2}x\right)^{-1} = \left(\dfrac{1}{2}\right)^{-1}(x)^{-1} = 2\left(\dfrac{1}{x}\right) = \dfrac{2}{x}$

   b. $\left(5x^{-1}\right)^2 = 5^2\left(x^{-1}\right)^2 = 5^2 x^{(-1)(2)}$
   $$= 25x^{-2} = \dfrac{25}{x^2}$$

   c. $\left(xy^2\right)^3 = x^3 y^{2(3)} = x^3 y^6$

   d. $\left(x^{-2}y\right)^{-3} = x^{(-2)(-3)}y^{-3} = \dfrac{x^6}{y^3}$

6. a. $\left(\dfrac{1}{3}\right)^2 = \dfrac{1^2}{3^2} = \dfrac{1}{9}$

   b. $\left(\dfrac{10}{7}\right)^{-2} = \left(\dfrac{7}{10}\right)^2 = \dfrac{7^2}{10^2} = \dfrac{49}{100}$

7. a. $\left(2x^4\right)^{-2} = \left(\dfrac{1}{2x^4}\right)^2 = \dfrac{1}{4x^8}$

   b. $\dfrac{x^2(-y)^3}{\left(xy^2\right)^3} = \dfrac{-x^2 y^3}{x^3 y^6} = -\dfrac{1}{xy^3}$

8. $732,000 = 7.32 \times 10^5$

9. $\dfrac{2.9 \times 10^{10}}{3.25 \times 10^8} = \dfrac{290}{325} \times 10^2 \approx 0.89 \times 10^2 \approx \$89$
   per person

P.2 Concepts and Vocabulary

1. In the expression 7^2, the number 2 is called the <u>exponent</u>.

3. The number $\dfrac{1}{4^{-2}}$ simplifies to be the positive integer <u>16</u>.

5. False. $(-11)^{10} = 11^{10}$

7. True

P.2 Building Skills

9. $3^{-2} = \left(\dfrac{1}{3}\right)^2 = \left(\dfrac{1}{3}\right)\left(\dfrac{1}{3}\right) = \dfrac{1}{9}$

11. $\left(\dfrac{1}{2}\right)^{-4} = 2^4 = 16$

13. $7^0 = 1$

15. $-(-7)^0 = -1$

17. $\left(2^3\right)^2 = 2^{3 \cdot 2} = 2^6 = 64$

19. $\left(3^2\right)^{-2} = 3^{2(-2)} = 3^{-4} = \dfrac{1}{3^4} = \dfrac{1}{81}$

21. $\left(5^{-2}\right)^3 = \left(\dfrac{1}{5^2}\right)^3 = \left(\dfrac{1}{25}\right)^3 = \dfrac{1}{15,625}$

23. $\left(4^{-3}\right) \cdot \left(4^5\right) = 4^{(-3+5)} = 4^2 = 16$

25. $\left(\sqrt{3}\right)^0 + 10^0 = 1 + 1 = 2$

27. $3^{-2} + \left(\dfrac{1}{3}\right)^2 = \left(\dfrac{1}{3}\right)^2 + \left(\dfrac{1}{3}\right)^2 = \dfrac{1}{9} + \dfrac{1}{9} = \dfrac{2}{9}$

29. $-2^{-3} = -\dfrac{1}{2^3} = -\dfrac{1}{8}$

31. $(-3)^{-2} = \dfrac{1}{(-3)^2} = \dfrac{1}{9}$

33. $\dfrac{2^{11}}{2^{10}} = 2^{(11-10)} = 2^1 = 2$

35. $\dfrac{\left(5^3\right)^4}{5^{12}} = \dfrac{5^{12}}{5^{12}} = 1$

37. $\dfrac{2^5 \cdot 3^{-2}}{2^4 \cdot 3^{-3}} = 2^{(5-4)} \cdot 3^{(-2-(-3))} = 2^1 \cdot 3^1 = 6$

39. $\dfrac{-5^{-2}}{2^{-1}} = -\dfrac{1}{5^2} \cdot 2^1 = -\dfrac{1}{25} \cdot 2 = -\dfrac{2}{25}$

41. $\left(\dfrac{2}{3}\right)^{-1} = \dfrac{3}{2}$

43. $\left(\dfrac{2}{3}\right)^{-2} = \left(\dfrac{3}{2}\right)^2 = \dfrac{3^2}{2^2} = \dfrac{9}{4}$

45. $\left(\dfrac{11}{7}\right)^{-2} = \dfrac{1}{\left(\dfrac{11}{7}\right)^2} = \dfrac{1}{\dfrac{121}{49}} = \dfrac{49}{121}$

47. $x^4 y^0 = x^4 \cdot 1 = x^4$

49. $x^{-1}y = \dfrac{1}{x} \cdot y = \dfrac{y}{x}$

51. $x^{-1}y^{-2} = \dfrac{1}{xy^2}$

53. $\left(x^{-3}\right)^4 = x^{-12} = \dfrac{1}{x^{12}}$

55. $\left(x^{-11}\right)^{-3} = x^{(-11)\cdot(-3)} = x^{33}$

57. $-3(xy)^5 = -3x^5 y^5$

59. $4\left(xy^{-1}\right)^2 = 4x^2 y^{-2} = \dfrac{4x^2}{y^2}$

61. $3\left(x^{-1}y\right)^{-5} = 3x^{(-1)\cdot(-5)}y^{-5} = \dfrac{3x^5}{y^5}$

63. $\dfrac{\left(x^3\right)^2}{\left(x^2\right)^5} = \dfrac{x^{3\cdot2}}{x^{2\cdot5}} = \dfrac{x^6}{x^{10}} = x^{6-10} = x^{-4} = \dfrac{1}{x^4}$

65. $\left(\dfrac{2xy}{x^2}\right)^3 = \dfrac{2^3 x^3 y^3}{x^{2\cdot3}} = \dfrac{8x^3 y^3}{x^6} = 8x^{3-6}y^3$

$= 8x^{-3}y^3 = \dfrac{8y^3}{x^3}$

67. $\left(\dfrac{-3x^2 y}{x}\right)^5 = \dfrac{(-3)^5 x^{2\cdot5} y^5}{x^5} = \dfrac{-243x^{10} y^5}{x^5}$

$= -243x^{10-5}y^5 = -243x^5 y^5$

69. $\left(\dfrac{-3x}{5}\right)^{-2} = \dfrac{(-3)^{-2}x^{-2}}{5^{-2}} = \dfrac{\dfrac{1}{9}\cdot\dfrac{1}{x^2}}{\dfrac{1}{25}} = \dfrac{\dfrac{1}{9x^2}}{\dfrac{1}{25}} = \dfrac{25}{9x^2}$

71. $\left(\dfrac{4x^{-2}}{xy^5}\right)^3 = \dfrac{4^3 x^{-2\cdot3}}{x^3 y^{5\cdot3}} = \dfrac{64x^{-6}}{x^3 y^{15}} = \dfrac{64}{x^9 y^{15}}$

73. $\dfrac{x^3 y^{-3}}{x^{-2}y} = x^{3-(-2)}y^{-3-1} = x^5 y^{-4} = \dfrac{x^5}{y^4}$

75. $\dfrac{27x^{-3}y^5}{9x^{-4}y^7} = 3x^{-3-(-4)}y^{5-7} = 3x^1 y^{-2} = \dfrac{3x}{y^2}$

77. $\dfrac{1}{x^3}\left(x^2\right)^3 x^{-4} = x^{-3}\left(x^{2\cdot3}\right)x^{-4} = x^{-3}\left(x^6\right)x^{-4}$

$= x^{-3+6-4} = x^{-1} = \dfrac{1}{x}$

79. $\left(-xy^2\right)^3\left(-2x^2 y^2\right)^{-4} = \dfrac{\left(-xy^2\right)^3}{\left(-2x^2 y^2\right)^4}$

$= \dfrac{(-x)^3\left(y^2\right)^3}{(-2)^4\left(x^2\right)^4\left(y^2\right)^4}$

$= \dfrac{-x^3 \cdot y^{2\cdot3}}{16x^{2\cdot4}y^{2\cdot4}} = \dfrac{-x^3 y^6}{16x^8 y^8}$

$= -\dfrac{1}{16}\left(x^{3-8}\right)\left(y^{6-8}\right)$

$= -\dfrac{1}{16}x^{-5}y^{-2} = -\dfrac{1}{16x^5 y^2}$

81. $\left(4x^3y^2z\right)^2\left(x^3y^2z\right)^{-7} = \dfrac{\left(4x^3y^2z\right)^2}{\left(x^3y^2z\right)^7}$

$= \dfrac{4^2\left(x^3\right)^2\left(y^2\right)^2 z^2}{\left(x^3\right)^7\left(y^2\right)^7 z^7}$

$= \dfrac{16x^{3\cdot2}y^{2\cdot2}z^2}{x^{3\cdot7}y^{2\cdot7}z^7}$

$= \dfrac{16x^6y^4z^2}{x^{21}y^{14}z^7}$

$= 16x^{6-21}y^{4-14}z^{2-7}$

$= 16x^{-15}y^{-10}z^{-5}$

$= \dfrac{16}{x^{15}y^{10}z^5}$

83. $\dfrac{5a^{-2}bc^2}{a^4b^{-3}c^2} = 5a^{-2-4}b^{1-(-3)}c^{2-2}$

$= 5a^{-6}b^4\cdot1 = \dfrac{5b^4}{a^6}$

85. $\left(\dfrac{xy^{-3}z^{-2}}{x^2y^{-4}z^3}\right)^{-3} = \dfrac{x^{-3}y^{(-3)(-3)}z^{(-2)(-3)}}{x^{2(-3)}y^{(-4)(-3)}z^{3(-3)}}$

$= \dfrac{x^{-3}y^9z^6}{x^{-6}y^{12}z^{-9}}$

$= x^{-3-(-6)}y^{9-12}z^{6-(-9)}$

$= x^3y^{-3}z^{15} = \dfrac{x^3z^{15}}{y^3}$

87. $125 = 1.25\times10^2$

89. $850,000 = 8.5\times10^5$

91. $0.007 = 7.0\times10^{-3}$

93. $0.00000275 = 2.75\times10^{-6}$

P.2 Applying the Concepts

95. $135\text{ ft}^3\times2^3 = 1080\text{ ft}^3$

97. a. $(2x)^2 = 4x^2 = 4A = 2^2A$

b. $(3x)^2 = 9x^2 = 9A = 3^2A$

99. $F = Sw^2 = 25,000(0.25)^2 = 1562.5\text{ lb}$

101. $365.25\text{ days}\times\dfrac{24\text{ hr}}{\text{day}}\times\dfrac{60\text{ min}}{\text{hr}}\times\dfrac{60\text{ sec}}{\text{min}}$

$= 365.25\times24\times60\times60\text{ sec}$

$= 31,557,600\text{ sec}$

$= 3.15576\times10^7\text{ sec}$

103.

Celestial Body	Equatorial Diameter (km)	Scientific Notation
Earth	12,700	1.27×10^4 km
Moon	3480	3.48×10^3 km
Sun	1,390,000	1.39×10^6 km
Jupiter	134,000	1.34×10^5 km
Mercury	4800	4.8×10^3 km

105. $602,000,000,000,000,000,000 = 6.02\times10^{20}$ atoms

107. $0.00000000000000000000167 = 1.67\times10^{-21}$ kg

109. $5,980,000,000,000,000,000,000,000 = 5.98\times10^{24}$ kg

P.2 Beyond the Basics

111. $2^x = 32$

a. $2^{x+2} = 2^x\cdot2^2 = 32\cdot4 = 128$

b. $2^{x-1} = 2^x\cdot2^{-1} = 32\cdot\dfrac{1}{2} = 16$

113. $5^x = 11$

a. $5^{x+1} = 5^x\cdot5^1 = 11\cdot5 = 55$

b. $5^{x-2} = 5^x\cdot5^{-2} = 11\cdot\left(\dfrac{1}{5}\right)^2 = 11\cdot\dfrac{1}{25} = \dfrac{11}{25}$

115. $\dfrac{3^{2-n}\cdot9^{2n-2}}{3^{3n}} = \dfrac{3^{2-n}\cdot\left(3^2\right)^{2n-2}}{3^{3n}}$

$= \dfrac{3^{2-n}\cdot3^{2(2n-2)}}{3^{3n}}$

$= \dfrac{3^{3n-2}}{3^{3n}} = 3^{(3n-2)-3n}$

$= 3^{-2} = \dfrac{1}{3^2} = \dfrac{1}{9}$

117. $\dfrac{2^{x(y-z)}}{2^{y(x-z)}} \div \left(\dfrac{2^y}{2^x}\right)^z = \dfrac{2^{xy-xz}}{2^{xy-yz}} \div \dfrac{2^{yz}}{2^{xz}}$

$\qquad = \dfrac{\dfrac{2^{xy}}{2^{xz}}}{\dfrac{2^{xy}}{2^{yz}}} \cdot \dfrac{2^{xz}}{2^{yz}}$

$\qquad = \dfrac{2^{yz}}{2^{xz}} \cdot \dfrac{2^{xz}}{2^{yz}} = 1$

P.2 Getting Ready for the Next Section

119. a. $x^2 \cdot x^5 = x^{2+5} = x^7$

     b. $(2x)(-5x^2) = -10x^3$

     c. $(2y^2)(3y^3)(4y^5) = 24y^{2+3+5} = 24y^{10}$

121. False. $5x^2$ and $3x^3$ are not like terms, so they cannot be combined.

P.3 Polynomials

P.3 Practice Problems

1. $16(7)^2 + 15(7) = 889$ ft

2. $(7x^3 + 2x^2 - 5) + (-2x^3 + 3x^2 + 2x + 1)$
$\qquad = 5x^3 + 5x^2 + 2x - 4$

3. $(3x^4 - 5x^3 + 2x^2 + 7) - (-2x^4 + 3x^2 + x - 5)$
$\qquad = 5x^4 - 5x^3 - x^2 - x + 12$

4. $-2x^3(4x^2 + 2x - 5) = -8x^5 - 4x^4 + 10x^3$

5. $(5x^2 + 2x)(-2x^2 + x - 7)$
$\qquad = 5x^2(-2x^2 + x - 7) + 2x(-2x^2 + x - 7)$
$\qquad = -10x^4 + 5x^3 - 35x^2 - 4x^3 + 2x^2 - 14x$
$\qquad = -10x^4 + x^3 - 33x^2 - 14x$

6. a. $(4x - 1)(x + 7) = 4x^2 + 28x - x - 7$
$\qquad\qquad = 4x^2 + 27x - 7$

     b. $(3x - 2)(2x - 5) = 6x^2 - 15x - 4x + 10$
$\qquad\qquad = 6x^2 - 19x + 10$

7. $(3x + 2)^2 = (3x)^2 + 2(3x)(2) + 2^2$
$\qquad = 9x^2 + 12x + 4$

8. $(1 - 2x)(1 + 2x) = 1^2 - (2x)^2 = 1 - 4x^2$

9. a. $(x + 2y)(x - 2y) = x^2 - (2y)^2 = x^2 - 4y^2$

     b. $(2x - y)^3 = (2x)^3 - 3(2x)^2 + 3(2x)y^2 - y^3$
$\qquad = 8x^3 - 3(4x^2) + 6xy^2 - y^3$
$\qquad = 8x^3 - 12x^2 + 6xy^2 - y^3$

P.3 Concepts and Vocabulary

1. The polynomial $-3x^7 + 2x^2 - 9x + 4$ has leading coefficient $\underline{-3}$ and degree $\underline{7}$.

3. When a polynomial in x of degree 3 is added to a polynomial in x of degree 4, the resulting polynomial has degree $\underline{4}$.

5. True

7. False. This is not a polynomial because the term $-3\sqrt{x}$ does not have an exponent that is either a positive integer or zero.

P.3 Building Skills

9. A polynomial; $x^2 + 2x + 1$

11. Not a polynomial

13. Not a polynomial

15. A polynomial; in standard form

17. Degree: 1; terms: $7x$, 3

19. Degree: 4; terms: $-x^4, x^2, 2x, -9$

21. $(x^3 + 2x^2 - 5x + 3) + (-x^3 + 2x - 4)$
$\qquad = (x^3 + (-x^3)) + 2x^2 + (-5x + 2x) + (3 - 4)$
$\qquad = 2x^2 - 3x - 1$

23. $(2x^3 - x^2 + x - 5) - (x^3 - 4x + 3)$
$\qquad = (2x^3 - x^3) - x^2 + (x - (-4x)) + (-5 - 3)$
$\qquad = x^3 - x^2 + 5x - 8$

25. $(-2x^4 + 3x^2 - 7x) - (8x^4 + 6x^3 - 9x^2 - 17)$
$\qquad = (-2x^4 - 8x^4) - 7x - 6x^3$
$\qquad\qquad + (3x^2 - (-9x^2)) - (-17)$
$\qquad = -10x^4 - 6x^3 + 12x^2 - 7x + 17$

27. $-2(3x^2 + x + 1) + 6(-3x^2 - 2x - 2)$
$\qquad = -6x^2 - 2x - 2 - 18x^2 - 12x - 12$
$\qquad = -24x^2 - 14x - 14$

29. $\left(3y^3-4y+2\right)+(2y+1)-\left(y^3-y^2+4\right)$
$=3y^3-4y+2+2y+1-y^3+y^2-4$
$=2y^3+y^2-2y-1$

31. $6x(2x+3)=12x^2+18x$

33. $(x+1)\left(x^2+2x+2\right)$
$=x\left(x^2+2x+2\right)+1\left(x^2+2x+2\right)$
$=x^3+2x^2+2x+x^2+2x+2$
$=x^3+3x^2+4x+2$

35. $(3x-2)\left(x^2-x+1\right)$
$=3x\left(x^2-x+1\right)-2\left(x^2-x+1\right)$
$=3x^3-3x^2+3x-2x^2+2x-2$
$=3x^3-5x^2+5x-2$

37. $(x+1)(x+2)=x(x+2)+1(x+2)$
$=x^2+2x+x+2=x^2+3x+2$

39. $(3x+2)(3x+1)=9x^2+3x+6x+2$
$=9x^2+9x+2$

41. $(-4x+5)(x+3)=-4x^2-12x+5x+15$
$=-4x^2-7x+15$

43. $(3x-2)(2x-1)=6x^2-3x-4x+2$
$=6x^2-7x+2$

45. $(2x-3a)(2x+5a)=4x^2+10ax-6ax-15a^2$
$=4x^2+4ax-15a^2$

47. $(x+2)^2-x^2=x^2+4x+4-x^2=4x+4$

49. $(4x+1)^2=16x^2+8x+1$

51. $\left(x+\frac{3}{4}\right)^2=x^2+2\left(\frac{3}{4}\right)x+\left(\frac{3}{4}\right)^2$
$=x^2+\frac{3}{2}x+\frac{9}{16}$

53. $(3x-4y)^2=(3x)^2-2(3x)(4y)+(4y)^2$
$=9x^2-24xy+16y^2$

55. $\left(x^2+2\right)^2=\left(x^2\right)^2+2\left(x^2\right)(2)+2^2$
$=x^4+4x^2+4$

57. $\left(\frac{x}{2}+\frac{2}{x}\right)^2=\left(\frac{x}{2}\right)^2+2\left(\frac{x}{2}\right)\left(\frac{2}{x}\right)+\left(\frac{2}{x}\right)^2$
$=\frac{x^2}{4}+2+\frac{4}{x^2}=\frac{x^2}{4}+\frac{4}{x^2}+2$

59. $(x+2)^3=x^3+3x^2(2)+3x(2)^2+2^3$
$=x^3+6x^2+12x+8$

61. $(x+3)^3-x^3$
Note that this is the difference of cubes, with $A=x+3$ and $B=x$.
$(x+3)^3-x^3$
$=\left[(x+3)-x\right]\left[(x+3)^2+(x+3)x+x^2\right]$
$=3\left(x^2+6x+9+x^2+3x+x^2\right)$
$=3\left(3x^2+9x+9\right)$
$=9x^2+27x+27$

63. $(x+2y)^3=x^3+3x^2(2y)+3x(2y)^2+2^3$
$=x^3+6x^2y+12xy^2+8$

65. $(5+2x)(5-2x)=5^2-(2x)^2=25-4x^2$

67. $(2x+3y)(2x-3y)=(2x)^2-(3y)^2$
$=4x^2-9y^2$

69. $\left(x+\frac{1}{x}\right)\left(x-\frac{1}{x}\right)=x^2-\left(\frac{1}{x}\right)^2=x^2-\frac{1}{x^2}$

71. $\left(x^2+3\right)\left(x^2-3\right)=\left(x^2\right)^2-3^2=x^4-9$

73. $(2x-3)\left(x^2-3x+5\right)$
$=2x\left(x^2-3x+5\right)-3\left(x^2-3x+5\right)$
$=2x^3-6x^2+10x-3x^2+9x-15$
$=2x^3-9x^2+19x-15$

75. $(1+y)\left(1-y+y^2\right)$
$=1\left(1-y+y^2\right)+y\left(1-y+y^2\right)$
$=1-y+y^2+y-y^2+y^3=y^3+1$

77. $(x-6)\left(x^2+6x+36\right)$
$=x\left(x^2+6x+36\right)-6\left(x^2+6x+36\right)$
$=x^3+6x^2+36x-6x^2-36x-216$
$=x^3-216$

79. $(x+2y)(3x+5y) = 3x^2 + 5xy + 6xy + 10y^2 = 3x^2 + 11xy + 10y^2$

81. $(2x-y)(3x+7y) = 6x^2 + 14xy - 3xy - 7y^2 = 6x^2 + 11xy - 7y^2$

83. $(x-y)^2 (x+y)^2 = (x-y)(x-y)(x+y)(x+y) = \left[(x-y)(x+y)\right]\left[(x-y)(x+y)\right]$
$$= \left(x^2 - y^2\right)\left(x^2 - y^2\right) = x^4 - 2x^2 y^2 + y^4$$

85. $(x+y)(x-2y)^2 = (x+y)\left(x^2 - 4xy + 4y^2\right) = x\left(x^2 - 4xy + 4y^2\right) + y\left(x^2 - 4xy + 4y^2\right)$
$$= x^3 - 4x^2 y + 4xy^2 + x^2 y - 4xy^2 + 4y^3 = x^3 - 3x^2 y + 4y^3$$

87. $(x-2y)^3 (x+2y) = \left(x^3 + 3x^2 (-2y) + 3x(-2y)^2 + (-2y)^3\right) \cdot (x+2y)$
$$= \left(x^3 - 6x^2 y + 12xy^2 - 8y^3\right)(x+2y)$$
$$= x\left(x^3 - 6x^2 y + 12xy^2 - 8y^3\right) + 2y\left(x^3 - 6x^2 y + 12xy^2 - 8y^3\right)$$
$$= x^4 - 6x^3 y + 12x^2 y^2 - 8xy^3 + 2x^3 y - 12x^2 y^2 + 24xy^3 - 16y^4$$
$$= x^4 - 4x^3 y + 16xy^3 - 16y^4$$

For exercises 89–93, use $a^2 + b^2 = (a+b)^2 - 2ab = (a-b)^2 + 2ab$.

89. $x + y = 4,\ xy = 3$
$$x^2 + y^2 = (x+y)^2 - 2xy = 4^2 - 2 \cdot 3 = 16 - 6 = 10$$

91. a. $x + \dfrac{1}{x} = 3$

Let $a = x$ and $b = \dfrac{1}{x}$. Then $ab = x\left(\dfrac{1}{x}\right) = 1$.

$$x^2 + \frac{1}{x^2} = \left(x + \frac{1}{x}\right)^2 - 2x\left(\frac{1}{x}\right) = 3^2 - 2 = 9 - 2 = 7$$

b. Let $a = x^2$ and $b = \dfrac{1}{x^2}$. Using the result from part a, we have

$$x^4 + \frac{1}{x^4} = \left(x^2 + \frac{1}{x^2}\right)^2 - 2x^2 \left(\frac{1}{x^2}\right) = 7^2 - 2 = 49 - 2 = 47$$

93. Let $a = 3x$ and $b = 2y$. Then $ab = 6xy$.
$$9x^2 + 4y^2 = (3x+2y)^2 - 2(6xy)$$
$$= 12^2 - 2 \cdot 6 \cdot 6 = 144 - 72 = 72$$

P.3 Applying the Concepts

95. $-0.025\left(6^2\right) + 0.44(6) + 4.28 = \$6.02$ in 2012 (six years after 2006)

97. $0.1\left(40^2\right) + 40 + 50 = \$250.00$

99. $d = 16\left(5^2\right) + 20(5) = 16(25) + 100 = 500$ feet

101. a. $-x + 22.50$

b. $(30 + 10x)(22.50 - x) = 30(22.50) - 30x + 225x - 10x^2 = 675 + 195x - 10x^2 = -10x^2 + 195x + 675$

P.3 Beyond the Basics

103.
$$(a+b+c)^2 = (a+b+c)(a+b+c)$$
$$= a(a+b+c)+b(a+b+c)+c(a+b+c)$$
$$= a^2 + ab + ac + ab + b^2 + bc + ac + bc + c^2$$
$$= a^2 + b^2 + c^2 + 2ab + 2bc + 2ac$$

105.
$$(2x+y-z)^2 = (2x)^2 + y^2 + (-z)^2 + 2(2x)y + 2y(-z) + 2(2x)(-z)$$
$$= 4x^2 + y^2 + z^2 + 4xy - 2yz - 4xz$$
$$(2x-y+z)^2 = (2x)^2 + (-y)^2 + z^2 + 2(2x)(-y) + 2(-y)z + 2(2x)z$$
$$= 4x^2 + y^2 + z^2 - 4xy - 2yz + 4xz$$
$$(2x+y-z)^2 - (2x-y+z)^2 = \left(4x^2 + y^2 + z^2 + 4xy - 2yz - 4xz\right) - \left(4x^2 + y^2 + z^2 - 4xy - 2yz + 4xz\right)$$
$$= 8xy - 8xz$$

Alternatively, recognize the difference of two squares.
$$(2x+y-z)^2 - (2x-y+z)^2 = \left[(2x+y-z)+(2x-y+z)\right]\left[(2x+y-z)-(2x-y+z)\right]$$
$$= (4x)(2y-2z) = 8xy - 8xz$$

107. $x+y+z = 12;\ x^2 + y^2 + z^2 = 44$
$$(x+y+z)^2 = x^2 + y^2 + z^2 + 2xy + 2yz + 2xz = x^2 + y^2 + z^2 + 2(xy+yz+xz)$$
$$12^2 = 44 + 2(xy+yz+xz)$$
$$144 = 44 + 2(xy+yz+xz)$$
$$100 = 2(xy+yz+xz)$$
$$50 = xy+yz+xz$$

109.
$$(a+b+c)\left(a^2+b^2+c^2-ab-bc-ca\right)$$
$$= a\left(a^2+b^2+c^2-ab-bc-ca\right)+b\left(a^2+b^2+c^2-ab-bc-ca\right)+c\left(a^2+b^2+c^2-ab-bc-ca\right)$$
$$= a^3 + ab^2 + ac^2 - a^2b - abc - ca^2 + a^2b + b^3 + bc^2 - ab^2 - b^2c - abc + a^2c + b^2c + c^3 - abc - bc^2 - c^2a$$
$$= a^3 + b^3 + c^3 - 3abc$$

111. Let $a = x - y,\ b = y - z,$ and $c = z - x.$ Then $a + b + c = (x-y)+(y-z)+(z-x) = 0.$

From exercise 110, if $a+b+c = 0,$ then $a^3 + b^3 + c^3 = 3abc.$ So,
$$(x-y)^3 + (y-z)^3 + (z-x)^3 = 3(x-y)(y-z)(z-x).$$

113. From exercise 109, $(a+b+c)\left(a^2+b^2+c^2-ab-bc-ca\right) = a^3 + b^3 + c^3 - 3abc.$

If $a+b+c = 8$ and $ab+bc+ca = 19,$ we have
$$(a+b+c)\left(a^2+b^2+c^2-ab-bc-ca\right) = (a+b+c)\left(a^2+b^2+c^2-(ab+bc+ca)\right) = a^3 + b^3 + c^3 - 3abc \Rightarrow$$
$$8\left(a^2+b^2+c^2-19\right) = a^3 + b^3 + c^3 - 3abc \quad (1)$$

From exercise 103,
$$(a+b+c)^2 = a^2+b^2+c^2+2ab+2bc+2ac = a^2+b^2+c^2+2(ab+bc+ca) \Rightarrow$$
$$8^2 = a^2+b^2+c^2+2(19) \Rightarrow 64 = a^2+b^2+c^2+38 \Rightarrow a^2+b^2+c^2 = 26$$

Substituting into equation (1) gives us $8(26-19) = a^3 + b^3 + c^3 - 3abc \Rightarrow 56 = a^3 + b^3 + c^3 - 3abc.$

P.3 Getting Ready for the Next Section

	a	b	$a+b$	ab
115.	3	4	<u>7</u>	<u>12</u>
117.	<u>4</u>	2	6	8
119.	<u>2</u>	<u>5</u>	7	10
121.	<u>−5</u>	<u>7</u>	2	−35
123.	<u>−2</u>	<u>−3</u>	−5	6

P.4   Factoring Polynomials

P.4 Practice Problems

1. a. $6x^5 + 14x^3 = 2x^3\left(3x^2 + 7\right)$

   b. $7x^5 + 21x^4 + 35x^2 = 7x^2\left(x^3 + 3x^2 + 5\right)$

   c. $5x^2(x-y) + 2(x-y) = (x-y)\left(5x^2 + 2\right)$

2. a $x^2 + 6x + 8 = (x+4)(x+2)$

   b. $x^2 - 3x - 10 = (x-5)(x+2)$

3. a. $x^2 + 4x + 4 = (x+2)^2$

   b. $9x^2 - 6x + 1 = (3x-1)^2$

4. a. $x^2 - 16 = (x-4)(x+4)$

   b. $4x^2 - 25 = (2x-5)(2x+5)$

5. $x^4 - 81 = \left(x^2 - 9\right)\left(x^2 + 9\right)$
   $$= (x-3)(x+3)\left(x^2 + 9\right)$$

6. a. $x^3 - 125 = (x-5)\left(x^2 + 5x + 25\right)$

   b. $27x^3 + 8 = (3x+2)\left(9x^2 - 6x + 4\right)$

7.   Following the reasoning in Example 7, in four years, the company will have invested the intial 12 million dollars, plus an additional 4 million dollars. Thus, the total investment is 16 million dollars. To find the profit or loss with 16 million dollars invested, let $x = 16$ in the profit-loss polynomial:
$$(0.012)(x-14)\left(x^2 + 14x + 196\right)$$
$$= (0.012)(16-14)\left(16^2 + 14\cdot 16 + 196\right)$$
$$= 16.224$$

The company will have made a profit of 16.224 million dollars in four years.

8. a. $x^3 + 3x^2 + x + 3 = x^2(x+3) + 1(x+3)$
   $$= (x+3)\left(x^2 + 1\right)$$

   b. $28x^3 - 20x^2 - 7x + 5$
   $$= 4x^2(7x-5) - (7x-5)$$
   $$= (7x-5)\left(4x^2 - 1\right)$$
   $$= (7x-5)(2x+1)(2x-1)$$

   c. $x^2 - y^2 - 2y - 1 = x^2 - \left(y^2 + 2y + 1\right)$
   $$= x^2 - (y+1)^2$$
   $$= (x-(y+1))(x+(y+1))$$
   $$= (x-y-1)(x+y+1)$$

9. $x^4 + 5x^2 + 9 = x^4 + 6x^2 - x^2 + 9$
   $$= x^4 + 6x^2 + 9 - x^2$$
   $$= \left(x^2 + 3\right)^2 - x^2$$
   $$= \left[\left(x^2 + 3\right) + x\right]\left[\left(x^2 + 3\right) - x\right]$$
   $$= \left(x^2 + x + 3\right)\left(x^2 - x + 3\right)$$

10. a. $5x^2 + 11x + 2 = 5x^2 + 10x + x + 2$
   $$= \left(5x^2 + 10x\right) + (x+2)$$
   $$= 5x(x+2) + 1(x+2)$$
   $$= (5x+1)(x+2)$$

   b. $4x^2 + x - 3 = 4x^2 + 4x - 3x - 3$
   $$= \left(4x^2 + 4x\right) - (3x+3)$$
   $$= 4x(x+1) - 3(x+1)$$
   $$= (4x-3)(x+1)$$

P.4 Concepts and Vocabulary

1.   The polynomials $x + 2$ and $x - 2$ are called <u>factors</u> of the polynomial $x^2 - 4$.

3.   The GCF of the polynomial $10x^3 + 30x^2$ is <u>$10x^2$</u>.

5.   True

7.   False. The polynomial $x^2 - 4$ can be factored as $(x+2)(x-2)$. Therefore, it is not irreducible.

P.4 Building Skills

9. $8x - 24 = 8(x - 3)$

11. $-6x^2 + 12x = -6x(x - 2)$

13. $7x^2 + 14x^3 = 7x^2(1 + 2x)$

15. $x^4 + 2x^3 + x^2 = x^2(x^2 + 2x + 1)$

17. $3x^3 - x^2 = x^2(3x - 1)$

19. $8ax^3 + 4ax^2 = 4ax^2(2x + 1)$

21. $x(x - y) + 3(x - y) = (x + 3)(x - y)$

23. $3x(2x + y) + 5y(2x + y) = (3x + 5y)(2x + y)$

25. $(y + 2)^2 - 3(y + 2) = (y + 2)(y + 2 - 3)$
$ = (y + 2)(y - 1)$

27. $x^3 + 3x^2 + x + 3 = x^2(x + 3) + 1(x + 3)$
$ = (x + 3)(x^2 + 1)$

29. $x^3 - 5x^2 + x - 5 = x^2(x - 5) + 1(x - 5)$
$ = (x - 5)(x^2 + 1)$

31. $6x^3 + 4x^2 + 3x + 2 = 2x^2(3x + 2) + 1(3x + 2)$
$ = (3x + 2)(2x^2 + 1)$

33. $12x^7 + 4x^5 + 3x^4 + x^2$
$= 4x^5(3x^2 + 1) + x^2(3x^2 + 1)$
$= (3x^2 + 1)(4x^5 + x^2)$
$= x^2(3x^2 + 1)(4x^3 + 1)$

35. $xy + ab + bx + ay = (xy + bx) + (ab + ay)$
$= x(y + b) + a(b + y)$
$= (x + a)(b + y)$

37. $x^2 + 7x + 12 = (x + 3)(x + 4)$

39. $x^2 - 6x + 8 = (x - 4)(x - 2)$

41. $x^2 - 3x - 4 = (x - 4)(x + 1)$

43. Irreducible

45. $2x^2 + x - 36 = (2x + 9)(x - 4)$

47. $6x^2 + 17x + 12 = (2x + 3)(3x + 4)$

49. $3x^2 - 11x - 4 = (3x + 1)(x - 4)$

51. Irreducible

53. $x^2 - xy - 20y^2 = (x + 4y)(x - 5y)$

55. $15x^2 - 28 + x = 15x^2 + x - 28$
$= (3x - 4)(5x + 7)$

57. $x^2 + 6x + 9 = (x + 3)^2$

59. $9x^2 + 6x + 1 = (3x + 1)^2$

61. $25x^2 - 20x + 4 = (5x - 2)^2$

63. $49x^2 + 42x + 9 = (7x + 3)^2$

65. $x^2 - 64 = (x - 8)(x + 8)$

67. $4x^2 - 1 = (2x - 1)(2x + 1)$

69. $16x^2 - 9 = (4x - 3)(4x + 3)$

71. $x^4 - 1 = (x^2 - 1)(x^2 + 1) = (x - 1)(x + 1)(x^2 + 1)$

73. $20x^4 - 5 = 5(4x^4 - 1) = 5(2x^2 - 1)(2x^2 + 1)$

75. $x^3 + 64 = x^3 + 4^3 = (x + 4)(x^2 - 4x + 16)$

77. $x^3 - 27 = x^3 - 3^3 = (x - 3)(x^2 + 3x + 9)$

79. $8 - x^3 = 2^3 - x^3 = (2 - x)(4 + 2x + x^2)$

81. $8x^3 - 27 = (2x)^3 - 3^3 = (2x - 3)(4x^2 + 6x + 9)$

83. $40x^3 + 5 = 5(8x^3 + 1) = 5((2x)^3 + 1^3)$
$= 5(2x + 1)(4x^2 - 2x + 1)$

85. $x^4 + x^2 + 25 = x^4 + 10x^2 - 9x^2 + 25$
$= (x^4 + 10x^2 + 25) - 9x^2$
$= (x^2 + 5)^2 - 9x^2$
$= [(x^2 + 5) + 3x][(x^2 + 5) - 3x]$
$= (x^2 + 3x + 5)(x^2 - 3x + 5)$

87. $x^4 + 15x^2 + 64 = x^4 + 16x^2 - x^2 + 64$
$$= \left(x^4 + 16x^2 + 64\right) - x^2$$
$$= \left(x^2 + 8\right)^2 - x^2$$
$$= \left[\left(x^2 + 8\right) + x\right]\left[\left(x^2 + 8\right) - x\right]$$
$$= \left(x^2 + x + 8\right)\left(x^2 - x + 8\right)$$

89. $x^4 - x^2 + 16 = x^4 + 8x^2 - 9x^2 + 16$
$$= \left(x^4 + 8x^2 + 16\right) - 9x^2$$
$$= \left(x^2 + 4\right)^2 - 9x^2$$
$$= \left[\left(x^2 + 4\right) + 3x\right]\left[\left(x^2 + 4\right) - 3x\right]$$
$$= \left(x^2 + 3x + 4\right)\left(x^2 - 3x + 4\right)$$

91. $x^4 + 4 = x^4 + 4x^2 - 4x^2 + 4$
$$= \left(x^4 + 4x^2 + 4\right) - 4x^2$$
$$= \left(x^2 + 2\right)^2 - 4x^2$$
$$= \left[\left(x^2 + 2\right) + 2x\right]\left[\left(x^2 + 2\right) - 2x\right]$$
$$= \left(x^2 + 2x + 2\right)\left(x^2 - 2x + 2\right)$$

93. $1 - 16x^2 = (1 + 4x)(1 - 4x)$

95. $x^2 - 6x + 9 = (x - 3)^2$

97. $4x^2 + 4x + 1 = (2x + 1)^2$

99. $2x^2 - 8x - 10 = 2\left(x^2 - 4x - 5\right)$
$$= 2(x - 5)(x + 1)$$

101. $2x^2 + 3x - 20 = (2x - 5)(x + 4)$

103. $x^2 - 24x + 36$ is irreducible.

105. $3x^5 + 12x^4 + 12x^3 = 3x^3(x^2 + 4x + 4)$
$$= 3x^3(x + 2)^2$$

107. $9x^2 - 1 = (3x - 1)(3x + 1)$

109. $16x^2 + 24x + 9 = (4x + 3)^2$

111. $x^2 + 15$ is irreducible.

113. $45x^3 + 8x^2 - 4x = x\left(45x^2 + 8x - 4\right)$
$$= x(5x + 2)(9x - 2)$$

115. $ax^2 - 7a^2x - 8a^3 = a\left(x^2 - 7ax - 8a^2\right)$
$$= a(x - 8a)(x + a)$$

117. $x^2 - 16a^2 + 8x + 16 = \left(x^2 + 8x + 16\right) - 16a^2$
$$= (x + 4)^2 - 16a^2$$
$$= (x + 4 - 4a)(x + 4 + 4a)$$

119. $3x^5 + 12x^4 y + 12x^3 y^2 = 3x^3\left(x^2 + 4xy + 4y^2\right)$
$$= 3x^3(x + 2y)^2$$

121. $x^2 - 25a^2 + 4x + 4 = \left(x^2 + 4x + 4\right) - 25a^2$
$$= (x + 2)^2 - 25a^2$$
$$= (x + 2 - 5a)(x + 2 + 5a)$$

123. $18x^6 + 12x^5 y + 2x^4 y^2 = 2x^4\left(9x^2 + 6xy + y^2\right)$
$$= 2x^4(3x + y)^2$$

P.4 Applying the Concepts

125.

If one side of the garden is x feet, and the perimeter is 16 feet, then $\dfrac{16 - 2x}{2} = 8 - x$ gives the other dimension of the rectangle. So, the area of the garden is $x(8 - x)$.

Use the figure below for exercises 127 and 128.

127. If x = the length of the cut corner, then $36 - 2x$ = the length of the box, and $16 - 2x$ = the width of the box. The height of the box is x. So,
$$v = x(36 - 2x)(16 - 2x) = x(2(18 - x))(2(8 - x))$$
$$= 4x(18 - x)(8 - x)$$

129. The area of the outside circle = $4\pi \text{ cm}^2$. The area of the inside circle = $\pi x^2 \text{ cm}^2$. So the area of the disk = area of outside circle – area of inside circle = $4\pi - \pi x^2 = \pi(4 - x^2) = \pi(2 - x)(2 + x) \text{ cm}^2$.

131. If one side of the fence is x feet, and the rancher needs a total of 2800 feet of fencing, then the width of the fence is $2800 - 2x$. So, the area of the pen is $x(2800 - 2x) = 2800x - 2x^2 = 2x(1400 - x)\,\text{ft}^2$.

P.4 Beyond the Basics

133. $x^3 + y^3 + x + y = (x+y)\left(x^2 - xy + y^2\right) + x + y$

$\qquad\qquad\qquad = (x+y)\left(x^2 - xy + y^2 + 1\right)$

135. $x^4 + y^4 + x^2 y^2 = x^4 + 2x^2 y^2 + y^4 - x^2 y^2 = \left(x^2 + y^2\right)^2 - x^2 y^2$

$\qquad\qquad\qquad = \left(x^2 + y^2 - xy\right)\left(x^2 + y^2 + xy\right) = \left(x^2 - xy + y^2\right)\left(x^2 + xy + y^2\right)$

137. $x^6 - 64 = \left(x^3\right)^2 - 8^2 = \left(x^3 - 8\right)\left(x^3 + 8\right)$

$\qquad\qquad = (x - 2)(x + 2x + 4)(x + 2)(x - 2x + 4)$

$\qquad\qquad = (x - 2)(x + 2)(x + 2x + 4)(x - 2x + 4)$

139. $\left(x^2 - 3x\right)^2 - 38\left(x^2 - 3x\right) - 80 = \left[\left(x^2 - 3x\right) - 40\right]\left[\left(x^2 - 3x\right) + 2\right]$

$\qquad\qquad\qquad\qquad = \left(x^2 - 3x - 40\right)\left(x^2 - 3x + 2\right)$

$\qquad\qquad\qquad\qquad = (x - 8)(x + 5)(x - 2)(x - 1)$

141. From exercise 109 in Section P.3, we have $(a + b + c)\left(a^2 + b^2 + c^2 - ab - bc - ca\right) = a^3 + b^3 + c^3 - 3abc$.

$\quad$ Then, $a^3 + b^3 + c^3 - 3abc = (a + b + c)\left(a^2 + b^2 + c^2 - ab - bc - ca\right)$

$\qquad\qquad\qquad = \frac{1}{2}(a + b + c) \cdot 2\left(a^2 + b^2 + c^2 - ab - bc - ca\right)$

$\qquad\qquad\qquad = \frac{1}{2}(a + b + c)\left(2a^2 + 2b^2 + 2c^2 - 2ab - 2bc - 2ca\right)$

$\qquad\qquad\qquad = \frac{1}{2}(a + b + c)\left(\left(a^2 - 2ab + b^2\right) + \left(b^2 - 2bc + c^2\right) + \left(c^2 - 2ca + a^2\right)\right)$

$\qquad\qquad\qquad = \frac{1}{2}(a + b + c)\left[(a - b)^2 + (b - c)^2 + (c - a)^2\right]$

$\quad$ If $a^3 + b^3 + c^3 - 3abc = 0$, then either $a + b + c = 0$ or $(a - b)^2 + (b - c)^2 + (c - a)^2 = 0 \Rightarrow a = b = c$.

143. Let $a = z(x - y) = xz - yz$, $b = x(y - z) = xy - xz$, and $c = y(z - x) = yz - xy$. Then

$\quad a + b + c = xz - yz + xy - xz + yz - xy = 0$, and we can apply Section P.3 exercise 110.

$\quad z^3(x - y)^3 + x^3(y - z)^3 + y^3(z - x)^3 = 3xyz(x - y)(y - z)(z - x)$

P.4 Getting Ready for the Next Section

145. $\dfrac{1}{4}+\dfrac{1}{6}=\dfrac{3}{12}+\dfrac{2}{12}=\dfrac{5}{12}$

147. $\dfrac{\overset{1}{\cancel{9}}}{\underset{5}{\cancel{20}}}\cdot\dfrac{\overset{4}{\cancel{16}}}{\underset{3}{\cancel{27}}}=\dfrac{4}{15}$

149. $\dfrac{2x}{4x+6}=\dfrac{2x}{2(2x+3)}=\dfrac{x}{2x+3}$

151. $\dfrac{x^2+3x+2}{x^2+4x+3}=\dfrac{(x+1)(x+2)}{(x+1)(x+3)}=\dfrac{x+2}{x+3}$

P.5   Rational Expressions

P.5 Practice Problems

1. $\dfrac{0.05(1-0.1)}{0.95(0.1)+(0.05)(1-0.1)}\approx 0.32$

The likelihood that a student who tests positive is a nonuser when the test that is used is 95% accurate is about 32%.

2. a. $\dfrac{2x^3+8x^2}{3x^2+12x}=\dfrac{2x^2(x+4)}{3x(x+4)}=\dfrac{2x}{3}$

b. $\dfrac{x^2-4}{x^2+4x+4}=\dfrac{(x-2)(x+2)}{(x+2)^2}=\dfrac{x-2}{x+2}$

3. $\dfrac{\dfrac{x^2-2x-3}{7x^3+28x^2}}{\dfrac{4x+4}{2x^4+8x^3}}=\dfrac{x^2-2x-3}{7x^3+28x^2}\div\dfrac{4x+4}{2x^4+8x^3}$

$=\dfrac{x^2-2x-3}{7x^3+28x^2}\cdot\dfrac{2x^4+8x^3}{4x+4}$

$=\dfrac{(x-3)(x+1)}{7x^2(x+4)}\cdot\dfrac{2x^3(x+4)}{4(x+1)}$

$=\dfrac{x(x-3)}{14}$

4. a. $\dfrac{5x+22}{x^2-36}+\dfrac{2(x+10)}{x^2-36}=\dfrac{5x+22+2x+20}{x^2-36}$

$=\dfrac{7x+42}{x^2-36}$

$=\dfrac{7(x+6)}{(x-6)(x+6)}=\dfrac{7}{x-6}$

b. $\dfrac{4x+1}{x^2+x-12}-\dfrac{3x+4}{x^2+x-12}=\dfrac{(4x+1)-(3x+4)}{x^2+x-12}$

$=\dfrac{x-3}{x^2+x-12}$

$=\dfrac{x-3}{(x+4)(x-3)}$

$=\dfrac{1}{x+4}$

5. a. $\dfrac{2x}{x+2}+\dfrac{3x}{x-5}$

$=\dfrac{2x(x-5)}{(x+2)(x-5)}+\dfrac{3x(x+2)}{(x+2)(x-5)}$

$=\dfrac{2x^2-10x+3x^2+6x}{(x+2)(x-5)}=\dfrac{5x^2-4x}{(x+2)(x-5)}$

$=\dfrac{x(5x-4)}{(x+2)(x-5)}$

b. $\dfrac{5x}{x-4}-\dfrac{2x}{x+3}$

$=\dfrac{5x(x+3)}{(x-4)(x+3)}-\dfrac{2x(x-4)}{(x-4)(x+3)}$

$=\dfrac{5x^2+15x-2x^2+8x}{(x-4)(x+3)}=\dfrac{3x^2+23x}{(x-4)(x+3)}$

$=\dfrac{x(3x+23)}{(x-4)(x+3)}$

6. a. $\dfrac{x^2+3x}{x^2(x+2)^2(x-2)},\dfrac{4x^2+1}{3x(x-2)^2}$

$\text{LCD}=3x^2(x+2)^2(x-2)^2$

b. $\dfrac{2x-1}{x^2-25},\dfrac{3-7x^2}{x^2+4x-5}$

$x^2-25=(x-5)(x+5)$

$x^2+4x-5=(x-1)(x+5)$

$\text{LCD}=(x-1)(x-5)(x+5)$

7. a. $\dfrac{4}{x^2-4x+4}+\dfrac{x}{x^2-4}$

$=\dfrac{4}{(x-2)^2}+\dfrac{x}{(x-2)(x+2)}$

$=\dfrac{4(x+2)}{(x-2)^2(x+2)}+\dfrac{x(x-2)}{(x-2)^2(x+2)}$

$=\dfrac{(4x+8)+(x^2-2x)}{(x-2)^2(x+2)}=\dfrac{x^2+2x+8}{(x-2)^2(x+2)}$

b. $\dfrac{2x}{3(x-5)^2} - \dfrac{6x}{2\left(x^2-5x\right)}$

$$= \dfrac{2x}{3(x-5)^2} - \dfrac{6x}{2x(x-5)}$$

$$= \dfrac{2x(2x)}{3\cdot 2x(x-5)^2} - \dfrac{6x(3)(x-5)}{3\cdot 2x(x-5)^2}$$

$$= \dfrac{4x^2-18x^2+90x}{6x(x-5)^2} = \dfrac{-14x^2+90x}{6x(x-5)^2}$$

$$= \dfrac{2x(-7x+45)}{6x(x-5)^2} = \dfrac{-7x+45}{3(x-5)^2}$$

8. $\dfrac{\dfrac{5}{3x}+\dfrac{1}{3}}{\dfrac{x^2-25}{3x}} = \dfrac{\left(\dfrac{5}{3x}+\dfrac{1}{3}\right)\cdot 3x}{\left(\dfrac{x^2-25}{3x}\right)\cdot 3x} = \dfrac{5+x}{x^2-25}$

$$= \dfrac{x+5}{(x-5)(x+5)} = \dfrac{1}{x-5}$$

9. $\dfrac{5x}{3x-\dfrac{4}{2-\dfrac{1}{3}}} = \dfrac{5x}{3x-\dfrac{4}{\dfrac{5}{3}}} = \dfrac{5x}{3x-4\left(\dfrac{3}{5}\right)} = \dfrac{5x}{3x-\dfrac{12}{5}}$

$$= \dfrac{5x\cdot 5}{\left(3x-\dfrac{12}{5}\right)\cdot 5} = \dfrac{25x}{15x-12}$$

P.5 Concepts and Vocabulary

1. The least common denominator of two rational expressions is the polynomial of least degree that contains <u>each denominator</u> as a factor.

3. If the denominators of two rational expressions are x^2-2x and x^2-x-2, then the LCD is $\underline{x(x-2)(x+1)}$.

5. False

7. False

P.5 Building Skills

9. $\dfrac{2x+2}{x^2+2x+1} = \dfrac{2(x+1)}{(x+1)^2} = \dfrac{2}{x+1}, x\neq -1$

11. $\dfrac{3x+3}{x^2-1} = \dfrac{3(x+1)}{(x-1)(x+1)} = \dfrac{3}{x-1}, x\neq -1, x\neq 1$

13. $\dfrac{2x-6}{9-x^2} = \dfrac{2(x-3)}{(3-x)(3+x)} = \dfrac{-2(3-x)}{(3-x)(3+x)}$

$$= -\dfrac{2}{3+x}, x\neq -3, x\neq 3$

15. $\dfrac{2x-1}{1-2x} = \dfrac{-1(1-2x)}{1-2x} = -1, x\neq \dfrac{1}{2}$

17. $\dfrac{x^2-6x+9}{4x-12} = \dfrac{(x-3)^2}{4(x-3)} = \dfrac{x-3}{4}, x\neq 3$

19. $\dfrac{7x^2+7x}{x^2+2x+1} = \dfrac{7x(x+1)}{(x+1)^2} = \dfrac{7x}{x+1}, x\neq -1$

21. $\dfrac{x^2-11x+10}{x^2+6x-7} = \dfrac{(x-10)(x-1)}{(x+7)(x-1)}$

$$= \dfrac{x-10}{x+7}, x\neq -7, x\neq 1$

23. $\dfrac{6x^4+14x^3+4x^2}{6x^4-10x^3-4x^2} = \dfrac{2x^2\left(3x^2+7x+2\right)}{2x^2\left(3x^2-5x-2\right)}$

$$= \dfrac{2x^2(3x+1)(x+2)}{2x^2(3x+1)(x-2)}$$

$$= \dfrac{x+2}{x-2}, x\neq -\dfrac{1}{3}, x\neq 0, x\neq 2$

25. $\dfrac{x-3}{2x+4}\cdot\dfrac{10x+20}{5x-15} = \dfrac{x-3}{2(x+2)}\cdot\dfrac{10(x+2)}{5(x-3)} = 1$

27. $\dfrac{2x+6}{4x-8}\cdot\dfrac{x^2+x-6}{x^2-9} = \dfrac{2(x+3)}{4(x-2)}\cdot\dfrac{(x+3)(x-2)}{(x+3)(x-3)}$

$$= \dfrac{x+3}{2(x-3)}$

29. $\dfrac{x^2-7x}{x^2-6x-7}\cdot\dfrac{x^2-1}{x^2}$

$$= \dfrac{x(x-7)}{(x+1)(x-7)}\cdot\dfrac{(x+1)(x-1)}{x^2} = \dfrac{x-1}{x}$

31. $\dfrac{x^2-x-6}{x^2+3x+2}\cdot\dfrac{x^2-1}{x^2-9}$

$$= \dfrac{(x-3)(x+2)}{(x+1)(x+2)}\cdot\dfrac{(x-1)(x+1)}{(x-3)(x+3)} = \dfrac{x-1}{x+3}$

33. $\dfrac{2-x}{x+1}\cdot\dfrac{x^2+3x+2}{x^2-4}$

$$= \dfrac{-1(x-2)}{x+1}\cdot\dfrac{(x+1)(x+2)}{(x-2)(x+2)} = -1$

35. $\dfrac{x+2}{6}\div\dfrac{4x+8}{9} = \dfrac{x+2}{6}\cdot\dfrac{9}{4(x+2)} = \dfrac{3}{8}$

37. $\dfrac{x^2-9}{x} \div \dfrac{2x+6}{5x^2} = \dfrac{(x-3)(x+3)}{x} \cdot \dfrac{5x^2}{2(x+3)} = \dfrac{5x(x-3)}{2}$

39. $\dfrac{x^2+2x-3}{x^2+8x+16} \div \dfrac{x-1}{3x+12} = \dfrac{(x-1)(x+3)}{(x+4)^2} \cdot \dfrac{3(x+4)}{x-1} = \dfrac{3(x+3)}{x+4}$

41. $\left(\dfrac{x^2-9}{x^3+8} \div \dfrac{x+3}{x^3+2x^2-x-2}\right)\left(\dfrac{1}{x^2-1}\right) = \dfrac{(x-3)(x+3)}{(x+2)(x^2-2x+4)} \cdot \dfrac{(x^2-1)(x+2)}{x+3} \cdot \dfrac{1}{x^2-1} = \dfrac{x-3}{x^2-2x+4}$

43. $\dfrac{x}{5} + \dfrac{3}{5} = \dfrac{x+3}{5}$

45. $\dfrac{x}{2x+1} + \dfrac{4}{2x+1} = \dfrac{x+4}{2x+1}$

47. $\dfrac{x^2}{x+1} - \dfrac{x^2-1}{x+1} = \dfrac{x^2-(x^2-1)}{x+1} = \dfrac{1}{x+1}$

49. $\dfrac{4}{3-x} + \dfrac{2x}{x-3} = \dfrac{4}{-1(x-3)} + \dfrac{2x}{x-3} = \dfrac{-4+2x}{x-3} = \dfrac{2(x-2)}{x-3}$

51. $\dfrac{5x}{x^2+1} + \dfrac{2x}{x^2+1} = \dfrac{7x}{x^2+1}$

53. $\dfrac{7x}{2(x-3)} + \dfrac{x}{2(x-3)} = \dfrac{8x}{2(x-3)} = \dfrac{4x}{x-3}$

55. $\dfrac{x}{x^2-4} - \dfrac{2}{x^2-4} = \dfrac{x-2}{(x-2)(x+2)} = \dfrac{1}{x+2}$

57. $\dfrac{x-2}{2x+1} - \dfrac{x}{2x-1} = \dfrac{(x-2)(2x-1)}{(2x+1)(2x-1)} - \dfrac{x(2x+1)}{(2x+1)(2x-1)} = \dfrac{(2x^2-5x+2)-(2x^2+x)}{(2x+1)(2x-1)}$

$= \dfrac{-6x+2}{(2x+1)(2x-1)} = \dfrac{2(1-3x)}{(2x+1)(2x-1)}$

59. $\dfrac{-x}{x+2} + \dfrac{x-2}{x} - \dfrac{x}{x-2} = \dfrac{-x^2(x-2)}{x(x+2)(x-2)} + \dfrac{(x-2)^2(x+2)}{x(x+2)(x-2)} - \dfrac{x^2(x+2)}{x(x+2)(x-2)}$

$= \dfrac{(-x^3+2x^2)}{x(x+2)(x-2)} + \dfrac{(x^3-2x^2-4x+8)}{x(x+2)(x-2)} + \dfrac{-(x^3+2x^2)}{x(x+2)(x-2)}$

$= \dfrac{-x^3-2x^2-4x+8}{x(x+2)(x-2)} = -\dfrac{x^3+2x^2+4x-8}{x(x+2)(x-2)}$

In exercises 61–68, to find the LCD, first factor each denominator and then multiply each prime factor the greatest number of times it appears as a factor.

61. The denominators are $3x-6 = 3(x-2)$ and $4x-8 = 4(x-2) \Rightarrow$ LCD $= 3\cdot4(x-2) = 12(x-2)$

63. The denominators are $4x^2-1 = (2x+1)(2x-1)$ and $(2x+1)^2 = (2x+1)(2x+1) \Rightarrow$ LCD $= (2x-1)(2x+1)^2$

65. The denominators are $x^2+3x+2 = (x+1)(x+2)$ and $x^2-1 = (x-1)(x+1) \Rightarrow$ LCD $= (x-1)(x+1)(x+2)$

67. The denominators are $x^2-5x+4 = (x-1)(x-4)$ and

$x^2+x-2 = (x-1)(x+2) \Rightarrow$ LCD $= (x-1)(x-4)(x+2)$

69. $\dfrac{5}{x-3}+\dfrac{2x}{x^2-9}=\dfrac{5}{x-3}+\dfrac{2x}{(x-3)(x+3)}=\dfrac{5(x+3)}{(x-3)(x+3)}+\dfrac{2x}{(x-3)(x+3)}=\dfrac{5x+15+2x}{(x-3)(x+3)}=\dfrac{7x+15}{(x-3)(x+3)}$

71. $\dfrac{2x}{x^2-4}-\dfrac{x}{x+2}=\dfrac{2x}{(x-2)(x+2)}-\dfrac{x}{x+2}=\dfrac{2x}{(x-2)(x+2)}-\dfrac{x(x-2)}{(x+2)(x-2)}$

$\qquad=\dfrac{2x-(x^2-2x)}{(x-2)(x+2)}=\dfrac{-x^2+4x}{(x-2)(x+2)}=\dfrac{x(4-x)}{(x-2)(x+2)}$

73. $\dfrac{x-2}{x^2+3x-10}+\dfrac{x+3}{x^2+x-6}=\dfrac{x-2}{(x-2)(x+5)}+\dfrac{x+3}{(x-2)(x+3)}=\dfrac{1}{x+5}+\dfrac{1}{x-2}$

$\qquad=\dfrac{x-2}{(x+5)(x-2)}+\dfrac{x+5}{(x+5)(x-2)}=\dfrac{2x+3}{(x-2)(x+5)}$

75. $\dfrac{2x-3}{9x^2-1}+\dfrac{4x-1}{(3x-1)^2}=\dfrac{2x-3}{(3x-1)(3x+1)}+\dfrac{4x-1}{(3x-1)^2}=\dfrac{(2x-3)(3x-1)}{(3x+1)(3x-1)^2}+\dfrac{(4x-1)(3x+1)}{(3x+1)(3x-1)^2}$

$\qquad=\dfrac{(6x^2-11x+3)+(12x^2+x-1)}{(3x+1)(3x-1)^2}=\dfrac{18x^2-10x+2}{(3x+1)(3x-1)^2}=\dfrac{2(9x^2-5x+1)}{(3x+1)(3x-1)^2}$

77. $\dfrac{x-3}{x^2-25}-\dfrac{x-3}{x^2+9x+20}=\dfrac{x-3}{(x-5)(x+5)}-\dfrac{x-3}{(x+4)(x+5)}=\dfrac{(x-3)(x+4)}{(x-5)(x+5)(x+4)}-\dfrac{(x-3)(x-5)}{(x+4)(x+5)(x-5)}$

$\qquad=\dfrac{(x^2+x-12)-(x^2-8x+15)}{(x-5)(x+5)(x+4)}=\dfrac{9x-27}{(x-5)(x+5)(x+4)}=\dfrac{9(x-3)}{(x-5)(x+5)(x+4)}$

79. $\dfrac{3}{x^2-4}+\dfrac{1}{2-x}-\dfrac{1}{2+x}=-\dfrac{3}{4-x^2}+\dfrac{1}{2-x}-\dfrac{1}{2+x}=-\dfrac{3}{(2-x)(2+x)}+\dfrac{1}{2-x}-\dfrac{1}{2+x}$

$\qquad=\dfrac{-3}{(2-x)(2+x)}+\dfrac{1(2+x)}{(2-x)(2+x)}-\dfrac{1(2-x)}{(2+x)(2-x)}=\dfrac{-3+(2+x)-(2-x)}{(2-x)(2+x)}=\dfrac{2x-3}{(2-x)(2+x)}$

81. $\dfrac{x+3a}{x-5a}-\dfrac{x+5a}{x-3a}=\dfrac{(x+3a)(x-3a)}{(x-5a)(x-3a)}-\dfrac{(x+5a)(x-5a)}{(x-3a)(x-5a)}=\dfrac{(x^2-9a^2)-(x^2-25a^2)}{(x-5a)(x-3a)}=\dfrac{16a^2}{(x-5a)(x-3a)}$

83. $\dfrac{1}{x+h}-\dfrac{1}{x}=\dfrac{x}{x(x+h)}-\dfrac{1(x+h)}{x(x+h)}$

$\qquad=\dfrac{x-(x+h)}{x(x+h)}=-\dfrac{h}{x(x+h)}$

85. $\dfrac{\dfrac{2}{x}}{\dfrac{3}{x^2}}=\dfrac{2}{x}\cdot\dfrac{x^2}{3}=\dfrac{2x}{3}$

87. $\dfrac{\dfrac{1}{x}}{1-\dfrac{1}{x}}=\dfrac{1}{x}\div\left(1-\dfrac{1}{x}\right)=\dfrac{1}{x}\div\left(\dfrac{x-1}{x}\right)$

$\qquad=\dfrac{1}{x}\cdot\dfrac{x}{x-1}=\dfrac{1}{x-1}$

We use Method 2 for exercises 89– 91.

89. $\dfrac{\dfrac{1}{x}-1}{\dfrac{1}{x}+1}=\dfrac{\left(\dfrac{1}{x}-1\right)x}{\left(\dfrac{1}{x}+1\right)x}=\dfrac{1-x}{1+x}$

91. $\dfrac{\dfrac{1}{x}-x}{1-\dfrac{1}{x^2}}=\dfrac{\left(\dfrac{1}{x}-x\right)x^2}{\left(1-\dfrac{1}{x^2}\right)x^2}=\dfrac{x-x^3}{x^2-1}=\dfrac{x\left(1-x^2\right)}{x^2-1}$

$\qquad=-x,\ x\neq0$

93. $x - \dfrac{x}{x+\dfrac{1}{2}} = x - \dfrac{x}{\dfrac{2x+1}{2}} = x - \left(x \cdot \dfrac{2}{2x+1}\right) = x - \dfrac{2x}{2x+1} = \dfrac{x(2x+1)}{2x+1} - \dfrac{2x}{2x+1}$

$$= \dfrac{2x^2 + x - 2x}{2x+1} = \dfrac{2x^2 - x}{2x+1} = \dfrac{x(2x-1)}{2x+1}$$

95. $\dfrac{\dfrac{1}{x+h} - \dfrac{1}{x}}{h} = \left(\dfrac{1}{x+h} - \dfrac{1}{x}\right) \div h = \left(\dfrac{x}{x(x+h)} - \dfrac{x+h}{x(x+h)}\right) \div h$

$$= \dfrac{x-(x+h)}{x(x+h)} \div h = \dfrac{-h}{x(x+h)} \div h = \dfrac{-h}{x(x+h)} \cdot \dfrac{1}{h} = -\dfrac{1}{x(x+h)}$$

97. $\dfrac{\dfrac{1}{x-a} + \dfrac{1}{x+a}}{\dfrac{1}{x-a} - \dfrac{1}{x+a}} = \left(\dfrac{1}{x-a} + \dfrac{1}{x+a}\right) \div \left(\dfrac{1}{x-a} - \dfrac{1}{x+a}\right) = \dfrac{(x+a)+(x-a)}{(x-a)(x+a)} \div \dfrac{(x+a)-(x-a)}{(x-a)(x+a)}$

$$= \dfrac{2x}{(x-a)(x+a)} \cdot \dfrac{(x-a)(x+a)}{2a} = \dfrac{x}{a}$$

P.5 Applying the Concepts

99. $d = 4 \Rightarrow r = 2$. Substituting 2 for r in the formula gives $\dfrac{125.6}{(3.14)(2^2)} = \dfrac{125.6}{12.56} = 10$ cm.

101. a. Originally the citrus extract is 3 out of 100 gallons or $\dfrac{3}{100}$. When x gallons of water are added to the mixture, the total number of gallons in the mixture is $100 + x$. There are still 3 gallons of citrus extract, so the fraction is $\dfrac{3}{100+x}$.

b. If $x = 50$, then $\dfrac{3}{100+x} = \dfrac{3}{150} = 0.02 = 2\%$.

103. a. The volume of a cylinder is given by $V = 120 = \pi r^2 h \Rightarrow h = \dfrac{120}{\pi r^2}$. The cost of each base $= 0.05\pi x^2$; since there are two bases, the total cost of the bases is $0.10\pi x^2$.

The cost of the side is given by $0.01(2\pi rh) = 0.02\pi rh$. Since $r = x$ and $h = \dfrac{120}{\pi r^2}$, this gives

$0.02\pi x\left(\dfrac{120}{\pi x^2}\right) = \dfrac{2.4}{x}$. So the total cost for the container is $0.1\pi x^2 + \dfrac{2.4}{x} = \dfrac{0.1\pi x^3 + 2.4}{x}$.

b. $\dfrac{0.1(3.14)(2^3 + 240)}{2} \approx \$2.46$

P.5 Beyond the Basics

105. $\dfrac{(3x+4)^2 - (2x+3)^2}{(3x+4)^2} = \dfrac{(9x^2 + 24x + 16) - (4x^2 + 12x + 9)}{(3x+4)^2} = \dfrac{5x^2 + 12x + 7}{(3x+4)^2} = \dfrac{(5x+7)(x+1)}{(3x+4)^2}$

107.
$$\frac{\left(x^2-2x+1\right)(2x+2)-\left(x^2+2x+1\right)(2x-2)}{\left(x^2-2x+1\right)^2}=\frac{(x-1)^2(2)(x+1)-(x+1)^2(2)(x-1)}{\left[(x-1)^2\right]^2}$$

$$=\frac{2(x-1)(x+1)\left[(x-1)-(x+1)\right]}{(x-1)^4}$$

$$=\frac{2(x+1)(-2)}{(x-1)^3}=-\frac{4(x+1)}{(x-1)^3}$$

109.
$$\frac{x^2-x-20}{x^2-25}\cdot\frac{x^2-x-2}{x^2+2x-8}\div\frac{x+1}{x^2+5x}=\frac{(x-5)(x+4)}{(x-5)(x+5)}\cdot\frac{(x-2)(x+1)}{(x+4)(x-2)}\cdot\frac{x(x+5)}{x+1}=x$$

111.
$$\frac{1}{a-\dfrac{1}{a+\dfrac{1}{a}}}-\frac{1}{a+\dfrac{1}{a-\dfrac{1}{a}}}=\frac{1}{a-\dfrac{1}{\dfrac{a^2+1}{a}}}-\frac{1}{a+\dfrac{1}{\dfrac{a^2-1}{a}}}=\frac{1}{a-\dfrac{a}{a^2+1}}-\frac{1}{a+\dfrac{a}{a^2-1}}$$

$$=\frac{1}{\dfrac{a^3+a-a}{a^2+1}}-\frac{1}{\dfrac{a^3-a+a}{a^2-1}}=\frac{1}{\dfrac{a^3}{a^2+1}}-\frac{1}{\dfrac{a^3}{a^2-1}}=\frac{a^2+1}{a^3}-\frac{a^2-1}{a^3}=\frac{2}{a^3}$$

113.
$$\left(\frac{x^2}{1-x^4}+\frac{2x^4}{1-x^8}\right)\div\frac{x^2+1}{x}=\left(\frac{x^2}{1-x^4}+\frac{2x^4}{\left(1-x^4\right)\left(1+x^4\right)}\right)\div\frac{x^2+1}{x}=\frac{x^2\left(1+x^4\right)+2x^4}{\left(1-x^4\right)\left(1+x^4\right)}\div\frac{x^2+1}{x}$$

$$=\frac{x^6+2x^4+x^2}{\left(1-x^2\right)\left(1+x^2\right)\left(1+x^4\right)}\cdot\frac{x}{x^2+1}=\frac{x^2\left(x^4+2x^2+1\right)}{\left(1-x^2\right)\left(1+x^2\right)\left(1+x^4\right)}\cdot\frac{x}{x^2+1}$$

$$=\frac{x^2\left(x^2+1\right)^2}{\left(1-x^2\right)\left(1+x^2\right)\left(1+x^4\right)}\cdot\frac{x}{x^2+1}=\frac{x^3}{\left(1-x^2\right)\left(1+x^4\right)}=-\frac{x^3}{\left(x^2-1\right)\left(1+x^4\right)}$$

P.5 Getting Ready for the Next Section

115. True

117. False

$$\sqrt{(-4)^2}=\sqrt{16}=4$$

119. True

P.6 Rational Exponents and Radicals

P.6 Practice Problems

1. a. $\sqrt{144}=\sqrt{12^2}=12$

b. $\sqrt{\dfrac{1}{49}}=\sqrt{\left(\dfrac{1}{7}\right)^2}=\dfrac{1}{7}$

c. $\sqrt{\dfrac{4}{64}}=\sqrt{\dfrac{2^2}{8^2}}=\sqrt{\left(\dfrac{2}{8}\right)^2}=\dfrac{2}{8}=\dfrac{1}{4}$

2. a. $\sqrt{20}=\sqrt{4\cdot5}=\sqrt{4}\sqrt{5}=2\sqrt{5}$

b. $\sqrt{6}\sqrt{8}=\sqrt{6\cdot8}=\sqrt{48}=\sqrt{16\cdot3}$
$=\sqrt{16}\sqrt{3}=4\sqrt{3}$

c. $\sqrt{12x^2}=\sqrt{4\cdot3\cdot x^2}=\sqrt{4}\sqrt{3}\sqrt{x^2}=2|x|\sqrt{3}$

d. $\sqrt{\dfrac{20y^3}{27x^2}}=\dfrac{\sqrt{20y^3}}{\sqrt{27x^2}}=\dfrac{\sqrt{4\cdot5\cdot y^2\cdot y}}{\sqrt{9\cdot3\cdot x^2}}$

$=\dfrac{\sqrt{4}\sqrt{5}\sqrt{y^2}\sqrt{y}}{\sqrt{9}\sqrt{3}\sqrt{x^2}}=\dfrac{2y\sqrt{5y}}{3|x|\sqrt{3}}$

3. Using Bernoulli's equation, the velocity of the water is

$$v=\sqrt{2gh}\approx\sqrt{2\left(10\,\text{m/s}^2\right)(45\,\text{m})}$$

$$=\sqrt{900\ \text{m}^2/\text{s}^2}=30\,\text{m/s}$$

$30\,\text{m/s}\times60\,\text{s/min}\times60\,\text{min/hr}=108{,}000\,\text{m/hr}$

$108\,\text{km/hr}\times0.6214\,\text{mi/km}\approx67\,\text{mi/hr}$

4. a. $\sqrt[3]{-8} = -2$ **b.** $\sqrt[5]{32} = 2$

   c. $\sqrt[4]{81} = 3$

   d. $\sqrt[6]{-4}$ is not a real number

5. a. $\sqrt[3]{72} = \sqrt[3]{8 \cdot 9} = \sqrt[3]{8} \cdot \sqrt[3]{9} = 2\sqrt[3]{9}$

   b. $\sqrt[4]{48a^2} = \sqrt[4]{16 \cdot 3a^2} = 2\sqrt[4]{3a^2}$

   c. $\sqrt[3]{\sqrt{64}} = \sqrt[3]{8} = 2$

6. a. $3\sqrt{12} + 7\sqrt{3} = 3\sqrt{4 \cdot 3} + 7\sqrt{3}$
   $\qquad = 6\sqrt{3} + 7\sqrt{3} = 13\sqrt{3}$

   b. $2\sqrt[3]{135x} - 3\sqrt[3]{40x} = 2\sqrt[3]{27 \cdot 5x} - 3\sqrt[3]{8 \cdot 5x}$
   $\qquad = 2 \cdot 3\sqrt[3]{5x} - 3 \cdot 2\sqrt[3]{5x}$
   $\qquad = 6\sqrt[3]{5x} - 6\sqrt[3]{5x} = 0$

7. a. $\sqrt{3}\sqrt[5]{2} = \sqrt[2]{3}\sqrt[5]{2} = \sqrt[2 \cdot 5]{3^5} \cdot \sqrt[5 \cdot 2]{2^2} = \sqrt[10]{3^5}\sqrt[10]{2^2}$
   $\qquad = \sqrt[10]{2^2 \cdot 3^5} = \sqrt[10]{972}$

   b. The least common multiple of the indices, 6 and 8, is 24.
   $\sqrt[6]{x^5} = \sqrt[6 \cdot 4]{x^{5 \cdot 4}} = \sqrt[24]{x^{20}}$
   $\sqrt[8]{y^7} = \sqrt[8 \cdot 3]{y^{7 \cdot 3}} = \sqrt[24]{y^{21}}$

8. a. $\dfrac{7}{\sqrt{8}} = \dfrac{7}{2\sqrt{2}} = \dfrac{7\sqrt{2}}{2\sqrt{2}\sqrt{2}} = \dfrac{7\sqrt{2}}{2 \cdot 2} = \dfrac{7\sqrt{2}}{4}$

   b. $\dfrac{\sqrt[3]{3}}{\sqrt[3]{4}} = \dfrac{\sqrt[3]{3} \cdot \sqrt[3]{2}}{\sqrt[3]{4} \cdot \sqrt[3]{2}} = \dfrac{\sqrt[3]{6}}{\sqrt[3]{8}} = \dfrac{\sqrt[3]{6}}{2}$

9. a. $\dfrac{2}{\sqrt{7} - \sqrt{5}} = \dfrac{2(\sqrt{7} + \sqrt{5})}{(\sqrt{7} - \sqrt{5})(\sqrt{7} + \sqrt{5})}$
   $\qquad = \dfrac{2(\sqrt{7} + \sqrt{5})}{7 - 5} = \dfrac{2(\sqrt{7} + \sqrt{5})}{2}$
   $\qquad = \sqrt{7} + \sqrt{5}$

   b. $\dfrac{x - 2}{\sqrt{x} + \sqrt{2}} = \dfrac{(x-2)(\sqrt{x} - \sqrt{2})}{(\sqrt{x} + \sqrt{2})(\sqrt{x} - \sqrt{2})}$
   $\qquad = \dfrac{(x-2)(\sqrt{x} - \sqrt{2})}{x - 2}$
   $\qquad = \sqrt{x} - \sqrt{2}$

10. a. $\left(\dfrac{1}{4}\right)^{1/2} = \sqrt{\dfrac{1}{4}} = \dfrac{1}{2}$

   b. $(125)^{1/3} = \sqrt[3]{125} = 5$

c. $(-32)^{1/5} = \sqrt[5]{-32} = -2$

11. a. $(25)^{2/3} = \sqrt[3]{25^2} = \sqrt[3]{5^4} = 5\sqrt[3]{5}$

   b. $-36^{3/2} = -\left(\sqrt{36}\right)^3 = -(6)^3 = -216$

   c. $16^{-5/2} = \dfrac{1}{\left(\sqrt{16}\right)^5} = \dfrac{1}{4^5} = \dfrac{1}{1024}$

   d. $(-36)^{-1/2} = \dfrac{1}{\sqrt{-36}}$
   Not a real number

12. a. $4x^{1/2} \cdot 3x^{1/5} = 12x^{1/2 + 1/5} = 12x^{7/10}$

   b. $\dfrac{25x^{-1/4}}{5x^{1/3}} = 5x^{-1/4 - 1/3} = 5x^{-7/12} = \dfrac{5}{x^{7/12}}$

   c. $\left(x^{2/3}\right)^{-1/5} = x^{-2/15} = \dfrac{1}{x^{2/15}}$

13. a. $\sqrt[6]{x^4} = x^{4/6} = x^{2/3} = \sqrt[3]{x^2}$

   b. $\sqrt[4]{25}\sqrt{5} = \sqrt[4]{5^2}\sqrt{5} = 5^{2/4} \cdot 5^{1/2} = 5^{1/2} \cdot 5^{1/2} = 5$

   c. $\sqrt{\sqrt[3]{x^{12}}} = \sqrt{x^{12/3}} = \sqrt{x^4} = \left(x^4\right)^{1/2} = x^2$

14. $x(x+3)^{-1/2} + (x+3)^{1/2}$
   $\qquad = \dfrac{x}{(x+3)^{1/2}} + (x+3)^{1/2}$
   $\qquad = \dfrac{x}{(x+3)^{1/2}} + \dfrac{(x+3)^{1/2}(x+3)^{1/2}}{(x+3)^{1/2}}$
   $\qquad = \dfrac{x + (x+3)}{(x+3)^{1/2}} = \dfrac{2x+3}{(x+3)^{1/2}}$
   $\qquad = \dfrac{(2x+3)(x+3)^{1/2}}{(x+3)^{1/2}(x+3)^{1/2}} = \dfrac{(2x+3)(x+3)^{1/2}}{x+3}$

P.6 Concepts and Vocabulary

1. Any positive number has <u>two</u> square roots.

3. Radicals that have the same index and the same radicand are called <u>like radicals</u>.

5. False. For all real x, $\sqrt{x^2} = |x|$.

7. True

P.6 Building Skills

9. $\sqrt{64} = \sqrt{8^2} = 8$

11. $\sqrt[3]{64} = \sqrt[3]{4^3} = 4$

13. $\sqrt[3]{-27} = \sqrt[3]{(-3)^3} = -3$

15. $\sqrt[3]{-\dfrac{1}{8}} = \sqrt[3]{\left(-\dfrac{1}{2}\right)^3} = -\dfrac{1}{2}$

17. $\sqrt{(-3)^2} = \sqrt{9} = 3$

19. There is no real number x such that $x^4 = -16$.

21. $-\sqrt[5]{-1} = -\sqrt[5]{(-1)^5} = -(-1) = 1$

23. $\sqrt[5]{(-7)^5} = -7$

25. $\sqrt{32} = \sqrt{16 \cdot 2} = \sqrt{16}\sqrt{2} = 4\sqrt{2}$

27. $\sqrt{18x^2} = \sqrt{9 \cdot 2x^2} = \sqrt{9}\sqrt{2}\sqrt{x^2} = 3x\sqrt{2}$

29. $\sqrt{9x^3} = \sqrt{9 \cdot x^2 \cdot x} = \sqrt{9}\sqrt{x^2}\sqrt{x} = 3x\sqrt{x}$

31. $\sqrt{6x}\sqrt{3x} = \sqrt{18x^2} = \sqrt{9 \cdot 2 \cdot x^2}$
$\qquad = \sqrt{9}\sqrt{2}\sqrt{x^2} = 3x\sqrt{2}$

33. $\sqrt{15x}\sqrt{3x^2} = \sqrt{45x^3} = \sqrt{9 \cdot 5 \cdot x^2 \cdot x}$
$\qquad = \sqrt{9}\sqrt{5}\sqrt{x^2}\sqrt{x} = 3x\sqrt{5x}$

35. $\sqrt[3]{-8x^3} = \sqrt[3]{-8}\sqrt[3]{x^3} = -2x$

37. $\sqrt[3]{-x^6} = -x^2$

38. $\sqrt[3]{-27x^6} = -3x^2$

39. $\sqrt{\dfrac{5}{32}} = \dfrac{\sqrt{5}}{\sqrt{32}} = \dfrac{\sqrt{5}}{\sqrt{16 \cdot 2}} = \dfrac{\sqrt{5}}{4\sqrt{2}}$
$\qquad = \dfrac{\sqrt{5}\sqrt{2}}{4\sqrt{2}\sqrt{2}} = \dfrac{\sqrt{10}}{8}$

41. $\sqrt[3]{\dfrac{8}{x^3}} = \dfrac{\sqrt[3]{8}}{\sqrt[3]{x^3}} = \dfrac{2}{x}$

43. $\sqrt[4]{\dfrac{2}{x^5}} = \dfrac{\sqrt[4]{2}}{\sqrt[4]{x^5}} = \dfrac{\sqrt[4]{2}}{\sqrt[4]{x^4 \cdot x}} = \dfrac{\sqrt[4]{2}}{\sqrt[4]{x^4}\sqrt[4]{x}} = \dfrac{\sqrt[4]{2}}{x\sqrt[4]{x}}$
$\qquad = \dfrac{\sqrt[4]{2}\sqrt[4]{x^3}}{x\sqrt[4]{x}\sqrt[4]{x^3}} = \dfrac{\sqrt[4]{2x^3}}{x\sqrt[4]{x^4}} = \dfrac{\sqrt[4]{2x^3}}{x^2}$

45. $\sqrt{\sqrt{x^8}} = \sqrt{x^4} = x^2$

47. $\sqrt{\dfrac{9x^6}{y^4}} = \dfrac{\sqrt{9x^6}}{\sqrt{y^4}} = \dfrac{3x^3}{y^2}$

49. $2\sqrt{3} + 5\sqrt{3} = 7\sqrt{3}$

51. $6\sqrt{5} - \sqrt{5} + 4\sqrt{5} = 9\sqrt{5}$

53. $\sqrt{98x} - \sqrt{32x} = 7\sqrt{2x} - 4\sqrt{2x} = 3\sqrt{2x}$

55. $\sqrt[3]{24} - \sqrt[3]{81} = \sqrt[3]{8 \cdot 3} - \sqrt[3]{27 \cdot 3}$
$\qquad = 2\sqrt[3]{3} - 3\sqrt[3]{3} = -\sqrt[3]{3}$

57. $\sqrt[3]{3x} - 2\sqrt[3]{24x} + \sqrt[3]{375x}$
$\qquad = \sqrt[3]{3x} - 2\sqrt[3]{8 \cdot 3x} + \sqrt[3]{125 \cdot 3x}$
$\qquad = \sqrt[3]{3x} - 2\sqrt[3]{8}\sqrt[3]{3x} + \sqrt[3]{125} \cdot \sqrt[3]{3x}$
$\qquad = \sqrt[3]{3x} - 2 \cdot 2\sqrt[3]{3x} + 5\sqrt[3]{3x}$
$\qquad = \sqrt[3]{3x} - 4\sqrt[3]{3x} + 5\sqrt[3]{3x} = 2\sqrt[3]{3x}$

59. $\sqrt{2x^5} - 5\sqrt{32x} + \sqrt{18x^3}$
$\qquad = \sqrt{2x^4 x} - 5\sqrt{16 \cdot 2x} + \sqrt{9 \cdot 2x^2 x}$
$\qquad = x^2\sqrt{2x} - 5 \cdot 4\sqrt{2x} + 3x\sqrt{2x}$
$\qquad = x^2\sqrt{2x} - 20\sqrt{2x} + 3x\sqrt{2x}$
$\qquad = \sqrt{2x}\left(x^2 + 3x - 20\right)$

61. $\sqrt{48x^5 y} - 4y\sqrt{3x^3 y} + y\sqrt{3xy^3}$
$\qquad = \sqrt{16 \cdot 3x^4 xy} - 4y\sqrt{3x^2 xy} + y\sqrt{3xy^2 y}$
$\qquad = 4x^2\sqrt{3xy} - 4xy\sqrt{3xy} + y^2\sqrt{3xy}$
$\qquad = \sqrt{3xy}\left(4x^2 - 4xy + y^2\right) = \sqrt{3xy}(2x - y)^2$

63. $\dfrac{2}{\sqrt{3}} = \dfrac{2\sqrt{3}}{\sqrt{3}\sqrt{3}} = \dfrac{2\sqrt{3}}{3}$

65. $\dfrac{7}{\sqrt{15}} = \dfrac{7\sqrt{15}}{\sqrt{15}\sqrt{15}} = \dfrac{7\sqrt{15}}{15}$

67. $\dfrac{1}{\sqrt{2} + x} = \dfrac{1\left(\sqrt{2} - x\right)}{\left(\sqrt{2} + x\right)\left(\sqrt{2} - x\right)} = \dfrac{\sqrt{2} - x}{2 - x^2}$

69. $\dfrac{3}{2 - \sqrt{3}} = \dfrac{3\left(2 + \sqrt{3}\right)}{\left(2 - \sqrt{3}\right)\left(2 + \sqrt{3}\right)} = \dfrac{3\left(2 + \sqrt{3}\right)}{2^2 - \sqrt{3}^2}$
$\qquad = \dfrac{3\left(2 + \sqrt{3}\right)}{4 - 3} = 3\left(2 + \sqrt{3}\right)$

71. $\dfrac{1}{\sqrt{3}+\sqrt{2}} = \dfrac{1\left(\sqrt{3}-\sqrt{2}\right)}{\left(\sqrt{3}+\sqrt{2}\right)\left(\sqrt{3}-\sqrt{2}\right)} = \dfrac{\sqrt{3}-\sqrt{2}}{\sqrt{3}^2-\sqrt{2}^2} = \dfrac{\sqrt{3}-\sqrt{2}}{3-2} = \sqrt{3}-\sqrt{2}$

73. $\dfrac{\sqrt{5}-\sqrt{2}}{\sqrt{5}+\sqrt{2}} = \dfrac{\left(\sqrt{5}-\sqrt{2}\right)\left(\sqrt{5}-\sqrt{2}\right)}{\left(\sqrt{5}+\sqrt{2}\right)\left(\sqrt{5}-\sqrt{2}\right)} = \dfrac{\sqrt{5}^2-2\sqrt{10}+\sqrt{2}^2}{\sqrt{5}^2-\sqrt{2}^2} = \dfrac{7-2\sqrt{10}}{3}$

75. $\dfrac{\sqrt{x+h}-\sqrt{x}}{\sqrt{x+h}+\sqrt{x}} = \dfrac{\left(\sqrt{x+h}-\sqrt{x}\right)\left(\sqrt{x+h}-\sqrt{x}\right)}{\left(\sqrt{x+h}+\sqrt{x}\right)\left(\sqrt{x+h}-\sqrt{x}\right)} = \dfrac{\left(\sqrt{x+h}\right)^2-2\sqrt{x}\sqrt{x+h}+\sqrt{x}^2}{\left(\sqrt{x+h}\right)^2-\sqrt{x}^2}$

$= \dfrac{x+h-2\sqrt{x(x+h)}+x}{x+h-x} = \dfrac{2x+h-2\sqrt{x(x+h)}}{h}$

77. $\dfrac{\sqrt{4+h}-2}{h} = \dfrac{\left(\sqrt{4+h}-2\right)\left(\sqrt{4+h}+2\right)}{h\left(\sqrt{4+h}+2\right)} = \dfrac{4+h-4}{h\left(\sqrt{4+h}+2\right)} = \dfrac{h}{h\left(\sqrt{4+h}+2\right)} = \dfrac{1}{\sqrt{4+h}+2}$

79. $\dfrac{2-\sqrt{4-x}}{x} = \dfrac{\left(2-\sqrt{4-x}\right)\left(2+\sqrt{4-x}\right)}{x\left(2+\sqrt{4-x}\right)} = \dfrac{4-(4-x)}{x\left(2+\sqrt{4-x}\right)} = \dfrac{x}{x\left(2+\sqrt{4-x}\right)} = \dfrac{1}{2+\sqrt{4-x}}$

81. $\dfrac{\sqrt{x}-2}{x-4} = \dfrac{\left(\sqrt{x}-2\right)\left(\sqrt{x}+2\right)}{(x-4)\left(\sqrt{x}+2\right)} = \dfrac{x-4}{(x-4)\left(\sqrt{x}+2\right)} = \dfrac{1}{\sqrt{x}+2}$

83. $\dfrac{\sqrt{x^2+4x}-x}{2} = \dfrac{\left(\sqrt{x^2+4x}-x\right)\left(\sqrt{x^2+4x}+x\right)}{2\left(\sqrt{x^2+4x}+x\right)} = \dfrac{x^2+4x-x^2}{2\left(\sqrt{x^2+4x}+x\right)} = \dfrac{4x}{2\left(\sqrt{x^2+4x}+x\right)} = \dfrac{2x}{\sqrt{x^2+4x}+x}$

85. $25^{1/2} = \sqrt{25} = 5$

87. $(-8)^{1/3} = \sqrt[3]{-8} = -2$

89. $8^{2/3} = \sqrt[3]{8^2} = \sqrt[3]{(2^3)^2} = \sqrt[3]{2^6} = 2^2 = 4$

91. $-25^{-3/2} = -\dfrac{1}{25^{3/2}} = -\dfrac{1}{\sqrt{25^3}} = -\dfrac{1}{\sqrt{(5^2)^3}}$

$= -\dfrac{1}{\sqrt{5^6}} = -\dfrac{1}{5^3} = -\dfrac{1}{125}$

93. $\left(\dfrac{9}{25}\right)^{-3/2} = \left(\dfrac{25}{9}\right)^{3/2} = \sqrt{\left(\dfrac{25}{9}\right)^3} = \dfrac{\sqrt{25^3}}{\sqrt{9^3}}$

$= \dfrac{\sqrt{(5^2)^3}}{\sqrt{(3^2)^3}} = \dfrac{\sqrt{5^6}}{\sqrt{3^6}} = \dfrac{5^3}{3^3} = \dfrac{125}{27}$

95. $x^{1/2} \cdot x^{2/5} = x^{1/2+2/5} = x^{9/10}$

97. $x^{3/5} \cdot x^{-1/2} = x^{3/5-1/2} = x^{1/10}$

99. $\left(8x^6\right)^{2/3} = 8^{2/3} x^{6(2/3)} = \left(2^3\right)^{2/3} x^{6(2/3)}$

$= 2^2 x^4 = 4x^4$

101. $\left(27x^6 y^3\right)^{-2/3} = 27^{-2/3} x^{6(-2/3)} y^{3(-2/3)}$

$= 3^{3(-2/3)} x^{6(-2/3)} y^{3(-2/3)}$

$= 3^{-2} x^{-4} y^{-2}$

$= \dfrac{1}{3^2 x^4 y^2} = \dfrac{1}{9x^4 y^2}$

103. $\dfrac{15x^{3/2}}{3x^{1/4}} = 5x^{3/2-1/4} = 5x^{5/4}$

105. $\left(\dfrac{x^{-1/4}}{y^{-2/3}}\right)^{-12} = \dfrac{x^{(-1/4)(-12)}}{y^{(-2/3)(-12)}} = \dfrac{x^3}{y^8}$

107. $\sqrt[4]{3^2} = 3^{2/4} = 3^{1/2}$

109. $\sqrt[3]{x^9} = x^{9/3} = x^3$

111. $\sqrt[3]{x^6 y^9} = x^{6/3} y^{9/3} = x^2 y^3$

113. $\sqrt[4]{9}\sqrt{3} = \sqrt[4]{3^2}\sqrt{3} = 3^{2/4} \cdot 3^{1/2}$

$= 3^{1/2} \cdot 3^{1/2} = 3^{1/2+1/2} = 3^1 = 3$

115. $\sqrt{\sqrt[3]{x^{10}}} = \sqrt{x^{10/3}} = \left(x^{10/3}\right)^{1/2} = x^{(10/3)(1/2)} = x^{5/3}$

117. $\sqrt[3]{2}\cdot\sqrt[5]{3} = \sqrt[3\cdot5]{2^5}\cdot\sqrt[3\cdot5]{3^3} = \sqrt[15]{32}\cdot\sqrt[15]{27} = \sqrt[15]{32\cdot27} = \sqrt[15]{864}$

119. $\sqrt[3]{x^2}\cdot\sqrt[4]{x^3} = \sqrt[3\cdot4]{\left(x^2\right)^4}\cdot\sqrt[3\cdot4]{\left(x^3\right)^3} = \sqrt[12]{x^8}\cdot\sqrt[12]{x^9} = \sqrt[12]{x^8\cdot x^9} = \sqrt[12]{x^{17}} = \sqrt[12]{x^{12}\cdot x^5} = \sqrt[12]{x^{12}}\cdot\sqrt[12]{x^5} = x\sqrt[12]{x^5}$

121. $\sqrt[9]{2m^2 n}\cdot\sqrt[3]{5m^5 n^2} = \sqrt[9]{2m^2 n}\cdot\sqrt[3\cdot3]{\left(5m^5 n^2\right)^3} = \sqrt[9]{2m^2 n}\cdot\sqrt[9]{5^3\left(m^5\right)^3\left(n^2\right)^3} = \sqrt[9]{2m^2 n}\cdot\sqrt[9]{125m^{15}n^6}$
$= \sqrt[9]{2m^2 n\cdot125m^{15}n^6} = \sqrt[9]{250m^{2+15}n^{1+6}} = \sqrt[9]{250m^{17}n^7} = m\sqrt[9]{250m^8 n^7}$

123. $\sqrt[5]{x^4 y^3}\cdot\sqrt[6]{x^3 y^5} = \sqrt[5\cdot6]{\left(x^4 y^3\right)^6}\cdot\sqrt[5\cdot6]{\left(x^3 y^5\right)^5} = \sqrt[30]{\left(x^4\right)^6\left(y^3\right)^6}\cdot\sqrt[30]{\left(x^3\right)^5\left(y^5\right)^5} = \sqrt[30]{x^{24}y^{18}}\cdot\sqrt[30]{x^{15}y^{25}}$
$= \sqrt[30]{x^{24}y^{18}\cdot x^{15}y^{25}} = \sqrt[30]{x^{24+15}y^{18+25}} = \sqrt[30]{x^{39}y^{43}} = \sqrt[30]{x^{30+9}y^{30+13}}$
$= \sqrt[30]{x^{30}x^9 y^{30}y^{13}} = xy\sqrt[30]{x^9 y^{13}}$

125. $\frac{4}{3}x^{1/3}(2x-3) + 2x^{4/3} = 2x^{1/3}\left(\frac{2}{3}(2x-3)+x\right) = 2x^{1/3}\left(\frac{4x}{3}-2+x\right) = 2x^{1/3}\left(\frac{7x}{3}-2\right) = \frac{2}{3}x^{1/3}(7x-6)$

127. $4(3x+1)^{1/3}(2x-1) + 2(3x+1)^{4/3} = 2(3x+1)^{1/3}\left(2(2x-1)+(3x+1)\right) = 2(3x+1)^{1/3}(4x-2+3x+1)$
$= 2(3x+1)^{1/3}(7x-1)$

129. $3x\left(x^2+1\right)^{1/2}(2x^2-x) + 2\left(x^2+1\right)^{3/2}(4x-1) = \left(x^2+1\right)^{1/2}\left[3x\left(2x^2-x\right)+2\left(x^2+1\right)(4x-1)\right]$
$= \left(x^2+1\right)^{1/2}\left(6x^3-3x^2+8x^3-2x^2+8x-2\right)$
$= \left(x^2+1\right)^{1/2}\left(14x^3-5x^2+8x-2\right)$

P.6 Applying the Concepts

131. Substituting 692 for A in $\sqrt{\frac{4A}{\sqrt{3}}}$ gives $\sqrt{\frac{4(692)}{1.73}} = \sqrt{\frac{2768}{1.73}} = \sqrt{1600} = 40$ cm.

133. Substituting 6 for w in $V = \sqrt{\frac{w}{1.5}}$ gives $V = \sqrt{\frac{6}{1.5}} = \sqrt{4} = 2$ cm/sec.

135. Substituting 1058 for W and 2 for R in $I = \sqrt{\frac{W}{R}}$ gives $I = \sqrt{\frac{1058}{2}} = \sqrt{529} = 23$ amp.

137. Substituting 16.25 for h in $P = 14.7(0.5)^{h/3.25}$ gives $P = 14.7(0.5)^{16.25/3.25} = 14.7(0.5)^5 =$
14.7(0.03125) ≈ 0.46 lb/sq in.

P.6 Beyond the Basics

139. $\dfrac{1}{3-\sqrt{8}} - \dfrac{1}{\sqrt{8}-\sqrt{7}} + \dfrac{1}{\sqrt{7}-\sqrt{6}} - \dfrac{1}{\sqrt{6}-\sqrt{5}} + \dfrac{1}{\sqrt{5}-2}$

$= \dfrac{1(3+\sqrt{8})}{(3-\sqrt{8})(3+\sqrt{8})} - \dfrac{1(\sqrt{8}+\sqrt{7})}{(\sqrt{8}-\sqrt{7})(\sqrt{8}+\sqrt{7})} + \dfrac{1(\sqrt{7}+\sqrt{6})}{(\sqrt{7}-\sqrt{6})(\sqrt{7}+\sqrt{6})} - \dfrac{1(\sqrt{6}+\sqrt{5})}{(\sqrt{6}-\sqrt{5})(\sqrt{6}+\sqrt{5})} + \dfrac{1(\sqrt{5}+2)}{(\sqrt{5}-2)(\sqrt{5}+2)}$

$= \dfrac{3+\sqrt{8}}{9-8} - \dfrac{\sqrt{8}+\sqrt{7}}{8-7} + \dfrac{\sqrt{7}+\sqrt{6}}{7-6} - \dfrac{\sqrt{6}+\sqrt{5}}{6-5} + \dfrac{\sqrt{5}+2}{5-4}$

$= 3+\sqrt{8} - \left(\sqrt{8}+\sqrt{7}\right) + \left(\sqrt{7}+\sqrt{6}\right) - \left(\sqrt{6}+\sqrt{5}\right) + \left(\sqrt{5}+2\right)$

$= 3+\sqrt{8}-\sqrt{8}-\sqrt{7}+\sqrt{7}+\sqrt{6}-\sqrt{6}-\sqrt{5}+\sqrt{5}+2 = 3+2 = 5$

141. a. $\left(2+\sqrt{3}\right) + \dfrac{1}{2+\sqrt{3}} = 2+\sqrt{3} + \dfrac{2-\sqrt{3}}{\left(2+\sqrt{3}\right)\left(2-\sqrt{3}\right)} = 2+\sqrt{3} + \dfrac{2-\sqrt{3}}{4-3} = 2+\sqrt{3}+2-\sqrt{3} = 4$

   b. $\left(2+\sqrt{3}\right)^2 + \dfrac{1}{\left(2+\sqrt{3}\right)^2} = 7+4\sqrt{3} + \dfrac{7-4\sqrt{3}}{\left(7+4\sqrt{3}\right)\left(7-4\sqrt{3}\right)} = 7+4\sqrt{3} + \dfrac{7-4\sqrt{3}}{49-48} = 7+4\sqrt{3}+7-4\sqrt{3} = 14$

143. Convert each radical to a radical with a common index. In this case, convert each radical to a radical with index 6.

$\sqrt{5} = \sqrt[6]{5^3} = \sqrt[6]{125},\ \sqrt[3]{11} = \sqrt[6]{11^2} = \sqrt[6]{121},\ \sqrt[6]{123}$

In increasing order of magnitude, the radicals are: $\sqrt[3]{11},\ \sqrt[6]{123},\ \sqrt{5}$

145. Convert each radical to a radical with a common index. In this case, convert each radical to a radical with index 12.

$\sqrt{3} = \sqrt[12]{3^6} = \sqrt[12]{729},\ \sqrt[3]{7} = \sqrt[12]{7^4} = \sqrt[12]{2401},$

$\sqrt[4]{10} = \sqrt[12]{10^3} = \sqrt[12]{1000}$

In increasing order of magnitude, the radicals are: $\sqrt{3},\ \sqrt[4]{10},\ \sqrt[3]{7}$

147. $\left(3\sqrt{5}+4\sqrt{3}\right)\left(3\sqrt{5}-4\sqrt{3}\right) = \left(3\sqrt{5}\right)^2 - \left(4\sqrt{3}\right)^2 = 45-48 = -3$

149. a. If $\sqrt{c}$ is an irrational number, $a+b\sqrt{c}=0$ only if $a=b=0$.

   b. $a+b\sqrt{c} = x+y\sqrt{c} \Rightarrow a+b\sqrt{c} - \left(x+y\sqrt{c}\right) = 0 \Rightarrow (a-x)+(b-y)\sqrt{c} = 0$

   From part (a), $(a-x)+(b-y)\sqrt{c}=0 \Rightarrow a-x=0$ and $b-y=0 \Rightarrow a=x$ and $b=y$.

151. $\sqrt{16+2\sqrt{55}} = \sqrt{a+\sqrt{c}} = \sqrt{x}+\sqrt{y}$

   From exercise 150, we have $x+y=a$ and $\sqrt{c}=2\sqrt{xy}$. Then $a=16=x+y$ and $\sqrt{c}=\sqrt{55}=\sqrt{xy}$.

   $\begin{cases} 16=x+y \\ 55=xy \end{cases} \Rightarrow \begin{cases} 16-x=y \\ 55=xy \end{cases} \Rightarrow 55 = x(16-x) \Rightarrow -x^2+16x-55=0 \Rightarrow$

   $x^2-16x+55=0 \Rightarrow (x-11)(x-5)=0 \Rightarrow x=11$ or $x=5$

   If $x=11$, $y=5$, and $\sqrt{a+\sqrt{c}} = \sqrt{x}+\sqrt{y} \Rightarrow \sqrt{16+2\sqrt{55}} = \sqrt{11}+\sqrt{5}$.

P.6 Getting Ready for the Next Section

153. $\dfrac{7}{3}+\dfrac{3}{2}=\dfrac{7(2)}{3(2)}+\dfrac{3(3)}{3(2)}=\dfrac{14}{6}+\dfrac{9}{6}=\dfrac{23}{6}$

155. $\dfrac{3}{2}\cdot\dfrac{12}{7}=\dfrac{3\cdot6}{7}=\dfrac{18}{7}$

157. $-7x+2x=-5x$

159. $(4x-5)+(7x-3)=11x-8$

Chapter P Review Exercises

Building Skills

1. a. $\{4\}$

b. $\{0,4\}$

c. $\{-5,0,4\}$

d. $\left\{-5,0,0.2,0.\overline{31},\dfrac{1}{2},4\right\}$

e. $\left\{\sqrt{7},\sqrt{12}\right\}$

f. $\left\{-5,0,0.2,0.\overline{31},\dfrac{1}{2},\sqrt{7},\sqrt{12},4\right\}$

3. Distributive property

5. Multiplicative identity

7. $x\le1$

9. $(0,\infty)$

11. $\big|2-|-3|\big|=|2-3|=|-1|=1$

13. $\big|1-\sqrt{15}\big|=\sqrt{15}-1$

15. $(-4)^3=-64$

17. $2^4-5\cdot3^2=16-5\cdot9=16-45=-29$

19. $-25^{1/2}=-5$

21. $8^{4/3}=\left(2^3\right)^{4/3}=2^{3(4/3)}=2^4=16$

23. $\left(\dfrac{25}{36}\right)^{-3/2}=\left(\dfrac{36}{25}\right)^{3/2}=\dfrac{\left(6^2\right)^{3/2}}{\left(5^2\right)^{3/2}}$
$=\dfrac{6^{2(3/2)}}{5^{2(3/2)}}=\dfrac{6^3}{5^3}=\dfrac{216}{125}$

25. $2\cdot5^3=2\cdot125=250$

27. $\dfrac{21\times10^6}{3\times10^7}=\dfrac{7}{10}=0.7$

29. $\sqrt{5^2}=5$

31. $\sqrt[3]{-125}=\sqrt[3]{(-5)^3}=-5$

33. $64^{-1/3}=\dfrac{1}{64^{1/3}}=\dfrac{1}{\left(4^3\right)^{1/3}}=\dfrac{1}{4}$

35. $\left(\sqrt{3}+2\right)\left(\sqrt{3}-2\right)=\left(\sqrt{3}\right)^2-2^2=3-4=-1$

37. $\dfrac{3^{-2}\cdot7^0}{18^{-1}}=\dfrac{\frac{1}{3^2}\cdot1}{\frac{1}{18}}=\dfrac{1}{9}\cdot18=2$

39. $\dfrac{2-6\sqrt{4}}{4(2)}=\dfrac{2-6(2)}{8}=\dfrac{2-12}{8}=\dfrac{-10}{8}=-\dfrac{5}{4}$

41. $16^{-1/2}=\dfrac{1}{16^{1/2}}=\dfrac{1}{4}$

43. $\left(x^{-2}\right)^{-5}=x^{(-2)(-5)}=x^{10}$

45. $\left(\dfrac{x^{-3}}{y^{-1}}\right)^{-2}=\dfrac{x^{-3(-2)}}{y^{-1(-2)}}=\dfrac{x^6}{y^2}$

47. $\left(16x^{-2/3}y^{-4/3}\right)^{3/2}$
$=16^{3/2}x^{(-2/3)(3/2)}y^{(-4/3)(3/2)}$
$=4^{2(3/2)}x^{-1}y^{-2}=\dfrac{4^3}{xy^2}=\dfrac{64}{xy^2}$

49. $\left(\dfrac{64y^{-9/2}}{x^{-3}}\right)^{-2/3}=\dfrac{4^{3(-2/3)}y^{(-9/2)(-2/3)}}{x^{-3(-2/3)}}$
$=\dfrac{4^{-2}y^3}{x^2}=\dfrac{y^3}{4^2x^2}=\dfrac{y^3}{16x^2}$

51. $\left(7x^{1/4}\right)\left(3x^{3/2}\right)=21x^{1/4+3/2}=21x^{7/4}$

53. $\dfrac{x^5(2x)^3}{4x^3} = \dfrac{x^5(2^3)x^3}{4x^3} = \dfrac{8x^{5+3}}{4x^3} = 2x^{8-3} = 2x^5$

55. $\left(\dfrac{x^2y^{4/3}}{x^{1/3}}\right)^6 = \dfrac{x^{2(6)}y^{(4/3)(6)}}{x^{(1/3)(6)}} = \dfrac{x^{12}y^8}{x^2}$
$\qquad = x^{12-2}y^8 = x^{10}y^8$

57. $\dfrac{\sqrt{64}}{\sqrt{11}} = \dfrac{8}{\sqrt{11}} = \dfrac{8\sqrt{11}}{\sqrt{11}\cdot\sqrt{11}} = \dfrac{8\sqrt{11}}{11}$

59. $7\sqrt{6} - 3\sqrt{24} = 7\sqrt{6} - 3\sqrt{4\cdot6}$
$\qquad = 7\sqrt{6} - 3\sqrt{4}\sqrt{6} = 7\sqrt{6} - 3(2)\sqrt{6}$
$\qquad = 7\sqrt{6} - 6\sqrt{6} = \sqrt{6}$

61. $\sqrt{2x}\sqrt{6x} = \sqrt{12x^2} = \sqrt{4\cdot3x^2}$
$\qquad = \sqrt{4}\sqrt{3}\sqrt{x^2} = 2x\sqrt{3}$

63. $\dfrac{\sqrt{100x^3}}{\sqrt{4x}} = \sqrt{\dfrac{100x^3}{4x}} = \sqrt{25x^2} = 5x$

65. $4\sqrt[3]{135} + \sqrt[3]{40} = 4\sqrt[3]{27\cdot5} + \sqrt[3]{8\cdot5}$
$\qquad = 4\sqrt[3]{27}\sqrt[3]{5} + \sqrt[3]{8}\sqrt[3]{5}$
$\qquad = 4(3)\sqrt[3]{5} + 2\sqrt[3]{5}$
$\qquad = 12\sqrt[3]{5} + 2\sqrt[3]{5} = 14\sqrt[3]{5}$

67. $\dfrac{7}{\sqrt{3}} = \dfrac{7\sqrt{3}}{\sqrt{3}\sqrt{3}} = \dfrac{7\sqrt{3}}{3}$

69. $\dfrac{1-\sqrt{3}}{1+\sqrt{3}} = \dfrac{\left(1-\sqrt{3}\right)\left(1-\sqrt{3}\right)}{\left(1+\sqrt{3}\right)\left(1-\sqrt{3}\right)} = \dfrac{1-2\sqrt{3}+3}{1^2-\left(\sqrt{3}\right)^2}$
$\qquad = \dfrac{4-2\sqrt{3}}{1-3} = \dfrac{4-2\sqrt{3}}{-2} = -2+\sqrt{3}$

71. $3.7 \times (6.23\times10^{12}) = 23.051\times10^{12}$
$\qquad = 2.3051\times10^{13}$

73. $\left(x^3-6x^2+4x-2\right)+\left(3x^3-6x^2+5x-4\right)$
$\qquad = 4x^3-12x^2+9x-6$

75. $\left(4x^4+3x^3-5x^2+9\right)+\left(5x^4+8x^3-7x^2+5\right)$
$\qquad = 9x^4+11x^3-12x^2+14$

77. $(x-12)(x-3) = x^2-3x-12x+36$
$\qquad = x^2-15x+36$

79. $\left(x^5-2\right)\left(x^5+2\right) = \left(x^5\right)^2 - 2^2 = x^{10}-4$

81. $(2x+5)(3x-11) = 6x^2-22x+15x-55$
$\qquad = 6x^2-7x-55$

83. $x^2-3x-10 = (x-5)(x+2)$

85. $24x^2-38x-11 = (6x-11)(4x+1)$

87. $x(x+11)+5(x+11) = (x+11)(x+5)$

89. $x^4-x^3+7x-7 = x^4+7x-x^3-7$
$\qquad = x(x^3+7)-1(x^3+7)$
$\qquad = (x^3+7)(x-1)$

91. $10x^2+23x+12 = (5x+4)(2x+3)$

93. $12x^2+7x-12 = (4x-3)(3x+4)$

95. $4x^2-49 = (2x-7)(2x+7)$

97. $x^2+12+36 = (x+6)^2$

99. $64x^2+48x+9 = (8x+3)^2$

101. $8x^3+27 = (2x+3)(4x^2-6x+9)$

103. $x^3+5x^2-16x-80 = x^3-16x+5x^2-80$
$\qquad = x(x^2-16)+5(x^2-16)$
$\qquad = (x^2-16)(x+5)$
$\qquad = (x-4)(x+4)(x+5)$

105. $\dfrac{4}{x-9}-\dfrac{10}{9-x} = \dfrac{4}{x-9}+\dfrac{10}{x-9} = \dfrac{14}{x-9}$

107. $\dfrac{3x+5}{x^2+14x+48}-\dfrac{3x-2}{x^2+10x+16}$
$\qquad = \dfrac{3x+5}{(x+6)(x+8)}-\dfrac{3x-2}{(x+2)(x+8)}$
$\qquad = \dfrac{(3x+5)(x+2)}{(x+6)(x+8)(x+2)}-\dfrac{(3x-2)(x+6)}{(x+2)(x+6)(x+8)}$
$\qquad = \dfrac{3x^2+11x+10-(3x^2+16x-12)}{(x+2)(x+6)(x+8)}$
$\qquad = \dfrac{-5x+22}{(x+2)(x+6)(x+8)}$

109. $\dfrac{x-1}{x+1}-\dfrac{x+1}{x-1} = \dfrac{(x-1)^2}{(x+1)(x-1)}-\dfrac{(x+1)^2}{(x+1)(x-1)}$
$\qquad = \dfrac{x^2-2x+1-(x^2+2x+1)}{(x+1)(x-1)}$
$\qquad = -\dfrac{4x}{(x+1)(x-1)}$

111. $\dfrac{x-1}{2x-3} \cdot \dfrac{4x^2-9}{2x^2-x-1}$

$= \dfrac{\cancel{x-1}}{\cancel{2x-3}} \cdot \dfrac{(2x+3)\cancel{(2x-3)}}{(2x+1)\cancel{(x-1)}} = \dfrac{2x+3}{2x+1}$

113. $\dfrac{x^2-4}{4x^2-9} \cdot \dfrac{2x^2-3x}{2x+4}$

$= \dfrac{(x-2)\cancel{(x+2)}}{\cancel{(2x-3)}(2x+3)} \cdot \dfrac{x\cancel{(2x-3)}}{2\cancel{(x+2)}} = \dfrac{x(x-2)}{2(2x+3)}$

115. $\dfrac{\dfrac{1}{x^2}-x}{\dfrac{1}{x^2}+x} = \dfrac{\dfrac{1-x^3}{x^2}}{\dfrac{1+x^3}{x^2}} = \dfrac{1-x^3}{x^2} \cdot \dfrac{x^2}{1+x^3}$

$= \dfrac{1-x^3}{1+x^3} = \dfrac{(1-x)(1+x+x^2)}{(1+x)(1-x+x^2)}$

117. $\dfrac{\dfrac{x}{x-3}+x}{\dfrac{x}{3-x}-x} = \dfrac{\dfrac{x+x(x-3)}{x-3}}{\dfrac{x-x(3-x)}{3-x}} = \dfrac{x^2-2x}{x-3} \cdot \dfrac{3-x}{x^2-2x}$

$= \dfrac{x^2-2x}{x-3} \cdot \left(-\dfrac{x-3}{x^2-2x}\right) = -1$

Applying the Concepts

119. Using the Pythagorean Theorem, we have
$c^2 = 20^2 + 21^2 \Rightarrow c^2 = 841 \Rightarrow c = 29$

121. First, find the length of the missing leg:
$20^2 = 12^2 + x^2 \Rightarrow 400 = 144 + x^2 \Rightarrow$
$256 = x^2 \Rightarrow 16 = x$.
Area $= \dfrac{1}{2}bh \Rightarrow A = \dfrac{1}{2}(12)(16) = 96$ in.^{2} .
Perimeter = the sum of the sides $\Rightarrow$
$P = 12 + 16 + 20 = 48$ in.

123. First, find the length of the missing side:
$10^2 = 6^2 + x^2 \Rightarrow 100 = 36 + x^2 \Rightarrow$
$64 = x^2 \Rightarrow 8 = x$. So the area $= (6)(8) = 48$
square feet.

125. $\sqrt{7920(0.7) + 0.7^2} = \sqrt{5544 + 0.49}$
≈ 74.46 miles

127. $\dfrac{36 - 3(2^2)}{8(2)} = \dfrac{36 - 3(4)}{16} = \dfrac{36 - 12}{16} = \dfrac{24}{16} = 1.5$ ft

Chapter P Practice Test

1. $\big|7 - |-3|\big| = |7 - 3| = |4| = 4$

2. $\big|\sqrt{2} - 100\big| = 100 - \sqrt{2}$

3. $\dfrac{5 - 3(-3)}{2} + (5)(-3) = \dfrac{5+9}{2} - 15 = -8$

4. $x \neq -9, x \neq 3 \Rightarrow (-\infty, -9) \cup (-9, 3) \cup (3, \infty)$

5. $\left(\dfrac{-3x^2 y}{x}\right)^3 = \dfrac{(-3)^3 x^{2(3)} y^3}{x^3} = \dfrac{-27x^6 y^3}{x^3}$

$= -27x^{6-3} y^3 = -27x^3 y^3$

6. $\sqrt[3]{-8x^6} = \sqrt[3]{(-2)^3 x^6} = -2x^2$

7. $\sqrt{75x} - \sqrt{27x} = \sqrt{3 \cdot 25x} - \sqrt{3 \cdot 9x}$
$= 5\sqrt{3x} - 3\sqrt{3x} = 2\sqrt{3x}$

8. $-16^{-3/2} = -\dfrac{1}{16^{3/2}} = -\dfrac{1}{(4^2)^{3/2}}$

$= -\dfrac{1}{4^{2(3/2)}} = -\dfrac{1}{4^3} = -\dfrac{1}{64}$

9. $\left(\dfrac{x^{-2} y^4}{25x^3 y^3}\right)^{-1/2}$

$= \left(\dfrac{25x^3 y^3}{x^{-2} y^4}\right)^{1/2} = \dfrac{25^{1/2}\left(x^3\right)^{1/2}\left(y^3\right)^{1/2}}{\left(x^{-2}\right)^{1/2}\left(y^4\right)^{1/2}}$

$= \dfrac{5x^{3/2} y^{3/2}}{x^{-1} y^2} = 5x^{(3/2)-(-1)} y^{(3/2)-2}$

$= 5x^{5/2} y^{-1/2} = \dfrac{5x^{5/2}}{y^{1/2}} = \dfrac{5x^{5/2} y^{1/2}}{y}$

10. $\dfrac{5}{1-\sqrt{3}} = \dfrac{5\left(1+\sqrt{3}\right)}{\left(1-\sqrt{3}\right)\left(1+\sqrt{3}\right)}$

$= \dfrac{5 + 5\sqrt{3}}{1 - 3} = -\dfrac{5 + 5\sqrt{3}}{2}$

11. $4(x^2 - 3x + 2) + 3(5x^2 - 2x + 1)$
$= 4x^2 - 12x + 8 + 15x^2 - 6x + 3$
$= 19x^2 - 18x + 11$

12. $(x-2)(5x-1) = 5x^2 - 11x + 2$

13. $\left(x^2 + 3y^2\right)^2 = x^4 + 6x^2 y^2 + 9y^4$

14.

The total area is πx^2 square feet. The area of the rug without the border is 25π square feet. So, the area of the border = the total area – the area of the rug =

$$\pi x^2 - 25\pi = \pi(x^2 - 25) = \pi(x - 5)(x + 5).$$

15. $x^2 - 5x + 6 = (x - 3)(x - 2)$

16. $9x^2 + 12x + 4 = (3x + 2)^2$

17. $8x^3 - 27 = (2x - 3)(4x^2 + 6x + 9)$

18. $\dfrac{6 - 3x}{x^2 - 4} - \dfrac{12}{2x + 4}$

$$= \dfrac{6 - 3x}{(x - 2)(x + 2)} - \dfrac{12}{2(x + 2)}$$

$$= \dfrac{2(6 - 3x)}{2(x - 2)(x + 2)} - \dfrac{12(x - 2)}{2(x - 2)(x + 2)}$$

$$= \dfrac{12 - 6x - (12x - 24)}{2(x - 2)(x + 2)} = \dfrac{-18x + 36}{2(x - 2)(x + 2)}$$

$$= \dfrac{-18(x - 2)}{2(x - 2)(x + 2)} = -\dfrac{9}{x + 2}$$

19. $\dfrac{\dfrac{1}{x^2}}{9 - \dfrac{1}{x^2}} = \dfrac{\dfrac{1}{x^2}}{\dfrac{9x^2 - 1}{x^2}} = \dfrac{1}{x^2} \cdot \dfrac{x^2}{(3x - 1)(3x + 1)}$

$$= \dfrac{1}{(3x - 1)(3x + 1)}$$

20. $c^2 = 35^2 + 12^2 \Rightarrow c^2 = 1369 \Rightarrow c = 37$

Chapter 1 Equations and Inequalities

1.1 Linear Equations in One Variable

1.1 Practice Problems

1. a. Both sides of the equation $\dfrac{x}{3} - 7 = 5$ are defined for all real numbers, so the domain is $(-\infty, \infty)$.

b. The left side of the equation $\dfrac{2}{2-x} = 4$ is not defined if $x = 2$. The right side of the equation is defined for all real numbers, so the domain is $(-\infty, 2) \cup (2, \infty)$.

c. The left side of the equation $\sqrt{x-1} = 0$ is not defined if $x < 1$. The right side of the equation is defined for all real numbers, so the domain is $[1, \infty)$.

2. $\dfrac{2}{3} - \dfrac{3}{2}x = \dfrac{1}{6} - \dfrac{7}{3}x$

To clear the fractions, multiply both sides of the equation by the LCD, 6.
$$4 - 9x = 1 - 14x$$
$$4 - 9x + 14x = 1 - 14x + 14x$$
$$4 + 5x = 1$$
$$4 + 5x - 4 = 1 - 4$$
$$5x = -3$$
$$\frac{5x}{5} = \frac{-3}{5}$$
$$x = -\frac{3}{5}$$

Solution set: $\left\{ -\dfrac{3}{5} \right\}$

3. $3x - \left[2x - 6(x+1) \right] = -1$
$$3x - (2x - 6x - 6) = -1$$
$$3x - (-4x - 6) = -1$$
$$3x + 4x + 6 = -1$$
$$7x + 6 = -1$$
$$7x + 6 - 6 = -1 - 6$$
$$7x = -7$$
$$\frac{7x}{7} = \frac{-7}{7}$$
$$x = -1$$

Solution set: $\{-1\}$

4. $3x - \left[2x - 6(x+1) \right] = 7x - 1$
$$3x - (2x - 6x - 6) = 7x - 1$$
$$3x - (-4x - 6) = 7x - 1$$
$$3x + 4x + 6 = 7x - 1$$
$$7x + 6 = 7x - 1$$
$$7x + 6 - 7x = 7x - 1 - 7x$$
$$6 = -1$$

Since $6 = -1$ is false, no number satisfies this equation. Thus, the equation is inconsistent, and the solution set is $\varnothing$.

5. $2(3x - 6) + 5 = 12 - (19 - 6x)$
$$6x - 12 + 5 = 12 - 19 + 6x$$
$$6x - 7 = -7 + 6x$$
$$6x - 7 - 6x = -7 + 6x - 6x$$
$$-7 = -7$$
$$-7 + 7 = -7 + 7$$
$$0 = 0$$

The equation $0 = 0$ is always true. Therefore, the original equation is an identity, and the solution set is $(-\infty, \infty)$.

6. Following the reasoning in example 6, we have $x + 2x = 3x$ is the maximum extended length (in feet) of the cord.
$$3x + 7 + 10 = 120$$
$$3x + 17 = 120$$
$$3x + 17 - 17 = 120 - 17$$
$$3x = 103$$
$$\frac{3x}{3} = \frac{103}{3} \Rightarrow x \approx 34.3$$

The cord should be no longer than 34.3 feet.

7. $F = \dfrac{9}{5}C + 32$
$$50 = \frac{9}{5}C + 32$$
$$50 - 32 = \frac{9}{5}C + 32 - 32$$
$$18 = \frac{9}{5}C$$
$$18 \cdot \frac{5}{9} = \frac{5}{9} \cdot \frac{9}{5}C$$
$$10 = C$$

Thus, 50°F converts to 10°C.

8. $P = 2\ell + 2w$

Subtract 2ℓ from both sides.
$$P - 2\ell = 2w$$

Now, divide both sides by 2.
$$\frac{P - 2\ell}{2} = w$$

9. Use the formula $V = hwd$, where V is the volume, h is the height, w is the width, and d is the depth of the suitcase. So, $h = 21$, $w = 13$, and $d = 9$.
$$V = hwd = 21 \cdot 13 \cdot 9 = 2457 \text{ in.}^3$$

1.1 Concepts and Vocabulary

1. The domain of the variable in an equation is the set of all real number for which both sides of the equation are <u>defined</u>.

3. Two equations with the same solution set are called <u>equivalent</u>.

5. False. A linear equation that is equivalent to the equation $0 = 4$ is an inconsistent equation.

7. True

1.1 Building Skills

9. **a.** Substitute 0 for x in the equation $x - 2 = 5x + 6$:
$$0 - 2 = 5(0) + 6 \Rightarrow -2 \neq 6$$
So, 0 is not a solution of the equation.

   b. Substitute -2 for x in the equation $x - 2 = 5x + 6$:
$$-2 - 2 = 5(-2) + 6 \Rightarrow -4 = -10 + 6 \Rightarrow$$
$$-4 = -4$$
So, -2 is a solution of the equation.

11. **a.** Substitute 4 for x in the equation $\dfrac{2}{x} = \dfrac{1}{3} + \dfrac{1}{x+2}$:
$$\frac{2}{4} = \frac{1}{3} + \frac{1}{4+2} \Rightarrow \frac{1}{2} = \frac{1}{3} + \frac{1}{6} \Rightarrow \frac{1}{2} = \frac{1}{2}$$
So, 4 is a solution of the equation.

   b. Substitute 1 for x in the equation $\dfrac{2}{x} = \dfrac{1}{3} + \dfrac{1}{x+2}$:
$$\frac{2}{1} = \frac{1}{3} + \frac{1}{1+2} \Rightarrow 2 = \frac{1}{3} + \frac{1}{3} \Rightarrow 2 \neq \frac{2}{3}$$
So, 1 is not a solution of the equation.

13. **a.** The equation $2x + 3x = 5x$ is an identity, so every real number is a solution of the equation. Thus 157 is a solution of the equation. This can be checked by substituting 157 for x in the equation:
$$2(157) + 3(157) = 5(157) \Rightarrow$$
$$314 + 471 = 785 \Rightarrow 785 = 785$$

   b. The equation $2x + 3x = 5x$ is an identity, so every real number is a solution of the equation. Thus -2046 is a solution of the equation. This can be checked by substituting -2046 for x in the equation:
$$2(-2046) + 3(-2046) = 5(-2046)$$
$$-4092 - 6138 = -10{,}230$$
$$-10{,}230 = -10{,}230$$

15. The left side of the equation $\dfrac{y}{y-1} = \dfrac{3}{y+2}$ is not defined if $y = 1$, and the right side of the equation is not defined if $y = -2$. The domain is $(-\infty, -2) \cup (-2, 1) \cup (1, \infty)$.

17. The left side of the equation $\dfrac{3x}{(x-3)(x-4)} = 2x + 9$ is not defined if $x = 3$ or $x = 4$. The right side is defined for all real numbers. So, the domain is $(-\infty, 3) \cup (3, 4) \cup (4, \infty)$.

19. Substitute 0 for x in $2x + 3 = 5x + 1$. Because $3 \neq 1$, the equation is not an identity.

21. When the terms on the left side of the equation $\dfrac{1}{x} + \dfrac{1}{2} = \dfrac{2+x}{2x}$ are collected, the equation becomes $\dfrac{2+x}{2x} = \dfrac{2+x}{2x}$, which is an identity.

In exercises 23–48, solve the equations using the procedures listed on page 84 in your text: eliminate fractions, simplify, isolate the variable term, combine terms, isolate the variable term, and check the solution.

23.
$$3x + 5 = 14$$
$$3x + 5 - 5 = 14 - 5$$
$$3x = 9$$
$$\frac{3x}{3} = \frac{9}{3}$$
$$x = 3$$
Solution set: $\{3\}$

25.
$$-10x + 12 = 32$$
$$-10x + 12 - 12 = 32 - 12$$
$$-10x = 20$$
$$\frac{-10x}{-10} = \frac{20}{-10}$$
$$x = -2$$
Solution set: $\{-2\}$

27.
$$3 - y = -4$$
$$3 - y - 3 = -4 - 3$$
$$-y = -7 \Rightarrow y = 7$$
Solution set: $\{7\}$

29.
$$7x + 7 = 2(x + 1)$$
$$7x + 7 = 2x + 2$$
$$7x + 7 - 7 = 2x + 2 - 7$$
$$7x = 2x - 5$$
$$7x - 2x = 2x - 5 - 2x$$
$$5x = -5$$
$$\frac{5x}{5} = \frac{-5}{5} \Rightarrow x = -1$$
Solution set: $\{-1\}$

31.
$$3(2 - y) + 5y = 3y$$
$$6 - 3y + 5y = 3y$$
$$6 + 2y = 3y$$
$$6 + 2y - 2y = 3y - 2y \Rightarrow 6 = y$$
Solution set: $\{6\}$

33. $4y - 3y + 7 - y = 2 - (7 - y)$
Distribute -1 to clear the parentheses.
$$7 = 2 - 7 + y$$
$$7 = -5 + y$$
$$7 + 5 = -5 + y + 5 \Rightarrow 12 = y$$
Solution set: $\{12\}$

35. $3(x - 2) + 2(3 - x) = 1$
$$3x - 6 + 6 - 2x = 1 \Rightarrow x = 1$$
Solution set: $\{1\}$

37.
$$2x + 3(x - 4) = 7x + 10$$
$$2x + 3x - 12 = 7x + 10$$
$$5x - 12 = 7x + 10$$
$$5x - 12 + 12 = 7x + 10 + 12$$
$$5x = 7x + 22$$
$$5x - 7x = 7x + 22 - 7x$$
$$-2x = 22$$
$$\frac{-2x}{-2} = \frac{22}{-2} \Rightarrow x = -11$$
Solution set: $\{-11\}$

39. $4[x + 2(3 - x)] = 2x + 1$
Distribute 2 to clear the inner parentheses.
$$4[x + 6 - 2x] = 2x + 1$$
Combine like terms within the brackets.
$$4[6 - x] = 2x + 1$$
Distribute 4 to clear the brackets.
$$24 - 4x = 2x + 1$$
$$24 - 4x - 24 = 2x + 1 - 24$$
$$-4x = 2x - 23$$
$$-4x - 2x = -23$$
$$-6x = -23$$
$$\frac{-6x}{-6} = \frac{-23}{-6} \Rightarrow x = \frac{23}{6}$$
Solution set: $\left\{\dfrac{23}{6}\right\}$

41. $3(4y - 3) = 4[y - (4y - 3)]$
Distribute 3 on the left side and -1 on the right side to clear parentheses.
$$12y - 9 = 4[y - 4y + 3]$$
Combine like terms in the brackets.
$$12y - 9 = 4[-3y + 3]$$
Distribute 4 to clear the brackets.
$$12y - 9 = -12y + 12$$
$$12y - 9 + 9 = -12y + 12 + 9$$
$$12y = -12y + 21$$
$$12y + 12y = -12y + 21 + 12y$$
$$24y = 21$$
$$\frac{24y}{24} = \frac{21}{24} \Rightarrow y = \frac{21}{24} = \frac{7}{8}$$
Solution set: $\left\{\dfrac{7}{8}\right\}$

43. $2x - 3(2 - x) = (x - 3) + 2x + 1$
Distribute -3 on the left to clear the parentheses.
$$2x - 6 + 3x = x - 3 + 2x + 1$$
$$5x - 6 = 3x - 2$$
$$5x - 6 + 6 = 3x - 2 + 6$$
$$5x = 3x + 4$$
$$5x - 3x = 3x + 4 - 3x$$
$$2x = 4$$
$$\frac{2x}{2} = \frac{4}{2} \Rightarrow x = 2$$
Solution set: $\{2\}$

45.
$$\frac{2x + 1}{9} - \frac{x + 4}{6} = 1$$
To clear the fractions, multiply both sides of the equation by the least common denominator, 36.
$$36\left(\frac{2x + 1}{9} - \frac{x + 4}{6}\right) = 36(1)$$
$$4(2x + 1) - 6(x + 4) = 36$$
$$8x + 4 - 6x - 24 = 36$$
$$2x - 20 = 36$$
$$2x - 20 + 20 = 36 + 20$$
$$2x = 56$$
$$\frac{2x}{2} = \frac{56}{2} \Rightarrow x = 28$$
Solution set: $\{28\}$

47.
$$\frac{1-x}{4}+\frac{5x+1}{2}=3-\frac{2(x+1)}{8}$$
To clear the fractions, multiply both sides by the least common denominator, 8.
$$8\left(\frac{1-x}{4}+\frac{5x+1}{2}\right)=8\left(3-\frac{2(x+1)}{8}\right)$$
Distribute the 8 on both sides.
$$8\left(\frac{1-x}{4}\right)+8\left(\frac{5x+1}{2}\right)=8(3)-8\left(\frac{2(x+1)}{8}\right)$$
$$2(1-x)+4(5x+1)=8(3)-2(x+1)$$
Simplify by collecting like terms and combining constants.
$$2-2x+20x+4=24-2x-2$$
$$18x+6=22-2x$$
$$18x+6+2x=22-2x+2x$$
$$20x+6=22$$
$$20x+6-6=22-6$$
$$20x=16$$
$$\frac{20x}{20}=\frac{16}{20}$$
$$x=\frac{16}{20}=\frac{4}{5}$$
Solution set: $\left\{\frac{4}{5}\right\}$

49. To solve $d=rt$ for r, divide both sides of the equation by t. $r=\frac{d}{t}$.

51. To solve $C=2\pi r$ for r, divide both sides of the equation by 2π. $r=\frac{C}{2\pi}$.

53. To solve $I=\frac{E}{R}$ for R, multiply both sides by R.
$$RI=R\left(\frac{E}{R}\right)\Rightarrow RI=E$$
Divide both sides by I.
$$\frac{RI}{I}=\frac{E}{I}\Rightarrow R=\frac{E}{I}$$

55. To solve $A=\frac{(a+b)h}{2}$ for h, multiply both sides by 2.
$$2A=(a+b)h$$
Divide both sides by $(a+b)$.
$$\frac{2A}{a+b}=\frac{(a+b)h}{a+b}\Rightarrow\frac{2A}{a+b}=h$$

57. To solve $\frac{1}{f}=\frac{1}{u}+\frac{1}{v}$ for u, clear the fractions by multiplying both sides by the least common denominator, fuv.
$$fuv\left(\frac{1}{f}\right)=fuv\left(\frac{1}{u}+\frac{1}{v}\right)$$
$$fuv\left(\frac{1}{f}\right)=fuv\left(\frac{1}{u}\right)+fuv\left(\frac{1}{v}\right)$$
Simplify.
$$uv=fv+fu$$
Subtract fu from both sides.
$$uv-fu=fv+fu-fu$$
$$uv-fu=fv$$
Factor the left side.
$$u(v-f)=fv$$
Divide both sides by $v-f$.
$$\frac{u(v-f)}{v-f}=\frac{fv}{v-f}\Rightarrow u=\frac{fv}{v-f}$$

59. To solve $y=mx+b$ for m, subtract b from both sides.
$$y-b=mx+b-b\Rightarrow y-b=mx$$
Divide both sides by x.
$$\frac{y-b}{x}=\frac{mx}{x}\Rightarrow\frac{y-b}{x}=m$$

1.1 Applying the Concepts

61. The formula for volume is $V=lwh$. Substitute 2808 for V, 18 for l, and 12 for h. Solve for w.

$$2808=18\cdot12\cdot w$$
$$2808=216w$$
$$\frac{2808}{216}=\frac{216w}{216}$$
$$13=w$$
The width of the pool is 13 ft.

63. Let $w=$ the width of the rectangle. Then $2w-5=$ the length of the rectangle.
$$2w+2(2w-5)=80$$
$$2w+4w-10=80$$
$$6w-10=80$$
$$6w=90$$
$$w=15,\ 2w-5=25$$
The width of the rectangle is 15 ft and its length is 25 feet.

65. The formula for circumference of a circle is $C = 2\pi r$. Substitute 114π for C. Solve for r.

$$114\pi = 2\pi r \Rightarrow \frac{114\pi}{2\pi} = \frac{2\pi r}{2\pi} \Rightarrow 57 = r$$

The radius is 57 cm.

67. The formula for surface area of a cylinder is $S = 2\pi rh + 2\pi r^2$. Substitute 6π for S and 1 for r. Solve for h.

$$6\pi = 2\pi(1)h + 2\pi(1^2)$$
$$6\pi = 2\pi h + 2\pi$$
$$6\pi - 2\pi = 2\pi h + 2\pi - 2\pi$$
$$4\pi = 2\pi h$$
$$\frac{4\pi}{2\pi} = \frac{2\pi h}{2\pi} \Rightarrow 2 = h$$

The height is 2 m.

69. The formula for area of a trapezoid is $A = \frac{1}{2}h(b_1 + b_2)$. Substitute 66 for A, 6 for h, and 3 for b_1. Solve for b_2.

$$66 = \frac{1}{2} \cdot 6(3 + b_2)$$
$$66 = 3(3 + b_2)$$
$$66 = 9 + 3b_2$$
$$66 - 9 = 9 + 3b_2 - 9$$
$$57 = 3b_2$$
$$\frac{57}{3} = \frac{3b_2}{3}$$
$$19 = b_2$$

The length of the second base is 19 ft.

71. $F = \frac{9}{5}C + 32 = \frac{9}{5}(20) + 32 = 36 + 32 = 68°$

$20°C$ is equivalent to $68°F$.

73. $C = \frac{5}{9}(F - 32) = \frac{5}{9}(57 - 32) = \frac{5}{9}(25) \approx 13.9°$

$57°F$ is equivalent to approximately $14°C$.

1.1 Beyond the Basics

75. First, solve $7x + 2 = 16$. Subtracting 2 from both sides, we have $7x = 14$. Then divide both sides by 7; we obtain $x = 2$. Now substitute 2 for x in $3x - 1 = k$. This becomes $3(2) - 1 = k$, so $k = 5$.

77. If $k = -2$ then the equation becomes $\frac{3}{y-2} = \frac{4}{y-2}$, which is inconsistent. Note that two fractions with the same denominator are not equal if the numerators are not equal.

In exercises 79–84, assume $a \neq 0$, $b \neq 0$.

79. To solve $a(a + x) = b^2 - bx$ for x, distribute a on the left side of the equation.

$$a^2 + ax = b^2 - bx$$

Add bx to both sides.

$$a^2 + ax + bx = b^2 - bx + bx$$
$$a^2 + ax + bx = b^2$$

Subtract a^2 from both sides.

$$a^2 + ax + bx - a^2 = b^2 - a^2$$
$$ax + bx = b^2 - a^2$$

Factor both sides. The right side is the difference of two squares.

$$x(a + b) = (b - a)(b + a)$$

Divide both sides by $(a + b)$.

$$\frac{x(a+b)}{(a+b)} = \frac{(b-a)(b+a)}{(a+b)}$$
$$x = b - a$$

81. To solve $\frac{ax}{b} - \frac{bx}{a} = \frac{(a+b)^2}{ab}$ for x, multiply both sides by the common denominator, ab.

$$ab\left(\frac{ax}{b} - \frac{bx}{a}\right) = ab\left(\frac{(a+b)^2}{ab}\right)$$
$$a^2x - b^2x = (a+b)^2$$

Factor the left side.

$$x(a^2 - b^2) = (a+b)^2$$

Divide both sides by $(a^2 - b^2)$.

$$\frac{x(a^2-b^2)}{(a^2-b^2)} = \frac{(a+b)^2}{(a^2-b^2)} \Rightarrow x = \frac{(a+b)^2}{(a^2-b^2)}$$

Factor the numerator and denominator on the right side. Then simplify.

$$x = \frac{(a+b)(a+b)}{(a-b)(a+b)}$$
$$x = \frac{a+b}{a-b}$$

83. To solve $\dfrac{b(bx-1)}{a} - \dfrac{a(1+ax)}{b} = 1$ for x,

multiply both sides by the common denominator, ab.

$$ab\left(\dfrac{b(bx-1)}{a} - \dfrac{a(1+ax)}{b}\right) = ab(1)$$

$$b(b(bx-1)) - a(a(1+ax)) = ab$$

$$b^2(bx-1) - a^2(1+ax) = ab$$

$$b^3x - b^2 - a^2 - a^3x = ab$$

Add $b^2 + a^2$ to both sides.

$$b^3x - b^2 - a^2 - a^3x + b^2 + a^2$$
$$= ab + b^2 + a^2$$
$$b^3x - a^3x = a^2 + ab + b^2$$

Factor both sides.

$$x(b-a)(b^2+ab+a^2) = a^2 + ab + b^2$$

Divide both sides by $(b-a)(b^2+ab+a^2)$ and then simplify.

$$\dfrac{x(b-a)(b^2+ab+a^2)}{(b-a)(b^2+ab+a^2)} = $$
$$\dfrac{a^2+ab+b^2}{(b-a)(b^2+ab+a^2)}$$
$$x = \dfrac{1}{b-a}$$

1.1 Critical Thinking/Discussion/Writing

85. a. The solution set of $x^2 = x$ is $\{0,1\}$, while the solution set of $x = 1$ is $\{1\}$. Therefore, the equations are not equivalent.

b. The solution set of $x^2 = 9$ is $\{-3,3\}$, while the solution set of $x = 3$ is $\{3\}$. Therefore, the equations are not equivalent.

c. The solution set of $x^2 - 1 = x - 1$ is $\{0, 1\}$, while the solution set of $x = 0$ is $\{0\}$. Therefore, the equations are not equivalent.

d. The equation $\dfrac{x}{x-2} = \dfrac{2}{x-2}$ is an inconsistent equation, so its solution set is $\varnothing$. The solution set of $x = 2$ is $\{2\}$. Therefore, the equations are not equivalent.

86. In the division step, there is division by zero because $x = 1$. The remainder of the argument is invalid due to the division by zero.

1.1 Getting Ready for the Next Section

87. $(3x+7) - x + 2(5-x) = 3x + 7 - x + 10 - 2x$
$$= 17$$

89. $-2x + 7(x-2) + 4 = -2x + 7x - 14 + 4$
$$= 5x - 10$$

91. $\dfrac{2x}{5} + \dfrac{x}{2} = \dfrac{4x}{10} + \dfrac{5x}{10} = \dfrac{9x}{10}$

93. $\dfrac{3x}{8} - \dfrac{2x}{3} = \dfrac{9x}{24} - \dfrac{16x}{24} = -\dfrac{7x}{24}$

95. 24 minutes $= \dfrac{24}{60} = \dfrac{2}{5} = 0.4$ hour

97. 0.25 hour $= \dfrac{1}{4} \times 60 = 15$ minutes

99. 5 yards $= \underline{5 \times 3 = 15}$ feet

101. 3 inches $= \underline{3 \times 2.54 = 7.62}$ centimeters

1.2 Applications of Linear Equations: Modeling

1.2 Practice Problems

1. Let $x =$ the original price.
Then the discounted price is $x - 0.15x$.
So we have
$$x - 0.15x = 29.75$$
$$0.85x = 29.75$$
$$x = 35$$
The original price of the backpack was \$35.

2. Let $w =$ the width of the rectangle. Then $2w + 5 =$ the length of the rectangle.
$P = 2l + 2w$, so we have
$$28 = 2(2w+5) + 2w$$
$$28 = 4w + 10 + 2w$$
$$28 = 6w + 10$$
$$18 = 6w$$
$$3 = w$$
The width of the rectangle is 3 m and the length is $2(3) + 5 = 11$ m

3. Let $x =$ the amount invested in stocks. Then $15,000 - x =$ the amount invested in bonds.
$$x = 3(15,000 - x)$$
$$x = 45,000 - 3x$$
$$4x = 45,000$$
$$x = 11,250$$
Tyrick invested \$11,250 in stocks and \$15,000 − \$11,250 = \$3,750 in bonds.

4. Let x = the amount of capital. Then $\dfrac{x}{5}$ = the amount invested at 5%, $\dfrac{x}{6}$ = the amount invested at 8%, and $x - \left(\dfrac{x}{5} + \dfrac{x}{6}\right) = \dfrac{19x}{30}$ = the amount invested at 10%.

Principal	Rate	Time	Interest
$\dfrac{x}{5}$	0.05	1	$0.05\left(\dfrac{x}{5}\right)$
$\dfrac{x}{6}$	0.08	1	$0.08\left(\dfrac{x}{6}\right)$
$\dfrac{19x}{30}$	0.1	1	$0.1\left(\dfrac{19x}{30}\right)$

The total interest is $130, so

$$0.05\left(\dfrac{x}{5}\right) + 0.08\left(\dfrac{x}{6}\right) + 0.1\left(\dfrac{19x}{30}\right) = 130$$

Multiply by the LCD, 30.

$$0.3x + 0.4x + 1.9x = 3900$$
$$2.6x = 3900$$
$$x = 1500$$

The total capital is $1500.

5. Let x = the length of the bridge.
Then $x + 130$ = the distance the train travels.
$rt = d$, so
$$25(21) = x + 130 \Rightarrow 525 = x + 130 \Rightarrow 395 = x$$
The bridge is 395 m long.

6. The initial separation between the ship and the aircraft is 955 miles.
Let t = time elapsed when ship and aircraft meet. Then $32t$ = the distance the ship traveled and $350t$ = distance the aircraft traveled.
$$32t + 350t = 955$$
$$382t = 955$$
$$t = 2.5$$
The aircraft and ship meet after 2.5 hours.

7. Let x = the amount of time they worked together to complete the job. Then $\dfrac{1}{45}x$ = the portion of the job done by Jim and $\dfrac{1}{30}x$ = the portion of the job done by Anita.

$$\dfrac{1}{45}x + \dfrac{1}{30}x = 1$$
$$2x + 3x = 90 \qquad \text{Multiply by the LCD, 90.}$$
$$5x = 90$$
$$x = 18$$
It took them 18 minutes to wash the car together.

8. Let x = the amount of 40% sulfuric acid.

Acid	Amount of acid	% acid	Amount of pure acid
40%	x	0.4	$0.4x$
20%	$50 - x$	0.2	$0.2(50 - x)$
25%	50	0.25	$0.25(50)$

Then, we have $0.4x + 0.2(50 - x) = 0.25(50)$.

$$0.4x + 0.2(50 - x) = 0.25(50)$$
$$0.4x + 10 - 0.2x = 12.5$$
$$0.2x + 10 = 12.5$$
$$0.2x = 2.5$$
$$x = 12.5$$

12.5 gallons of 40% sulfuric acid solution should be added.

1.2 Concepts and Vocabulary

1. If the sides of a rectangle are a and b units, the perimeter of the rectangle is $\underline{2a + 2b}$ units.

3. The distance d traveled by an object moving at rate r for time t is given by $d = \underline{rt}$.

5. False. The original price can be found by solving the equation $x - 0.3x = 238$ or $0.7x = 238$.

7. False. The interest $I = (100)(0.05)(3)$.

1.2 Building Skills

9. $0.065x$ 11. $\$22{,}000 - x$

13. a. Natasha: $\dfrac{t}{4}$ b. Natasha's brother: $\dfrac{t}{5}$

15. a. $10 - x$ b. $4.60x$

c. $7.40(10 - x)$

17. a. $8 + x$ b. $0.8 + x$

1.2 Applying the Concepts

19. Let x = the original price. Then,
$x - 0.2x = 1192 \Rightarrow 0.8x = 1192 \Rightarrow x = 1490$
The original price was $1490.

21. Let x = the fourth test score. Then,
$$\frac{81 + 75 + 77 + x}{4} = 80$$
$$233 + x = 320 \Rightarrow x = 87$$
Kylee needs an 87 on the next test to raise the average to 80.

23. Let w = the width of the rectangle.
Then $2w - 5$ = the length of the rectangle.
$$2w + 2(2w - 5) = 80$$
$$2w + 4w - 10 = 80$$
$$6w - 10 = 80$$
$$6w = 90$$
$$w = 15, \ 2w - 5 = 25$$
The width of the rectangle is 15 ft and its length is 25 feet.

25. Substitute 920 for A, 500 for P and 0.06 for r into the formula $A = P + Prt$. Solve for t.
$$920 = 500 + 500 \cdot 0.06t$$
$$920 = 500 + 30t$$
$$920 - 500 = 500 + 30t - 500$$
$$420 = 30t$$
$$\frac{420}{30} = \frac{30t}{30} \Rightarrow 14 = t$$
It will take 14 years until the amount of money available is $920.

27. Substitute 427 for M and 302 for b into the formula $M = \dfrac{m - b}{4}$. Solve for m.
$$427 = \frac{m - 302}{4}$$
$$4(427) = 4\left(\frac{m - 302}{4}\right)$$
$$1708 = m - 302$$
$$1708 + 302 = m - 302 + 302 \Rightarrow 2010 = m$$
The gross monthly income must be $2010.

29. Substitute 170 for P into the formula $P = 200 - 0.02q$. Solve for q.
$$170 = 200 - 0.02q$$
$$170 - 200 = 200 - 0.02q - 200$$
$$-30 = -0.02q$$
$$\frac{-30}{-0.02} = \frac{-0.02q}{-0.02} \Rightarrow 1500 = q$$
Note that the solution must fall between 100 and 2000 cameras. The retailer must order 1500 cameras.

31. Let x = the cost of the less expensive land. Then $x + 23,000$ = the cost of the more expensive land. Together they cost $147,000, so
$$x + (x + 23,000) = 147,000$$
$$2x + 23,000 = 147,000$$
$$2x = 124,000 \Rightarrow x = 62,000$$
The less expensive piece of land costs $62,000 and the more expensive piece of land costs $62,000 + $23,000 = $85,000.

33. Let x = the lottery ticket sales in July. Then $1.10x$ = the lottery ticket sales in August. A total of 1113 tickets were sold, so
$$x + 1.10x = 1113$$
$$2.10x = 1113 \Rightarrow x = 530$$
530 tickets were sold in July, and $1.10(530) = 583$ tickets were sold in August.

35. Let x = the amount the younger son receives. Then $4x$ = the amount the older son receives. Together they receive $225,000, so
$$x + 4x = 225,000 \Rightarrow 5x = 225,000 \Rightarrow$$
$$x = 45,000$$
The younger son will received $45,000, and the older son will receive $4($45,000) = $180,000.

37. a. Let x = the number of points needed to average 75.
$$\frac{87 + 59 + 73 + x}{4} = 75$$
$$219 + x = 300$$
$$x = 81$$
You need to score 81 in order to average 75.

b.
$$\frac{87 + 59 + 73 + 2x}{5} = 75$$
$$219 + 2x = 375$$
$$2x = 156$$
$$x = 78$$
You need to score 78 in order to average 75 if the final carries double weight.

39. Let x = the amount invested in a tax shelter. Then $7000 - x$ = the amount invested in a bank.

Investment	Principal	Rate	Time	Interest
Tax shelter	x	0.09	1	$0.09x$
Bank	$7000 - x$	0.06	1	$0.06(7000 - x)$

The total interest was $540, so
$$0.09x + 0.06(7000 - x) = 540$$
$$0.09x + 420 - 0.06x = 540$$
$$0.03x + 420 = 540$$
$$0.03x = 120 \Rightarrow x = 4000$$
Mr. Mostafa invested $4000 in a tax shelter and $7000 - 4000 = \$3000$ in a bank.

41. Let x = the amount to be invested at 8%.

Principal	Rate	Time	Interest
5000	0.05	1	250
x	0.08	1	$0.08x$
$5000 + x$	0.06	1	$0.06(5000 + x)$

The amount of interest for the total investment is the sum of the interest earned on the individual investments, so
$$0.06(5000 + x) = 250 + 0.08x$$
$$300 + 0.06x = 250 + 0.08x$$
$$50 + 0.06x = 0.08x$$
$$50 = 0.02x \Rightarrow 2500 = x$$
So, $2500 must be invested at 8%.

43. There is a profit of $2 on each shaving set. They want to earn $40,000 + \$30,000 = \$70,000$. Let x = the number of shaving sets to be sold. Then $2x$ = the amount of profit for x shaving sets. So, $2x = 70,000 \Rightarrow x = 35,000$
They must sell 35,000 shaving sets.

45. Let x = the time the second car travels. Then $1 + x$ = the time the first car travels. So,

	Rate	Time	Distance
First car	50	$1 + x$	$50(1 + x)$
Second car	70	x	$70x$

The distances are equal, so
$$50(1 + x) = 70x$$
$$50 + 50x = 70x$$
$$50 = 20x \Rightarrow 2.5 = x$$
So, it will take the second car 2.5 hours to overtake the first car.

47. At 20 miles per hour, it will take Lucas two minutes to bike the remaining 2/3 of a mile.
$$\left(\frac{20\,\text{mi}}{1\,\text{hr}} = \frac{20\,\text{mi}}{60\,\text{min}} = \frac{1\,\text{mi}}{3\,\text{min}} \right)$$ So his brother will have to bike 1 mile in 2 minutes:
$$\frac{1\,\text{mi}}{2\,\text{min}} = \frac{30\,\text{mi}}{60\,\text{min}} = \frac{30\,\text{mi}}{1\,\text{hr}}$$

49. Let x = the rate the slower car travels. Then $x + 7$ = the rate the faster car travels. So,

	Rate	Time	Distance
First car	x	3	$3x$
Second car	$x + 7$	3	$3(x + 7)$

The planes are 621 miles apart, so
$$3x + 3(x + 7) = 621$$
$$3x + 3x + 21 = 621$$
$$6x + 21 = 621 \Rightarrow 6x = 600 \Rightarrow x = 100$$
One car is traveling at 100 miles per hour, and the other car is traveling at 107 miles per hour.

51. Let x = the amount of time it takes both pumps to drain the pool together. The old pump drains 1/6 of the pool in one hour, so it will drain $x/6$ of the pool. The new pump drains 1/4 of the pool in one hour, so it will drain $x/4$ of the pool. So,
$$\frac{x}{6} + \frac{x}{4} = 1$$
$$12\left(\frac{x}{6} + \frac{x}{4} \right) = 12(1)$$
$$2x + 3x = 12$$
$$5x = 12 \Rightarrow x = \frac{12}{5} = 2.4$$
It will take the two pumps 2.4 hours or 2 hours and 24 minutes to drain the pool working together.

53. Let x = the amount of time it takes both blowers to fill the blimp together. The first blower fills 1/6 of the blimp in one hour, so it will fill $x/6$ of the blimp. The second blower fills 1/9 of the blimp in one hour, so it will fill $x/9$ of the blimp. So,
$$\frac{x}{6} + \frac{x}{9} = 1$$
$$36\left(\frac{x}{6} + \frac{x}{9} \right) = 36(1)$$
$$6x + 4x = 36$$
$$10x = 36 \Rightarrow x = 3.6$$

(*continued on next page*)

(*continued*)

It will take the two blowers 3.6 hours or 3 hours and 36 minutes to fill the blimp working together.

55. Let x = the time for the new sorter to complete the job alone. Then $2x$ = the time for the old sorter to complete the job alone.. So, $8/x$ is the portion of the job done by the new sorter and $8/(2x) = 4/x$ is the portion of the job done by the old sorter.

$$\frac{8}{x} + \frac{4}{x} = 1 \Rightarrow x\left(\frac{8}{x} + \frac{4}{x}\right) = x(1) \Rightarrow$$
$$8 + 4 = x \Rightarrow 12 = x$$

It will take 12 hours for the new sorter to complete the job working alone.

57. Let x = the amount of time they worked together to complete the job. Then $x/9$ = the portion of the job done by a professor and $x/6$ = the portion of the job done by a student.

$$3\left(\frac{1}{9}x\right) + 2\left(\frac{1}{6}x\right) = 1$$
$$3\left(\frac{x}{3} + \frac{x}{3}\right) = 1(3)$$
$$2x = 3 \Rightarrow x = 1.5$$

The job will take 1.5 hours.

59. Let x = the number of grams of pure gold to be added. Then $120 + x$ = the number of grams in the new alloy.

Number of grams of gold in the new alloy	=	Number of grams of gold in the original alloy	+	Number of grams of pure gold added.

$$0.8(120 + x) = 0.75(120) + x \Rightarrow 96 + 0.8x = 90 + x \Rightarrow 6 + 0.8x = x \Rightarrow 6 = 0.2x \Rightarrow 30 = x$$

So, 30 grams of pure gold must be added.

61. Let x = the amount of 60-40 solder to be added.

Solder	Amount of solder	% tin	Amount of tin
60-40	x	0.6	$0.6x$
40-60	$600 - x$	0.4	$0.4(600 - x)$
55-45	600	0.55	$0.55(600)$

$$0.6x + 0.4(600 - x) = 0.55(600)$$
$$0.6x + 240 - 0.4x = 330$$
$$0.2x + 240 = 330$$
$$0.2x = 90$$
$$x = 450$$

450 g of 60-40 solder must be added to 150 g of 40-60 solder.

63. Let x = the amount of 60% boric acid

Solution	Amount of solution	% boric acid	Amount of boric acid
60%	x	0.6	$0.6x$
8%	7.5	0.08	$0.08(7.5)$
20%	$x + 7.5$	0.2	$0.2(x + 7.5)$

$$0.6x + 0.08(7.5) = 0.2(x + 7.5)$$
$$0.6x + 0.6 = 0.2x + 1.5$$
$$0.4x = 0.9 \Rightarrow x = 2.25$$

2.25 liters of 60% boric acid solution are needed.

65. Let x = the number of dimes. Then $3x$ = the number of nickels, and $4x$ = the number of quarters.

Coin	Number of coins	Value of each coin	Total
Nickels	$3x$	0.05	$0.15x$
Dimes	x	0.10	$0.10x$
Quarters	$4x$	0.25	x

$$0.15x + 0.10x + x = 96.25$$
$$1.25x = 96.25 \Rightarrow x = 77$$

So, there are 77 dimes, $3(77) = 231$ nickels, and $4(77) = 308$ quarters.

67. Let x = Eric's grandfather's age now. Then $x - 57 =$ Eric's age now. $x + 5 =$ Eric's grandfather's age five years from now, and $(x - 57) + 5 = x - 52 =$ Eric's age five years from now. So,

$$x + 5 = 4(x - 52)$$
$$x + 5 = 4x - 208$$
$$x + 213 = 4x \Rightarrow 213 = 3x \Rightarrow 71 = x$$

Eric's grandfather is 71 years old now.

69. Let x = the length of the tin rectangle. Then $x - 2$ = the length of the box.

The formula for volume is $V = lwh$, so
$$(x - 2)(1)(1) = 2 \Rightarrow x - 2 = 2 \Rightarrow x = 4$$
The length of the tin rectangle is 4 m.

1.2 Beyond the Basics

71. Let x = the average speed for the second half of the trip.

	Rate	Distance	Time
1st half	75	D	$\frac{D}{75}$
2nd half	x	D	$\frac{D}{x}$
Whole trip	60	$2D$	$\frac{2D}{60}$

So,
$$\frac{D}{75} + \frac{D}{x} = \frac{2D}{60}$$
$$300x\left(\frac{D}{75} + \frac{D}{x}\right) = 300x\left(\frac{2D}{60}\right)$$
$$4Dx + 300D = 5x(2D)$$
$$4Dx + 300D = 10Dx$$
$$300D = 6Dx \Rightarrow 50 = x$$

The average speed for the second half of the drive is 50 mph.

73. Let x = the number of liters of water in the original mixture. Then $5x$ = the number of liters of alcohol in the original mixture, and $6x$ = the total number of liters in the original mixture.

$x + 5$ = the number of liters of water in the new mixture. Then $6x + 5$ = the total number of liters in the new mixture. Since the ratio of alcohol to water in the new mixture is 5:2, then the amount of alcohol in the new mixture is 5/7 of the total mixture or $\frac{5}{7}(6x + 5)$.

There was no alcohol added, so the amount of alcohol in the original mixture equals the amount of alcohol in the new mixture. This gives

$$\frac{5}{7}(6x + 5) = 5x$$
$$5(6x + 5) = 35x$$
$$30x + 25 = 35x \Rightarrow 25 = 5x \Rightarrow 5 = x$$

So, there were 5 liters of water in the original mixture and 25 liters of alcohol.

75. Let x = Democritus' age now. Then $x/6$ = the number of years as a boy, $x/8$ = the number of years as a youth, and $x/2$ = the number of years as a man. He has spent 15 years as a mature adult. So,

$$\frac{x}{6} + \frac{x}{8} + \frac{x}{2} + 15 = x$$
$$24\left(\frac{x}{6} + \frac{x}{8} + \frac{x}{2} + 15\right) = 24x$$
$$4x + 3x + 12x + 360 = 24x$$
$$19x + 360 = 24x$$
$$360 = 5x \Rightarrow 72 = x$$

Democritus is 72 years old.

77. There are 180 minutes from 3 p.m. to 6 p.m. So, the number of minutes before 6 p.m. plus 50 minutes plus 4 × the number of minutes before 6 p.m. equals 180 minutes. Let x = the number of minutes before 6 p.m. So,
$$x + 50 + 4x = 180 \Rightarrow 5x + 50 = 180 \Rightarrow$$
$$5x = 130 \Rightarrow x = 26$$
So it is 26 minutes before 6 p.m. or 5:34 p.m. Check this by verifying that $26 + 50 = 76$ minutes before 6 p.m. is the same time as $4(26) = 104$ minutes after 3 p.m. Seventy-six minutes before 6 p.m. is 4:44 p.m., while 104 minutes after 3 p.m. is also 4:44 p.m.

79. a. Because of the head wind, the plane flies at 140 mph from Atlanta to Washington and 160 mph from Washington to Atlanta. Let x = the distance the plane flew before turning back. So,

	Rate	Distance	Time
to	140	x	$x/140$
from	160	x	$x/160$

$$\frac{x}{140} + \frac{x}{160} = 1.5$$
$$160x + 140x = 1.5(140)(160)$$
$$300x = 33,600 \Rightarrow x = 112$$

The plane flew 112 miles before turning back.

b. The plane traveled 224 miles in 1.5 hours, so the average speed is $\frac{224}{1.5} = 149.33$ mph.

81. The ends of the trains are 440 feet apart when the trains first meet. Because the speed is in miles per hour, but the distance is measured in feet, convert the speed to feet per hour. The distance is relatively short, so we need to convert the feet per hour to feet per second.

$$\frac{50 \text{ mi}}{\text{hr}} \cdot \frac{5280 \text{ ft}}{\text{mi}} \cdot \frac{1 \text{ hr}}{60 \text{ min}} \cdot \frac{1 \text{ min}}{60 \text{ sec}} = \frac{220}{3} \text{ feet per second.}$$

$$\frac{60 \text{ mi}}{\text{hr}} \cdot \frac{5280 \text{ ft}}{\text{mi}} \cdot \frac{1 \text{ hr}}{60 \text{ min}} \cdot \frac{1 \text{ min}}{60 \text{ sec}} = 88 \text{ feet per second.}$$

	Rate	Time	Distance
Train A	$\frac{220}{3}$	t	$\frac{220t}{3}$
Train B	88	t	$88t$

230 ft   210 ft
50 mph   60 mph
$\frac{220}{3}$ ft/sec   88 ft/sec

The total distance is 440 feet, so we have

$$\frac{220t}{3} + 88t = 440$$
$$220t + 264t = 1320$$
$$484t = 1320$$
$$t = 2.7$$

The ends of the trains will pass each other after 2.7 seconds.

1.2 Critical Thinking/Discussion/Writing

83. If x represents the amount the pawn shop owner paid for the first watch and the owner made a profit of 10%, then $1.1x = 499$, so $x = 453.64$. If y represents the amount the pawn shop owner paid for the second watch and the owner lost 10%, then $0.9y = 499$, so $y = 554.44$. Together the two watches cost $\$453.64 + \$554.44 = \$1008.08$. But the pawn shop owner sold the two watches for $998, so there was a loss. The amount of loss is $(1008.08 - 998)/1008.08 = 10.08/1008.08 \approx 0.01 = 1\%$. The answer is (C).

84. Let x represent the amount of gasoline used in July. Then $0.8x$ represents the amount of gasoline used in August. Let y represent the price of gasoline in July. Then $1.2y$ represents the cost of gasoline in August. The cost of gasoline used in July is xy (amount × price), and the cost of gasoline used in August is $0.8x \times 1.2y = 0.96xy$. So the cost of gasoline used in August is 96% of the cost of gasoline used in July, which is a decrease of 4%. The answer is (D).

1.2 Getting Ready for the Next Section

85. $\sqrt{8} = \sqrt{4 \cdot 2} = \sqrt{4}\sqrt{2} = 2\sqrt{2}$

87. $\sqrt{12} = \sqrt{4 \cdot 3} = \sqrt{4}\sqrt{3} = 2\sqrt{3}$

89. $\dfrac{2 + 3\sqrt{8}}{2} = \dfrac{2 + 3 \cdot 2\sqrt{2}}{2} = \dfrac{2}{2} + \dfrac{3 \cdot 2\sqrt{2}}{2} = 1 + 3\sqrt{2}$

91. $\dfrac{15 - \sqrt{75}}{5} = \dfrac{15 - 5\sqrt{3}}{5} = \dfrac{15}{5} - \dfrac{5\sqrt{3}}{5} = 3 - \sqrt{3}$

93. $x^2 + x = x(x+1)$

95. $x^2 - 4 = (x-2)(x+2)$

97. $x^2 + 4x + 4 = (x+2)^2$

99. $x^2 - 8x + 7 = (x-1)(x-7)$

101. $6x^2 - x - 1 = (3x+1)(2x-1)$

103. $-5x^2 + 3x + 2 = (5x+2)(-x+1)$
$$= (5x+2)(1-x)$$

1.3 Quadratic Equations

1.3 Practice Problems

1.
$$x^2 + 25x = -84$$
$$x^2 + 25x + 84 = 0$$
$$(x+4)(x+21) = 0$$

$$x + 4 = 0 \quad | \quad x + 21 = 0$$
$$x = -4 \quad | \quad x = -21$$

Solution set: $\{-21, -4\}$

2.
$$2m^2 = 5m$$
$$2m^2 - 5m = 0$$
$$m(2m - 5) = 0$$

$$m = 0 \quad | \quad 2m - 5 = 0$$
$$\quad | \quad 2m = 5$$
$$\quad | \quad m = \frac{5}{2}$$

Solution set: $\left\{0, \frac{5}{2}\right\}$

3.
$$x^2 - 6x = -9$$
$$x^2 - 6x + 9 = 0$$
$$(x - 3)^2 = 0$$
$$x - 3 = 0$$
$$x = 3$$

Solution set: $\{3\}$

4.
$$(x + 2)^2 = 5$$
$$x + 2 = \pm\sqrt{5}$$
$$x = -2 \pm \sqrt{5}$$

Solution set: $\left\{-2 - \sqrt{5}, -2 + \sqrt{5}\right\}$

5.
$$x^2 - 6x + 7 = 0$$
$$x^2 - 6x = -7$$
$$x^2 - 6x + 9 = -7 + 9$$
$$(x - 3)^2 = 2$$
$$x - 3 = \pm\sqrt{2}$$
$$x = 3 \pm \sqrt{2}$$

Solution set: $\left\{3 - \sqrt{2}, 3 + \sqrt{2}\right\}$

6.
$$4x^2 - 24x + 25 = 0$$
$$4x^2 - 24x = -25$$
$$x^2 - 6x = -\frac{25}{4}$$
$$x^2 - 6x + 9 = -\frac{25}{4} + 9$$

Solution set: $\left\{3 - \dfrac{\sqrt{11}}{2}, \ 3 + \dfrac{\sqrt{11}}{2}\right\}$

7.
$$6x^2 - x - 2 = 0$$
$$a = 6, \ b = -1, \ c = -2$$
$$x = \frac{-b \pm \sqrt{b^2 - 4ac}}{2a}$$
$$= \frac{-(-1) \pm \sqrt{(-1)^2 - 4(6)(-2)}}{2(6)}$$
$$= \frac{1 \pm \sqrt{49}}{12} = \frac{1 \pm 7}{12} = \frac{-6}{12} = -\frac{1}{2} \text{ or } \frac{8}{12} = \frac{2}{3}$$

Solution set: $\left\{-\dfrac{1}{2}, \dfrac{2}{3}\right\}$

8. a. $3x^2 - 2x - 5 = 0$
$$a = 3, \ b = -2, \ c = -5$$
$$b^2 - 4ac = (-2)^2 - 4(3)(-5) = 64$$

Because 64 is a perfect square, there are two distinct rational solutions.

b. $x^2 + 4 = 4x$

Rewrite the equation in standard form.
$$x^2 - 4x + 4 = 0$$
$$a = 1, \ b = -4, \ c = 4$$
$$b^2 - 4ac = (-4)^2 - 4(1)(4) = 0$$

The discriminant is 0, so there is 1 rational solution (a double solution).

c. $3x^2 + 6x = -6$

Rewrite the equation in standard form.
$$3x^2 + 6x + 6 = 0$$
$$a = 3, \ b = 6, \ c = 6$$
$$b^2 - 4ac = 6^2 - 4(3)(6) = -36$$

The discriminant is negative, so there are no real solutions.

9. Let r = the radius of the battery. Then $2r$ = the width of the square, and $(2r)^2$ the area of the square. The area of the battery is $\pi r^2 = 64\pi$, so $r^2 = 64 \Rightarrow r = 8$ mm. (Note that the radius must be positive.) Therefore the area of the square is $(2r)^2 = (2 \cdot 8)^2 = 256$ sq mm. The area of the square not covered by the battery is $256 - 64\pi \approx 256 - 64(3.14) \approx 55$ sq mm.

10. Let x = the frontage of the building. Then $5x$ = the depth of the building and $5x - 45$ = the depth of the rear portion.
$$x(5x - 45) = 2100$$
$$5x^2 - 45x = 2100$$
$$5x^2 - 45x - 2100 = 0$$
$$a = 5,\ b = -45,\ c = -2100$$
$$x = \frac{-(-45) \pm \sqrt{(-45)^2 - 4(5)(-2100)}}{2(5)}$$
$$= \frac{45 \pm \sqrt{44{,}025}}{10} \approx -16.48 \text{ or } 25.48$$
Reject the negative solution.
$5x = 5 \cdot 25.482 = 127.41$
The building is approximately 25.48 ft by 127.41 ft.

11. $\Phi = \dfrac{\text{length}}{\text{width}} \Rightarrow \dfrac{1 + \sqrt{5}}{2} = \dfrac{x}{36}$
$$x = 36\left(\frac{1 + \sqrt{5}}{2}\right) = 18 + 18\sqrt{5} \approx 58.25 \text{ ft}$$

1.3 Concepts and Vocabulary

1. Any equation of the form $ax^2 + bx + c = 0$ with $a \neq 0$, is called a <u>quadratic</u> equation.

3. From the Square Root Property, we know that if $u^2 = 5$, then $\underline{u = \pm\sqrt{5}}$.

5. True

7. True

1.3 Building Skills

9. $(-6)^2 + 4(-6) - 12 = 36 - 24 - 12 = 0$
   –6 is a solution of the equation.

11. $3\left(\dfrac{2}{3}\right)^2 + 7\left(\dfrac{2}{3}\right) - 6 = 3\left(\dfrac{4}{9}\right) + \dfrac{14}{3} - 6$
$$= \frac{4}{3} + \frac{14}{3} - 6 = 0$$
2/3 is a solution of the equation.

13. $\left(2 - \sqrt{3}\right)^2 - 4\left(2 - \sqrt{3}\right) + 1$
$$= \left(4 - 4\sqrt{3} + 3\right) - 8 + 4\sqrt{3} + 1 = 0$$
$2 - \sqrt{3}$ is a solution of the equation.

15. $4\left(2 + \sqrt{3}\right)^2 - 8\left(2 + \sqrt{3}\right) + 13$
$$= 4\left(7 + 4\sqrt{3}\right) - 8\left(2 + \sqrt{3}\right) + 13$$
$$= 28 + 16\sqrt{3} - 16 - 8\sqrt{3} + 13 = 25 + 8\sqrt{3} \neq 0$$
$2 + \sqrt{3}$ is not a solution of the equation.

17. $k(1)^2 + 1 - 3 = 0 \Rightarrow k - 2 = 0 \Rightarrow k = 2$

19. $x^2 - 5x = 0 \Rightarrow x(x - 5) = 0 \Rightarrow$
    $x = 0$ or $x - 5 = 0 \Rightarrow x = 0$ or $x = 5$

21. $x^2 + 5x = 14$
    $x^2 + 5x - 14 = 0$
    $(x + 7)(x - 2) = 0$
    $\qquad x + 7 = 0$ or $x - 2 = 0$
    $\qquad\qquad x = -7$ or $x = 2$

23. $x^2 = 5x + 6$
    $x^2 - 5x - 6 = 0$
    $(x - 6)(x + 1) = 0$
    $\qquad x - 6 = 0$ or $x + 1 = 0$
    $\qquad\qquad x = 6$ or $x = -1$

25. $3x^2 = 48 \Rightarrow x^2 = 16 \Rightarrow x = \pm4$

27. $x^2 + 1 = 5 \Rightarrow x^2 = 4 \Rightarrow x = \pm2$

29. $(x - 1)^2 = 16$
    $x - 1 = -4$ or $x - 1 = 4 \Rightarrow x = -3$ or $x = 5$

31. To complete the square, find 1/2 of the coefficient of the x-term, $4/2 = 2$, and then square the answer. $2^2 = 4$.

33. To complete the square, find 1/2 of the coefficient of the x-term, $6/2 = 3$, and then square the answer. $3^2 = 9$.

35. To complete the square, find $1/2$ of the coefficient of the x-term and then square the answer. $\left(\dfrac{-7}{2}\right)^2 = \dfrac{49}{4}$.

37. To complete the square, find $1/2$ of the coefficient of the x-term, $\dfrac{1}{2} \cdot \dfrac{1}{3} = \dfrac{1}{6}$ and then square the answer. $\left(\dfrac{1}{6}\right)^2 = \dfrac{1}{36}$.

39. To complete the square, find $1/2$ of the coefficient of the x-term and then square the answer. $(a/2)^2 = a^2/4$.

41. $x^2 + 2x - 5 = 0 \Rightarrow x^2 + 2x = 5$
Now, complete the square.
$x^2 + 2x + 1 = 5 + 1 \Rightarrow (x+1)^2 = 6 \Rightarrow$
$x + 1 = \pm\sqrt{6} \Rightarrow x = -1 \pm \sqrt{6}$

43. $x^2 - 3x - 1 = 0$
$x^2 - 3x = 1$
Now, complete the square.
$x^2 - 3x + \dfrac{9}{4} = 1 + \dfrac{9}{4}$
$\left(x - \dfrac{3}{2}\right)^2 = \dfrac{13}{4}$
$x - \dfrac{3}{2} = \pm\dfrac{\sqrt{13}}{2}$
$x = \dfrac{3}{2} \pm \dfrac{\sqrt{13}}{2} = \dfrac{3 \pm \sqrt{13}}{2}$

45. $2r^2 + 3r = 9$
$r^2 + \dfrac{3}{2}r = \dfrac{9}{2}$
$r^2 + \dfrac{3}{2}r + \dfrac{9}{16} = \dfrac{9}{2} + \dfrac{9}{16}$
$\left(r + \dfrac{3}{4}\right)^2 = \dfrac{81}{16}$
$r + \dfrac{3}{4} = \pm\dfrac{9}{4} \Rightarrow r = \dfrac{-3 \pm 9}{4} = \dfrac{3}{2}$ or -3

In exercises 47–52, use the quadratic formula $x = \dfrac{-b \pm \sqrt{b^2 - 4ac}}{2a}$.

47. $x^2 + 2x - 4 = 0 \Rightarrow a = 1, b = 2, c = -4$
$x = \dfrac{-2 \pm \sqrt{2^2 - 4(1)(-4)}}{2(1)} = \dfrac{-2 \pm \sqrt{4 + 16}}{2}$
$= \dfrac{-2 \pm \sqrt{20}}{2} = \dfrac{-2 \pm 2\sqrt{5}}{2} = -1 \pm \sqrt{5}$

49. $6x^2 = 7x + 5 \Rightarrow 6x^2 - 7x - 5 = 0 \Rightarrow$
$a = 6, b = -7, c = -5$
$x = \dfrac{-(-7) \pm \sqrt{(-7)^2 - 4(6)(-5)}}{2(6)}$
$= \dfrac{7 \pm \sqrt{49 + 120}}{12} = \dfrac{7 \pm \sqrt{169}}{12} = \dfrac{7 \pm 13}{12}$
$x = \dfrac{7 + 13}{12} = \dfrac{20}{12} = \dfrac{5}{3}$ or
$x = \dfrac{7 - 13}{12} = \dfrac{-6}{12} = -\dfrac{1}{2}$

51. $3z^2 - 2z = 7 \Rightarrow 3z^2 - 2z - 7 = 0 \Rightarrow$
$a = 3, b = -2, c = -7$
$z = \dfrac{-(-2) \pm \sqrt{(-2)^2 - 4(3)(-7)}}{2(3)}$
$= \dfrac{2 \pm \sqrt{4 + 84}}{6} = \dfrac{2 \pm \sqrt{88}}{6} = \dfrac{2 \pm 2\sqrt{22}}{6}$
$= \dfrac{1 \pm \sqrt{22}}{3}$

53. $2x^2 + 5x - 3 = 0$
$(2x - 1)(x + 3) = 0$
$2x - 1 = 0$ or $x + 3 = 0$
$x = \dfrac{1}{2}$ or $x = -3$

55. $(3x - 2)^2 - 16 = 0$
$(3x - 2)^2 = 16$
$3x - 2 = \pm 4$
$3x - 2 = -4$ or $3x - 2 = 4$
$3x = -2$ $3x = 6$
$x = -\dfrac{2}{3}$ or $x = 2$

57.
$$5x^2 - 6x = 4x^2 + 6x - 3$$
$$x^2 - 12x = -3$$
Now, complete the square.
$$x^2 - 12x + 36 = -3 + 36$$
$$(x-6)^2 = 33$$
$$x - 6 = \pm\sqrt{33} \Rightarrow x = 6 \pm \sqrt{33}$$

59. $3p^2 + 8p + 4 = 0 \Rightarrow a = 3, b = 8, c = 4$
$$p = \frac{-8 \pm \sqrt{8^2 - 4(3)(4)}}{2(3)}$$
$$= \frac{-8 \pm \sqrt{64 - 48}}{6} = \frac{-8 \pm \sqrt{16}}{6} = \frac{-8 \pm 4}{6}$$
$$p = \frac{-4}{6} = -\frac{2}{3} \text{ or } p = \frac{-12}{6} = -2$$

61.
$$3y^2 + 5y + 2 = 0$$
$$(3y+2)(y+1) = 0$$
$$3y + 2 = 0 \text{ or } y + 1 = 0$$
$$y = -\frac{2}{3} \text{ or } y = -1$$

63.
$$5x^2 + 12x + 4 = 0$$
$$(5x+2)(x+2) = 0$$
$$5x + 2 = 0 \text{ or } x + 2 = 0$$
$$x = -\frac{2}{5} \text{ or } x = -2$$

65.
$$5y^2 + 10y + 4 = 2y^2 + 3y + 1$$
$$3y^2 + 7y = -3$$
$$y^2 + \frac{7}{3}y = -1$$
Now, complete the square.
$$y^2 + \frac{7}{3}y + \frac{49}{36} = -1 + \frac{49}{36}$$
$$\left(y + \frac{7}{6}\right)^2 = \frac{13}{36}$$
$$y + \frac{7}{6} = \pm\sqrt{\frac{13}{36}} = \pm\frac{\sqrt{13}}{6}$$
$$y = -\frac{7}{6} \pm \frac{\sqrt{13}}{6} = \frac{-7 \pm \sqrt{13}}{6}$$

67.
$$2x^2 + x = 15$$
$$2x^2 + x - 15 = 0$$
$$(2x-5)(x+3) = 0$$
$$2x - 5 = 0 \text{ or } x + 3 = 0 \Rightarrow x = \frac{5}{2} \text{ or } x = -3$$

69.
$$12x^2 - 10x = 12$$
$$12x^2 - 10x - 12 = 0$$
$$2(6x^2 - 5x - 6) = 0$$
$$6x^2 - 5x - 6 = 0$$
$$(3x+2)(2x-3) = 0$$
$$3x + 2 = 0 \text{ or } 2x - 3 = 0$$
$$x = -\frac{2}{3} \text{ or } x = \frac{3}{2}$$

71. $(x+13)(x+5) = -2 \Rightarrow x^2 + 18x + 65 = -2 \Rightarrow$
$$x^2 + 18x + 67 = 0 \Rightarrow a = 1, b = 18, c = 67$$
$$x = \frac{-18 \pm \sqrt{18^2 - 4(1)(67)}}{2(1)}$$
$$= \frac{-18 \pm \sqrt{324 - 268}}{2} = \frac{-18 \pm \sqrt{56}}{2}$$
$$= \frac{-18 \pm 2\sqrt{14}}{2} = -9 \pm \sqrt{14}$$

73.
$$18x^2 - 45x = -7$$
$$18x^2 - 45x + 7 = 0$$
$$(3x-7)(6x-1) = 0$$
$$3x - 7 = 0 \text{ or } 6x - 1 = 0 \Rightarrow x = \frac{7}{3} \text{ or } x = \frac{1}{6}$$

75. $2t^2 - 5 = 0 \Rightarrow a = 2, b = 0, c = -5$
$$t = \frac{0 \pm \sqrt{0^2 - 4(2)(-5)}}{2(2)} = \frac{\pm\sqrt{40}}{4}$$
$$= \pm\frac{2\sqrt{10}}{4} = \pm\frac{\sqrt{10}}{2}$$

77.
$$4x^2 - 10x - 750 = 0$$
$$2(2x^2 - 5x - 375) = 0$$
$$2x^2 - 5x - 375 = 0$$
$$(2x+25)(x-15) = 0$$
$$2x + 25 = 0 \text{ or } x - 15 = 0$$
$$x = -\frac{25}{2} \text{ or } x = 15$$

79. $x^2 + 4x - 2 = 0$
$a = 1, b = 4, c = -2$
$$b^2 - 4ac = 4^2 - 4(1)(-2) = 24$$
Because the discriminant, 24, is not a perfect square, there are two distinct irrational solutions.

81. $-3x^2 = 4$

Write the equation in standard form.
Write the equation in standard form.

$-3x^2 - 4 = 0$

$a = -3, b = 0, c = -4$

$b^2 - 4ac = (0)^2 - 4(-3)(-4) = -48$

Because the discriminant is -48, there are no real solutions.

83. $6x^2 - 3x - 2 = 0$

$a = 6, b = -3, c = -2$

$b^2 - 4ac = (-3)^2 - 4(6)(-2) = 57$

Because the discriminant, 57, is not a perfect square, there are two distinct irrational solutions.

85. $2x^2 - 3x + 1 = 0$

$a = 2, b = -3, c = 1$

$b^2 - 4ac = (-3)^2 - 4(2)(1) = 1$

Because the discriminant, 1, is a perfect square, there are two distinct rational solutions.

87. $\Phi = \dfrac{\text{length}}{\text{width}} \Rightarrow \dfrac{1+\sqrt{5}}{2} = \dfrac{x}{14.72}$

$x = 14.72\left(\dfrac{1+\sqrt{5}}{2}\right) \approx 23.82$ in.

89. $\Phi = \dfrac{\text{length}}{\text{width}} \Rightarrow \dfrac{1+\sqrt{5}}{2} = \dfrac{8.46}{x}$

$x = 8.46\left(\dfrac{2}{1+\sqrt{5}}\right) \approx 5.23$ cm

1.3 Applying the Concepts

91. Let $x =$ the width of the plot. Then $3x =$ the length of the plot. So, $3x^2 = 10,800 \Rightarrow$

$x^2 = 3600 \Rightarrow x = 60$.
The plot is 60 ft by 180 ft.

93. Let $x =$ the first integer. Then $28 - x =$ the second integer. So, $x(28 - x) = 147 \Rightarrow$

$28x - x^2 = 147 \Rightarrow 0 = x^2 - 28x + 147 \Rightarrow$

$0 = (x - 7)(x - 21) \Rightarrow x = 7$ or $x = 21$.

The sum of the two numbers is 28. So, the numbers are 7 and 21.

95. $p = 0.003x^2 + 0.2573x + 83.975$

a. The year 2014 corresponds to $x = 14$.

$p = 0.003(14)^2 + 0.2573(14) + 83.975$
$= 88.1652$

In 2014, about 88% of people age 25 and older obtained a bachelor's degrees.

b. The year 2020 corresponds to $x = 20$.

$p = 0.003(20)^2 + 0.2573(20) + 83.975$
$= 90.321$

In 2020, about 90% of people age 25 and older will obtain a bachelor's degrees.

97. Let $x =$ the width of the rectangle. Then $x + 5 =$ the length of the rectangle. So,

$x(x + 5) = 500 \Rightarrow x^2 + 5x = 500 \Rightarrow$

$x^2 + 5x - 500 = 0 \Rightarrow (x + 25)(x - 20) = 0 \Rightarrow$

$x = -25$ or $x = 20$. The length cannot be negative, so reject that solution.
$x = 20 \Rightarrow x + 5 = 25$. The rectangle is 25 cm by 20 cm.

99. Let $x =$ one piece of the wire. Then $16 - x =$ the other piece of the wire. Each piece is bent into a square, so the sides of the squares are $\dfrac{x}{4}$ and $\dfrac{16-x}{4}$, respectively.

$\left(\dfrac{x}{4}\right)^2 + \left(\dfrac{16-x}{4}\right)^2 = 10$

$\dfrac{x^2}{16} + \dfrac{256 - 32x + x^2}{16} = 10$

$x^2 + 256 - 32x + x^2 = 160$

$2x^2 - 32x + 96 = 0$

$x^2 - 16x + 48 = 0$

$(x - 12)(x - 4) = 0 \Rightarrow x = 12$ or $x = 4$

If one piece of the wire is 12, then the other piece is $16 - 12 = 4$. The pieces are 12 in. and 4 in.

101. Let r = the radius of the can. Then
$$32\pi = 2\pi r(6) + 2\pi r^2$$
$$32\pi = 12\pi r + 2\pi r^2$$
Divide both sides by 2π
$$16 = 6r + r^2$$
$$0 = r^2 + 6r - 16$$
$$0 = (r-2)(r+8)$$
$$r = 2 \text{ or } r = -8$$
The answer cannot be negative, so we reject −8. The radius of the can is 2 inches.

103. Let x = the length of the piece of tin. Then $x - 10$ = the length of the box.

$$(x-10)(x-10)(5) = 480$$
$$5(x^2 - 20x + 100) = 480$$
$$x^2 - 20x + 100 = 96$$
$$x^2 - 20x + 4 = 0$$
Solve using the quadratic formula with $a = 1$, $b = -20$, and $c = 4$.
$$x = \frac{-(-20) \pm \sqrt{(-20)^2 - 4(1)(4)}}{2(1)}$$
$$= \frac{20 \pm \sqrt{400 - 16}}{2} = \frac{20 \pm \sqrt{384}}{2}$$
$$\approx \frac{20 + 19.6}{2} \text{ or } \frac{20 - 19.6}{2} \approx 19.8 \text{ or } 0.2$$
The length cannot be 0.2 inches, so we reject that solution. The tin is 19.8 in. by 19.8 in.

105. Let x = the time the buses travel. So the distance the first bus travels = $52x$ mi, and the distance the second bus travels = $39x$ mi.

Using the Pythagorean theorem, we have
$$(52x)^2 + (39x)^2 = 390^2$$
$$2704x^2 + 1521x^2 = 152,100$$
$$4225x^2 = 152,100$$
$$x^2 = 36 \Rightarrow x = \pm 6$$
Time cannot be negative, so we reject −6. The buses will be 390 miles apart after 6 hours.

107. Let x = the width of the border. Then length of the garden with the border is $25 + 2x$, and the width of the garden with the border is $15 + 2x$. The area of the border = the area of the garden with the border − the area of the garden.

$$A_b = 624 = (25 + 2x)(15 + 2x) - (15)(25)$$
$$= 375 + 80x + 4x^2 - 375$$
$$= 80x + 4x^2$$
$$624 = 80x + 4x^2$$
$$0 = 4x^2 + 80x - 624$$
$$0 = x^2 + 20x - 156$$
$$0 = (x + 26)(x - 6)$$
$$x = -26 \text{ or } x = 6$$
We reject the negative solution. The width of the border is 6 feet.

109. a. $h = -16(2^2) + 112(2) = 160$ feet

b. $96 = -16t^2 + 112t \Rightarrow 16t^2 - 112t + 96 = 0$
$\Rightarrow t^2 - 7t + 6 = 0 \Rightarrow (t-1)(t-6) = 0 \Rightarrow$
$t = 1 \text{ or } t = 6$
The ball will be at a height of 96 feet at 1 second and at 6 seconds.

c. $0 = -16t^2 + 112t \Rightarrow 16t^2 - 112t = 0 \Rightarrow$
$t^2 - 7t = 0 \Rightarrow t(t-7) = 0 \Rightarrow t = 0 \text{ or }$
$t = 7$
The ball will return to the ground at 7 seconds.

111. a. $-16t^2 + 96t + 480 = 592$

$-16t^2 + 96t - 112 = 0$

$a = -16,\ b = 96,\ c = -112$

$$x = \frac{-96 \pm \sqrt{96^2 - 4(-16)(-112)}}{2(-16)}$$

≈ 1.59 or 4.41

The projectile will be at a height of 592 ft at about 1.59 sec and 4.41 sec.

b. $-16t^2 + 96t + 480 = 0$

$a = -16,\ b = 96,\ c = 480$

$$x = \frac{-96 \pm \sqrt{96^2 - 4(-16)(480)}}{2(-16)}$$

≈ -3.24 (reject this) or 9.24

The projectile will crash on the ground after approximately 9.24 seconds.

113. a. $-16t^2 + 256t + 480 = 592$

$-16t^2 + 256t - 112 = 0$

$a = -16,\ b = 256,\ c = -112$

$$x = \frac{-256 \pm \sqrt{256^2 - 4(-16)(-112)}}{2(-16)}$$

≈ 0.45 or 15.55

The projectile will be at a height of 592 ft at about 0.45 sec and 15.55 sec.

b. $-16t^2 + 256t + 480 = 0$

$a = -16,\ b = 256,\ c = 480$

$$x = \frac{-256 \pm \sqrt{256^2 - 4(-16)(480)}}{2(-16)}$$

≈ -1.70 (reject this) or 17.70

The projectile will crash on the ground after 17.70 seconds.

1.3 Beyond the Basics

In exercises 115–120, the equation has equal roots if the discriminant equals 0.

115. $x^2 - kx + 3 = 0$

$a = 1,\ b = -k,\ c = 3$

$b^2 - 4ac = (-k)^2 - 4(1)(3) = k^2 - 12$

Now solve $k^2 - 12 = 0$.

$k^2 - 12 = 0 \Rightarrow k^2 = 12 \Rightarrow k = \pm\sqrt{12} = \pm 2\sqrt{3}$

117. $2x^2 + kx + k = 0$

$a = 2,\ b = k,\ c = k$

$b^2 - 4ac = k^2 - 4(2)k = k^2 - 8k$

Now solve $k^2 - 8k = 0$.

$k^2 - 8k = 0 \Rightarrow k(k - 8) = 0 \Rightarrow k = 0$ or $k = 8$

119.   Since r and s are roots of the equation, then

$$ar^2 + br + c = 0 = as^2 + bs + c \Rightarrow$$
$$ar^2 + br + c - c = as^2 + bs + c - c$$
$$ar^2 + br = as^2 + bs$$
$$ar^2 - as^2 = bs - br$$
$$a(r^2 - s^2) = b(s - r)$$
$$a(r^2 - s^2) = -b(r - s)$$
$$\frac{r^2 - s^2}{r - s} = -\frac{b}{a} \Rightarrow r + s = -\frac{b}{a}$$

To find $r \cdot s$, first divide both sides of

$ar^2 + br + c = 0$ by a. We have

$r^2 + \dfrac{b}{a}r + \dfrac{c}{a} = 0$. Now use the results from

the first part of the problem and substitute

$-(r + s)$ for $\dfrac{b}{a}$. This gives

$$r^2 - (r + s)r + \frac{c}{a} = 0$$
$$r^2 - r^2 - rs + \frac{c}{a} = 0 \Rightarrow \frac{c}{a} = rs$$

Use the results from exercise 119 to solve exercises 121 and 122. The sum of the roots is $-\dfrac{b}{a}$, and the product of the roots is $\dfrac{c}{a}$.

121. $-\dfrac{k-3}{2} = \dfrac{3k-5}{2} \Rightarrow -k + 3 = 3k - 5 \Rightarrow$
$-4k = -8 \Rightarrow k = 2$

123.   From exercise 119, we have

$r + s = -\dfrac{b}{a} \Rightarrow b = -a(r + s)$ and

$rs = \dfrac{c}{a} \Rightarrow c = ars$. Substitute these values

into the equation:

$$ax^2 + bx + c = ax^2 - a(r + s)x + ars$$
$$= a(x^2 - (r + s)x + rs)$$
$$= a(x^2 - rx - sx + rs)$$
$$= a[x(x - r) - s(x - r)]$$
$$= a(x - r)(x - s)$$

125. a. $(x-(-3))(x-4)=0 \Rightarrow (x+3)(x-4)=0 \Rightarrow$
$x^2-x-12=0$

b. $(x-5)(x-5)=0 \Rightarrow x^2-10x+25=0$

c. $(x-(3+\sqrt{2}))(x-(3-\sqrt{2}))=0 \Rightarrow$
$x^2-3x-x\sqrt{2}-3x+x\sqrt{2}+7=0 \Rightarrow$
$x^2-6x+7=0$

1.3 Critical Thinking/Discussion/Writing

127. $(x-a)(x-b)=k^2 \Rightarrow$
$x^2-(a+b)x+ab-k^2=0$
$\left[-(a+b)\right]^2-4(1)\left(ab-k^2\right)$
$=a^2+2ab+b^2-4ab+4k^2$
$=a^2-2ab+b^2+4k^2$
$=(a-b)^2+4k^2 \geq 0$
Thus, the solutions of the equation are real.

128. $ax(1-x)=1 \Rightarrow ax-ax^2=1 \Rightarrow$
$-ax^2+ax-1=0$
Examining the discriminant, we have
$a^2-4(-a)(-1)=a^2-4a$
$a^2-4a<0 \Rightarrow ax(1-x)=1$ has no real
solutions for $0<a<4$.

1.3 Getting Ready for the Next Section

129. $(3+\sqrt{2})(3-\sqrt{2})=3^2-(\sqrt{2})^2=9-2=7$

131. $(2+\sqrt{5})^2=2^2+2\cdot2\cdot\sqrt{5}+(\sqrt{5})^2$
$=4+4\sqrt{5}+5=9+4\sqrt{5}$

133. $\frac{5+5\sqrt{20}}{5}=1+\sqrt{20}=1+\sqrt{4\cdot5}=1+2\sqrt{5}$

135. $\frac{16-\sqrt{100}}{2}=\frac{16-10}{2}=\frac{6}{2}=3$

137. $(3x+2)+(x-7)=4x-5$

139. $(9x+4)-(2x+12)=9x+4-2x-12$
$=7x-8$

141. $(3x+2)(x-9)=3x^2-27x+2x-18$
$=3x^2-25x-18$

143. $(x-3)(x+3)=x^2-9$

Solve each equation in exercises 145–148 using the quadratic formula, $x=\dfrac{-b\pm\sqrt{b^2-4ac}}{2a}$.

145. $x^2+4x+1=0$
$a=1, b=4, c=1$
$x=\dfrac{-4\pm\sqrt{4^2-4\cdot1\cdot1}}{2\cdot1}=\dfrac{-4\pm\sqrt{12}}{2}$
$=\dfrac{-4\pm2\sqrt{3}}{2}=-2\pm\sqrt{3}$

147. $5x^2+8x+2=0$
$a=5, b=8, c=2$
$x=\dfrac{-8\pm\sqrt{8^2-4\cdot5\cdot2}}{2\cdot5}=\dfrac{-8\pm\sqrt{24}}{10}$
$=\dfrac{-8\pm2\sqrt{6}}{10}=\dfrac{-4\pm\sqrt{6}}{5}$

1.4 Complex Numbers: Quadratic Equations with Complex Solutions

1.4 Practice Problems

1. a. $-1+2i$
real part: -1, imaginary part: 2

b. $-\dfrac{1}{3}-6i$
real part: $-\dfrac{1}{3}$, imaginary part: -6

c. $8=8+0i$
real part: 8, imaginary part: 0

2. a. $(1-4i)+(3+2i)=4-2i$

b. $(4+3i)-(5-i)=-1+4i$

c. $(3-\sqrt{-9})-(5-\sqrt{-64})=(3-3i)-(5-8i)$
$=-2+5i$

3. a. $(2-6i)(1+4i)=2+8i-6i-24i^2$
$=2+2i+24=26+2i$

b. $-3i(7-5i)=-21i+15i^2=-15-21i$

4. a. $(-3+\sqrt{-4})^2=(-3+2i)^2$
$=(-3)^2+2(-3)(2i)+(2i)^2$
$=9-12i+4i^2=9-12i-4$
$=5-12i$

b. $\left(5+\sqrt{-2}\right)\left(4+\sqrt{-8}\right)=\left(5+i\sqrt{2}\right)\left(4+2i\sqrt{2}\right)$
$$=20+10\sqrt{2}i+4\sqrt{2}i+4i^2$$
$$=20+14\sqrt{2}i-4$$
$$=16+14\sqrt{2}i$$

5. a. $z=1+6i\Rightarrow \overline{z}=1-6i$
$$z\overline{z}=(1+6i)(1-6i)=1-36i^2=1+36=37$$

b. $z=-2i\Rightarrow \overline{z}=2i$
$$z\overline{z}=(-2i)(2i)=-4i^2=4$$

6. a. $\dfrac{2}{1-i}=\dfrac{2}{1-i}\cdot\dfrac{1+i}{1+i}=\dfrac{2+2i}{1-i^2}=\dfrac{2+2i}{1+1}$
$$=\dfrac{2+2i}{2}=1+i$$

b. $\dfrac{-3i}{4+\sqrt{-25}}=\dfrac{-3i}{4+5i}=\dfrac{-3i}{4+5i}\cdot\dfrac{4-5i}{4-5i}$
$$=\dfrac{-12i+15i^2}{16-25i^2}=\dfrac{-15-12i}{16+25}$$
$$=\dfrac{-15-12i}{41}=-\dfrac{15}{41}-\dfrac{12i}{41}$$

7. $Z_t=\dfrac{Z_1Z_2}{Z_1+Z_2}=\dfrac{(1+2i)(2-3i)}{(1+2i)+(2-3i)}$
$$=\dfrac{2-3i+4i-6i^2}{3-i}=\dfrac{2+i+6}{3-i}=\dfrac{8+i}{3-i}$$
$$=\dfrac{8+i}{3-i}\cdot\dfrac{3+i}{3+i}=\dfrac{24+8i+3i+i^2}{9-i^2}$$
$$=\dfrac{24+11i-1}{9+1}=\dfrac{23+11i}{10}=\dfrac{23}{10}+\dfrac{11}{10}i$$

8. a. $4x^2+9=0$
$$4x^2=-9$$
$$x^2=-\dfrac{9}{4}\Rightarrow x=\pm\sqrt{-\dfrac{9}{4}}=\pm\dfrac{3}{2}i$$
Solution set: $\left\{-\dfrac{3}{2}i,\dfrac{3}{2}i\right\}$

b. $x^2=4x-13$
$$x^2-4x+13=0$$
$$x=\dfrac{-(-4)\pm\sqrt{(-4)^2-4(1)(13)}}{2(1)}$$
$$=\dfrac{4\pm\sqrt{-36}}{2}=\dfrac{4\pm6i}{2}=2\pm3i$$
Solution set: $\{2-3i,2+3i\}$

9. a. $9x^2-6x+1=0\Rightarrow a=9, b=-6, c=1$
So, $D=(-6)^2-4(9)(1)=36-36=0$.
Therefore, there is one real root.

b. $x^2-5x+3=0\Rightarrow a=1, b=-5, c=3$
So, $D=(-5)^2-4(1)(3)=25-12=13>0$
Therefore, there are two unequal real roots.

c. $2x^2-3x+4=0\Rightarrow a=2, b=-3, c=4$
So, $D=(-3)^2-4(2)(4)=9-32=-23<0$
Therefore, there are two nonreal complex solutions.

1.4 Concepts and Vocabulary

1. We define $i=\sqrt{-1}$, so that $i^2=\underline{-1}$.

3. For $b>0$, $\sqrt{-b}=\underline{i\sqrt{b}}$.

5. True

7. False. Every imaginary number has a conjugate.

9. $(5+2i)+(3+i)=(5+3)+(2+1)i=8+3i$

11. $(4-3i)-(5+3i)=(4-5)+(-3-3)i$
$$=-1-6i$$

13. $(-2-3i)+(-3-2i)=[-2+(-3)]+(-3-2)i$
$$=-5-5i$$

15. $3(5+2i)=3(5)+3(2i)=15+6i$

17. $-4(2-3i)=-4(2)-4(-3i)=-8+12i$

19. $3i(5+i)=3i(5)+3i(i)=15i+3i^2$
Because $i^2=-1$, $3i^2=-3$.
So, $15i+3i^2=15i-3=-3+15i$.

21. $4i(2-5i)=4i(2)+4i(-5i)=8i-20i^2$.
Because $i^2=-1$, $-20i^2=(-20)(-1)=20$.
So, $8i-20i^2=8i+20=20+8i$.

23. $(3+i)(2+3i)=3\cdot2+3\cdot3i+i\cdot2+i\cdot3i$
$$=6+9i+2i+3i^2$$
$$=6+11i+3(-1)$$
$$=3+11i$$

25. $(2-3i)(2+3i)=2\cdot2+2\cdot3i+(-3i)\cdot2+(-3i)\cdot3i$
$$=4+6i-6i-9i^2$$
$$=4-9(-1)=4+9=13$$

27. $(3+4i)(4-3i)$
$$=3\cdot4+3\cdot(-3i)+4i\cdot4+(4i)\cdot(-3i)$$
$$=12-9i+16i-12i^2$$
$$=12+7i-12(-1)$$
$$=12+7i+12=24+7i$$

29. $\left(\sqrt{3} - 12i\right)^2$

$= \left(\sqrt{3} - 12i\right)\left(\sqrt{3} - 12i\right)$

$= \sqrt{3} \cdot \sqrt{3} + \sqrt{3} \cdot (-12i)$

$\qquad - 12i \cdot \sqrt{3} + (-12i) \cdot (-12i)$

$= 3 - 12\sqrt{3}i - 12\sqrt{3}i + 144i^2$

$= 3 - 24\sqrt{3}i + 144(-1) = 3 - 24\sqrt{3}i - 144$

$= -141 - 24\sqrt{3}i$

31. $\left(2 - \sqrt{-16}\right)(3 + 5i)$

$= (2 - 4i)(3 + 5i)$

$= 2 \cdot 3 + 2 \cdot 5i - 4i \cdot 3 - 4i \cdot 5i$

$= 6 + 10i - 12i - 20i^2$

$= 6 - 2i - 20(-1)$

$= 6 - 2i + 20$

$= 26 - 2i$

33. If $z = 2 - 3i$ then $\overline{z} = 2 + 3i$, and

$z\overline{z} = (2 - 3i)(2 + 3i) = 4 - 9i^2 = 4 + 9 = 13$.

35. If $z = \dfrac{1}{2} - 2i$ then $\overline{z} = \dfrac{1}{2} + 2i$, and

$z\overline{z} = \left(\dfrac{1}{2} - 2i\right)\left(\dfrac{1}{2} + 2i\right) = \dfrac{1}{4} - 4i^2 = \dfrac{1}{4} + 4 = \dfrac{17}{4}$

37. If $z = \sqrt{2} - 3i$ then $\overline{z} = \sqrt{2} + 3i$, and

$z\overline{z} = \left(\sqrt{2} - 3i\right)\left(\sqrt{2} + 3i\right) = 2 - 9i^2 = 2 + 9 = 11$

39. The denominator is $-i$, so its conjugate is i. Multiply the numerator and denominator by i.

$\dfrac{5}{-i} = \dfrac{5i}{-i \cdot i} = \dfrac{5i}{1} = 5i$

41. The denominator is $1 + i$, so its conjugate is $1 - i$. Multiply the numerator and denominator by $1 - i$.

$\dfrac{-1}{1 + i} = \dfrac{-1(1 - i)}{(1 + i)(1 - i)} = \dfrac{-1 + i}{1 + 1}$

$= \dfrac{-1 + i}{2} = -\dfrac{1}{2} + \dfrac{1}{2}i$

43. The denominator is $2 + i$, so its conjugate is $2 - i$. Multiply the numerator and denominator by $2 - i$.

$\dfrac{5i}{2 + i} = \dfrac{5i(2 - i)}{(2 + i)(2 - i)} = \dfrac{10i - 5i^2}{4 + 1}$

$= \dfrac{10i + 5}{5} = 1 + 2i$

45. The denominator is $1 + i$, so its conjugate is $1 - i$. Multiply the numerator and denominator by $1 - i$.

$\dfrac{2 + 3i}{1 + i} = \dfrac{(2 + 3i)(1 - i)}{(1 + i)(1 - i)} = \dfrac{2 - 2i + 3i - 3i^2}{1 + 1}$

$= \dfrac{2 + i + 3}{2} = \dfrac{5 + i}{2} = \dfrac{5}{2} + \dfrac{1}{2}i$

47. The denominator is $4 - 7i$, so its conjugate is $4 + 7i$. Multiply the numerator and denominator by $4 + 7i$.

$\dfrac{2 - 5i}{4 - 7i} = \dfrac{(2 - 5i)(4 + 7i)}{(4 - 7i)(4 + 7i)} = \dfrac{8 + 14i - 20i - 35i^2}{16 + 49}$

$= \dfrac{8 - 6i + 35}{65} = \dfrac{43 - 6i}{65} = \dfrac{43}{65} - \dfrac{6}{65}i$

49. The denominator is $1 + i$, so its conjugate is $1 - i$. Multiply the numerator and denominator by $1 - i$.

$\dfrac{2 + \sqrt{-4}}{1 + i} = \dfrac{(2 + 2i)}{(1 + i)} = \dfrac{(2 + 2i)(1 - i)}{(1 + i)(1 - i)}$

$= \dfrac{2 - 2i + 2i - 2i^2}{1 + 1} = \dfrac{2 - 2i^2}{2} = \dfrac{2 + 2}{2} = 2$

51. The denominator is $2 - 3i$, so its conjugate is $2 + 3i$. Multiply the numerator and denominator by $2 + 3i$.

$\dfrac{-2 + \sqrt{-25}}{2 - 3i} = \dfrac{-2 + 5i}{2 - 3i} = \dfrac{(-2 + 5i)(2 + 3i)}{(2 - 3i)(2 + 3i)}$

$= \dfrac{-4 - 6i + 10i + 15i^2}{4 + 9}$

$= \dfrac{-4 + 4i - 15}{13} = \dfrac{-19 + 4i}{13}$

$= -\dfrac{19}{13} + \dfrac{4}{13}i$

53. $x^2 + 5 = 1 \Rightarrow x^2 = -4 \Rightarrow x = \pm\sqrt{-4} = \pm 2i$

55. $z^2 - 2z + 2 = 0$

$z^2 - 2z = -2$

Now, complete the square.

$z^2 - 2z + 1 = -2 + 1$

$(z - 1)^2 = -1 \Rightarrow z - 1 = \pm i \Rightarrow z = 1 \pm i$

57. $2x^2 - 20x + 49 = -7$

$2x^2 - 20x = -56$

$x^2 - 10x = -28$

Now, complete the square.

$x^2 - 10x + 25 = -28 + 25 \Rightarrow (x - 5)^2 = -3$

$x - 5 = \pm i\sqrt{3} \Rightarrow x = 5 \pm i\sqrt{3}$

59. $8(x^2 - x) = x^2 - 3 \Rightarrow 8x^2 - 8x = x^2 - 3 \Rightarrow$
$7x^2 - 8x + 3 = 0 \Rightarrow a = 7, b = -8, c = 3$

$$x = \frac{-(-8) \pm \sqrt{(-8)^2 - 4(7)(3)}}{2(7)}$$

$$= \frac{8 \pm \sqrt{64 - 84}}{14} = \frac{8 \pm \sqrt{-20}}{14}$$

$$= \frac{8 \pm 2i\sqrt{5}}{14} = \frac{4 \pm i\sqrt{5}}{7}$$

$$x = \frac{4}{7} + \frac{\sqrt{5}}{7}i \text{ or } x = \frac{4}{7} - \frac{\sqrt{5}}{7}i$$

61. $9k^2 + 25 = 0 \Rightarrow a = 9, b = 0, c = 25$

$$t = \frac{0 \pm \sqrt{0^2 - 4(9)(25)}}{2(9)} = \frac{\pm\sqrt{-900}}{18}$$

$$= \pm\frac{30i}{18} = \pm\frac{5}{3}i$$

1.4 Applying the Concepts

63. $Z_1 = 4 + 3i$ and $Z_2 = 5 - 2i$.
So, $Z_1 + Z_2 = (4 + 3i) + (5 - 2i) = 9 + i$.

65. $Z = \frac{V}{I}$, $I = 7 + 5i$, $V = 35 + 70i$. Then,

$Z = \frac{35 + 70i}{7 + 5i}$. Simplify the fraction by

multiplying the numerator and denominator
by $7 - 5i$.

$$\frac{35 + 70i}{7 + 5i} = \frac{(35 + 70i)(7 - 5i)}{(7 + 5i)(7 - 5i)}$$

$$= \frac{245 - 175i + 490i - 350i^2}{49 + 25}$$

$$= \frac{245 + 315i + 350}{74} = \frac{595 + 315i}{74}$$

$$= \frac{595}{74} + \frac{315}{74}i$$

67. $Z = \frac{V}{I}$, $Z = 5 - 7i$, $I = 2 + 5i$. Then,

$V = ZI = (5 - 7i)(2 + 5i)$

$= 10 + 25i - 14i - 35i^2$

$= 10 + 11i + 35 = 45 + 11i$

69. $Z = \frac{V}{I}$, $V = 12 + 10i$, $Z = 12 + 6i$. Then,

$I = \frac{V}{Z} = \frac{12 + 10i}{12 + 6i}$. Simplify the fraction by

multiplying the numerator and denominator
by $12 - 6i$.

$$\frac{12 + 10i}{12 + 6i} = \frac{(12 + 10i)(12 - 6i)}{(12 + 6i)(12 - 6i)}$$

$$= \frac{144 - 72i + 120i - 60i^2}{144 + 36}$$

$$= \frac{144 + 48i + 60}{180}$$

$$= \frac{204 + 48i}{180} = \frac{17}{15} + \frac{4}{15}i$$

1.4 Beyond the Basics

71. To find i^{17}, first divide 17 by 4. The
remainder is 1, so $i^{17} = i^1 = i$.

73. To find i^{-7}, first rewrite it as $\frac{1}{i^7}$. Then

divide 7 by 4. The remainder is 3, so $\frac{1}{i^7} = \frac{1}{i^3}$.

Simplify the fraction by multiplying the
numerator and denominator by i.

$\frac{1}{i^3} \cdot \frac{i}{i} = \frac{i}{i^4} = i$.

75. To find i^{10}, first divide 10 by 4. The
remainder is 2, so $i^{10} = i^2 = -1$. So
$i^{10} + 7 = -1 + 7 = 6$.

77. To find i^5, first divide 5 by 4. The remainder
is 1, so $i^5 = i$. So $3i^5 = 3i$. $i^3 = -i$, so
$-2i^3 = 2i$. Then, $3i^5 - 2i^3 = 3i + 2i = 5i$.

79. $i^3 = -i$, so $2i^3 = -2i$. $i^4 = 1$, so $1 + i^4$
$= 1 + 1 = 2$. Then $2i^3(1 + i^4) = -2i(2) = -4i$.

81. $\frac{1}{a + bi} = \frac{1(a - bi)}{(a + bi)(a - bi)} = \frac{a - bi}{a^2 + b^2}$

$$= \frac{a}{a^2 + b^2} - \frac{b}{a^2 + b^2}i$$

83. $\frac{z - \bar{z}}{2i} = \frac{(a + bi) - (a - bi)}{2i} = \frac{2bi}{2i} = b$

85. $\frac{1 - 2i}{5 - 5i} = \frac{1 - 2i}{5 - 5i} \cdot \frac{5 + 5i}{5 + 5i} = \frac{5 + 5i - 10i - 10i^2}{25 - 25i^2}$

$$= \frac{5 - 5i + 10}{25 + 25} = \frac{15 - 5i}{50} = \frac{3}{10} - \frac{1}{10}i$$

$\frac{2 - 3i}{9 - 7i} = \frac{2 - 3i}{9 - 7i} \cdot \frac{9 + 7i}{9 + 7i} = \frac{18 + 14i - 27i - 21i^2}{81 - 49i^2}$

$$= \frac{18 - 13i + 21}{81 + 49} = \frac{39 - 13i}{130} = \frac{3}{10} - \frac{1}{10}i$$

(*continued on next page*)

(*continued*)

$$\frac{1-2i}{5-5i} = \frac{3}{10} - \frac{1}{10}i = \frac{2-3i}{9-7i} \Rightarrow \frac{1-2i}{5-5i} = \frac{2-3i}{9-7i}$$

87. $(1+3i)z + (2+4i) = 7-3i$

$$(1+3i)z = 7-3i-(2+4i) = 5-7i$$

$$z = \frac{5-7i}{1+3i} = \frac{5-7i}{1+3i} \cdot \frac{1-3i}{1-3i}$$

$$= \frac{5-15i-7i+21i^2}{1-9i^2}$$

$$= \frac{5-22i-21}{1+9} = \frac{-16-22i}{10}$$

$$= -\frac{8}{5} - \frac{11}{5}i$$

1.4 Critical Thinking/Discussion/Writing

89. a. True. Every real number a can be written as a complex number $a + 0i$.

  b. False.

  c. False. A complex number with the form $a + 0i$ does not have an imaginary component.

  d. True

  e. True. $(a+bi)(a-bi) = a^2 + b^2$. There is no imaginary component.

  f. True

90. $\dfrac{1+i}{1-i} = \dfrac{1+i}{1-i} \cdot \dfrac{1+i}{1+i} = \dfrac{1+2i-1}{2} = \dfrac{2i}{2} = i$

$i^n = 1 \Rightarrow n = 4$

91. If $i = 0$, then $i^2 = 0^2 \Rightarrow -1 = 0$, which is a contradiction. If $i < 0$, then $i \cdot i > 0 \cdot i$ (since i is negative) $\Rightarrow i^2 > 0 \Rightarrow -1 > 0$, a contradiction.

If $i > 0$, then $i \cdot i > 0 \cdot i \Rightarrow i^2 > 0 \Rightarrow -1 > 0$, a contradiction. Thus, the set of complex numbers does not have the ordering properties of the set of real numbers.

92. Answers may vary. Sample answer:

$zw = 0 \Rightarrow (a+bi)(c+di) = 0 \Rightarrow$

$ac + (bd)i = 0 \Rightarrow ac = 0, bd = 0$

It is given that $z \neq 0$ and $w \neq 0$, so only one of a and b can be zero, and only one of c and d can be zero. If $a \neq 0$, then $b = 0$, $c = 0$, and $d \neq 0$. One example is $z = a + 0i$, $w = 0 + di$.

1.4 Getting Ready for the Next Section

93. $x^3 - x^2 - 9x + 9 = (x^3 - x^2) - (9x - 9)$

$$= x^2(x-1) - 9(x-1)$$

$$= (x^2 - 9)(x-1)$$

$$= (x-3)(x+3)(x-1)$$

95. $\dfrac{1}{x-1}, \dfrac{1}{x}$

LCD $= x(x-1)$

97. $\dfrac{15}{x+3}, \dfrac{2x+1}{x-3}$

LCD $= (x+3)(x-3)$

99. $\dfrac{x+1}{(x-3)(2-x)}, \dfrac{12}{(x-2)(x+1)}$

Note that $x - 2 = -(2-x)$.

The LCD is $(x-3)(x-2)(x+1)$.

101. $27^{2/3} = \left(\sqrt[3]{27}\right)^2 = 3^2 = 9$

103. $4^{5/2} = \left(\sqrt{4}\right)^5 = 2^5 = 32$

105. $\left(\sqrt{3+x^2}\right)^2 = 3 + x^2$

107. $\left(\left(3x^2+7\right)^{1/3}\right)^3 = \left(3x^2+7\right)^{(1/3)(3)} = 3x^2 + 7$

1.5 Solving Other Types of Equations

1.5 Practice Problems

1. $x^3 + 2x^2 - x - 2 = 0$

$x^2(x+2) - (x+2) = 0$

$(x^2 - 1)(x+2) = 0$

$(x-1)(x+1)(x+2) = 0$

$x - 1 = 0 \Rightarrow x = 1$ or $x + 1 = 0 \Rightarrow x = -1$ or $x + 2 = 0 \Rightarrow x = -2$

Solution set: $\{-2, -1, 1\}$

2.
$$x^4 = 4x^2$$
$$x^4 - 4x^2 = 0$$
$$x^2(x^2 - 4) = 0$$
$$x^2(x-2)(x+2) = 0$$
$x^2 = 0 \Rightarrow x = 0$ or $x - 2 = 0 \Rightarrow x = 2$ or
$x + 2 = 0 \Rightarrow x = -2$
Solution set: $\{-2, 0, 2\}$

3.
$$\frac{1}{x} - \frac{12}{5x+10} = \frac{1}{5}$$
$$\frac{1}{x} - \frac{12}{5(x+2)} = \frac{1}{5}$$
Multiply by the LCD, $5x(x+2)$.
$$5(x+2) - 12x = x(x+2)$$
$$5x + 10 - 12x = x^2 + 2x$$
$$0 = x^2 + 9x - 10$$
$$0 = (x+10)(x-1)$$
$x + 10 = 0 \Rightarrow x = -10$ or $x - 1 = 0 \Rightarrow x = 1$
Be sure to check the solutions in the original
equation.
Solution set: $\{-10, 1\}$

4.
$$\frac{x}{x-2} - \frac{2}{x+2} = \frac{4x}{x^2-4}$$
$$\frac{x}{x-2} - \frac{2}{x+2} = \frac{4x}{(x-2)(x+2)}$$
Multiply by the common denominator
$(x-2)(x+2)$.
$$x(x+2) - 2(x-2) = 4x$$
$$x^2 + 2x - 2x + 4 = 4x$$
$$x^2 - 4x + 4 = 0$$
$$(x-2)^2 = 0 \Rightarrow x = 2$$
Since neither $\frac{x}{x-2}$ nor $\frac{4x}{x^2-4}$ is defined for
$x = 2$, this is an extraneous solution.
Solution set: $\varnothing$

5.
$$x = \sqrt{x^3 - 6x}$$
$$x^2 = \left(\sqrt{x^3 - 6x}\right)^2 = x^3 - 6x$$
$$0 = x^3 - x^2 - 6x$$
$$= x(x^2 - x - 6)$$
$$= x(x-3)(x+2)$$
$x = 0$ or $x - 3 = 0 \Rightarrow x = 3$ or
$x + 2 = 0 \Rightarrow x = -2$

Now check to see if -2, 0, or 3 are extraneous
solutions. If $x = -2$, then
$\sqrt{(-2)^3 - 6(-2)} = \sqrt{4} = 2 \neq -2$, so -2 is an
extraneous solution. If $x = 0$, then
$\sqrt{0^3 - 6(0)} = \sqrt{0} = 0$, so 0 is a solution.
If $x = 3$, then $\sqrt{3^3 - 6(3)} = \sqrt{9} = 3$, so 3 is a
solution.
Solution set: $\{0, 3\}$

6.
$$\sqrt{6x+4} + 2 = x$$
$$\sqrt{6x+4} = x - 2$$
$$\left(\sqrt{6x+4}\right)^2 = (x-2)^2$$
$$6x + 4 = x^2 - 4x + 4$$
$$0 = x^2 - 10x = x(x-10)$$
$x = 0$ or $x - 10 = 0 \Rightarrow x = 10$
Now check to see if 0 or 10 are extraneous
solutions. If $x = 0$, then
$\sqrt{6(0)+4} + 2 = \sqrt{4} + 2 = 2 + 2 = 4 \neq 0$, so 0 is
an extraneous solution. If $x = 10$, then
$\sqrt{6(10)+4} + 2 = \sqrt{64} + 2 = 8 + 2 = 10$, so 10
is a solution.
Solution set: $\{10\}$

7.
$$\sqrt{x-5} + \sqrt{x} = 5$$
$$\sqrt{x-5} = 5 - \sqrt{x}$$
$$\left(\sqrt{x-5}\right)^2 = \left(5 - \sqrt{x}\right)^2$$
$$x - 5 = 25 - 10\sqrt{x} + x$$
$$-30 = -10\sqrt{x}$$
$$3 = \sqrt{x} \Rightarrow 9 = x$$
Now check to see if 9 is an extraneous
solution.
$\sqrt{9-5} + \sqrt{9} = \sqrt{4} + \sqrt{9} = 2 + 3 = 5$, so 9 is a
solution.
Solution set: $\{9\}$

8. a.
$$3(x-2)^{3/5} + 4 = 7$$
$$3(x-2)^{3/5} = 3$$
$$(x-2)^{3/5} = 1$$
$$\left[(x-2)^{3/5}\right]^{5/3} = 1^{5/3}$$
$$x - 2 = 1$$
$$x = 3$$
Be sure to check that 3 satisfies the original
equation.
Solution set: $\{3\}$

b. $(2x+1)^{4/3} - 7 = 9$

$$(2x+1)^{4/3} = 16$$

$$\left[(2x+1)^{4/3}\right]^{3/4} = \pm\left(16^{3/4}\right)$$

$$2x+1 = \pm 8$$

$2x+1 = -8$	$2x+1 = 8$
$2x = -9$	$2x = 7$
$x = -\dfrac{9}{2}$	$x = \dfrac{7}{2}$

Check:

$$\left[2\left(-\frac{9}{2}\right)+1\right]^{4/3} - 7 \overset{?}{=} 9$$

$$(-9+1)^{4/3} - 7 \overset{?}{=} 9$$

$$(-8)^{4/3} - 7 \overset{?}{=} 9$$

$$16 - 7 = 9$$

$$\left[2\left(\frac{7}{2}\right)+1\right]^{4/3} - 7 \overset{?}{=} 9$$

$$(7+1)^{4/3} - 7 \overset{?}{=} 9$$

$$(8)^{4/3} - 7 \overset{?}{=} 9$$

$$16 - 7 = 9$$

Solution set: $\left\{-\dfrac{9}{2}, \dfrac{7}{2}\right\}$

9. $x^{2/3} - 7x^{1/3} + 6 = 0$

Let $u = x^{1/3}$. Then the original equation becomes

$$u^2 - 7u + 6 = 0 \Rightarrow (u-1)(u-6) = 0 \Rightarrow$$
$$u = 1, 6$$

Now solve for x:

$$1 = x^{1/3} \Rightarrow 1^3 = \left(x^{1/3}\right)^3 \Rightarrow 1 = x$$

$$6 = x^{1/3} \Rightarrow 6^3 = \left(x^{1/3}\right)^3 \Rightarrow 216 = x$$

Be sure to check that both solutions satisfy the original equation.
Solution set: $\{1, 216\}$

10. $\left(1+\dfrac{1}{x}\right)^2 - 6\left(1+\dfrac{1}{x}\right) + 8 = 0$

Let $u = \left(1+\dfrac{1}{x}\right)$. Then the original equation becomes

$$u^2 - 6u + 8 = 0 \Rightarrow (u-2)(u-4) = 0 \Rightarrow$$
$$u = 2, 4$$

Now solve for x:

$$1 + \frac{1}{x} = 2 \Rightarrow x+1 = 2x \Rightarrow x = 1$$

$$1 + \frac{1}{x} = 4 \Rightarrow x+1 = 4x \Rightarrow 1 = 3x \Rightarrow \frac{1}{3} = x$$

Be sure to check that both solutions satisfy the original equation.

Solution set: $\left\{\dfrac{1}{3}, 1\right\}$

11. $t_0 = t\sqrt{1 - \dfrac{v^2}{c^2}}$, $t_0 = 20$, $t = 25$

$$20 = 25\sqrt{1 - \frac{v^2}{c^2}} \Rightarrow \frac{4}{5} = \sqrt{1 - \frac{v^2}{c^2}} \Rightarrow$$

$$\frac{16}{25} = 1 - \frac{v^2}{c^2} \Rightarrow \frac{v^2}{c^2} = 1 - \frac{16}{25} = \frac{9}{25} \Rightarrow$$

$$\frac{v}{c} = \frac{3}{5} \Rightarrow v = 0.6c$$

The spacecraft must have been traveling at 60% of the speed of light.

1.5 Concepts and Vocabulary

1. If an apparent solution does not satisfy the equation, it is called an <u>extraneous</u> solution.

3. If $x^{3/4} = 8$, then $x = \underline{16}$.

5. False

7. False. The solution for the equation $x^2 = 9$ is $x = \pm 3$.

1.5 Building Skills

9. $x^3 = 2x^2 \Rightarrow x^3 - 2x^2 = 0 \Rightarrow x^2(x-2) = 0 \Rightarrow$
$x^2 = 0$ or $x - 2 = 0 \Rightarrow x = 0$ or $x = 2$

11. $\left(\sqrt{x}\right)^3 = \sqrt{x} \Rightarrow \left(\sqrt{x}\right)^2\sqrt{x} = \sqrt{x} \Rightarrow$
$x\sqrt{x} = \sqrt{x} \Rightarrow x\sqrt{x} - \sqrt{x} = 0 \Rightarrow$
$\sqrt{x}(x-1) = 0 \Rightarrow \sqrt{x} = 0$ or $(x-1) = 0 \Rightarrow$
$x = 0$ or $x = 1$

13. $x^3 + x = 0 \Rightarrow x(x^2 + 1) = 0 \Rightarrow$
$x = 0$ or $x^2 + 1 = 0 \Rightarrow x = 0$ or $x^2 = -1 \Rightarrow$
$x = 0$ or $x = i$
We are looking for real roots only, so we reject $x = i$. Solution set: $\{0\}$

15.
$$x^4 - x^3 = x^2 - x$$
$$x^4 - x^3 - x^2 + x = 0$$
Factor by grouping.
$$x^3(x-1) - x(x-1) = 0$$
$$(x^3 - x)(x-1) = 0$$
$$x(x^2 - 1)(x-1) = 0$$
$$x(x+1)(x-1)(x-1) = 0$$
$$x = 0 \text{ or } x = -1 \text{ or } x = 1$$
Solution set: $\{-1, 0, 1\}$

17.
$$x^4 = 27x$$
$$x^4 - 27x = 0$$
$$x(x^3 - 27) = 0$$
$$x(x^3 - 3^3) = 0$$
$$x(x-3)(x^2 + 3x + 9) = 0$$
$$x = 0 \text{ or } x = 3 \text{ or } x^2 + 3x + 9 = 0$$
Solving $x^2 + 3x + 9 = 0$ gives
$$x = -\frac{3}{2} \pm \frac{3i\sqrt{3}}{2}.$$ However, we are looking for
the real roots only, so we reject these roots.
Solution set: $\{0, 3\}$

19.
$$\frac{x+1}{3x-2} = \frac{5x-4}{3x+2}$$
$$(x+1)(3x+2) = (5x-4)(3x-2)$$
$$3x^2 + 5x + 2 = 15x^2 - 22x + 8$$
$$-12x^2 + 27x - 6 = 0$$
$$-3(4x^2 - 9x + 2) = 0$$
$$4x^2 - 9x + 2 = 0 \Rightarrow (4x-1)(x-2) = 0$$
$$4x - 1 = 0 \text{ or } x - 2 = 0 \Rightarrow x = \frac{1}{4} \text{ or } x = 2$$

21.
$$\frac{1}{x} + \frac{2}{x+1} = 1$$
$$(x)(x+1)\left(\frac{1}{x} + \frac{2}{x+1}\right) = (x)(x+1)(1)$$
$$(x+1) + 2x = x^2 + x$$
$$3x + 1 = x^2 + x$$
$$0 = x^2 - 2x - 1$$
Solve using the quadratic formula.
$$x = \frac{-(-2) \pm \sqrt{(-2)^2 - 4(1)(-1)}}{2}$$
$$x = \frac{2 \pm \sqrt{4+4}}{2} = \frac{2 \pm \sqrt{8}}{2} = \frac{2 \pm 2\sqrt{2}}{2} = 1 \pm \sqrt{2}$$

23.
$$\frac{6x-7}{x} - \frac{1}{x^2} = 5$$
$$x^2\left(\frac{6x-7}{x} - \frac{1}{x^2}\right) = 5x^2$$
$$x(6x-7) - 1 = 5x^2$$
$$6x^2 - 7x - 1 = 5x^2$$
$$x^2 - 7x - 1 = 0$$
Solve using the quadratic formula.
$$x = \frac{-(-7) \pm \sqrt{(-7)^2 - 4(1)(-1)}}{2}$$
$$x = \frac{7 \pm \sqrt{49+4}}{2} = \frac{7 \pm \sqrt{53}}{2} = \frac{7}{2} \pm \frac{\sqrt{53}}{2}$$

25.
$$\frac{1}{x} + \frac{1}{x-3} = \frac{7}{3x-5}$$
$$x(x-3)(3x-5)\left(\frac{1}{x} + \frac{1}{x-3} = \frac{7}{3x-5}\right)$$
$$(x-3)(3x-5) + x(3x-5) = 7x(x-3)$$
$$3x^2 - 14x + 15 + 3x^2 - 5x = 7x^2 - 21x$$
$$6x^2 - 19x + 15 = 7x^2 - 21x$$
$$-x^2 + 2x + 15 = 0$$
$$-1(x^2 - 2x - 15) = 0$$
$$x^2 - 2x - 15 = 0$$
$$(x-5)(x+3) = 0 \Rightarrow$$
$$x - 5 = 0 \text{ or } x + 3 = 0 \Rightarrow x = 5 \text{ or } x = -3$$

27.
$$\frac{5}{x+1} - \frac{4}{2x+2} + \frac{2}{2x-1} = \frac{13}{18}$$
$$\frac{5}{x+1} - \frac{4}{2(x+1)} + \frac{2}{2x-1} = \frac{13}{18}$$
$$18(x+1)(2x-1)\left(\frac{5}{x+1} - \frac{4}{2(x+1)} + \frac{2}{2x-1}\right)$$
$$= 18(x+1)(2x-1)\left(\frac{13}{18}\right)$$
$$90(2x-1) - 36(2x-1) + 36(x+1)$$
$$= 13(x+1)(2x-1)$$
$$180x - 90 - 72x + 36 + 36x + 36$$
$$= 13(2x^2 + x - 1)$$
$$144x - 18 = 26x^2 + 13x - 13$$
$$0 = 26x^2 - 131x + 5$$
$$0 = (26x - 1)(x - 5)$$
$$26x - 1 = 0 \text{ or } x - 5 = 0$$
$$x = \frac{1}{26} \text{ or } x = 5$$

29.
$$\frac{x}{x-5} - \frac{5}{x+5} = \frac{10x}{x^2-25}$$
$$x(x+5) - 5(x-5) = 10x$$
$$x^2 + 5x - 5x + 25 = 10x$$
$$x^2 - 10x + 25 = 0$$
$$(x-5)^2 = 0 \Rightarrow x = 5$$

Since $x = 5$ makes the denominator in the first fraction equal zero, there is no solution. Solution set: $\varnothing$

31.
$$\frac{x}{x-3} + \frac{3}{x+3} = \frac{6x}{x^2-9}$$
$$x(x+3) + 3(x-3) = 6x$$
$$x^2 + 3x + 3x - 9 = 6x$$
$$x^2 - 9 = 0$$
$$(x+3)(x-3) = 0 \Rightarrow x = -3, 3$$

Since $x = -3$ makes the denominator in the second fraction equal zero and $x = 3$ makes the denominator in the first fraction equal zero, there is no solution. Solution set: $\varnothing$

33.
$$\frac{1}{x-1} + \frac{x}{x+3} = \frac{4}{x^2+2x-3}$$
$$\frac{1}{x-1} + \frac{x}{x+3} = \frac{4}{(x-1)(x+3)}$$
$$(x+3) + x(x-1) = 4$$
$$x+3 + x^2 - x = 4 \Rightarrow x^2 = 1 \Rightarrow x = \pm 1$$

If $x = 1$, then the denominator in the first fraction equals zero, so 1 is an extraneous solution. Solution set: $\{-1\}$

35.
$$\frac{2x}{x+3} - \frac{x}{x-1} = \frac{14}{x^2+2x-3}$$
$$\frac{2x}{x+3} - \frac{x}{x-1} = \frac{14}{(x+3)(x-1)}$$
$$2x(x-1) - x(x+3) = 14$$
$$2x^2 - 2x - x^2 - 3x = 14$$
$$x^2 - 5x - 14 = 0$$
$$(x-7)(x+2) = 0 \Rightarrow x = 7, -2$$

Solution set: $\{-2, 7\}$

37. $\sqrt[3]{3x-1} = 2 \Rightarrow \left(\sqrt[3]{3x-1}\right)^3 = 2^3 \Rightarrow$
$3x - 1 = 8 \Rightarrow 3x = 9 \Rightarrow x = 3$

39. There is no solution for $\sqrt{x-1} = -2$ because the square root is not negative. The solution set is $\varnothing$.

41. $x + \sqrt{x+6} = 0 \Rightarrow x = -\sqrt{x+6} \Rightarrow$
$x^2 = \left(-\sqrt{x+6}\right)^2 \Rightarrow x^2 = x + 6 \Rightarrow$
$x^2 - x - 6 = 0 \Rightarrow (x-3)(x+2) = 0 \Rightarrow$
$x = 3$ or $x = -2$
Now check to see if 3 or −2 are extraneous roots. If $x = 3$, then $x + \sqrt{x+6} = 3 + \sqrt{3+6} = $
$3 + 3 \neq 0$. So 3 is extraneous. If $x = -2$, then $x + \sqrt{x+6} = -2 + \sqrt{-2+6} = -2 + \sqrt{4} = $
$-2 + 2 = 0$. Solution set: $\{-2\}$

43. $\sqrt{y+6} = y \Rightarrow \left(\sqrt{y+6}\right)^2 = y^2 \Rightarrow y + 6 = y^2 \Rightarrow$
$0 = y^2 - y - 6 \Rightarrow 0 = (y-3)(y+2) \Rightarrow$
$y = 3$ or $y = -2$
Now check to see if 3 or −2 are extraneous roots. If $y = 3$, then $\sqrt{y+6} = y \Rightarrow$
$\sqrt{3+6} = 3 \Rightarrow \sqrt{9} = 3 \Rightarrow 3 = 3$. So 3 is a solution. If $y = -2$, then $\sqrt{y+6} = y \Rightarrow$
$\sqrt{-2+6} = -2 \Rightarrow \sqrt{4} = -2 \Rightarrow 2 \neq -2$. So −2 is extraneous. Solution set: $\{3\}$

45.
$$\sqrt{6y-11} = 2y - 7$$
$$\left(\sqrt{6y-11}\right)^2 = (2y-7)^2$$
$$6y - 11 = 4y^2 - 28y + 49$$
$$0 = 4y^2 - 34y + 60$$
$$0 = 2y^2 - 17y + 30$$
$$0 = (2y-5)(y-6) \Rightarrow y = \frac{5}{2} \text{ or } y = 6$$

Now check to see if 5/2 or 6 are extraneous roots. If $y = \frac{5}{2}$, then $\sqrt{6y-11} = 2y - 7 \Rightarrow$
$\sqrt{6\left(\frac{5}{2}\right)-11} = 2\left(\frac{5}{2}\right) - 7 \Rightarrow \sqrt{15-11} = 5 - 7 \Rightarrow$
$\sqrt{4} = -2 \Rightarrow 2 \neq -2$. So 5/2 is extraneous.
If $y = 6$, then $\sqrt{6y-11} = 2y - 7 \Rightarrow$
$\sqrt{6(6)-11} = 2(6) - 7 \Rightarrow \sqrt{36-11} = 12 - 7 \Rightarrow$
$\sqrt{25} = 5 \Rightarrow 5 = 5$. Solution set: $\{6\}$

47.
$$t - \sqrt{3t+6} = -2$$
$$t + 2 = \sqrt{3t+6}$$
$$(t+2)^2 = \left(\sqrt{3t+6}\right)^2$$
$$t^2 + 4t + 4 = 3t + 6$$
$$t^2 + t - 2 = 0$$
$$(t+2)(t-1) = 0 \Rightarrow t = -2 \text{ or } t = 1$$

(continued on next page)

(*continued*)

Now check to see if −2 or 1 are extraneous roots. If $t = -2$, then $t - \sqrt{3t+6} = -2 \Rightarrow$
$-2 - \sqrt{3(-2)+6} = -2 \Rightarrow -2 - \sqrt{-6+6} \Rightarrow$
$-2 - 0 = -2$. So −2 is a solution.
If $t = 1$, then $t - \sqrt{3t+6} = -2 \Rightarrow$
$1 - \sqrt{3(1)+6} = -2 \Rightarrow 1 - \sqrt{9} = -2 \Rightarrow$
$1 - 3 = -2 \Rightarrow -2 = -2$. So 1 is a solution.
Solution set: $\{-2, 1\}$

49. $\sqrt{x-3} = \sqrt{2x-5} - 1$
$$\left(\sqrt{x-3}\right)^2 = \left(\sqrt{2x-5}-1\right)^2$$
$$x - 3 = (2x-5) - 2\sqrt{2x-5} + 1$$
$$x - 3 = 2x - 4 - 2\sqrt{2x-5}$$
$$2\sqrt{2x-5} = x - 1 \Rightarrow \left(2\sqrt{2x-5}\right)^2 = (x-1)^2$$
$$4(2x-5) = x^2 - 2x + 1$$
$$8x - 20 = x^2 - 2x + 1$$
$$0 = x^2 - 10x + 21$$
$$0 = (x-7)(x-3) \Rightarrow x = 7 \text{ or } x = 3$$

Now check to see if 3 or 7 are extraneous roots. If $x = 3$, then $\sqrt{x-3} = \sqrt{2x-5} - 1 \Rightarrow$
$\sqrt{3-3} = \sqrt{2(3)-5} - 1 \Rightarrow 0 = 0$. So 3 is a solution.
If $x = 7$, then $\sqrt{x-3} = \sqrt{2x-5} - 1 \Rightarrow$
$\sqrt{7-3} = \sqrt{2(7)-5} - 1 \Rightarrow \sqrt{4} = \sqrt{9} - 1 \Rightarrow$
$2 = 2$. So 7 is a solution.
Solution set: $\{3, 7\}$

51. $\sqrt{2y+9} = 2 + \sqrt{y+1}$
$$\left(\sqrt{2y+9}\right)^2 = \left(2 + \sqrt{y+1}\right)^2$$
$$2y + 9 = 4 + 4\sqrt{y+1} + y + 1$$
$$y + 4 = 4\sqrt{y+1}$$
$$(y+4)^2 = \left(4\sqrt{y+1}\right)^2$$
$$y^2 + 8y + 16 = 16(y+1)$$
$$y^2 + 8y + 16 = 16y + 16$$
$$y^2 - 8y = 0$$
$$y(y-8) = 0 \Rightarrow y = 0 \text{ or } y = 8$$

Now check to see if 0 or 8 are extraneous roots. If $y = 0$, then $\sqrt{2y+9} = 2 + \sqrt{y+1} \Rightarrow$
$\sqrt{2(0)+9} = 2 + \sqrt{0+1} \Rightarrow \sqrt{9} = 2 + 1 \Rightarrow 3 = 3$.

So 0 is a solution. If $y = 8$, then
$\sqrt{2y+9} = 2 + \sqrt{y+1} \Rightarrow \sqrt{2(8)+9} =$
$2 + \sqrt{8+1} \Rightarrow \sqrt{25} = 2 + \sqrt{9} \Rightarrow 5 = 5$. So 8 is a solution. Solution set: $\{0, 8\}$

53. $\sqrt{7z+1} - \sqrt{5z+4} = 1$
$$\sqrt{7z+1} = \sqrt{5z+4} + 1$$
$$\left(\sqrt{7z+1}\right)^2 = \left(\sqrt{5z+4}+1\right)^2$$
$$7z + 1 = 5z + 4 + 2\sqrt{5z+4} + 1$$
$$2z - 4 = 2\sqrt{5z+4}$$
$$(2z-4)^2 = \left(2\sqrt{5z+4}\right)^2$$
$$4z^2 - 16z + 16 = 4(5z+4)$$
$$4z^2 - 16z + 16 = 20z + 16$$
$$4z^2 - 36z = 0 \Rightarrow 4z(z-9) = 0 \Rightarrow$$
$$z = 0 \text{ or } z = 9$$

Now check to see if 0 or 9 are extraneous roots. If $z = 0$, then $\sqrt{7z+1} - \sqrt{5z+4} = 1 \Rightarrow$
$\sqrt{7(0)+1} - \sqrt{5(0)+4} = 1 \Rightarrow \sqrt{1} - \sqrt{4} = 1 \Rightarrow$
$1 - 2 \neq 1$. So 0 is extraneous. If $z = 9$, then
$\sqrt{7z+1} - \sqrt{5z+4} = 1 \Rightarrow \sqrt{7(9)+1} - \sqrt{5(9)+4}$
$= 1 \Rightarrow \sqrt{64} - \sqrt{49} \Rightarrow 8 - 7 = 1 \Rightarrow 1 = 1$. So 9 is a solution. Solution set: $\{9\}$

55. $\sqrt{2x+5} + \sqrt{x+6} = 3$
$$\sqrt{2x+5} = 3 - \sqrt{x+6}$$
$$\left(\sqrt{2x+5}\right)^2 = \left(3 - \sqrt{x+6}\right)^2$$
$$2x + 5 = 9 - 6\sqrt{x+6} + x + 6$$
$$x - 10 = -6\sqrt{x+6}$$
$$(x-10)^2 = \left(-6\sqrt{x+6}\right)^2$$
$$x^2 - 20x + 100 = 36x + 216$$
$$x^2 - 56x - 116 = 0$$
$$(x-58)(x+2) = 0 \Rightarrow x = 58 \text{ or } x = -2$$

Now check to see if 58 or −2 are extraneous roots. If $x = 58$, then $\sqrt{2x+5} + \sqrt{x+6} = 3 \Rightarrow$
$\sqrt{2(58)+5} + \sqrt{58+6} = 3 \Rightarrow \sqrt{121} + \sqrt{64} = 3 \Rightarrow$
$11 + 8 \neq 3$. So 58 is extraneous. If $x = -2$,
then $\sqrt{2x+5} + \sqrt{x+6} = 3 \Rightarrow \sqrt{2(-2)+5} +$
$\sqrt{-2+6} = 3 \Rightarrow \sqrt{1} + \sqrt{4} \Rightarrow 1 + 2 = 3$.
So −2 is a solution. Solution set: $\{-2\}$

57.
$$\sqrt{2x-5} - \sqrt{x-3} = 1$$
$$\sqrt{2x-5} = 1 + \sqrt{x-3}$$
$$\left(\sqrt{2x-5}\right)^2 = \left(1 + \sqrt{x-3}\right)^2$$
$$2x-5 = 1 + 2\sqrt{x-3} + x - 3$$
$$x - 3 = 2\sqrt{x-3}$$
$$(x-3)^2 = \left(2\sqrt{x-3}\right)^2$$
$$x^2 - 6x + 9 = 4x - 12$$
$$x^2 - 10x + 21 = 0$$
$$(x-7)(x-3) = 0 \Rightarrow x = 7 \text{ or } x = 3$$

Now check to see if 7 or 3 are extraneous roots. If $x = 7$, then $\sqrt{2x-5} - \sqrt{x-3} = 1 \Rightarrow$ $\sqrt{2(7)-5} - \sqrt{7-3} = 1 \Rightarrow \sqrt{9} - \sqrt{4} = 1 \Rightarrow$ $3 - 2 = 1$. So 7 is a solution. If $x = 3$, then $\sqrt{2x-5} - \sqrt{x-3} = 1 \Rightarrow \sqrt{2(3)-5} -$ $\sqrt{3-3} = 1 \Rightarrow \sqrt{1} - \sqrt{0} = 1 \Rightarrow 1 = 1$. So 3 is a solution. Solution set: $\{3, 7\}$

59.
$$(x-4)^{3/2} = 27$$
$$\left[(x-4)^{3/2}\right]^{2/3} = 27^{2/3}$$
$$x - 4 = 9 \Rightarrow x = 13$$

Verify that 13 satisfies the original equation. Solution set: $\{13\}$

61.
$$(5x-3)^{2/3} - 5 = 4$$
$$(5x-3)^{2/3} = 9$$
$$\left[(5x-3)^{2/3}\right]^{3/2} = \pm 9^{3/2}$$
$$5x - 3 = \pm 27$$

$$\begin{array}{c|c} 5x-3 = -27 & 5x-3 = 27 \\ 5x = -24 & 5x = 30 \\ x = -\dfrac{24}{5} & x = 6 \end{array}$$

If $x = -24/5$, then

$$\left(5\left(-\frac{24}{5}\right) - 3\right)^{2/3} - 5 = (-24-3)^{2/3} - 5$$
$$= (-27)^{2/3} - 5$$
$$= 9(-1)^{2/3} - 5 = 4,$$

so $x = -\dfrac{24}{5}$ is a solution. Verify that 6 satisfies the original equation.

Solution set: $\left\{-\dfrac{24}{5}, 6\right\}$

63. Let $u = \sqrt{x}$. Then $u^2 = x$. So the equation becomes $u^2 - 5u + 6 = 0 \Rightarrow$ $(u-3)(u-2) = 0 \Rightarrow u = 3$ or $u = 2$. Solve for x: $3 = \sqrt{x} \Rightarrow x = 9$ or $2 = \sqrt{x} \Rightarrow 4 = x$. Check to make sure that neither solution is extraneous. If $x = 9$, then $9 - 5\sqrt{9} + 6 = 0 \Rightarrow$ $9 - 15 + 6 = 0 \Rightarrow 0 = 0$. If $x = 4$, then $4 - 5\sqrt{4} + 6 = 0 \Rightarrow 4 - 10 + 6 = 0 \Rightarrow 0 = 0$. Solution set: $\{4, 9\}$

65. Let $u = \sqrt{y}$. Then $u^2 = y$. So the equation becomes $2u^2 - 15u = -7 \Rightarrow$ $2u^2 - 15u + 7 = 0 \Rightarrow (2u-1)(u-7) = 0 \Rightarrow$ $u = \dfrac{1}{2}$ or $u = 7$. Solve for y: $\dfrac{1}{2} = \sqrt{y} \Rightarrow$ $\dfrac{1}{4} = y$ or $7 = \sqrt{y} \Rightarrow 49 = y$. Now check to make sure that neither solution is extraneous.

If $y = \dfrac{1}{4}$, then $2\left(\dfrac{1}{4}\right) - 15\sqrt{\dfrac{1}{4}} = -7 \Rightarrow$ $\dfrac{1}{2} - \dfrac{15}{2} = -7 \Rightarrow -7 = -7$.

If $y = 49$, then $2(49) - 15\sqrt{49} = -7 \Rightarrow 98 - 105 = -7 \Rightarrow$ $-7 = -7$. Solution set: $\left\{\dfrac{1}{4}, 49\right\}$

67. $x^{-2} - x^{-1} - 42 = 0$

Let $u = x^{-1}$. Then the equation becomes $u^2 - u - 42 = 0 \Rightarrow (u-7)(u+6) = 0 \Rightarrow$ $u = 7, -6$. Now solve for x:

$7 = x^{-1} \Rightarrow x = \dfrac{1}{7}$; $-6 = x^{-1} \Rightarrow x = -\dfrac{1}{6}$

Be sure to check that neither of the solutions are extraneous. Solution set: $\left\{-\dfrac{1}{6}, \dfrac{1}{7}\right\}$

69. $x^{2/3} - 6x^{1/3} + 8 = 0$

Let $u = x^{1/3}$. Then the equation becomes $u^2 - 6x + 8 = 0 \Rightarrow (u-4)(u-2) = 0 \Rightarrow$ $u = 4, 2$. Now solve for x:

$x^{1/3} = 4 \Rightarrow x = 64$; $x^{1/3} = 2 \Rightarrow x = 8$

Be sure to check that neither of the solutions are extraneous. Solution set: $\{8, 64\}$

71. $2x^{1/2} + 3x^{1/4} - 2 = 0$

Let $u = x^{1/4}$. Then the equation becomes

$2u^2 + 3u - 2 = 0 \Rightarrow (2u - 1)(u + 2) = 0 \Rightarrow$

$u = \dfrac{1}{2}, -2$. Now solve for x:

$x^{1/4} = \dfrac{1}{2} \Rightarrow x = \dfrac{1}{16}$; $x^{1/4} = -2 \Rightarrow x$ is not a

real number. Be sure to check that $\dfrac{1}{16}$ is not

extraneous. Solution set: $\left\{\dfrac{1}{16}\right\}$

73. Let $u = x^2$. Then $x^4 = u^2$. So the equation

becomes $u^2 - 13u + 36 = 0 \Rightarrow$

$(u - 4)(u - 9) = 0 \Rightarrow u = 4$ or $u = 9$. Now

solve for x: $4 = x^2 \Rightarrow \pm 2 = x$ or

$9 = x^2 \Rightarrow \pm 3 = x$. Now check to make sure

that none of the solutions are extraneous.

If $x = \pm 2$, then $(\pm 2)^4 - 13(\pm 2)^2 + 36 = 0 \Rightarrow$

$16 - 52 + 36 = 0 \Rightarrow 0 = 0$. If $x = \pm 3$, then

$(\pm 3)^4 - 13(\pm 3)^2 + 36 = 0 \Rightarrow 81 - 117 + 36 = 0$

$\Rightarrow 0 = 0$. The solutions are ± 2 and ± 3.

75. Let $u = t^2$. Then $t^4 = u^2$. So the equation

becomes $2u^2 + u - 1 = 0 \Rightarrow (2u - 1)(u + 1) = 0$

$\Rightarrow u = \dfrac{1}{2}$ or $u = -1$. Now solve for t:

$\dfrac{1}{2} = t^2 \Rightarrow \pm\sqrt{\dfrac{1}{2}} = \pm\dfrac{\sqrt{2}}{2} = t$ or $-1 = t^2 \Rightarrow$

$\pm i = t$. Now check to make sure that none of

the solutions are extraneous.

If $t = \pm\dfrac{\sqrt{2}}{2}$, then $2\left(\pm\dfrac{\sqrt{2}}{2}\right)^4 + \left(\dfrac{\sqrt{2}}{2}\right)^2 - 1 = 0 \Rightarrow$

$\dfrac{1}{2} + \dfrac{1}{2} - 1 = 0 \Rightarrow 0 = 0$. If $x = \pm i$, then

$2(\pm i)^4 + (\pm i)^2 - 1 = 0 \Rightarrow 2 - 1 - 1 = 0 \Rightarrow 0 = 0$.

The solutions are $\pm i$ and $\pm\dfrac{\sqrt{2}}{2}$.

77. Let $u = \sqrt{p^2 - 3}$. Then $p^2 - 3 = u^2$. So the

equation becomes $u^2 + 4u - 5 = 0 \Rightarrow$

$(u + 5)(u - 1) = 0 \Rightarrow u = -5$ or $u = 1$.

Now solve for p: $-5 = \sqrt{p^2 - 3}$, which is not

possible, or $1 = \sqrt{p^2 - 3} \Rightarrow$

$1 = p^2 - 3 \Rightarrow 1 = p^2 - 3 \Rightarrow 4 = p^2 \Rightarrow \pm 2 = p$.

Now check to make sure that none of the

solutions are extraneous. If $p = \pm 2$, then

$(\pm 2)^2 - 3 + 4\sqrt{(\pm 2)^2 - 3} - 5 = 0 \Rightarrow$

$4 - 3 + 4\sqrt{1} - 5 = 0 \Rightarrow 0 = 0$.

Solution set: $\{\pm 2\}$.

79. Let $u = 3t + 1$. Then the equation becomes

$u^2 - 3u + 2 = 0 \Rightarrow (u - 2)(u - 1) = 0 \Rightarrow$

$u = 2$ or $u = 1$. Now solve for t: $2 = 3t + 1 \Rightarrow$

$\dfrac{1}{3} = t$ or $1 = 3t + 1 \Rightarrow 0 = t$. Now check to

make sure that neither solution is extraneous.

If $t = 0$,

$(3 \bullet 0 + 1)^2 - 3(3 \bullet 0 + 1) + 2 = 0 \Rightarrow$

$1 - 3 + 2 = 0 \Rightarrow 0 = 0$, so 0 is a solution. If

$t = \dfrac{1}{3}$, then $\left(3\left(\dfrac{1}{3}\right) + 1\right)^2 - 3\left(3\left(\dfrac{1}{3}\right) + 1\right) + 2 = 0 \Rightarrow$

$4 - 6 + 2 = 0 \Rightarrow 0 = 0$. So $1/3$ is a solution.

Solution set: $\left\{0, \dfrac{1}{3}\right\}$.

81. Let $u = \sqrt{y} + 5$. Then the equation becomes

$u^2 - 9u + 20 = 0 \Rightarrow (u - 5)(u - 4) = 0 \Rightarrow$

$u = 5$ or $u = 4$. Now solve for y:

$5 = \sqrt{y} + 5 \Rightarrow y = 0$ or $4 = \sqrt{y} + 5 \Rightarrow$

$-1 = \sqrt{y}$, which is not defined. Now check to

see if $y = 0$ is extraneous.

$\left(\sqrt{0} + 5\right)^2 - 9\left(\sqrt{0} + 5\right) + 20 = 0 \Rightarrow$

$25 - 45 + 20 = 0 \Rightarrow 0 = 0$.

Solution set: $\{0\}$

83. Let $u = x^2 - 4$. So the equation becomes
$$u^2 - 3u - 4 = 0 \Rightarrow (u-4)(u+1) = 0 \Rightarrow$$
$u = 4$ or $u = -1$. Now solve for x:
$$4 = x^2 - 4 \Rightarrow x = \pm 2\sqrt{2} \text{ or}$$
$-1 = x^2 - 4 \Rightarrow x = \pm\sqrt{3}$. Now check to make sure that neither solution is extraneous. If $x = \pm 2\sqrt{2}$, then
$$\left[(\pm 2\sqrt{2})^2 - 4\right]^2 - 3\left[(\pm 2\sqrt{2})^2 - 4\right] - 4 = 0 \Rightarrow$$
$16 - 12 - 4 = 0 \Rightarrow 0 = 0$. So $\pm 2\sqrt{2}$ are solutions.
If $x = \pm\sqrt{3}$, then
$$\left[(\pm\sqrt{3})^2 - 4\right]^2 - 3\left[(\pm\sqrt{3})^2 - 4\right] - 4 = 0 \Rightarrow$$
$1 + 3 - 4 = 0 \Rightarrow 0 = 0$. So $\pm\sqrt{3}$ are solutions.
Solution set: $\left\{\pm\sqrt{3}, \pm 2\sqrt{2}\right\}$.

85. Let $u = x^2 - 3x$. So the equation becomes
$$u^2 - 2u - 8 = 0 \Rightarrow (u+2)(u-4) = 0 \Rightarrow$$
$u = -2$ or $u = 4$. Now solve for x:
$$-2 = x^2 - 3x \Rightarrow x^2 - 3x + 2 = 0 \Rightarrow$$
$(x-2)(x-1) = 0 \Rightarrow x = 2$ or $x = 1$ or
$$4 = x^2 - 3x \Rightarrow x^2 - 3x - 4 = 0 \Rightarrow$$
$(x-4)(x+1) = 0 \Rightarrow x = 4$ or $x = -1$.

Now check to make sure that none of the solutions are extraneous. If $x = 2$, then
$$(2^2 - 3(2))^2 - 2(2^2 - 3(2)) - 8 = 0 \Rightarrow$$
$4 - (-4) - 8 = 0 \Rightarrow 0 = 0$. So 2 is a solution.
If $x = 1$, then
$$(1^2 - 3(1))^2 - 2(1^2 - 3(1)) - 8 = 0 \Rightarrow$$
$4 + 4 - 8 = 0 \Rightarrow 0 = 0$. So 1 is a solution. If $x = 4$, then
$$(4^2 - 3(4))^2 - 2(4^2 - 3(4)) - 8 = 0 \Rightarrow$$
$16 - 8 - 8 = 0 \Rightarrow 0 = 0$. So 4 is a solution.
If $x = -1$, then
$$\left[(-1)^2 - 3(-1)\right]^2 - 2\left[(-1)^2 - 3(-1)\right] - 8 = 0 \Rightarrow$$
$16 - 8 - 8 = 0 \Rightarrow 0 = 0$. So -1 is a solution.
Solution set: $\{-1, 1, 2, 4\}$.

1.5 Applying the Concepts

87. Let $x =$ the amount of time it takes Todrick to do the job alone. Then, $x + 3 =$ the amount of time it takes his brother to do the job alone.

$$\frac{2}{x} + \frac{2}{x+3} = 1$$
$$x(x+3)\left(\frac{2}{x} + \frac{2}{x+3}\right) = 1 \cdot x(x+3)$$
$$2(x+3) + 2x = x^2 + 3x$$
$$4x + 6 = x^2 + 3x$$
$$0 = x^2 - x - 6$$
$$0 = (x-3)(x+2)$$
$$x = 3 \text{ or } x = -2$$
Disregard the negative solution. Todrick can do the job in 3 days alone; his brother can complete the job in $3+3 = 6$ days alone.

89. $63 + \sqrt{t+40} = 70$
$$\sqrt{t+40} = 7$$
$$t + 40 = 49 \Rightarrow t = 9$$
The pressure will be 70 psi at time $t = 9$.

91. Let $x =$ the denominator. Then $x - 2 =$ the numerator.
$$\frac{x-2}{x} + \frac{x}{x-2} = \frac{25}{12}$$
$$12x(x-2)\left(\frac{x-2}{x} + \frac{x}{x-2}\right) = 12x(x-2)\left(\frac{25}{12}\right)$$
$$12(x-2)^2 + 12x^2 = 25x(x-2)$$
$$24x^2 - 48x + 48 = 25x^2 - 50x$$
$$0 = x^2 - 2x - 48$$
$$0 = (x-8)(x+6)$$
$$x = 8 \text{ or } x = -6$$
The solution must be positive. So the fraction is $\frac{6}{8}$.

93. Let $x =$ the number of shares of stock. Then $1800/x =$ the price of each share of stock.
$$\left(\frac{1800}{x} - 18\right)(x+5) = 1800$$
$$\frac{1800(x+5)}{x} - 18(x+5) = 1800$$
$$\frac{1800x + 9000}{x} - 18x - 90 = 1800$$
$$1800x + 9000 - 18x^2 - 90x = 1800x$$
$$-18x^2 - 90x + 9000 = 0$$
$$-18(x^2 + 5x - 500) = 0$$
$$(x-20)(x+25) = 0$$
$$x = 20 \text{ or } x = -25$$
The answer must be positive, so we reject -25.
a. Latasha bought 20 shares of stock.

b. She paid $1800/20 = \$90$ for each share.

95. Let d = the depth of the well and let t_1 = the time it takes for the stone to hit the water. Using Galileo's formula, we have

$d = 16t_1^2 \Rightarrow \dfrac{\sqrt{d}}{4} = t_1$. Let t_2 = the time it

takes for the sound of the splash to return to

the top of the well. Then $t_2 = \dfrac{d}{1100}$, and

$$t_1 + t_2 = \dfrac{\sqrt{d}}{4} + \dfrac{d}{1100} = 4$$

$$1100\left(\dfrac{\sqrt{d}}{4} + \dfrac{d}{1100}\right) = 1100(4)$$

$$275\sqrt{d} + d = 4400$$

Now let $u = \sqrt{d}$ and $u^2 = d$. This gives $u^2 + 275u = 4400 \Rightarrow u^2 + 275u - 4400 = 0$. Now solve for u using the quadratic formula with $a = 1$, $b = 275$, and $c = -4400$:

$$u = \dfrac{-275 \pm \sqrt{275^2 - 4(1)(-4400)}}{2(1)} \approx 15.2 \ .$$

Note that only the positive answer has meaning, so we reject the negative answer. $u^2 = d \approx 230$. So the well is approximately 230 feet deep.

97. Let x = the speed of the current. Then $10 + x$ = the speed of the motorboat going downstream and $10 - x$ = the speed of the motorboat going

upstream. $\dfrac{12}{10 + x}$ = the time the boat took to

go downstream and $\dfrac{12}{10 - x}$ = the time the

boat took to go upstream. So, we have

$$\dfrac{12}{10 - x} = \dfrac{12}{10 + x} + \dfrac{1}{2}$$

$$2(10 + x)(10 - x)\left(\dfrac{12}{10 - x} = \dfrac{12}{10 + x} + \dfrac{1}{2}\right)$$

$$240 + 24x = 240 - 24x + 100 - x^2$$

$$x^2 + 48x - 100 = 0$$

$$(x + 50)(x - 2) = 0 \Rightarrow x = -50 \text{ or } x = 2$$

Only the positive answer has meaning, so we reject –50. The rate of the current was 2 mph.

99. Let x = the time it took the wife to wash the car alone. Then $x + 20$ = the time it took the husband to wash the car alone. So we have

$$\dfrac{1}{x} + \dfrac{1}{x + 20} = \dfrac{1}{24}$$

$$24x(x + 20)\left(\dfrac{1}{x} + \dfrac{1}{x + 20} = \dfrac{1}{24}\right)$$

$$24x + 480 + 24x = x^2 + 20x$$

$$0 = x^2 - 28x - 480$$

$$0 = (x - 40)(x + 12)$$

$$x = 40 \text{ or } x = -12$$

Only the positive answer has meaning, so we reject –12. The wife took 40 minutes to wash the car alone.

101. Let x = the length of the shorter side. Then $100 + x$ = the length of the longer side. Use the Pythagorean theorem to find the length of the

diagonal $= \sqrt{x^2 + (x + 100)^2}$. So we have

$$\sqrt{x^2 + (x + 100)^2} + (x + 100) = 3x \Rightarrow$$

$$\sqrt{x^2 + (x + 100)^2} = 2x - 100 \Rightarrow$$

$$\left(\sqrt{x^2 + (x + 100)^2}\right)^2 = (2x - 100)^2 \Rightarrow$$

$$x^2 + (x + 100)^2 = 4x^2 - 400x + 10,000 \Rightarrow$$

$$2x^2 + 200x + 10,000 = 4x^2 - 400x + 10,000 \Rightarrow$$

$$0 = 2x^2 - 600x \Rightarrow 0 = 2x(x - 300) \Rightarrow$$

$$x = 0 \text{ or } x = 300$$

The shorter side is 300 feet and the longer side is 400 feet. So the area is 120,000 square feet.

103. Use the Pythagorean theorem to find the

lengths of AE and BE. $AE = \sqrt{64 + x^2}$ and

$BE = \sqrt{(18 - x)^2 + 25}$.

So we have $AE + BE = 23$ or

$$\sqrt{64 + x^2} + \sqrt{(18 - x)^2 + 25} = 23 \Rightarrow$$

$$\sqrt{(18 - x)^2 + 25} = 23 - \sqrt{64 + x^2} \Rightarrow$$

$$\left(\sqrt{(18 - x)^2 + 25}\right)^2 = \left(23 - \sqrt{64 + x^2}\right)^2 \Rightarrow$$

$$(18 - x)^2 + 25 = 529 - 46\sqrt{64 + x^2} + 64 + x^2 \Rightarrow$$

$$324 - 36x + x^2 + 25 = 593 - 46\sqrt{64 + x^2} + x^2 \Rightarrow$$

$$-244 - 36x = -46\sqrt{64 + x^2} \Rightarrow$$

$$(244 + 36x)^2 = \left(46\sqrt{64 + x^2}\right)^2$$

$$1296x^2 + 17,568x + 59,536 =$$
$$135,424 + 2116x^2$$

$$-820x^2 + 17,568x - 75,888 = 0$$

$$-4(205x^2 - 4392x + 18,972) = 0$$

$$(x - 6)(205x - 3162) = 0 \Rightarrow$$

$$x = 6 \text{ or } x = \dfrac{3162}{205} \approx 15.4$$

Check for extraneous solutions. Neither is extraneous, so CE = 6 mi or $CE \approx 15.4$ mi.

1.5 Beyond the Basics

105. $\left(\dfrac{x^2-3}{x}\right)^2 - 6\left(\dfrac{x^2-3}{x}\right) + 8 = 0$

Let $u = \dfrac{x^2-3}{x}$, so we have

$u^2 - 6u + 8 = 0 \Rightarrow (u-4)(u-2) = 0 \Rightarrow$
$u = 4$ or $u = 2$.

Now solve for x.

$\dfrac{x^2-3}{x} = 4 \Rightarrow x^2 - 3 = 4x \Rightarrow$

$x^2 - 4x - 3 = 0 \Rightarrow$

$x = 2 \pm 2\sqrt{7}$ (using the quadratic formula) or

$\dfrac{x^2-3}{x} = 2 \Rightarrow x^2 - 3 = 2x \Rightarrow$

$x^2 - 2x - 3 = 0 \Rightarrow$

$(x-3)(x+1) = 0 \Rightarrow x = 3$ or $x = -1$

Be sure to check to make sure that none of the solutions are extraneous. None are extraneous.

Solution set: $\left\{-1,\ 2 \pm 2\sqrt{7},\ 3\right\}$.

107. $\sqrt{\sqrt{3x-2} + 2\sqrt{4x+1}} = 2\sqrt{2}$

$\sqrt{3x-2} + 2\sqrt{4x+1} = 8$

$\sqrt{3x-2} = 8 - 2\sqrt{4x+1}$

$3x - 2 = 64 - 32\sqrt{4x+1} + 4(4x+1)$

$3x - 2 = 16x + 68 - 32\sqrt{4x+1}$

$-13x - 70 = -32\sqrt{4x+1}$

$169x^2 + 1820x + 4900 = 4096x + 1024$

$169x^2 - 2276x + 3876 = 0$

Solve for x using the quadratic formula.

$x = \dfrac{2276 \pm \sqrt{2276^2 - 4(169)(3876)}}{2(169)}$

$x = 2$ or $x \approx 11.47$

Check to make sure that neither solution is extraneous. 11.47 is extraneous.
Solution set: $\{2\}$.

109. $\dfrac{x+2}{x} + \dfrac{2x+6}{x^2+4x} = -\dfrac{1}{x+4}$

$x(x+4)\left(\dfrac{x+2}{x} + \dfrac{2x+6}{x^2+4x} = -\dfrac{1}{x+4}\right)$

$(x+4)(x+2) + 2x + 6 = -x$

$x^2 + 9x + 14 = 0$

$(x+7)(x+2) = 0$

$x = -7$ or $x = -2$

Solution set: $\{-7, -2\}$

111. $\left(\dfrac{t}{t+2}\right)^2 + \dfrac{2t}{t+2} - 15 = 0$

Let $u = \dfrac{t}{t+2}$. This gives

$u^2 + 2u - 15 = 0 \Rightarrow (u+5)(u-3) = 0 \Rightarrow$
$u = -5$ or $u = 3$.

Now solve for t.

$-5 = \dfrac{t}{t+2} \Rightarrow -5t - 10 = t \Rightarrow t = -\dfrac{5}{3}$ or

$3 = \dfrac{t}{t+2} \Rightarrow 3t + 6 = t \Rightarrow t = -3$.

Be sure to check to make sure that neither solution is extraneous. Neither is extraneous.

Solution set: $\left\{-3,\ -\dfrac{5}{3}\right\}$.

113. $x^{4/3} - 4x^{2/3} + 3 = 0$

Let $u = x^{2/3}$. The equation becomes

$u^2 - 4u + 3 = 0 \Rightarrow (u-1)(u-3) = 0 \Rightarrow$
$u = 1, 3$

$\begin{array}{c|c} x^{2/3} = 1 & x^{2/3} = 3 \\ \left(x^{2/3}\right)^{3/2} = \pm\left(1^{3/2}\right) & \left(x^{2/3}\right)^{3/2} = \pm\left(3^{3/2}\right) \\ x = \pm 1 & x = \pm\sqrt{27} = \pm 3\sqrt{3} \end{array}$

Verify that none of the solutions are extraneous. Solution set: $\left\{\pm 1,\ \pm 3\sqrt{3}\right\}$.

115. $\dfrac{x+b}{x-b} = \dfrac{x-5b}{2x-5b}$

$(2x-5b)(x+b) = (x-b)(x-5b)$

$2x^2 - 3bx - 5b^2 = x^2 - 6bx + 5b^2$

$x^2 + 3bx - 10b^2 = 0$

$(x+5b)(x-2b) = 0 \Rightarrow x = -5b$ or $x = 2b$

Solution set: $\{-5b, 2b\}$

1.5 Critical Thinking/Discussion/Writing

117. To find the value of x, we can let $x = \sqrt{1+x}$ because the square root contains a copy of itself. Now solve the equation:

$x = \sqrt{1+x} \Rightarrow x^2 = 1 + x \Rightarrow x^2 - x - 1 = 0$.

Now use the quadratic formula to solve for x:

$x = \dfrac{1 \pm \sqrt{1+4}}{2} = \dfrac{1 \pm \sqrt{5}}{2}$.

The original expression is positive, so

$x = \dfrac{1 - \sqrt{5}}{2}$ cannot be the solution.

Solution set: $\left\{\dfrac{1+\sqrt{5}}{2}\right\}$.

118. To find the value of x, we can let $x = \sqrt{20 + x}$ because the square root contains a copy of itself. Now solve the equation:
$x = \sqrt{20 + x} \Rightarrow x^2 = 20 + x \Rightarrow$
$x^2 - x - 20 = 0$. Now use the quadratic

formula. $x = \dfrac{1 \pm \sqrt{1 + 80}}{2} = \dfrac{1 \pm 9}{2} \Rightarrow x = 5$

or $x = -4$. The original expression is positive, so $x = -4$ cannot be a solution.
Solution set: $\{5\}$.

119. To find the value of x, we can let $x = \sqrt{n + x}$ because the square root contains a copy of itself. Now solve the equation:
$x = \sqrt{n + x} \Rightarrow x^2 = n + x \Rightarrow$
$x^2 - x - n = 0$. Now use the quadratic

formula. $x = \dfrac{1 \pm \sqrt{1 + 4n}}{2} = \dfrac{1}{2} \pm \dfrac{\sqrt{1 + 4n}}{2}$.

The original expression is positive, so we reject the negative root.

Solution set: $\left\{ \dfrac{1}{2} + \dfrac{\sqrt{1 + 4n}}{2} \right\}$.

120. $\sqrt{x + 2\sqrt{x - 1}} + \sqrt{x - 2\sqrt{x - 1}} = 4$
Note that $x + 2\sqrt{x - 1} = (x - 1) + 1 + 2\sqrt{x - 1}$
$= \left(\sqrt{x - 1} + 1\right)^2$. Similarly, $x - 2\sqrt{x - 1} =$
$(x - 1) - 2\sqrt{x - 1} + 1 = \left(\sqrt{x - 1} - 1\right)^2$.
Substitute these expressions into the original equation:
$\sqrt{\left(\sqrt{x - 1} + 1\right)^2} + \sqrt{\left(\sqrt{x - 1} - 1\right)^2} = 4 \Rightarrow$
$\sqrt{x - 1} + 1 + \sqrt{x - 1} - 1 = 4 \Rightarrow$
$2\sqrt{x - 1} = 4 \Rightarrow \sqrt{x - 1} = 2 \Rightarrow x - 1 = 4 \Rightarrow$
$x = 5$.
Solution set: $\{5\}$

1.5 Getting Ready for the Next Section

121. $5 > -1$

123. $7 \geq 7$

125. $3 > 3x + 4$

127. $x \geq 0$

129. $2(x - 3) + 17 = 4x + 1$
$2x - 6 + 17 = 4x + 1$
$2x + 11 = 4x + 1$
$-2x = -10$
$x = 5$
Solution set: $\{5\}$

131. $2x - 3 - (3x - 1) = 6$
$2x - 3 - 3x + 1 = 6$
$-x - 2 = 6$
$-x = 8 \Rightarrow x = -8$
Solution set: $\{-8\}$

133. $\dfrac{5}{2x} + 3 = \dfrac{5 + 3(2x)}{2x} = \dfrac{5 + 6x}{2x}$

135. $\dfrac{3}{x + 1} - \dfrac{x - 2}{x + 3} = \dfrac{3(x + 3) - (x + 1)(x - 2)}{(x + 1)(x + 3)}$

 $= \dfrac{3x + 9 - \left(x^2 - x - 2\right)}{(x + 1)(x + 3)}$

$= \dfrac{-x^2 + 4x + 11}{(x + 1)(x + 3)}$

1.6 Inequalities

1.6 Practice Problems

1. a. $3x + 1 \leq 10 \Rightarrow 3x \leq 9 \Rightarrow x \leq 3$
Solution set: $(-\infty, 3]$

b. $4x + 9 > 2(x + 6) + 1 \Rightarrow 4x + 9 > 2x + 13 \Rightarrow$
$2x > 4 \Rightarrow x > 2$
Solution set: $(2, \infty)$

c. $7 - 2x \geq -3 \Rightarrow -2x \geq -10 \Rightarrow x \leq 5$
Solution set: $(-\infty, 5]$

2. Let t = time elapsed since the plane went on autopilot. Then $340t$ = distance the plane has flown in t hours, and $185 + 340t$ = the plane's distance from Miami after t hours.
$185 + 340t \geq 1035 \Rightarrow 340t \geq 850 \Rightarrow t \geq 2.5$
The tower will suspect trouble if the plane has not arrived in 2.5 hours.

3. a. $2(4 - x) + 6x < 4(x + 1) + 7$
$8 - 2x + 6x < 4x + 4 + 7$
$8 + 4x < 4x + 11$
$0 < 3$
The last inequality is always true, so the solution set is $(-\infty, \infty)$.

b. $3(x-2)+5 \geq 7(x-1)-4(x-2)$
$$3x-6+5 \geq 7x-7-4x+8$$
$$3x-1 \geq 3x+1 \Rightarrow -1 \geq 1$$

The last inequality is always false, so the solution set is $\varnothing$.

4. $3x-5 \geq 7$ or $5-2x \geq 1$
$$3x \geq 12 \quad \text{or} \quad -2x \geq -4$$
$$x \geq 4 \quad \text{or} \quad x \leq 2$$

Solution set: $(-\infty, 2] \cup [4, \infty)$

5. $2(3-x)-3 < 5$ and $2(x-5)+7 \leq 3$
$$6-2x-3 < 5 \quad \text{and} \quad 2x-10+7 \leq 3$$
$$3-2x < 5 \quad \text{and} \quad 2x-3 \leq 3$$
$$-2x < 2 \quad \text{and} \quad 2x \leq 6$$
$$x > -1 \quad \text{and} \quad x \leq 3$$

Solution set: $(-1, 3]$

6. $-6 \leq 4x-2 < 4 \Rightarrow -4 \leq 4x < 6 \Rightarrow -1 \leq x < \dfrac{3}{2}$

Solution set: $\left[-1, \dfrac{3}{2}\right)$

7. Start with the interval for x.
$$-3 \leq x \leq 2$$
$$3(-3) \leq 3x \leq 3(2)$$
$$-9 \leq 3x \leq 6$$
$$-9+5 \leq 3x+5 \leq 6+5$$
$$-4 \leq 3x+5 \leq 11$$

Therefore, $a = -4$ and $b = 11$.

8. $15 < C < 25$
$$\left(\dfrac{9}{5}\right)15 < \dfrac{9}{5}C < \left(\dfrac{9}{5}\right)25$$
$$27 < \dfrac{9}{5}C < 45$$
$$27+32 < \dfrac{9}{5}C+32 < 45+32$$
$$59 < \dfrac{9}{5}C+32 < 77$$

The temperature range from $15°C$ to $25°C$ corresponds to a range from $59°F$ to $77°F$.

9. $x^2+2 < 3x+6$
$$x^2-3x-4 < 0 \Rightarrow (x+1)(x-4) < 0$$
The factors equal 0 at $x = -1$ and $x = 4$.

Interval	Test point	Value of $x^2 - 3x - 4$	Result
$(-\infty, -1)$	-2	6	$+$
$(-1, 4)$	0	-4	$-$
$(4, \infty)$	5	6	$+$

The solution set is $(-1, 4)$.

10. $\dfrac{2x+5}{x-1} \leq 1$
$$\dfrac{2x+5}{x-1}-1 \leq 0$$
$$\dfrac{2x+5-x+1}{x-1} \leq 0 \Rightarrow \dfrac{x+6}{x-1} \leq 0$$
$$x+6 = 0 \Rightarrow x = -6$$
$$x-1 = 0 \Rightarrow x = 1$$
The test points are -6 and 1.

Interval	Test point	Value of $\dfrac{x+6}{x-1}$	Result
$(-\infty, -6)$	-9	$\dfrac{3}{10}$	$+$
$(-6, 1)$	0	-6	$-$
$(1, \infty)$	2	8	$+$

Note that the expression is undefined for $x = 1$. The solution set is $[-6, 1)$.

1.6 Concepts and Vocabulary

1. $<$ **3.** $<$

5. True   **7.** True

1.6 Building Skills

9. $(-2, 5)$

11. $(0, 4]$

13. $[-1, \infty)$

15. $(-\infty, -2]$

17. $2x-3 > -5 \Rightarrow 2x > -2 \Rightarrow x > -1$
$(-1, \infty)$

19. $x+3 < 6 \Rightarrow x < 3$
$(-\infty, 3)$

21. $1 - x \le 4 \Rightarrow -x \le 3 \Rightarrow x \ge -3$
$[-3, \infty)$

23. $2x + 5 < 9 \Rightarrow 2x < 4 \Rightarrow x < 2$
$(-\infty, 2)$

25. $3 - 3x > 15 \Rightarrow -3x > 12 \Rightarrow x < -4$
$(-\infty, -4)$

26. $8 - 4x \ge 12 \Rightarrow -4x \ge 4 \Rightarrow x \le -1$
$(-\infty, -1]$

29. $3(x - 3) \le 3 - x \Rightarrow 3x - 9 \le 3 - x \Rightarrow$
$4x \le 12 \Rightarrow x \le 3$
$(-\infty, 3]$

31. $6x + 4 > 3x + 10 \Rightarrow 3x > 6 \Rightarrow x > 2$
$(2, \infty)$

33. $8(x - 1) - x \le 7x - 12$
$8x - 8 - x \le 7x - 12$
$7x - 8 \le 7x - 12 \Rightarrow -8 \le -12$ False
The solution set is $\varnothing$.

35. $5(x + 2) \le 3(x + 1) + 10 \Rightarrow$
$5x + 10 \le 3x + 3 + 10 \Rightarrow 5x + 10 \le 3x + 13 \Rightarrow$
$2x \le 3 \Rightarrow x \le \dfrac{3}{2}$
$\left(-\infty, \dfrac{3}{2}\right]$

37. $2(x + 1) + 3 \ge 2(x + 2) - 1$
$2x + 2 + 3 \ge 2x + 4 - 1$
$2x + 5 \ge 2x + 3 \Rightarrow 5 \ge 3$ True
$(-\infty, \infty)$

39. $2(x + 1) - 2 \le 3(2 - x) + 9$
$2x + 2 - 2 \le 6 - 3x + 9$
$2x \le 15 - 3x$
$5x \le 15 \Rightarrow x \le 3$
$(\infty, 3]$

41. $9x - 6 \ge \dfrac{3}{2} x + 9$
$18x - 12 \ge 3x + 18$
$15x \ge 30 \Rightarrow x \ge 2$
Solution set: $[2, \infty)$

43. $\dfrac{x - 3}{3} \le 2 + \dfrac{x}{2} \Rightarrow 2x - 6 \le 12 + 3x \Rightarrow$
$-18 \le x \Rightarrow x \ge -18$
Solution set: $[-18, \infty)$

45. $\dfrac{3x + 1}{2} < x - 1 + \dfrac{x}{2}$
$3x + 1 < 2x - 2 + x$
$3x + 1 < 3x - 2 \Rightarrow 1 < -2$ False
Solution set: $\varnothing$

47. $\dfrac{x - 3}{2} \ge \dfrac{x}{3} + 1$
$3x - 9 \ge 2x + 6 \Rightarrow x \ge 15$
Solution set: $[15, \infty)$

49. $\dfrac{3x + 1}{3} - \dfrac{x}{2} \le \dfrac{x + 2}{2}$
$6x + 2 - 3x \le 3x + 6$
$3x + 2 \le 3x + 6 \Rightarrow 2 \le 6$ True
Solution set: $(-\infty, \infty)$

51. $2x + 5 < 1$ or $2 + x > 4$
$2x < -4$ or $x > 2$
$x < -2$ or
Solution set: $(-\infty, -2) \cup (2, \infty)$

53. $\dfrac{2x - 3}{4} \le 2$ or $\dfrac{4 - 3x}{2} \ge 2$
$2x - 3 \le 8$ or $4 - 3x \ge 4$
$2x \le 11$ or $-3x \ge 0$
$x \le \dfrac{11}{2}$ or $x \le 0$
Solution set: $\left(-\infty, \dfrac{11}{2}\right]$

55. $\dfrac{2x + 1}{3} \ge x + 1$ or $\dfrac{x}{2} - 1 > \dfrac{x}{3}$
$2x + 1 \ge 3x + 3$ or $3x - 6 > 2x$
$-2 \ge x$ or $-6 > -x$
$x \le -2$ or $6 < x \Rightarrow x > 6$
Solution set: $(-\infty, -2] \cup (6, \infty)$

57. $\dfrac{x - 1}{2} > \dfrac{x}{3} - 1$ or $\dfrac{2x + 5}{3} \le \dfrac{x + 1}{6}$
$3x - 3 > 2x - 6$ or $4x + 10 \le x + 1$
$x > -3$ or $3x \le -9 \Rightarrow x \le -3$
Solution set: $(-\infty, \infty)$

59. $3 - 2x \le 7$ and $2x - 3 \le 7$
$\quad\quad -2x \le 4$ and $\quad 2x \le 10$
$\quad\quad\quad x \ge -2$ and $\quad\quad x \le 5$

Solution set: $[-2, 5]$

61. $2(x+1)+3 \ge 1$ and $2(2-x) > -6$
$\quad 2x+2+3 \ge 1$ and $\quad 4-2x > -6$
$\quad\quad 2x+5 \ge 1$ and $\quad\quad -2x > -10$
$\quad\quad\quad 2x \ge -4$ and $\quad\quad\quad x < 5$
$\quad\quad\quad\quad x \ge -2$

Solution set: $[-2, 5)$

63. $2(x+1)-3 > 7$ and $3(2x+1)+1 < 10$
$\quad 2x+2-3 > 7$ and $\quad 6x+3+1 < 10$
$\quad\quad 2x-1 > 7$ and $\quad\quad 6x+4 < 10$
$\quad\quad\quad 2x > 8$ and $\quad\quad\quad 6x < 6$
$\quad\quad\quad\quad x > 4$ and $\quad\quad\quad\quad x < 1$

Since x cannot be less than 1 and greater than 4 at the same time, the solution set is $\varnothing$.

65. $5+3(x-1) < 3+3(x+1)$ and $3x-7 \le 8$
$\quad 5+3x-3 < 3+3x+3$ and $\quad 3x \le 15$
$\quad\quad 2+3x < 6+3x$ and $\quad\quad x \le 5$
$\quad\quad\quad 2 < 6$ True

Solution set: $(-\infty, 5]$

67. $3 < x+5 < 4 \Rightarrow -2 < x < -1$ or $(-2,-1)$

69. $-4 \le x-2 < 2 \Rightarrow -2 \le x < 4$ or $[-2,4)$

71. $-9 \le 2x+3 \le 5 \Rightarrow -12 \le 2x \le 2 \Rightarrow$
$\quad -6 \le x \le 1$ or $[-6,1]$

73. $0 \le 1-\dfrac{x}{3} < 2 \Rightarrow 0 \le 3-x < 6 \Rightarrow$
$\quad -3 \le -x < 3 \Rightarrow 3 \ge x > -3$ or $(-3,3]$

75. $-1 < \dfrac{2x-3}{5} \le 0 \Rightarrow -5 < 2x-3 \le 0 \Rightarrow$
$\quad -2 < 2x \le 3 \Rightarrow -1 < x \le \dfrac{3}{2}$ or $\left(-1, \dfrac{3}{2}\right]$

77. $5x \le 3x+1 < 4x-2 \Rightarrow 2x \le 1 < x-2 \Rightarrow$
$\quad x \le \dfrac{1}{2}$ and $3 < x \Rightarrow x > 3$

x cannot be less than or equal to $1/2$ at the same time that 3 is less than x, so the solution set is $\varnothing$.

79. $-2 < x < 1 \Rightarrow -2+7 < x+7 < 1+7 \Rightarrow$
$\quad 5 < x+7 < 8 \Rightarrow a = 5, b = 8$

81. $-1 < x < 1 \Rightarrow 1 > -x > -1 \Rightarrow$
$\quad 2+1 > 2-x > 2-1 \Rightarrow 3 > 2-x > 1 \Rightarrow$
$\quad 1 < 2-x < 3 \Rightarrow a = 1, b = 3$

83. $0 < x < 4 \Rightarrow 0 < 5x < 20 \Rightarrow -1 < 5x-1 < 19 \Rightarrow$
$\quad a = -1, b = 19$

85. $x^2 + 4x - 12 \le 0 \Rightarrow (x+6)(x-2) \le 0$.
Now solve the associated equation:
$(x+6)(x-2) = 0 \Rightarrow x = -6$ or $x = 2$.
So, the intervals are
$(-\infty, -6], [-6, 2],$ and $[2, \infty)$.

Interval	Test point	Value of $x^2 + 4x - 12$	Result
$(-\infty, -6]$	-10	48	$+$
$[-6, 2]$	0	-12	$-$
$[2, \infty)$	3	9	$+$

The solution set is $[-6, 2]$.

87. $6x^2 + 7x - 3 \ge 0 \Rightarrow (3x-1)(2x+3) \ge 0$.
Now solve the associated equation:
$(3x-1)(2x+3) = 0 \Rightarrow x = 1/3$ or $x = -3/2$.
The intervals are $(-\infty, -3/2], [-3/2, 1/3],$
and $[1/3, \infty)$.

Interval	Test point	Value of $6x^2 + 7x - 3$	Result
$(-\infty, -3/2]$	-2	7	$+$
$[-3/2, 1/3]$	0	-3	$-$
$[1/3, \infty)$	1	10	$+$

The solution set is $(-\infty, -3/2] \cup [1/3, \infty)$.

89. $(x+3)(x+1)(x-1) \ge 0$.
Now solve the associated equation:
$(x+3)(x+1)(x-1) = 0 \Rightarrow$
$x = -3, x = -1$ or $x = 1$.
The intervals are
$(-\infty, -3], [-3, -1], [-1, 1],$ and $[1, \infty)$.

(continued on next page)

(continued)

Interval	Test point	Value of $(x+3)(x+1)(x-1)$	Result
$(-\infty,-3]$	-10	-693	$-$
$[-3,-1]$	-2	3	$+$
$[-1,1]$	0	-3	$-$
$[1,\infty)$	10	1287	$+$

The solution set is $[-3,-1]\cup[1,\infty)$.

91. $x^3-4x^2-12x>0$

Solve the associated equation
$x^3-4x^2-12x=0$.
$x^3-4x^2-12x=0 \Rightarrow x(x^2-4x-12)=0 \Rightarrow$
$x(x+2)(x-6)=0 \Rightarrow x=0,-2,6$
The intervals are $(-\infty,-2),(-2,0),(0,6),$
and $(6,\infty)$.

Interval	Test point	Value of x^3-4x^2-12x	Result
$(-\infty,-2)$	-10	-1280	$-$
$(-2,0)$	-1	7	$+$
$(0,6)$	1	-15	$-$
$(6,\infty)$	10	480	$+$

The solution set is $(-2,0)\cup(6,\infty)$.

93. $x^2+2x<-1 \Rightarrow x^2+2x+1<0$.

Now solve the associated equation:
$x^2+2x+1=0 \Rightarrow (x+1)^2=0 \Rightarrow x=-1$.
The intervals are $(-\infty,-1)$ and $(-1,\infty)$.

Interval	Test point	Value of x^2+2x+1	Result
$(\infty,-1)$	-2	1	$+$
$(-1,\infty)$	0	1	$+$

Neither interval has a negative solution, so the solution set is $\varnothing$.

95. $x^3-x^2\ge0$.

Now solve the associated equation:
$x^3-x^2=0 \Rightarrow x^2(x-1)=0 \Rightarrow$
$x=0$ or $x=1$.
The intervals are $(-\infty,0],[0,1]$ and $[1,\infty)$.

Interval	Test point	Value of x^3-x^2	Result
$(-\infty,0]$	-1	-2	$-$
$[0,1]$	$\frac{1}{2}$	$-\frac{1}{8}$	$-$
$[1,\infty)$	2	4	$+$

Note that 0 is a solution. The solution set is $\{0\}\cup[1,\infty)$.

97. $x^2\ge1 \Rightarrow x^2-1\ge0$.

Now solve the associated equation:
$x^2-1=0 \Rightarrow (x-1)(x+1)=0 \Rightarrow$
$x=-1$ or $x=1$.
The intervals are $(-\infty,-1],[-1,1],$ and $[1,\infty)$.

Interval	Test point	Value of x^2-1	Result
$(-\infty,-1]$	-2	3	$+$
$[-1,1]$	0	-1	$-$
$[1,\infty)$	2	3	$+$

The solution set is $(-\infty,-1]\cup[1,\infty)$.

99. $x^3<-8 \Rightarrow x^3+8<0$.

Now solve the associated equation:
$x^3+8=0 \Rightarrow (x+2)(x^2-2x+4)=0 \Rightarrow$
$x=-2$ or $x^2-2x+4=0$.
Solve $x^2-2x+4=0$ using the quadratic formula:
$$x=\frac{-(-2)\pm\sqrt{(-2)^2-4(1)(4)}}{2(1)}=1\pm i\sqrt{3}.$$
We cannot use complex intervals, so the only intervals we examine are $(-\infty,-2)$ and $(-2,\infty)$.

Interval	Test point	Value of x^3+8	Result
$(-\infty,-2)$	-10	-992	$-$
$(-2,\infty)$	10	1008	$+$

The solution set is $(-\infty,-2)$.

Copyright © 2019 Pearson Education Inc.

101. $\dfrac{x+2}{x-5} < 0.$

Now solve $x+2=0 \Rightarrow x=-2$ and
$x-5=0 \Rightarrow x=5$.

So the intervals are
$(-\infty,-2),(-2,5),$ and $(5,\infty)$.

Interval	Test point	Value of $\dfrac{x+2}{x-5}$	Result
$(-\infty,-2)$	-10	$8/15$	$+$
$(-2,5)$	0	$-2/5$	$-$
$(5,\infty)$	10	$12/5$	$+$

Note that the fraction is undefined if $x=5$.
The solution set is $(-2,5)$.

103. $\dfrac{x+4}{x} < 0.$

Solve $x+4=0 \Rightarrow x=-4$ and $x=0$.
The fraction is undefined if $x=0$. So the
intervals are $(-\infty,-4)$, $(-4,0)$, and $(0,\infty)$.

Interval	Test point	Value of $\dfrac{x+4}{x}$	Result
$(-\infty,-4)$	-10	$3/5$	$+$
$(-4,0)$	-2	-1	$-$
$(0,\infty)$	2	3	$+$

The solution set is $(-4,0)$.

105. $\dfrac{x+1}{x+2} \le 3 \Rightarrow \dfrac{x+1}{x+2}-3 \le 0 \Rightarrow$

$\dfrac{x+1-3(x+2)}{x+2} \le 0 \Rightarrow \dfrac{-2x-5}{x+2} \le 0.$

Now solve $-2x-5=0 \Rightarrow x=-5/2$ and
$x+2=0 \Rightarrow x=-2$. So the intervals are
$(-\infty,-5/2], [-5/2,-2),$ and $(-2,\infty)$. The
original fraction is not defined if $x=-2$, so -2
is not included in the intervals.

Interval	Test point	Value of $\dfrac{-2x-5}{x+2}$	Result
$(-\infty,-5/2]$	-3	-1	$-$
$[-5/2,-2)$	$-9/4$	2	$+$
$(-2,\infty)$	3	$-11/5$	$-$

The solution set is $(-\infty,-5/2] \cup (-2,\infty)$.

107. $\dfrac{(x-2)(x+2)}{x} > 0.$

Now we have $x=0$ and
$(x-2)(x+2)=0 \Rightarrow x=2$ and $x=-2$.

So the intervals are $(-\infty,-2),(-2,0),(0,2),$
and $(2,\infty)$. Note that the original fraction is
not defined if $x=0$, so 0 is not included in the
intervals.

Interval	Test point	Value of $\dfrac{(x-2)(x+2)}{x}$	Result
$(-\infty,-2)$	-3	$-5/3$	$-$
$(-2,0)$	-1	3	$+$
$(0,2)$	1	-3	$-$
$(2,\infty)$	3	$5/3$	$+$

The solution set is $(-2,0) \cup (2,\infty)$.

109. $\dfrac{(x-2)(x+1)}{(x-3)(x+5)} \ge 0.$

Set the numerator and denominator equal to
zero and solve for x.

$(x-2)(x+1)=0 \Rightarrow x=2,-1$
$(x-3)(x+5)=0 \Rightarrow x=3,-5$

The intervals are $(-\infty,-5), (-5,-1],$
$[-1,2], [2,3),$ and $(3,\infty)$.

Interval	Test point	Value of $\dfrac{(x-2)(x+1)}{(x-3)(x+5)}$	Result
$(-\infty,-5)$	-6	$40/9$	$+$
$(-5,-1]$	-2	$-4/15$	$-$
$[-1,2]$	0	$2/15$	$+$
$[2,3)$	$5/2$	$-5/15$	$-$
$(3,\infty)$	5	$9/10$	$+$

The solution set is
$(-\infty,-5) \cup [-1,2] \cup (3,\infty)$.

111. $\dfrac{x^2-1}{x^2-4} \le 0$.

Set the numerator and denominator equal to zero and solve for x.

$x^2-1=0 \Rightarrow x=\pm 1;\ x^2-4=0 \Rightarrow x=\pm 2$

The intervals are $(-\infty,-2)$, $(-2,-1]$, $[-1,1]$, $[1,2)$, and $[2,\infty)$.

Interval	Test point	Value of $\dfrac{x^2-1}{x^2-4}$	Result
$(-\infty,-2)$	-3	$8/5$	$+$
$(-2,-1]$	$-3/2$	$-5/7$	$-$
$[-1,1]$	0	$1/4$	$+$
$[1,2)$	$3/2$	$-5/7$	$-$
$[2,\infty)$	3	$8/5$	$+$

The solution set is $(-2,-1]\cup[1,2)$.

113. $\dfrac{x+4}{3x-2} \ge 1 \Rightarrow \dfrac{x+4}{3x-2}-1 \ge 0 \Rightarrow$

$\dfrac{x+4-3x+2}{3x-2} \ge 0 \Rightarrow \dfrac{-2x+6}{3x-2} \ge 0$.

Now we have $-2x+6=0 \Rightarrow x=3$ and $3x-2=0 \Rightarrow x=2/3$. So the intervals are $(-\infty,2/3)$, $(2/3,3]$, and $[3,\infty)$. Note that the original fraction is not defined if $x=2/3$, so $2/3$ is not included in the intervals.

Interval	Test point	Value of $\dfrac{x+4}{3x-2}-1$	Result
$(-\infty,2/3)$	0	-3	$-$
$(2/3,3]$	1	4	$+$
$[3,\infty)$	4	$-1/5$	$-$

The solution set is $(2/3, 3]$.

115. $3 \le \dfrac{2x+6}{2x+1} \Rightarrow \dfrac{2x+6}{2x+1}-3 \ge 0 \Rightarrow$

$\dfrac{2x+6-3(2x+1)}{2x+1} \ge 0 \Rightarrow \dfrac{-4x+3}{2x+1} \ge 0$.

This gives $-4x+3=0 \Rightarrow x=3/4$ and $2x+1=0 \Rightarrow x=-1/2$. The intervals are $(-\infty,-1/2)$, $(-1/2,3/4)$, and $[3/4,\infty)$. The original fraction is not defined if $x=-1/2$, so $-1/2$ is not included in the intervals.

Interval	Test point	Value of $\dfrac{2x+6}{2x+1}-3$	Result
$(-\infty,-1/2)$	-1	-7	$-$
$(-1/2,3/4)$	0	3	$+$
$[3/4,\infty)$	1	$-1/3$	$-$

The solution set is $(-1/2, 3/4]$.

117. $\dfrac{x+2}{x-3} \ge \dfrac{x-1}{x+3} \Rightarrow \dfrac{x+2}{x-3}-\dfrac{x-1}{x+3} \ge 0 \Rightarrow$

$\dfrac{(x+2)(x+3)-(x-1)(x-3)}{(x-3)(x+3)} \ge 0 \Rightarrow$

$\dfrac{\left(x^2+5x+6\right)-\left(x^2-4x+3\right)}{(x-3)(x+3)} \ge 0 \Rightarrow$

$\dfrac{9x+3}{(x-3)(x+3)} \ge 0$

Set the numerator and the denominator equal to zero and solve for x. $9x+3=0 \Rightarrow x=-1/3$ and $(x-3)(x+3)=0 \Rightarrow x=3$ or $x=-3$. The intervals are $(-\infty,-3)$, $(-3,-1/3]$, $[-1/3,3)$, and $(3,\infty)$. Note that the original fractions are not defined if $x=-3$ or $x=3$, so -3 and 3 are not included in the intervals.

Interval	Test point	Value of $\dfrac{x+2}{x-3}-\dfrac{x-1}{x+3}$	Result
$(-\infty,-3)$	-6	$-17/9$	$-$
$(-3,-1/3]$	-1	$3/4$	$+$
$[-1/3,3)$	1	$-3/2$	$-$
$(3,\infty)$	6	$19/9$	$+$

The solution set is $(-3,-1/3]\cup(3,\infty)$.

119. $\dfrac{x-1}{x+1} \le \dfrac{x+2}{x-3} \Rightarrow \dfrac{x-1}{x+1} - \dfrac{x+2}{x-3} \le 0 \Rightarrow$

$\dfrac{(x-1)(x-3)-(x+2)(x+1)}{(x+1)(x-3)} \le 0 \Rightarrow$

$\dfrac{\left(x^2-4x+3\right)-\left(x^2+3x+2\right)}{(x+1)(x-3)} \le 0 \Rightarrow$

$\dfrac{-7x+1}{(x+1)(x-3)} \le 0$

Set the numerator and the denominator equal to zero and solve for x. $-7x+1 = 0 \Rightarrow x = 1/7$ and $(x+1)(x-3) = 0 \Rightarrow x = -1$ or $x = 3$. The intervals are $(-\infty, -1)$, $(-1, 1/7]$, $[1/7, 3)$, and $(3, \infty)$. Note that the original fractions are not defined if $x = -1$ or $x = 3$, so -1 and 3 are not included in the intervals.

Interval	Test point	Value of $\dfrac{x-1}{x+1} - \dfrac{x+2}{x-3}$	Result
$(-\infty, -1)$	-2	3	$+$
$(-1, 1/7]$	0	$-1/3$	$-$
$[1/7, 3)$	1	$3/2$	$+$
$(3, \infty)$	4	$-27/5$	$-$

The solution set is $(-1, 1/7] \cup (3, \infty)$.

1.6 Applying the Concepts

121. Let $x =$ the selling price of the refrigerator. Then
$1750 + 0.15(1750) \le x \le 1750 + 0.20(1750) \Rightarrow$
$2012.50 \le x \le 2100$
The refrigerator's selling price ranges from \$2012.50 to \$2100.

123. Let $x =$ the amount of gasoline in the car at the start of the trip. Then
$300 \le 40x \le 480 \Rightarrow 7.5 \le x \le 12$
The car had between 7.5 gallons and 12 gallons of gas at the start of the trip.

125. Let $x =$ the amount of cream.
Then $270 - x =$ the amount of milk. So,
$0.3x + 0.03(270 - x) \ge 0.045(270) \Rightarrow$
$0.3x + 8.1 - 0.03x \ge 12.15 \Rightarrow$
$0.27x \ge 4.05 \Rightarrow x \ge 15$
At least 15 quarts of cream must be added.

127. Let $x =$ the number of 4-door sedans. Then $3x =$ the number of SUV's and $2x =$ the number of convertibles.
$x + 3x + 2x \ge 48 \Rightarrow 6x \ge 48 \Rightarrow x \ge 8$
There are at least 8 four-door sedans.

129. $132t - t^2 < 3200 \Rightarrow -t^2 + 132t - 3200 > 0 \Rightarrow$
$t^2 - 132t + 3200 < 0 \Rightarrow (t-100)(t-32) < 0$
Solving the associated equation gives $t = 100$ or $t = 32$. The intervals to be checked are $(-\infty, 32), (32, 100),$ and $(100, \infty)$. Checking a number in each interval shows that the temperature must fall in the range $(32°F, 100°F)$ for the number of mites to exceed 3200.

131. Probability must be less than 1, so we have
$0.5 < \dfrac{64 - 0.2x}{208 - x} \le 1 \Rightarrow$
$0.5 < \dfrac{64 - 0.2x}{208 - x}$ and $\dfrac{64 - 0.2x}{208 - x} \le 1$
Solving each inequality independently gives
$0.5 < \dfrac{64 - 0.2x}{208 - x} \Rightarrow 0 < \dfrac{64 - 0.2x}{208 - x} - 0.5 \Rightarrow$
$0 < \dfrac{64 - 0.2x - 0.5(208 - x)}{208 - x} \Rightarrow$
$0 < \dfrac{-40 + 0.3x}{208 - x}$
Now we have $-40 + 0.3x = 0 \Rightarrow x = 400/3$ and $208 - x = 0 \Rightarrow 208 = x$. The original fraction is not defined if $x = 208$, so 208 is not in the solution set. So there need to be more than 133 cards. (Note that the value of x is rounded up to account for the partial value.)
$\dfrac{64 - 0.2x}{208 - x} \le 1 \Rightarrow \dfrac{64 - 0.2x}{208 - x} - 1 \le 0 \Rightarrow$
$\dfrac{64 - 0.2x - 208 + x}{208 - x} \le 0 \Rightarrow \dfrac{0.8x - 144}{208 - x} \le 0 \Rightarrow$
$0.8x - 144 \le 0 \Rightarrow 0.8x \le 144 \Rightarrow x \le 180$
So, the likelihood that the next card dealt would be a jack, queen, king, or ace is greater than 50% if more than 133 cards and less than 180 cards are dealt.

1.6 Beyond the Basics

133. $2x^2 + kx + 2 = 0$ has two real solutions if the discriminant is greater than zero. So
$k^2 - 4(2)(2) > 0 \Rightarrow k^2 - 16 > 0$. Solving the associated equation gives $k = -4$ or $k = 4$. The intervals to be tested are $(-\infty, -4), (-4, 4),$ and $(4, \infty)$. $k^2 - 16 > 0$ for $(-\infty, -4) \cup (4, \infty)$.

135. $x^2 + kx + k = 0$ has two real solutions if the discriminant is greater than zero. So $k^2 - 4k > 0 \Rightarrow k(k-4) > 0$. Solving the associated equation gives $k = 0$ or $k = 4$. The intervals to be tested are $(-\infty, 0)$, $(0, 4)$, and $(4, \infty)$. $k^2 - 4k > 0$ for $(-\infty, 0) \cup (4, \infty)$.

137. $\dfrac{x}{2x+1} \ge \dfrac{1}{4}$ and $\dfrac{6x}{4x-1} < \dfrac{1}{2}$

We will solve each inequality independently and then determine where the solution sets intersect.

$\dfrac{x}{2x+1} \ge \dfrac{1}{4} \Rightarrow \dfrac{x}{2x+1} - \dfrac{1}{4} \ge 0 \Rightarrow$

$\dfrac{4x - (2x+1)}{4(2x+1)} \ge 0 \Rightarrow \dfrac{2x-1}{8x+4} \ge 0$

Now, we have $2x - 1 = 0 \Rightarrow x = 1/2$ and $8x + 4 = 0 \Rightarrow x = -1/2$. Note that the original fraction $\dfrac{x}{2x+1}$ is not defined if $x = -1/2$, so the intervals to be tested are $(-\infty, -1/2)$, $(-1/2, 1/2]$, and $[1/2, \infty)$.

Interval	Test point	Value of $\dfrac{x}{2x+1} - \dfrac{1}{4}$	Result
$(-\infty, -1/2)$	-1	$3/4$	$+$
$(-1/2, 1/2]$	0	$-1/4$	$-$
$[1/2, \infty)$	1	$1/12$	$+$

The solution set is $(-\infty, -1/2) \cup [1/2, \infty)$.

$\dfrac{6x}{4x-1} < \dfrac{1}{2} \Rightarrow \dfrac{6x}{4x-1} - \dfrac{1}{2} < 0 \Rightarrow$

$\dfrac{12x - (4x-1)}{2(4x-1)} < 0 \Rightarrow \dfrac{8x+1}{8x-2} < 0$

The original fractions $\dfrac{6x}{4x-1}$ and $\dfrac{8x+1}{8x-2}$ are not defined if $x = 1/4$. Also, $8x + 1 = 0 \Rightarrow x = -1/8$. The intervals to be tested are $(-\infty, -1/8)$, $(-1/8, 1/4)$, and $(1/4, \infty)$.

Interval	Test point	Value of $\dfrac{6x}{4x-1} - \dfrac{1}{2}$	Result
$(-\infty, -1/8)$	-1	$7/10$	$+$
$(-1/8, 1/4)$	0	$-1/2$	$-$
$(1/4, \infty)$	1	$3/2$	$+$

The solution set is $(-1/8, 1/4)$.

The two solution sets do not intersect, so the solution set of $\dfrac{x}{2x+1} \ge \dfrac{1}{4}$ and $\dfrac{6x}{4x-1} < \dfrac{1}{2}$ is $\varnothing$.

139. Let $x =$ one number. Then $c - x =$ the other number. So we have

$x(c - x) = 36 \Rightarrow cx - x^2 = 36 \Rightarrow$

$x^2 - cx + 36 = 0$

$x^2 - cx + 36 = 0$ has one or two real solutions if the discriminant is greater than or equal to zero. So

$(-c)^2 - 4(1)(36) \ge 0 \Rightarrow c^2 - 144 \ge 0 \Rightarrow$

$c^2 \ge 144 \Rightarrow c \ge 12$ or $c \le -12$

Solution set: $(-\infty, -12] \cup [12, \infty)$

141. Let $x =$ the total number of radios imported. Then $x - 1000 =$ the number of radios subject to the penalty tax. So we have

$10x + 0.05x(x - 1000) \le 640,000$

$10x + 0.05x^2 - 50x \le 640,000$

$0.05x^2 - 40x - 640,000 \le 0$

$x^2 - 800x - 12,800,000 \le 0$

$(x - 4000)(x + 3200) \le 0$

Solving the associated equation gives $x = 4000$ or $x = -3200$. Checking the intervals $[0, 4000]$ and $[4000, \infty)$, we find that the solution is in the range $[0, 4000]$. So they can import no more than 4000 radios.

1.6 Critical Thinking/Discussion/Writing

In Exercises 143 and 144, answers may vary. Sample responses are given.

143. a. $(x + 4)(x - 5) < 0$

b. $(x + 2)(x - 6) \le 0$

c. $x^2 \ge 0$ **d.** $x^2 < 0$

e. $(x - 3)^2 \le 0$ **f.** $(x - 2)^2 > 0$

144. a. $\dfrac{x-4}{x+2} \le 0$ **b.** $\dfrac{x-3}{x-5} \le 0$

    c. It is not possible to have a quadratic inequality with solution set (2, 5). If we try $(x-2)(x-5) < 0$, then the solution set will be (2, 5). If we try $(x-2)(x-5) \le 0$, then the solution set will be [2, 5].

1.6 Getting Ready for the Next Section

145. $|-3| = 3$ **147.** $|6-4| = |2| = 2$

149. $|0| = 0$

151. $d = |5-(-2)| = |7| = 7$

153. $d = |5.7 - 2.3| = |3.4| = 3.4$

155. $|x-(-2)| = 5$ or $|x+2| = 5$

157. $|x-4| \le 2$ **159.** $|x-5| \le 3$

1.7 Equations and Inequalities Involving Absolute Value

1.7 Practice Problems

   1. a. $|x-2| = 0 \Rightarrow x-2 = 0 \Rightarrow x = 2$
      Solution set: $\{2\}$

     b. $|6x-3| - 8 = 1$
$$|6x-3| = 9$$
$$6x-3 = 9 \quad \text{or} \quad 6x-3 = -9$$
$$6x = 12 \qquad\qquad 6x = -6$$
$$x = 2 \qquad\qquad x = -1$$
      Solution set: $\{-1, 2\}$

   2. $|x+2| = |x-3|$
$$x+2 = x-3 \qquad \text{or} \quad x+2 = -(x-3)$$
$$0 = -5 \text{ False} \qquad x+2 = -x+3$$
$$2x = 1 \Rightarrow x = \dfrac{1}{2}$$
      Solution set: $\left\{\dfrac{1}{2}\right\}$

   3. $|3x-4| = 2(x-1)$
$$3x-4 = 2(x-1) \quad \text{or} \quad 3x-4 = -2(x-1)$$
$$3x-4 = 2x-2 \qquad\qquad 3x-4 = -2x+2$$
$$x = 2 \qquad\qquad 5x = 6 \Rightarrow x = \dfrac{6}{5}$$
      Solution set: $\left\{\dfrac{6}{5}, 2\right\}$

   4. $|3x+3| \le 6 \Rightarrow -6 \le 3x+3 \le 6 \Rightarrow$
$$-9 \le 3x \le 3 \Rightarrow -3 \le x \le 1$$
     Solution set: $[-3, 1]$

   5. Let x = the actual speed of the search plane, in miles per hour. Then,
$$|x-115| \le 25 \Rightarrow -25 \le x-115 \le 25 \Rightarrow$$
$$90 \le x \le 140$$
     Thus, the actual speed of the plane is between 90 and 140 miles per hour. Since the plane uses 10 gallons of fuel per hour and has 30 gallons of fuel, it can fly for 3 hours. The actual number of miles the search plane can fly is $3x$:
$$3(90) \le 3x \le 3(140) \Rightarrow 270 \le x \le 420$$
     The plane can fly between 270 miles and 420 miles, inclusive.

   6. $|2x+3| \ge 6 \Rightarrow 2x+3 \le -6$ or $2x+3 \ge 6$

$2x+3 \le -6$	$2x+3 \ge 6$
$2x \le -9$	$2x \ge 3$
$x \le -\dfrac{9}{2}$	$x \ge \dfrac{3}{2}$

     Solution set: $\left(-\infty, -\dfrac{9}{2}\right] \cup \left[\dfrac{3}{2}, \infty\right)$

   7. a. $|5-9x| > -3$

      Since absolute value is always nonnegative, $|5-9x| > -3$ is true for all real numbers.

      Solution set: $(-\infty, \infty)$

     b. $|7x-4| \le -1$

      Since absolute value is always nonnegative, $|7x-4| \le -1$ is false for all real numbers.

      Solution set: $\varnothing$

8. $|x-2| < 4|x+4| \Rightarrow \left|\dfrac{x-2}{x+4}\right| < 4 \Rightarrow$

$$-4 < \dfrac{x-2}{x+4} < 4$$

$$0 < \dfrac{x-2}{x+4} + 4 \qquad\qquad \dfrac{x-2}{x+4} - 4 < 0$$

$$0 < \dfrac{x-2+4(x+4)}{x+4} \quad \dfrac{x-2-4(x+4)}{x+4} < 0$$

$$0 < \dfrac{x-2+4x+16}{x+4} \quad \dfrac{x-2-4x-16}{x+4} < 0$$

$$0 < \dfrac{5x+14}{x+4} \qquad\qquad \dfrac{-3x-18}{x+4} < 0$$

$$5x+14 = 0 \Rightarrow x = -\dfrac{14}{5};\ x+4 = 0 \Rightarrow x = -4$$

$$-3x-18 = 0 \Rightarrow x = -6$$

Interval	Test point	Value of $\dfrac{5x+14}{x+4}$	Result
$(-\infty, -6)$	-10	6	$+$
$\left(-4, -\frac{14}{5}\right)$	-3	-1	$-$
$\left(-\frac{14}{5}, \infty\right)$	-1	3	$+$

Note that the expression is undefined for $x = -4$. The solution set is for this part of the original inequality is $\left(-\infty, -6\right) \cup \left(-\frac{14}{5}, \infty\right)$.

Interval	Test point	Value of $\dfrac{-3x-18}{x+4}$	Result
$(-\infty, -6)$	-7	-1	$-$
$(-6, -4)$	-5	3	$+$
$(-4, \infty)$	-1	-5	$-$

Note that the expression is undefined for $x = 4$. The solution set is for this part of the original inequality is $\left(-\infty, -6\right) \cup \left(-4, \infty\right)$.

Interval	Test point	Value of $\dfrac{-3x-18}{x+4}$	Result
$(-\infty, -6)$	-7	-1	$-$
$(-6, -4)$	-5	3	$+$
$(-4, \infty)$	-1	-5	$-$

Note that the expression is undefined for $x = 4$. The solution set is for this part of the original inequality is $\left(-\infty, -6\right) \cup \left(-4, \infty\right)$.

The figure shows that both inequalities are true on $\left(-\infty, -6\right) \cup \left(-\frac{14}{5}, \infty\right)$, so the solution set is $\left(-\infty, -6\right) \cup \left(-\frac{14}{5}, \infty\right)$.

1.7 Concepts and Vocabulary

1. The solution set of the equation $|x| = a$ is $\{-a, a\}$.

3. The solution set of the inequality $|x| \geq a$ is $(-\infty, -a] \cup [a, \infty)$.

5. True

7. True

1.7 Building Skills

9. $|3x| = 9 \Rightarrow 3x = 9$ or $3x = -9 \Rightarrow$ $x = 3$ or $x = -3$

11. $|-2x| = 6 \Rightarrow -2x = 6$ or $-2x = -6 \Rightarrow$ $x = -3$ or $x = 3$

13. $|x+3| = 2 \Rightarrow x+3 = 2$ or $x+3 = -2 \Rightarrow$ $x = -1$ or $x = -5$

15. $|6-2x| = 8 \Rightarrow 6-2x = 8$ or $6-2x = -8 \Rightarrow$ $x = -1$ or $x = 7$

17. $|6x-2| = 9 \Rightarrow 6x-2 = 9$ or $6x-2 = -9 \Rightarrow$ $x = \dfrac{11}{6}$ or $x = -\dfrac{7}{6}$

19. $|2x+3| - 1 = 0 \Rightarrow |2x+3| = 1 \Rightarrow 2x+3 = 1$ or $2x+3 = -1 \Rightarrow x = -1$ or $x = -2$

21. $\dfrac{1}{2}|x| = 3 \Rightarrow |x| = 6 \Rightarrow x = -6$ or $x = 6$

23. $\left|\dfrac{1}{4}x + 2\right| = 3 \Rightarrow \dfrac{1}{4}x + 2 = -3$ or $\dfrac{1}{4}x + 2 = 3 \Rightarrow$ $x = -20$ or $x = 4$

25. $6|1-2x|-8=10 \Rightarrow 6|1-2x|=18 \Rightarrow$
$|1-2x|=3 \Rightarrow 1-2x=3 \text{ or } 1-2x=-3 \Rightarrow$
$x=-1 \text{ or } x=2$

27. $2|3x-4|+9=7 \Rightarrow 2|3x-4|=-2 \Rightarrow$
$|3x-4|=-1$
The solution set is $\varnothing$ because an absolute value cannot be negative.

29. $|2x+1|=-1$
The solution set is $\varnothing$ because an absolute value cannot be negative.

31. $|x^2-4|=0 \Rightarrow x^2-4=0 \Rightarrow x=\pm 2$
Solution set: $\{-2, 2\}$

33. $|1-2x|=3 \Rightarrow 1-2x=3 \text{ or } 1-2x=-3 \Rightarrow$
$x=-1 \text{ or } x=2$
Solution set: $\{-1, 2\}$

35. $\left|\dfrac{1}{3}-x\right|=\dfrac{2}{3} \Rightarrow \dfrac{1}{3}-x=\dfrac{2}{3} \text{ or } \dfrac{1}{3}-x=-\dfrac{2}{3} \Rightarrow$

$x=-\dfrac{1}{3} \text{ or } x=1$

Solution set: $\left\{-\dfrac{1}{3}, 1\right\}$

In exercises 37–46, be sure to check answers to eliminate extraneous solutions.

37. $|x+3|=|x+5| \Rightarrow x+3=x+5$ (impossible)
or $x+3=-(x+5) \Rightarrow x+3=-x-5 \Rightarrow x=-4$
The solution set is $\{-4\}$.

39. $|3x-2|=|6x+7| \Rightarrow 3x-2=6x+7 \Rightarrow$
$x=-3 \text{ or } 3x-2=-(6x+7) \Rightarrow$
$3x-2=-6x-7 \Rightarrow x=-\dfrac{5}{9}$

The solution set is $\left\{-3, -\dfrac{5}{9}\right\}$.

41. $|2x-1|=x+1 \Rightarrow$
$2x-1=x+1 \text{ or } 2x-1=-(x+1)$
    $x=2$ $2x-1=-x-1$
                         $3x=0 \Rightarrow x=0$
Solution set: $\{0, 2\}$

43. $|4-3x|=x-1 \Rightarrow$
$4-3x=x-1 \text{ or } 4-3x=-(x-1)$
   $5=4x$ $4-3x=-x+1$
   $\dfrac{5}{4}=x$ $3=2x \Rightarrow \dfrac{3}{2}=x$

Solution set: $\left\{\dfrac{5}{4}, \dfrac{3}{2}\right\}$

45. $|3x+2|=2(x-1) \Rightarrow$
$3x+2=2(x-1) \text{ or } 3x+2=-2(x-1)$
$3x+2=2x-2$ $3x+2=-2x+2$
   $x=-4$ $5x=0 \Rightarrow x=0$
If $x=-4$, then $|3x+2|=|3(-4)+2|=|-10|=$
10, while $2(x-1)=2(-4-1)=2(-5)=-10$.
Therefore, -4 is not a solution.
If $x=0$, then $|3x+2|=|3(0)+2|=|2|=2$,
while $2(x-1)=2(0-1)=2(-1)=-2$.
Therefore, 0 is not a solution.
Solution set: $\varnothing$

47. $|3x|<12 \Rightarrow -12<3x<12 \Rightarrow -4<x<4$
The solution set is $(-4, 4)$.

49. $|4x|>16 \Rightarrow 4x<-16 \text{ or } 4x>16 \Rightarrow$
$x<-4 \text{ or } x>4$.
The solution set is $(-\infty,-4)\cup(4,\infty)$.

51. $|x+1|<3 \Rightarrow -3<x+1<3 \Rightarrow -4<x<2$.
The solution set is $(-4,2)$.

53. $|x|+2\geq-1 \Rightarrow |x|\geq-3$
Since absolute value is always nonnegative, the inequality is true for all real numbers.
Solution set: $(-\infty,\infty)$

55. $|2x-3|<4 \Rightarrow -4<2x-3<4 \Rightarrow$
$-1<2x<7 \Rightarrow -\dfrac{1}{2}<x<\dfrac{7}{2}$.

The solution set is $\left(-\dfrac{1}{2}, \dfrac{7}{2}\right)$.

57. $|5-2x|>3 \Rightarrow 5-2x<-3 \text{ or } 5-2x>3 \Rightarrow$
$x>4 \text{ or } x<1$.
The solution set is $(-\infty,1)\cup(4,\infty)$.

59. $|3x+4|\leq19 \Rightarrow -19\leq3x+4\leq19 \Rightarrow$
$-23\leq3x\leq15 \Rightarrow -\dfrac{23}{3}\leq x\leq5$.

The solution set is $\left[-\dfrac{23}{3}, 5\right]$.

61. $|2x-15| < 0$. The solution set is $\varnothing$ because an absolute value cannot be negative.

63. $\left|\dfrac{x-2}{x+3}\right| < 1 \Rightarrow -1 < \dfrac{x-2}{x+3} < 1$

$0 < \dfrac{x-2}{x+3}+1$ ⎪ $\dfrac{x-2}{x+3}-1 < 0$

$0 < \dfrac{x-2+(x+3)}{x+3}$ ⎪ $\dfrac{x-2-(x+3)}{x+3} < 0$

$0 < \dfrac{2x+1}{x+3}$ ⎪ $\dfrac{-5}{x+3} < 0$

$2x+1 = 0 \Rightarrow x = -\dfrac{1}{2}; \quad x+3 = 0 \Rightarrow x = -3$

Interval	Test point	Value of $\dfrac{2x+1}{x+3}$	Result
$(-\infty, -3)$	-4	7	$+$
$\left(-3, -\dfrac{1}{2}\right)$	-2	-3	$-$
$\left(-\dfrac{1}{2}, \infty\right)$	2	1	$+$

Note that the expression is undefined for $x = -3$. The solution set is for this part of the original inequality is $(-\infty, -3) \cup \left(-\dfrac{1}{2}, \infty\right)$.

Interval	Test point	Value of $\dfrac{-5}{x+3}$	Result
$(-\infty, -3)$	-4	5	$+$
$\left(-3, -\dfrac{1}{2}\right)$	-2	-5	$-$
$\left(-\dfrac{1}{2}, \infty\right)$	2	-1	$-$

Note that the expression is undefined for $x = -3$. The solution set is for this part of the original inequality is $\left(-3, -\dfrac{1}{2}\right) \cup \left(-\dfrac{1}{2}, \infty\right)$.

Both inequalities are true on $\left(-\dfrac{1}{2}, \infty\right)$, so the solution set is $\left(-\dfrac{1}{2}, \infty\right)$.

65. $\left|\dfrac{2x-3}{x+1}\right| \le 3 \Rightarrow -3 \le \dfrac{2x-3}{x+1} \le 3$

$0 \le \dfrac{2x-3}{x+1}+3$ ⎪ $\dfrac{2x-3}{x+1}-3 \le 0$

$0 \le \dfrac{2x-3+3(x+1)}{x+1}$ ⎪ $\dfrac{2x-3-3(x+1)}{x+1} \le 0$

$0 \le \dfrac{5x}{x+1}$ ⎪ $\dfrac{-x-6}{x+1} \le 0$

$5x = 0 \Rightarrow x = 0; \quad x+1 = 0 \Rightarrow x = -1$
$-x-6 = 0 \Rightarrow x = -6$

Interval	Test point	Value of $\dfrac{5x}{x+1}$	Result
$(-\infty, -6]$	-11	$\dfrac{11}{2}$	$+$
$[-6, -1)$	-2	10	$+$
$(-1, 0]$	$-\dfrac{1}{2}$	-5	$-$
$[0, \infty)$	4	4	$+$

Note that the expression is undefined for $x = -1$. The solution set is for this part of the original inequality is $(-\infty, -6] \cup [-6, -1) \cup [0, \infty)$.

Interval	Test point	Value of $\dfrac{-x-6}{x+1}$	Result
$(-\infty, -6]$	-11	$-\dfrac{1}{2}$	$-$
$[-6, -1)$	-2	4	$+$
$(-1, 0]$	$-\dfrac{1}{2}$	-11	$-$
$[0, \infty)$	4	-2	$-$

Note that the expression is undefined for $x = -1$. The solution set is for this part of the original inequality is $(-\infty, -6] \cup (-1, 0] \cup [0, \infty)$.

Both inequalities are true on $(-\infty, -6] \cup [0, \infty)$, so the solution set is $(-\infty, -6] \cup [0, \infty)$.

67. $\left|\dfrac{x-1}{x+2}\right| \ge 2 \Rightarrow \dfrac{x-1}{x+2} \le -2$ or $\dfrac{x-1}{x+2} \ge 2$

$$\dfrac{x-1}{x+2} \le -2 \qquad\qquad \dfrac{x-1}{x+2} \ge 2$$

$$\dfrac{x-1}{x+2} + 2 \le 0 \qquad\qquad \dfrac{x-1}{x+2} - 2 \ge 0$$

$$\dfrac{x-1+2(x+2)}{x+2} \le 0 \qquad \dfrac{x-1-2(x+2)}{x+2} \ge 0$$

$$\dfrac{3x+3}{x+2} \le 0 \qquad\qquad \dfrac{-x-5}{x+2} \ge 0$$

$3x+3 = 0 \Rightarrow x = -1$; $x+2 = 0 \Rightarrow x = -2$
$-x-5 = 0 \Rightarrow x = -5$

Interval	Test point	Value of $\dfrac{3x+3}{x+2}$	Result
$(-\infty, -5]$	-6	$\dfrac{15}{4}$	$+$
$[-5, -2)$	-3	6	$+$
$(-2, -1]$	$-\dfrac{3}{2}$	-3	$-$
$[-1, \infty)$	1	2	$+$

Note that the expression is undefined for $x = -2$. The solution set is for this part of the original inequality is $(-2, -1]$.

Interval	Test point	Value of $\dfrac{-x-5}{x+2}$	Result
$(-\infty, -5]$	-6	$-\dfrac{1}{4}$	$-$
$[-5, -2)$	-3	2	$+$
$(-2, -1]$	$-\dfrac{3}{2}$	-7	$-$
$[-1, \infty)$	1	-2	$-$

Note that the expression is undefined for $x = -2$. The solution set is for this part of the original inequality is $[-5, -2)$. Since the original inequality is an "or" inequality, the solution set of the original inequality is the union of the two solution sets.
The solution set is $(-2, -1] \cup [-5, -2)$.

69. $\left|\dfrac{2x+1}{x-1}\right| > 4 \Rightarrow \dfrac{2x+1}{x-1} < -4$ or $\dfrac{2x+1}{x-1} > 4$

$$\dfrac{2x+1}{x-1} < -4 \qquad\qquad \dfrac{2x+1}{x-1} > 4$$

$$\dfrac{2x+1}{x-1} + 4 < 0 \qquad\qquad \dfrac{2x+1}{x-1} - 4 > 0$$

$$\dfrac{2x+1+4(x-1)}{x-1} < 0 \qquad \dfrac{2x+1-4(x-1)}{x-1} > 0$$

$$\dfrac{6x-3}{x-1} < 0 \qquad\qquad \dfrac{-2x+5}{x-1} > 0$$

$6x-3 = 0 \Rightarrow x = \dfrac{1}{2}$; $x-1 = 0 \Rightarrow x = 1$

$-2x+5 = 0 \Rightarrow x = \dfrac{5}{2}$

Interval	Test point	Value of $\dfrac{6x-3}{x-1}$	Result
$\left(-\infty, \dfrac{1}{2}\right)$	0	3	$+$
$\left(\dfrac{1}{2}, 1\right)$	$\dfrac{3}{4}$	-6	$-$
$\left(1, \dfrac{5}{2}\right)$	2	9	$+$
$\left(\dfrac{5}{2}, \infty\right)$	4	7	$+$

Note that the expression is undefined for $x = -2$. The solution set is for this part of the original inequality is $\left(\dfrac{1}{2}, 1\right)$.

Interval	Test point	Value of $\dfrac{-2x+5}{x-1}$	Result
$\left(-\infty, \dfrac{1}{2}\right)$	0	-5	$-$
$\left(\dfrac{1}{2}, 1\right)$	$\dfrac{3}{4}$	-14	$-$
$\left(1, \dfrac{5}{2}\right)$	2	1	$+$
$\left(\dfrac{5}{2}, \infty\right)$	4	-1	$-$

Note that the expression is undefined for $x = -2$. The solution set is for this part of the original inequality is $\left(1, \dfrac{5}{2}\right)$. Since the original inequality is an "or" inequality, the solution set of the original inequality is the union of the two solution sets. The solution set is $\left(\dfrac{1}{2}, 1\right) \cup \left(1, \dfrac{5}{2}\right)$.

71. $|x-1| \leq 2|2x-5| \Rightarrow \left|\dfrac{x-1}{2x-5}\right| \leq 2 \Rightarrow$

$-2 \leq \dfrac{x-1}{2x-5} \leq 2$

$0 \leq \dfrac{x-1}{2x-5} + 2$ $\qquad$ $\dfrac{x-1}{2x-5} - 2 \leq 0$

$0 \leq \dfrac{x-1+2(2x-5)}{2x-5}$ $\qquad$ $\dfrac{x-1-2(2x-5)}{2x-5} \leq 0$

$0 \leq \dfrac{x-1+4x-10}{2x-5}$ $\qquad$ $\dfrac{x-1-4x+10}{2x-5} \leq 0$

$0 \leq \dfrac{5x-11}{2x-5}$ $\qquad$ $\dfrac{-3x+9}{2x-5} \leq 0$

$5x-11 = 0 \Rightarrow x = \dfrac{11}{5}$; $\ 2x-5 = 0 \Rightarrow x = \dfrac{5}{2}$

$-3x+9 = 0 \Rightarrow x = 3$

Interval	Test point	Value of $\dfrac{5x-11}{2x-5}$	Result
$\left(-\infty, \frac{11}{5}\right]$	0	$\frac{11}{5}$	$+$
$\left[\frac{11}{5}, \frac{5}{2}\right)$	$\frac{23}{10}$	$-\frac{5}{4}$	$-$
$\left(\frac{5}{2}, \infty\right)$	3	4	$+$

Note that the expression is undefined for $x = \frac{5}{2}$. The solution set is for this part of the original inequality is $S_1 = \left(-\infty, \frac{11}{5}\right] \cup \left(\frac{5}{2}, \infty\right)$.

Interval	Test point	Value of $\dfrac{-3x+9}{2x-5}$	Result
$\left(-\infty, \frac{5}{2}\right)$	0	$-\frac{9}{5}$	$-$
$\left(\frac{5}{2}, 3\right]$	$\frac{11}{4}$	$\frac{3}{2}$	$+$
$[3, \infty)$	5	$-\frac{6}{5}$	$-$

Note that the expression is undefined for $x = \frac{5}{2}$. The solution set is for this part of the original inequality is $S_2 = \left(-\infty, \frac{5}{2}\right) \cup [3, \infty)$.

The figure shows that both inequalities are true on $\left(-\infty, \frac{11}{5}\right] \cup [3, \infty)$, so the solution set is $\left(-\infty, \frac{11}{5}\right] \cup [3, \infty)$.

1.7 Applying the Concepts

73. $|T - 75| = 20 \Rightarrow T - 75 = -20$ or
$T - 75 = 20 \Rightarrow T = 55$ or $T = 95$
The temperatures in Tampa during December are between 55°F and 95°F, inclusive.

75. $|x - 700| \leq 50$

77. $|x - 120| \leq 6.75$

79. Let x = the number of people at a party. Then
$|120 - x| \leq 15 \Rightarrow -15 \leq 120 - x \leq 15 \Rightarrow$
$-135 \leq -x \leq -105 \Rightarrow 105 \leq x \leq 135$. So, between 105 and 135 people will be at the party. The total spent on food will be between $48(105) = \$5040$ and $48(135) = \$6480$, inclusive.

81. Let x = the actual weight of Sarah's catch.
Then $|32 - x| \leq \dfrac{1}{2} \Rightarrow -\dfrac{1}{2} \leq 32 - x \leq \dfrac{1}{2} \Rightarrow$
$-1 \leq 64 - 2x \leq 1 \Rightarrow -65 \leq -2x \leq -63 \Rightarrow$.
$32.5 \geq x \geq 31.5$.
So, her catch is between 31.5 pounds and 32.5 pounds. She will be paid between
$0.60(31.5) = \$18.90$ and $0.60(32.5) = \$19.50$, inclusive.

1.7 Beyond the Basics

83. a. $|x^2 - 9| = x - 3 \Rightarrow x^2 - 9 = -(x - 3)$ or
$x^2 - 9 = x - 3$
$x^2 - 9 = -(x - 3) \Rightarrow x^2 - 9 = -x + 3 \Rightarrow$
$x^2 + x - 12 = 0 \Rightarrow (x + 4)(x - 3) = 0 \Rightarrow$
$x = -4, 3$

$x^2 - 9 = x - 3 \Rightarrow x^2 - x - 6 = 0 \Rightarrow$
$(x + 2)(x - 3) = 0 \Rightarrow x = -2, 3$

Checking $x = -4$, $x = -2$, and $x = 3$ in the original equation shows that $x = -4$ and $x = -2$ are extraneous solutions.
The solution set is $\{3\}$.

b. $\left|x^2 - 8\right| = -2x \Rightarrow x^2 - 8 = -(-2x)$ or
$x^2 - 8 = -2x$

$x^2 - 8 = -(-2x) \Rightarrow x^2 - 8 = 2x \Rightarrow$
$x^2 - 2x + 8 = 0 \Rightarrow (x + 2)(x - 4) = 0 \Rightarrow$
$x = -2, 4$

$x^2 - 8 = -2x \Rightarrow x^2 + 2x - 8 = 0 \Rightarrow$
$(x + 4)(x - 2) = 0 \Rightarrow x = -4, 2$

Checking $x = -4$, $x = -2$, $x = 2$, and $x = 4$ in the original equation shows that $x = 2$, and $x = 4$ are extraneous solution.

The solution set is $\{-4, -2\}$.

c. $\left|x^2 - 5x\right| = 6 \Rightarrow x^2 - 5x = -6$ or
$x^2 - 5x = 6$

$x^2 - 5x = -6 \Rightarrow x^2 - 5x + 6 = 0 \Rightarrow$
$(x - 2)(x - 3) = 0 \Rightarrow x = 2, 3$

$x^2 - 5x = 6 \Rightarrow x^2 - 5x - 6 = 0 \Rightarrow$
$(x + 1)(x - 6) = 0 \Rightarrow x = -1, 6$

Checking $x = -1$, $x = 2$, $x = 3$, and $x = 6$ in the original equation shows that all values are solutions.

The solution set is $\{-1, 2, 3, 6\}$.

d. $\left|x^2 + 3x - 2\right| = 2 \Rightarrow x^2 + 3x - 2 = -2$ or
$x^2 + 3x - 2 = 2$. If $x^2 + 3x - 2 = -2$, then
$x^2 + 3x = 0 \Rightarrow x(x + 3) = 0 \Rightarrow x = 0$ or
$x = -3$.

If $x^2 + 3x - 2 = 2$, then $x^2 + 3x - 4 = 0 \Rightarrow$
$(x + 4)(x - 1) = 0 \Rightarrow x = 1$ or $x = -4$.

Checking $x = -4$, $x = -3$, $x = 0$, and $x = 1$ in the original equation shows that all values are solutions.

The solution set is $\{-4, -3, 0, 1\}$.

85. $|2x - 3| + |x - 2| = 4$

The points $x = \frac{3}{2}$ and $x = 2$ are the points where the absolute value expressions equal zero. Since these expressions must be negative or positive for other x-values, then these points divide the number line into intervals each of which should be considered separately. Thus, we will consider the intervals $\left(-\infty, \frac{3}{2}\right]$, $\left(\frac{3}{2}, 2\right]$, and $(2, \infty)$.

Interval	Test point	Sign of $2x - 3$	Sign of $x - 2$
$\left(-\infty, \frac{3}{2}\right]$	0	$-$	$-$
$\left(\frac{3}{2}, 2\right]$	$\frac{7}{4}$	$+$	$-$
$(2, \infty)$	3	$+$	$+$

On the first interval, both absolute-value expressions will have negative values, so change the signs on both of them when taking the bars off.

$-(2x - 3) - (x - 2) = 4$
$-2x + 3 - x + 2 = 4$
$-3x + 5 = 4$
$-3x = -1 \Rightarrow x = \frac{1}{3}$

Because $x = \frac{1}{3}$ lies in the interval $\left(-\infty, \frac{3}{2}\right]$, this is a valid solution.

On the second interval, $|2x - 3|$ is positive, so just take the bars off. But $|x - 2|$ is negative, change the sign when taking the bars off.

$(2x - 3) - (x - 2) = 4$
$2x - 3 - x + 2 = 4$
$x - 1 = 4 \Rightarrow x = 5$

Since $x = 5$ does not lie in the interval $\left(\frac{3}{2}, 2\right]$, it is not a valid solution of the original equation.

On the third interval, the arguments of both absolute values expressions are positive, so just remove the absolute value bars.

$(2x - 3) + (x - 2) = 4$
$2x - 3 + x - 2 = 4$
$3x - 5 = 4$
$3x = 9 \Rightarrow x = 3$

Because $x = 3$ lies in the interval $(2, \infty)$, the solution is valid. Thus, the solution set for the original equation is $\left\{\frac{1}{3}, 3\right\}$.

87. Let $u = |x|$. Then $|x|^2 - 4|x| - 7 = 5 \Rightarrow$
$u^2 - 4u - 7 = 5 \Rightarrow u^2 - 4u - 12 = 0 \Rightarrow$
$(u - 6)(u + 2) = 0 \Rightarrow u = -2, 6$.

Now solve for x: $-2 = |x|$ is not possible.

$6 = |x| \Rightarrow x = 6$ or $x = -6$

Solution set: $\{-6, 6\}$.

89. $0 < a < b \Rightarrow b - a > 0$ and
$0 < c < d \Rightarrow d - c > 0.$ The product of two positive numbers is positive, so
$(b - a)c > 0 \Leftrightarrow bc > ac$ (1) and
$(d - c)a > 0 \Leftrightarrow ad > ac$ (2). We know that
$bc > ac$ and $bd > ad,$ so
$bd - bc > ad - ac \Rightarrow bd + ac > ad + bc.$
Substituting (1) and (2) into the inequality, we have $bd + ac > ad + bc > ac + ac \Rightarrow bd > ac.$

91. $x^2 < a \Rightarrow x^2 - a < 0 \Rightarrow \left(x - \sqrt{a}\right)\left(x + \sqrt{a}\right) < 0$
$\left(x - \sqrt{a}\right)\left(x + \sqrt{a}\right) < 0 \Rightarrow$
$\left(x - \sqrt{a}\right) > 0$ and $\left(x + \sqrt{a}\right) < 0 \Rightarrow$
$x > \sqrt{a}$ and $x < -\sqrt{a} \Rightarrow \sqrt{a} < x < -\sqrt{a} \Rightarrow$
$\sqrt{a} < -\sqrt{a},$ a contradiction
or
$\left(x - \sqrt{a}\right) < 0$ and $\left(x + \sqrt{a}\right) > 0 \Rightarrow$
$x > -\sqrt{a}$ and $x < \sqrt{a} \Rightarrow -\sqrt{a} < x < \sqrt{a}$
Thus, the solution set is $\left(-\sqrt{a}, \sqrt{a}\right).$

In Exercises 93–104, answers may vary. Sample responses are given.

93. $|x - 4| < 3$ **95.** $|x - 4| \le 6$

97. $|x - 7| > 4$ **99.** $|2x - 5| \ge 15$

101. $\left|x - \dfrac{a+b}{2}\right| < \dfrac{b-a}{2} \Rightarrow |2x - a - b| < b - a$

103. $|2x - a - b| > b - a$

105. $|x - 39| \le 31$

107. $1 \le |x - 2| \le 3 \Rightarrow$
$1 \le x - 2 \le 3$ or $1 \le -(x - 2) \le 3$
$3 \le x \le 5$ $1 \le -x + 2 \le 3$
                        $-1 \le -x \le 1$
                        $1 \ge x \ge -1 \Rightarrow -1 \le x \le 1$
The union of the two solution sets is the solution set of the original inequality.
Solution set: $[-1, 1] \cup [3, 5]$

109. $|x - 2| + |x - 4| \ge 8.$
Solve the equation $|x - 2| + |x - 4| = 8$ to find the critical values for the inequality.
The points $x = 2$ and $x = 4$ are the points where the absolute value expressions equal zero. Because these expressions must be negative or positive for other x-values, then these points divide the number line into intervals each of which should be considered separately. Thus, we will consider the intervals $(-\infty, 2], (2, 4],$ and $(4, \infty).$

Interval	Test point	Sign of $x - 2$	Sign of $x - 4$
$(-\infty, 2]$	0	–	–
$(2, 4]$	3	+	–
$(4, \infty)$	5	+	+

On the first interval, both absolute-value expressions will have negative values, so change the signs on both of them when taking the bars off.
$-(x - 2) + [-(x - 4)] = 8$
$-2x + 6 = 8$
$-2x = 2 \Rightarrow x = -1$
Since $x = -1$ lies in the interval $(-\infty, 2],$ $x = -1$ is a valid solution.
On the second interval, $x - 2$ is positive, so just take the bars off. But $x - 4$ is negative, change the sign when taking the bars off.
$(x - 2) + [-(x - 4)] = 8$
$2 = 8$
This is a false statement, so there are no values of x in the interval $(2, 4]$ that are valid solutions of the equation.
On the third interval, the arguments of both absolute values expressions are positive, so just remove the absolute value bars.
$x - 2 + x - 4 = 8$
$2x - 6 = 8$
$2x = 14 \Rightarrow x = 7$
Because 7 lies in the interval $(2, \infty),$ $x = 7$ is a valid solution. So $x = -1, 2, 4,$ and 7 are critical values. Test values in the intervals $(-\infty, -1], [-1, 2], [2, 4], [4, 7],$ and $[7, \infty)$ to see where the original inequality is true.

(continued on next page)

(*continued*)

Interval	Test point	Value of $\lvert x-2 \rvert + \lvert x-4 \rvert$	Value ≥ 8?
$(-\infty, -1]$	-2	10	Yes
$[-1, 2]$	0	6	No
$[2, 4]$	3	2	No
$[4, 7]$	5	4	No
$[7, \infty)$	10	14	Yes

Thus, the solution set is $(-\infty, -1] \cup [7, \infty)$.

Verify this by graphing $Y_1 = \lvert x-2 \rvert + \lvert x-4 \rvert$ and $Y_2 = 8$.

111. $\dfrac{\lvert x-1 \rvert - (x+2)}{x+2} \geq 0$.

The value of $x - 1$ changes from negative to positive at $x = 1$, where it is 0. So, we will consider two cases,

$$\begin{cases} \dfrac{(x-1)-(x+2)}{x+2} \geq 0 & \text{if } x \geq 1 \\ \dfrac{-(x-1)-(x+2)}{x+2} \geq 0 & \text{if } x < 1 \end{cases}$$

Test the case for $x \geq 1$.

$$\dfrac{(x-1)-(x+2)}{x+2} \geq 0 \Rightarrow \dfrac{-3}{x+2} \geq 0$$

$x + 2 = 0 \Rightarrow x = -2$

Since this test case is for $x \geq 1$, it is only necessary to test the interval $[1, \infty)$.

Interval	Test point	Value of $\dfrac{-3}{x+2}$	Result
$[1, \infty)$	2	$-\dfrac{3}{4}$	$-$

$-\dfrac{3}{4}$ is not ≥ 1, so there is no solution in this interval.

Now test the case $x < 1$.

$$\dfrac{-(x-1)-(x+2)}{x+2} \geq 0 \Rightarrow \dfrac{-2x-1}{x+2} \geq 0$$

$-2x - 1 = 0 \Rightarrow x = -\dfrac{1}{2}$; $x + 2 = 0 \Rightarrow x = -2$

The intervals to be tested are $(-\infty, -2)$, $\left(-2, -\dfrac{1}{2}\right]$, and $\left[-\dfrac{1}{2}, 1\right]$.

Interval	Test point	Value of $\dfrac{-2x-1}{x+2}$	Result
$(-\infty, -2)$	-3	-5	$-$
$\left(-2, -\dfrac{1}{2}\right]$	-1	1	$+$
$\left[-\dfrac{1}{2}, 1\right]$	0	$-\dfrac{1}{2}$	$-$

The only interval where the value of $\dfrac{-2x-1}{x+2} \geq 0$ is $\left(-2, -\dfrac{1}{2}\right]$, so the solution set is $\left(-2, -\dfrac{1}{2}\right]$. Verify this by graphing

$$Y_1 = \dfrac{\lvert x-1 \rvert - (x+2)}{x+2}.$$

1.7 Critical Thinking/Discussion/Writing

112. To find what values make $\sqrt{(x-3)^2} = x-3$ true, solve $x - 3 = 0 \Rightarrow x = 3$, so we check the intervals $(-\infty, 3]$ and $[3, \infty)$ to see which makes the equation true. The equation is true for $[3, \infty)$.

113. To find what values make

$$\sqrt{\left(x^2 - 6x + 8\right)^2} = x^2 - 6x + 8 \text{ true, solve}$$

$x^2 - 6x + 8 = 0 \Rightarrow (x-2)(x-4) = 0 \Rightarrow x = 2$ or $x = 4$. Then check the intervals $(-\infty, 2]$, $[2, 4]$, and $[4, \infty)$ to see which make the equation true. The equation is true for $(-\infty, 2] \cup [4, \infty)$.

114. To solve $|x-3|^2 - 7|x-3| + 10 = 0$, let
$u = |x-3|$. So we have $u^2 - 7u + 10 = 0 \Rightarrow$
$(u-5)(u-2) = 0 \Rightarrow u = 5$ or $u = 2$. Now solve
for x. $5 = |x-3| \Rightarrow -5 = x-3$ or $5 = x-3 \Rightarrow$
$x = -2$ or $x = 8$. $2 = |x-3| \Rightarrow -2 = x-3$ or
$2 = x-3 \Rightarrow x = 1$ or $x = 5$. So the solution set
is $\{-2, 1, 5, 8\}$.

1.7 Getting Ready for the Next Section

115. $\dfrac{2+5}{2} = \dfrac{7}{2} = 3.5$

116. $\dfrac{-3+7}{2} = \dfrac{4}{2} = 2$

117. $\dfrac{-3-7}{2} = \dfrac{-10}{2} = -5$

118. $\dfrac{(a-b)-(a+b)}{2} = \dfrac{a-b-a-b}{2}$
$\phantom{\dfrac{(a-b)-(a+b)}{2}} = \dfrac{-2b}{2} = -b$

119. $\sqrt{(5-2)^2 + (3-7)^2} = \sqrt{3^2 + (-4)^2}$
$\phantom{\sqrt{(5-2)^2 + (3-7)^2}} = \sqrt{9+16} = \sqrt{25} = 5$

121. $\sqrt{(2-5)^2 + (8-6)^2} = \sqrt{(-3)^2 + 2^2} = \sqrt{9+4}$
$\phantom{\sqrt{(2-5)^2 + (8-6)^2}} = \sqrt{13}$

123. $x^2 + 4x + \underline{\ 4\ } = (x+2)^2$

125. $x^2 - 5x + \left(\dfrac{-5}{2}\right)^2 = x^2 - 5x + \dfrac{25}{4} = \left(x - \dfrac{5}{2}\right)^2$

127. $x^2 + \dfrac{3}{2}x + \left(\dfrac{\frac{3}{2}}{2}\right)^2 = x^2 + \dfrac{3}{2}x + \left(\dfrac{3}{4}\right)^2$
$ = x^2 + \dfrac{3}{2}x + \dfrac{9}{16} = \left(x + \dfrac{3}{4}\right)^2$

Chapter 1 Review Exercises

Building Skills

1. $5x - 4 = 11 \Rightarrow 5x = 15 \Rightarrow x = 3$

3. $3(2x-4) = 9 - (x+7) \Rightarrow 6x - 12 = 2 - x \Rightarrow$
$7x = 14 \Rightarrow x = 2$

5. $3x + 8 = 3(x+2) + 2 \Rightarrow 3x + 8 = 3x + 8$
This is an identity, so the solution set is
$(-\infty, \infty)$.

7. $x - (5x-2) = 7(x-1) - 2 \Rightarrow$
$-4x + 2 = 7x - 9 \Rightarrow -11x = -11 \Rightarrow x = 1$

9. $\dfrac{2}{x+3} = \dfrac{5}{11x-1} \Rightarrow 2(11x-1) = 5(x+3) \Rightarrow$
$22x - 2 = 5x + 15 \Rightarrow 17x = 17 \Rightarrow x = 1$

11. $\dfrac{y+5}{2} + \dfrac{y-1}{3} = \dfrac{7y+3}{8} + \dfrac{4}{3} \Rightarrow$
$12(y+5) + 8(y-1) = 3(7y+3) + 8(4) \Rightarrow$
$20y + 52 = 21y + 41 \Rightarrow 11 = y$

13. $|2x-3| = |4x+5| \Rightarrow 2x - 3 = 4x + 5 \Rightarrow$
$-2x = 8 \Rightarrow x = -4$
or $2x - 3 = -(4x+5) \Rightarrow$
$2x - 3 = -4x - 5 \Rightarrow 6x = -2 \Rightarrow x = -1/3$

The solution set is $\left\{-4, -\dfrac{1}{3}\right\}$.

15. $|2x-1| = |2x+7|$
$2x - 1 = 2x + 7$ or $2x - 1 = -(2x+7)$
$ -2 = 7$ False $ 2x - 1 = -2x - 7$
$ 4x = -6 \Rightarrow x = -\dfrac{3}{2}$

Solution set: $\left\{-\dfrac{3}{2}\right\}$

17. $|3x-2| = 2x + 1$
$3x - 2 = 2x + 1$ or $3x - 2 = -(2x+1)$
$ x = 3 3x - 2 = -2x - 1$
$ 5x = 1$
$ x = \dfrac{1}{5}$

Solution set: $\left\{\dfrac{1}{5}, 3\right\}$

19. $p = k + gt \Rightarrow p - k = gt \Rightarrow \dfrac{p-k}{t} = g$

21. $T = \dfrac{2B}{B-1} \Rightarrow TB - T = 2B \Rightarrow TB - 2B = T \Rightarrow$
$B(T-2) = T \Rightarrow B = \dfrac{T}{T-2}$

23. $x^2 - 7x = 0 \Rightarrow x(x-7) = 0 \Rightarrow$
$x = 0$ or $x = 7$

25. $x^2 - 3x - 10 = 0 \Rightarrow (x-5)(x+2) = 0 \Rightarrow$
$x = 5$ or $x = -2$

27. $(x-1)^2 = 2x^2 + 3x - 5 \Rightarrow$
$x^2 - 2x + 1 = 2x^2 + 3x - 5 \Rightarrow$
$0 = x^2 + 5x - 6 \Rightarrow 0 = (x+6)(x-1) \Rightarrow$
$x = -6$ or $x = 1$

29. $\dfrac{x^2}{4} + x = \dfrac{5}{4} \Rightarrow x^2 + 4x = 5 \Rightarrow$
$x^2 + 4x - 5 = 0 \Rightarrow (x+5)(x-1) = 0 \Rightarrow$
$x = -5$ or $x = 1$

31. $3x(x+1) = 2x + 2 \Rightarrow 3x^2 + 3x = 2x + 2 \Rightarrow$
$3x^2 + x - 2 = 0 \Rightarrow (3x-2)(x+1) = 0 \Rightarrow$
$x = \dfrac{2}{3}$ or $x = -1$

33. Use the quadratic formula to solve
$x^2 - 3x - 1 = 0$:
$$x = \frac{-(-3) \pm \sqrt{(-3)^2 - 4(1)(-1)}}{2(1)}$$
$$= \frac{3 \pm \sqrt{13}}{2} = \frac{3}{2} \pm \frac{\sqrt{13}}{2}$$

35. $2x^2 + x - 1 = 0 \Rightarrow (2x-1)(x+1) = 0 \Rightarrow$
$x = \dfrac{1}{2}$ or $x = -1$

37. Divide $3x^2 - 12x - 24 = 0$ by 3 to obtain
$x^2 - 4x - 8 = 0$. Now use the quadratic formula:
$$x = \frac{-(-4) \pm \sqrt{(-4)^2 - 4(1)(-8)}}{2(1)}$$
$$= \frac{4 \pm \sqrt{48}}{2} = \frac{4 \pm 4\sqrt{3}}{2} = 2 \pm 2\sqrt{3}$$

39. Use the quadratic formula to solve
$2x^2 - x - 2 = 0$:
$$x = \frac{-(-1) \pm \sqrt{(-1)^2 - 4(2)(-2)}}{2(2)}$$
$$= \frac{1 \pm \sqrt{17}}{4} = \frac{1}{4} \pm \frac{\sqrt{17}}{4}$$

41. Use the quadratic formula to solve
$x^2 - x + 1 = 0$:
$$x = \frac{-(-1) \pm \sqrt{(-1)^2 - 4(1)(1)}}{2(1)}$$
$$= \frac{1 \pm \sqrt{-3}}{2} = \frac{1 \pm i\sqrt{3}}{2} = \frac{1}{2} \pm \frac{\sqrt{3}}{2}i$$

43. Use the quadratic formula to solve
$x^2 - 6x + 13 = 0$:
$$x = \frac{-(-6) \pm \sqrt{(-6)^2 - 4(1)(13)}}{2(1)}$$
$$= \frac{6 \pm \sqrt{-16}}{2} = \frac{6 \pm 4i}{2} = 3 \pm 2i$$

45. Use the quadratic formula to solve
$4x^2 - 8x + 13 = 0$:
$$x = \frac{-(-8) \pm \sqrt{(-8)^2 - 4(4)(13)}}{2(4)}$$
$$= \frac{8 \pm \sqrt{-144}}{8} = \frac{8 \pm 12i}{8} = 1 \pm \frac{3}{2}i$$

47. The discriminant is $(-11)^2 - 4(3)(6) = 49 > 0$, so there are 2 real unequal roots.

49. The discriminant is $2^2 - 4(5)(1) = -16 < 0$, so there are 2 unequal complex roots.

51. $\sqrt{x^2 - 16} = 0 \Rightarrow x^2 - 16 = 0 \Rightarrow$
$(x+4)(x-4) = 16 \Rightarrow x = \pm 4$

53. $\sqrt{4 - 7x} = \sqrt{2}x \Rightarrow 4 - 7x = 2x^2 \Rightarrow$
$2x^2 + 7x - 4 = 0 \Rightarrow (2x-1)(x+4) = 0 \Rightarrow$
$x = \dfrac{1}{2}$ or $x = -4$
If $x = -4$, then the equation becomes
$\sqrt{4 - 7(-4)} = \sqrt{2}(-4) \Rightarrow \sqrt{32} = -4\sqrt{2}$, which is not true, so we reject that root. The solution set is $\left\{\dfrac{1}{2}\right\}$.

55. $y - 2\sqrt{y} - 3 = 0$. Let $u = \sqrt{y}$, so the equation becomes $u^2 - 2u - 3 = 0 \Rightarrow$
$(u-3)(u+1) = 0 \Rightarrow u = 3$ or $u = -1$.
Now solve for y. $3 = \sqrt{y} \Rightarrow 9 = y$ or $-1 = \sqrt{y}$ (reject this). The solution set is $\{9\}$.

57. $\sqrt{x} - 1 = \sqrt{5 + \sqrt{x}}$. Let $u = \sqrt{x}$. The equation becomes $u - 1 = \sqrt{5 + u} \Rightarrow (u-1)^2 = 5 + u \Rightarrow$
$u^2 - 2u + 1 = 5 + u \Rightarrow u^2 - 3u - 4 \Rightarrow$
$(u-4)(u+1) = 0 \Rightarrow u = 4$ or $u = -1$.
Now solve for x.
$4 = \sqrt{x} \Rightarrow 16 = x$ or $-1 = \sqrt{x}$
(not possible). The solution set is $\{16\}$.

59. $(7x+5)^2 + 2(7x+5) - 15 = 0$. Let $u = 7x+5$.
Then the equation becomes
$u^2 + 2u - 15 = 0 \Rightarrow$
$(u+5)(u-3) = 0 \Rightarrow u = -5$ or $u = 3$.

Now solve for x: $-5 = 7x+5 \Rightarrow x = -\dfrac{10}{7}$ or

$3 = 7x+5 \Rightarrow x = -\dfrac{2}{7}$.
The solution set is $\{-10/7, -2/7\}$.

61. $x^{2/3} + 3x^{1/3} - 4 = 0$. Let $u = x^{1/3}$. So the
equation becomes $u^2 + 3u - 4 = 0 \Rightarrow$
$(u+4)(u-1) = 0 \Rightarrow u = -4$ or $u = 1$. Now
solve for u: $-4 = x^{1/3} \Rightarrow -64 = x$ or
$1 = x^{1/3} \Rightarrow 1 = x$. The solution set is $\{-64,\ 1\}$.

63. $\left(\sqrt{t}+5\right)^2 - 9\left(\sqrt{t}+5\right) + 20 = 0$.

Let $u = \sqrt{t}+5$, so the equation becomes
$u^2 - 9u + 20 = 0 \Rightarrow (u-5)(u-4) = 0 \Rightarrow$
$u = 5$ or $u = 4$. Now solve for t. $5 = \sqrt{t}+5 \Rightarrow$
$0 = t$ or $4 = \sqrt{t}+5 \Rightarrow -1 = \sqrt{t}$. (reject this)
The solution set is $\{0\}$.

65. $4x^4 - 37x^2 + 9 = 0$. Let $u = x^2$. So the
equation becomes $4u^2 - 37u + 9 = 0 \Rightarrow$
$(4u-1)(u-9) = 0 \Rightarrow u = \dfrac{1}{4}$ or $u = 9$. Now

solve for x: $\dfrac{1}{4} = x^2 \Rightarrow x = \pm\dfrac{1}{2}$ or $9 = x^2 \Rightarrow$

$x = \pm 3$. The solution set is $\left\{-3, -\dfrac{1}{2}, \dfrac{1}{2}, 3\right\}$.

67. $\dfrac{2x+1}{2x-1} = \dfrac{x-1}{x+1}$
$(2x+1)(x+1) = (2x-1)(x-1)$
$2x^2 + 3x + 1 = 2x^2 - 3x + 1$
$6x = 0 \Rightarrow x = 0$
The solution set is $\{0\}$.

69. $\left(\dfrac{7x}{x+1}\right)^2 - 3\left(\dfrac{7x}{x+1}\right) = 18$. Let $u = \dfrac{7x}{x+1}$. So,

the equation becomes $u^2 - 3u = 18 \Rightarrow$
$u^2 - 3u - 18 = 0 \Rightarrow (u-6)(u+3) = 0 \Rightarrow$
$u = 6$ or $u = -3$. Now solve for x:

$-3 = \dfrac{7x}{x+1} \Rightarrow -3x-3 = 7x \Rightarrow x = -\dfrac{3}{10}$ or

$6 = \dfrac{7x}{x+1} \Rightarrow 6x+6 = 7x \Rightarrow x = 6$.

The solution set is $\left\{-\dfrac{3}{10}, 6\right\}$.

71. $x^2 + 2yx - 3y^2 = 0$
$(x+3y)(x-y) = 0$
$x+3y = 0 \Rightarrow x = -3y$
$x-y = 0 \Rightarrow x = y$
Solution set: $\{-3y, y\}$

73. $x^2 + (3-2y)x + y^2 - 3y + 2 = 0$
Use the quadratic formula with $a = 1$,
$b = 3-2y$ and $c = y^2 - 3y + 2$.

$x = \dfrac{-(3-2y) \pm \sqrt{(3-2y)^2 - 4(1)\left(y^2 - 3y + 2\right)}}{2(1)}$

$x = \dfrac{2y-3 \pm \sqrt{4y^2 - 12y + 9 - \left(4y^2 - 12y + 8\right)}}{2}$

$= \dfrac{(2y-3)\pm 1}{2} = \dfrac{2y-4}{2} = y-2$ or

$\dfrac{2y-2}{2} = y-1$
Solution set: $\{y-2, y-1\}$

75. $x+5 < 3 \Rightarrow x < -2$
The solution set is $(-\infty, -2)$.

77. $3(x-3) \le 8 \Rightarrow 3x-9 \le 8 \Rightarrow 3x \le 17 \Rightarrow x \le \dfrac{17}{3}$

The solution set is $\left(-\infty, \dfrac{17}{3}\right]$.

79. $x+2 \ge \dfrac{2}{3}x - 2x \Rightarrow 3x+6 \ge 2x - 6x \Rightarrow$

$3x+6 \ge -4x \Rightarrow 6 \ge -7x \Rightarrow -\dfrac{6}{7} \le x$

The solution set is $\left[-\dfrac{6}{7}, \infty\right)$.

81. $\dfrac{1}{6} > \dfrac{4-3x}{3} \Rightarrow 1 > 2(4-3x) \Rightarrow 1 > 8 - 6x \Rightarrow$

$-7 > -6x \Rightarrow \dfrac{7}{6} < x$

Solution set: $\left(\dfrac{7}{6}, \infty\right)$

83. $\dfrac{x-3}{3} - 2 \le \dfrac{x}{6} + \dfrac{1}{2} \Rightarrow 2(x-3) - 2(6) \le x+3 \Rightarrow$

$2x - 6 - 12 \le x+3 \Rightarrow 2x - 18 \le x+3 \Rightarrow x \le 21$

Solution set: $(-\infty, 21]$

85. $\dfrac{3-2x}{4} + 1 > \dfrac{x-5}{3}$

$3(3-2x) + 12 > 4(x-5)$

$9 - 6x + 12 > 4x - 20$

$-6x + 21 > 4x - 20$

$-10x > -41 \Rightarrow x < \dfrac{41}{10}$

Solution set: $\left(-\infty, \dfrac{41}{10}\right)$

87. $3x - 1 < 2 \text{ or } 11 - 2x < 5$

$\quad 3x < 3 \text{ or } \quad -2x < -6$

$\quad\; x < 1 \text{ or } \quad\;\; x > 3$

Solution set: $(-\infty, 1) \cup (3, \infty)$

89. $4x - 5 < 7 \text{ and } 7 - 3x < 1$

$\quad 4x < 12 \text{ and } \quad -3x < -6$

$\quad\; x < 3 \text{ and } \quad\;\; x > 2$

Solution set: $(2, 3)$

91. $-3 \le 2x + 1 < 7 \Rightarrow -4 \le 2x < 6 \Rightarrow -2 \le x < 3$

Solution set: $[-2, 3)$

93. $-3 < 3 - 2x \le 97 \Rightarrow -6 < -2x \le 94 \Rightarrow$

$3 > x \ge -47 \Rightarrow -47 \le x < 3$

Solution set: $[-47, 3)$

95. $x^2 + x - 6 \ge 0 \Rightarrow (x+3)(x-2) \ge 0$

Solve the associated equation:

$(x+3)(x-2) = 0 \Rightarrow x = -3 \text{ or } x = 2.$

So, the intervals are $(-\infty, -3]$, $[-3, 2]$, and and $[2, \infty)$.

Interval	Test point	Value of $x^2 + x - 6$	Result
$(-\infty, -3]$	-4	6	$+$
$[-3, 2]$	0	-6	$-$
$[2, \infty)$	3	6	$+$

The solution set is $(-\infty, -3] \cup [2, \infty)$.

97. $\dfrac{(x-1)(x+3)}{(x+2)(x+5)} \ge 0$

Set the numerator and denominator equal to zero and solve for x.

$(x-1)(x+3) = 0 \Rightarrow x = 1, -3$

$(x+2)(x+5) = 0 \Rightarrow x = -2, -5$

The intervals are $(-\infty, -5)$, $(-5, -3]$, $[-3, -2)$, $(-2, 1]$, and $(1, \infty)$.

Interval	Test point	Value of $\dfrac{(x-1)(x+3)}{(x+2)(x+5)}$	Result
$(-\infty, -5)$	-6	$\dfrac{21}{4}$	$+$
$(-5, -3]$	-4	$-\dfrac{5}{2}$	$-$
$[-3, -2)$	$-\dfrac{5}{2}$	$\dfrac{7}{5}$	$+$
$(-2, 1]$	0	$-\dfrac{3}{10}$	$-$
$(1, \infty)$	2	$\dfrac{5}{28}$	$+$

The solution set is $(-\infty, -5) \cup [-3, -2) \cup (1, \infty)$.

99. $|3x + 2| \le 7 \Rightarrow -7 \le 3x + 2 \le 7 \Rightarrow$

$-9 \le 3x \le 5 \Rightarrow -3 \le x \le \dfrac{5}{3}$

The solution set is $\left[-3, \dfrac{5}{3}\right]$.

101. $4|x-2| + 8 > 12 \Rightarrow 4|x-2| > 4 \Rightarrow$

$|x-2| > 1 \Rightarrow x - 2 < -1 \text{ or } x - 2 > 1 \Rightarrow$

$x < 1 \text{ or } x > 3$

The solution set is $(-\infty, 1) \cup (3, \infty)$.

103. $\left|\dfrac{4-x}{5}\right| \ge 1 \Rightarrow \dfrac{4-x}{5} \le -1 \text{ or } \dfrac{4-x}{5} \ge 1 \Rightarrow$

$4 - x \le -5 \Rightarrow x \ge 9 \text{ or } 4 - x \ge 5 \Rightarrow x \le -1$

The solution set is $(-\infty, -1] \cup [9, \infty)$.

105. $\left|\dfrac{x-1}{x+2}\right| \le 3 \Rightarrow -3 \le \dfrac{x-1}{x+2} \le 3$

$$0 \le \dfrac{x-1}{x+2}+3 \qquad\qquad \dfrac{x-1}{x+2}-3 \le 0$$

$$0 \le \dfrac{x-1+3(x+2)}{x+2} \qquad \dfrac{x-1-3(x+2)}{x+2} \le 0$$

$$0 \le \dfrac{4x+5}{x+2} \qquad\qquad \dfrac{-2x-7}{x+2} \le 0$$

$$4x+5=0 \Rightarrow x=-\dfrac{5}{4};\ x+2=0 \Rightarrow x=-2$$

Interval	Test point	Value of $\dfrac{4x+5}{x+2}$	Result
$(-\infty, -2)$	-4	$\dfrac{11}{2}$	$+$
$\left(-2, -\dfrac{5}{4}\right]$	$-\dfrac{3}{2}$	-2	$-$
$\left[-\dfrac{5}{4}, \infty\right)$	0	$\dfrac{5}{2}$	$+$

Note that the expression is undefined for $x=-2$. The solution set for this part of the original inequality is $(-\infty, -2)\cup\left[-\dfrac{5}{4}, \infty\right)$.

$$-2x-7=0 \Rightarrow x=-\dfrac{7}{2};\ x+2=0 \Rightarrow x=-2$$

Interval	Test point	Value of $\dfrac{-2x-7}{x+2}$	Result
$\left(-\infty, -\dfrac{7}{2}\right]$	-4	$-\dfrac{1}{2}$	$-$
$\left[-\dfrac{7}{2}, -2\right)$	-3	1	$+$
$(-2, \infty)$	0	$-\dfrac{7}{2}$	$-$

Note that the expression is undefined for $x=-2$. The solution set is for this part of the original inequality is $\left(-\infty, -\dfrac{7}{2}\right]\cup(-2, \infty)$.

Both inequalities are true on $\left(-\infty, -\dfrac{7}{2}\right]$ and $\left[-\dfrac{5}{4}, \infty\right)$, so the solution set is $\left(-\infty, -\dfrac{7}{2}\right]\cup\left[-\dfrac{5}{4}, \infty\right)$.

107. $\left|\dfrac{2x-3}{x+2}\right| \ge 2 \Rightarrow \dfrac{2x-3}{x+2} \le -2$ or $\dfrac{2x-3}{x+2} \ge 2$

$$\dfrac{2x-3}{x+2} \le -2 \qquad\qquad \dfrac{2x-3}{x+2} \ge 2$$

$$\dfrac{2x-3}{x+2}+2 \le 0 \qquad\qquad \dfrac{2x-3}{x+2}-2 \ge 0$$

$$\dfrac{2x-3+2(x+2)}{x+2} \le 0 \qquad \dfrac{2x-3-2(x+2)}{x+2} \ge 0$$

$$\dfrac{4x+1}{x+2} \le 0 \qquad\qquad \dfrac{-7}{x+2} \ge 0$$

$$4x+1=0 \Rightarrow x=-\dfrac{1}{4};\ x+2=0 \Rightarrow x=-2$$

Interval	Test point	Value of $\dfrac{4x+1}{x+2}$	Result
$(-\infty, -2)$	-3	11	$+$
$\left(-2, -\dfrac{1}{4}\right]$	-1	-3	$-$
$\left[-\dfrac{1}{4}, \infty\right)$	0	$\dfrac{1}{2}$	$+$

Note that the expression is undefined for $x=-2$. The solution set is for this part of the original inequality is $\left(-2, -\dfrac{1}{4}\right]$.

Interval	Test point	Value of $\dfrac{-7}{x+2}$	Result
$(-\infty, -2)$	-3	$\dfrac{7}{5}$	$+$
$(-2, \infty)$	0	$-\dfrac{7}{2}$	$-$

Note that the expression is undefined for $x=-2$. The solution set is for this part of the original inequality is $(-\infty, -2)$. Since the original inequality is an "or" inequality, the solution set of the original inequality is the union of the two solution sets.

The solution set is $(-\infty, -2)\cup\left(-2, -\dfrac{1}{4}\right]$.

Applying the Concepts

109. $C=2\pi r \Rightarrow 22=2\pi r \Rightarrow \dfrac{11}{\pi}=r$

The radius is $11/\pi$ cm.

111. $A=\dfrac{1}{2}h(b_1+b_2) \Rightarrow 32=\dfrac{1}{2}(8)(5+b_2) \Rightarrow$
$32=4(5+b_2) \Rightarrow 8=5+b_2 \Rightarrow 3=b_2$
The other base is 3 meters.

113. $V = lwh \Rightarrow 4212 = 27(12)h \Rightarrow 13 = h$
The box is 13 cm high.

115. $V = \pi r^2 h \Rightarrow 8750\pi = 5^2 \pi h \Rightarrow 350 = h$
The height is 350 cm.

117. Let x = measure or the base angle.
Then $40 + 2x$ = the measure of the third angle.
So,
$x + x + 40 + 2x = 180 \Rightarrow 4x + 40 = 180 \Rightarrow$
$x = 35$.
The three angles are $35°$, $35°$, and $110°$.

119. Let x = amount invested at 6%.
Then $30,000 - x$ = amount invested at 8%. So,
we have $0.06x + 0.08(30,000 - x) = 2160 \Rightarrow$
$-0.02x + 2400 = 2160 \Rightarrow -0.02x = -240 \Rightarrow$
$x = 12,000$. \$12,000 was invested at 6%, and
\$18,000 was invested at 8%.

121. Let x = amount of 4 1/2% solution.
Then $10 - x$ = the amount of 12% solution.
So, we have
$0.045x + 0.12(10 - x) = 0.06(10) \Rightarrow$
$-0.075x + 1.2 = 0.6 \Rightarrow -0.075x = -0.6 \Rightarrow$
$x = 8$. There are 8 liters of the 4 1/2%
solution and 2 liters of the 12% solution.

123. Let x = the number of people at the party.
Each person shook $x - 1$ hands. So, there are
$x(x - 1)$ handshakes. However, each
handshake is counted twice (if A shakes hands
with B, that is the same as B shaking hands
with A).
So, we have $\dfrac{x(x-1)}{2} = 28 \Rightarrow x^2 - x = 56 \Rightarrow$
$x^2 - x - 56 = 0 \Rightarrow (x - 8)(x + 7) = 0 \Rightarrow$
$x = 8$ or $x = -7$. We reject the negative
answer. There were 8 people at the party.

125. Let x = the number of shares that Lavina
bought. She spent \$18,040 for the stock, or
$18,040/x$ per share.
She sold $x - 20$ shares at $\dfrac{18,040}{x} + 18$ per
share for a total of \$20,000. Therefore,
$(x - 20)\left(\dfrac{18,040}{x} + 18\right) = 20,000$
$\dfrac{18,040 + 18x}{x} = \dfrac{20,000}{x - 20}$
$(x - 20)(18,040 + 18x) = 20,000x$

$18x^2 + 17,680x - 360,800 = 20,000x$
$18x^2 - 2320x - 360,800 = 0$
$2(9x + 820)(x - 220) = 0$

$9x + 820 = 0 \Rightarrow x = -\dfrac{820}{9}$ (reject this)

$x - 220 = 0 \Rightarrow x = 220$
Lavina bought 220 shares of stock.

127. Let x = the number of horses in the herd. Then
$x/4$ = the number of horses in the forest, and
$2\sqrt{x}$ = the number of horses in the
mountains.
$\dfrac{x}{4} + 2\sqrt{x} + 15 = x \Rightarrow x + 8\sqrt{x} + 60 = 4x \Rightarrow$
$8\sqrt{x} = 3x - 60 \Rightarrow 64x = (3x - 60)^2 \Rightarrow$
$64x = 9x^2 - 360x + 3600 \Rightarrow$
$9x^2 - 424x + 3600 = 0 \Rightarrow$
$(x - 36)(9x - 100) = 0 \Rightarrow x = 36$ or $x = 100/9$
Reject the fractional solution. There were 36
horses in the herd.

129. Let x = the original number of members going
on the trip. Then $x + 4$ = the number of members
going on the trip. The cost per member for the
original number going on the trip was $\dfrac{324}{x}$, and
the cost for the final number going on the trip is
$\dfrac{324}{x} - 0.9$. Therefore,
$\dfrac{324}{x + 4} = \dfrac{324}{x} - 0.9$
$324x = 324(x + 4) - 0.9x(x + 4)$
$324x = 324x + 1296 - 0.9x^2 - 3.6x$
$0 = -0.9x^2 - 3.6x + 1296$
$0 = x^2 + 4x - 1440$
$0 = (x - 36)(x + 40) \Rightarrow x = 36$ or $x = -40$
(Reject the negative solution). So, 36 people
originally signed up for the trip, and 40 people
went on the trip.

Chapter 1 Practice Test A

1. $5x - 9 = 3x - 5 \Rightarrow 2x = 4 \Rightarrow x = 2$

2. $\dfrac{7}{24} = \dfrac{x}{8} + \dfrac{1}{6} \Rightarrow 7 = 3x + 4 \Rightarrow x = 1$

3. $\dfrac{1}{x-2}-5=\dfrac{1}{x+2}\Rightarrow\dfrac{1-5x+10}{x-2}=\dfrac{1}{x+2}\Rightarrow$

    $\dfrac{-5x+11}{x-2}=\dfrac{1}{x+2}\Rightarrow$

    $(-5x+11)(x+2)=x-2$

    $\quad -5x^2+x+22=x-2$

    $\quad\quad -5x^2+24=0\Rightarrow -5x^2=-24\Rightarrow$

    $x^2=\dfrac{24}{5}\Rightarrow x=\pm\dfrac{2\sqrt{6}}{\sqrt{5}}=\pm\dfrac{2\sqrt{30}}{5}$

4. Let $x=$ the length of the rectangle.
    Then $x-3=$ the width of the rectangle.

    $x(x-3)=54\Rightarrow x^2-3x-54=0\Rightarrow$

    $(x+6)(x-9)=0\Rightarrow x=-6$ or $x=9$

    We reject the negative answer. The rectangle
    is 9 cm by 6 cm.

5. $x^2+36=-13x\Rightarrow x^2+13x+36=0\Rightarrow$
    $(x+9)(x+4)=0\Rightarrow x=-9$ or $x=-4$

6. Let $x=$ the amount to be invested at 8%.
    Then the total amount invested is $x+8200$.

    $0.06(8200)+0.08x=0.07(8200+x)$

    $\quad 492+0.08x=574+0.07x$

    $\quad\quad\quad 0.01x=82\Rightarrow x=8200$

    Fran must invest \$8200 at 7%.

7. Let $x=$ the width of the border. Then the
    length of the border is $2x+25$, and the width
    of the border is $2x+11$.

    $(2x+25)(2x+11)=351$

    $\quad 4x^2+72x+275=351$

    $\quad\quad 4x^2+72x-76=0$

    $\quad 4(x+19)(x-1)=0\Rightarrow x=-19$ or $x=1$

    The border is 1 inch wide.

8. $\quad\quad\quad -6x-15=(2x+5)^2$

    $\quad\quad\quad -6x-15=4x^2+20x+25$

    $\quad 4x^2+26x+40=0$

    $\quad 2(2x+5)(x+4)=0\Rightarrow x=-\dfrac{5}{2}$ or $x=-4$

    The solution set is $\left\{-4,-5/2\right\}$.

9. To find the constant term, find 1/2 of 2/3 = 1/3
    and then square the answer: $(1/3)^2=1/9$.

    The trinomial is $x^2+\dfrac{2}{3}x+\dfrac{1}{9}$, which factors

    into $\left(x+\dfrac{1}{3}\right)^2$.

10. $x=\dfrac{-(-5)\pm\sqrt{(-5)^2-4(3)(-1)}}{2(3)}$

    $\quad =\dfrac{5\pm\sqrt{37}}{6}=\dfrac{5}{6}\pm\dfrac{\sqrt{37}}{6}$

11. Let $x=$ the length of the side of the original
    piece of cardboard. Then $x-4=$ the length of
    the side of the box.

    $2(x-4)^2=50\Rightarrow (x-4)^2=25\Rightarrow$

    $x-4=5\Rightarrow x=9$ (Reject the negative
    solution.) The side of the original cardboard
    square is 9 inches.

12. $x=\dfrac{-12\pm\sqrt{12^2-4(1)(33)}}{2(1)}$

    $\quad =\dfrac{-12\pm\sqrt{12}}{2}=\dfrac{-12\pm2\sqrt{3}}{2}=-6\pm\sqrt{3}$

13. $3x^4-75x^2=0\Rightarrow 3x^2(x^2-25)=0\Rightarrow$
    $3x^2(x-5)(x+5)=0\Rightarrow 3x^2=0\Rightarrow x=0$
    or $x-5=0\Rightarrow x=5$
    or $x+5=0\Rightarrow x=-5$
    The solution set is $\{-5,0,5\}$.

14. Let $u=\sqrt{x}$. The equation becomes
    $3u^2-2-5u=0\Rightarrow 3u^2-5u-2=0\Rightarrow$
    $(3u+1)(u-2)=0\Rightarrow u=-\dfrac{1}{3}$ or $u=2$

    Now solve for x: $2=\sqrt{x}\Rightarrow x=4$ (We reject
    the negative solution). The solution set is $\{4\}$.

15. $\left|\frac{1}{3}x+5\right|=\left|\frac{2}{3}x+7\right| \Rightarrow \frac{1}{3}x+5=\frac{2}{3}x+7$

or $\frac{1}{3}x+5=-\left(\frac{2}{3}x+7\right)$.

$\frac{1}{3}x+5=\frac{2}{3}x+7 \Rightarrow x+15=2x+21 \Rightarrow x=-6$.

$\frac{1}{3}x+5=-\left(\frac{2}{3}x+7\right) \Rightarrow \frac{1}{3}x+5=-\frac{2}{3}x-7 \Rightarrow$

$x+15=-2x-21 \Rightarrow 3x=-36 \Rightarrow x=-12$

The solution set is $\{-12, -6\}$.

16. $\frac{x}{2}-5 \geq \frac{4x}{9} \Rightarrow 9x-90 \geq 8x \Rightarrow x \geq 90$

The solution set is $[90, \infty)$.

17. $-4 < 2x-3 < 4 \Rightarrow -1 < 2x < 7 \Rightarrow$

$-\frac{1}{2} < x < \frac{7}{2}$

The solution set is $\left(-\frac{1}{2}, \frac{7}{2}\right)$.

18. $\left|\frac{2}{3}x-1\right|-2 > \frac{1}{3} \Rightarrow \left|\frac{2}{3}x-1\right| > \frac{7}{3} \Rightarrow$

$\frac{2}{3}x-1 < -\frac{7}{3}$ or $\frac{2}{3}x-1 > \frac{7}{3}$

$\frac{2}{3}x-1 < -\frac{7}{3} \Rightarrow 2x-3 < -7 \Rightarrow 2x < -4 \Rightarrow$

$x < -2$

$\frac{2}{3}x-1 > \frac{7}{3} \Rightarrow 2x-3 > 7 \Rightarrow 2x > 10 \Rightarrow x > 5$

The solution set is $(-\infty, -2) \cup (5, \infty)$.

19. $\frac{2}{5}y-13 \leq -\left(7+\frac{13}{5}y\right)$

$\frac{2}{5}y-13 \leq -7-\frac{13}{5}y$

$2y-65 \leq -35-13y \Rightarrow 15y \leq 30 \Rightarrow y \leq 2$

The solution set is $(-\infty, 2]$.

20. $0 \leq 5x-2 \leq 8 \Rightarrow 2 \leq 5x \leq 10 \Rightarrow \frac{2}{5} \leq x \leq 2$

The solution set is $\left[\frac{2}{5}, 2\right]$.

Chapter 1 Practice Test B

1. $2x-2=5x+34 \Rightarrow -3x=36 \Rightarrow x=-12$.
The answer is D.

2. $\frac{z}{2}=2z+35 \Rightarrow z=4z+70 \Rightarrow -3z=70 \Rightarrow$

$z=-\frac{70}{3}$. The answer is D.

3. $\frac{1}{t-2}-\frac{1}{2}=\frac{-2t}{4t-1} \Rightarrow$
$2(4t-1)-(t-2)(4t-1)=-2t(2)(t-2) \Rightarrow$
$8t-2-4t^2+9t-2=-4t^2+8t \Rightarrow$

$9t-4=0 \Rightarrow 9t=4 \Rightarrow t=\frac{4}{9}$.

The answer is A.

4. Let $w=$ the width of the rectangle.
Then $w+4=$ the length of the rectangle.

$w(w+4)=77 \Rightarrow w^2+4w-77=0 \Rightarrow$
$(w+11)(w-7)=0 \Rightarrow w=-11$ or $w=7$

We reject the negative solution. The rectangle is 7 cm by 11 cm. The answer is C.

5. $x^2+12=-7x \Rightarrow x^2+7x+12=0 \Rightarrow$
$(x+4)(x+3)=0 \Rightarrow x=-4$ or $x=-3$

The answer is B.

6. Let $x=$ the amount to be invested at 12%.
Then the total amount invested is $7500+x$.
$0.07(7500)+0.12x=0.10(7500+x)$
$525+0.12x=750+0.10x$
$0.02x=225 \Rightarrow x=11,250$
Rena must invest \$11,250 at 12%.
The answer is A.

7. Let $x=$ the width of the border. Then the length of the border is $2x+20$, and the width of the border is $2x+13$.

$(2x+20)(2x+13)=368$

$4x^2+66x+260=368$

$4x^2+66x-108=0$

$2(2x-3)(x+18)=0 \Rightarrow x=\frac{3}{2}$ or $x=-18$

The border is 1.5 inch wide. The answer is D.

8.
$$-6x - 2 = (3x + 1)^2$$
$$-6x - 2 = 9x^2 + 6x + 1$$
$$9x^2 + 12x + 3 = 0$$
$$3x^2 + 4x + 1 = 0$$
$$(3x + 1)(x + 1) = 0 \Rightarrow x = -\frac{1}{3} \text{ or } x = -1$$
The answer is D.

9. To find the constant term, find
$1/2$ of $1/6 = 1/12$ and then square the answer:
$(1/12)^2 = 1/144$.

The trinomial is $x^2 + \frac{1}{6}x + \frac{1}{144}$, which

factors into $\left(x + \frac{1}{12}\right)^2$. The answer is B.

10.
$$x = \frac{-10 \pm \sqrt{10^2 - 4(7)(2)}}{2(7)}$$
$$= \frac{-10 \pm \sqrt{44}}{14} = \frac{-10 \pm 2\sqrt{11}}{14} = -\frac{5}{7} \pm \frac{\sqrt{11}}{7}$$
The answer is A.

11. Let $x =$ the length of the side of the original piece of cardboard. Then $x - 6 =$ the length of the side of the box.

$$3(x - 6)^2 = 675 \Rightarrow (x - 6)^2 = 225 \Rightarrow$$
$x - 6 = 15 \Rightarrow x = 21$ (Note that we reject the negative solution.) The side of the original cardboard square is 21 inches.
The answer is B.

12.
$$x = \frac{-14 \pm \sqrt{14^2 - 4(1)(38)}}{2(1)}$$
$$= \frac{-14 \pm \sqrt{44}}{2} = \frac{-14 \pm 2\sqrt{11}}{2}$$
$$= -7 \pm \sqrt{11}$$
The answer is D.

13. $5x^4 - 45x^2 = 0 \Rightarrow 5x^2(x^2 - 9) = 0 \Rightarrow$
$5x^2(x - 3)(x + 3) = 0 \Rightarrow 5x^2 = 0 \Rightarrow x = 0$
or $x - 3 = 0 \Rightarrow x = 3$
or $x + 3 = 0 \Rightarrow x = -3$
The solution set is $\{-3, 0, 3\}$.
The answer is A.

14. Let $u = \sqrt{x}$. The equation becomes
$$u^2 - 2048 - 32u = 0 \Rightarrow$$
$$u^2 - 32u - 2048 = 0 \Rightarrow$$
$$(u - 64)(u + 32) = 0 \Rightarrow u = 64 \text{ or } u = -32$$
Now solve for x: $64 = \sqrt{x} \Rightarrow x = 4096$ (We reject the negative solution). The answer is A.

15. $\left|\frac{1}{2}x + 2\right| = \left|\frac{3}{4}x - 2\right| \Rightarrow \frac{1}{2}x + 2 = \frac{3}{4}x - 2$

or $\frac{1}{2}x + 2 = -\left(\frac{3}{4}x - 2\right)$.

$\frac{1}{2}x + 2 = \frac{3}{4}x - 2 \Rightarrow 2x + 8 = 3x - 8 \Rightarrow x = 16$

$\frac{1}{2}x + 2 = -\left(\frac{3}{4}x - 2\right) \Rightarrow \frac{1}{2}x + 2 = -\frac{3}{4}x + 2 \Rightarrow$

$2x + 8 = -3x + 8 \Rightarrow 5x = 0 \Rightarrow x = 0$
The answer is D.

16. $\frac{x}{6} - \frac{1}{3} \le \frac{x}{3} + 1 \Rightarrow x - 2 \le 2x + 6 \Rightarrow -8 \le x$
The answer is D.

17. $-13 \le -3x + 2 < -4 \Rightarrow -15 \le -3x < -6 \Rightarrow$
$5 \ge x > 2$
The answer is B.

18. $8 + \left|1 - \frac{x}{2}\right| \ge 10 \Rightarrow \left|1 - \frac{x}{2}\right| \ge 2 \Rightarrow$

$1 - \frac{x}{2} \le -2 \Rightarrow -\frac{x}{2} \le -3 \Rightarrow x \ge 6$ or

$1 - \frac{x}{2} \ge 2 \Rightarrow -\frac{x}{2} \ge 1 \Rightarrow x \le -2$

The solution set is $(-\infty, -2] \cup [6, \infty)$.
The answer is C.

19. $\frac{2}{3}x - 2 < \frac{5}{3}x \Rightarrow 2x - 6 < 5x \Rightarrow -6 < 3x \Rightarrow$
$-2 < x$
The solution set is $(-2, \infty)$. The answer is A.

20. $0 \le 7x - 1 \le 13 \Rightarrow 1 \le 7x \le 14 \Rightarrow \frac{1}{7} \le x \le 2$

The solution set is $\left[\frac{1}{7}, 2\right]$. The answer is A.

Chapter 2 Graphs and Functions

2.1 The Coordinate Plane

2.1 Practice Problems

1.

2.

3. $(x_1, y_1) = (-5, 2);\ (x_2, y_2) = (-4, 1)$

$$d = \sqrt{(x_2 - x_1)^2 + (y_2 - y_1)^2}$$
$$= \sqrt{(-4 - (-5))^2 + (1 - 2)^2}$$
$$= \sqrt{1^2 + (-1)^2} = \sqrt{2} \approx 1.4$$

4. $(x_1, y_1) = (6, 2);\ (x_2, y_2) = (-2, 0)$
$(x_3, y_3) = (1, 5)$

$$d_1 = \sqrt{(x_2 - x_1)^2 + (y_2 - y_1)^2}$$
$$= \sqrt{(-2 - 6)^2 + (0 - 2)^2}$$
$$= \sqrt{(-8)^2 + (-2)^2} = \sqrt{68}$$

$$d_2 = \sqrt{(x_3 - x_1)^2 + (y_3 - y_1)^2}$$
$$= \sqrt{(1 - 6)^2 + (5 - 2)^2}$$
$$= \sqrt{(-5)^2 + (3)^2} = \sqrt{34}$$

$$d_3 = \sqrt{(x_3 - x_2)^2 + (y_3 - y_2)^2}$$
$$= \sqrt{(1 - (-2))^2 + (5 - 0)^2}$$
$$= \sqrt{(3)^2 + (5)^2} = \sqrt{34}$$

Yes, the triangle is an isosceles triangle.

5.

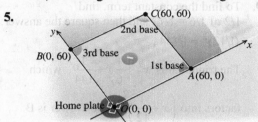

We are asked to find the distance between the points $A(60, 0)$ and $B(0, 60)$.

$$d(A,\ B) = \sqrt{(60 - 0)^2 + (0 - 60)^2}$$
$$= \sqrt{(60)^2 + (-60)^2} = \sqrt{2(60)^2}$$
$$= 60\sqrt{2} \approx 84.85 \text{ ft}$$

6. $M = \left(\dfrac{5 + 6}{2}, \dfrac{-2 + (-1)}{2}\right) = \left(\dfrac{11}{2}, -\dfrac{3}{2}\right)$

2.1 Concepts and Vocabulary

1. A point with a negative first coordinate and a positive second coordinate lies in the <u>second</u> quadrant.

3. The distance between the points $P(x_1, y_1)$ and $Q(x_2, y_2)$ is given by the formula
$$d(P, Q) = \sqrt{(x_2 - x_1)^2 + (y_2 - y_1)^2}.$$

5. True

7. False. Every point in quadrant II has a negative x-coordinate.

2.1 Building Skills

9.

(2, 2): Q1; (3, −1): Q4; (−1, 0): *x*-axis
(−2, −5): Q3; (0, 0): origin; (−7, 4): Q2
(0, 3): *y*-axis; (−4, 2): Q2

11. a. If the *x*-coordinate of a point is 0, the point lies on the *y*-axis.

b.

The set of all points of the form (−1, *y*) is a vertical line that intersects the *x*-axis at −1.

13. a. $y > 0$ **b.** $y < 0$

 c. $x < 0$ **d.** $x > 0$

In Exercises 15−24, use the distance formula,
$d = \sqrt{(x_2 - x_1)^2 + (y_2 - y_1)^2}$ and the midpoint
formula, $(x, y) = \left(\dfrac{x_1 + x_2}{2}, \dfrac{y_1 + y_2}{2} \right)$.

15. a. $d = \sqrt{(2-2)^2 + (5-1)^2} = \sqrt{4^2} = 4$

 b. $M = \left(\dfrac{2+2}{2}, \dfrac{1+5}{2} \right) = (2, 3)$

17. a. $d = \sqrt{(2-(-1))^2 + (-3-(-5))^2}$
 $= \sqrt{3^2 + 2^2} = \sqrt{13}$

 b. $M = \left(\dfrac{-1+2}{2}, \dfrac{-5+(-3)}{2} \right) = (0.5, -4)$

19. a. $d = \sqrt{(3-(-1))^2 + (-6.5-1.5)^2}$
 $= \sqrt{4^2 + (-8)^2} = \sqrt{80} = 4\sqrt{5}$

 b. $M = \left(\dfrac{-1+3}{2}, \dfrac{1.5+(-6.5)}{2} \right) = (1, -2.5)$

21. a. $d = \sqrt{\left(\sqrt{2} - \sqrt{2}\right)^2 + (5-4)^2} = \sqrt{1^2} = 1$

 b. $M = \left(\dfrac{\sqrt{2} + \sqrt{2}}{2}, \dfrac{4+5}{2} \right) = \left(\sqrt{2}, 4.5 \right)$

23. a. $d = \sqrt{(k-t)^2 + (t-k)^2}$
 $= \sqrt{\left(k^2 - 2tk + t^2\right) + \left(t^2 - 2kt + k^2\right)}$
 $= \sqrt{2t^2 - 4tk + 2k^2} = \sqrt{2\left(t^2 - 2tk + k^2\right)}$
 $= \sqrt{2(t-k)^2} = |t-k|\sqrt{2}$

 b. $M = \left(\dfrac{t+k}{2}, \dfrac{k+t}{2} \right)$

25. $P = (-1, -2), Q = (0, 0), R = (1, 2)$
$d(P, Q) = \sqrt{(0-(-1))^2 + (0-(-2))^2} = \sqrt{5}$
$d(Q, R) = \sqrt{(1-0)^2 + (2-0)^2} = \sqrt{5}$
$d(P, R) = \sqrt{(1-(-1))^2 + (2-(-2))^2}$
$\qquad = \sqrt{2^2 + 4^2} = \sqrt{20} = 2\sqrt{5}$
Because $d(P, Q) + d(Q, R) = d(P, R)$, the points are collinear.

27. $P = (4, -2), Q = (1, 3), R = (-2, 8)$
$d(P, Q) = \sqrt{(1-4)^2 + (3-(-2))^2} = \sqrt{34}$
$d(Q, R) = \sqrt{(-2-1)^2 + (8-3)^2} = \sqrt{34}$

$d(P, R) = \sqrt{(-2-4)^2 + (8-(-2))^2}$
$\qquad = \sqrt{(-6)^2 + 10^2} = \sqrt{136} = 2\sqrt{34}$
Because $d(P, Q) + d(Q, R) = d(P, R)$, the points are collinear.

29. $P = (-1, 4), Q = (3, 0), R = (11, -8)$
$d(P, Q) = \sqrt{(3-(-1))^2 + (0-4)^2} = 4\sqrt{2}$
$d(Q, R) = \sqrt{(11-3)^2 + ((-8)-0)^2} = 8\sqrt{2}$
$d(P, R) = \sqrt{(11-(-1))^2 + (-8-4)^2}$
$\qquad = \sqrt{(12)^2 + (-12)^2} = \sqrt{288} = 12\sqrt{2}$
Because $d(P, Q) + d(Q, R) = d(P, R)$, the points are collinear.

31. It is not possible to arrange the points in such a way so that $d(P, Q) + d(Q, R) = d(P, R)$, so the points are not collinear.

33. First, find the midpoint M of PQ.

$$M = \left(\frac{-4+0}{2}, \frac{0+8}{2}\right) = (-2, 4)$$

Now find the midpoint R of PM.

$$R = \left(\frac{-4+(-2)}{2}, \frac{0+4}{2}\right) = (-3, 2)$$

Finally, find the midpoint S of MQ.

$$S = \left(\frac{-2+0}{2}, \frac{4+8}{2}\right) = (-1, 6)$$

Thus, the three points are $(-3, 2)$, $(-2, 4)$, and $(-1, 6)$.

35. $d(P,Q) = \sqrt{(-1-(-5))^2 + (4-5)^2} = \sqrt{17}$

$d(Q,R) = \sqrt{(-4-(-1))^2 + (1-4)^2} = 3\sqrt{2}$

$d(P,R) = \sqrt{(-4-(-5))^2 + (1-5)^2} = \sqrt{17}$

The triangle is isosceles.

37. $d(P,Q) = \sqrt{(0-(-4))^2 + (7-8)^2} = \sqrt{17}$

$d(Q,R) = \sqrt{(-3-0)^2 + (5-7)^2} = \sqrt{13}$

$d(P,R) = \sqrt{(-3-(-4))^2 + (5-8)^2} = \sqrt{10}$

The triangle is scalene.

39. $d(P,Q) = \sqrt{(9-0)^2 + (-9-(-1))^2} = \sqrt{145}$

$d(Q,R) = \sqrt{(5-9)^2 + (1-(-9))^2} = 2\sqrt{29}$

$d(P,R) = \sqrt{(5-0)^2 + (1-(-1))^2} = \sqrt{29}$

The triangle is scalene.

41. $d(P,Q) = \sqrt{(-1-1)^2 + (1-(-1))^2} = 2\sqrt{2}$

$d(Q,R) = \sqrt{(-\sqrt{3}-(-1))^2 + (-\sqrt{3}-1)^2}$

$\quad = \sqrt{(3-2\sqrt{3}+1) + (3+2\sqrt{3}+1)}$

$\quad = \sqrt{8} = 2\sqrt{2}$

$d(P,R) = \sqrt{(-\sqrt{3}-1)^2 + (-\sqrt{3}-(-1))^2}$

$\quad = \sqrt{(3+2\sqrt{3}+1) + (3-2\sqrt{3}+1)}$

$\quad = \sqrt{8} = 2\sqrt{2}$

The triangle is equilateral.

43. First find the lengths of the sides:

$d(P,Q) = \sqrt{(-1-7)^2 + (3-(-12))^2} = 17$

$d(Q,R) = \sqrt{(14-(-1))^2 + (11-3)^2} = 17$

$d(R,S) = \sqrt{(22-14)^2 + (-4-11)^2} = 17$

$d(S,P) = \sqrt{(22-7)^2 + (-4-(-12))^2} = 17$

All the sides are equal, so the quadrilateral is either a square or a rhombus. Now find the length of the diagonals:

$d(P,R) = \sqrt{(14-7)^2 + (11-(-12))^2} = 17\sqrt{2}$

$d(Q,S) = \sqrt{(22-(-1))^2 + (-4-3)^2} = 17\sqrt{2}$

The diagonals are equal, so the quadrilateral is a square.

45. $5 = \sqrt{(x-2)^2 + (2-(-1))^2}$

$\quad = \sqrt{x^2 - 4x + 4 + 9} \Rightarrow$

$5 = \sqrt{x^2 - 4x + 13} \Rightarrow 25 = x^2 - 4x + 13 \Rightarrow$

$0 = x^2 - 4x - 12 \Rightarrow 0 = (x-6)(x+2) \Rightarrow$

$x = -2$ or $x = 6$

47. $P = (-5, 2)$, $Q = (2, 3)$, $R = (x, 0)$ (R is on the x-axis, so the y-coordinate is 0).

$$d(P,R) = \sqrt{(x-(-5))^2 + (0-2)^2}$$
$$d(Q,R) = \sqrt{(x-2)^2 + (0-3)^2}$$

$\sqrt{(x-(-5))^2 + (0-2)^2} = \sqrt{(x-2)^2 + (0-3)^2}$

$(x+5)^2 + (0-2)^2 = (x-2)^2 + (0-3)^2$

$x^2 + 10x + 25 + 4 = x^2 - 4x + 4 + 9$

$10x + 29 = -4x + 13$

$14x = -16$

$x = -\dfrac{8}{7}$

The coordinates of R are $\left(-\dfrac{8}{7}, 0\right)$.

2.1 Applying the Concepts

49.

51.

53.

55. 2014 is the midpoint of the initial range, so
$$M = \left(\frac{2012+2016}{2}, \frac{326+425}{2}\right)$$
$$= (2014, 375.5)$$
Americans spent about \$376 billion on prescription drugs in 2014.

57. Percentage of Android sales in June 2013: 51.5%

59. Android sales were at a maximum in June 2014.

61. Denote the diagonal connecting the endpoints of the edges a and b by d. Then a, b, and d form a right triangle. By the Pythagorean theorem, $a^2 + b^2 = d^2$. The edge c and the diagonals d and h also form a right triangle, so $c^2 + d^2 = h^2$. Substituting d^2 from the first equation, we obtain $a^2 + b^2 + c^2 = h^2$.

63. First, find the initial length of the rope using the Pythagorean theorem:
$$c = \sqrt{24^2 + 10^2} = 26.$$
After t seconds, the length of the rope is $26 - 3t$. Now find the distance from the boat to the dock, x, using the Pythagorean theorem again and solving for x:
$$(26-3t)^2 = x^2 + 10^2$$
$$676 - 156t + 9t^2 = x^2 + 100$$
$$576 - 156t + 9t^2 = x^2$$
$$\sqrt{576 - 156t + 9t^2} = x$$

2.1 Beyond the Basics

65. a. If AB is one of the diagonals, then DC is the other diagonal, and both diagonals have the same midpoint. The midpoint of AB is
$$\left(\frac{2+5}{2}, \frac{3+4}{2}\right) = (3.5, 3.5).$$ The midpoint of
$$DC = (3.5, 3.5) = \left(\frac{x+3}{2}, \frac{y+8}{2}\right).$$
So we have $3.5 = \frac{x+3}{2} \Rightarrow x = 4$ and $3.5 = \frac{y+8}{2} \Rightarrow y = -1$.
The coordinates of D are $(4, -1)$.

b. If AC is one of the diagonals, then DB is the other diagonal, and both diagonals have the same midpoint. The midpoint of AC is
$$\left(\frac{2+3}{2}, \frac{3+8}{2}\right) = (2.5, 5.5).$$ The midpoint of
$$DB = (2.5, 5.5) = \left(\frac{x+5}{2}, \frac{y+4}{2}\right).$$
So we have $2.5 = \frac{x+5}{2} \Rightarrow x = 0$ and $5.5 = \frac{y+4}{2} \Rightarrow y = 7$.
The coordinates of D are $(0, 7)$.

c. If BC is one of the diagonals, then DA is the other diagonal, and both diagonals have the same midpoint. The midpoint of BC is
$$\left(\frac{5+3}{2}, \frac{4+8}{2}\right) = (4, 6).$$ The midpoint of DA
is $(4, 6) = \left(\frac{x+2}{2}, \frac{y+3}{2}\right)$. So we have
$$4 = \frac{x+2}{2} \Rightarrow x = 6 \text{ and } 6 = \frac{y+3}{2} \Rightarrow y = 9.$$
The coordinates of D are $(6, 9)$.

67. a. The midpoint of the diagonal connecting
$(1, 2)$ and $(5, 8)$ is $\left(\frac{1+5}{2}, \frac{2+8}{2}\right) = (3, 5)$.
The midpoint of the diagonal connecting $(-2, 6)$ and $(8, 4)$ is
$\left(\frac{-2+8}{2}, \frac{6+4}{2}\right) = (3, 5)$. Because the midpoints are the same, the figure is a parallelogram.

b. The midpoint of the diagonal connecting $(3, 2)$ and (x, y) is $\left(\dfrac{3+x}{2}, \dfrac{2+y}{2} \right)$. The midpoint of the diagonal connecting $(6, 3)$ and $(6, 5)$ is $(6, 4)$. So $\dfrac{3+x}{2} = 6 \Rightarrow x = 9$ and $\dfrac{2+y}{2} = 4 \Rightarrow y = 6$.

69. Let $P(0, 0)$, $Q(a, 0)$, and $R(0, b)$ be the vertices of the right triangle. The midpoint M of the hypotenuse is $\left(\dfrac{a}{2}, \dfrac{b}{2} \right)$.

$$d(Q, M) = \sqrt{\left(a - \dfrac{a}{2} \right)^2 + \left(0 - \dfrac{b}{2} \right)^2}$$

$$= \sqrt{\left(\dfrac{a}{2} \right)^2 + \left(-\dfrac{b}{2} \right)^2} = \dfrac{\sqrt{a^2 + b^2}}{2}$$

$$d(R, M) = \sqrt{\left(0 - \dfrac{a}{2} \right)^2 + \left(b - \dfrac{b}{2} \right)^2}$$

$$= \sqrt{\left(-\dfrac{a}{2} \right)^2 + \left(\dfrac{b}{2} \right)^2} = \dfrac{\sqrt{a^2 + b^2}}{2}$$

$$d(P, M) = \sqrt{\left(0 - \dfrac{a}{2} \right)^2 + \left(0 - \dfrac{b}{2} \right)^2}$$

$$= \sqrt{\left(-\dfrac{a}{2} \right)^2 + \left(-\dfrac{b}{2} \right)^2} = \dfrac{\sqrt{a^2 + b^2}}{2}$$

71. Since ABC is an equilateral triangle and O is the midpoint of AB, then the coordinates of A are $(-a, 0)$.

$AB = AC = AB = 2a$. Using triangle BOC and the Pythagorean theorem, we have

$BC^2 = OB^2 + OC^2 \Rightarrow (2a)^2 = a^2 + OC^2 \Rightarrow$

$4a^2 = a^2 + OC^2 \Rightarrow 3a^2 = OC^2 \Rightarrow OC = \sqrt{3}a$

Thus, the coordinates of C are $\left(0, \sqrt{3}a \right)$ and the coordinates of D are $\left(0, -\sqrt{3}a \right)$.

2.1 Critical Thinking/Discussion/Writing

73. a. y-axis

b. x-axis

74. a. The union of the x- and y-axes

b. The plane without the x- and y-axes

75. a. Quadrants I and III

b. Quadrants II and IV

76. a. The origin

b. The plane without the origin

77. a. Right half-plane

b. Upper half-plane

78. Let (x, y) be the point.

The point lies in	if
Quadrant I	$x > 0$ and $y > 0$
Quadrant II	$x < 0$ and $y > 0$
Quadrant III	$x < 0$ and $y < 0$
Quadrant IV	$x > 0$ and $y < 0$

2.1 Getting Ready for the Next Section

79. a. $x^2 + y^2 = \left(\dfrac{1}{2} \right)^2 + \left(\dfrac{1}{2} \right)^2 = \dfrac{1}{4} + \dfrac{1}{4} = \dfrac{2}{4} = \dfrac{1}{2}$

b. $x^2 + y^2 = \left(\dfrac{\sqrt{2}}{2} \right)^2 + \left(\dfrac{\sqrt{2}}{2} \right)^2 = \dfrac{2}{4} + \dfrac{2}{4} = 1$

81. a. $\dfrac{x}{|x|} + \dfrac{|y|}{y} = \dfrac{2}{|2|} + \dfrac{|-3|}{-3} = \dfrac{2}{2} + \dfrac{3}{-3} = 1 - 1 = 0$

b. $\dfrac{x}{|x|} + \dfrac{|y|}{y} = \dfrac{-4}{|-4|} + \dfrac{|3|}{3} = \dfrac{-4}{4} + \dfrac{3}{3} = -1 + 1 = 0$

83. $x^2 - 6x + \left(\dfrac{-6}{2} \right)^2 = x^2 - 6x + 3^2$

$$= x^2 - 6x + \underline{9}$$

85. $y^2 + 3y = y^2 + 3y + \left(\dfrac{3}{2}\right)^2 = y^2 + 3y + \dfrac{9}{4}$

87. $x^2 - ax + \left(\dfrac{-a}{2}\right)^2 = x^2 - ax + \dfrac{a^2}{4}$

2.2 Graphs of Equations

2.2 Practice Problems

1. $y = -x^2 + 1$

x	$y = -x^2 + 1$	(x, y)
-2	$y = -(-2)^2 + 1$	$(-2, -3)$
-1	$y = -(-1)^2 + 1$	$(-1, 0)$
0	$y = -(0)^2 + 1$	$(0, 1)$
1	$y = -(1)^2 + 1$	$(1, 0)$
2	$y = -(2)^2 + 1$	$(2, -3)$

2. To find the x-intercept, let $y = 0$, and solve the equation for x: $0 = 2x^2 + 3x - 2 \Rightarrow$

$0 = (2x - 1)(x + 2) \Rightarrow x = \dfrac{1}{2}$ or $x = -2$. To find the y-intercept, let $x = 0$, and solve the equation for y:

$y = 2(0)^2 + 3(0) - 2 \Rightarrow y = -2$.

The x-intercepts are $\frac{1}{2}$ and -2; the y-intercept is -2.

3. To test for symmetry about the y-axis, replace x with $-x$ to determine if $(-x, y)$ satisfies the equation.

$(-x)^2 - y^2 = 1 \Rightarrow x^2 - y^2 = 1$, which is the same as the original equation. So the graph is symmetric about the y-axis.

4. x-axis: $x^2 = (-y)^3 \Rightarrow x^2 = -y^3$, which is not the same as the original equation, so the equation is not symmetric with respect to the x-axis.

y-axis: $(-x)^2 = y^3 \Rightarrow x^2 = y^3$, which is the same as the original equation, so the equation is symmetric with respect to the y-axis.

origin: $(-x)^2 = (-y)^3 \Rightarrow x^2 = -y^3$, which is not the same as the original equation, so the equation is not symmetric with respect to the origin.

5. $y = -t^4 + 77t^2 + 324$

a. First, find the intercepts. If $t = 0$, then $y = 324$, so the y-intercept is $(0, 324)$. If $y = 0$, then we have

$$0 = -t^4 + 77t^2 + 324$$
$$t^4 - 77t^2 - 324 = 0$$
$$\left(t^2 - 81\right)\left(t^2 + 4\right) = 0$$
$$(t + 9)(t - 9)\left(t^2 + 4\right) = 0 \Rightarrow t = -9, 9, \pm 2i$$

So, the t-intercepts are $(-9, 0)$ and $(9, 0)$. Next, check for symmetry.

t-axis: $-y = -t^4 + 77t^2 + 324$ is not the same as the original equation, so the equation is not symmetric with respect to the t-axis.

y-axis: $y = -(-t)^4 + 77(-t)^2 + 324 \Rightarrow$

$y = -t^4 + 77t^2 + 324$, which is the same as the original equation. So the graph is symmetric with respect to the y-axis.

origin: $-y = -(-t)^4 + 77(-t)^2 + 324 \Rightarrow$

$-y = -t^4 + 77t^2 + 324$, which is not the same as the original equation. So the graph is not symmetric with respect to the origin. Now, make a table of values. Since the graph is symmetric with respect to the y-axis, if (t, y) is on the graph, then so is $(-t, y)$. However, the graph pertaining to the physical aspects of the problem consists only of those values for $t \geq 0$.

t	$y = -t^4 + 77t^2 + 324$	(t, y)
0	324	$(0, 324)$
1	400	$(1, 400)$
2	616	$(2, 616)$
3	936	$(3, 936)$
4	1300	$(4, 1300)$
5	1624	$(5, 1624)$
6	1800	$(6, 1800)$
7	1696	$(7, 1696)$
8	1156	$(8, 1156)$
9	0	$(9, 0)$

(continued on next page)

(continued)

b.

c. The population becomes extinct after 9 years.

6. The standard form of the equation of a circle is $(x-h)^2 + (y-k)^2 = r^2$
$(h, k) = (3, -6)$ and $r = 10$
The equation of the circle is
$(x-3)^2 + (y+6)^2 = 100$.

7. $(x-2)^2 + (y+1)^2 = 36 \Rightarrow (h,k) = (2,-1)$, $r = 6$
This is the equation of a circle with center $(2, -1)$ and radius 6.

8. $x^2 + y^2 + 4x - 6y - 12 = 0 \Rightarrow$
$x^2 + 4x + y^2 - 6y = 12$
Now complete the square:
$x^2 + 4x + 4 + y^2 - 6y + 9 = 12 + 4 + 9 \Rightarrow$
$(x+2)^2 + (y-3)^2 = 25$
This is a circle with center $(-2, 3)$ and radius 5.

2.2 Concepts and Vocabulary

1. The graph of an equation in two variables, such as x and y, is the set of all ordered pairs (a, b) that satisfy the equation.

3. If $(0, -5)$ is a point of a graph, then -5 is a <u>y-</u> intercept of the graph.

5. False. The equation of a circle has both an x^2-term and a y^2-term. The given equation does not have a y^2-term.

7. False. The center of the circle with equation $(x+3)^2 + (y+4)^2 = 9$ is $(-3, -4)$.

2.2 Building Skills

In exercises 9–14, to determine if a point lies on the graph of the equation, substitute the point's coordinates into the equation to see if the resulting statement is true.

9. on the graph: $(-3, -4)$, $(1, 0)$, $(4, 3)$; not on the graph: $(2, 3)$

11. on the graph: $(3, 2)$, $(0, 1)$, $(8, 3)$; not on the graph: $(8, -3)$

13. on the graph: $(1, 0)$, $\left(2, \sqrt{3}\right)$, $\left(2, -\sqrt{3}\right)$; not on the graph: $(0, -1)$

15.

17.

19.

21.

23.

25.

27.

29.

31.

33.

35.

For exercises 37–46, read the answers directly from the given graphs.

37. *x*-intercepts: –1, 1
y-intercepts: none
symmetries: *y*-axis

39. *x*-intercepts: $-\pi$, 0, π
y-intercepts: 0
symmetries: origin

41. *x*-intercepts: –3, 3
y-intercepts: –2, 2
symmetries: *x*-axis, *y*-axis, origin

43. *x*-intercepts: –2, 0, 2
y-intercepts: 0
symmetries: origin

45. *x*-intercepts: –2, 0, 2
y-intercepts: 0, 3
symmetries: *y*-axis

47.

49.

51. To find the x-intercept, let $y = 0$, and solve the equation for x: $3x + 4(0) = 12 \Rightarrow x = 4$. To find the y-intercept, let $x = 0$, and solve the equation for y: $3(0) + 4y = 12 \Rightarrow y = 3$. The x-intercept is 4; the y-intercept is 3.

53. To find the x-intercept, let $y = 0$, and solve the equation for x: $\dfrac{x}{5} + \dfrac{0}{3} = 1 \Rightarrow x = 5$. To find the y-intercept, let $x = 0$, and solve the equation for y: $\dfrac{0}{5} + \dfrac{y}{3} = 1 \Rightarrow y = 3$. The x-intercept is 5; the y-intercept is 3.

55. To find the x-intercept, let $y = 0$, and solve the equation for x: $0 = \dfrac{x+2}{x-1} \Rightarrow x = -2$. To find the y-intercept, let $x = 0$, and solve the equation for y: $y = \dfrac{0+2}{0-1} = -2$. The x-intercept is -2; the y-intercept is -2.

57. To find the x-intercept, let $y = 0$, and solve the equation for x: $0 = x^2 - 6x + 8 \Rightarrow x = 4$ or $x = 2$. To find the y-intercept, let $x = 0$, and solve the equation for y: $y = 0^2 - 6(0) + 8 \Rightarrow y = 8$. The x-intercepts are 2 and 4; the y-intercept is 8.

59. To find the x-intercept, let $y = 0$, and solve the equation for x: $x^2 + 0^2 = 4 \Rightarrow x = \pm 2$. To find the y-intercept, let $x = 0$, and solve the equation for y: $0^2 + y^2 = 4 \Rightarrow y = \pm 2$. The x-intercepts are -2 and 2; the y-intercepts are -2 and 2.

61. To find the x-intercept, let $y = 0$, and solve the equation for x: $0 = \sqrt{9 - x^2} \Rightarrow x = \pm 3$. To find the y-intercept, let $x = 0$, and solve the equation for y: $y = \sqrt{9 - 0^2} \Rightarrow y = 3$. The x-intercepts are -3 and 3; the y-intercept is 3.

63. To find the x-intercept, let $y = 0$, and solve the equation for x: $x(0) = 1 \Rightarrow$ no solution. To find the y-intercept, let $x = 0$, and solve the equation for y: $(0)y = 1 \Rightarrow$ no solution. There is no x-intercept; there is no y-intercept.

In exercises 65–74, to test for symmetry with respect to the x-axis, replace y with $-y$ to determine if $(x, -y)$ satisfies the equation. To test for symmetry with respect to the y-axis, replace x with $-x$ to determine if $(-x, y)$ satisfies the equation. To test for symmetry with respect to the origin, replace x with $-x$ and y with $-y$ to determine if $(-x, -y)$ satisfies the equation.

65. $-y = x^2 + 1$ is not the same as the original equation, so the equation is not symmetric with respect to the x-axis.

$y = (-x)^2 + 1 \Rightarrow y = x^2 + 1$, so the equation is symmetric with respect to the y-axis.

$-y = (-x)^2 + 1 \Rightarrow -y = x^2 + 1$, is not the same as the original equation, so the equation is not symmetric with respect to the origin.

67. $-y = x^3 + x$ is not the same as the original equation, so the equation is not symmetric with respect to the x-axis.

$y = (-x)^3 - x \Rightarrow y = -x^3 - x \Rightarrow$

$y = -(x^3 + x)$ is not the same as the original equation, so the equation is not symmetric with respect to the y-axis.

$-y = (-x)^3 - x \Rightarrow -y = -x^3 - x \Rightarrow$

$-y = -(x^3 + x) \Rightarrow y = x^3 + x$, so the equation is symmetric with respect to the origin.

69. $-y = 5x^4 + 2x^2$ is not the same as the original equation, so the equation is not symmetric with respect to the x-axis.

$y = 5(-x)^4 + 2(-x)^2 \Rightarrow y = 5x^4 + 2x^2$, so the equation is symmetric with respect to the y-axis.

$-y = 5(-x)^4 + 2(-x) \Rightarrow -y = 5x^4 + 2x^2$ is not the same as the original equation, so the equation is not symmetric with respect to the origin.

71. $-y = -3x^5 + 2x^3$ is not the same as the original equation, so the equation is not symmetric with respect to the x-axis.

$y = -3(-x)^5 + 2(-x)^3 \Rightarrow y = 3x^5 - 2x^3$ is not the same as the original equation, so the equation is not symmetric with respect to the y-axis.

$-y = -3(-x)^5 + 2(-x)^3 \Rightarrow -y = 3x^5 - 2x^3 \Rightarrow$

$-y = -(-3x^5 + 2x^3) \Rightarrow y = -3x^5 + 2x^3$, so the equation is symmetric with respect to the origin.

73. $x^2(-y)^2 + 2x(-y) = 1 \Rightarrow x^2 y^2 - 2xy = 1$ is not the same as the original equation, so the equation is not symmetric with respect to the x-axis.

$(-x)^2 y^2 + 2(-x)y = 1 \Rightarrow x^2 y^2 - 2xy = 1$ is not the same as the original equation, so the equation is not symmetric with respect to the y-axis.

$(-x)^2(-y)^2 + 2(-x)(-y) = 1 \Rightarrow$

$x^2 y^2 + 2xy = 1$, so the equation is symmetric with respect to the origin.

For exercises 75–78, use the standard form of the equation of a circle, $(x - h)^2 + (y - k)^2 = r^2$.

75. Center (2, 3); radius = 6

77. Center (−2, −3); radius = $\sqrt{11}$

79. $x^2 + (y - 1)^2 = 4$

81. $(x + 1)^2 + (y - 2)^2 = 2$

83. Find the radius by using the distance formula:

$d = \sqrt{(-1-3)^2 + (5-(-4))^2} = \sqrt{97}$.

The equation of the circle is

$(x - 3)^2 + (y + 4)^2 = 97$.

85. The circle touches the x-axis, so the radius is 2. The equation of the circle is

$(x - 1)^2 + (y - 2)^2 = 4$.

87. Find the diameter by using the distance formula:

$d = \sqrt{(-3-7)^2 + (6-4)^2} = \sqrt{104} = 2\sqrt{26}$.

So the radius is $\sqrt{26}$. Use the midpoint formula to find the center:

$M = \left(\dfrac{7 + (-3)}{2}, \dfrac{4 + 6}{2} \right) = (2, 5)$. The equation of the circle is $(x - 2)^2 + (y - 5)^2 = 26$.

89. a. $x^2 + y^2 - 2x - 2y - 4 = 0 \Rightarrow$

$x^2 - 2x + y^2 - 2y = 4$

Now complete the square:

$x^2 - 2x + 1 + y^2 - 2y + 1 = 4 + 1 + 1 \Rightarrow$

$(x - 1)^2 + (y - 1)^2 = 6$. This is a circle with center (1, 1) and radius $\sqrt{6}$.

b. To find the *x*-intercepts, let *y* = 0 and solve for *x*:

$$(x-1)^2 + (0-1)^2 = 6 \Rightarrow (x-1)^2 + 1 = 6 \Rightarrow$$
$$(x-1)^2 = 5 \Rightarrow x - 1 = \pm\sqrt{5} \Rightarrow x = 1 \pm \sqrt{5}$$

Thus, the *x*-intercepts are $\left(1+\sqrt{5}, 0\right)$ and $\left(1-\sqrt{5}, 0\right)$.

To find the *y*-intercepts, let *x* = 0 and solve for *y*:

$$(0-1)^2 + (y-1)^2 = 6 \Rightarrow 1 + (y-1)^2 = 6 \Rightarrow$$
$$(y-1)^2 = 5 \Rightarrow y - 1 = \pm\sqrt{5} \Rightarrow y = 1 \pm \sqrt{5}$$

Thus, the *y*-intercepts are $\left(0, 1+\sqrt{5}\right)$ and $\left(0, 1-\sqrt{5}\right)$.

91. a. $2x^2 + 2y^2 + 4y = 0 \Rightarrow$

$$2(x^2 + y^2 + 2y) = 0 \Rightarrow x^2 + y^2 + 2y = 0.$$

Now complete the square:

$$x^2 + y^2 + 2y + 1 = 0 + 1 \Rightarrow x^2 + (y+1)^2 = 1.$$

This is a circle with center (0, –1) and radius 1.

b. To find the *x*-intercepts, let *y* = 0 and solve for *x*: $x^2 + (0+1)^2 = 1 \Rightarrow x^2 = 0 \Rightarrow x = 0$

Thus, the *x*-intercept is (0, 0).

To find the *y*-intercepts, let *x* = 0 and solve for *y*:

$$0^2 + (y+1)^2 = 1 \Rightarrow y + 1 = \pm 1 \Rightarrow y = 0, -2$$

Thus, the *y*-intercepts are (0, 0) and (0, –2).

93. a. $x^2 + y^2 - x = 0 \Rightarrow x^2 - x + y^2 = 0.$

Now complete the square:

$$x^2 - x + \frac{1}{4} + y^2 = 0 + \frac{1}{4} \Rightarrow$$

$$\left(x - \frac{1}{2}\right)^2 + y^2 = \frac{1}{4}.$$ This is a circle with

center $\left(\frac{1}{2}, 0\right)$ and radius $\frac{1}{2}$.

b. To find the *x*-intercepts, let *y* = 0 and solve

for *x*: $\left(x - \frac{1}{2}\right)^2 + 0^2 = \frac{1}{4} \Rightarrow x - \frac{1}{2} = \pm\frac{1}{2} \Rightarrow$

x = 0, 1. Thus, the *x*-intercepts are (0, 0) and (1, 0). To find the *y*-intercepts, let *x* = 0 and solve for *y*.

$$\left(0 - \frac{1}{2}\right)^2 + y^2 = \frac{1}{4} \Rightarrow y^2 + \frac{1}{4} = \frac{1}{4} \Rightarrow$$
$$y^2 = 0 \Rightarrow y = 0.$$

Thus, the *y*-intercept is (0, 0).

2.2 Applying the Concepts

95. The distance from *P*(*x*, *y*) to the *x*-axis is $|x|$ while the distance from *P* to the *y*-axis is $|y|$. So the equation of the graph is $|x| = |y|$.

97. If you save $100 each month, it will take 24 months (or two years) to save $2400. So, the graph starts at (0, 0) and increases to (2, 2400). It will take another 30 months (or 2.5) years to withdraw $80 per month until the $2400 is gone. Thus, the graph passes through (4.5, 0).

99. a. July 2018 is represented by *t* = 0, so March 2018 is represented by *t* = –4. The monthly profit for March is determined by

$$P = -0.5(-4)^2 - 3(-4) + 8 = \$12 \text{ million.}$$

b. July 2018 is represented by *t* = 0, so October 2018 is represented by *t* = 3. So the monthly profit for October is determined by

$$P = -0.5(3)^2 - 3(3) + 8 = -\$5.5 \text{ million.}$$

This is a loss.

c.

Because *t* = 0 represents July 2018, *t* = –6 represents January 2018, and *t* = 5 represents December 2018.

d. To find the *t*-intercept, set $P = 0$ and solve

for *t*: $0 = -0.5t^2 - 3t + 8 \Rightarrow$

$$t = \frac{3 \pm \sqrt{(-3)^2 - 4(-0.5)(8)}}{2(-0.5)} = \frac{3 \pm \sqrt{25}}{-1}$$

$$= 2 \text{ or } -8$$

The *t*-intercepts represent the months with no profit and no loss. In this case, $t = -8$ makes no sense in terms of the problem, so we disregard this solution. $t = 2$ represents Sept 2018.

e. To find the *P*-intercept, set $t = 0$ and solve

for *P*: $P = -0.5(0)^2 - 3(0) + 8 \Rightarrow P = 8$.

The *P*-intercept represents the profit in July 2018.

101. a.

t	Height = $-16t^2 + 128t + 320$
0	320 feet
1	432 feet
2	512 feet
3	560 feet
4	576 feet
5	560 feet
6	512 feet

b.

c. $0 \le t \le 10$

d. To find the *t*-intercept, set $y = 0$ and solve for *t*:

$0 = -16t^2 + 128t + 320 \Rightarrow$

$0 = -16(t^2 - 8t - 20) \Rightarrow$

$0 = (t - 10)(t + 2) \Rightarrow t = 10$ or $t = -2$.

The graph does not apply if $t < 0$, so the *t*-intercept is 10. This represents the time when the object hits the ground. To find the *y*-intercept, set $t = 0$ and solve for *y*:

$y = -16(0)^2 + 128(0) + 320 \Rightarrow y = 320$.

This represents the height of the building.

2.2 Beyond the Basics

103. $x^2 + y^2 - 4x + 2y - 20 = 0 \Rightarrow$

$x^2 - 4x + y^2 + 2y = 20 \Rightarrow$

$x^2 - 4x + 4 + y^2 + 2y + 1 = 20 + 4 + 1 \Rightarrow$

$(x - 2)^2 + (y + 1)^2 = 25$

So this is the graph of a circle with center $(2, -1)$ and radius 5. The area of this circle is 25π. $x^2 + y^2 - 4x + 2y - 31 = 0 \Rightarrow$

$x^2 - 4x + y^2 + 2y = 31 \Rightarrow$

$x^2 - 4x + 4 + y^2 + 2y + 1 = 31 + 4 + 1 \Rightarrow$

$(x - 2)^2 + (y + 1)^2 = 36$

So, this is the graph of a circle with center $(2, -1)$ and radius 6. The area of this circle is 36π. Both circles have the same center, so the area of the region bounded by the two circles equals $36\pi - 25\pi = 11\pi$.

2.2 Critical Thinking/Discussion/Writing

105. The graph of $y^2 = 2x$ is the union of the graphs of $y = \sqrt{2x}$ and $y = -\sqrt{2x}$.

106. Let (x, y) be a point on the graph. The graph is symmetric with regard to the x-axis, so the point $(x, -y)$ is also on the graph. Because the graph is symmetric with regard to the y-axis, the point $(-x, y)$ is also on the graph. Therefore the point $(-x, -y)$ is on the graph, and the graph is symmetric with respect to the origin. The graph of $y = x^3$ is an example of a graph that is symmetric with respect to the origin but is not symmetric with respect to the x- and y-axes.

107. a. First find the radius of the circle:

$$d(A, B) = \sqrt{(6-0)^2 + (8-1)^2} = \sqrt{85} \Rightarrow$$

$$r = \frac{\sqrt{85}}{2}.$$

The center of the circle is

$$\left(\frac{6+0}{2}, \frac{1+8}{2}\right) = \left(3, \frac{9}{2}\right).$$

So the equation of the circle is

$$(x-3)^2 + \left(y - \frac{9}{2}\right)^2 = \frac{85}{4}.$$

To find the x-intercepts, set $y = 0$, and solve for x:

$$(x-3)^2 + \left(0 - \frac{9}{2}\right)^2 = \frac{85}{4} \Rightarrow$$

$$(x-3)^2 + \frac{81}{4} = \frac{85}{4} \Rightarrow x^2 - 6x + 9 = 1 \Rightarrow$$

$$x^2 - 6x + 8 = 0$$

The x-intercepts are the roots of this equation.

b. First find the radius of the circle:

$$d(A, B) = \sqrt{(a-0)^2 + (b-1)^2}$$

$$= \sqrt{a^2 + (b-1)^2} \Rightarrow$$

$$r = \frac{\sqrt{a^2 + (b-1)^2}}{2}.$$

The center of the circle is

$$\left(\frac{a+0}{2}, \frac{b+1}{2}\right) = \left(\frac{a}{2}, \frac{b+1}{2}\right)$$

So the equation of the circle is

$$\left(x - \frac{a}{2}\right)^2 + \left(y - \frac{b+1}{2}\right)^2 = \frac{a^2 + (b-1)^2}{4}.$$

To find the x-intercepts, set $y = 0$ and solve for x:

$$\left(x - \frac{a}{2}\right)^2 + \left(0 - \frac{b+1}{2}\right)^2 = \frac{a^2 + (b-1)^2}{4}$$

$$x^2 - ax + \frac{a^2}{4} + \frac{(b+1)^2}{4} = \frac{a^2 + (b-1)^2}{4}$$

$$4x^2 - 4ax + a^2 + b^2 + 2b + 1 = a^2 + b^2 - 2b + 1$$

$$4x^2 - 4ax + 4b = 0$$

$$x^2 - ax + b = 0$$

The x-intercepts are the roots of this equation.

c. $a = 3$ and $b = 1$. Approximate the roots of the equation by drawing a circle whose diameter has endpoints $A(0, 1)$ and $B(3, 1)$.

The center of the circle is $\left(\frac{3}{2}, 1\right)$ and the

radius is $\frac{3}{2}$. The roots are approximately

$(0.4, 0)$ and $(2.6, 0)$.

108. a. The coordinates of the center of each circle are (r, r) and $(3r, r)$.

b. To find the area of the shaded region, first find the area of the rectangle shown in the figure below, and then subtract the sum of the areas of the two sectors, A and B.

(continued on next page)

(*continued*)

$$A_{\text{rect}} = r(2r) = 2r^2$$

$$A_{\text{sector } R} = A_{\text{sector } S} = \frac{1}{4}\pi r^2$$

$$A_{\text{shaded region}} = 2r^2 - \left(\frac{1}{4}\pi r^2 + \frac{1}{4}\pi r^2\right)$$

$$= 2r^2 - \frac{1}{2}\pi r^2 = \left(2 - \frac{\pi}{2}\right)r^2$$

2.2 Getting Ready for the Next Section

109. $\dfrac{5-3}{6-2} = \dfrac{2}{4} = \dfrac{1}{2}$

111. $\dfrac{2-(-3)}{3-13} = \dfrac{5}{-10} = -\dfrac{1}{2}$

113. $\dfrac{\dfrac{1}{2} - \dfrac{1}{4}}{\dfrac{3}{8} - \left(-\dfrac{1}{4}\right)} = \dfrac{\dfrac{1}{4}}{\dfrac{5}{8}} = \dfrac{1}{4} \cdot \dfrac{8}{5} = \dfrac{2}{5}$

115. $-\dfrac{1}{2}$ **117.** $\dfrac{3}{2}$

119. $-\dfrac{2 + \dfrac{3}{4}}{1 - \dfrac{1}{2}} = -\dfrac{\dfrac{11}{4}}{\dfrac{1}{2}} = -\dfrac{11}{4} \cdot 2 = -\dfrac{11}{2}$

121. $2x + 3y = 6 \Rightarrow 3y = -2x + 6 \Rightarrow y = -\dfrac{2}{3}x + 2$

123. $y - 2 - \dfrac{2}{3}(x+1) = 0 \Rightarrow y - 2 = \dfrac{2}{3}(x+1) \Rightarrow$

$y = \dfrac{2}{3}x + \dfrac{2}{3} + 2 = \dfrac{2}{3}x + \dfrac{8}{3}$

2.3   Lines

2.3 Practice Problems

1. $m = \dfrac{-3 - 5}{-6 - (-7)} = -\dfrac{8}{13}$

A slope of $-\frac{8}{13}$ means that the value of y decreases 8 units for every 13 units increase in x.

2. $P(-2, -3),\ m = -\dfrac{2}{3}$

$$y - (-3) = -\frac{2}{3}\left[x - (-2)\right]$$

$$y + 3 = -\frac{2}{3}(x + 2)$$

$$y + 3 = -\frac{2}{3}x - \frac{4}{3} \Rightarrow y = -\frac{2}{3}x - \frac{13}{3}$$

3. $m = \dfrac{-4 - 6}{-3 - (-1)} = \dfrac{-10}{-2} = 5$

Use either point to determine the equation of the line. Using $(-3, -4)$, we have

$y - (-4) = 5\left[x - (-3)\right] \Rightarrow y + 4 = 5(x + 3) \Rightarrow$
$y + 4 = 5x + 15 \Rightarrow y = 5x + 11$

Using $(-1, 6)$, we have

$y - 6 = 5\left[x - (-1)\right] \Rightarrow y - 6 = 5(x + 1) \Rightarrow$
$y - 6 = 5x + 5 \Rightarrow y = 5x + 11$

4. $y - y_1 = m(x - x_1) \Rightarrow y - (-3) = 2(x - 0)$
                              point-slope form

$y - (-3) = 2(x - 0) \Rightarrow$
$y + 3 = 2x \Rightarrow y = 2x - 3$

5. The slope is $-\dfrac{2}{3}$ and the y-intercept is 4. The line goes through $(0, 4)$, so locate a second point by moving two units down and three units right. Thus, the line goes through $(3, 2)$.

6. $x = -3$. The slope is undefined, and there is no y-intercept. The x-intercept is -3.
    $y = 7$. The slope is 0, and the y-intercept is 7.

7. First, solve for y to write the equation in slope-intercept form:
    $3x + 4y = 24 \Rightarrow 4y = -3x + 24 \Rightarrow$
    $y = -\dfrac{3}{4}x + 6$. The slope is $-\dfrac{3}{4}$, and the y-intercept is 6. Find the x-intercept by setting $y = 0$ and solving the equation for x:
    $0 = -\dfrac{3}{4}x + 6 \Rightarrow 6 = \dfrac{3}{4}x \Rightarrow 8 = x$. Thus, the graph passes through the points $(0, 6)$ and $(8, 0)$.

8. Use the equation $H = 2.6x + 65$.
    $H_1 = 2.6(43) + 65 = 176.8$
    $H_2 = 2.6(44) + 65 = 179.4$
    The person is between 176.8 cm and 179.4 cm tall, or 1.768 m and 1.794 m.

9. a. Parallel lines have the same slope, so the slope of the line is $m = \dfrac{3-7}{2-5} = \dfrac{-4}{-3} = \dfrac{4}{3}$.
    Using the point-slope form, we have
    $y - 5 = \dfrac{4}{3}\left[x - (-2)\right] \Rightarrow 3y - 15 = 4(x+2) \Rightarrow$
    $3y - 15 = 4x + 8 \Rightarrow 4x - 3y + 23 = 0$

b. The slopes of perpendicular lines are negative reciprocals. Write the equation $4x + 5y + 1 = 0$ in slope-intercept form to find its slope: $4x + 5y + 1 = 0 \Rightarrow$
    $5y = -4x - 1 \Rightarrow y = -\dfrac{4}{5}x - \dfrac{1}{5}$. The slope of a line perpendicular to this lines is $\dfrac{5}{4}$.
    Using the point-slope form, we have
    $y - (-4) = \dfrac{5}{4}(x - 3) \Rightarrow 4(y + 4) = 5(x - 3) \Rightarrow$
    $4y + 16 = 5x - 15 \Rightarrow 5x - 4y - 31 = 0$

10. Because 2016 is 10 years after 2006, set $x = 10$. Then $y = 0.44(10) + 6.70 = 11.1$
    There were 11.1 million registered motorcycles in the U.S. in 2016.

2.3 Concepts and Vocabulary

1. The slope of a horizontal line is _0_; the slope of a vertical line is _undefined_.

3. Every line parallel to the line $y = 3x - 2$ has slope, m, equal to _3_.

5. False. The slope of the line $y = -\frac{1}{4}x + 5$ is equal to $-\frac{1}{4}$.

7. True

2.3 Building Skills

9. $m = \dfrac{7-3}{4-1} = \dfrac{4}{3}$; the graph is rising.

11. $m = \dfrac{-2 - (-2)}{-2 - 6} = \dfrac{0}{-8} = 0$; the graph is horizontal.

13. $m = \dfrac{-3.5 - 2}{3 - 0.5} = \dfrac{-5.5}{2.5} = -2.2$; the graph is falling.

15. $m = \dfrac{5-1}{(1+\sqrt{2}) - \sqrt{2}} = \dfrac{4}{1} = 4$; the graph is rising.

17. ℓ_3 19. ℓ_4

21. ℓ_1 passes through the points $(2, 3)$ and $(-5, 4)$. $m_{\ell_1} = \dfrac{-4 - 3}{-5 - 2} = \dfrac{-7}{-7} = 1$.

23. ℓ_3 passes through the points (2, 3) and (0, −1). $m_{\ell_3} = \dfrac{-1-3}{0-2} = 2$.

25. $(0, 5)$; $m = 3$

$y = 3x + 5$

27. $y = \dfrac{1}{2}x + 4$

29. $y - 1 = -\dfrac{3}{2}(x - 2) \Rightarrow y - 1 = -\dfrac{3}{2}x + 3 \Rightarrow$

$y = -\dfrac{3}{2}x + 4$

31. $y + 4 = 0(x - 5) \Rightarrow y + 4 = 0 \Rightarrow y = -4$

33. $m = \dfrac{0-1}{1-0} = -1$. The y-intercept is (0, 1), so

the equation is $y = -x + 1$.

35. $m = \dfrac{3-3}{3-(-1)} = 0$ Because the slope = 0, the

line is horizontal. Its equation is $y = 3$.

37. $m = \dfrac{1-(-1)}{1-(-2)} = \dfrac{2}{3}$. Now write the equation in

point-slope form, and then solve for y to write
the equation in slope-intercept form.

$y + 1 = \dfrac{2}{3}(x + 2) \Rightarrow y + 1 = \dfrac{2}{3}x + \dfrac{4}{3} \Rightarrow$

$y = \dfrac{2}{3}x + \dfrac{1}{3}$

39. $m = \dfrac{2-\dfrac{1}{4}}{0-\dfrac{1}{2}} = \dfrac{\dfrac{7}{4}}{-\dfrac{1}{2}} = -\dfrac{7}{2}$. Now write the

equation in point-slope form, and then solve
for y to write the equation in slope-intercept

form. $y - 2 = -\dfrac{7}{2}x \Rightarrow y = -\dfrac{7}{2}x + 2$

41. $x = 5$ 43. $y = 0$

45. $y = 14$ 47. $y = -\dfrac{2}{3}x - 4$

49. $m = \dfrac{4-0}{0-(-3)} = \dfrac{4}{3}$; $y = \dfrac{4}{3}x + 4$

51. $y = 7$ 53. $y = -5$

55. $y = 3x - 2$

The slope is 3 and the y-intercept is $(0, -2)$.

$0 = 3x - 2 \Rightarrow 3x = 2 \Rightarrow x = \dfrac{2}{3}$

The x-intercept is $\left(\dfrac{2}{3}, 0\right)$.

57. $x + 2y - 4 = 0 \Rightarrow 2y = -x + 4 \Rightarrow y = -\dfrac{1}{2}x + 2$.

The slope is $-1/2$, and the y-intercept is $(0, 2)$. To find the x-intercept, set $y = 0$ and solve for x: $x + 2(0) - 4 = 0 \Rightarrow x = 4$.

59. $3x - 2y + 6 = 0 \Rightarrow 3x + 6 = 2y \Rightarrow \dfrac{3}{2}x + 3 = y$.

The slope is $3/2$, and the y-intercept is $(0, 3)$. To find the x-intercept, set $y = 0$ and solve for x: $3x - 2(0) + 6 = 0 \Rightarrow 3x = -6 \Rightarrow x = -2$.

61. $x - 5 = 0 \Rightarrow x = 5$. The slope is undefined, and there is no y-intercept. The x-intercept is 5.

63. $x = 0$. The slope is undefined, and the y-intercepts are the y-axis. This is a vertical line whose x-intercept is 0.

For exercises 65–68, the two-interecepts form of the equation of a line is $\dfrac{x}{a} + \dfrac{y}{b} = 1$.

65. $\dfrac{x}{4} + \dfrac{y}{3} = 1$

67. $2x + 3y = 6 \Rightarrow \dfrac{2x}{6} + \dfrac{3y}{6} = \dfrac{6}{6} \Rightarrow \dfrac{x}{3} + \dfrac{y}{2} = 1$;

x-intercept = 3; y-intercept = 2

69. $2x - 3y = 12 \Rightarrow \dfrac{2x}{12} - \dfrac{3y}{12} = 1 \Rightarrow \dfrac{x}{6} - \dfrac{y}{4} = 1$

The x-intercept is 6 and the y-intercept is –4.

71. $-5x + 2y = 10 \Rightarrow -\dfrac{5x}{10} + \dfrac{2y}{10} = 1 \Rightarrow -\dfrac{x}{2} + \dfrac{y}{5} = 1$

The *x*-intercept is –2 and the *y*-intercept is 5.

73. $m = \dfrac{9-4}{7-2} = \dfrac{5}{5} = 1.$ The equation of the line

through (2, 4) and (7, 9) is $y - 4 = 1(x - 2) \Rightarrow$

$y = x + 2.$ Check to see if (–1, 1) satisfies the

equation by substituting $x = -1$ and $y = 1$:

$1 = -1 + 2 \Rightarrow 1 = 1.$ So (–1, 1) lies on the line.

75. The given line passes through the points (0, 3) and (4, 0), so its slope is $-\dfrac{3}{4}.$ Any line

parallel to this line will have the same slope. The line that passes through the origin and is parallel to the given line has equation

$y = -\dfrac{3}{4}x.$

77. The red line passes through the points (–2, 0) and (0, 3), so its slope is $\dfrac{3}{2}.$ The blue line

passes through (4, 2) and has the same slope, so its equation is

$y - 2 = \dfrac{3}{2}(x - 4) \Rightarrow 2y - 4 = 3x - 12 \Rightarrow$

$2y = 3x - 8 \Rightarrow y = \dfrac{3}{2}x - 4$

79. The slope of $y = 3x - 1$ is 3. The slope of

$y = 3x + 2$ is also 3. The lines are parallel.

81. The slope of $y = 2x - 4$ is 2. The slope of

$y = -\dfrac{1}{2}x + 4$ is $-\dfrac{1}{2}.$ The lines are

perpendicular.

83. The slope of $3x + 8y = 7$ is $-3/8$, while the

slope of $5x - 7y = 0$ is $5/7.$ The lines are

neither parallel nor perpendicular.

85. The slope of $x = 4y + 8$ is $1/4.$ The slope of

$y = -4x + 1$ is -4, so the lines are

perpendicular.

87. Both lines are vertical lines. The lines are parallel.

89. The equation of the line through (2, –3) with slope 3 is

$y + 3 = 3(x - 2) \Rightarrow y + 3 = 3x - 6 \Rightarrow$

$y = 3x - 9.$

91. A line perpendicular to a line with slope $-\dfrac{1}{2}$

has slope 2. The equation of the line through

(–1, 2) with slope 2 is $y - 2 = 2(x - (-1)) \Rightarrow$

$y - 2 = 2(x + 1) \Rightarrow y - 2 = 2x + 2 \Rightarrow$

$y = 2x + 4.$

93. The slope of the line joining (1, –2) and (–3,

2) is $\dfrac{2 - (-2)}{-3 - 1} = -1.$ The equation of the line

through (–2, –5) with slope –1 is

$y - (-5) = -(x - (-2)) \Rightarrow y + 5 = -(x + 2) \Rightarrow$

$y + 5 = -x - 2 \Rightarrow y = -x - 7.$

95. The slope of the line joining (–3, 2) and

(–4, –1) is

$\dfrac{-1 - 2}{-4 - (-3)} = 3.$

A line perpendicular to this line has slope $-\dfrac{1}{3}.$

The equation of the line through (1, –2) with

slope $-\dfrac{1}{3}$ is

$y - (-2) = -\dfrac{1}{3}(x - 1) \Rightarrow 3(y + 2) = -(x - 1) \Rightarrow$

$3y + 6 = -x + 1 \Rightarrow 3y = -x - 5 \Rightarrow$

$y = -\dfrac{1}{3}x - \dfrac{5}{3}.$

97. The slope of the line $y = 6x + 5$ is 6. The

lines are parallel, so the slope of the new line

is also 6. The equation of the line with slope 6

and *y*-intercept 4 is $y = 6x + 4.$

99. The slope of the line $y = 6x + 5$ is 6. The

lines are perpendicular, so the slope of the

new line is $-\dfrac{1}{6}.$ The equation of the line with

slope $-\dfrac{1}{6}$ and *y*-intercept 4 is $y = -\dfrac{1}{6}x + 4.$

101. The slope of $x + y = 1$ is –1. The lines are

parallel, so they have the same slope. The

equation of the line through (1, 1) with slope

–1 is $y - 1 = -(x - 1) \Rightarrow y - 1 = -x + 1 \Rightarrow$

$y = -x + 2.$

103. The slope of $3x - 9y = 18$ is $1/3$. The lines are perpendicular, so the slope of the new line is -3. The equation of the line through $(-2, 4)$ with slope -3 is $y - 4 = -3(x - (-2)) \Rightarrow$ $y - 4 = -3x - 6 \Rightarrow y = -3x - 2$.

2.3 Applying the Concepts

105. a. The y-intercept represents the initial expenses.

  b. The x-intercept represents the point at which the teacher breaks even, i.e., the expenses equal the income.

  c. The teacher's profit if there are 16 students in the class is $640.

  d. The slope of the line is
  $$\frac{640 - (-750)}{16 - 0} = \frac{1390}{16} = \frac{695}{8}$$
  The equation of the line is
  $$P = \frac{695}{8}n - 750.$$

107. slope = $\dfrac{\text{rise}}{\text{run}} \Rightarrow \dfrac{4}{40} = \dfrac{1}{10}$

109. 8 in. in two weeks $\Rightarrow$ the plant grows 4 in. per week. John wants to trim the hedge when it grows 6 in., so he should trim it every $\frac{6}{4} = 1.5$ weeks ≈ 10 days.

111. a. x = the number of weeks; y = the amount of money in the account after x weeks; $y = 7x + 130$

  b. The slope is the amount of money deposited each week; the y-intercept is the initial deposit.

113. a. x = the number of months owed to pay off the refrigerator; y = the amount owed; $y = -15x + 600$

  b. The slope is the amount that the balance due changes per month; the y-intercept is the initial amount owed.

115. a. x = the number of years after 2010; y = the life expectancy of a female born in the year $2010 + x$; $y = 0.17x + 80.8$

  b. The slope is the rate of increase in life expectancy; the y-intercept is the current life expectancy.

117. There are 30 days in June. For the first 13 days, you used data at a rate of $\dfrac{435}{13} \approx 33.5$ MB/day. At the same rate, you will use $33.5(17) = 569.5$ MB for the rest of the month.
$435 + 569.5 = 1004.5$
So, you don't need to buy extra data. You will have about 20 MB left.

119. $y = 5x + 40{,}000$

121. a. The two points are $(100, 212)$ and $(0, 32)$. So the slope is $\dfrac{212 - 32}{100 - 0} = \dfrac{180}{100} = \dfrac{9}{5}$. The equation is
  $$F - 32 = \frac{9}{5}(C - 0) \Rightarrow F = \frac{9}{5}C + 32$$

  b. One degree Celsius change in the temperature equals $9/5$ degrees change in degrees Fahrenheit.

  c.

C	$F = \dfrac{9}{5}C + 32$
40°C	104°F
25°C	77°F
–5°C	23°F
–10°C	14°F

  d. $100°F = \dfrac{9}{5}C + 32 \Rightarrow C = 37.78°C$

  $90°F = \dfrac{9}{5}C + 32 \Rightarrow C = 32.22°C$

  $75°F = \dfrac{9}{5}C + 32 \Rightarrow C = 23.89°C$

  $-10°F = \dfrac{9}{5}C + 32 \Rightarrow C = -23.33°C$

  $-20°F = \dfrac{9}{5}C + 32 \Rightarrow C = -28.89°C$

  e. $97.6°F = \dfrac{9}{5}C + 32 \Rightarrow C = 36.44°C$;

  $99.6°F = \dfrac{9}{5}C + 32 \Rightarrow C = 37.56°C$

  f. Let $x = °F = °C$. Then $x = \dfrac{9}{5}x + 32 \Rightarrow$

  $-\dfrac{4}{5}x = 32 \Rightarrow x = -40$. At $-40°$, $°F = °C$.

123. a. The year 2005 is represented by $t = 0$, and the year 2011 is represented by $t = 6$. The points are $(0, 2425)$ and $(6, 4026)$. So the slope is $\dfrac{4026 - 2425}{6} \approx 266.8$. The equation is $y - 2425 = 266.8(t - 0) \Rightarrow$
$y = 266.8t + 2425$

b.

c. The year 2008 is represented by $t = 3$. So $y = 266.8(3) + 2425 \Rightarrow y = 3225.4$. Note that there cannot be a fraction of a person, so. there were 3225 women prisoners in 2008.

d. The year 2017 is represented by $t = 12$. So $y = 266.8(12) + 2425 \Rightarrow y = 5626.6$. There will be 5627 women prisoners in 2017.

125. The independent variable t represents the number of years after 2000, with $t = 0$ representing 2000. The two points are $(0, 11.7)$ and $(5, 12.7)$. So the slope is $\dfrac{12.7 - 11.7}{5} = 0.2$. The equation is $p - 11.7 = 0.2(t - 0) \Rightarrow p = 0.2t + 11.7$. The year 2010 is represented by $t = 10$. $p = 0.2(10) + 11.7 \Rightarrow p = 13.7\%$.

127. a.

$y = -2x + 12.4$

b.

[0, 6, 1] by [0, 12, 1]

c. The price in the table is given as the number of nickels. $35¢ = 7$ nickels, so let $x = 7$. $y = -2(7) + 12.4 = -1.6$
Thus, no newspapers will be sold if the price per copy is $35¢$. Note that this is also clear from the graph, which appears to cross the x-axis at approximately $x = 6$.

2.3 Beyond the Basics

129. $3 = \dfrac{c - 3}{1 - (-2)} \Rightarrow 9 = c - 3 \Rightarrow c = 12$

131. a. Let $A = (0, 1)$, $B = (1, 3)$, $C = (-1, -1)$.
$m_{AB} = \dfrac{3 - 1}{1 - 0} = 2; m_{BC} = \dfrac{-1 - 3}{-1 - 1} = \dfrac{-4}{-2} = 2$
$m_{AC} = \dfrac{-1 - 1}{-1 - 0} = 2$
The slopes of the three segments are the same, so the points are collinear.

b. $d(A, B) = \sqrt{(1 - 0)^2 + (3 - 1)^2} = \sqrt{5}$
$d(B, C) = \sqrt{(-1 - 1)^2 + (-1 - 3)^2} = 2\sqrt{5}$
$d(A, C) = \sqrt{(-1 - 0)^2 + (-1 - 1)^2} = \sqrt{5}$
Because $d(B, C) = d(A, B) + d(A, C)$, the three points are collinear.

133. a. $m_{AB} = \dfrac{4 - 1}{-1 - 1} = -\dfrac{3}{2}; m_{BC} = \dfrac{8 - 4}{5 - (-1)} = \dfrac{2}{3}$.
The product of the slopes $= -1$, so $AB \perp BC$.

b. $d(A, B) = \sqrt{(-1 - 1)^2 + (4 - 1)^2} = \sqrt{13}$
$d(B, C) = \sqrt{(5 - (-1))^2 + (8 - 4)^2} = \sqrt{52}$
$d(A, C) = \sqrt{(5 - 1)^2 + (8 - 1)^2} = \sqrt{65}$
$\left(d(A, B)\right)^2 + \left(d(B, C)\right)^2 = \left(d(A, C)\right)^2$, so the triangle is a right triangle.

For exercises 135 and 136, refer to the figures accompanying the exercises in your text.

135. $\overline{AD}$ and $\overline{BC}$ are parallel because they lie on parallel lines l_1 and l_2. $\overline{AB}$ and $\overline{CD}$ are parallel because they are parallel to the x-axis. Therefore, $ABCD$ is a parallelogram. $\overline{AB} \cong \overline{CD}$ and $\overline{AD} \cong \overline{CB}$ because opposite sides of a parallelogram are congruent. $\triangle ABD \cong \triangle CDB$ by SSS. Then

$$m_1 = \frac{\text{rise}}{\text{run}} = \frac{BD}{CD} \text{ and } m_2 = \frac{\text{rise}}{\text{run}} = \frac{BD}{AB}. \text{ Since}$$

$AB = CD$, $m_1 = \dfrac{BD}{CD} = \dfrac{BD}{AB} = m_2$.

137. Let the quadrilateral $ABCD$ be such that $AB \cong CD$ and $AB \parallel CD$. Locate the points as shown in the figure.

Because $AB \parallel CD$, the y-coordinates of C and D are equal. Because $AB \cong CD$, the x-coordinates of the points are as shown in the figure. The slope of AD is d/c. The slope of

BC is $\dfrac{d-0}{b+c-b} = \dfrac{d}{c}$. So $AD \parallel BC$.

$d(A, D) = \sqrt{d^2 + c^2}$.

$d(B, C) = \sqrt{d^2 + ((b+c) - b)^2} = \sqrt{d^2 + c^2}$.

So $AD \cong BC$.

139. Let (x, y) be the coordinates of point B. Then

$d(A, B) = 12.5 = \sqrt{(x-2)^2 + (y-2)^2} \Rightarrow$

$(x-2)^2 + (y-2)^2 = 156.25$ and

$m_{AB} = \dfrac{4}{3} = \dfrac{y-2}{x-2} \Rightarrow 4(x-2) = 3(y-2) \Rightarrow$

$y = \dfrac{4}{3}x - \dfrac{2}{3}$. Substitute this into the first equation and solve for x:

$$(x-2)^2 + \left(\left(\frac{4}{3}x - \frac{2}{3}\right) - 2\right)^2 = 156.25$$

$$(x-2)^2 + \left(\frac{4}{3}x - \frac{8}{3}\right)^2 = 156.25$$

$$x^2 - 4x + 4 + \frac{16}{9}x^2 - \frac{64}{9}x + \frac{64}{9} = 156.25$$

$$9x^2 - 36x + 36 + 16x^2 - 64x + 64 = 1406.25$$

$$25x^2 - 100x - 1306.25 = 0$$

Solve this equation using the quadratic formula:

$$x = \frac{100 \pm \sqrt{100^2 - 4(25)(-1306.25)}}{2(25)}$$

$$= \frac{100 \pm \sqrt{10,000 + 130,625}}{50}$$

$$= \frac{100 \pm \sqrt{140,625}}{50} = \frac{100 \pm 375}{50}$$

$$= 9.5 \text{ or } -5.5$$

Now find y by substituting the x-values into

the slope formula: $\dfrac{4}{3} = \dfrac{y-2}{9.5-2} \Rightarrow y = 12$ or

$\dfrac{4}{3} = \dfrac{y-2}{-5.5-2} \Rightarrow y = -8$. So the coordinates of B are (9.5, 12) or (−5.5, −8).

141.

$$m_{\overline{OP}} = \frac{12-0}{-5-0} = -\frac{12}{5}$$

Since the tangent line ℓ is perpendicular to $\overline{OP}$, the slope of ℓ is the negative reciprocal of $-\dfrac{12}{5}$ or $\dfrac{5}{12}$. Using the point-slope form, we have

$$y - 12 = \frac{5}{12}\left[x - (-5)\right] \Rightarrow y - 12 = \frac{5}{12}(x+5) \Rightarrow$$

$$y - 12 = \frac{5}{12}x + \frac{25}{12} \Rightarrow y = \frac{5}{12}x + \frac{169}{12}$$

143.

The family of lines has slope 2. The lines have different y-intercepts.

145.

The lines pass through (1, 0). The lines have different slopes.

2.3 Critical Thinking/Discussion/Writing

147. a.

This is a family of lines parallel to the line $y = -2x$. They all have slope –2.

b.

This is a family of lines that passes through the point (0, –4). Their y-intercept is –4.

148. $\left.\begin{array}{l} y = m_1 x + b_1 \\ y = m_2 x + b_2 \end{array}\right\} \Rightarrow m_1 x + b_1 = m_2 x + b_2 \Rightarrow$

$m_1 x - m_2 x = b_2 - b_1 \Rightarrow x(m_1 - m_2) = b_2 - b_1 \Rightarrow$

$x = \dfrac{b_2 - b_1}{m_1 - m_2}$

a. If $m_1 > m_2 > 0$ and $b_1 > b_2$, then

$x = \dfrac{b_2 - b_1}{m_1 - m_2} = -\dfrac{b_1 - b_2}{m_1 - m_2}$.

b. If $m_1 > m_2 > 0$ and $b_1 < b_2$, then

$x = \dfrac{b_2 - b_1}{m_1 - m_2}$.

c. If $m_1 < m_2 < 0$ and $b_1 \geq b_2$, then

$x = \dfrac{b_2 - b_1}{m_1 - m_2} = \dfrac{b_1 - b_2}{m_2 - m_1}$.

d. If $m_1 < m_2 < 0$ and $b_1 < b_2$, then

$x = \dfrac{b_2 - b_1}{m_1 - m_2} = -\dfrac{b_2 - b_1}{m_2 - m_1}$.

2.3 Getting Ready for the Next Section

149. $x^2 - 4 = 0 \Rightarrow (x + 2)(x - 2) = 0 \Rightarrow$

$x + 2 = 0 \Rightarrow x = -2$ or

$x - 2 = 0 \Rightarrow x = 2$

Solution: $\{-2, 2\}$

151. $x^2 - x - 2 = 0$

$(x + 1)(x - 2) = 0$

$x + 1 = 0 \Rightarrow x = -1$ or $x - 2 = 0 \Rightarrow x = 2$

Solution: $\{-1, 2\}$

153. $\left(3(a + h) + 1\right) - (3a + 1) = (3a + 3h + 1) - (3a + 1)$

$= 3h$

155. $\dfrac{-(a + h)^2 + a^2}{h} = \dfrac{-\left(a^2 + 2ah + h^2\right) + a^2}{h}$

$= \dfrac{-2ah - h^2}{h} = \dfrac{h(-2a - h)}{h}$

$= -2a - h$

157. $(x - 1)(x - 3) < 0$

Solve the associated equation:

$(x - 1)(x - 3) = 0 \Rightarrow x = 1$ or $x = 3$.

So, the intervals are

$(-\infty, 1), (1, 3),$ and $(1, \infty)$.

Interval	Test point	Value of $(x-1)(x-3)$	Result
$(-\infty, 1)$	0	3	+
$(1, 3)$	2	–2	–
$(3, \infty)$	4	3	+

The solution set is $(1, 3)$.

2.4 Functions

2.4 Practice Problems

1. a. The domain of R is $\{2, -2, 3\}$ and its range is $\{1, 2\}$. The relation R is a function because no two ordered pairs in R have the same first component.

b. The domain of S is $\{2, 3\}$ and its range is $\{5, -2\}$. The relation S is not a function because the ordered paired $(3, -2)$ and $(3, 5)$ have the same first component.

2. Solve each equation for y.

a. $2x^2 - y^2 = 1 \Rightarrow 2x^2 - 1 = y^2 \Rightarrow$
$\pm\sqrt{2x^2 - 1} = y$; not a function

b. $x - 2y = 5 \Rightarrow x - 5 = 2y \Rightarrow \frac{1}{2}(x - 5) = y$;
a function

3. $g(x) = -2x^2 + 5x$

a. $g(0) = -2(0)^2 + 5(0) = 0$

b. $g(-1) = -2(-1)^2 + 5(-1) = -7$

c. $g(x + h) = -2(x + h)^2 + 5(x + h)$
$= -2(x^2 + 2xh + h^2) + 5x + 5h$
$= -2x^2 - 4hx + 5x - 2h^2 + 5h$

4. $A_{TLMS} = (\text{length})(\text{height}) = (|3 - 1|)(22)$
$= (2)(22) = 44$ sq. units

5. a. $f(x) = \dfrac{1}{\sqrt{1 - x}}$ is not defined when
$1 - x = 0 \Rightarrow x = 1$ or
when $1 - x < 0 \Rightarrow 1 < x.$. Thus, the domain of f is $(-\infty, 1)$.

b. $g(x) = \sqrt{x^2 + 2x - 3}$ is not defined when
$x^2 + 2x - 3 < 0.$
Use the test point method to see that
$x^2 + 2x - 3 < 0$ on the interval $(-3, 1)$.
Thus, the domain of g is
$(-\infty, -3] \cup [1, \infty)$.

$x^2 + 2x - 3 = (x + 3)(x - 1) \geq 0$ or $(-\infty, -3] \cup [1, \infty)$

6. $f(x) = x^2$, domain $X = [-3, 3]$

a. $f(x) = 10 \Rightarrow x^2 = 10 \Rightarrow x = \pm\sqrt{10} \approx \pm 3.16$
Since $\sqrt{10} > 3$ and $-\sqrt{10} < -3$, neither solution is in the interval $X = [-3, 3]$. Therefore, 10 is not in the range of f.

b. $f(x) = 4 \Rightarrow x^2 = 4 \Rightarrow x = \pm 2$
Since $-3 < -2 < 2 < 3$, 4 is in the range of f.

c. The range of f is the interval $[0, 9]$ because for each number y in this interval, the number $x = \sqrt{y}$ is in the interval $[-3, 3]$.

7.

The graph is not a function because a vertical line can be drawn through three points, as shown.

8.

Domain: $(-3, \infty)$; range: $(-2, 2] \cup \{3\}$

9. $y = f(x) = x^2 + 4x - 5$

   a.   Check whether the ordered pair (2, 7) satisfies the equation:

$$7 \overset{?}{=} 2^2 + 4(2) - 5$$
$$7 = 7 \checkmark$$

     The point (2, 7) is on the graph.

   b.   Let $y = -8$, then solve for x:

$$-8 = x^2 + 4x - 5 \Rightarrow 0 = x^2 + 4x + 3 \Rightarrow$$
$$0 = (x+3)(x+1) \Rightarrow x = -3 \text{ or } x = -1$$

     The points $(-3, -8)$ and $(-1, -8)$ lie on the graph.

   c.   Let $x = 0$, then solve for y:

$$y = 0^2 + 4(0) - 5 = -5$$

     The y-intercept is -5.

   d.   Let $y = 0$, then solve for x:

$$0 = x^2 + 4x - 5 \Rightarrow 0 = (x+5)(x-1) \Rightarrow$$
$$x = -5 \text{ or } x = 1$$

     The x-intercepts are -5 and 1.

10.   The range of $C(t)$ is [6, 12].
$$C(11) = \tfrac{1}{2}C(10) + 6 = \tfrac{1}{2}(11.989) + 6 \approx 11.995.$$

11.   From Example 11, we have

$$AP = \sqrt{500^2 + x^2} \text{ and } \overline{PD} = 1200 - x \text{ feet. If}$$
$c =$ the cost on land, the total cost C is given by $C = 1.3c(PD) + c(AP)$

$$= 1.3c\sqrt{500^2 + x^2} + c(1200 - x)$$
$$= c\left[1.3\sqrt{500^2 + x^2} + 1200 - x\right]$$

12. a. $C(x) = 1200x + 100,000$

    b. $R(x) = 2500x$

    c. $P(x) = R(x) - C(x)$
$$= 2500x - (1200x + 100,000)$$
$$= 1300x - 100,000$$

    d.   The break-even point occurs when $C(x) = R(x)$.

$$1200x + 100,000 = 2500x$$
$$100,000 = 1300x \Rightarrow x \approx 77$$

    Metro needs 77 shows to break even.

2.4 Basic Concepts and Skills

1.   In the functional notation $y = f(x)$, x is the independent variable.

3.   If the point $(9, -14)$ is on the graph of a function f, then $\underline{f(9) = -14}$.

5.   False.

7.   True. $-x = 7$ and the square root function is defined for all positive numbers.

2.4 Building Skills

9.   Domain: $\{a, b, c\}$; range: $\{d, e\}$; function

11.   Domain: $\{a, b, c\}$; range: $\{1, 2\}$; function

13.   Domain: $\{0, 3, 8\}$; range: $\{-3, -2, -1, 1, 2\}$; not a function

15. $x + y = 2 \Rightarrow y = -x + 2$; a function

17. $y = \dfrac{1}{x}$; a function

19. $y^2 = x^2 \Rightarrow y = \pm\sqrt{x^2} \Rightarrow y = \pm x$; not a function

21. $y = \dfrac{1}{\sqrt{2x - 5}}$; a function

23. $2 - y = 3x \Rightarrow y = 2 - 3x$; a function

25. $x + y^2 = 8 \Rightarrow y = \pm\sqrt{8 - x}$; not a function

27. $x^2 + y^3 = 5 \Rightarrow y = \sqrt[3]{5 - x^2}$; a function

In exercises 29–32, $f(x) = x^2 - 3x + 1$, $g(x) = \dfrac{2}{\sqrt{x}}$, and $h(x) = \sqrt{2 - x}$.

29. $f(0) = 0^2 - 3(0) + 1 = 1$

$g(0) = \dfrac{2}{\sqrt{0}} \Rightarrow g(0)$ is undefined

$h(0) = \sqrt{2 - 0} = \sqrt{2}$

$f(a) = a^2 - 3a + 1$

$f(-x) = (-x)^2 - 3(-x) + 1 = x^2 + 3x + 1$

31. $f(-1) = (-1)^2 - 3(-1) + 1 = 5$;

$g(-1) = \dfrac{2}{\sqrt{-1}} \Rightarrow g(-1)$ is undefined;

$h(-1) = \sqrt{2 - (-1)} = \sqrt{3}; h(c) = \sqrt{2 - c}$;

$h(-x) = \sqrt{2 - (-x)} = \sqrt{2 + x}$

33. a. $f(0) = \dfrac{2(0)}{\sqrt{4-0^2}} = 0$

b. $f(1) = \dfrac{2(1)}{\sqrt{4-1^2}} = \dfrac{2}{\sqrt{3}} = \dfrac{2\sqrt{3}}{3}$

c. $f(2) = \dfrac{2(2)}{\sqrt{4-2^2}} = \dfrac{4}{0} \Rightarrow f(2)$ is undefined

d. $f(-2) = \dfrac{2(-2)}{\sqrt{4-(-2)^2}} = \dfrac{-4}{0} \Rightarrow f(-2)$ is undefined

e. $f(-x) = \dfrac{2(-x)}{\sqrt{4-(-x)^2}} = \dfrac{-2x}{\sqrt{4-x^2}}$

35. The width of each rectangle is 1. The height of the left rectangle is $f(1) = 1^2 + 2 = 3$. The height of the right rectangle is
$f(2) = 2^2 + 2 = 6$.
$A = (1)(f(1)) + (1)(f(2))$
$\quad = 1(3) + (1)(6) = 9$ sq. units

37. $(-\infty, \infty)$

39. The denominator is not defined for $x = 9$. The domain is $(-\infty, 9) \cup (9, \infty)$

41. The denominator is not defined for $x = -1$ or $x = 1$. The domain is
$(-\infty, -1) \cup (-1, 1) \cup (1, \infty)$.

43. The numerator is not defined for $x < 3$, and the denominator is not defined for $x = -2$. The domain is $[3, \infty)$

45. The denominator equals 0 if $x = -1$ or $x = -2$. The domain is $(-\infty, -2) \cup (-2, -1) \cup (-1, \infty)$.

47. The denominator is not defined for $x = 0$. The domain is $(-\infty, 0) \cup (0, \infty)$

49.

a function

51.

a function

53.

not a function

55. $f(-4) = -2; f(-1) = 1; f(3) = 5; f(5) = 7$

57. $h(-2) = -5; h(-1) = 4; h(0) = 3; h(1) = 4$

59. $h(x) = 7$, so solve the equation
$7 = x^2 - x + 1$
$x^2 - x - 6 = 0 \Rightarrow (x-3)(x+2) = 0 \Rightarrow x = -2$
or $x = 3$.

61. a. $1 = -2(1+1)^2 + 7 \Rightarrow 1 = -1$, which is false. Therefore, $(1, 1)$ does not lie on the graph of f.

b. $1 = -2(x+1)^2 + 7 \Rightarrow 2(x+1)^2 = 6 \Rightarrow$
$(x+1)^2 = 3 \Rightarrow x+1 = \pm\sqrt{3} \Rightarrow x = -1 \pm \sqrt{3}$
The points $\left(-1-\sqrt{3}, 1\right)$ and $\left(-1+\sqrt{3}, 1\right)$ lie on the graph of f.

c. $y = -2(0+1)^2 + 7 \Rightarrow y = 5$
The y-intercept is $(0, 5)$.

d. $0 = -2(x+1)^2 + 7 \Rightarrow -7 = -2(x+1)^2 \Rightarrow$
$\dfrac{7}{2} = (x+1)^2 \Rightarrow \pm\sqrt{\dfrac{7}{2}} = \pm\dfrac{\sqrt{14}}{2} = x+1 \Rightarrow$
$x = -1 \pm \dfrac{\sqrt{14}}{2}$

The x-intercepts are $\left(-1 - \dfrac{\sqrt{14}}{2}, 0\right)$ and

$\left(-1 + \dfrac{\sqrt{14}}{2}, 0\right)$.

63. Domain: $[-3, 2]$; range: $[-3, 3]$

65. Domain: $[-4, \infty)$; range: $[-2, 3]$

67. Domain: $[-3, \infty)$; range: $[-1, 4] \cup \{-3\}$

69. Domain: $(-\infty, 4] \cup [-2, 2] \cup [4, \infty)$
Range: $[-2, 2] \cup \{3\}$

71. $[-9, \infty)$

73. $-3, 4, 7, 9$

75. $f(-7) = 4, \ f(1) = 5, \ f(5) = 2$

77. $\{-3.75, -2.25, 3\} \cup [12, \infty)$

79. $[-9, \infty)$

81. $g(-4) = -1, \ g(1) = 3, \ g(3) = 4$

83. $[-9, -5)$

2.4 Applying the Concepts

85. A function because there is only one high temperature per day.

87. Not a function because there are several states that begin with N (i.e., New York, New Jersey, New Mexico, Nevada, North Carolina, North Dakota); there are also several states that begin with T and S.

89. $A(x) = x^2; \ A(4) = 16; \ A(4)$ represents the area of a tile with side 4.

91. It is a function. $S(x) = 6x^2; \ S(3) = 54$

93. a. The domain is $[0, 8]$.

b. $h(2) = 128(2) - 16(2^2) = 192$
$h(4) = 128(4) - 16(4^2) = 256$
$h(6) = 128(6) - 16(6^2) = 192$

c. $0 = 128t - 16t^2 \Rightarrow 0 = 16t(8 - t) \Rightarrow$
$t = 0$ or $t = 8$. It will take 8 seconds for the stone to hit the ground.

d.

95. $x + y = 28 \Rightarrow y = 28 - x$
$P = x(28 - x) = 28x - x^2$

97. Note that the length of the base = the width of the base = x.
$V = lwh = x^2h = 64 \Rightarrow h = \dfrac{64}{x^2}$
$S = 2lw + 2lh + 2wh$
$\quad = 2x^2 + 2x\left(\dfrac{64}{x^2}\right) + 2x\left(\dfrac{64}{x^2}\right)$
$\quad = 2x^2 + \dfrac{128}{x} + \dfrac{128}{x} = 2x^2 + \dfrac{256}{x}$

99. The piece with length x is formed into a circle, so $C = x = 2\pi r \Rightarrow r = \dfrac{x}{2\pi}$. Thus, the area of the circle is $A = \pi r^2 = \pi\left(\dfrac{x}{2\pi}\right)^2 = \dfrac{x^2}{4\pi}$.
The piece with length $20 - x$ is formed into a square, so $P = 20 - x = 4s \Rightarrow s = \dfrac{1}{4}(20 - x)$.
Thus, the area of the square is
$s^2 = \left[\dfrac{1}{4}(20 - x)\right]^2 = \dfrac{1}{16}(20 - x)^2$.
The sum of the areas is $A = \dfrac{x^2}{4\pi} + \dfrac{1}{16}(20 - x)^2$

101. The volume of the pool is
$V = 288 = x^2h \Rightarrow h = \dfrac{288}{x^2}$.

The total area to be tiled is
$4xh = 4x\left(\dfrac{288}{x^2}\right) = \dfrac{1152}{x}$
The cost of the tile is $6\left(\dfrac{1152}{x}\right) = \dfrac{6912}{x}$.
The area of the bottom of the pool is x^2, so the cost of the cement is $2x^2$. Therefore, the total cost is $C = 2x^2 + \dfrac{6912}{x}$.

103.

Using the distance formula we have

$$d = \sqrt{(x-2)^2 + (y-1)^2}$$

$$= \sqrt{(x-2)^2 + \left[\left(x^3 - 3x + 6\right) - 1\right]^2}$$

$$= \sqrt{(x-2)^2 + \left(x^3 - 3x + 5\right)^2}$$

$$= \left[(x-2)^2 + \left(x^3 - 3x + 5\right)^2\right]^{1/2}$$

105. a. $p(5) = 1275 - 25(5) = 1150$. If 5000 TVs can be sold, the price per TV is \$1150.
$p(15) = 1275 - 25(15) = 900$. If 15,000 TVs can be sold, the price per TV is \$900.
$p(30) = 1275 - 25(30) = 525$. If 30,000 TVs can be sold, the price per TV is \$525.

b.

c. $650 = 1275 - 25x \Rightarrow -625 = -25x \Rightarrow x = 25$
25,000 TVs can be sold at \$650 per TV.

107. a. $C(x) = 5.5x + 75,000$

b. $R(x) = 0.6(15)x = 9x$

c. $P(x) = R(x) - C(x) = 9x - (5.5x + 75,000)$
$= 3.5x - 75,000$

d. The break-even point is when the profit is zero: $3.5x - 75,000 = 0 \Rightarrow x = 21,429$

e. $P(46,000) = 3.5(46,000) - 75,000$
$= \$86,000$
The company's profit is \$86,000 when 46,000 copies are sold.

2.4 Beyond the Basics

109. $x = \dfrac{2}{y-4} \Rightarrow xy - 4x = 2 \Rightarrow xy = 2 + 4x \Rightarrow$

$y = \dfrac{4x+2}{x} \Rightarrow f(x) = \dfrac{4x+2}{x};$

Domain: $(-\infty, 0) \cup (0, \infty)$. $f(4) = \dfrac{9}{2}$.

111. $\left(x^2 + 1\right)y + x = 2 \Rightarrow y = \dfrac{2-x}{x^2+1} \Rightarrow$

$f(x) = \dfrac{2-x}{x^2+1};$ Domain: $(-\infty, \infty); f(4) = -\dfrac{2}{17}$

113. $f(x) \neq g(x)$ because they have different domains.

115. $f(x) \neq g(x)$ because they have different domains. $g(x)$ is not defined for $x = -1$, while $f(x)$ is defined for all real numbers.

117. $f(x) = g(x)$ because $f(3) = 10 = g(3)$ and $f(5) = 26 = g(5)$.

119. $f(2) = 15 = a(2^2) + 2a - 3 \Rightarrow 15 = 6a - 3 \Rightarrow$
$a = 3$.

121. $h(6) = 0 = \dfrac{3(6) + 2a}{2(6) - b} \Rightarrow 0 = 18 + 2a \Rightarrow a = -9$

$h(3)$ is undefined $\Rightarrow \dfrac{3(3) + 2(-9)}{2(3) - b}$ has a zero

in the denominator. So $6 - b = 0 \Rightarrow b = 6$.

123. $g(x) = x^2 - \dfrac{1}{x^2} \Rightarrow g\left(\dfrac{1}{x}\right) = \dfrac{1}{x^2} - \dfrac{1}{\dfrac{1}{x^2}} = \dfrac{1}{x^2} - x^2$

$g(x) + g\left(\dfrac{1}{x}\right) = \left(x^2 - \dfrac{1}{x^2}\right) + \left(\dfrac{1}{x^2} - x^2\right) = 0$

125. $f(x) = \dfrac{x+3}{4x-5} \Rightarrow$

$f(t) = \dfrac{\dfrac{3+5x}{4x-1} + 3}{4\left(\dfrac{3+5x}{4x-1}\right) - 5} = \dfrac{\dfrac{(3+5x) + 3(4x-1)}{4x-1}}{\dfrac{(12+20x) - (5(4x-1))}{4x-1}}$

$= \dfrac{(3+5x) + (12x-3)}{(12+20x) - (20x-5)} = \dfrac{17x}{17} = x$

2.4 Critical Thinking/Discussion/Writing

126. Answers may vary. Sample answers are given

    a. $y = \sqrt{x-2}$ **b.** $y = \dfrac{1}{\sqrt{x-2}}$

    c. $y = \sqrt{2-x}$ **d.** $y = \dfrac{1}{\sqrt{2-x}}$

127. a. $ax^2 + bx + c = 0$

    b. $y = c$

    c. The equation will have no x-intercepts if $b^2 - 4ac < 0$.

    d. It is not possible for the equation to have no y-intercepts because $y = f(x)$.

128. a. $f(x) = |x|$

    b. $f(x) = 0$

    c. $f(x) = x$

    d. $f(x) = \sqrt{-x^2}$ (Note: the point is the origin.)

    e. $f(x) = 1$

    f. A vertical line is not a function.

129.a.

$\{(a, 1), (b, 1)\}$	$\{(a, 2), (b, 1)\}$
$\{(a, 1), (b, 2)\}$	$\{(a, 2), (b, 2)\}$
$\{(a, 1), (b, 3)\}$	$\{(a, 2), (b, 3)\}$

    $\{(a, 3), (b, 1)\}$
    $\{(a, 3), (b, 2)\}$
    $\{(a, 3), (b, 3)\}$

    There are nine functions from X to Y.

    b. $\{(1, a)\}, \{(2, a)\}, \{(3, a)\}$
        $\{(1, a)\}, \{(2, a)\}, \{(3, b)\}$
        $\{(1, a)\}, \{(2, b)\}, \{(3, a)\}$
        $\{(1, b)\}, \{(2, a)\}, \{(3, a)\}$
        $\{(1, b)\}, \{(2, a)\}, \{(3, a)\}$
        $\{(1, b)\}, \{(2, b)\}, \{(3, a)\}$
        $\{(1, b)\}, \{(2, a)\}, \{(3, b)\}$
        $\{(1, b)\}, \{(2, b)\}, \{(3, b)\}$

    There are eight functions from Y to X.

130. If a set X has m elements and a set of Y has n elements, there are n^m functions that can be defined from X to Y. This is true since a function assigns each element of X to an element of Y. There are m possibilities for each element of X, so there are

$$\underbrace{n \cdot n \cdot n \cdots n}_{m} = n^m \text{ possible functions.}$$

2.4 Getting Ready for the Next Section

131. $2x - 4 < 12 \Rightarrow 2x < 16 \Rightarrow x < 8$
The solution set is $(-\infty, 8)$.

133. $x^2 > 0$
Solve the associated equation:
$x^2 = 0 \Rightarrow x = 0$.
So, the intervals are $(-\infty, 0)$ and $(0, \infty)$.

Interval	Test point	Value of x^2	Result
$(-\infty, 0)$	-1	1	$+$
$(0, \infty)$	1	1	$+$

The solution set is $(-\infty, 0) \cup (0, \infty)$

For exercises 135–140, $f(x) = 3 - 2x^2$,
$g(x) = \sqrt{x+3}$, and $h(x) = \dfrac{2}{x^2+1}$.

135. $f(0) = 3 - 2(0)^2 = 3$
$g(0) = \sqrt{0+3} = \sqrt{3}$
$h(0) = \dfrac{2}{0^2+1} = 2$

137. $f(-2) = 3 - 2(-2)^2 = 3 - 8 = -5$
$g(-2) = \sqrt{-2+3} = \sqrt{1} = 1$
$h(-2) = \dfrac{2}{(-2)^2+1} = \dfrac{2}{5}$

139. $f(-x) = 3 - 2(-x)^2 = 3 - x^2$
$g(-x) = \sqrt{-x+3}$
$h(-x) = \dfrac{2}{(-x)^2+1} = \dfrac{2}{x^2+1}$

2.5 Properties of Functions

2.5 Practice Problems

1. The function is decreasing on $(0, 3)$, $(12, 13)$, and $(15, 24)$; increasing on $(3, 12)$ and $(13, 15)$

2. Relative maxima of 3640 at $x = 12$ and 4070 at $x = 15$; relative minima of 40 at $x = 3$ and 3490 at $x = 13$.

3.

   Relative minimum of 0 at $x = 0$
   Relative maximum of 1 at $x = 1$

4. $v = (11 - r)r^2$

   $[0, 13, 1]$ by $[0, 250, 25]$

   Mrs. Osborn's windpipe should be contracted to a radius of 7.33 mm for maximizing the airflow velocity.

5. $f(x) = -x^2$

   Replace x with $-x$:

   $f(-x) = -(-x)^2 = -x^2 = f(x)$

   Thus, the function is even.

6. $f(x) = -x^3$

   Replace x with $-x$:

   $f(-x) = -(-x)^3 = x^3 = -f(x)$

   Thus, the function is odd.

7. a. $g(-x) = 3(-x)^4 - 5(-x)^2$
      $= 3x^4 - 5x^2 = f(x) \Rightarrow$
      $g(x)$ is even.

   b. $f(-x) = 4(-x)^5 + 2(-x)^3 = -4x^5 - 2x^3$
      $= -(4x^5 + 2x^3) = -f(x) \Rightarrow$
      $f(x)$ is odd.

   c. $h(-x) = 2(-x) + 1 = -2x + 1$
      $\neq h(x)$
      $\neq h(-x) \Rightarrow h$ is neither even nor odd.

8. $f(x) = 1 - x^2; \ a = 2, b = 4$
   $f(2) = 1 - 2^2 = -3; \ f(4) = 1 - 4^2 = -15$
   $\dfrac{f(b) - f(a)}{b - a} = \dfrac{-15 - (-3)}{4 - 2} = \dfrac{-12}{2} = -6$
   The average rate of change is -6.

9. $f(t) = 1 - t; \ a = 2, \ b = x, \ x \neq 2$
   $f(a) = f(2) = 1 - 2 = -1$
   $f(b) = f(x) = 1 - x$
   $\dfrac{f(b) - f(a)}{b - a} = \dfrac{(1 - x) - (-1)}{x - 2} = \dfrac{2 - x}{x - 2}$
   $\qquad\qquad\quad = \dfrac{-1(x - 2)}{x - 2} = -1$
   The average rate of change is -1.

10. $f(x) = 100x^2 - 800x + 2000$
    $f(0) = 100(0)^2 - 800(0) + 2000 = 2000$
    $f(3) = 100(3)^2 - 800(3) + 2000 = 500$
    $\dfrac{f(3) - f(0)}{3 - 0} = \dfrac{500 - 2000}{3} = \dfrac{-1500}{3} = -500$

(continued on next page)

(continued)

The average rate of change is −500, so the number of bacteria per cubic centimeter decreases by 500 each day after adding the chlorine.

11. $f(x) = -x^2 + x - 3$

$f(x+h) = -(x+h)^2 + (x+h) - 3$
$\quad\quad = -x^2 - 2xh - h^2 + x + h - 3$

$\dfrac{f(x+h) - f(x)}{h}$

$= \dfrac{\left(-x^2 - 2xh - h^2 + x + h - 3\right) - \left(-x^2 + x - 3\right)}{h}$

$= \dfrac{-2xh - h^2 + h}{h} = -2x - h + 1$

2.5 Concepts and Vocabulary

1. A function f is decreasing if $x_1 < x_2$ implies that $\underline{f(x_1) > f(x_2)}$.

3. A function f is even if $\underline{f(-x) = f(x)\text{ for all }x\text{ in the domain of }f}$.

5. True

7. True

2.5 Building Skills

9. Increasing on $(-\infty, \infty)$

11. Increasing on $(-\infty, 2)$, decreasing on $(2, \infty)$

13. Increasing on $(-\infty, -2)$, constant on $(-2, 2)$, increasing on $(2, \infty)$

15. Increasing on $(-\infty, -3)$ and $\left(-\frac{1}{2}, 2\right)$, decreasing on $\left(-3, -\frac{1}{2}\right)$ and $(2, \infty)$

17. No relative extrema

19. $(2, 10)$ is a relative maximum point and a turning point.

21. Any point on $(x, 2)$ is a relative maximum and a relative minimum point on the interval $(-2, 2)$. Relative maximum at $(-2, 2)$; relative minimum at $(2, 2)$. None of these points are turning points.

23. $(-3, 4)$ and $(2, 5)$ are relative maxima points and turning points. $\left(-\frac{1}{2}, -2\right)$ is a relative minimum and a turning point.

For exercises 25−34, recall that the graph of an even function is symmetric about the y-axis, and the graph of an odd function is symmetric about the origin.

25. The graph is symmetric with respect to the origin. The function is odd.

27. The graph has no symmetries, so the function is neither odd nor even.

29. The graph is symmetric with respect to the origin. The function is odd.

31. The graph is symmetric with respect to the y-axis. The function is even.

33. The graph is symmetric with respect to the origin. The function is odd.

For exercises 35−48, $f(-x) = f(x) \Rightarrow f(x)$ is even and $f(-x) = -f(x) \Rightarrow f(x)$ is odd.

35. $f(-x) = 2(-x)^4 + 4 = 2x^4 + 4 = f(x) \Rightarrow$ $f(x)$ is even.

37. $f(-x) = 5(-x)^3 - 3(-x) = -5x^3 + 3x$
$\quad\quad = -(5x^3 - 3x) = -f(x) \Rightarrow$
$f(x)$ is odd.

39. $f(-x) = 2(-x) + 4 = -2x + 4$
$\quad\quad \neq -f(x) \neq f(x) \Rightarrow$
$f(x)$ is neither even nor odd.

41. $f(-x) = \dfrac{1}{(-x)^2 + 4} = \dfrac{1}{x^2 + 4} = f(x) \Rightarrow$ $f(x)$ is even.

43. $f(-x) = \dfrac{(-x)^3}{(-x)^2 + 1} = -\dfrac{x^3}{x^2 + 1} = -f(x) \Rightarrow$ $f(x)$ is odd.

45. $f(-x) = \dfrac{-x}{(-x)^5 - 3(-x)^3} = \dfrac{-x}{-x^5 + 3x^3}$

$\quad\quad = \dfrac{(-1)(-x)}{(-1)\left(-x^5 + 3x^3\right)} = \dfrac{x}{x^5 - 3x^3} = f(x)$

Thus, $f(x)$ is even.

47. $f(-x) = \dfrac{(-x)^2 - 2(-x)}{5(-x)^4 + 4(-x)^2 + 7} = \dfrac{x^2 + 2x}{5x^4 + 4x^2 + 7}$

$\quad\quad \neq -f(x) \neq f(x)$

Thus, $f(x)$ is neither even nor odd.

49. a. domain: $(-\infty, \infty)$; range: $(-\infty, 3]$

   b. x-intercepts: $(-3, 0), (3, 0)$
   y-intercept: $(0, 3)$

   c. increasing on $(-\infty, 0)$, decreasing on
       $(0, \infty)$

   d. relative maximum at $(0, 3)$

   e. even

51. a. domain: $(-3, 4)$; range: $[-2, 2]$

   b. x-intercept: $(1, 0)$; y-intercept: $(0, -1)$

   c. constant on $(-3, -1)$ and $(3, 4)$
       increasing on $(-1, 3)$

   d. Since the function is constant on $(-3, -1)$,
       any point $(x, -2)$ is both a relative
       maximum and a relative minimum on that
       interval. Since the function is constant on
       $(3, 4)$, any point $(x, 2)$ is both a relative
       maximum and a relative minimum on that
       interval.

   e. neither even not odd

53. a. domain: $(-2, 4)$; range: $[-2, 3]$

   b. x-intercept: $(0, 0)$; y-intercept: $(0, 0)$

   c. decreasing on $(-2, -1)$ and $(3, 4)$
       increasing on $(-1, 3)$

   d. relative maximum: $(3, 3)$
       relative minimum: $(-1, -2)$

   e. neither even nor odd

55. a. domain: $(-\infty, \infty)$; range: $(0, \infty)$

   b. no x-intercept; y-intercept: $(0, 1)$

   c. increasing on $(-\infty, \infty)$

   d. no relative minimum or relative maximum

   e. neither even nor odd

57. $f(x) = -2x + 7$; $a = -1$, $b = 3$
       $f(3) = -2(3) + 7 = 1$; $f(-1) = -2(-1) + 7 = 9$
       average rate of change $= \dfrac{f(3) - f(-1)}{3 - (-1)}$
       $= \dfrac{1 - 9}{4} = -2$

59. $f(x) = 3x + c$; $a = 1$, $b = 5$
       $f(5) = 3 \cdot 5 + c = 15 + c$; $f(1) = 3 \cdot 1 + c = 3 + c$
       average rate of change $= \dfrac{f(5) - f(1)}{5 - 1}$
       $= \dfrac{15 + c - (3 + c)}{4}$
       $= \dfrac{12}{4} = 3$

61. $h(x) = x^2 - 1$; $a = -2$, $b = 0$
       $h(0) = 0^2 - 1 = -1$; $h(-2) = (-2)^2 - 1 = 3$
       average rate of change $= \dfrac{h(0) - h(-2)}{0 - (-2)}$
       $= \dfrac{-1 - 3}{2} = -2$

63. $f(x) = (3 - x)^2$; $a = 1$, $b = 3$
       $f(4) = (3 - 3)^2 = 0$; $f(1) = (3 - 1)^2 = 4$
       average rate of change $= \dfrac{f(3) - f(1)}{3 - 1}$
       $= \dfrac{0 - 4}{2} = -2$

65. $g(x) = x^3$; $a = -1$, $b = 3$
       $g(3) = 3^3 = 27$; $g(-1) = (-1)^3 = -1$
       average rate of change $= \dfrac{g(3) - g(-1)}{3 - (-1)}$
       $= \dfrac{27 - (-1)}{4} = 7$

67. $h(x) = \dfrac{1}{x}$; $a = 2$, $b = 6$
       $h(2) = \dfrac{1}{2}$; $h(6) = \dfrac{1}{6}$
       average rate of change $= \dfrac{h(6) - h(2)}{6 - 2}$
       $= \dfrac{\frac{1}{6} - \frac{1}{2}}{4} = -\dfrac{1}{12}$

69. $f(x + h) = x + h$
       $f(x + h) - f(x) = x + h - x = h$
       $\dfrac{f(x + h) - f(x)}{h} = \dfrac{h}{h} = 1$

71. $f(x + h) = -2(x + h) + 3 = -2x - 2h + 3$
       $f(x + h) - f(x) = -2x - 2h + 3 - (-2x + 3)$
       $= -2h$
       $\dfrac{f(x + h) - f(x)}{h} = \dfrac{-2h}{h} = -2$

73. $f(x + h) = m(x + h) + b = mx + mh + b$
       $f(x + h) - f(x) = mx + mh + b - (mx + b)$
       $= mh$
       $\dfrac{f(x + h) - f(x)}{h} = \dfrac{mh}{h} = m$

75. $f(x+h) = (x+h)^2 = x^2 + 2xh + h^2$

$f(x+h) - f(x) = x^2 + 2xh + h^2 - x^2$
$= 2xh + h^2$

$\dfrac{f(x+h) - f(x)}{h} = \dfrac{2xh + h^2}{h} = 2x + h$

77. $f(x+h) = 2(x+h)^2 + 3(x+h)$
$= 2x^2 + 4xh + 2h^2 + 3x + 3h$
$= 2x^2 + 4xh + 3x + 2h^2 + 3h$

$f(x+h) - f(x)$
$= 2x^2 + 4xh + 3x + 2h^2 + 3h - (2x^2 + 3x)$
$= 4xh + 2h^2 + 3h$

$\dfrac{f(x+h) - f(x)}{h} = \dfrac{4xh + 2h^2 + 3h}{h}$
$= 4x + 2h + 3$

79. $f(x+h) = 4$

$f(x+h) - f(x) = 4 - 4 = 0$

$\dfrac{f(x+h) - f(x)}{h} = \dfrac{0}{h} = 0$

81. $f(x+h) = \dfrac{1}{x+h}$

$f(x+h) - f(x) = \dfrac{1}{x+h} - \dfrac{1}{x}$

$= \dfrac{x}{x(x+h)} - \dfrac{x+h}{x(x+h)}$

$= -\dfrac{h}{x(x+h)}$

$\dfrac{f(x+h) - f(x)}{h} = \dfrac{-\dfrac{h}{x(x+h)}}{h} = -\dfrac{1}{x(x+h)}$

2.5 Applying the Concepts

83. a. Increasing: (2006, 2009), (2011, 2012), (2013, 2014)
Decreasing: (2009, 2011), (2012, 2013)

b. Relative maxima: 251.1 at $x = 2009$, 293.2 at $x = 2012$
Relative minima: 21.5 at $x = 2011$, 187.0 at $x = 2013$

85. domain: $[0, \infty)$
The particle's motion is tracked indefinitely from time $t = 0$.

87. The graph is above the t-axis on the intervals (0, 9) and (21, 24). This means that the particle was moving forward between 0 and 9 seconds and between 21 and 24 seconds.

89. The function is increasing on (0, 3), (5, 6), (16, 19), and (21, 23). However, the speed $|v|$ of the particle is increasing on (0, 3), (5, 6), (11, 15), and (21, 23). Note that the particle is moving forward on (0, 3), (5, 6), and (21, 23), and moving backward on (11, 15).

91. The maximum speed is between times $t = 15$ and $t = 16$.

93. The particle is moving forward with increasing velocity.

95.

a. $V = lwh = (12 - 2x)(12 - 2x)x$
$= (144 - 48x + 4x^2)x$
$= 4x^3 - 48x^2 + 144x$

b. The length of the squares in the corners must be greater than 0 and less than 6, so the domain of V is (0, 6).

c.

[0, 6, 1] by [−25, 150, 25]
range: [0, 128]

d. V is at its maximum when $x = 2$.

97. a. $C(x) = 210x + 10,500$

b. $C(50) = 210(50) + 10,500 = \$21,000$
It costs \$21,000 to produce 50 notebooks per day.

c. average cost $= \dfrac{\$21,000}{50} = \$420$

d. $\dfrac{210x + 10,500}{x} = 315$
$210x + 10,500 = 315x$
$10,500 = 105x \Rightarrow x = 100$

The average cost per notebook will be \$315 when 100 notebooks are produced.

99. average rate of increase $= \dfrac{f(2014) - f(2000)}{2014 - 2000}$

$$= \dfrac{20.2 - 15.3}{14} = 0.35$$

The average rate of increase was 0.35 million, or 350,000, students per year.

101. a. $f(0) = 0^2 + 3(0) + 4 = 4$

The particle is 4 ft to the right from the origin.

b. $f(4) = 4^2 + 3(4) + 4 = 32$

The particle started 4 ft from the origin, so it traveled $32 - 4 = 28$ ft in four seconds.

c. $f(3) = 3^2 + 3(3) + 4 = 22$

The particle started 4 ft from the origin, so it traveled $22 - 4 = 18$ ft in three seconds. The average velocity is $18/3 = 6$ ft/sec

d. $f(2) = 2^2 + 3(2) + 4 = 14$

$f(5) = 5^2 + 3(5) + 4 = 44$

The particle traveled $44 - 14 = 30$ ft between the second and fifth seconds. The average velocity is $30/(5-2) = 10$ ft/sec

2.5 Beyond the Basics

103. $f(x) = \dfrac{x-1}{x+1}$

$f(2x) = \dfrac{2x-1}{2x+1}$

$\dfrac{3f(x)+1}{f(x)+3} = \dfrac{3\left(\dfrac{x-1}{x+1}\right)+1}{\dfrac{x-1}{x+1}+3} = \dfrac{\dfrac{3x-3}{x+1}+1}{\dfrac{x-1+3(x+1)}{x+1}}$

$= \dfrac{\dfrac{3x-3+x+1}{x+1}}{\dfrac{x-1+3(x+1)}{x+1}} = \dfrac{4x-2}{4x+2}$

$= \dfrac{2(2x-1)}{2(2x+1)} = \dfrac{2x-1}{2x+1} = f(2x)$

105. In order to find the relative maximum, first observe that the relative maximum of $-(x+1)^2 \le 0$. Then $-(x+1)^2 \le 0 \Rightarrow$

$(x+1)^2 \ge 0 \Rightarrow x \ge -1$.

Thus, the x-coordinate of the relative maximum is -1. $f(-1) = -(-1+1)^2 + 5 = 5$
The relative maximum is $(-1, 5)$.
There is no relative minimum.

107. $f(x) = \sqrt{x}$

$\dfrac{f(x+h) - f(x)}{h} = \dfrac{\sqrt{x+h} - \sqrt{x}}{h} \cdot \dfrac{\sqrt{x+h} + \sqrt{x}}{\sqrt{x+h} + \sqrt{x}}$

$= \dfrac{x+h-x}{h\left(\sqrt{x+h} + \sqrt{x}\right)}$

$= \dfrac{h}{h\left(\sqrt{x+h} + \sqrt{x}\right)}$

$= \dfrac{1}{\sqrt{x+h} + \sqrt{x}}$

109. $f(x) = -\dfrac{1}{\sqrt{x}}$

$\dfrac{f(x+h) - f(x)}{h}$

$= \dfrac{-\dfrac{1}{\sqrt{x+h}} + \dfrac{1}{\sqrt{x}}}{h} = \dfrac{\dfrac{\sqrt{x+h} - \sqrt{x}}{\sqrt{x}\sqrt{x+h}}}{h}$

$= \dfrac{\dfrac{\sqrt{x+h} - \sqrt{x}}{\sqrt{x(x+h)}}}{h} = \dfrac{\sqrt{x+h} - \sqrt{x}}{h\sqrt{x(x+h)}}$

$= \dfrac{\sqrt{x+h} - \sqrt{x}}{h\sqrt{x(x+h)}} \cdot \dfrac{\sqrt{x+h} + \sqrt{x}}{\sqrt{x+h} + \sqrt{x}}$

$= \dfrac{(x+h) - x}{h\sqrt{x(x+h)}\left(\sqrt{x+h} + \sqrt{x}\right)}$

$= \dfrac{1}{\sqrt{x(x+h)}\left(\sqrt{x} + \sqrt{x+h}\right)}$

2.5 Critical Thinking/Discussion/Writing

111. f has a relative maximum at $x = a$ if there is an interval $[a, x_1]$ with $a < x_1 < b$ such that

$f(a) \ge f(x)$, or $f(x) \le f(a)$, for every x in the interval $(x_1, b]$.

112. f has a relative minimum at $x = b$ if there is x_1 in $[a, b]$ such that $f(x) \ge f(b)$ for every x in the interval $(x_1, b]$.

113. Answers will vary. Sample answers are given.

a. $f(x) = x$ on the interval $[-1, 1]$

b. $f(x) = \begin{cases} x & \text{if } 0 \le x < 1 \\ 0 & \text{if } x = 1 \end{cases}$

c. $f(x) = \begin{cases} x & \text{if } 0 < x \le 1 \\ 1 & \text{if } x = 0 \end{cases}$

d. $f(x) = \begin{cases} 0 & \text{if } x = 0 \text{ or } x = 1 \\ 1 & \text{if } 0 < x < 1 \text{ and } x \text{ is rational} \\ -1 & \text{if } 0 < x < 1 \text{ and } x \text{ is irrational} \end{cases}$

2.5 Getting Ready for the Next Section

115. $m = \dfrac{-1-2}{7-6} = \dfrac{-3}{1} = -3$

   $y - 2 = -3(x - 6) \Rightarrow y - 2 = -3x + 18 \Rightarrow$
   $y = -3x + 20$

117.

119. If $x = -3$, then $|x + 1| = \underline{2}$.

121. $f(x) = x^{3/2}$

   (i) $f(2) = 2^{3/2} = \left(\sqrt{2}\right)^3 = \sqrt{8} = 2\sqrt{2}$

   (ii) $f(4) = 4^{3/2} = \left(\sqrt{4}\right)^3 = 2^3 = 8$

   (iii) $f(-4) = (-4)^{3/2} = \left(\sqrt{-4}\right)^3 = (2i)^3 = -8i$

2.6 A Library of Functions

2.6 Practice Problems

1. Because $g(-2) = 2$ and $g(1) = 8$, the line passes through the points $(-2, 2)$ and $(1, 8)$.

   $m = \dfrac{8-2}{1-(-2)} = \dfrac{6}{3} = 2$

   Use the point-slope form:
   $y - 8 = 2(x - 1) \Rightarrow y - 8 = 2x - 2 \Rightarrow$
   $y = 2x + 6 \Rightarrow g(x) = 2x + 6$

2. Using the formula
   Shark length = (0.96)(tooth height) − 0.22,
   gives:
   Shark length = (0.96)(16.4) − 0.22 = 15.524 m

3.

   Domain: $(-\infty, 0]$; range: $[0, \infty)$

4.

   Domain: $(-\infty, \infty)$; range: $(-\infty, \infty)$

5. $f(x) = \begin{cases} x^2 & \text{if } x \le -1 \\ 2x & \text{if } x > -1 \end{cases}$

   $f(-2) = (-2)^2 = 4; \quad f(3) = 2(3) = 6$

6. a. $f(x) = \begin{cases} 50 + 4(x - 55) & 56 \le x < 75 \\ 200 + 5(x - 75) & x \ge 75 \end{cases}$

   b. The fine for driving 60 mph is
      $50 + 4(60 - 55) = \$70$.

   c. The fine for driving 90 mph is
      $200 + 5(90 - 75) = \$275$.

7. The graph of f is made up of two parts: a line segment passing through $(1, 5)$ and $(3, 2)$ on the interval $[1, 3]$, and a line segment passing through $(3, 2)$ and $(5, 4)$ on the interval $[3, 5]$.

(continued on next page)

(*continued*)

For the first line segment:

$$m = \frac{2-5}{3-1} = -\frac{3}{2}$$

$$y - 5 = -\frac{3}{2}(x-1) \Rightarrow 2y - 10 = -3(x-1) \Rightarrow$$

$$2y - 10 = -3x + 3 \Rightarrow 2y = -3x + 13 \Rightarrow$$

$$y = -\frac{3}{2}x + \frac{13}{2}$$

For the second line segment:

$$m = \frac{4-2}{5-3} = 1$$

$$y - 4 = x - 5 \Rightarrow y = x - 1$$

The piecewise function is

$$g(x) = \begin{cases} -\dfrac{3}{2}x + \dfrac{13}{2} & \text{if } 1 \le x \le 3 \\ x - 1 & \text{if } 3 < x \le 5 \end{cases}$$

8. $f(x) = \begin{cases} -3x & \text{if } x \le -1 \\ 2x & \text{if } x > -1 \end{cases}$

Graph $f(x) = -3x$ on the interval $(-\infty, -1]$, and graph $f(x) = 2x$ on the interval $(-1, \infty)$.

9. $f(x) = [\![x]\!]$
$f(-3.4) = -4;$ $f(4.7) = 4$

2.6 Concepts and Vocabulary

1. The graph of the linear function $f(x) = b$ is a <u>horizontal</u> line.

3. The graph of the function

$$f(x) = \begin{cases} x^2 + 2 & \text{if } x \le 1 \\ ax & \text{if } x > 1 \end{cases}$$

will have a break at $x = 1$ unless $a = \underline{3}$.

$$f(x) = \begin{cases} x^2 + 2 & \text{if } x \le 1 \\ 3x & \text{if } x > 1 \end{cases}$$

5. True. The equation of the graph of a vertical line has the format $x = a$.

7. True

2.6 Building Skills

In exercises 9–18, first find the slope of the line using the two points given. Then substitute the coordinates of one of the points into the slope-intercept form of the equation to solve for b.

9. The two points are $(0, 1)$ and $(-1, 0)$.

$$m = \frac{0-1}{-1-0} = 1.\quad 1 = 1(0) + b \Rightarrow b = 1.$$

$$f(x) = x + 1$$

11. The two points are $(-1, 1)$ and $(2, 7)$.

$$m = \frac{7-1}{2-(-1)} = 2.\quad 1 = 2(-1) + b \Rightarrow 3 = b.$$

$$f(x) = 2x + 3$$

13. The two points are $(1, 1)$ and $(2, -2)$.

$m = \dfrac{-2-1}{2-1} = -3$. $1 = -3(1) + b \Rightarrow b = 4$.

$f(x) = -3x + 4$.

b.

15. The two points are $(-2, 2)$ and $(2, 4)$.

$m = \dfrac{4-2}{2-(-2)} = \dfrac{1}{2}$. $4 = \dfrac{1}{2}(2) + b \Rightarrow b = 3$.

$f(x) = \dfrac{1}{2}x + 3$.

21. $g(x) = \begin{cases} 1 & \text{if } x > 0 \\ -1 & \text{if } x < 0 \end{cases}$

a. $f(-15) = -1$; $f(12) = 1$

b.

c. domain: $(-\infty, 0) \cup (0, \infty)$

range: $\{-1, 1\}$

17. The two points are $(0, -1)$ and $(3, -3)$.

$m = \dfrac{-3-(-1)}{3-0} = -\dfrac{2}{3}$.

$-1 = -\dfrac{2}{3}(0) + b \Rightarrow b = -1$.

$f(x) = -\dfrac{2}{3}x - 1$.

23. $f(x) = \begin{cases} 2x & \text{if } x < 0 \\ x^2 & \text{if } x \geq 0 \end{cases}$

Domain: $(-\infty, \infty)$; range: $(-\infty, \infty)$

19. $f(x) = \begin{cases} x & \text{if } x \geq 2 \\ 2 & \text{if } x < 2 \end{cases}$

a. $f(1) = 2$; $f(2) = 2$; $f(3) = 3$

25. $g(x) = \begin{cases} \dfrac{1}{x} & \text{if } x < 0 \\ \sqrt{x} & \text{if } x \geq 0 \end{cases}$

Domain: $(-\infty, \infty)$; range: $(-\infty, \infty)$

27. $f(x) = \begin{cases} [\![x]\!] & \text{if } x < 1 \\ \sqrt[3]{x} & \text{if } x \geq 1 \end{cases}$

Domain: $(-\infty, \infty)$;

range: $\{\ldots, -3, -2, -1, 0\} \cup [1, \infty)$

29. $g(x) = \begin{cases} |x| & \text{if } x < 1 \\ x^3 & \text{if } x \geq 1 \end{cases}$

Domain: $(-\infty, \infty)$; range: $[0, \infty)$

31. $f(x) = \begin{cases} |x| & \text{if } -2 \leq x < 1 \\ \sqrt{x} & \text{if } x \geq 1 \end{cases}$

Domain: $[-2, \infty)$; range: $[0, \infty)$

33. $f(x) = \begin{cases} [\![x]\!] & \text{if } x < 1 \\ -2x & \text{if } 1 \leq x \leq 4 \end{cases}$

Domain: $(-\infty, 4]$;

range: $\{\ldots, -3, -2, -1, 0\} \cup [-8, -2]$

35. $f(x) = \begin{cases} 2x+3 & \text{if } x < -2 \\ x+1 & -2 \leq x < 1 \\ -x+3 & \text{if } x \geq 1 \end{cases}$

Domain: $(-\infty, \infty)$; range: $(-\infty, 2]$

37. The graph of f is made up of two parts. For $x < 2$, the graph is made up of the half-line passing through the points $(-1, 0)$ and $(2, 3)$.

$$m = \frac{3-0}{2-(-1)} = \frac{3}{3} = 1$$
$$y - 0 = x - (-1) \Rightarrow y = x + 1$$

For $x \geq 2$, the graph is a half-line passing through the points $(2, 3)$ and $(3, 0)$.

$$m = \frac{0-3}{3-2} = -3$$
$$y - 0 = -3(x - 3) \Rightarrow y = -3x + 9$$

Combining the two parts, we have

$$f(x) = \begin{cases} x+1 & \text{if } x < -2 \\ -3x+9 & \text{if } x \geq 2 \end{cases}$$

39. The graph of f is made up of three parts. For $x < -2$, the graph is the half-line passing through the points $(-2, 2)$ and $(-3, 0)$.

$$m = \frac{0-2}{-3-(-2)} = \frac{-2}{-1} = 2$$
$$y - 0 = 2(x - (-3)) \Rightarrow y = 2(x + 3) \Rightarrow$$
$$y = 2x + 6$$

For $-2 \leq x < 2$, the graph is a horizontal line segment passing through the points $(-2, 4)$ and $(2, 4)$, so the equation is $y = 4$.
For $x \geq 2$, the graph is the half-line passing through the points $(2, 1)$ and $(3, 0)$.

$$m = \frac{0-1}{3-2} = -1$$
$$y - 0 = -(x - 3) \Rightarrow y = -x + 3$$

Combining the three parts, we have

$$f(x) = \begin{cases} 2x+6 & \text{if } x < -2 \\ 4 & \text{if } -2 \leq x < 2 \\ -x+3 & \text{if } x \geq 2 \end{cases}$$

2.6 Applying the Concepts

41. a. $f(x) = \dfrac{x}{33.81}$

Domain: $[0, \infty)$; range: $[0, \infty)$.

b. $f(3) = \dfrac{3}{33.81} \approx 0.0887$

This means that 3 oz ≈ 0.0887 liter.

c. $f(12) = \dfrac{12}{33.81} \approx 0.3549$ liter.

43. a. $P(0) = \dfrac{1}{33}(0) + 1 = 1$. The y-intercept is 1.

This means that the pressure at sea level ($d = 0$) is 1 atm.

$$0 = \frac{1}{33}d + 1 \Rightarrow d = -33.$$

d can't be negative, so there is no d-intercept.

b. $P(0) = 1$ atm; $P(10) = \dfrac{1}{33}(10) + 1 \approx 1.3$ atm;

$$P(33) = \frac{1}{33}(33) + 1 = 2 \text{ atm;}$$

$$P(100) = \frac{1}{33}(100) + 1 \approx 4.03 \text{ atm.}$$

c. $5 = \dfrac{1}{33}d + 1 \Rightarrow d = 132$ feet

The pressure is 5 atm at 132 feet.

45. a. $C(x) = 50x + 6000$

b. The y-intercept is the fixed overhead cost.

c. $11,500 = 50x + 6000 \Rightarrow 110$

110 printers were manufactures on a day when the total cost was $11,500.

47. a. $R = 900 - 30x$

b. $R(6) = 900 - 30(6) = 720$

If you move in 6 days after the first of the month, the rent is $720.

c. $600 = 900 - 30x \Rightarrow x = 10$

You moved in ten days after first of the month.

49. The rate of change (slope) is $\dfrac{100-40}{20-80} = -1$.

Use the point $(20, 100)$ to find the equation of the line: $100 = -20 + b \Rightarrow b = 120$. The equation of the line is $y = -x + 120$. Now solve $50 = -x + 120 \Rightarrow x = 70$.

Age 70 corresponds to 50% capacity.

51. a. The rate of change (slope) is

$$\frac{50-30}{420-150} = \frac{2}{27}.$$

The equation of the line is
$$y - 30 = \frac{2}{27}(x - 150) \Rightarrow$$
$$y = \frac{2}{27}(x - 150) + 30.$$

b. $y = \frac{2}{27}(350 - 150) + 30 \Rightarrow y = \frac{1210}{27} \approx 44.8$

There can't be a fractional number of deaths, so round up. There will be about 45 deaths when $x = 350$ milligrams per cubic meter.

c. $70 = \frac{2}{27}(x - 150) + 30 \Rightarrow x = 690$

If the number of deaths per month is 70, the concentration of sulfur dioxide in the air is 690 mg/m^3.

53. a.

b. **(i)** $T(12,000) = 0.04(12,000) = \$480$

(ii)
$$T(20,000) = 800 + 0.06(20,000 - 20,000)$$
$$= \$800$$

(iii)
$$T(50,000) = 800 + 0.06(50,000 - 20,000)$$
$$= \$2600$$

c. **(i)** $600 = 0.04x \Rightarrow x = \$15,000$

(ii) $1200 = 0.04x \Rightarrow x = \$30,000$, which is outside of the domain. Try
$$1200 = 800 + 0.06(x - 20,000) \Rightarrow$$
$$x \approx \$26,667$$

(iii) $2300 = 800 + 0.06(x - 20,000) \Rightarrow$
$$x = \$45,000$$

2.6 Beyond the Basics

55. $2(3) - 1 = a - 3(3) \Rightarrow 5 = a - 9 \Rightarrow a = 14$

57. a. Domain: $(-\infty, \infty)$; range: $[0,1)$

b. The function is increasing on $(n, n+1)$ for every integer n.

c. $f(-x) = -x - [\![-x]\!] \neq -f(x) \neq f(x)$, so the function is neither even nor odd.

59. a. **(i)** $WCI(2) = 40$

(ii)
$$WCI(16)$$
$$= 91.4 + (91.4 - 40)$$
$$\cdot \left(0.0203(16) - 0.304\sqrt{16} - 0.474\right)$$
$$\approx 21$$

(iii) $WCI(50) = 1.6(40) - 55 = 9$

b. **(i)** $-58 = 91.4 + (91.4 - T) \bullet$
$$(0.0203(36) - 0.304\sqrt{36} - 0.474$$
$$-58 = 91.4 + (91.4 - T)(-1.5672)$$
$$-58 = 91.4 - 143.24 + 1.5672T$$
$$-58 = -51.84 + 1.5672T \Rightarrow T \approx -4°F$$

(ii) $-10 = 1.6T - 55 \Rightarrow T \approx 28°F$

61. $C(x) = 2[\![x]\!] + 4$

2.6 Critical Thinking/Discussion/Writing

63. D   **64.** C

65. a. If f is even, then f is increasing on $(-\infty, -1)$ and decreasing on $(-1, 0)$.

b. If f is odd, then f is decreasing on $(-\infty, -1)$ and increasing on $(-1, 0)$.

66. a. If f is even, then f has a relative maximum at $x = -1$ and a relative minimum at $x = -3$.

b. If f is odd, then f has a relative minimum at $x = -1$ and a relative maximum at $x = -3$.

2.6 Getting Ready for the Next Section

67. If we add 3 to each y-coordinate of the graph of f, we will obtain the graph of $y = \underline{f(x) + 3}$.

69. If we replace each x-coordinate with its opposite in the graph of f, we will obtain the graph of $y = \underline{f(-x)}$.

71.

73.

For exercises 75–78, refer to section 1.3 in your text for help on completing the square.

75. $x^2 + 8x + \underline{16}$; $(x+4)^2$

77. $x^2 - \dfrac{2}{3}x + \dfrac{1}{9}$; $\left(x - \dfrac{1}{3}\right)^2$

2.7 Transformations of Functions

2.7 Practice Problems

1.

The graph of g is the graph of f shifted one unit up. The graph of h is the graph of f shifted two units down.

2.

The graph of g is the graph of f shifted one unit to the right. The graph of h is the graph of f shifted two units to the left.

3. The graph of $f(x) = \sqrt{x-2} + 3$ is the graph of $g(x) = \sqrt{x}$ shifted two units to the right and three units up.

4. The graph of $y = -(x-1)^2 + 2$ can be obtained from the graph of $y = x^2$ by first shifting the graph of $y = x^2$ one unit to the right. Reflect the resulting graph about the x-axis, and then shift the graph two units up

5. The graph of $y = 2x - 4$ is obtained from the graph of $y = 2x$ by shifting it down by four units. We know that
$$|y| = \begin{cases} y & \text{if } y \geq 0 \\ -y & \text{if } y < 0. \end{cases}$$

This means that the portion of the graph on or above the x-axis $(y \geq 0)$ is unchanged while the portion of the graph below the x-axis $(y < 0)$ is reflected above the x-axis. The graph of $y = |f(x)| = |2x - 4|$ is given on the right.

6.

The graph of g is the graph of f stretched vertically by multiplying each of its y-coordinates by 2.

7. a.

b.

8. Start with the graph of $y = \sqrt{x}$. Shift the graph one unit to the left, then stretch the graph vertically by a factor of three. Shift the resulting graph down two units.

9.

Shift the graph one unit right to graph $y = f(x-1)$.

Compress horizontally by a factor of 2. Multiply each x-coordinate by $\frac{1}{2}$ to graph $y = f(2x-1)$.

Compress vertically by a factor of $\frac{1}{2}$. Multiply each y-coordinate by $\frac{1}{2}$ to graph $y = \frac{1}{2}f(2x-1)$.

Shift the graph up three units to graph $y = \frac{1}{2}f(2x-1)+3$.

$$y = f(x) \rightarrow y = f(x-1) \rightarrow y = f(2x-1) \rightarrow$$
$$y = \frac{1}{2}f(2x-1) \Rightarrow y = \frac{1}{2}f(2x-1)+3$$

2.7 Concepts and Vocabulary

1. The graph of $y = f(x) - 3$ is found by vertically shifting the graph of $y = f(x)$ three units <u>down</u>.

3. The graph of $y = f(bx)$ is a horizontal compression of the graph of $y = f(x)$ is b <u>is greater than 1</u>.

5. False. The graphs are the same if the function is an even function.

7. False. The graph on the left shows $y = x^2$ first shifted up two units and then reflected about the x-axis, while the graph on the right shows $y = x^2$ reflected about the x-axis and then shifted up two units.

2.7 Building Skills

9. a. The graph of g is the graph of f shifted two units up.

   b. The graph of h is the graph of f shifted one unit down.

11. a. The graph of g is the graph of f shifted one unit to the left.

   b. The graph of h is the graph of f shifted two units to the right.

13. a. The graph of g is the graph of f shifted one unit left, then two units down.

   b. The graph of h is the graph of f shifted one unit right, then three units up.

15. a. The graph of g is the graph of f reflected about the x-axis.

   b. The graph of h is the graph of f reflected about the y-axis.

17. a. The graph of g is the graph of f vertically stretched by a factor of 2.

   b. The graph of h is the graph of f horizontally compressed by a factor of 2.

19. a. The graph of g is the graph of f reflected about the x-axis and then shifted one unit up.

   b. The graph of h is the graph of f reflected about the y-axis and then shifted one unit up.

21. a. The graph of g is the graph of f shifted one unit up.

   b. The graph of h is the graph of f shifted one unit to the left.

23. e   **25.** g

27. i   **29.** b

31. l   **33.** d

35.

$f(x) = x^2 - 2$

37.

$g(x) = \sqrt{x} + 1$

39. $f(x) = |x| + 2$

41. $f(x) = x^3 + 2$

43. $f(x) = \dfrac{1}{x} + 1$

45.

$f(x) = (x - 3)^3$

47. $f(x) = \sqrt{x-1}$

49.

$h(x) = |x+1|$

51. $f(x) = (x+1)^3$

53. $f(x) = \dfrac{1}{x-3}$

55. $f(x) = \sqrt{-x}$

57. $f(x) = -x^2$

59. $f(x) = 2x^2$

61. $f(x) = 2|x|$

63. $f(x) = (x-2)^2 + 1$

Start with the graph of $f(x) = x^2$, then shift it two units right and one unit up.

65. $f(x) = 5 - (x-3)^2$

Start with the graph of $f(x) = x^2$, then shift it three units right. Reflect the graph across the x-axis. Shift it five units up.

67. $f(x) = \sqrt{x+1} - 3$

Start with the graph of $f(x) = \sqrt{x}$, then shift it one unit left and three units down.

69. $f(x) = \sqrt{1-x} + 2$

Start with the graph of $f(x) = \sqrt{x}$, then shift it one unit left. Reflect the graph across the y-axis, and then shift it two units up.

71. $f(x) = |x-1| - 2$

Start with the graph of $f(x) = |x|$, then shift it one unit right and two units down.

73. $f(x) = \dfrac{1}{x-1} + 3$

Start with the graph of $f(x) = \frac{1}{x}$, then shift it one unit right and three units up.

75. $f(x) = 2(x+1)^2 - 1$

Start with the graph of $f(x) = x^2$, then shift it one unit left. Stretch the graph vertically by a factor of 2, then shift it one unit down.

77. $f(x) = 2 - \frac{1}{2}(x-3)^2$

Start with the graph of $f(x) = x^2$, then shift it three units right. Compress the graph vertically by a factor of 1/2, reflect it across the *x*-axis, then shift it two units up.

79. $f(x) = 2\sqrt{x+1} - 3$

Start with the graph of $f(x) = \sqrt{x}$, then shift it one unit left. Stretch the graph vertically by a factor of 2, and then shift it three units down.

81. $f(x) = -2|x-1| + 2$

Start with the graph of $f(x) = |x|$, then shift it one unit right. Stretch the graph vertically by a factor of 2, then reflect it across the *x*-axis. Shift the graph up two units.

83. $y = x^3 + 2$

85. $y = -|x|$

87. $y = (x-3)^2 + 2$

89. $y = -\sqrt{x+3} - 2$

91. $y = 3(-x+4)^3 + 2$

92. $y = -(-x+1)^3 + 1$

93. $y = -2|x-4| - 3$

95.

97.

99.

101.

103.

105.

107.

109. a.

b.

111. a.

b.

113. a.

b.

115. a.

b.

117. a.

b.

2.7 Applying the Concepts

119. $g(x) = f(x) + 800$

121. $p(x) = 1.02 \big(f(x) + 500 \big)$

123. a.   Shift one unit right, stretch vertically by a factor of 10, and shift 5000 units up.

b. $C(400) = 5000 + 10\sqrt{400 - 1} = \$5199.75$

125. a.   Shift one unit left, reflect across the x-axis, and shift up 109,561 units.

b. $69,160 = 109,561 - (p+1)^2$
$40,401 = (p+1)^2$
$\quad 201 = p + 1 \Rightarrow p = 200\text{¢} = \$2.00$

c. $\quad 0 = 109,561 - (p+1)^2$
$109,561 = (p+1)^2$
$\quad\quad 331 = p + 1 \Rightarrow p = 330\text{¢} = \$3.30$

127.   The first coordinate gives the month; the second coordinate gives the hours of daylight. From March to September, there is daylight more than half of the day each day. From September to March, more than half of the day is dark each day.

2.7 Beyond the Basics

129.   The graph is shifted one unit right, then reflected about the x-axis, and finally reflected about the y-axis. The equation is
$g(x) = -\sqrt{1 - x}$.

131.   Shift two units left, then 4 units down.

133. $f(x) = -x^2 + 2x = -\big(x^2 - 2x + 1 \big) + 1$
$\quad\quad\quad = -(x-1)^2 + 1$

Shift one unit right, reflect about the x-axis, then shift one unit up.

135. $f(x) = 2x^2 - 4x = 2\left(x^2 - 2x + 1\right) - 2$

$\qquad = 2(x-1)^2 - 2$

Shift one unit right, stretch vertically by a factor of 2, then shift two units down.

137. $f(x) = -2x^2 - 8x + 3 = -2\left(x^2 + 4x\right) + 3$

$\qquad = -2\left(x^2 + 4x + 4\right) + 3 + 2(4)$

$\qquad = -2(x+2)^2 + 11$

Shift two units left, stretch vertically by a factor of 2, reflect across the x-axis, then shift eleven units up.

139.

141.

143.

2.7 Critical Thinking/Discussion/Writing

145. a. $y = f(x+2)$ is the graph of $y = f(x)$ shifted two units left. So the x-intercepts are $-1 - 2 = -3$, $0 - 2 = -2$, and $2 - 2 = 0$.

b. $y = f(x-2)$ is the graph of $y = f(x)$ shifted two units right. So the x-intercepts are $-1 + 2 = 1$, $0 + 2 = 2$, and $2 + 2 = 4$.

c. $y = -f(x)$ is the graph of $y = f(x)$ reflected across the x-axis. The x-intercepts are the same, -1, 0, 2.

d. $y = f(-x)$ is the graph of $y = f(x)$ reflected across the y-axis. The x-intercepts are the opposites, 1, 0, -2.

e. $y = f(2x)$ is the graph of $y = f(x)$ compressed horizontally by a factor of 1/2. The x-intercepts are $-\frac{1}{2}$, 0, 1.

f. $y = f\left(\frac{1}{2}x\right)$ is the graph of $y = f(x)$ stretched horizontally by a factor of 2. The x-intercepts are -2, 0, 4.

146. a. $y = f(x) + 2$ is the graph of $y = f(x)$ shifted two units up. The y-intercept is $2 + 2 = 4$.

b. $y = f(x) - 2$ is the graph of $y = f(x)$ shifted two units down. The y-intercept is $2 - 2 = 0$.

c. $y = -f(x)$ is the graph of $y = f(x)$ reflected across the x-axis. The y-intercept is the opposite, -2.

d. $y = f(-x)$ is the graph of $y = f(x)$ reflected across the y-axis. The y-intercept is the same, 2.

e. $y = 2f(x)$ is the graph of $y = f(x)$ stretched vertically by a factor of 2. The y-intercept is 4.

f. $y = \frac{1}{2}f(x)$ is the graph of $y = f(x)$ compressed horizontally by a factor of 1/2. The y-intercept is 1.

147. a. $y = f(x+2)$ is the graph of $y = f(x)$ shifted two units left. The domain is $[-1-2, 3-2] = [-3, 1]$. The range is the same, $[-2, 1]$.

b. $y = f(x) - 2$ is the graph of $y = f(x)$ shifted two units down. The domain is the same, $[-1, 3]$. The range is $[-2-2, 1-2] = [-4, -1]$.

c. $y = -f(x)$ is the graph of $y = f(x)$ reflected across the x-axis. The domain is the same, $[-1, 3]$. The range is the opposite, $[-1, 2]$.

d. $y = f(-x)$ is the graph of $y = f(x)$ reflected across the y-axis. The domain is the opposite, $[-3, 1]$. The range is the same, $[-2, 1]$.

e. $y = 2f(x)$ is the graph of $y = f(x)$ stretched vertically by a factor of 2. The domain is the same, $[-1, 3]$. The range is $[2(-2), 2(1)] = [-4, 2]$.

f. $y = \frac{1}{2}f(x)$ is the graph of $y = f(x)$ compressed horizontally by a factor of 1/2. The domain is the same, $[-1, 3]$. The range is $\left[\frac{1}{2}(-2), \frac{1}{2}(1)\right] = \left[-1, \frac{1}{2}\right]$.

148. a. $y = f(x+2)$ is the graph of $y = f(x)$ shifted two units left. So the relative maximum is at $x = 1 - 2 = -1$, and the relative minimum is at $x = 2 - 2 = 0$.

b. $y = f(x) - 2$ is the graph of $y = f(x)$ shifted two units down. The locations of the relative maximum and minimum do not change. Relative maximum at $x = 1$, relative minimum at $x = 2$.

c. $y = -f(x)$ is the graph of $y = f(x)$ reflected across the x-axis. The relative maximum and relative minimum switch. The relative maximum occurs at $x = 2$, and the relative minimum occurs at $x = 1$.

d. $y = f(-x)$ is the graph of $y = f(x)$ reflected across the y-axis. The relative maximum and relative minimum occur at their opposites. The relative maximum occurs at $x = -1$, and the relative minimum occurs at $x = -2$.

e. $y = 2f(x)$ is the graph of $y = f(x)$ stretched vertically by a factor of 2. The locations of the relative maximum and minimum do not change. Relative maximum at $x = 1$, relative minimum at $x = 2$.

f. $y = \frac{1}{2}f(x)$ is the graph of $y = f(x)$ compressed horizontally by a factor of 1/2. The locations of the relative maximum and minimum do not change. Relative maximum at $x = 1$, relative minimum at $x = 2$.

2.7 Getting Ready for the Next Section

149. $(5x^2 + 5x + 7) + (x^2 + 9x - 4) = 6x^2 + 14x + 3$

151. $(5x^2 + 6x - 2) - (3x^2 - 9x + 1) = 2x^2 + 15x - 3$

153. $(x-2)(x^2 + 2x + 4)$
$= x^3 + 2x^2 + 4x - 2x^2 - 4x - 8$
$= x^3 - 8$

155. $f(x) = \dfrac{2x-3}{x^2 - 5x + 6}$
The function is not defined when the denominator is zero.
$x^2 - 5x + 6 = 0 \Rightarrow (x-2)(x-3) = 0 \Rightarrow x = 2, 3$
The domain is $(-\infty, 2) \cup (2, 3) \cup (3, \infty)$.

157. $f(x) = \sqrt{2x-3}$
The function is defined only if $2x - 3 \geq 0$.
$2x - 3 \geq 0 \Rightarrow 2x \geq 3 \Rightarrow x \geq \dfrac{3}{2}$

The domain is $\left[\dfrac{3}{2}, \infty\right)$.

159. $\dfrac{x-1}{x-10} < 0$

First, solve $x-1=0 \Rightarrow x=1$ and
$x-10=0 \Rightarrow x=10$.
So the intervals are
$(-\infty, 1), (1, 10),$ and $(10, \infty)$.

Interval	Test point	Value of $\dfrac{x-1}{x-10}$	Result
$(-\infty, 1)$	0	$\dfrac{1}{10}$	+
$(1, 10)$	5	$-\dfrac{4}{5}$	−
$(10, \infty)$	15	$\dfrac{14}{5}$	+

Note that the fraction is undefined if $x = 10$.
The solution set is $(1, 10)$.

161. $\dfrac{-2x+8}{x^2+1} \leq 0$

Set the numerator and denominator equal to
zero and solve for x.
$-2x+8=0 \Rightarrow -2x=-8 \Rightarrow x=4$
$x^2+1=0$ has no real solution.
The intervals are $(-\infty, 4]$ and $[4, \infty)$.

Interval	Test point	Value of $\dfrac{-2x+8}{x^2+1}$	Result
$(-\infty, 4]$	0	8	+
$[4, \infty)$	5	$-\dfrac{1}{13}$	−

The solution set is $[4, \infty)$.

Interval	Test point	Value of $\dfrac{(x-3)(x-1)}{(x-5)(x+1)}$	Result
$(-\infty, -1)$	−2	$\dfrac{15}{7}$	+
$(-1, 1]$	0	$-\dfrac{3}{5}$	−
$[1, 3]$	2	$\dfrac{1}{9}$	+
$[3, 5)$	4	$-\dfrac{3}{5}$	−
$(5, \infty)$	6	$\dfrac{15}{7}$	+

The solution set is $(-\infty, -1)\cup[1, 3]\cup(5, \infty)$.

2.8 Combining Functions; Composite Functions

2.8 Practice Problems

1. $f(x)=3x-1, \ g(x)=x^2+2$

$(f+g)(x)=f(x)+g(x)$
$\qquad = 3x-1+x^2+2 = x^2+3x+1$
$(f-g)(x)=f(x)-g(x)$
$\qquad = (3x-1)-(x^2+2) = -x^2+3x-3$

$(fg)(x)=f(x)\cdot g(x)$
$\qquad = (3x-1)(x^2+2) = 3x^3-x^2+6x-2$
$\left(\dfrac{f}{g}\right)(x)=\dfrac{f(x)}{g(x)}=\dfrac{3x-1}{x^2+2}$

2. $f(x)=\sqrt{x-1}, \ g(x)=\sqrt{3-x}$

The domain of f is $[1, \infty)$ and the domain of g
is $(-\infty, 3]$. The intersection of D_f and D_g,
$D_f \cap D_g = [1, 3]$.
The domain of fg is $[1, 3]$.
The domain of $\dfrac{f}{g}$ is $[1, 3)$.
The domain of $\dfrac{g}{f}$ is $(1, 3]$.

3. $f(x)=-5x, \ g(x)=x^2+1$

  a. $(f \circ g)(0)=f(g(0))$
  $\qquad = f(0^2+1)=f(1)=-5$

  b. $(g \circ f)(0)=g(f(0))=g(-5\cdot0)=g(0)=1$

4. $f(x)=2-x, \ g(x)=2x^2+1$

  a. $(g \circ f)(x)=g(f(x))=g(2-x)$
  $\qquad = 2(2-x)^2+1$
  $\qquad = 2(4-4x+x^2)+1$
  $\qquad = 8-8x+2x^2+1=2x^2-8x+9$

  b. $(f \circ g)(x)=f(g(x))=f(2x^2+1)$
  $\qquad = 2-(2x^2+1)=1-2x^2$

c. $(g \circ g)(x) = g(g(x)) = g(2x^2 + 1)$

$\qquad = 2(2x^2 + 1)^2 + 1$

$\qquad = 2(4x^4 + 4x^2 + 1) + 1$

$\qquad = 8x^4 + 8x^2 + 3$

5. $f(x) = \sqrt{x+1}, \ g(x) = \dfrac{2}{x-3}$

Let $A = \{x \mid g(x) \text{ is defined}\}$.

$g(x)$ is not defined if $x = 3$, so

$A = (-\infty, 3) \cup (3, \infty)$.

Let $B = \{x \mid f(g(x)) \text{ is defined}\}$.

$f(g(x)) = \sqrt{\dfrac{2}{x-3} + 1} = \sqrt{\dfrac{2+x-3}{x-3}} = \sqrt{\dfrac{x-1}{x-3}}$

$f(g(x))$ is not defined if $x = 3$ or if

$\dfrac{x-1}{x-3} < 0.$

$x - 1 = 0 \Rightarrow x = 1$

Interval	Test point	Value of $\dfrac{x-1}{x-3}$	Result
$(-\infty, 1]$	0	$\frac{1}{3}$	+
$[1, 3)$	2	-1	$-$
$(3, \infty)$	4	3	+

$f(g(x))$ is not defined for $[1, 3)$, so

$B = (-\infty, 1] \cup (3, \infty)$.

The domain of $f \circ g$ is

$A \cap B = (-\infty, 1] \cup (3, \infty)$.

6. $f(x) = \sqrt{x}; \ g(x) = 3 - x$

$f \circ g = f(3 - x) = \sqrt{3-x}$

$f \circ g$ is not defined if $3 - x < 0 \Rightarrow 3 < x$ or

$x > 3$. $f \circ g$ is defined for $x < 3$. Thus, the

domain of $f \circ g$ is $(-\infty, 3]$.

7. $f(x) = \sqrt{x-1}, \ g(x) = \sqrt{4 - x^2}$

a. $(f \circ g)(x) = f(g(x)) = f\left(\sqrt{4-x^2}\right)$

$\qquad = \sqrt{\sqrt{4-x^2} - 1}$

The function $g(x) = \sqrt{4 - x^2}$ is defined

for $4 - x^2 \geq 0 \Rightarrow x^2 \leq 4 \Rightarrow -2 \leq x \leq 2$. So,

$A = [-2, 2]$.

The function $f(g(x))$ is defined for

$\sqrt{4-x^2} - 1 \geq 0 \Rightarrow \sqrt{4-x^2} \geq 1 \Rightarrow$

$4 - x^2 \geq 1 \Rightarrow -x^2 \geq -3 \Rightarrow x^2 \leq 3 \Rightarrow$

$-\sqrt{3} \leq x \leq \sqrt{3}$

So, $B = \left[-\sqrt{3}, \sqrt{3}\right]$.

The domain of $f \circ g$ is

$A \cap B = \left[-\sqrt{3}, \sqrt{3}\right]$.

b. $(g \circ f)(x) = g(f(x)) = \sqrt{4 - \left(\sqrt{x-1}\right)^2}$

$\qquad = \sqrt{4 - (x-1)} = \sqrt{5 - x}$

The function $f(x) = \sqrt{x-1}$ is defined for

$x - 1 \geq 0 \Rightarrow x \geq 1$. So, $A = [1, \infty)$.

The function $g(f(x))$ is defined for

$5 - x \geq 0 \Rightarrow 5 \geq x$, or $x \leq 5$. So,

$B = (-\infty, 5]$.

The domain of $g \circ f$ is $A \cap B = [1, 5]$.

8. $H(x) = \dfrac{1}{\sqrt{2x^2 + 1}}, \ g(x) = \sqrt{2x^2 + 1}$

If $f(x) = \dfrac{1}{x}$, then

$H(x) = (f \circ g)(x) = f\left(\sqrt{2x^2 + 1}\right) = \dfrac{1}{\sqrt{2x^2 + 1}}.$

9. a. $A = f\big(g(t)\big) = f\big(g(3)\big) = f(3t)$
$= \pi(3t)^2 = 9\pi t^2$

b. $A = 9\pi t^2 = 9\pi(6)^2 = 324\pi$
The area covered by the oil slick is $324\pi \approx 1018$ square miles.

10. a. $r(x) = x - 4500$

b. $d(x) = x - 0.06x = 0.94x$

c. i. $(r \circ d)(x) = r(0.94x) = 0.94x - 4500$

ii. $(d \circ r)(x) = d(x - 4500)$
$= 0.94(x - 4500)$
$= 0.94x - 4230$

d. $(d \circ r)(x) - (r \circ d)(x)$
$= (0.94x - 4230) - (0.94x - 4500)$
$= 270$

2.8 Concepts and Vocabulary

1. $(f \cdot g)(x) = \underline{f(x) \cdot g(x)}$.

3. The composition of the function f with the function g is written as $f \circ g$ and is defined by $f \circ g(x) = \underline{f\big(g(x)\big)}$.

5. False. For example, if $f(x) = 2x$ and $g(x) = x^2$, then
$(f \circ g)(x) = f\big(g(x)\big) = f\big(x^2\big) = 2x^2$, while
$(g \circ f)(x) = g\big(f(x)\big) = g(2x) = 4x^2$.

7. False. The domain of $f \cdot g$ may include $g(x) = 0$, but the domain of $\frac{f}{g}$ cannot include $g(x) = 0$.

2.8 Building Skills

9. $(f + g)(-2) = f(-2) + g(-2) = 1 + 2 = 3$

11. $(f - g)(4) = f(4) - g(4) = -2 - 1 = -3$

13. $(f \cdot g)(-1) = f(-1) \cdot g(-1) = 1 \cdot (-4) = -4$

15. $\left(\dfrac{f}{g}\right)(-2) = \dfrac{f(-2)}{g(-2)} = \dfrac{1}{2}$

17. $(f \circ g)(1) = f\big(g(1)\big) = f(-2) = 1$

19. $(f \circ g)(-3) = f\big(g(-3)\big) = f(0) = 0$

21. a. $(f + g)(-1) = f(-1) + g(-1)$
$= 2(-1) + -(-1) = -2 + 1 = -1$

b. $(f - g)(0) = f(0) - g(0) = 2(0) - (-0) = 0$

c. $(f \cdot g)(2) = f(2) \cdot g(2) = 2(2) \cdot (-2) = -8$

d. $\left(\dfrac{f}{g}\right)(1) = \dfrac{f(1)}{g(1)} = \dfrac{2(1)}{-1} = -2$

23. a. $(f + g)(-1) = f(-1) + g(-1)$
$= \dfrac{1}{\sqrt{-1+2}} + (2(-1)+1) = 0$

b. $(f - g)(0) = f(0) - g(0)$
$= \dfrac{1}{\sqrt{0+2}} - (2(0)+1) = \dfrac{\sqrt{2}}{2} - 1$

c. $(f \cdot g)(2) = f(2) \cdot g(2)$
$= \dfrac{1}{\sqrt{2+2}} \cdot (2(2)+1) = \dfrac{5}{2}$

d. $\left(\dfrac{f}{g}\right)(1) = \dfrac{f(1)}{g(1)} = \dfrac{\frac{1}{\sqrt{1+2}}}{2(1)+1} = \dfrac{1}{3\sqrt{3}} = \dfrac{\sqrt{3}}{9}$

25. a. $f + g = x^2 + x - 3$; domain: $(-\infty, \infty)$

b. $f - g = x - 3 - x^2 = -x^2 + x - 3$;
domain: $(-\infty, \infty)$

c. $f \cdot g = (x-3)x^2 = x^3 - 3x^2$;
domain: $(-\infty, \infty)$

d. $\dfrac{f}{g} = \dfrac{x-3}{x^2}$; domain: $(-\infty, 0) \cup (0, \infty)$

e. $\dfrac{g}{f} = \dfrac{x^2}{x-3}$; domain: $(-\infty, 3) \cup (3, \infty)$

27. a. $f + g = \big(x^3 - 1\big) + \big(2x^2 + 5\big) = x^3 + 2x^2 + 4$;
domain: $(-\infty, \infty)$

b. $f - g = \big(x^3 - 1\big) - \big(2x^2 + 5\big) = x^3 - 2x^2 - 6$;
domain: $(-\infty, \infty)$

c. $f \cdot g = \big(x^3 - 1\big)\big(2x^2 + 5\big)$
$= 2x^5 + 5x^3 - 2x^2 - 5$;
domain: $(-\infty, \infty)$

d. $\dfrac{f}{g}=\dfrac{x^3-1}{2x^2+5}$; domain: $(-\infty,\infty)$

e. $\dfrac{g}{f}=\dfrac{2x^2+5}{x^3-1}$; domain: $(-\infty,1)\cup(1,\infty)$

29. a. $f+g=2x+\sqrt{x}-1$; domain: $[0,\infty)$

b. $f-g=2x-\sqrt{x}-1$; domain: $[0,\infty)$

c. $f\cdot g=(2x-1)\sqrt{x}=2x\sqrt{x}-\sqrt{x}$; domain: $[0,\infty)$

d. $\dfrac{f}{g}=\dfrac{2x-1}{\sqrt{x}}$; domain: $(0,\infty)$

e. $\dfrac{g}{f}=\dfrac{\sqrt{x}}{2x-1}$; the numerator is defined only for $x\ge 0$, while the denominator $=0$ when $x=\dfrac{1}{2}$, so the domain is $\left[0,\dfrac{1}{2}\right)\cup\left(\dfrac{1}{2},\infty\right)$.

31. a. $f+g=x-6+\sqrt{x-3}$; domain: $[3,\infty)$

b. $f-g=x-6-\sqrt{x-3}$; domain: $[3,\infty)$

c. $f\cdot g=(x-6)\sqrt{x-3}$; domain: $[3,\infty)$

d. $\dfrac{f}{g}=\dfrac{x-6}{\sqrt{x-3}}$; domain: $(3,\infty)$

e. $\dfrac{g}{f}=\dfrac{\sqrt{x-3}}{x-6}$; the numerator is defined only for $x\ge 3$, while the denominator $=0$ when $x=6$, so the domain is $[3,6)\cup(6,\infty)$.

33. a. $f+g=1-\dfrac{2}{x+1}+\dfrac{1}{x}$
$=\dfrac{x(x+1)-2x+(x+1)}{x(x+1)}$
$=\dfrac{x^2+x-2x+x+1}{x(x+1)}=\dfrac{x^2+1}{x(x+1)}$
domain: $(-\infty,-1)\cup(-1,0)\cup(0,\infty)$

b. $f-g=1-\dfrac{2}{x+1}-\dfrac{1}{x}$
$=\dfrac{x(x+1)-2x-(x+1)}{x(x+1)}$
$=\dfrac{x^2+x-2x-x-1}{x(x+1)}=\dfrac{x^2-2x-1}{x(x+1)}$
domain: $(-\infty,-1)\cup(-1,0)\cup(0,\infty)$

c. $f\cdot g=\left(1-\dfrac{2}{x+1}\right)\dfrac{1}{x}=\left(\dfrac{x+1-2}{x+1}\right)\dfrac{1}{x}$
$=\left(\dfrac{x-1}{x+1}\right)\dfrac{1}{x}=\dfrac{x-1}{x(x+1)}$
domain: $(-\infty,-1)\cup(-1,0)\cup(0,\infty)$

d. $\dfrac{f}{g}=\dfrac{1-\frac{2}{x+1}}{\frac{1}{x}}=\left(1-\dfrac{2}{x+1}\right)(x)$
$=\left(\dfrac{x+1-2}{x+1}\right)x=\dfrac{x(x-1)}{x+1}$
domain: $(-\infty,-1)\cup(-1,0)\cup(0,\infty)$

e. $\dfrac{g}{f}=\dfrac{\frac{1}{x}}{1-\frac{2}{x+1}}=\dfrac{\frac{1}{x}(x+1)}{\left(1-\frac{2}{x+1}\right)(x+1)}=\dfrac{\frac{x+1}{x}}{x+1-2}$
$=\dfrac{\frac{x+1}{x}}{x-1}=\dfrac{x+1}{x(x-1)}$
The denominator equals zero when $x=0$ or $x=1$, so the domain is $(-\infty,-1)\cup(-1,0)\cup(0,1)\cup(1,\infty)$.

35. a. $f+g=\dfrac{2}{x+1}+\dfrac{x}{x+1}=\dfrac{2+x}{x+1}$
Neither f nor g is defined for $x=-1$, so the domain is $(-\infty,-1)\cup(-1,\infty)$.

b. $f-g=\dfrac{2}{x+1}-\dfrac{x}{x+1}=\dfrac{2-x}{x+1}$.
domain: $(-\infty,-1)\cup(-1,\infty)$.

c. $f\cdot g=\left(\dfrac{2}{x+1}\right)\left(\dfrac{x}{x+1}\right)=\dfrac{2x}{(x+1)^2}$.
domain: $(-\infty,-1)\cup(-1,\infty)$.

d. $\dfrac{f}{g}=\dfrac{\frac{2}{x+1}}{\frac{x}{x+1}}=\dfrac{2}{x}$. Neither f nor g is defined for $x=-1$, and f/g is not defined for $x=0$, so the domain is $(-\infty,-1)\cup(-1,0)\cup(0,\infty)$.

e. $\dfrac{g}{f}=\dfrac{\frac{x}{x+1}}{\frac{2}{x+1}}=\dfrac{x}{2}$. Neither f nor g is defined for $x=-1$, so the domain is $(-\infty,-1)\cup(-1,\infty)$.

37. $f(x) = \dfrac{x^2}{x+1}$; $g(x) = \dfrac{2x}{x^2-1}$

    a. $f+g = \dfrac{x^2}{x+1} + \dfrac{2x}{x^2-1} = \dfrac{x^2(x-1)}{x^2-1} + \dfrac{2x}{x^2-1}$

         $= \dfrac{x^3-x^2+2x}{x^2-1}$

       f is not defined for $x = -1$, g is not defined for $x = \pm 1$, and $f+g$ is not defined for either -1 or 1, so the domain is $(-\infty, -1) \cup (-1, 1) \cup (1, \infty)$.

    b. $f-g = \dfrac{x^2}{x+1} - \dfrac{2x}{x^2-1} = \dfrac{x^2(x-1)}{x^2-1} - \dfrac{2x}{x^2-1}$

         $= \dfrac{x^3-x^2-2x}{x^2-1} = \dfrac{x(x^2-x-2)}{x^2-1}$

         $= \dfrac{x(x-2)(x+1)}{(x-1)(x+1)} = \dfrac{x^2-2x}{x-1}$

       f is not defined for $x = -1$, g is not defined for $x = \pm 1$, and $f-g$ is not defined for 1, so the domain is $(-\infty, -1) \cup (-1, 1) \cup (1, \infty)$.

    c. $f \cdot g = \dfrac{x^2}{x+1} \cdot \dfrac{2x}{x^2-1} = \dfrac{2x^3}{x^3+x^2-x-1}$

       f is not defined for $x = -1$, g is not defined for $x = \pm 1$, and fg is not defined for either -1 or 1, so the domain is $(-\infty, -1) \cup (-1, 1) \cup (1, \infty)$.

    d. $\dfrac{f}{g} = \dfrac{\dfrac{x^2}{x+1}}{\dfrac{2x}{x^2-1}} = \dfrac{x^2}{x+1} \cdot \dfrac{x^2-1}{2x} = \dfrac{x(x-1)}{2}$

       f is not defined for $x = -1$, g is not defined for $x = \pm 1$, and f/g is not defined for either -1, 0, or 1, so the domain is $(-\infty, -1) \cup (-1, 0) \cup (0, 1) \cup (1, \infty)$.

    e. $\dfrac{g}{f} = \dfrac{\dfrac{2x}{x^2-1}}{\dfrac{x^2}{x+1}} = \dfrac{2x}{x^2-1} \cdot \dfrac{x+1}{x^2} = \dfrac{2}{x(x-1)}$

       Neither f nor g is defined for $x = -1$ and g/f is not defined for $x = 0$ or $x = 1$, so the domain is $(-\infty, -1) \cup (-1, 0) \cup (0, 1) \cup (1, \infty)$.

39. $f(x) = \sqrt{x-1}$; $g(x) = \sqrt{5-x}$

    a. $f \cdot g = \sqrt{x-1} \cdot \sqrt{5-x}$

       f is not defined for $x < 1$, g is not defined for $x > 5$. The domain is $[1, 5]$.

    b. $\dfrac{f}{g} = \dfrac{\sqrt{x-1}}{\sqrt{5-x}}$

       f is not defined for $x < 1$, g is not defined for $x > 5$. The denominator is zero when $x = 5$. The domain is $[1, 5)$.

41. $f(x) = \sqrt{x+2}$; $g(x) = \sqrt{9-x^2}$

    a. $f \cdot g = \sqrt{x+2} \cdot \sqrt{9-x^2}$

       f is not defined for $x < -2$, g is defined for $[-3, 3]$. The domain is $[-2, 3]$.

    b. $\dfrac{f}{g} = \dfrac{\sqrt{x+2}}{\sqrt{9-x^2}}$

       f is not defined for $x < -2$, g is defined for $[-3, 3]$. The denominator is zero when $x = -3$ or $x = 3$. The domain is $[-2, 3)$.

43. $(g \circ f)(x) = 2(x^2-1) + 3 = 2x^2 + 1$;

    $(g \circ f)(2) = 2(2^2-1) + 3 = 9$;

    $(g \circ f)(-3) = 2((-3)^2-1) + 3 = 19$

45. $(f \circ g)(2) = 2(2(2^2)-3) + 1 = 11$

47. $(f \circ g)(-3) = 2(2(-3)^2-3) + 1 = 31$

49. $(f \circ g)(0) = 2(2(0^2)-3) + 1 = -5$

51. $(f \circ g)(-c) = 2(2(-c)^2-3) + 1 = 4c^2 - 5$

53. $(g \circ f)(a) = 2(2a+1)^2 - 3$

             $= 2(4a^2+4a+1) - 3$

             $= 8a^2 + 8a - 1$

55. $(f \circ f)(1) = 2(2(1)+1) + 1 = 7$

57. $f(x) = \dfrac{1}{x}$; $g(x) = 10 - 5x$

    $(f \circ g)(x) = \dfrac{1}{10-5x}$

    The domain of $f \circ g$ is the set of all real numbers such that $10 - 5x \neq 0$, or $x \neq 2$. The domain of $f \circ g$ is $(-\infty, 2) \cup (2, \infty)$.

59. $f(x) = \sqrt{x};\ g(x) = 2x - 8$
$(f \circ g)(x) = \sqrt{2x - 8}$
The domain of $f \circ g$ is the set of all real numbers such that $2x - 8 \geq 0,$ or $x \geq 4.$ The domain of $f \circ g$ is $[4, \infty).$

61. $(f \circ g)(x) = \dfrac{2}{\dfrac{1}{x} + 1} = \dfrac{2}{\dfrac{x+1}{x}} = \dfrac{2x}{x+1}$
The domain of g is $(-\infty, 0) \cup (0, \infty).$ Since -1 is not in the domain of f, we must exclude those values of x that make $g(x) = -1.$
$\dfrac{1}{x} = -1 \Rightarrow x = -1$
Thus, the domain of $f \circ g$ is
$(-\infty, -1) \cup (-1, 0) \cup (0, \infty).$

63. $(f \circ g)(x) = \sqrt{(2 - 3x) - 3} = \sqrt{-1 - 3x}.$
The domain of g is $(-\infty, \infty).$ Since f is not defined for $(-\infty, 3),$ we must exclude those values of x that make $g(x) < 3.$
$2 - 3x < 3 \Rightarrow -3x < 1 \Rightarrow x > -\dfrac{1}{3}$
Thus, the domain of $f \circ g$ is $\left(-\infty, -\dfrac{1}{3}\right].$

65. $(f \circ g)(x) = \left| x^2 - 1 \right|;$ domain: $(-\infty, \infty)$

67. a. $(f \circ g)(x) = 2(x + 4) - 3 = 2x + 5;$
domain: $(-\infty, \infty)$

 b. $(g \circ f)(x) = (2x - 3) + 4 = 2x + 1;$
domain: $(-\infty, \infty)$

 c. $(f \circ f)(x) = 2(2x - 3) - 3 = 4x - 9;$
domain: $(-\infty, \infty)$

 d. $(g \circ g)(x) = (x + 4) + 4 = x + 8;$
domain: $(-\infty, \infty)$

69. a. $(f \circ g)(x) = 1 - 2(1 + x^2) = -2x^2 - 1;$
domain: $(-\infty, \infty)$

 b. $(g \circ f)(x) = 1 + (1 - 2x)^2 = 4x^2 - 4x + 2;$
domain: $(-\infty, \infty)$

 c. $(f \circ f)(x) = 1 - 2(1 - 2x) = 4x - 1;$
domain: $(-\infty, \infty)$

 d. $(g \circ g)(x) = 1 + (1 + x^2)^2 = x^4 + 2x^2 + 2;$
domain: $(-\infty, \infty)$

71. a. $(f \circ g)(x) = 2(2x - 1)^2 + 3(2x - 1)$
$\qquad = 2(4x^2 - 4x + 1) + 6x - 3$
$\qquad = 8x^2 - 2x - 1;$ domain: $(-\infty, \infty)$

 b. $(g \circ f)(x) = 2(2x^2 + 3x) - 1 = 4x^2 + 6x - 1;$
domain: $(-\infty, \infty)$

 c. $(f \circ f)(x) = 2(2x^2 + 3x)^2 + 3(2x^2 + 3x)$
$\qquad = 2(4x^4 + 12x^3 + 9x^2) + 6x^2 + 9x$
$\qquad = 8x^4 + 24x^3 + 24x^2 + 9x;$
domain: $(-\infty, \infty)$

 d. $(g \circ g)(x) = 2(2x - 1) - 1 = 4x - 3;$
domain: $(-\infty, \infty)$

73. a. $(f \circ g)(x) = \left(\sqrt{x}\right)^2 = x;$ domain: $[0, \infty)$

 b. $(g \circ f)(x) = \sqrt{x^2} = |x|;$ domain: $(-\infty, \infty)$

 c. $(f \circ f)(x) = \left(x^2\right)^2 = x^4;$ domain: $(-\infty, \infty)$

 d. $(g \circ g)(x) = \sqrt{\sqrt{x}} = \sqrt[4]{x};$ domain: $[0, \infty)$

75. a. $(f \circ g)(x) = \dfrac{1}{2\left(\dfrac{1}{x^2}\right) - 1} = \dfrac{1}{\dfrac{2 - x^2}{x^2}}$
$\qquad = \dfrac{x^2}{2 - x^2} = -\dfrac{x^2}{x^2 - 2}.$
The domain of g is $(-\infty, 0) \cup (0, \infty).$ Since $\frac{1}{2}$ is not in the domain of f, we must find those values of x that make $g(x) = \frac{1}{2}.$
$\dfrac{1}{x^2} = \dfrac{1}{2} \Rightarrow x^2 = 2 \Rightarrow x = \pm\sqrt{2}$
Thus, the domain of $f \circ g$ is
$\left(-\infty, -\sqrt{2}\right) \cup \left(-\sqrt{2}, 0\right) \cup \left(0, \sqrt{2}\right) \cup \left(\sqrt{2}, \infty\right).$

 b. $(g \circ f) = \dfrac{1}{\left(\dfrac{1}{2x - 1}\right)^2} = (2x - 1)^2$
The domain of f is $\left(-\infty, \frac{1}{2}\right) \cup \left(\frac{1}{2}, \infty\right).$ Since 0 is not in the domain of g, we must find those values of x that make $f(x) = 0.$ However, there are no such values, so the domain of $g \circ f$ is $\left(-\infty, \frac{1}{2}\right) \cup \left(\frac{1}{2}, \infty\right).$

c. $(f \circ f)(x) = \dfrac{1}{2\left(\dfrac{1}{2x-1}\right)-1} = \dfrac{1}{\dfrac{2-2x+1}{2x-1}}$

$\qquad = \dfrac{2x-1}{3-2x} = -\dfrac{2x-1}{2x-3}$

The domain of f is $\left(-\infty, \dfrac{1}{2}\right) \cup \left(\dfrac{1}{2}, \infty\right)$.

$-\dfrac{2x-1}{2x-3}$ is defined for $\left(-\infty, \dfrac{3}{2}\right) \cup \left(\dfrac{3}{2}, \infty\right)$,

so the domain of $f \circ f$ is

$\left(-\infty, \dfrac{1}{2}\right) \cup \left(\dfrac{1}{2}, \dfrac{3}{2}\right) \cup \left(\dfrac{3}{2}, \infty\right)$.

d. $(g \circ g)(x) = \dfrac{1}{\left(\dfrac{1}{x^2}\right)} = x^4$.

The domain of g is $(-\infty, 0) \cup (0, \infty)$, while $g \circ g = x^4$ is defined for all real numbers. Thus, the domain of $g \circ g$ is $(-\infty, 0) \cup (0, \infty)$.

77. a. $(f \circ g)(x) = \sqrt{\sqrt{4-x}-1}$; domain: $(-\infty, 3]$

b. $(g \circ f)(x) = \sqrt{4 - \sqrt{x-1}}$; domain: $[1, 17]$

c. $(f \circ f)(x) = \sqrt{\sqrt{x-1}-1}$; domain: $[2, \infty)$

d. $(g \circ g)(x) = \sqrt{4 - \sqrt{4-x}}$; domain: $[-12, 4]$

79. a. $(f \circ g)(x) = \dfrac{1 - \dfrac{x+3}{x-4}}{\dfrac{x+3}{x-4}+2} = \dfrac{\left(1 - \dfrac{x+3}{x-4}\right)(x-4)}{\left(\dfrac{x+3}{x-4}+2\right)(x-4)}$

$\qquad = \dfrac{(x-4)-(x+3)}{(x+3)+2(x-4)} = -\dfrac{7}{3x-5}$

The domain of g is $(-\infty, 4) \cup (4, \infty)$. The denominator of $f \circ g$ is 0 when $x = \dfrac{5}{3}$, so the domain of $f \circ g$ is

$\left(-\infty, \dfrac{5}{3}\right) \cup \left(\dfrac{5}{3}, 4\right) \cup (4, \infty)$.

b. $(g \circ f)(x) = \dfrac{\dfrac{1-x}{x+2}+3}{\dfrac{1-x}{x+2}-4} = \dfrac{\left(\dfrac{1-x}{x+2}+3\right)(x+2)}{\left(\dfrac{1-x}{x+2}-4\right)(x+2)}$

$\qquad = \dfrac{(1-x)+3(x+2)}{(1-x)-4(x+2)} = \dfrac{2x+7}{-5x-7}$

$\qquad = -\dfrac{2x+7}{5x+7}$

The domain of f is $(-\infty, -2) \cup (-2, \infty)$. The denominator of $g \circ f$ is 0 when $x = -\dfrac{7}{5}$, so, the domain of $g \circ f$ is

$\left(-\infty, -2\right) \cup \left(-2, -\dfrac{7}{5}\right) \cup \left(-\dfrac{7}{5}, \infty\right)$.

c. $(f \circ f)(x) = \dfrac{1 - \dfrac{1-x}{x+2}}{\dfrac{1-x}{x+2}+2} = \dfrac{\left(1 - \dfrac{1-x}{x+2}\right)(x+2)}{\left(\dfrac{1-x}{x+2}+2\right)(x+2)}$

$\qquad = \dfrac{(x+2)-(1-x)}{(1-x)+2(x+2)} = \dfrac{2x+1}{x+5}$

The domain of f is $(-\infty, -2) \cup (-2, \infty)$. The denominator of $f \circ f$ is 0 when $x = -5$, so, the domain of $f \circ f$ is

$(-\infty, -5) \cup (-5, -2) \cup (-2, \infty)$.

d. $(g \circ g)(x) = \dfrac{\dfrac{x+3}{x-4}+3}{\dfrac{x+3}{x-4}-4} = \dfrac{\left(\dfrac{x+3}{x-4}+3\right)(x-4)}{\left(\dfrac{x+3}{x-4}-4\right)(x-4)}$

$\qquad = \dfrac{(x+3)+3(x-4)}{(x+3)-4(x-4)} = \dfrac{4x-9}{-3x+19}$

$\qquad = -\dfrac{4x-9}{3x-19}$

The domain of g is $(-\infty, 4) \cup (4, \infty)$. The denominator of $g \circ g$ is 0 when $x = \dfrac{19}{3}$, so the domain of $g \circ g$ is

$\left(-\infty, 4\right) \cup \left(4, \dfrac{19}{3}\right) \cup \left(\dfrac{19}{3}, \infty\right)$.

In exercises 81–90, sample answers are given. Other answers are possible.

81. $H(x) = \sqrt{x+2} \Rightarrow f(x) = \sqrt{x}, g(x) = x+2$

83. $H(x) = \left(x^2 - 3\right)^{10} \Rightarrow f(x) = x^{10}, g(x) = x^2 - 3$

85. $H(x) = \dfrac{1}{3x-5} \Rightarrow f(x) = \dfrac{1}{x}, g(x) = 3x-5$

87. $H(x) = \sqrt[3]{x^2 - 7} \Rightarrow f(x) = \sqrt[3]{x}, g(x) = x^2 - 7$

89. $H(x) = \dfrac{1}{\left|x^3 - 1\right|} \Rightarrow f(x) = \dfrac{1}{|x|}, g(x) = x^3 - 1$

2.8 Applying the Concepts

91. a. $f(x)$ is the cost function.

b. $g(x)$ is the revenue function.

c. $h(x)$ is the selling price of x shirts including sales tax.

d. $P(x)$ is the profit function.

93. a. $P(x) = R(x) - C(x) = 25x - (350 + 5x)$
$$= 20x - 350$$

b. $P(20) = 20(20) - 350 = 50.$ This represents the profit when 20 radios are sold.

c. $P(x) = 20x - 350; 500 = 20x - 350 \Rightarrow x = 43$

d. $C = 350 + 5x \Rightarrow x = \dfrac{C - 350}{5} = x(C).$
$$(R \circ x)(C) = 25\left(\dfrac{C - 350}{5}\right) = 5C - 1750.$$
This function represents the revenue in terms of the cost C.

95. a. $f(x) = 0.7x$

b. $g(x) = x - 5$

c. $(g \circ f)(x) = 0.7x - 5$

d. $(f \circ g)(x) = 0.7(x - 5)$

e. $(f \circ g) - (g \circ f) = 0.7(x - 5) - (0.7x - 5)$
$$= 0.7x - 3.5 - 0.7x + 5$$
$$= \$1.50$$

97. a. $f(x) = 1.1x; g(x) = x + 8$

b. $(f \circ g)(x) = 1.1(x + 8) = 1.1x + 8.8$
. This represents a final test score computed by first adding 8 points to the original score and then increasing the total by 10%.

c. $(g \circ f)(x) = 1.1x + 8$
This represents a final test score computed by first increasing the original score by 10% and then adding 8 points.

d. $(f \circ g)(70) = 1.1(70 + 8) = 85.8;$
$(g \circ f)(70) = 1.1(70) + 8 = 85.0;$

e. $(f \circ g)(x) \neq (g \circ f)(x)$

f. (i) $(f \circ g)(x) = 1.1x + 8.8 \geq 90 \Rightarrow x \geq 73.82$

(ii) $(g \circ f)(x) = 1.1x + 8 \geq 90 \Rightarrow x \geq 74.55$

99. a. $f(x) = \pi x^2$

b. $g(x) = \pi(x + 30)^2$

c. $g(x) - f(x)$ represents the area between the fountain and the fence.

d. The circumference of the fence is $2\pi(x + 30)$.
$$10.5(2\pi(x + 30)) = 4200 \Rightarrow$$
$$\pi(x + 30) = 200 \Rightarrow$$
$$\pi x + 30\pi = 200 \Rightarrow \pi x = 200 - 30\pi.$$

$$g(x) - f(x) = \pi(x + 30)^2 - \pi x^2$$
$$= \pi(x^2 + 60x + 900) - \pi x^2$$
$$= 60\pi x + 900\pi. \text{ Now substitute}$$
$200 - 30\pi$ for πx to compute the estimate:
$$1.75[60(200 - 30\pi) + 900\pi]$$
$$= 1.75(12,000 - 900\pi) \approx \$16,052.$$

101. a. $(f \circ g)(t) = \pi(2t + 1)^2$

b. $A(t) = f(2t + 1) = \pi(2t + 1)^2$

c. They are the same.

2.8 Beyond the Basics

103. a. When you are looking for the domain of the sum of two functions that are given as sets, you are looking for the intersection of their domains. Since the x-values that f and g have in common are -2, 1, and 3, the domain of $f + g$ is $\{-2, 1, 3\}$. Now add the y-values.
$$(f + g)(-2) = 3 + 0 = 3$$
$$(f + g)(1) = 2 + (-2) = 0$$
$$(f + g)(3) = 0 + 2 = 2$$
Thus, $f + g = \{(-2, 3), (1, 0), (3, 2)\}$.

b. When you are looking for the domain of the product of two functions that are given as sets, you are looking for the intersection of their domains. Since the x-values that f and g have in common are -2, 1, and 3, the domain of $f + g$ is $\{-2, 1, 3\}$. Now multiply the y-values.
$$(fg)(-2) = 3 \cdot 0 = 0$$
$$(fg)(1) = 2 \cdot (-2) = -4$$
$$(fg)(3) = 0 \cdot 2 = 0$$
Thus, $fg = \{(-2, 0), (1, -4), (3, 0)\}$.

c. When you are looking for the domain of the quotient of two functions that are given as sets, you are looking for the intersection of their domains and values of x that do not cause the denominator to equal zero. The x-values that f and g have in common are -2, 1, and 3; however, $g(-2) = 0$, so the domain is $\{1, 3\}$. Now divide the y-values.

$$\left(\frac{f}{g}\right)(1) = \frac{2}{-2} = -1$$

$$\left(\frac{f}{g}\right)(3) = \frac{0}{2} = 0$$

Thus, $\dfrac{f}{g} = \{(1, -1), (3, 0)\}$.

d. When you are looking for the domain of the composition of two functions that are given as sets, you are looking for values that come from the domain of the inside function and when you plug those values of x into the inside function, the output is in the domain of the outside function.

$f(g(-2)) = f(0)$, which is undefined

$f(g(0)) = f(2) = 1$,

$f(g(1)) = f(-2) = 3$,

$f(g(3)) = f(2) = 1$

Thus, $f \circ g = \{(0, 1), (1, 3), (3, 1)\}$.

105. a. $f(-x) = h(-x) + h(-(-x)) = h(-x) + h(x)$
$= f(x) \Rightarrow f(x)$ is an even function.

b. $g(-x) = h(-x) - h(-(-x)) = h(-x) - h(x)$
$= -g(x) \Rightarrow g(x)$ is an odd function.

c. $\begin{cases} f(x) = h(x) + h(-x) \\ g(x) = h(x) - h(-x) \end{cases} \Rightarrow$

$f(x) + g(x) = 2h(x) \Rightarrow$

$h(x) = \dfrac{f(x) + g(x)}{2} = \dfrac{f(x)}{2} + \dfrac{g(x)}{2} \Rightarrow$

$h(x)$ is the sum of an even function and an odd function.

107. $f(x) = \sqrt{\dfrac{1 - |x|}{2 - |x|}}$

$f(x)$ is defined if $\dfrac{1 - |x|}{2 - |x|} \geq 0$ and $2 - |x| \neq 0$.

$2 - |x| = 0 \Rightarrow 2 = |x| \Rightarrow x = \pm 2$

Thus, the values -2 and 2 are not in the domain of f.

$\dfrac{1 - |x|}{2 - |x|} \geq 0$ if $1 - |x| \geq 0$ and $2 - |x| > 0$, or if

$1 - |x| \leq 0$ and $2 - |x| < 0$.

Case 1: $1 - |x| \geq 0$ and $2 - |x| > 0$.

$1 - |x| \geq 0 \Rightarrow 1 \geq |x| \Rightarrow -1 \leq x \leq 1$

$2 - |x| > 0 \Rightarrow 2 > |x| \Rightarrow -2 < x < 2$

Thus, $1 - |x| \geq 0$ and $2 - |x| > 0 \Rightarrow -1 \leq x \leq 1$.

Case 2: $1 - |x| \leq 0$ and $2 - |x| < 0$.

$1 - |x| \leq 0 \Rightarrow 1 \leq |x| \Rightarrow (-\infty, -1] \cup [1, \infty)$

$2 - |x| < 0 \Rightarrow 3 \leq |x| \Rightarrow (-\infty, -2) \cup (2, \infty)$

Thus, $1 - |x| \leq 0$ and $2 - |x| < 0 \Rightarrow$

$(-\infty, -2) \cup (2, \infty)$.

The domain of f is

$(-\infty, -2) \cup [-1, 1] \cup (2, \infty)$.

2.8 Critical Thinking/Discussion/Writing

109. a. The domain of $f(x)$ is $(-\infty, 0) \cup [1, \infty)$.

b. The domain of $g(x)$ is $[0, 2]$.

c. The domain of $f(x) + g(x)$ is $[1, 2]$.

d. The domain of $\dfrac{f(x)}{g(x)}$ is $[1, 2)$.

110. a. The domain of f is $(-\infty, 0)$. The domain of $f \circ f$ is $\varnothing$ because $f \circ f = \dfrac{1}{\sqrt{-\dfrac{1}{\sqrt{-x}}}}$ and

the denominator is the square root of a negative number.

b. The domain of f is $(-\infty, 1)$. The domain of $f \circ f$ is $(-\infty, 0)$ because

$f \circ f = \dfrac{1}{\sqrt{1 - \dfrac{1}{\sqrt{1 - x}}}}$ and the denominator

must be greater than 0. If $x = 0$, then the denominator $= 0$.

111. a. The sum of two even functions is an even function.

$f(x) = f(-x)$ and $g(x) = g(-x) \Rightarrow$

$(f + g)(x) = f(x) + g(x) = f(-x) + g(-x)$

$= (f + g)(-x)$.

b. The sum of two odd functions is an odd function.

$f(-x) = -f(x)$ and $g(-x) = -g(x) \Rightarrow$

$(f+g)(-x) = f(-x) + g(-x) = -f(x) - g(x)$

$= -(f+g)(x).$

c. The sum of an even function and an odd function is neither even nor odd.

$f(x)$ even $\Rightarrow f(x) = f(-x)$ and $g(x)$ odd $\Rightarrow$

$g(-x) = -g(x) \Rightarrow f(-x) + g(-x) =$

$f(x) + (-g(x))$, which is neither even nor odd.

d. The product of two even functions is an even function.

$f(x) = f(-x)$ and $g(x) = g(-x) \Rightarrow$

$(f \cdot g)(x) = f(x) \cdot g(x) = f(-x) \cdot (g(-x))$

$= (f \cdot g)(-x).$

e. The product of two odd functions is an even function.

$f(-x) = -f(x)$ and $g(-x) = -g(x) \Rightarrow$

$(f \cdot g)(-x) = f(-x) \cdot g(-x) = -f(x) \cdot (-g(x))$

$= (f \cdot g)(x).$

f. The product of an even function and an odd function is an odd function.

$f(x)$ even $\Rightarrow f(x) = f(-x)$ and $g(x)$ odd $\Rightarrow$

$g(-x) = -g(x) \Rightarrow$

$f(-x) \cdot g(-x) = f(x) \cdot (-g(x)) = -(f \cdot g)(x)$

112. a. $f(-x) = -f(x)$ and $g(-x) = -g(x) \Rightarrow$

$(f \circ g)(-x) = f(g(-x)) = f(-g(x)) =$

$-f(g(x)) \Rightarrow (f \circ g)(x)$ is odd.

b. $f(x) = f(-x)$ and $g(x) = g(-x) \Rightarrow$

$(f \circ g)(-x) = f(g(-x)) = f(g(x)) \Rightarrow$

$(f \circ g)(x)$ is even.

c. $f(x)$ odd $\Rightarrow f(-x) = -f(x)$ and

$g(x)$ even $\Rightarrow g(x) = g(-x) \Rightarrow (f \circ g)(-x)$

$f(g(x)) = f(g(-x)) \Rightarrow (f \circ g)(x)$ is even.

d. $f(x)$ even $\Rightarrow f(x) = f(-x)$ and $g(x)$ odd $\Rightarrow$

$g(-x) = -g(x) \Rightarrow (f \circ g)(-x) = f(-g(x))$

$= f(g(x)) = (f \circ g)(-x) \Rightarrow (f \circ g)(x)$ is even.

2.8 Getting Ready for the Next Section

113. a. Yes, R defines a function.

b. $S = \{(2, -3), (1, -1), (3, 1), (1, 2)\}$

No, S does not define a function since the first value 1 maps to two different second values, −1 and 2.

115. $x = 2y + 3 \Rightarrow x - 3 = 2y \Rightarrow \dfrac{x-3}{2} = y$

117. $x^2 + y^2 = 4,\ x \le 0 \Rightarrow x^2 = 4 - y^2 \Rightarrow$

$x = -\sqrt{4 - y^2}$

2.9 Inverse Functions

2.9 Practice Problems

1. $f(x) = (x-1)^2$ is not one-to-one because the horizontal line $y = 1$ intersects the graph at two different points.

2. a. $f^{-1}(12) = -3$

b. $f(9) = 4$

3. $f(x) = 3x - 1,\ g(x) = \dfrac{x+1}{3}$

$(f \circ g)(x) = f\left(\dfrac{x+1}{3}\right) = 3\left(\dfrac{x+1}{3}\right) - 1 = x$

$(g \circ f)(x) = g(3x - 1) = \dfrac{3x - 1 + 1}{3} = x$

Because $f(g(x)) = g(f(x)) = x$, the two functions are inverses.

4. The graph of f^{-1} is the reflection of the graph of f about the line $y = x$.

5. $f(x) = -2x + 3$ is a one-to-one function, so the function has an inverse. Interchange the variables and solve for y:

$f(x) = y = -2x + 3 \Rightarrow x = -2y + 3 \Rightarrow$

$\dfrac{x-3}{-2} = y \Rightarrow y = f^{-1}(x) = \dfrac{3-x}{2}$.

6. Interchange the variables and solve for y:

$f(x) = y = \dfrac{x}{x+3}, x \neq -3$

$x = \dfrac{y}{y+3} \Rightarrow xy + 3x = y \Rightarrow 3x = y - xy \Rightarrow$

$3x = y(1-x) \Rightarrow \dfrac{3x}{1-x} = y \Rightarrow$

$f^{-1}(x) = \dfrac{3x}{1-x}, x \neq 1$

7. $f(x) = \dfrac{x}{x+3}$

The function is not defined if the denominator is zero, so the domain is $(-\infty, -3) \cup (-3, \infty)$.

The range of the function is the same as the domain of the inverse, thus the range is $(-\infty, 1) \cup (1, \infty)$.

8. G is one-to-one because the domain is restricted, so an inverse exists.

$G(x) = y = x^2 - 1, x \leq 0$. Interchange the variables and solve for y:

$x = y^2 - 1, y \leq 0 \Rightarrow y = G^{-1}(x) = -\sqrt{x+1}$.

9. From the text, we have $d = \dfrac{11p}{5} - 33$.

$d = \dfrac{11 \cdot 1650}{5} - 33 = 3597$

The bell was 3597 feet below the surface when the gauge failed.

2.9 Concepts and Vocabulary

1. If no horizontal line intersects the graph of a function f in more than one point, the f is a <u>one-to-one</u> function.

3. If $f(x) = 3x$, then $f^{-1}(x) = \dfrac{1}{3}x$.

5. True

7. False. $f^{-1}(x)$ means the inverse of f.

2.9 Building Skills

9. One-to-one   **11.** Not one-to-one

13. Not one-to-one   **15.** One-to-one

17. $f(2) = 7 \Rightarrow f^{-1}(7) = 2$

19. $f(-1) = 2 \Rightarrow f^{-1}(2) = -1$

21. $f(a) = b \Rightarrow f^{-1}(b) = a$

23. $\left(f^{-1} \circ f\right)(337) = f^{-1}\left(f(337)\right) = 337$

25. a. $f(3) = 2(3) - 3 = 3$

   b. Using the result from part (a), $f^{-1}(3) = 3$.

   c. $\left(f \circ f^{-1}\right)(19) = f\left(f^{-1}(19)\right) = 19$

   d. $\left(f \circ f^{-1}\right)(5) = f\left(f^{-1}(5)\right) = 5$

27. a. $f(1) = 1^3 + 1 = 2$

   b. Using the result from part (a), $f^{-1}(2) = 1$.

   c. $\left(f \circ f^{-1}\right)(269) = f\left(f^{-1}(269)\right) = 269$

29. $f(g(x)) = 3\left(\dfrac{x-1}{3}\right) + 1 = x - 1 + 1 = x$

$g(f(x)) = \dfrac{(3x+1)-1}{3} = \dfrac{3x}{3} = x$

31. $f(g(x)) = \left(\sqrt[3]{x}\right)^3 = x$

$g(f(x)) = \sqrt[3]{x^3} = x$

33. $f\big(g(x)\big) = \dfrac{\dfrac{1+2x}{1-x} - 1}{\dfrac{1+2x}{1-x} + 2} = \dfrac{\dfrac{1+2x-(1-x)}{1-x}}{\dfrac{1+2x+2(1-x)}{1-x}}$

$= \dfrac{3x}{3} = x$

$g\big(f(x)\big) = \dfrac{1+2\left(\dfrac{x-1}{x+2}\right)}{1-\dfrac{x-1}{x+2}} = \dfrac{1+\dfrac{2x-2}{x+2}}{1-\dfrac{x-1}{x+2}}$

$= \dfrac{\dfrac{x+2+2x-2}{x+2}}{\dfrac{x+2-(x-1)}{x+2}} = \dfrac{3x}{3} = x$

35.

37.

39.

41.

43. a. One-to-one

b. $f(x) = y = 15 - 3x.$ Interchange the variables and solve for y: $x = 15 - 3y \Rightarrow$
$y = f^{-1}(x) = \dfrac{15-x}{3} = 5 - \dfrac{1}{3}x.$

c.

d. Domain of f: $(-\infty, \infty)$; x-intercept of f: 5; y-intercept of f: 15
domain of f^{-1}: $(-\infty, \infty)$; x-intercept of f^{-1}: 15; y-intercept of f^{-1}: 5

45. a. Not one-to-one

47. a. One-to-one

b. $f(x) = y = \sqrt{x} + 3$ Interchange the variables and solve for y: $x = \sqrt{y} + 3 \Rightarrow$
$x - 3 = \sqrt{y} \Rightarrow y = f^{-1}(x) = (x-3)^2.$

c.

d. Domain of f: $[0, \infty)$; x-intercept of f: none; y-intercept of f: 3
domain of f^{-1}: $[3, \infty)$; x-intercept of f^{-1}: 3; y-intercept of f^{-1}: none

49. a. One-to-one

b. $g(x) = y = \sqrt[3]{x+1}.$ Interchange the variables and solve for y: $x = \sqrt[3]{y+1} \Rightarrow$
$x^3 = y+1 \Rightarrow y = g^{-1}(x) = x^3 - 1$

c.

c.

d. Domain of g: $(-\infty, \infty)$; x-intercept of g: -1;

y-intercept of g: 1

domain of g^{-1}: $(-\infty, \infty)$; x-intercept of g^{-1}: 1; y-intercept of g^{-1}: -1

d. Domain of f: $[-1, \infty)$; x-intercept of f: none; y-intercept of f: 3

Domain of f^{-1}: $[2, \infty)$;

x-intercept of f^{-1}: 3

y-intercept of f^{-1}: none

51. a. One-to-one

b. $f(x) = y = \dfrac{1}{x-1}$. Interchange the variables

and solve for y: $x = \dfrac{1}{y-1} \Rightarrow x(y-1) = 1 \Rightarrow$

$\dfrac{1}{x} = y - 1 \Rightarrow y = f^{-1}(x) = \dfrac{1}{x} + 1 = \dfrac{1+x}{x}$.

c.

d. Domain of f: $(-\infty, 1) \cup (1, \infty)$

x-intercept of f: none; y-intercept of f: -1

domain of f^{-1}: $(-\infty, 0) \cup (0, \infty)$

x-intercept of f^{-1}: -1

y-intercept of f^{-1}: none

53. a. One-to-one

b. $f(x) = y = 2 + \sqrt{x+1}$. Interchange the variables and solve for y: $x = 2 + \sqrt{y+1} \Rightarrow$

$x - 2 = \sqrt{y+1} \Rightarrow (x-2)^2 = y+1 \Rightarrow$

$y = f^{-1}(x) = (x-2)^2 - 1 = x^2 - 4x + 3$

55. The range of f is the same as the domain of f^{-1}.

Domain: $(-\infty, -2) \cup (-2, \infty)$

Range: $(-\infty, 1) \cup (1, \infty)$

57. $f(x) = y = \dfrac{x+1}{x-2}$. Interchange the variables

and solve for y: $x = \dfrac{y+1}{y-2} \Rightarrow xy - 2x = y + 1 \Rightarrow$

$xy - y = 2x + 1 \Rightarrow y(x-1) = 2x + 1 \Rightarrow$

$y = f^{-1}(x) = \dfrac{2x+1}{x-1}$.

Domain of f: $(-\infty, 2) \cup (2, \infty)$

Range of f: $(-\infty, 1) \cup (1, \infty)$.

59. $f(x) = y = \dfrac{1-2x}{1+x}$. Interchange the variables

and solve for y: $x = \dfrac{1-2y}{1+y} \Rightarrow$

$x + xy = 1 - 2y \Rightarrow xy + 2y = 1 - x \Rightarrow$

$y(x+2) = 1 - x \Rightarrow y = f^{-1}(x) = \dfrac{1-x}{x+2}$.

Domain of f: $(-\infty, -1) \cup (-1, \infty)$

Range of f: $(-\infty, -2) \cup (-2, \infty)$.

61. f is one-to-one since the domain is restricted, so an inverse exists.

$f(x) = y = -x^2, x \geq 0$. Interchange the variables and solve for y:

$x = -y^2 \Rightarrow y = \sqrt{-x}, x \leq 0$.

(continued on next page)

(*continued*)

$f^{-1}(x) = \sqrt{-x}$

63. f is one-to-one since the domain is restricted, so an inverse exists.

$f(x) = y = |x| = x$, $x \geq 0$. Interchange the variables and solve for y: $y = x$, $x \geq 0$.

65. f is one-to-one since the domain is restricted, so an inverse exists.

$f(x) = y = x^2 + 1$, $x \leq 0$. Interchange the variables and solve for y:

$x = y^2 + 1 \Rightarrow y = -\sqrt{x-1}$, $x \geq 1$.

67. f is one-to-one since the domain is restricted, so an inverse exists.

$f(x) = y = -x^2 + 2$, $x \leq 0$. Interchange the variables and solve for y:

$x = -y^2 + 2 \Rightarrow y = -\sqrt{2-x}$, $x \leq 2$.

$f(x) = -x^2 + 2$
$f^{-1}(x) = -\sqrt{2-x}$

2.9 Applying the Concepts

69. **a.** $K(C) = C + 273 \Rightarrow$
$C(K) = K - 273 = K^{-1}(C)$.
This represents the Celsius temperature corresponding to a given Kelvin temperature.

b. $C(300) = 300 - 273 = 27°C$

c. $K(22) = 22 + 273 = 295°K$

71. **a.** $F(K(C)) = \dfrac{9}{5}(C + 273) - \dfrac{2297}{5}$

$= \dfrac{9}{5}C + \dfrac{9(273)}{5} - \dfrac{2297}{5}$

$= \dfrac{9}{5}C + \dfrac{160}{5} = \dfrac{9}{5}C + 32$

b. $C(K(F)) = \dfrac{5}{9}F + \dfrac{2297}{9} - 273$

$= \dfrac{5}{9}F + \dfrac{2297 - 2457}{9}$

$= \dfrac{5}{9}F - \dfrac{160}{9}$

73. **a.** $E(x) = 0.75x$, where x represents the number of dollars
$D(x) = 1.25x$, where x represents the number of euros.

b. $E(D(x)) = 0.75(1.25x) = 0.9375x \neq x$.
Therefore, the two functions are not inverses.

c. She loses money either way.

75. **a.** $7 = 4 + 0.05x \Rightarrow x = \$60$. This means that if food sales $\leq \$60$, he will receive the minimum hourly wage. If food sales $> \$60$, his wages will be based on food sales.

$w = \begin{cases} 4 + 0.05x & \text{if } x > 60 \\ 7 & \text{if } x \leq 60 \end{cases}$

b. The function does not have an inverse because it is constant on (0, 60), and it is not one-to-one.

c. If the domain is restricted to [60, ∞), the function has an inverse.

77. a. $V = 8\sqrt{x} \Rightarrow \dfrac{V}{8} = \sqrt{x} \Rightarrow \dfrac{1}{64}V^2 = x = V^{-1}(x)$

This represents the height of the water in terms of the velocity.

b. **(i)** $x = \dfrac{1}{64}\left(30^2\right) = 14.0625$ ft

(ii) $x = \dfrac{1}{64}\left(20^2\right) = 6.25$ ft

79. a. The function represents the amount she still owes after x months.

b. $y = 36,000 - 600x$. Interchange the variables and solve for y: $x = 36,000 - 600y \Rightarrow$

$600y = 36,000 - x \Rightarrow y = 60 - \dfrac{x}{600} \Rightarrow$

$f^{-1}(x) = 60 - \dfrac{1}{600}x$.

This represents the number of months that have passed from the first payment until the balance due is \$$x$.

c. $y = 60 - \dfrac{1}{600}(22,000) = 23.33 \approx 24$ months

There are 24 months remaining.

2.9 Beyond the Basics

81. $f(g(3)) = f(1) = 3, f(g(5)) = f(3) = 5,$ and $f(g(2)) = f(4) = 2 \Rightarrow f(g(x)) = x$ for each x. $g(f(1)) = g(3) = 1, g(f(3)) = g(5) = 3,$ and $g(f(4)) = g(2) = 4 \Rightarrow g(f(x)) = x$ for each x. So, f and g are inverses.

83. a.

b. f is not one-to-one

c. Domain: [–2, 2]; range: [0, 2]

85. a. f satisfies the horizontal line test.

b. $y = 1 - \dfrac{1}{x+1}$. Interchange the variables

and solve for y: $x = 1 - \dfrac{1}{y+1} \Rightarrow$

$\dfrac{1}{y+1} = 1 - x \Rightarrow 1 = y + 1 - xy - x \Rightarrow$

$xy - y = -x \Rightarrow y(x-1) = -x \Rightarrow$

$y = f^{-1}(x) = -\dfrac{x}{x-1} = \dfrac{x}{1-x}$

c. Domain of f: $(-\infty, -1) \cup (-1, \infty)$; range of f: $(-\infty, 1) \cup (1, \infty)$.

87. a. **(i)** $f(x) = y = 2x - 1$. Interchange the variables and solve for y: $x = 2y - 1 \Rightarrow$

$y = f^{-1}(x) = \dfrac{1}{2}x + \dfrac{1}{2}$

(ii) $g(x) = y = 3x + 4$. Interchange the variables and solve for y: $x = 3y + 4 \Rightarrow$

$y = g^{-1}(x) = \dfrac{1}{3}x - \dfrac{4}{3}$

(iii) $(f \circ g)(x) = 2(3x + 4) - 1 = 6x + 7$

(iv) $(g \circ f)(x) = 3(2x - 1) + 4 = 6x + 1$

(v) $(f \circ g)(x) = y = 6x + 7$. Interchange the variables and solve for y:

$x = 6y + 7 \Rightarrow (f \circ g)^{-1}(x) = \dfrac{1}{6}x - \dfrac{7}{6}$

(vi) $(g \circ f)(x) = y = 6x + 1$. Interchange the variables and solve for y:

$x = 6y + 1 \Rightarrow (f \circ g)^{-1}(x) = \dfrac{1}{6}x - \dfrac{1}{6}$

(vii) $\left(f^{-1} \circ g^{-1}\right)(x) = \dfrac{1}{2}\left(\dfrac{1}{3}x - \dfrac{4}{3}\right) + \dfrac{1}{2}$

$= \dfrac{1}{6}x - \dfrac{2}{3} + \dfrac{1}{2} = \dfrac{1}{6}x - \dfrac{1}{6}$

(viii) $\left(g^{-1} \circ f^{-1}\right)(x) = \frac{1}{3}\left(\frac{1}{2}x + \frac{1}{2}\right) - \frac{4}{3}$

$$= \frac{1}{6}x + \frac{1}{6} - \frac{4}{3} = \frac{1}{6}x - \frac{7}{6}$$

b. (i) $(f \circ g)^{-1}(x) = \frac{1}{6}x - \frac{7}{6}$
$$= \left(g^{-1} \circ f^{-1}\right)(x)$$

(ii) $(g \circ f)^{-1}(x) = \frac{1}{6}x - \frac{1}{6}$
$$= \left(f^{-1} \circ g^{-1}\right)(x)$$

2.9 Critical Thinking/Discussion/Writing

89. No. For example, $f(x) = x^3 - x$ is odd, but it does not have an inverse, because $f(0) = f(1)$, so it is not one-to-one.

90. Yes. The function $f = \{(0,1)\}$ is even, and it has an inverse: $f^{-1} = \{(1,0)\}$.

91. Yes, because increasing and decreasing functions are one-to-one.

92. a. $R = \{(-1,1), (0,0), (1,1)\}$

   b. $R = \{(-1,1), (0,0), (1,2)\}$

2.9 Getting Ready for the Next Section

93. $x^2 - x - 12 = (x+3)(x-4)$

95. $x^2 + 2x - 8 = (x+4)(x-2)$

97. $x^2 - 7x + 12 = 0$
$(x-3)(x-4) = 0$
$x - 3 = 0 \mid x - 4 = 0$
$\quad x = 3 \mid \quad x = 4$
Solution: $\{3, 4\}$

99. $3x^2 + 7x + 2 = 0$
$(3x+1)(x+2) = 0$
$3x + 1 = 0 \mid x + 2 = 0$
$\quad x = -\frac{1}{3} \mid \quad x = -2$
Solution: $\left\{-2, -\frac{1}{3}\right\}$

101. $y = (x+2)^2 - 3$

Start with the graph of $f(x) = x^2$, then shift it two units left and three units down.

103. $y = -(x+1)^2 + 2$

Start with the graph of $f(x) = x^2$, then shift it one unit left. Reflect the graph across the *x*-axis and then shift it two units up

Chapter 2 Review Exercises

Building Skills

1. False. The midpoint is
$$\left(\frac{-3+3}{2}, \frac{1+11}{2}\right) = (0,6).$$

3. True

5. False.
The slope is 4/3 and the *y*-intercept is 3.

7. True

9. a. $d(P,Q) = \sqrt{(-1-3)^2 + (3-5)^2} = 2\sqrt{5}$

   b. $M = \left(\frac{3+(-1)}{2}, \frac{5+3}{2}\right) = (1,4)$

   c. $m = \frac{3-5}{-1-3} = \frac{1}{2}$

11. a. $d(P,Q) = \sqrt{(9-4)^2 + (-8-(-3))^2} = 5\sqrt{2}$

   b. $M = \left(\frac{4+9}{2}, \frac{-3+(-8)}{2}\right) = \left(\frac{13}{2}, -\frac{11}{2}\right)$

   c. $m = \frac{-8-(-3)}{9-4} = -1$

13. a. $D(P,Q) = \sqrt{(5-2)^2 + (-2-(-7))^2} = \sqrt{34}$

   b. $M = \left(\frac{2+5}{2}, \frac{-7+(-2)}{2}\right) = \left(\frac{7}{2}, -\frac{9}{2}\right)$

c. $m = \dfrac{-2-(-7)}{5-2} = \dfrac{5}{3}$

15. $d(A,B) = \sqrt{(-2-0)^2 + (-3-5)^2} = \sqrt{68}$

$d(A,C) = \sqrt{(3-0)^2 + (0-5)^2} = \sqrt{34}$

$d(B,C) = \sqrt{(3-(-2))^2 + (0-(-3))^2} = \sqrt{34}$

Using the Pythagorean theorem, we have

$AC^2 + BC^2 = (\sqrt{34})^2 + (\sqrt{34})^2$

$\qquad = 68 = (\sqrt{68})^2 = AB^2$

Alternatively, we can show that AC and CB are perpendicular using their slopes.

$m_{AC} = \dfrac{0-5}{3-0} = -\dfrac{5}{3}; m_{CB} = \dfrac{0-(-3)}{3-(-2)} = \dfrac{3}{5}$

$m_{AC} \cdot m_{CB} = -1 \Rightarrow AC \perp CB$, so $\triangle ABC$ is a right triangle.

17. $A = (-6,3), B = (4,5)$

$d(A,O) = \sqrt{(-6-0)^2 + (3-0)^2} = \sqrt{45}$

$d(B,O) = \sqrt{(4-0)^2 + (5-0)^2} = \sqrt{41}$

(4, 5) is closer to the origin.

19. $A = (-5,3), B = (4,7), C = (x,0)$

$d(A,C) = \sqrt{(x-(-5))^2 + (0-3)^2}$

$\qquad = \sqrt{(x+5)^2 + 9}$

$d(B,C) = \sqrt{(x-4)^2 + (0-7)^2}$

$\qquad = \sqrt{(x-4)^2 + 49}$

$d(A,C) = d(B,C) \Rightarrow$

$\sqrt{(x+5)^2 + 9} = \sqrt{(x-4)^2 + 49}$

$(x+5)^2 + 9 = (x-4)^2 + 49$

$x^2 + 10x + 34 = x^2 - 8x + 65$

$x = \dfrac{31}{18} \Rightarrow$ The point is $\left(\dfrac{31}{18}, 0\right)$.

21. Not symmetric with respect to the x-axis; symmetric with respect to the y-axis; not symmetric with respect to the origin.

23. Symmetric with respect to the x-axis; not symmetric with respect to the y-axis; not symmetric with respect to the origin.

25. x-intercept: 4; y-intercept: 2; not symmetric with respect to the x-axis; not symmetric with respect to the y-axis; not symmetric with respect to the origin.

27. x-intercept: 0; y-intercept: 0; not symmetric with respect to the x-axis; symmetric with respect to the y-axis; not symmetric with respect to the origin.

29. x-intercept: 0; y-intercept: 0; not symmetric with respect to the x-axis; not symmetric with respect to the y-axis; symmetric with respect to the origin.

31. No x-intercept; y-intercept: 2; not symmetric with respect to the x-axis; symmetric with respect to the y-axis; not symmetric with respect to the origin.

33. *x*-intercepts: –4, 4; *y*-intercepts: –4, 4;
    symmetric with respect to the *x*-axis;
    symmetric with respect to the *y*-axis;
    symmetric with respect to the origin.

35. $(x-2)^2 + (y+3)^2 = 25$

37. The radius is 2, so the equation of the circle is
    $(x+2)^2 + (y+5)^2 = 4$.

39. $\dfrac{x}{2} - \dfrac{y}{5} = 1 \Rightarrow 5x - 2y = 10 \Rightarrow \dfrac{5}{2}x - 5 = y$. Line
    with slope 5/2 and *y*-intercept –5.

41. $x^2 + y^2 - 2x + 4y - 4 = 0 \Rightarrow$
    $x^2 - 2x + 1 + y^2 + 4y + 4 = 4 + 1 + 4 \Rightarrow$
    $(x-1)^2 + (y+2)^2 = 9$.
    Circle with center (1, –2) and radius 3.

43. $y - 2 = -2(x-1) \Rightarrow y = -2x + 4$

45. $m = \dfrac{7-3}{-1-1} = -2; 3 = -2(1) + b \Rightarrow 5 = b \Rightarrow$
    $y = -2x + 5$

47. **a.** $y = 3x - 2 \Rightarrow m = 3; y = 3x + 2 \Rightarrow m = 3$
    The slopes are equal, so the lines are
    parallel.

    b. $3x - 5y + 7 \Rightarrow m = 3/5;$
    $5x - 3y + 2 = 0 \Rightarrow m = 5/3$
    The slopes are neither equal nor negative
    reciprocals, so the lines are neither parallel
    nor perpendicular.

    c. $ax + by + c = 0 \Rightarrow m = -a/b;$
    $bx - ay + d = 0 \Rightarrow m = b/a$
    The slopes are negative reciprocals, so the
    lines are perpendicular.

    d. $y + 2 = \dfrac{1}{3}(x-3) \Rightarrow m = \dfrac{1}{3};$
    $y - 5 = 3(x-3) \Rightarrow m = 3$
    The slopes are neither equal nor negative
    reciprocals, so the lines are neither parallel
    nor perpendicular.

49. Domain: {–1, 0, 1, 2}; range: {–1, 0, 1, 2}.
    This is a function.

51. Domain: $(-\infty, \infty)$; range: $(-\infty, \infty)$.
    This is a function.

53. Domain: $[-0.2, 0.2]$; range: $[-0.2, 0.2]$.
This is not a function.

55. Domain: $\{1\}$; range: $(-\infty, \infty)$.
This is not a function.

57. Domain: $(-\infty, \infty)$; range: $[0, \infty)$.
This is a function.

59. $f(-2) = 3(-2) + 1 = -5$

61. $f(x) = 4 \Rightarrow 3x + 1 = 4 \Rightarrow x = 1$

63. $(f+g)(1) = f(1) + g(1)$
$= (3(1) + 1) + (1^2 - 2) = 3$

65. $(f \cdot g)(-2) = f(-2) \cdot g(-2)$
$= (3(-2) + 1) \cdot ((-2)^2 - 2) = -10$

67. $(f \circ g)(3) = 3(3^2 - 2) + 1 = 22$

69. $(f \circ g)(x) = 3(x^2 - 2) + 1 = 3x^2 - 5$

71. $(f \circ f)(x) = 3(3x + 1) + 1 = 9x + 4$

73. $f(a + h) = 3(a + h) + 1 = 3a + 3h + 1$

75. $\dfrac{f(x+h) - f(x)}{h} = \dfrac{(3(x+h)+1) - (3x+1)}{h}$
$= \dfrac{3x + 3h + 1 - 3x - 1}{h} = \dfrac{3h}{h} = 3$

77. Domain: $(-\infty, \infty)$; range: $\{-3\}$.
Constant on $(-\infty, \infty)$.

79. Domain: $\left[\dfrac{2}{3}, \infty\right)$; range: $[0, \infty)$
Increasing on $\left(\dfrac{2}{3}, \infty\right)$.

81. Domain: $(-\infty, \infty)$; range: $[1, \infty)$. Decreasing on $(-\infty, 0)$; increasing on $(0, \infty)$.

83. The graph of g is the graph of f shifted one unit left.

85. The graph of g is the graph of f shifted two units right, and then reflected in the x-axis.

87. $f(-x) = (-x)^2 - (-x)^4 = x^2 - x^4 = f(x) \Rightarrow$ $f(x)$ is even. Not symmetric with respect to the x-axis; symmetric with respect to the y-axis; not symmetric with respect to the origin.

89. $f(-x) = |-x| + 3 = |x| + 3 = f(x) \Rightarrow$ $f(x)$ is even. Not symmetric with respect to the x-axis; symmetric with respect to the y-axis; not symmetric with respect to the origin.

91. $f(-x) = \sqrt{-x} \neq f(x)$ or $f(-x) \Rightarrow f(x)$ is neither even nor odd. Not symmetric with respect to the x-axis; not symmetric with respect to the y-axis; not symmetric with respect to the origin.

93. $f(x) = \sqrt{x^2 - 4} \Rightarrow f(x) = (g \circ h)(x)$ where $g(x) = \sqrt{x}$ and $h(x) = x^2 - 4$.

95. $h(x) = \sqrt{\dfrac{x-3}{2x+5}} \Rightarrow h(x) = (f \circ g)(x)$ where $f(x) = \sqrt{x}$ and $g(x) = \dfrac{x-3}{2x+5}$.

97. $f(x)$ is one-to-one. $f(x) = y = x + 2$. Interchange the variables and solve for y: $x = y + 2 \Rightarrow y = x - 2 = f^{-1}(x)$.

99. $f(x)$ is one-to-one. $f(x) = y = \sqrt[3]{x-2}$. Interchange the variables and solve for y: $x = \sqrt[3]{y-2} \Rightarrow y = x^3 + 2 = f^{-1}(x)$.

101. $f(x) = y = \dfrac{x-1}{x+2}, x \neq -2$.

Interchange the variables and solve for y.

$x = \dfrac{y-1}{y+2} \Rightarrow xy + 2x = y - 1 \Rightarrow$

$xy - y = -2x - 1 \Rightarrow y(x-1) = -2x - 1 \Rightarrow$

$y = \dfrac{-2x-1}{x-1} \Rightarrow y = f^{-1}(x) = \dfrac{2x+1}{1-x}$

Domain of f: $(-\infty, -2) \cup (-2, \infty)$

Range of f: $(-\infty, 1) \cup (1, \infty)$

103. a. $A = (-3, -3), B = (-2, 0), C = (0, 1), D = (3, 4)$. Find the equation of each segment:

$m_{AB} = \dfrac{0 - (-3)}{-2 - (-3)} = 3.\ 0 = 3(-2) + b \Rightarrow b = 6.$

The equation of AB is $y = 3x + 6$.

$m_{BC} = \dfrac{1 - 0}{0 - (-2)} = \dfrac{1}{2}; b = 1.$

The equation of BC is $y = \dfrac{1}{2}x + 1$.

$m_{CD} = \dfrac{4-1}{3-0} = 1; b = 1.$

The equation of CD is $y = x + 1$.

So,

$$f(x) = \begin{cases} 3x + 6 & \text{if } -3 \leq x \leq -2 \\ \dfrac{1}{2}x + 1 & \text{if } -2 < x < 0 \\ x + 1 & \text{if } 0 \leq x \leq 3 \end{cases}$$

b. Domain: $[-3, 3]$; range: $[-3, 4]$

c. x-intercept: -2; y-intercept: 1

d.

e.

f.

g.

h.

i.

j.

k. *f* is one-to-one because it satisfies the horizontal line test.

l.

Applying the Concepts

105. a. rate of change (slope) $= \dfrac{173,000 - 54,000}{223,000 - 87,000}$
$$= 0.875$$
$54,000 = 0.875(87,000) + b \Rightarrow$
$b = -22,125.$
The equation is $C = 0.875w - 22,125.$

b. The slope represents the cost to dispose of one pound of waste. The *x*-intercept represents the amount of waste that can be disposed with no cost. The *y*-intercept represents the fixed cost.

c. $C = 0.875(609,000) - 22,125 = \$510,750$

d. $1,000,000 = 0.875w - 22,125 \Rightarrow$
$w = 1,168,142.86$ pounds

107. a. $f(2) = 100 + 55(2) - 3(2)^2 = \$198.$
She started with $100, so she won $98.

b. She was winning at a rate of $49/hour.

c. $0 = 100 + 55t - 3t^2 \Rightarrow (-t + 20)(3t + 5) \Rightarrow$
$t = 20, t = -5/3.$ Since t represents the amount of time, we reject $t = -5/3.$
Chloe will lose all her money after playing for 20 hours.

d. $100/20 = $5/hour.

109. a. $(L \circ x)(t) = 0.5\sqrt{\left(1 + 0.002t^2\right)^2 + 4}$
$= 0.5\sqrt{0.000004t^4 + 0.004t^2 + 5}$

b. $(L \circ x)(5) = 0.5\sqrt{\left(1 + 0.002(5^2)\right)^2 + 4}$
$= 0.5\sqrt{(1.05)^2 + 4} = 0.5\sqrt{5.1025}$
≈ 1.13

111. a.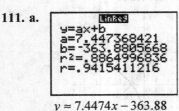
$y \approx 7.4474x - 363.88$

b.
[70, 90, 2] by [150, 300, 25]

c. $y \approx 7.4474(76) - 363.88 \approx 202$
A player whose height is 76 inches weighs about 202 pounds.

Chapter 2 Practice Test A

1. The endpoints of the diameter are $(-2, 3)$ and $(-4, 5)$, so the center of the circle is
$C = \left(\dfrac{-2 + (-4)}{2}, \dfrac{3 + 5}{2}\right) = (-3, 4).$
The length of the diameter is
$\sqrt{\left(-4 - (-2)\right)^2 + \left(5 - 3\right)^2} = \sqrt{8} = 2\sqrt{2}.$
Therefore, the length of the radius is $\sqrt{2}.$
The equation of the circle is
$(x + 3)^2 + (y - 4)^2 = 2.$

2. To test if the graph is symmetric with respect to the y-axis, replace x with $-x$:
$3(-x) + 2(-x)y^2 = 1 \Rightarrow -3x - 2xy^2 = 1,$ which is not the same as the original equation, so the graph is not symmetric with respect to the y-axis. To test if the graph is symmetric with respect to the x-axis, replace y with $-y$:
$3x + 2x(-y)^2 = 1 \Rightarrow 3x + 2xy^2 = 1,$ which is the same as the original equation, so the graph is symmetric with respect to the x-axis. To test if the graph is symmetric with respect to the origin, replace x with $-x$ and y with $-y$:
$3(-x) + 2(-x)(-y)^2 = 1 \Rightarrow -3x - 2xy^2 = 1,$
which is not the same as the original equation, so the graph is not symmetric with respect to the origin.

3. $0 = x^2(x - 3)(x + 1) \Rightarrow x = 0$ or $x = 3$ or $x = -1$
$y = 0^2(0 - 3)(0 + 1) \Rightarrow y = 0.$ The x-intercepts are 0, 3, and -1; the y-intercept is 0.

4.

Intercepts:
$y^2 - 2y = 2 \Rightarrow y = 1 \pm \sqrt{3}$
$x^2 - 2x = 2 \Rightarrow x = 1 \pm \sqrt{3}$

5. $7 = -1(2) + b \Rightarrow 9 = b$
The equation is $y = -x + 9.$

6. $8x - 2y = 7 \Rightarrow y = 4x - \dfrac{7}{2} \Rightarrow$ the slope of the line is 4. $-1 = 4(2) + b \Rightarrow b = -9.$ So the equation is $y = 4x - 9.$

7. $(fg)(2) = f(2) \cdot g(2)$
$= (-2(2) + 1)\left(2^2 + 3(2) + 2\right)$
$= (-3)(12) = -36$

8. $g(f(2)) = g(2(2) - 3) = g(1) = 1 - 2(1)^2 = -1$

9. $(f \circ f)(x) = (x^2 - 2x)^2 - 2(x^2 - 2x)$
$= x^4 - 4x^3 + 4x^2 - 2x^2 + 4x$
$= x^4 - 4x^3 + 2x^2 + 4x$

10. a. $f(-1) = (-1)^3 - 2 = -3$

b. $f(0) = 0^3 - 2 = -2$

c. $f(1) = 1 - 2(1)^2 = -1$

11. $1 - x > 0 \Rightarrow x < 1$; x must also be greater than or equal to 0, so the domain is [0, 1).

12. $\dfrac{f(4) - f(1)}{4 - 1} = \dfrac{(2(4) + 7) - (2(1) + 7)}{3} = 2$

13. $f(-x) = 2(-x)^4 - \dfrac{3}{(-x)^2} = 2x^4 - \dfrac{3}{x^2} = f(x) \Rightarrow$

$f(x)$ is even.

14. Increasing on $(-\infty, 0)$ and $(2, \infty)$; decreasing on $(0, 2)$.

15. Shift the graph of $y = \sqrt{x}$ three units to the right, then stretch the graph vertically by a factor of 2, and then shift the resulting graph four units up.

16. $25 = 25 - (2t - 5)^2 \Rightarrow 0 = -(2t - 5)^2 \Rightarrow$
$0 = 2t - 5 \Rightarrow t = 5/2 = 2.5$ seconds

17. $f(2) = 7 \Rightarrow f^{-1}(7) = 2$

18. $f(x) = y = \dfrac{2x}{x - 1}$. Interchange the variables

and solve for y: $x = \dfrac{2y}{y - 1} \Rightarrow$

$xy - x = 2y \Rightarrow xy - 2y = x \Rightarrow$

$y(x - 2) = x \Rightarrow y = f^{-1}(x) = \dfrac{x}{x - 2}$

19. $A(x) = 100x + 1000$

20. a. $C(230) = 0.25(230) + 30 = \$87.50$

b. $57.50 = 0.25m + 30 \Rightarrow m = 110$ miles

Chapter 2 Practice Test B

1. To test if the graph is symmetric with respect to the y-axis, replace x with $-x$:
$|-x| + 2|y| = 2 \Rightarrow |x| + 2|y| = 2$, which is the same as the original equation, so the graph is symmetric with respect to the y-axis. To test if the graph is symmetric with respect to the x-axis, replace y with $-y$:

$|x| + 2|-y| = 2 \Rightarrow |x| + 2|y| = 2$, which is the same as the original equation, so the graph is symmetric with respect to the x-axis. To test if the graph is symmetric with respect to the origin, replace x with $-x$, and y with $-y$:
$|-x| + 2|-y| = 2 \Rightarrow |x| + 2|y| = 2$, which is the same as the original equation, so the graph is symmetric with respect to the origin. The answer is D.

2. $0 = x^2 - 9 \Rightarrow x = \pm 3; y = 0^2 - 9 \Rightarrow y = -9$.
The x-intercepts are ± 3; the y-intercept is -9.
The answer is B.

3. D   **4.** D   **5.** C

6. Suppose the coordinates of the second point are (a, b). Then $-\dfrac{1}{2} = \dfrac{b - 2}{a - 3}$. Substitute each of the points given into this equation to see which makes it true. The answer is C.

7. Find the slope of the original line:
$6x - 3y = 5 \Rightarrow y = 2x - \dfrac{5}{3}$. The slope is 2.
The equation of the line with slope 2, passing through $(-1, 2)$ is $y - 2 = 2(x + 1)$.
The answer is D.

8. $(f \circ g)(x) = 3(2 - x^2) - 5 = 1 - 3x^2$.
The answer is B.

9. $(f \circ f)(x) = 2(2x^2 - x)^2 - (2x^2 - x)$
$= 8x^4 - 8x^3 + x$
The answer is A.

10. $g(a - 1) = \dfrac{1 - (a - 1)}{1 + (a - 1)} = \dfrac{2 - a}{a}$.
The answer is C.

11. $1 - x \geq 0 \Rightarrow x \leq 1$; x must also be greater than or equal to 0, so the domain is [0, 1].
The answer is A.

12. $x^2 + 3x - 4 = 6 \Rightarrow x^2 + 3x - 10 = 0 \Rightarrow$
$(x + 5)(x - 2) = 0 \Rightarrow x = -5, 2$
The answer is D.

13. A   **14.** A   **15.** B

16. D   **17.** C

18. $f(x) = y = \dfrac{-x}{3x+2}$. Interchange the variables

and solve for y: $x = \dfrac{-y}{3y+2} \Rightarrow$

$3xy + 2x = -y \Rightarrow 3xy - y = -2x \Rightarrow$

$y(3x - 1) = -2x \Rightarrow y = f^{-1}(x) = -\dfrac{2x}{3x-1} \Rightarrow$

$f^{-1}(x) = \dfrac{2x}{1-3x}$

The answer is C.

19. $w = 5x - 190; w = 5(70) - 190 = 160.$
The answer is B.

20. $50 = 0.2m + 25 \Rightarrow m = 125.$ The answer is A.

Cumulative Review Exercises (Chapters P–2)

1. a. $\left(\dfrac{x^3}{y^2}\right)^2 \left(\dfrac{y^2}{x^3}\right)^3 = \left(\dfrac{x^6}{y^4}\right)\left(\dfrac{y^6}{x^9}\right) = \dfrac{y^2}{x^3}$

b. $\dfrac{x^{-1}y^{-1}}{x^{-1}+y^{-1}} = \dfrac{\dfrac{1}{x} \cdot \dfrac{1}{y}}{\dfrac{1}{x} + \dfrac{1}{y}} = \dfrac{\dfrac{1}{xy}}{\dfrac{y+x}{xy}} = \dfrac{1}{x+y}$

2. a. $2x^2 + x - 15 = (2x-5)(x+3)$

b. $x^3 - 2x^2 + 4x - 8 = x^2(x-2) + 4(x-2)$
$= (x^2+4)(x-2)$

3. a. $\sqrt{75} + \sqrt{108} - \sqrt{192} = 5\sqrt{3} + 6\sqrt{3} - 8\sqrt{3}$
$= 3\sqrt{3}$

b. $\dfrac{x-1}{x+1} - \dfrac{x-2}{x+2} = \dfrac{(x-1)(x+2) - (x-2)(x+1)}{(x+1)(x+2)}$
$= \dfrac{(x^2+x-2) - (x^2-x-2)}{(x+1)(x+2)}$
$= \dfrac{2x}{(x+1)(x+2)}$

4. a. $\dfrac{1}{2+\sqrt{3}} = \dfrac{1}{2+\sqrt{3}} \cdot \dfrac{2-\sqrt{3}}{2-\sqrt{3}} = \dfrac{2-\sqrt{3}}{4-3} = 2-\sqrt{3}$

b. $\dfrac{1}{\sqrt{5}-2} = \dfrac{1}{\sqrt{5}-2} \cdot \dfrac{\sqrt{5}+2}{\sqrt{5}+2} = \dfrac{\sqrt{5}+2}{5-4} = \sqrt{5}+2$

5. a. $3x - 7 = 5 \Rightarrow 3x = 12 \Rightarrow x = 4$

b. $\dfrac{1}{x-1} = \dfrac{3}{x-1} \Rightarrow$ There is no solution.

6. a. $x^2 - 3x = 0 \Rightarrow x(x-3) = 0 \Rightarrow$
$x = 0$ or $x = 3$

b. $x^2 + 3x - 10 = 0 \Rightarrow (x+5)(x-2) = 0 \Rightarrow$
$x = -5$ or $x = 2$

7. a. $2x^2 - x + 3 = 0 \Rightarrow x = \dfrac{1 \pm \sqrt{1 - 4(2)(3)}}{2(2)} \Rightarrow$
$x = \dfrac{1 \pm \sqrt{-23}}{4} \Rightarrow x = \dfrac{1 \pm i\sqrt{23}}{4}$

b. $4x^2 - 12x + 9 = 0 \Rightarrow (2x-3)^2 = 0 \Rightarrow x = \dfrac{3}{2}$

8. a. $x - 6\sqrt{x} + 8 = 0 \Rightarrow \left(\sqrt{x} - 4\right)\left(\sqrt{x} - 2\right) = 0 \Rightarrow$
$\sqrt{x} = 4 \Rightarrow x = 16$ or $\sqrt{x} = 2 \Rightarrow x = 4$

b. $\left(x - \dfrac{1}{x}\right)^2 - 10\left(x - \dfrac{1}{x}\right) + 21 = 0.$

Let $u = x - \dfrac{1}{x}$.

$u^2 - 10u + 21 = 0 \Rightarrow$
$(u-7)(u-3) = 0 \Rightarrow u = 7$ or $u = 3$;

$x - \dfrac{1}{x} = 7 \Rightarrow x^2 - 1 = 7x \Rightarrow$

$x^2 - 7x - 1 = 0 \Rightarrow x = \dfrac{7 \pm \sqrt{7^2 - (4)(-1)}}{2} \Rightarrow$

$x = \dfrac{7 \pm \sqrt{53}}{2}; x - \dfrac{1}{x} = 3 \Rightarrow x^2 - 1 = 3x \Rightarrow$

$x^2 - 3x - 1 = 0 \Rightarrow x = \dfrac{3 \pm \sqrt{3^2 - 4(-1)}}{2} \Rightarrow$

$x = \dfrac{3 \pm \sqrt{13}}{2}$

The solution set is
$\left\{ \dfrac{7-\sqrt{53}}{2}, \dfrac{7+\sqrt{53}}{2}, \dfrac{3-\sqrt{13}}{2}, \dfrac{3+\sqrt{13}}{2} \right\}.$

9. a. $\sqrt{3x-1}=2x-1\Rightarrow 3x-1=(2x-1)^2\Rightarrow$

$3x-1=4x^2-4x+1\Rightarrow 4x^2-7x+2=0\Rightarrow$

$x=\dfrac{7\pm\sqrt{(-7)^2-4(4)(2)}}{2(4)}=\dfrac{7\pm\sqrt{17}}{8}$. If

$x=\dfrac{7-\sqrt{17}}{8},\ \sqrt{3\left(\dfrac{7-\sqrt{17}}{8}\right)-1}\approx 0.281$

while $2\left(\dfrac{7-\sqrt{17}}{8}\right)-1\approx -0.281$, so the

solution set is $\left\{\dfrac{7+\sqrt{17}}{8}\right\}$.

b.
$$\sqrt{1-x}=2-\sqrt{2x+1}$$
$$\left(\sqrt{1-x}\right)^2=\left(2-\sqrt{2x+1}\right)^2$$
$$1-x=4-4\sqrt{2x+1}+2x+1$$
$$-4-3x=-4\sqrt{2x+1}$$
$$(-4-3x)^2=\left(-4\sqrt{2x+1}\right)^2$$
$$16+24x+9x^2=16(2x+1)$$
$$16+24x+9x^2=32x+16$$
$$9x^2-8x=0\Rightarrow x(9x-8)=0$$
$$x=0 \text{ or } x=\frac{8}{9}.$$

Check to make sure that neither solution is extraneous. The solution set is $\{0,8/9\}$.

10. a. $2x-5<11\Rightarrow x<8\Rightarrow(-\infty,8)$

b. $-3x+4>-5\Rightarrow x<3\Rightarrow(-\infty,3)$

11. a. $-3<2x-3<5\Rightarrow 0<2x<8\Rightarrow 0<x<4$.
The solution set is $(0,4)$.

b. $5\le 1-2x\le 7\Rightarrow 4\le -2x\le 6\Rightarrow -2\ge x\ge -3$.
The solution set is $[-3,-2]$.

12. a. $|2x-1|\le 7\Rightarrow -7\le 2x-1\le 7\Rightarrow$
$-6\le 2x\le 8\Rightarrow -3\le x\le 4$
The solution set is $[-3,4]$.

b. $|2x-3|\ge 5\Rightarrow 2x-3\ge 5\Rightarrow x\ge 4$ or
$2x-3\le -5\Rightarrow x\le -1$.
The solution set is $(-\infty,-1]\cup[4,\infty)$.

13. $d(A,C)=\sqrt{(2-5)^2+(2-(-2))^2}=5$

$d(B,C)=\sqrt{(2-6)^2+(2-5)^2}=5$

Since the lengths of the two sides are equal, the triangle is isosceles.

14.

15. First, find the equation of the circle with center $(2,-1)$ and radius determined by $(2,-1)$ and $(-3,-1)$:
$r=\sqrt{2-(-3))^2+(-1-(-1))^2}=5$.

The equation is $(x-2)^2+(y+1)^2=5^2$. Now check to see if the other three points satisfy the equation:
$(2-2)^2+(4+1)^2=5^2\Rightarrow 5^2=5^2$,
$(5-2)^2+(3+1)^2=5^2\Rightarrow 3^2+4^2=5^2$ (true because 3, 4, 5 is a Pythagorean triple), and
$(6-2)^2+(2+1)^2=5^2\Rightarrow 4^2+3^2=5^2$.
Since all the points satisfy the equation, they lie on the circle.

16. $x^2+y^2-6x+4y+9=0\Rightarrow$
$x^2-6x+y^2+4y=-9$.
Now complete both squares:
$x^2-6x+9+y^2+4y+4=-9+9+4\Rightarrow$
$(x-3)^2+(y+2)^2=4$.
The center is $(3,-2)$ and the radius is 2.

17. $y=-3x+5$

18. The x-intercept is 4, so $(4,0)$ satisfies the equation. To write the equation in slope-intercept form, find the y-intercept:
$0=2(4)+b\Rightarrow -8=b$
The equation is $y=2x-8$.

19. The slope of the perpendicular line is the negative reciprocal of the slope of the original line. The slope of the original line is 2, so the slope of the perpendicular is $-1/2$. Now find the y-intercept of the perpendicular:
$-1=-\dfrac{1}{2}(2)+b\Rightarrow b=0$. The equation of the

perpendicular is $y=-\dfrac{1}{2}x$.

20. The slope of the parallel line is the same as the slope of the original line, 2. Now find the y-intercept of the parallel line: $-1 = 2(2) + b \Rightarrow$ $b = -5$. The equation of the parallel line is $y = 2x - 5$.

21. The slope of the perpendicular line is the negative reciprocal of the slope of the original line. The slope of the original line is $\dfrac{7-(-1)}{5-3} = 4$, so the slope of the perpendicular is $-1/4$. The perpendicular bisector passes through the midpoint of the original segment. The midpoint is $\left(\dfrac{3+5}{2}, \dfrac{-1+7}{2}\right) = (4,3)$. Use this point and the slope to find the y-intercept: $3 = -\dfrac{1}{4}(4) + b \Rightarrow b = 4$. The equation of the perpendicular bisector is $y = -\dfrac{1}{4}x + 4$.

22. The slope is undefined because the line is vertical. Because it passes through $(5, 7)$, the equation of the line is $x = 5$.

23. Use the slope formula to solve for x:
$2 = \dfrac{5-11}{x-5} \Rightarrow 2(x-5) = -6 \Rightarrow 2x - 10 = -6 \Rightarrow$
$x = 2$

24. The line through $(x, 3)$ and $(3, 7)$ has slope -2 because it is perpendicular to a line with slope $1/2$. Use the slope formula to solve for x:
$-2 = \dfrac{3-7}{x-3} \Rightarrow -2(x-3) = -4 \Rightarrow x - 3 = 2 \Rightarrow$
$x = 5$

25.

26.

27.

28.

29. Let x = the number of books initially purchased, and $\dfrac{1650}{x}$ = the cost of each book. Then $x - 16$ = the number of books sold, and $\dfrac{1650}{x-16}$ = the selling price of each book. The profit = the selling price – the cost, so
$\dfrac{1650}{x-16} - \dfrac{1650}{x} = 10 \Rightarrow$
$1650x - 1650(x-16) = 10x(x-16) \Rightarrow$
$1650x - 1650x + 26,400 = 10x^2 - 160x \Rightarrow$
$10x^2 - 160x - 26,400 = 0 \Rightarrow$
$x^2 - 16x - 2640 = 0 \Rightarrow (x-60)(x+44) = 0 \Rightarrow$
$x = 60$, $x = -44$. Reject -44 because there cannot be a negative number of books. So she bought 60 books.

30. Let x = the monthly note on the 1.5 year lease, and $1.5(12)x = 18x$ = the total expense for the 1.5 year lease. Then $x - 250$ = the monthly note on the 2 year lease, and $2(12)(x - 250) = 24x - 6000$ the total expenses for the 2 year lease. Then
$18x + 24x - 6000 = 21,000 \Rightarrow$
$42x = 27,000 \Rightarrow x = 642.86.$ So the monthly note for the 1.5 year lease is \$642.86, and the monthly note for the 2 year lease is
$\$642.86 - 250 = \$392.86.$

31. a. The domain of f is the set of all values of x which make $x + 1 \geq 0$ (because the square root of a negative number is not a real value.) So $x \geq -1$ or $[-1, \infty)$ in interval notation is the domain.

  b. $y = \sqrt{0+1} - 3 \Rightarrow y = -2; 0 = \sqrt{x+1} - 3 \Rightarrow$
  $3 = \sqrt{x+1} \Rightarrow 9 = x + 1 \Rightarrow 8 = x.$ The x-intercept is 8, and the y-intercept is –2.

  c. $f(-1) = \sqrt{-1+1} - 3 = -3$

  d. $f(x) > 0 \Rightarrow \sqrt{x+1} - 3 > 0 \Rightarrow \sqrt{x+1} > 3 \Rightarrow$
  $x + 1 > 9 \Rightarrow x > 8.$ In interval notation, this is $(8, \infty)$.

32. a. $f(-2) = -(-2) = 2; f(0) = 0^2 = 0;$
  $f(2) = 2^2 = 4$

  b. f decreases on $(-\infty, 0)$ and increases on $(0, \infty)$.

33. a. $(f \circ g)(x) = \dfrac{1}{\dfrac{2}{x} - 2} = \dfrac{1}{\dfrac{2 - 2x}{x}} = \dfrac{x}{2 - 2x}.$

  Because 0 is not in the domain of g, it must be excluded from the domain of $(f \circ g)$.

  Because 2 is not in the domain of f, any values of x for which $g(x) = 2$ must also be excluded from the domain of
  $(f \circ g) : \dfrac{2}{x} = 2 \Rightarrow x = 1,$ so 1 is excluded also. The domain of $(f \circ g)$ is
  $(-\infty, 0) \cup (0,1) \cup (1, \infty).$

  b. $(g \circ f)(x) = \dfrac{2}{\dfrac{1}{x - 2}} = 2(x - 2) = 2x - 4.$

  Because 2 is not in the domain of f, it must be excluded from the domain of $(g \circ f)$.

  Because 0 is not in the domain of g, any values of x for which $f(x) = 0$ must also be excluded from the domain of $(g \circ f)$.

  However, there is no value for x which makes $f(x) = 0$. So the domain of $(g \circ f)$ is
  $(-\infty, 2) \cup (2, \infty).$

Chapter 3 Polynomial and Rational Functions

3.1 Quadratic Functions

3.1 Practice Problems

1. Substitute 1 for h, -5 for k, 3 for x, and 7 for y in the standard form for a quadratic equation to solve for a:

   $7 = a(3-1)^2 - 5 \Rightarrow 7 = 4a - 5 \Rightarrow a = 3$

   The equation is $y = 3(x-1)^2 - 5$.

   Since $a = 3 > 0$, f has a minimum value of -5 at $x = 1$.

2. The graph of $f(x) = -2(x+1)^2 + 3$ is a parabola with $a = -2$, $h = -1$ and $k = 3$. Thus, the vertex is $(-1, 3)$, and the maximum value of the function is 3. The parabola opens down because $a < 0$. Now, find the x-intercepts:

   $0 = -2(x+1)^2 + 3 \Rightarrow 2(x+1)^2 = 3 \Rightarrow$

   $(x+1)^2 = \dfrac{3}{2} \Rightarrow x + 1 = \pm\sqrt{\dfrac{3}{2}} \Rightarrow x = \pm\sqrt{\dfrac{3}{2}} - 1 \Rightarrow$

   $x \approx 0.22$ or $x \approx -2.22$. Next, find the y-intercept: $f(0) = -2(0+1)^2 + 3 = 1$

   Plot the vertex, the x-intercepts, and the y-intercept, and join them with a parabola.

3. The graph of $f(x) = 3x^2 - 3x - 6$ is a parabola with $a = 3$, $b = -3$ and $c = -6$. The parabola opens up because $a > 0$. Now, find the vertex:

   $h = -\dfrac{b}{2a} = -\dfrac{-3}{2(3)} = \dfrac{1}{2}$

   $k = f(h) = f\left(\dfrac{1}{2}\right) = 3\left(\dfrac{1}{2}\right)^2 - 3\left(\dfrac{1}{2}\right) - 6 = -\dfrac{27}{4}$

   Thus, the vertex (h, k) is $\left(\dfrac{1}{2}, -\dfrac{27}{4}\right)$, and the

   minimum value of the function is $-\dfrac{27}{4}$.

Next, find the x-intercepts:

$3x^2 - 3x - 6 = 0 \Rightarrow 3\left(x^2 - x - 2\right) = 0 \Rightarrow$

$(x-2)(x+1) = 0 \Rightarrow x = 2$ or $x = -1$

Now, find the y-intercept:

$f(0) = 3(0)^2 - 3(0) - 6 = -6$.

Thus, the intercepts are $(-1, 0)$, $(2, 0)$ and $(0, -6)$. Use the fact that the parabola is

symmetric with respect to its axis, $x = \dfrac{1}{2}$, to

locate additional points. Plot the vertex, the x-intercepts, the y-intercept, and any additional points, and join them with a parabola.

4. The graph of $f(x) = 3x^2 - 6x - 1$ is a parabola with $a = 3$, $b = -6$ and $c = -1$. The parabola opens up because $a > 0$. Complete the square to write the equation in standard form:

   $f(x) = 3x^2 - 6x - 1 = 3\left(x^2 - 2x\right) - 1$

   $= 3\left(x^2 - 2x + 1\right) - 1 - 3 = 3(x-1)^2 - 4$

   Thus, the vertex is $(1, -4)$. The domain of f is $(-\infty, \infty)$ and the range is $[-4, \infty)$.

   Next, find the x-intercepts:

   $0 = 3(x-1)^2 - 4 \Rightarrow \dfrac{4}{3} = (x-1)^2 \Rightarrow$

   $\pm\dfrac{2\sqrt{3}}{3} = x - 1 \Rightarrow 1 \pm \dfrac{2\sqrt{3}}{3} = x \Rightarrow x \approx 2.15$ or

   $x \approx -0.15$. Now, find the

   y-intercept: $f(0) = 3(0)^2 - 6(0) - 1 = -1$.

   (continued on next page)

(*continued*)

Use the fact that the parabola is symmetric with respect to its axis, $x = 1$, to locate additional points. Plot the vertex, the x-intercepts, the y-intercept, and any additional points, and join them with a parabola.

The graph of f is below the x-axis between the x-intercepts, so the solution set for
$f(x) = 3x^2 - 6x - 1 \le 0$ is

$$\left[1 - \frac{2\sqrt{3}}{3}, 1 + \frac{2\sqrt{3}}{3}\right] \text{ or } \left[\frac{3 - 2\sqrt{3}}{3}, \frac{3 + 2\sqrt{3}}{3}\right].$$

5. $h(t) = -\dfrac{g_E}{2}t^2 + v_0 t + h_0$

    $g_E = 32 \text{ ft/s}^2$, $h_0 = 100 \text{ ft}$, max height $= 244 \text{ ft}$

    a. Using the given values, we have
    $$h(t) = -\frac{32}{2}t^2 + v_0 t + 100$$
    $$= -16t^2 + v_0 t + 100$$
    Using the formula for the vertex of a
    parabola gives $t = -\dfrac{v_0}{2(-16)} = \dfrac{v_0}{32}$. This is

    the time at which the maximum height $h(t) = 244 \text{ ft}$ is attained. Thus,
    $$h(t) = 244 = h\left(\frac{v_0}{32}\right)$$
    $$= -16\left(\frac{v_0}{32}\right)^2 + v_0\left(\frac{v_0}{32}\right) + 100$$
    $$= -\frac{v_0^2}{64} + \frac{v_0^2}{32} + 100 = \frac{v_0^2}{64} + 100$$
    Solving for v_0 yields
    $$244 = \frac{v_0^2}{64} + 100 \Rightarrow 144 = \frac{v_0^2}{64} \Rightarrow$$
    $$v_0^2 = 144 \cdot 64 \Rightarrow v_0 = \sqrt{144 \cdot 64} = 12 \cdot 8 = 96$$
    Thus, $h(t) = -16t^2 + 96t + 100$ feet.

b. Using the formula for the vertex of a
    parabola gives $t = -\dfrac{96}{2(-16)} = \dfrac{96}{32} = 3$.

    The ball reached its highest point 3 seconds after it was released.

6. Let $x =$ the length of the playground and $y =$ the width of the playground.
    Then $2(x + y) = 1000 \Rightarrow x + y = 500 \Rightarrow$
    $y = 500 - x$.
    The area of the playground is
    $A(x) = xy = x(500 - x) = 500x - x^2$.
    The vertex for the parabola is (h, k) where
    $$h = -\frac{500}{2(-1)} = 250 \text{ and}$$
    $$k = 500(250) - 250^2 = 62,500.$$
    Thus, the maximum area that can be enclosed is $62,500 \text{ ft}^2$. The playground is a square with side length 250 ft.

3.1 Concepts and Vocabulary

1. A point where the axis meets the parabola is called the <u>vertex</u>.

3. The vertical line passing through the vertex of a parabola is called the <u>axis or axis of symmetry</u>.

5. False. $a = 1 > 0$.

7. True. The x-coordinate of the vertex of the parabola is $h = -\dfrac{b}{2a}$, so the y-coordinate of the vertex is $f\left(-\dfrac{b}{2a}\right)$.

3.1 Building Skills

9. f   11. a

13. h   15. g

17. Substitute -8 for y and 2 for x to solve for a:
    $-8 = a(2)^2 \Rightarrow -2 = a.$
    The equation is $y = -2x^2$.

19. Substitute 20 for y and 2 for x to solve for a:
    $20 = a(2)^2 \Rightarrow 5 = a.$
    The equation is $y = 5x^2$.

21. Substitute 0 for h, 0 for k, 8 for y, and -2 for x in the standard form for a quadratic equation to solve for a: $8 = a(-2-0)^2 + 0 \Rightarrow 8 = 4a \Rightarrow$ $2 = a$. The equation is $y = 2x^2$.

23. Substitute -3 for h, 0 for k, -4 for y, and -5 for x in the standard form for a quadratic equation to solve for a:

$-4 = a(-5-(-3))^2 + 0 \Rightarrow -1 = a$.

The equation is $y = -(x+3)^2$.

25. Substitute 2 for h, 5 for k, 7 for y, and 3 for x in the standard form for a quadratic equation to solve for a: $7 = a(3-2)^2 + 5 \Rightarrow 2 = a$.

The equation is $y = 2(x-2)^2 + 5$.

27. Substitute 2 for h, -3 for k, 8 for y, and -5 for x in the standard form for a quadratic equation to solve for a: $8 = a(-5-2)^2 - 3 \Rightarrow \dfrac{11}{49} = a$.

The equation is $y = \dfrac{11}{49}(x-2)^2 - 3$.

29. Substitute $\dfrac{1}{2}$ for h, $\dfrac{1}{2}$ for k, $-\dfrac{1}{4}$ for y, and $\dfrac{3}{4}$ for x in the standard form for a quadratic equation to solve for a:

$-\dfrac{1}{4} = a\left(\dfrac{3}{4} - \dfrac{1}{2}\right)^2 + \dfrac{1}{2} \Rightarrow -12 = a$.

The equation is $y = -12\left(x - \dfrac{1}{2}\right)^2 + \dfrac{1}{2}$.

31. The vertex is $(-2, 0)$, and the graph passes through $(0, 3)$. Substitute -2 for h, 0 for k, 3 for y, and 0 for x in the standard form for a quadratic equation to solve for a:

$3 = a(0-(-2))^2 + 0 \Rightarrow \dfrac{3}{4} = a$.

The equation is $y = \dfrac{3}{4}(x+2)^2$.

33. The vertex is $(3, -1)$, and the graph passes through $(5, 2)$. Substitute 3 for h, -1 for k, 2 for y, and 5 for x in the standard form for a quadratic equation to solve for a:

$2 = a(5-3)^2 - 1 \Rightarrow \dfrac{3}{4} = a$.

The equation is $y = \dfrac{3}{4}(x-3)^2 - 1$.

35. $f(x) = 3x^2$

Stretch the graph of $y = x^2$ vertically by a factor of 3.

37. $g(x) = (x-4)^2$

Shift the graph of $y = x^2$ right four units.

39. $f(x) = -2x^2 - 4$

Stretch the graph of $y = x^2$ vertically by a factor of 2, reflect the resulting graph in the x-axis, then shift the resulting graph down 4 units.

41. $g(x) = (x-3)^2 + 2$

Shift the graph of $y = x^2$ right three units, then shift the resulting graph up two units.

43. $f(x) = -3(x-2)^2 + 4$

Shift the graph of $y = x^2$ right two units, stretch the resulting graph vertically by a factor of 3, reflect the resulting graph about the x-axis, and then shift the resulting graph up four units.

45. Complete the square to write the equation in standard form: $y = x^2 + 4x \Rightarrow$

$y + 4 = x^2 + 4x + 4 \Rightarrow y = (x+2)^2 - 4$. This is the graph of $y = x^2$ shifted two units left and four units down. The vertex is $(-2, -4)$. The axis of symmetry is $x = -2$. To find the x-intercepts, let $y = 0$ and solve

$0 = (x+2)^2 - 4 \Rightarrow (x+2)^2 = 4 \Rightarrow$

$x + 2 = \pm 2 \Rightarrow x = -4$ or $x = 0$

To find the y-intercept, let $x = 0$ and solve

$y = (0+2)^2 - 4 \Rightarrow y = 0$.

47. Complete the square to write the equation in standard form:

$y = 6x - 10 - x^2 \Rightarrow y = -(x^2 - 6x + 10) \Rightarrow$

$y - 9 = -(x^2 - 6x + 9) - 10 \Rightarrow y = -(x-3)^2 - 1$.

This is the graph of $y = x^2$ shifted three units right, reflected about the x-axis, and then shifted one unit down. The vertex is $(3, -1)$. The axis of symmetry is $x = 3$. To find the x-intercepts, let $y = 0$ and solve

$0 = -(x-3)^2 - 1 \Rightarrow -1 = (x-3)^2 \Rightarrow$ there is no x-intercept. To find the y-intercept, let $x = 0$ and solve $y = -(0-3)^2 - 1 \Rightarrow y = -10$.

49. Complete the square to write the equation in standard form:

$y = 2x^2 - 8x + 9 \Rightarrow y = 2(x^2 - 4x) + 9 \Rightarrow$

$y + 8 = 2(x^2 - 4x + 4) + 9 \Rightarrow y = 2(x-2)^2 + 1$.

This is the graph of $y = x^2$ shifted 2 units right, stretched vertically by a factor of 2, and then shifted one unit up. The vertex is $(2, 1)$. The axis of symmetry is $x = 2$. To find the x-intercepts, let $y = 0$ and solve

$0 = 2(x-2)^2 + 1 \Rightarrow -\dfrac{1}{2} = (x-2)^2 \Rightarrow$ there is no x-intercept. To find the y-intercept, let $x = 0$ and solve $y = 2(0-2)^2 + 1 \Rightarrow y = 9$.

(continued on next page)

(*continued*)

51. Complete the square to write the equation in standard form:

$$y = -3x^2 + 18x - 11 \Rightarrow y = -3(x^2 - 6x) - 11 \Rightarrow$$

$$y - 27 = -3(x^2 - 6x + 9) - 11 \Rightarrow$$

$$y = -3(x - 3)^2 + 16.$$ This is the graph of

$y = x^2$ shifted three units right, stretched vertically by a factor of three, reflected about the *x*-axis, and then shifted 16 units up. The vertex is (3, 16). The axis of symmetry is $x = 3$. To find the *x*-intercepts, let $y = 0$ and

solve $y = 0 = -3(x - 3)^2 + 16 \Rightarrow$

$$\frac{16}{3} = (x - 3)^2 \Rightarrow \pm \frac{4\sqrt{3}}{3} = x - 3 \Rightarrow$$

$3 \pm \frac{4}{3}\sqrt{3} = x.$ To find the *y*-intercept, let $x = 0$

and solve $y = -3(0 - 3)^2 + 16 \Rightarrow y = -11.$

53. Complete the square to write the equation in standard form:

$$y = -x^2 - 4x \Rightarrow y = -(x^2 + 4x) \Rightarrow$$

$$y - 4 = -(x^2 + 4x + 4) \Rightarrow y = -(x + 2)^2 + 4$$

This is the graph of $y = x^2$ shifted two units left, reflected about the *x*-axis, and then shifted four units up. The vertex is (–2, 4). The axis is $x = -2$. To find the *x*-intercepts, let $y = 0$ and

solve $0 = -(x + 2)^2 + 4 \Rightarrow 4 = (x + 2)^2 \Rightarrow$

$\pm 2 = x + 2 \Rightarrow x = -4,\ x = 0.$ To find the

y-intercept, let $x = 0$ and solve

$$y = -(0 + 2)^2 + 4 \Rightarrow y = 0.$$

55. a. $a = 1 > 0$, so the graph opens up.

b. The vertex is

$$\left(-\frac{-8}{2(1)}, f\left(-\frac{-8}{2(1)} \right) \right) = (4, -1).$$

c. The axis of symmetry is $x = 4$.

d. To find the *x*-intercepts, let $y = 0$ and solve

$0 = x^2 - 8x + 15 \Rightarrow 0 = (x - 3)(x - 5) \Rightarrow$

$x = 3$ or $x = 5$. To find the *y*-intercept, let

$x = 0$ and solve $y = 0^2 - 8(0) + 15 \Rightarrow$

$y = 15.$

e.

57. a. $a = 1 > 0$, so the graph opens up.

b. The vertex is

$$\left(-\frac{-1}{2(1)}, f\left(-\frac{-1}{2(1)} \right) \right) = \left(\frac{1}{2}, -\frac{25}{4} \right).$$

c. The axis of symmetry is $x = \frac{1}{2}$.

d. To find the *x*-intercepts, let $y = 0$ and solve
$0 = x^2 - x - 6 \Rightarrow 0 = (x-3)(x+2) \Rightarrow x = 3$
or $x = -2$. To find the *y*-intercept, let $x = 0$
and solve $y = 0^2 - (0) - 6 \Rightarrow y = -6$.

e.

59. a. $a = 1 > 0$, so the graph opens up.

b. The vertex is $\left(-\dfrac{-2}{2(1)}, f\left(-\dfrac{-2}{2(1)}\right)\right) = (1,3)$.

c. The axis of symmetry is $x = 1$.

d. To find the *x*-intercepts, let $y = 0$ and solve
$0 = x^2 - 2x + 4 \Rightarrow$
$x = \dfrac{-(-2) \pm \sqrt{(-2)^2 - 4(1)(4)}}{2(1)} \Rightarrow$
$x = \dfrac{2 \pm \sqrt{-12}}{2} \Rightarrow$ there are no *x*-intercepts.
To find the *y*-intercept, let $x = 0$ and solve
$y = 0^2 - 2(0) + 4 \Rightarrow y = 4$.

e.

61. a. $a = -1 < 0$, so the graph opens down.

b. The vertex is
$\left(-\dfrac{-2}{2(-1)}, f\left(-\dfrac{-2}{2(-1)}\right)\right) = (-1,7)$.

c. The axis of symmetry is $x = -1$.

d. To find the *x*-intercepts, let $y = 0$ and solve
$0 = 6 - 2x - x^2 \Rightarrow$
$x = \dfrac{-(-2) \pm \sqrt{(-2)^2 - 4(-1)(6)}}{2(-1)} \Rightarrow x = \dfrac{2 \pm \sqrt{28}}{-2} \Rightarrow$
$x = -1 \pm \sqrt{7}$.
To find the *y*-intercept, let $x = 0$ and solve
$y = 6 - 2(0) - 0^2 \Rightarrow y = 6$.

e.

63. a. $a = 1 > 0$, so the graph opens up and has a
minimum value. Find the minimum value
by finding the vertex:
$\left(-\dfrac{-4}{2(1)}, f\left(-\dfrac{-4}{2(1)}\right)\right) = (2,-1)$
The minimum value is -1.

b. The range of *f* is $[-1, \infty)$.

65. a. $a = -1 < 0$, so the graph opens down and
has a maximum value. Find the maximum
value by finding the vertex:
$\left(-\dfrac{4}{2(-1)}, f\left(-\dfrac{4}{2(-1)}\right)\right) = (2,0)$
The maximum value is 0.

b. The range of *f* is $(-\infty, 0]$.

67. a. $a = 2 > 0$, so the graph opens up and has a
minimum value. Find the minimum value
by finding the vertex:
$\left(-\dfrac{-8}{2(2)}, f\left(-\dfrac{-8}{2(2)}\right)\right) = (2,-5)$
The minimum value is -5.

b. The range of *f* is $[-5, \infty)$.

69. a. $a = -4 < 0$, so the graph opens down and has a maximum value. Find the maximum value by finding the vertex:

$$\left(-\frac{12}{2(-4)}, f\left(-\frac{12}{2(-4)}\right)\right) = \left(\frac{3}{2}, 16\right)$$

The maximum value is 16.

b. The range of f is $(-\infty, 16]$.

In exercises 71–76, the x-intercepts are the boundaries of the intervals.

71. Solution: $[-2, 2]$

73. Solution:
$(-\infty, 1) \cup (3, \infty)$

75. Solution: $\left(-1, \frac{7}{6}\right)$

3.1 Applying the Concepts

77. $h(x) = -\frac{32}{132^2}x^2 + x + 3$

a. Using the graph, we see that the ball traveled approximately 550 ft horizontally.

b. Using the graph, we see that the ball went approximately 140 high.

c. $0 = -\frac{32}{132^2}x^2 + x + 3$

$$x = \frac{-1 \pm \sqrt{1^2 - 4\left(-\frac{32}{132^2}\right)(3)}}{2\left(-\frac{32}{132^2}\right)} \approx -3 \text{ or } 547$$

Thus, the ball traveled approximately 547 ft horizontally. The ball reached its maximum height at the vertex of the function,

$$\left(-\frac{b}{2a}, h\left(-\frac{b}{2a}\right)\right).$$

$$-\frac{b}{2a} = -\frac{1}{2\left(-32/\left(132^2\right)\right)} \approx 272.25$$

$$h(272.25) = -\frac{32}{132^2}(272.25)^2 + 272.25 + 3$$
$$\approx 139$$

The ball reached approximately 139 ft

79. The vertex of the function is

$$\left(-\frac{114}{2(-3)}, f\left(-\frac{114}{2(-3)}\right)\right) = (19, 1098).$$

The revenue is at its maximum when $x = 19$.

81. $\left(-\frac{-50}{2(1)}, f\left(-\frac{-50}{2(1)}\right)\right) = (25, -425).$

The total cost is minimum when $x = 25$.

83. Let x = the length of the rectangle. Then

$$\frac{80-2x}{2} = 40 - x = \text{ the width of the rectangle.}$$

The area of the rectangle = $x(40-x)$

$= 40x - x^2$. Find the vertex to find the maximum value:

$$\left(-\frac{40}{2(-1)}, f\left(-\frac{40}{2(-1)}\right)\right) = (20, 400).$$

The rectangle with the maximum area is a square with sides of length 20 units. Its area is 400 square units.

85. Let x = the width of the fields. Then,

$$\frac{600-3x}{2} = 300 - \frac{3}{2}x = \text{ the length of the two}$$

fields together. (Note that there is fencing between the two fields, so there are three "widths.")

The total area = $x\left(300 - \frac{3}{2}x\right) = 300x - \frac{3}{2}x^2$.

Find the vertex to find the dimensions and maximum value:

$$\left(-\frac{300}{2(-3/2)}, f\left(-\frac{300}{2(-3/2)}\right)\right) = (100, 15,000).$$

So the width of each field is 100 meters. The length of the two fields together is $300 - 1.5(100) = 150$ meters, so the length of each field is $150/2 = 75$ meters. The area of each field is $100(75) = 7500$ square meters.

87. The yield per tree is modeled by the equation of a line passing through (26, 500) where the x-coordinate represents the number of trees planted, and the y-coordinate represents the number of apples per tree. The rate of change is -10; that is, for each tree planted the yield decreases by 10. So, the yield per tree is $y - 500 = -10(x - 26) \Rightarrow y = -10x + 760$. Since there are x trees, the total yield =

$x(-10x + 760) = -10x^2 + 760x$.

Use the vertex to find the number of trees that will maximize the yield:

$$\left(-\frac{760}{2(-10)}, f\left(-\frac{760}{2(-10)}\right)\right) = (38, 14440).$$

So the maximum yield occurs when 38 trees are planted per acre.

89. If 20 students or less go on the trip, the cost is \$72 per students. If more than 20 students go on the trip, the cost is reduced by \$2 per the number of students over 20. So the cost per student is a piecewise function based on the number of students, n, going on the trip:

$$f(n) = \begin{cases} 72 & \text{if } n \le 20 \\ 72 - 2(n-20) = 112 - 2n & \text{if } n > 20 \end{cases}$$

The total revenue is

$$nf(n) = \begin{cases} 72n & \text{if } n \le 20 \\ n(112 - 2n) = 112n - 2n^2 & \text{if } n > 20 \end{cases}$$

The maximum revenue is either 1440 (the revenue if 20 students go on the trip) or the maximum of $112n - 2n^2$. Find this by using the vertex:

$$\left(-\frac{112}{(2)(-2)}, f\left(-\frac{112}{(2)(-2)}\right)\right) = (28, 1568)$$

The maximum revenue is \$1568 when 28 students go on the trip.

91. $h(t) = -\frac{g_M}{2}t^2 + v_0 t + h_0$

$g_M = 1.6\,\text{m/s}^2$, $h_0 = 5$ m, max height = 25 m

a. Using the given values, we have

$$h(t) = -\frac{1.6}{2}t^2 + v_0 t + 5 = -0.8t^2 + v_0 t + 5$$

Using the formula for the vertex of a parabola gives $t = -\dfrac{v_0}{2(-0.8)} = \dfrac{v_0}{1.6}$. This is the time at which the height $h(t) = 25$ m is attained. Thus,

$$h(t) = 25 = h\left(\frac{v_0}{1.6}\right)$$

$$= -0.8\left(\frac{v_0}{1.6}\right)^2 + v_0\left(\frac{v_0}{1.6}\right) + 5$$

$$= -\frac{v_0^2}{3.2} + \frac{v_0^2}{1.6} + 5 = \frac{v_0^2}{3.2} + 5$$

Solving for v_0 yields

$$25 = \frac{v_0^2}{3.2} + 5 \Rightarrow 20 = \frac{v_0^2}{3.2} \Rightarrow$$

$$v_0^2 = 64 \Rightarrow v_0 = 8$$

Thus, $h(t) = -0.8t^2 + 8t + 5$.

b. Using the formula for the vertex of a parabola gives $t = -\dfrac{8}{2(-0.8)} = \dfrac{8}{1.6} = 5.$

The stone reached its highest point 5 seconds after it was released.

b. Using the formula for the vertex of a parabola gives $t = -\dfrac{12}{2(-0.8)} = \dfrac{12}{1.6} = 7.5.$

The stone reached its highest point 7.5 seconds after it was released.

93. a. The maximum height occurs at the vertex:

$$\left(-\dfrac{64}{2(-16)}, f\left(-\dfrac{64}{2(-16)}\right)\right) = (2, 64), \text{ so the}$$

maximum height is 64 feet.

b. When the projectile hits the ground, $h = 0$, so solve

$$0 = -16t^2 + 64t \Rightarrow -16t(t-4) = 0 \Rightarrow$$

$t = 0$ or $t = 4$.

The projectile hits the ground at 4 seconds.

95. Let x = the radius of the semicircle. Then the length of the rectangle is $2x$. The circumference of the semicircle is πx, so the perimeter of the rectangular portion of the window is $18 - \pi x$.

The width of the rectangle $= \dfrac{18 - \pi x - 2x}{2}.$

The area of the semicircle is $\pi x^2/2$, and the

area of the rectangle is $2x\left(\dfrac{18 - \pi x - 2x}{2}\right) =$

$18x - \pi x^2 - 2x^2.$ So the total area is

$$18x - \pi x^2 - 2x^2 + \dfrac{\pi x^2}{2} = 18x - 2x^2 - \dfrac{\pi x^2}{2} \Rightarrow$$

$$18x - \left(2 + \dfrac{\pi}{2}\right)x^2.$$

The maximum area occurs at the x-coordinate of the vertex:

$$-\dfrac{18}{2\left(-2 - \dfrac{\pi}{2}\right)} = -\dfrac{18}{-4 - \pi} = \dfrac{18}{4 + \pi}.$$

This is the radius of the semicircle. The length

of the rectangle is $2\left(\dfrac{18}{4 + \pi}\right) = \dfrac{36}{4 + \pi}$ ft.

The width of the rectangle is

$$\dfrac{18 - \pi\left(\dfrac{18}{4 + \pi}\right) - 2\left(\dfrac{18}{4 + \pi}\right)}{2} = \dfrac{18}{4 + \pi} \text{ ft.}$$

97. a. $V = \left(-\dfrac{1.18}{2(-0.01)}, f\left(-\dfrac{1.18}{2(-0.01)}\right)\right)$

$= (59, 36.81)$

b. $0 = -0.01x^2 + 1.18x + 2 \Rightarrow$

$$x = \dfrac{-1.18 \pm \sqrt{1.18^2 - 4(-0.01)(2)}}{2(-0.01)}$$

$$= \dfrac{-1.18 \pm \sqrt{1.4724}}{-0.02} \approx \dfrac{-1.18 \pm 1.21}{-0.02}$$

≈ -1.5 or 119.5.

Reject the negative answer, and round the positive answer to the nearest whole number. The ball hits the ground approximately 120 feet from the punter.

c. The maximum height occurs at the vertex. (See part (a).) The maximum height is approximately 37 ft.

d. The player is at $x = 6$ feet.

$f(6) = -0.01(6)^2 + 1.18(6) + 2 = 8.72.$

The player must reach approximately 9 feet to block the ball.

e. $7 = -0.01x^2 + 1.18x + 2$

$0 = -0.01x^2 + 1.18x - 5$

$$x = \dfrac{-1.18 \pm \sqrt{1.18^2 - 4(-0.01)(-5)}}{2(-0.01)}$$

$$= \dfrac{-1.18 \pm \sqrt{1.1924}}{-0.02} \Rightarrow$$

$x \approx 4.4$ feet or $x \approx 113.6$ feet

99. $f(x) = 1.51 - 0.22x + 0.029x^2$

    a. We must complete the square to write the function in standard form.

$$0.029x^2 - 0.22x + 1.51$$

$$= 0.029\left(x^2 - \frac{0.22}{0.029}x\right) + 1.51$$

$$= 0.029\left(x^2 - \frac{0.22}{0.029}x + \frac{0.22^2}{4 \cdot 0.029^2}\right)$$

$$\qquad + 1.51 - \left(\frac{0.22^2}{4 \cdot 0.029^2}\right)(0.029)$$

$$= 0.029\left(x - \frac{0.22}{2 \cdot 0.029}\right)^2 + 1.09$$

$$= 0.029(x - 3.79)^2 + 1.09$$

    b. Sales were at a minimum at $x \approx 3.79$, or during year 4 (2010). This fits the original data.

    c. The year 2011 is represented by $x = 5$.

$$f(5) = 1.51 - 0.22(5) + 0.029(5)^2 = 1.135$$

    In 2011, there were about 1.14 million vehicles sold.

3.1 Beyond the Basics

101. $y = 3(x+2)^2 + 3 \Rightarrow y = 3(x^2 + 4x + 4) + 3 \Rightarrow$
$y = 3x^2 + 12x + 15$

103. The y-intercept is 4, so the graph passes through $(0, 4)$. Substitute the coordinates $(0, 4)$ for x and y and $(1, -2)$ for h and k into the standard form $y = a(x-h)^2 + k$ to solve for a: $4 = a(0-1)^2 - 2 \Rightarrow 6 = a$.
The x-coordinate of the vertex is

$$1 = -\frac{b}{2a} = -\frac{b}{2(6)} \Rightarrow b = -12.$$ The y-intercept

$= c$, so the equation is $f(x) = 6x^2 - 12x + 4$.

105. The x-coordinate of the vertex is 3. Substitute the coordinates of the x-intercept and the x-coordinate of the vertex into the standard form $y = a(x-h)^2 + k$ to find an expression relating a and k:
$0 = a(7-3)^2 + k \Rightarrow -16a = k.$

Now substitute the coordinates of the y-intercept, the x-coordinate of the vertex, and the expression for k into the standard form $y = a(x-h)^2 + k$ to solve for a:

$14 = a(0-3)^2 - 16a \Rightarrow -2 = a.$ Use this value to find k: $k = -16(-2) = 32.$ The equation is $y = -2(x-3)^2 + 32 \Rightarrow y = -2x^2 + 12x + 14.$

107. The x-coordinate of the vertex is $\frac{-2+6}{2} = 2.$

Substitute the coordinates of one of the x-intercepts and the x-coordinate of the vertex into the standard form $y = a(x-h)^2 + k$ to find an expression relating a and k:
$0 = a(-2-2)^2 + k \Rightarrow -16a = k.$

Now substitute the coordinates of the other x-intercept, the x-coordinate of the vertex, and the expression for k into the standard form $y = a(x-h)^2 + k$ to solve for a:

$0 = a(6-2)^2 - 16a \Rightarrow 1 = a.$ Use this value to find k: $-16(1) = -16 = k.$

So, one equation is

$y = (x-2)^2 - 16 = x^2 - 4x - 12.$ The graph of this equation opens upward. To find the equation of the graph that opens downward, multiply the equation by -1:

$y = -x^2 + 4x + 12.$

109. The x-coordinate of the vertex is $\frac{-7-1}{2} = -4.$

Substitute the coordinates of one of the x-intercepts and the x-coordinate of the vertex into the standard form $y = a(x-h)^2 + k$ to find an expression relating a and k:
$0 = a(-7+4)^2 + k \Rightarrow -9a = k.$

Now substitute the coordinates of the other x-intercept, the x-coordinate of the vertex, and the expression for k into the standard form $y = a(x-h)^2 + k$ to solve for a:

$0 = a(-1+4)^2 - 9a \Rightarrow 1 = a.$ Use this value to find k: $-9(1) = -9 = k.$ So one equation is

$y = (x+4)^2 - 9 = x^2 + 8x + 7.$ The graph of this equation opens upward. To find the equation of the graph that opens downward, multiply the equation by -1:

$y = -x^2 - 8x - 7.$

3.1 Critical Thinking/Discussion/Writing

111. $f(h+p) = a(h+p)^2 + b(h+p) + c$

$f(h-p) = a(h-p)^2 + b(h-p) + c$

$f(h+p) = f(h-p) \Rightarrow a(h+p)^2 + b(h+p) + c = a(h-p)^2 + b(h-p) + c \Leftrightarrow$

$ah^2 + 2ahp + ap^2 + bh + bp + c = ah^2 - 2ahp + ap^2 + bh - bp + c \Rightarrow 4ahp = -2bp \Rightarrow 2ah = -b.$

(We can divide by p because $p \neq 0$.) Since $a \neq 0$, $2ah = -b \Rightarrow h = -\dfrac{b}{2a}$.

112. Write $y = 2x^2 - 8x + 9$ in standard form to find the axis of symmetry:

$y = 2x^2 - 8x + 9 \Rightarrow y + 8 = 2(x^2 - 4x + 4) + 9 \Rightarrow y = 2(x-2)^2 + 1.$

The axis of symmetry is $x = 2$.

Using the results of exercise 111, we know that $f(2+p) = f(-1) = f(2-p)$.

$2 + p = -1 \Rightarrow p = -3$, so $2 - p = 2 - (-3) = 5$.

The point symmetric to the point $(-1, 19)$ across the axis of symmetry is $(5, 19)$.

113. Answers will vary. If $a > 0$, then the parabola $f(x) = ax^2 + bx + c$ opens upward and the solution of the inequality is the portion of the graph that is above the x-axis. If the x-intercepts are represented by x_0 and x_1 with $x_0 \leq x_1$, then the solution of the inequality is $(-\infty, x_0) \cup (x_1, \infty)$. If $a < 0$, then the parabola $f(x) = ax^2 + bx + c$ opens downward and the solution of the inequality is the portion of the graph that is below the x-axis. If the x-intercepts are represented by x_0 and x_1 with $x_0 \leq x_1$, then the solution of the inequality is (x_0, x_1).

114. a. $(f \circ g)(x) = f(mx+b) = a\big[(mx+b) - h\big]^2 + k$

$\qquad = a\big[(mx+b)^2 - 2h(mx+b) + h^2\big] + k$

$\qquad = a\big[m^2x^2 + 2bmx + b^2 - 2hmx - 2hb + h^2\big] + k$

$\qquad = am^2x^2 + 2abmx + ab^2 - 2ahmx - 2ahb + ah^2 + k$

$\qquad = am^2x^2 + (2abm - 2ahm)x + \big(ab^2 - 2ahb + ah^2 + k\big)$

This is the equation of a parabola. The x-coordinate of the vertex is

$-\dfrac{2abm - 2ahm}{2am^2} = -\dfrac{2am(b-h)}{2am^2} = -\dfrac{b-h}{m}$ or $\dfrac{h-b}{m}$.

The y-coordinate of the vertex is

$am^2\left(\dfrac{h-b}{m}\right)^2 + (2abm - 2ahm)\left(\dfrac{h-b}{m}\right) + \big(ab^2 - 2ahb + k\big)$

$\qquad = a(h-b)^2 + 2a(b-h)(h-b) + \big(ab^2 - 2ahb + ah^2 + k\big)$

$\qquad = ah^2 - 2ahb + ab^2 - 2ab^2 + 4abh - 2ah^2 + ab^2 - 2ahb + ah^2 + k$

$\qquad = k$

The vertex is $\left(\dfrac{h-b}{m}, k\right)$.

b. $(g \circ f)(x) = g\left[a(x-h)^2 + k\right] = m\left[a(x-h)^2 + k\right] + b$

$\qquad = m\left(ax^2 - 2ahx + ah^2 + k\right) + b = max^2 - 2mahx + mah^2 + mk + b$

This is the equation of a parabola. The *x*-coordinate of the vertex is $-\dfrac{-2mah}{2ma} = h.$

The *y*-coordinate of the vertex is $mah^2 - 2mah^2 + mah^2 + mk + b = mk + b.$
Thus, the vertex is $(h, mk + b).$

115. If the discriminant equals zero, there is exactly one real solution. Thus, the vertex of
$y = f(x)$ lies on the *x*-axis at $x = -\dfrac{b}{2a}.$
If the discriminant > 0, there are two unequal real solutions. This means that the graph of
$y = f(x)$ crosses the *x*-axis in two places. If
$a > 0$, then the vertex lies below the *x*-axis and the parabola crosses the *x*-axis; if $a < 0$, then the vertex lies above the *x*-axis and the parabola crosses the *x*-axis. If the discriminant < 0, there are two nonreal complex solutions. If $a > 0$, then the vertex lies above the *x*-axis and the parabola does not cross the *x*-axis (it opens upward); if $a < 0$, then the vertex lies below the *x*-axis and the parabola does not cross the *x*-axis (it opens downward).

116. Let $g(x) = x^2$ and $k(x) = x - h.$ Then
$f(x) = g(k(x)) = g(x - h) = (x - h)^2,$ which is a horizontal translation of *g*.

3.1 Getting Ready for the Next Section

117. $-5^0 = -\left(5^0\right) = -1$

119. $-4^{-2} = -(4)^{-2} = -\dfrac{1}{(4)^2} = -\dfrac{1}{16}$

121. $x^7 \cdot x^{-7} = x^{7+(-7)} = x^0 = 1$

123. $x^2\left(3 - \dfrac{3}{4}\right) = x^2\left(\dfrac{12}{4} - \dfrac{3}{4}\right) = \dfrac{9}{4}x^2$

125. $4x^2 - 9 = (2x + 3)(2x - 3)$

127. $15x^2 + 11x - 12 = (3x + 4)(5x - 3)$

129. $x^2(x - 1) - 4(x - 1) = \left(x^2 - 4\right)(x - 1)$
$\qquad\qquad\qquad\qquad = (x + 2)(x - 2)(x - 1)$

131. $x^3 + 4x^2 + 3x + 12 = \left(x^3 + 4x^2\right) + (3x + 12)$
$\qquad\qquad\qquad\qquad = x^2(x + 4) + 3(x + 4)$
$\qquad\qquad\qquad\qquad = \left(x^2 + 3\right)(x + 4)$

133. $2x + 9 = 5x + 3 \Rightarrow -3x = -6 \Rightarrow x = 2$
Solution set: $\{2\}$

135. $6x^2 - x - 2 = 0 \Rightarrow (3x - 2)(2x + 1) = 0 \Rightarrow$
$x = \dfrac{2}{3}, \; x = -\dfrac{1}{2}$
Solution set: $\left\{-\dfrac{1}{2}, \dfrac{2}{3}\right\}$

3.2 Polynomial Functions

3.2 Practice Problems

1. a. $f(x) = \dfrac{x^2 + 1}{x - 1}$ is not a polynomial function because its domain is not $(-\infty, \infty).$

b. $g(x) = 2x^7 + 5x^2 - 17$ is a polynomial function. Its degree is 7, the leading term is $2x^7$, and the leading coefficient is 2.

2. $P(x) = 4x^3 + 2x^2 + 5x - 17$
$\qquad = x^3\left[4 + \dfrac{2}{x} + \dfrac{5}{x^2} - \dfrac{17}{x^3}\right]$

When |*x*| is large, the terms $\dfrac{2}{x}$, $\dfrac{5}{x^2}$, and $-\dfrac{17}{x^3}$ are close to 0. Therefore,
$P(x) = x^3(4 + 0 + 0 - 0) \approx 4x^3.$

3. Use the leading-term test to determine the end behavior of $y = f(x) = -2x^4 + 5x^2 + 3$. Here $n = 4$ and $a_n = -2 < 0$. Thus, Case 2 applies. The end behavior is described as
$y \to -\infty$ as $x \to -\infty$ and $y \to -\infty$ as $x \to \infty.$

4. First group the terms, then factor and solve $f(x) = 0$:

$$f(x) = 2x^3 - 3x^2 + 4x - 6$$
$$= 2x^3 + 4x - 3x^2 - 6$$
$$= 2x(x^2 + 2) - 3(x^2 + 2)$$
$$= (2x - 3)(x^2 + 2)$$

$$0 = (2x - 3)(x^2 + 2)$$

$$0 = 2x - 3 \Rightarrow x = \frac{3}{2} \text{ or}$$

$$0 = x^2 + 2 \text{ (no real solution)}$$

The only real zero is $\frac{3}{2}$.

5. $f(x) = 2x^3 - 3x - 6$

$f(1) = -7$ and $f(2) = 4$. Since $f(1)$ and $f(2)$ have opposite signs, by the Intermediate Value Theorem, f has a real zero between 1 and 2.

6. $f(x) = (x+1)^2(x-3)(x+5) = 0 \Rightarrow$
$(x+1)^2 = 0 \text{ or } x - 3 = 0 \text{ or } x + 5 = 0 \Rightarrow$
$x = -1 \text{ or } x = 3 \text{ or } x = -5$
$f(x)$ has three distinct zeros.

7. $f(x) = (x-1)^2(x+3)(x+5) = 0 \Rightarrow$
$(x-1)^2 = 0 \text{ or } x + 3 = 0 \text{ or } x + 5 = 0 \Rightarrow$
$x = 1$ (multiplicity 2) or $x = -3$ (multiplicity 1) or $x = -5$ (multiplicity 1)

8. $f(x) = -x^4 + 3x^2 - 2$ has at most three turning points. Using a graphing calculator, we see that there are indeed, three turning points.

9. $f(x) = -x^4 + 5x^2 - 4$

Since the degree, 4, is even and the leading coefficient is -1, the end behavior is as shown:

$y \rightarrow -\infty$ as $x \rightarrow -\infty$ and $y \rightarrow -\infty$ as $x \rightarrow \infty$.

Now find the zeros of the function:

$$0 = -x^4 + 5x^2 - 4 \Rightarrow 0 = -(x^4 - 5x^2 + 4)$$
$$= -(x^2 - 4)(x^2 - 1) \Rightarrow$$
$$0 = x^2 - 4 \text{ or } 0 = x^2 - 1 \Rightarrow$$
$$x = \pm 2 \quad \text{or} \quad x = \pm 1$$

There are four zeros, each of multiplicity 1, so the graph crosses the x-axis at each zero. Next, find the y-intercept:

$$f(0) = -x^4 + 5x^2 - 4 = -4$$

Now find the intervals on which the graph lies above or below the x-axis. The four zeros divide the x-axis into five intervals, $(-\infty, -2)$, $(-2, -1), (-1, 1), (1, 2),$ and $(2, \infty)$. Determine the sign of a test value in each interval

Interval	Test point	Value of $f(x)$	Above/below x-axis
$(-\infty, -2)$	-3	-40	below
$(-2, -1)$	-1.5	2.1875	above
$(-1, 1)$	0	-4	below
$(1, 2)$	1.5	2.1875	above
$(2, \infty)$	3	-40	below

Plot the zeros, y-intercepts, and test points, and then join the points with a smooth curve.

10. $f(x) = x(x+2)^3(x-1)^2$

First, find the zeros of the function.

$x(x+2)^3(x-1)^2 = 0 \Rightarrow$

$\begin{array}{c|c|c} x = 0 & x+2 = 0 & x-1 = 0 \\ & x = -2 & x = 1 \end{array}$

The function has three distinct zeros, –2, 1, and 0. The zero $x = -2$ has multiplicity 3, so the graph crosses the x-axis at –2, and near –2 the function looks like

$f(x) = -2(x+2)^3(-2-1)^2 = -18(x+2)^3$.

The zero $x = 0$ has multiplicity 1, so the graph crosses the x-axis at 0, and near 0 the function looks like $f(x) = x(0+2)^3(0-1)^2 = 8x$.

The zero $x = 1$ has multiplicity 2, so the graph touches the x-axis at 1, and near 1 the function looks like

$f(x) = 1(1+2)^3(x-1)^2 = 27(x-1)^2$.

We can determine the end behavior of the polynomial by finding its leading term as a product of the leading terms of each factor.

$f(x) = x(x+2)^3(x-1)^2 \approx x(x^3)(x^2) = x^6$

Therefore, the end behavior of $f(x)$ is Case 1:

Now find the intervals on which the graph lies above or below the x-axis. The three zeros divide the x-axis into four intervals, $(-\infty, -2)$, $(-2, 0), (0, 1)$, and $(1, \infty)$. Determine the sign of a test value in each interval.

Interval	Test point	Value of $f(x)$	Above/below x-axis
$(-\infty, -2)$	–3	48	above
$(-2, 0)$	–1	–4	below
$(0, 1)$	0.5	1.95	above
$(1, \infty)$	2	128	above

Plot the zeros, y-intercepts, and test points, and then join the points with a smooth curve.

11. $V = \dfrac{\pi}{3\sqrt{3}} x^3 = \dfrac{\pi}{3\sqrt{3}} \cdot 7^3 \approx 207.378 \text{ dm}^3$
$\approx 207.378 \text{ L}$

3.2 Concepts and Vocabulary

1. Consider the polynomial $2x^5 - 3x^4 + x - 6$. The degree of this polynomial is <u>5</u>, its leading term is $\underline{2x^5}$, its leading coefficient is <u>2</u>, and its constant term is <u>–6</u>.

3. If c is a zero of even multiplicity for a polynomial function f, then the graph of f <u>touches</u> the x-axis at c.

5. False. The graph of a polynomial function is a smooth curve. That means it has no corners or cusps.

7. False. The polynomial function of degree n has, at most, n zeros.

3.2 Building Skills

9. Polynomial function; degree: 5;
leading term: $2x^5$; leading coefficient: 2

11. Polynomial function; degree: 3;
leading term: $\dfrac{2}{3}x^3$; leading coefficient: $\dfrac{2}{3}$

13. Polynomial function; degree: 4; leading term: πx^4; leading coefficient: π

15. Not a polynomial function because the graph has sharp corners. It is not a smooth curve. There is a presence of $|x|$.

17. Not a polynomial function because the domain is not $(-\infty, \infty)$.

19. Not a polynomial function because of the presence of $\sqrt{x}$.

21. Not a polynomial function because the domain is not $(-\infty, \infty)$.

23. Not a polynomial function because the graph is not continuous.

25. Not a polynomial function because the graph is not continuous.

27. Not a polynomial function because it is not the graph of a function.

29. c   **31.** a   **33.** d

35. $f(x) = x - x^3$

The leading term is $-x^3$, so $f(x) \to \infty$ as $x \to -\infty$ and $f(x) \to -\infty$ as $x \to \infty$.

37. $f(x) = 4x^4 + 2x^3 + 1$

The leading term is $4x^4$, so $f(x) \to \infty$ as $x \to -\infty$ and $f(x) \to \infty$ as $x \to \infty$.

39. $f(x) = (x+2)^2 (2x-1)$

The leading term is $x^2 (2x) = 2x^3$, so $f(x) \to -\infty$ as $x \to -\infty$ and $f(x) \to \infty$ as $x \to \infty$.

41. $f(x) = (x+2)^2 (4-x)$

The leading term is $x^2 (-x) = -x^3$, so $f(x) \to \infty$ as $x \to -\infty$ and $f(x) \to -\infty$ as $x \to \infty$.

43. $f(x) = 3(x-1)(x+2)(x-3)$
Zeros: -2, 1, 3
$x = -2$: multiplicity: 1, crosses the x-axis
$x = 1$: multiplicity: 1, crosses the x-axis
$x = 3$: multiplicity: 1, crosses the x-axis

45. $f(x) = (x+2)^2 (2x-1)$
Zeros: -2, $\frac{1}{2}$
$x = -2$: multiplicity 2, touches but does not cross the x-axis
$x = \frac{1}{2}$: multiplicity 1, crosses the x-axis

47. $f(x) = x^2 (x^2 - 9)(3x+2)^3$
Zeros: $-3, -\frac{2}{3}, 0, 3$
$x = -3$: multiplicity 1, crosses the x-axis
$x = -\frac{2}{3}$: multiplicity 3, crosses the x-axis
$x = 0$: multiplicity 2, touches but does not cross the x-axis
$x = 3$: multiplicity 1, crosses the x-axis

49. $f(x) = (x^2 + 1)(3x-2)^2$

Zero: $\frac{2}{3}$, multiplicity 2, touches but does not cross the x-axis

51. $f(2) = 2^4 - 2^3 - 10 = -2$;
$f(3) = 3^4 - 3^3 - 10 = 44$.

Because the sign changes, there is a real zero between 2 and 3. The zero is approximately 2.09.

53. $f(2) = 2^5 - 9(2)^2 - 15 = -19$;
$f(3) = 3^5 - 9(3)^2 - 15 = 147$.

Because the sign changes, there is a real zero between 2 and 3. The zero is approximately 2.28.

55. The least possible degree is 3. The zeros are -2, 1, and 3, each with multiplicity 1. The end behavior is $f(x) \to -\infty$ as $x \to -\infty$ and $f(x) \to \infty$ as $x \to \infty$, so the leading coefficient is 1. The polynomial with smallest possible degree is $(x+2)(x-1)(x-3)$.

57. The least possible degree is 3. The zeros are -3 (multiplicity 1) and 2 (multiplicity 2). The end behavior is $f(x) \to -\infty$ as $x \to -\infty$ and $f(x) \to \infty$ as $x \to \infty$, so the leading coefficient is 1. The polynomial with smallest possible degree is $(x+3)(x-2)^2$.

59. The least possible degree is 4. The zeros are -1 (multiplicity 2) and 2 (multiplicity 2). The end behavior is $f(x) \to \infty$ as $x \to -\infty$ and $f(x) \to \infty$ as $x \to \infty$, so the leading coefficient is 1. The polynomial with smallest possible degree is $(x+1)^2 (x-2)^2$.

61. The zeros are –2 (multiplicity 2), 2 (multiplicity 1), and 3 (multiplicity 1), so the least possible degree is 4. The end behavior is $f(x) \to \infty$ as $x \to -\infty$ and $f(x) \to \infty$ as $x \to \infty$, so the leading coefficient is 1. The polynomial with smallest possible degree is $(x+2)^2 (x-2)(x-3)$.

For exercises 63–74, use the procedures shown in Examples 9 and 10 in the text to graph the function.

63. $f(x) = 2(x+1)(x-2)(x+4)$

65. $f(x) = (x-1)^2 (x+3)(x-4)$

67. $f(x) = -x^2 (x-3)^2$

69. $f(x) = (x-1)^2 (x+2)^3 (x-3)$

71. $f(x) = -x^2 (x^2 -1)(x+1)$

73. $f(x) = x^2 (x^2 +1)(x-2)$

3.2 Applying the Concepts

75. a. $P = \dfrac{0.04825}{746}(40)^3 \approx 4.14$ hp

b. $P = \dfrac{0.04825}{746}(65)^3 \approx 17.76$ hp

c. $P = \dfrac{0.04825}{746}(80)^3 \approx 33.12$ hp

d. $\dfrac{P(2v)}{P(v)} = \dfrac{\frac{0.04825}{746}(2v)^3}{\frac{0.04825}{746}v^3} = \dfrac{8v^3}{v^3} = 8$

77. a. $3x^2(4-x) = 0 \Rightarrow x = 0$ or $x = 4$. $x = 0$, multiplicity 2; $x = 4$, multiplicity 1.

b.

c. There are 2 turning points.

d. Domain: [0, 4]. The portion between the x-intercepts is the graph of $R(x)$.

79. a. Domain [0, 1]

b.

81. a. $R(x) = x\left(27 - \left(\dfrac{x}{300}\right)^2\right) = 27x - \dfrac{x^3}{90{,}000}$

b. The domain of $R(x)$ is the same as the domain of p. $p \geq 0$ when $p \leq 900\sqrt{3}$.
The domain is $\left[0, 900\sqrt{3}\right]$.

c.

83. a. $N(x) = (x+12)\left(400 - 2x^2\right)$

b. The low end of the domain is 0. (There cannot be fewer than 0 workers.) The upper end of the domain is the value where productivity is 0, so solve $N(x) = 0$ to find the upper end of the domain.
$(x+12)(400-2x^2) = 0 \Rightarrow x = -12$ (reject this) or $400-2x^2 = 0 \Rightarrow 200 = x^2 \Rightarrow$
$x = \pm 10\sqrt{2}$ (reject the negative solution).
The domain is $[0, 10\sqrt{2}]$.

c.

85. a. $V(x) = x(8-2x)(15-2x)$

b.

87. a. $V(x) = x^2(108 - 4x)$

b.

89. a. $V(x) = x^2(62 - 2x)$

b.

91. a. $V(x) = 2x^2(45 - 3x)$

b.

93.

95. $f(x) = 0.963 + 0.88x - 0.192x^2 + 0.0117x^3$

   a. The year 2015 is represented by $x = 7$.
      $f(7) \approx 1.73$ trillion

   b.

      [–2, 12] by [–2, 2.5]

      It appears that there is one zero and two
      turning points.

   c.

      [0, 8] by [–0.5, 2.5]

      The maximum occurs at $x \approx 3.27$, which is
      during 2011. This fits the original data.

3.2 Beyond the Basics

97. The graph of $f(x) = (x-1)^4$ is the graph of
$y = x^4$ shifted one unit right.
$(x-1)^4 = 0 \Rightarrow x = 1$. The zero is $x = 1$ with
multiplicity 4.

99. The graph of $f(x) = x^4 + 2$ is the graph of
$y = x^4$ shifted two units up.
$x^4 + 2 = 0 \Rightarrow x = \sqrt[4]{-2}$. There are no zeros.

101. The graph of $f(x) = -(x-1)^4$ is the graph of
$y = x^4$ shifted one unit right and then
reflected across the x-axis.
$-(x-1)^4 = 0 \Rightarrow x = 1$. The zero is $x = 1$ with
multiplicity 4.

103. The graph of $f(x) = x^5 + 1$ is the graph of
$y = x^5$ shifted one unit up. $x^5 + 1 = 0 \Rightarrow$
$x = \sqrt[5]{-1} \Rightarrow x = -1$. The zero is -1 with
multiplicity 1.

105. The graph of $f(x) = 8 - \dfrac{(x+1)^5}{4}$ is the graph

of $y = x^5$ shifted one unit left, compressed vertically by one-fourth, reflected in the x-axis, and then shifted up eight units.

$$8 - \frac{(x+1)^5}{4} = 0 \Rightarrow -\frac{(x+1)^5}{4} = -8 \Rightarrow$$

$(x+1)^5 = 32 \Rightarrow x+1 = 2 \Rightarrow x = 1.$ The zero is 1 with multiplicity 1.

107. If $0 < x < 1$, then $x^2 < 1.$ Multiplying both sides of the inequality by x^2, we obtain $x^4 < x^2.$ Thus, $0 < x^4 < x^2.$

109.

The graph of $y = x^2 - x^4$ represents the difference between the two functions $y = x^2$ and $y = x^4$, so the maximum distance between the graphs occurs at the local maximum of $y = x^2 - x^4.$ The maximum vertical distance is 0.25. It occurs at $x \approx 0.71.$

Answers will vary in exercises 111–114. Sample answers are given.

111. $f(x) = x^2(x+1)$

113. $f(x) = 1 - x^4$

115. The smallest possible degree is 5, because the graph has five x-intercepts and four turning points.

117. The smallest possible degree is 6, because the graph has five turning points.

119. $f(x) = 10x^4 + x + 1$

For large values of x, $f(x) \approx 10x^4.$ For small values of x, $f(x) \approx x + 1.$

121. $f(x) = x^4 - x^2 + 1$

For large values of x, $f(x) \approx x^4.$ For small values of x, $f(x) \approx -x^2 + 1.$

3.2 Critical Thinking/Discussion/Writing

123. If a human is scaled up by a factor of 20, then the total weight will increase $20^3 = 8000$ times but the bone strength will increase only $20^2 = 400$ times. The bones would probably break under the increased weight.

124. If a human is scaled down by a factor of 70, then the body surface area will decrease $70^2 = 4900$ times while the volume will decrease $70^3 = 343,000$ times. The body will rapidly lose heat and will have a very hard time maintaining body temperature unless its metabolic rate increases drastically.

125. It is not possible for a polynomial function to have no y-intercepts because the domain of any polynomial function is $(-\infty, \infty)$, which includes the point $x = 0.$

126. It is possible for a polynomial function to have no x-intercepts because the function can be shifted above the x-axis. An example is the function $y = x^2 + 1$.

127. It is not possible for the graph of a polynomial function of degree 3 to have exactly one local maximum and no local minimum because the graph of a function of degree 3 rises in one direction and falls in the other. This requires an even number of turning points. Since the degree is 3, there can be only zero or two turning points. Therefore, if there is a local maximum, there must also be another turning point, which will be a local minimum.

128. It is not possible for the graph of a polynomial function of degree 4 to have exactly one local maximum and exactly one local minimum because the graph rises in both directions or falls in both directions. This requires an odd number of turning points. Since the degree is 4, there can be only one or three turning points. If there were exactly one local maximum and exactly one local minimum, there would be two turning points.

129. If $P(x) = a_m x^m + a_{m-1} x^{m-1} + \cdots + a_1 x + a_0$ and $Q(x) = b_n x^n + b_{n-1} x^{n-1} + \cdots + b_1 x + b_0$, then

$$(P \circ Q)(x) = a_m \left(b_n x^n + b_{n-1} x^{n-1} + \cdots + b_1 x + b_0\right)^m + a_{m-1}\left(b_n x^n + b_{n-1} x^{n-1} + \cdots + b_1 x + b_0\right)^{m-1} + \cdots$$
$$+ a_1\left(b_n x^n + b_{n-1} x^{n-1} + \cdots + b_1 x + b_0\right) + a_0,$$

which is a polynomial of degree mn.

130. If $P(x) = a_m x^m + a_{m-1} x^{m-1} + \cdots + a_1 x + a_0$ and $Q(x) = b_n x^n + b_{n-1} x^{n-1} + \cdots + b_1 x + b_0$, then

$$(P \circ Q)(x) = a_m \left(b_n x^n + b_{n-1} x^{n-1} + \cdots + b_1 x + b_0\right)^m + a_{m-1}\left(b_n x^n + b_{n-1} x^{n-1} + \cdots + b_1 x + b_0\right)^{m-1} + \cdots$$
$$+ a_1\left(b_n x^n + b_{n-1} x^{n-1} + \cdots + b_1 x + b_0\right) + a_0,$$

and the leading coefficient is $a_m \left(b_n\right)^m$.

$$(Q \circ P)(x) = b_n \left(a_m x^m + a_{m-1} x^{m-1} + \cdots + a_1 x + a_0\right)^n + b_{n-1}\left(a_m x^m + a_{m-1} x^{m-1} + \cdots + a_1 x + a_0\right)^{n-1} + \cdots$$
$$+ b_1\left(a_m x^m + a_{m-1} x^{m-1} + \cdots + a_1 x + a_0\right) + b_0,$$

and the leading coefficient is $b_n \left(a_m\right)^n$.

3.2 Getting Ready for the Next Section

131. $f(x) = x^3 - 3x^2 + 2x - 9$
$f(-2) = (-2)^3 - 3(-2)^2 + 2(-2) - 9 = -33$

133. $g(x) = \left(-7x^5 + 2x\right)(x - 13)$
$g(13) = \left[-7(13)^5 + 2(13)\right]\left[(13) - 13\right] = 0$

135. $x^2 - 3x - 10 = (x - 5)(x + 2)$

137. $(2x + 3)\left(x^2 + 6x - 7\right) = (2x + 3)(x + 7)(x - 1)$

139. $3x^3 - 6x^2 + 5x - 10 = \left(3x^3 - 6x^2\right) + (5x - 10)$
$= 3x^2(x - 2) + 5(x - 2)$
$= \left(3x^2 + 5\right)(x - 2)$

3.3   Dividing Polynomials

3.3 Practice Problems

1.

$$
\begin{array}{r}
3x + 4 \\
x^2 + 0x + 1 \overline{)\,3x^3 + 4x^2 + x + 7} \\
(-)\underline{3x^3 + 0x^2 + 3x} \\
4x^2 - 2x + 7 \\
(-)\underline{4x^2 + 0x + 4} \\
-2x + 3
\end{array}
$$

Quotient: $3x + 4$; remainder: $-2x + 3$

2.
$$x^2 - x + 3 \overline{\smash{\big)}\ x^4 + 0x^3 + 5x^2 + 2x + 6}$$

quotient: $x^2 + x + 3$

$$\underline{x^4 - x^3 + 3x^2}$$
$$x^3 + 2x^2 + 2x$$
$$\underline{x^3 - x^2 + 3x}$$
$$3x^2 - x + 6$$
$$\underline{3x^2 - 3x + 9}$$
$$2x - 3$$

Quotient: $x^2 + x + 3$; remainder: $2x - 3$ or
$$x^2 + x + 3 + \frac{2x-3}{x^2-x+3}$$

3. $(2x^3 - 7x^2 + 5) \div (x - 3)$

$$\begin{array}{r|rrrr} 3 & 2 & -7 & 0 & 5 \\ & & 6 & -3 & -9 \\ \hline & 2 & -1 & -3 & -4 \end{array}$$

The quotient is $2x^2 - x - 3$ remainder -4 or
$$2x^2 - x - 3 - \frac{4}{x-3}.$$

4. $(2x^3 + x^2 - 18x - 7) \div (x + 3)$

$$\begin{array}{r|rrrr} -3 & 2 & 1 & -18 & -7 \\ & & -6 & 15 & 9 \\ \hline & 2 & -5 & -3 & 2 \end{array}$$

The quotient is $2x^2 - 5x - 3$ remainder 2 or
$$2x^2 - 5x - 3 + \frac{2}{x+3}.$$

5. $F(1) = 1^{110} - 2 \cdot 1^{57} + 5 = 1 - 2 + 5 = 4$, so the remainder when $F(x) = x^{110} - 2x^{57} + 5$ is divided by $x - 1$ is 4.

6.
$$\begin{array}{r|rrrrr} -2 & 1 & 0 & 10 & 2 & -20 \\ & & -2 & 4 & -28 & 52 \\ \hline & 1 & -2 & 14 & -26 & 32 \end{array}$$
The remainder is 32, so $f(-2) = 32$.

7. Because -2 is a zero of the function $3x^3 - x^2 - 20x - 12$, $x + 2$ is a factor. Use synthetic division to find the depressed equation.
$$\begin{array}{r|rrrr} -2 & 3 & -1 & -20 & -12 \\ & & -6 & 14 & 12 \\ \hline & 3 & -7 & -6 & 0 \end{array}$$
Thus,
$$3x^3 - x^2 - 20x - 12 = (x+2)(3x^2 - 7x - 6).$$

Now solve $3x^2 - 7x - 6 = 0$ to find the two remaining zeros:
$$3x^2 - 7x - 6 = 0 \Rightarrow (x-3)(3x+2) = 0 \Rightarrow$$
$$x - 3 = 0 \text{ or } 3x + 2 = 0 \Rightarrow x = 3 \text{ or } x = -\frac{2}{3}$$
The solution set is $\left\{-2, -\frac{2}{3}, 3\right\}$.

8. $C(x) = 0.23x^3 - 4.255x^2 + 0.345x + 41.05$
$C(3) = 10 \Rightarrow C(x) = (x-3)Q(x) + 10 \Rightarrow$
$C(x) - 10 = (x-3)Q(x)$
So, 3 is zero of $C(x) - 10 =$
$0.23x^3 - 4.255x^2 + 0.345x + 31.05$.
We need to find another positive zero of $C(x) - 10$. Use synthetic division to find the depressed equation.
$$\begin{array}{r|rrrr} 3 & 0.23 & -4.255 & 0.345 & 31.05 \\ & & 0.69 & -10.695 & -31.05 \\ \hline & 0.23 & -3.565 & -10.35 & 0 \end{array}$$
Solve the depressed equation
$0.23x^2 - 3.565x - 10.35 = 0$ using the quadratic formula:
$$x = \frac{3.565 \pm \sqrt{(-3.565)^2 - 4(0.23)(-10.35)}}{2(0.23)}$$
≈ 18 or -2.5
The positive zero is 18.
Check by verifying that $C(18) = 10$.

3.3 Concepts and Vocabulary

1. In the division
$$\frac{x^4 - 2x^3 + 5x^2 - 2x + 1}{x^2 - 2x + 3} = x^2 + 2 + \frac{2x-5}{x^2-2x+3},$$
the dividend is $x^4 - 2x^3 + 5x^2 - 2x + 1$, the divisor is $x^2 - 2x + 3$, the quotient is $x^2 + 2$, and the remainder is $2x - 5$.

3. The Remainder Theorem states that if a polynomial $F(x)$ is divided by $(x-a)$, then the remainder $R = F(a)$.

5. True

7. False. If the polynomial $F(x)$ has $(x+2)$ as a factor, then $F(-2) = 0$.

3.3 Building Skills

9.

$$
\begin{array}{r}
3x - 2 \\
2x+1\overline{)6x^2 - x - 2} \\
\underline{-(6x^2 + 3x)} \\
-4x - 2 \\
\underline{-(-4x - 2)} \\
0
\end{array}
$$

In exercises 11–16, insert zero coefficients for missing terms.

11.

$$
\begin{array}{r}
3x^3 - 3x^2 - 3x + 6 \\
x+1\overline{)3x^4 + 0x^3 - 6x^2 + 3x - 7} \\
\underline{-(3x^4 + 3x^3)} \\
-3x^3 - 6x^2 \\
\underline{-(-3x^3 - 3x^2)} \\
-3x^2 + 3x \\
\underline{-(-3x^2 - 3x)} \\
6x - 7 \\
\underline{-(6x + 6)} \\
-13
\end{array}
$$

13.

$$
\begin{array}{r}
2x - 1 \\
2x^2 - x - 5\overline{)4x^3 - 4x^2 - 9x + 5} \\
\underline{-(4x^3 - 2x^2 - 10x)} \\
-2x^2 + x + 5 \\
\underline{-(-2x^2 + x + 5)} \\
0
\end{array}
$$

15.

$$
\begin{array}{r}
z^2 + 2z + 1 \\
z^2 - 2z + 1\overline{)z^4 + 0z^3 - 2z^2 + 0z + 1} \\
\underline{-(z^4 - 2z^3 + z^2)} \\
2z^3 - 3z^2 + 0z \\
\underline{-(2z^3 - 4z^2 + 2z)} \\
z^2 - 2z + 1 \\
\underline{-(z^2 - 2z + 1)} \\
0
\end{array}
$$

17.

$$
\begin{array}{r|rrrr}
1\!\!\downarrow & 1 & -1 & -7 & 2 \\
  &   & 1 & 0 & -7 \\
\hline
  & 1 & 0 & -7 & -5
\end{array}
$$

The quotient is $x^2 - 7$ and the remainder is -5.

19.

$$
\begin{array}{r|rrrr}
-2\!\!\downarrow & 1 & 4 & -7 & -10 \\
  &   & -2 & -4 & 22 \\
\hline
  & 1 & 2 & -11 & 12
\end{array}
$$

The quotient is $x^2 + 2x - 11$ and the remainder is 12.

21.

$$
\begin{array}{r|rrrrr}
2\!\!\downarrow & 1 & -3 & 2 & 4 & 5 \\
  &   & 2 & -2 & 0 & 8 \\
\hline
  & 1 & -1 & 0 & 4 & 13
\end{array}
$$

The quotient is $x^3 - x^2 + 4$ and the remainder is 13.

23.

$$
\begin{array}{r|rrrr}
\frac{1}{2}\!\!\downarrow & 2 & 4 & -3 & 1 \\[4pt]
  &   & 1 & \frac{5}{2} & -\frac{1}{4} \\[4pt]
\hline
  & 2 & 5 & -\frac{1}{2} & \frac{3}{4}
\end{array}
$$

The quotient is $2x^2 + 5x - \dfrac{1}{2}$ and the remainder is $\dfrac{3}{4}$.

25.

$$
\begin{array}{r|rrrr}
-\frac{1}{2}\!\!\downarrow & 2 & -5 & 3 & 2 \\[4pt]
  &   & -1 & 3 & -3 \\[4pt]
\hline
  & 2 & -6 & 6 & -1
\end{array}
$$

The quotient is $2x^2 - 6x + 6$ and the remainder is -1.

27.

$$
\begin{array}{r|rrrrrr}
1\!\!\downarrow & 1 & 1 & -7 & 2 & 1 & -1 \\
  &   & 1 & 2 & -5 & -3 & -2 \\
\hline
  & 1 & 2 & -5 & -3 & -2 & -3
\end{array}
$$

The quotient is $x^4 + 2x^3 - 5x^2 - 3x - 2$ and the remainder is -3.

29.

$$
\begin{array}{r|rrrrrr}
-1\!\!\downarrow & 1 & 0 & 0 & 0 & 0 & 1 \\
  &   & -1 & 1 & -1 & 1 & -1 \\
\hline
  & 1 & -1 & 1 & -1 & 1 & 0
\end{array}
$$

The quotient is $x^4 - x^3 + x^2 - x + 1$ remainder 0.

31.

$$
\begin{array}{r|rrrrrr}
-1\!\!\downarrow & 1 & 0 & 2 & -1 & 0 & 5 \\
  &   & -1 & 1 & -3 & 4 & -4 \\
\hline
  & 1 & -1 & 3 & -4 & 4 & -4 & 9
\end{array}
$$

The quotient is $x^5 - x^4 + 3x^3 - 4x^2 + 4x - 4$ remainder 9.

33. a.

$$
\begin{array}{r|rrrr}
1\!\!\downarrow & 1 & 3 & 0 & 1 \\
  &   & 1 & 4 & 4 \\
\hline
  & 1 & 4 & 4 & 5
\end{array}
$$

The remainder is 5, so $f(1) = 5$.

b.
$$
\begin{array}{r|rrrr}
-1 & 1 & 3 & 0 & 1 \\
   &   & -1 & -2 & 2 \\
\hline
   & 1 & 2 & -2 & 3
\end{array}
$$
The remainder is 3, so $f(-1) = 3$.

c.
$$
\begin{array}{r|rrrr}
\frac{1}{2} & 1 & 3 & 0 & 1 \\
   &   & \frac{1}{2} & \frac{7}{4} & \frac{7}{8} \\
\hline
   & 1 & \frac{7}{2} & \frac{7}{4} & \frac{15}{8}
\end{array}
$$
The remainder is $\frac{15}{8}$, so $f\left(\frac{1}{2}\right) = \frac{15}{8}$.

d.
$$
\begin{array}{r|rrrr}
10 & 1 & 3 & 0 & 1 \\
   &   & 10 & 130 & 1300 \\
\hline
   & 1 & 13 & 130 & 1301
\end{array}
$$
The remainder is 1301, so $f(10) = 1301$.

35. a.
$$
\begin{array}{r|rrrrr}
1 & 1 & 5 & -3 & 0 & -20 \\
  &   & 1 & 6 & 3 & 3 \\
\hline
  & 1 & 6 & 3 & 3 & -17
\end{array}
$$
The remainder is -17, so $f(1) = -17$.

b.
$$
\begin{array}{r|rrrrr}
-1 & 1 & 5 & -3 & 0 & -20 \\
   &   & -1 & -4 & 7 & -7 \\
\hline
   & 1 & 4 & -7 & 7 & -27
\end{array}
$$
The remainder is -27, so $f(-1) = -27$.

c.
$$
\begin{array}{r|rrrrr}
-2 & 1 & 5 & -3 & 0 & -20 \\
   &   & -2 & -6 & 18 & -36 \\
\hline
   & 1 & 3 & -9 & 18 & -56
\end{array}
$$
The remainder is -56, so $f(-2) = -56$.

d.
$$
\begin{array}{r|rrrrr}
2 & 1 & 5 & -3 & 0 & -20 \\
  &   & 2 & 14 & 22 & 44 \\
\hline
  & 1 & 7 & 11 & 22 & 24
\end{array}
$$
The remainder is 24, so $f(2) = 24$.

37. $f(1) = 2(1)^3 + 3(1)^2 - 6(1) + 1 = 0 \Rightarrow x - 1$ is a factor of $2x^3 + 3x^2 - 6x + 1$.
Check as follows:
$$
\begin{array}{r|rrrr}
1 & 2 & 3 & -6 & 1 \\
  &   & 2 & 5 & -1 \\
\hline
  & 2 & 5 & -1 & 0
\end{array}
$$

39. $f(-1) = 5(-1)^4 + 8(-1)^3 + (-1)^2$
$$+ 2(-1) + 4 = 0 \Rightarrow x + 1 \text{ is a}$$
factor of $5x^4 + 8x^3 + x^2 + 2x + 4$.
Check as follows:
$$
\begin{array}{r|rrrrr}
-1 & 5 & 8 & 1 & 2 & 4 \\
   &   & -5 & -3 & 2 & -4 \\
\hline
   & 5 & 3 & -2 & 4 & 0
\end{array}
$$

41. $f(2) = 2^4 + 2^3 - 2^2 - 2 - 18 = 0 \Rightarrow x - 2$ is a factor of $x^4 + x^3 - x^2 - x - 18$.
Check as follows:
$$
\begin{array}{r|rrrrr}
2 & 1 & 1 & -1 & -1 & -18 \\
  &   & 2 & 6 & 10 & 18 \\
\hline
  & 1 & 3 & 5 & 9 & 0
\end{array}
$$

43. $f(-2) = (-2)^6 - (-2)^5 - 7(-2)^4 + (-2)^3$
$$+ 8(-2)^2 + 5(-2) + 2 = 0 \Rightarrow x + 2$$
is a factor of
$x^6 - x^5 - 7x^4 + x^3 + 8x^2 + 5x + 2$.
Check as follows:
$$
\begin{array}{r|rrrrrrr}
-2 & 1 & -1 & -7 & 1 & 8 & 5 & 2 \\
   &   & -2 & 6 & 2 & -6 & -4 & -2 \\
\hline
   & 1 & -3 & -1 & 3 & 2 & 1 & 0
\end{array}
$$

45. $f(-1) = 0 = (-1)^3 + 3(-1)^2 + (-1) + k \Rightarrow$
$0 = 1 + k \Rightarrow k = -1$

47. $f(2) = 0 = 2(2)^3 + (2^2)k - 2k - 2 \Rightarrow$
$14 + 2k = 0 \Rightarrow k = -7$

In exercises 49–52, use synthetic division to find the remainder.

49.
$$
\begin{array}{r|rrrr}
2 & -2 & 4 & -4 & 9 \\
  &   & -4 & 0 & -8 \\
\hline
  & -2 & 0 & -4 & 1
\end{array}
$$
The remainder is 1, so $x - 2$ is not a factor of $-2x^3 + 4x^2 - 4x + 9$.

51.
$$
\begin{array}{r|rrrrr}
-2 & 4 & 9 & 3 & 1 & 4 \\
   &   & -8 & -2 & -2 & 2 \\
\hline
   & 4 & 1 & 1 & -1 & 6
\end{array}
$$
The remainder is 6, so $x + 2$ is not a factor of $4x^4 + 9x^3 + 3x^2 + x + 4$.

3.3 Applying the Concepts

53. $A = lw \Rightarrow l = \dfrac{A}{w} \Rightarrow$

$$
\begin{array}{r}
2x^2 + 1 \\
x^2 - x + 2\,\overline{\big)\,2x^4 - 2x^3 + 5x^2 - x + 2} \\
-(2x^4 - 2x^3 + 4x^2) \\
\hline
x^2 - x + 2 \\
-(x^2 - x + 2) \\
\hline
0
\end{array}
$$

The width is $2x^2 + 1$.

55. a. $R(40) = 3000 \Rightarrow R(40) - 3000 = 0$ and $R(60) = 3000 \Rightarrow R(60) - 3000 = 0$. Thus, 40 and 60 are zeros of $R(x) - 3000$. Therefore, $R(x) - 3000 = a(x - 40)(x - 60) = a(x^2 - 100x + 2400)$

Since $(30, 2400)$ lies on $R(x)$, we have
$2400 - 3000 = a(30 - 40)(30 - 60) \Rightarrow -600 = a(-10)(-30) \Rightarrow a = -2.$

Thus, $R(x) - 3000 = -2(x - 40)(x - 60) = -2x^2 + 200x - 4800.$

b. $R(x) - 3000 = -2x^2 + 200x - 4800 \Rightarrow R(x) = -2x^2 + 200x - 1800$

c. The maximum weekly revenue occurs at the vertex of the function, $\left(-\dfrac{b}{2a},\ R\left(-\dfrac{b}{2a}\right)\right)$.

$-\dfrac{b}{2a} = -\dfrac{200}{2(-2)} = 50$

$R(50) = -2(50)^2 + 200(50) - 1800 = 3200$

The maximum revenue is $3200 if the phone is priced at $50.

57. Since $t = 11$ represents 2002, we have $C(11) = 97.6 \Rightarrow C(x) = (x - 11)Q(x) + 97.6 \Rightarrow$
$C(t) - 97.6 = (t - 11)Q(t)$. So

$-0.0006t^3 - 0.0613t^2 + 2.0829t + 82.904 - 97.6 = -0.0006t^3 - 0.0613t^2 + 2.0829t - 14.6960$
$$= (x - 11)Q(x).$$

Use synthetic division to find $Q(x)$:

$$
\begin{array}{r|rrrr}
11 & -0.0006 & -0.0613 & 2.0829 & -14.6960 \\
 & & -0.0066 & -0.7469 & 14.6960 \\
\hline
 & -0.0006 & -0.0679 & 1.3360 & 0
\end{array}
$$

Now solve the depressed equation to find another zero:

$-0.0006^2 - 0.0679t + 1.3360 = 0 \Rightarrow t = \dfrac{0.0679 \pm \sqrt{0.0679^2 - 4(-0.0006)(1.3360)}}{2(-0.0006)} \Rightarrow$

$t = \dfrac{0.0679 \pm \sqrt{0.00781681}}{-0.0012} = \dfrac{0.0679 \pm 0.0884}{-0.0012} \Rightarrow t \approx -130.26$ or $t \approx 17.0939.$

Since we must find t greater than 0, $t = 17$, and the year is $1991 + 17 = 2008$.

59. $M(t) = -0.0027t^3 + 0.3681t^2 - 5.8645t + 195.2782$

Since $M(2) = 191.736$, $M(t) = (t-2)Q(t) + 195.2782 \Rightarrow M(t) - 195.2782 = (t-2)Q(t)$

We must find two other zeros of

$F(t) = M(t) - 185 = -0.0027t^3 + 0.3681t^2 - 5.8645t + 195.2782 - 185$

$\quad\quad = -0.0027t^3 + 0.3681t^2 - 5.8645t + 10.2782$

Because 2 is a zero of $F(t)$, use synthetic division to find $Q(t)$.

$$\begin{array}{r|rrrr} 2 & -0.0027 & 0.3681 & -5.8645 & 10.2782 \\ & & -0.0054 & 0.7254 & -10.2782 \\ \hline & -0.0027 & 0.3627 & -5.1391 & 0 \end{array}$$

Now use the quadratic formula to solve the depressed equation.

$$t = \frac{-0.3627 \pm \sqrt{0.3627^2 - 4(-0.0027)(-5.1391)}}{2(-0.0027)} = \frac{-0.3627 \pm \sqrt{0.0760}}{-0.0054} \Rightarrow t \approx 16.1 \text{ or } t \approx 118$$

Thus, the model shows that the Marine Corps had about 186,000 when $t \approx 16$, or in the year $1990 + 16 = 2006$.

3.3 Beyond the Basics

61. Divide $4x^3 + 8x^2 - 11x + 3$ by $\left(x - \dfrac{1}{2}\right)$:

$$\begin{array}{r|rrrr} \frac{1}{2} & 4 & 8 & -11 & 3 \\ & & 2 & 5 & -3 \\ \hline & 4 & 10 & -6 & 0 \end{array}$$

Now divide $4x^2 + 10x - 6$ by $\left(x - \dfrac{1}{2}\right)$:

$$\begin{array}{r|rrr} \frac{1}{2} & 4 & 10 & -6 \\ & & 2 & 6 \\ \hline & 4 & 12 & 0 \end{array}$$

Since $\left(x - \dfrac{1}{2}\right)$ does not divide $4x + 12$,

$\left(x - \dfrac{1}{2}\right)$ is a root of multiplicity 2 of

$4x^3 + 8x^2 - 11x + 3$.

63. a. $x + a$ is a factor of $x^n + a^n$ if n is an odd integer. The possible rational zeros of $x^n + a^n$ are $\{\pm a, \pm a^2, \pm a^3, \ldots, \pm a^n\}$. Since $x + a$ is a factor means that $-a$ is a root, then $(-a)^n + a^n = 0 \Rightarrow (-a)^n = -a^n$ only for odd values of n.

b. $x + a$ is a factor of $x^n - a^n$ if n is an even integer. The possible rational zeros of $x^n - a^n$ are $\{\pm a, \pm a^2, \pm a^3, \ldots, \pm a^n\}$. Since $x + a$ is a factor means that $-a$ is a root, then $(-a)^n - a^n = 0 \Rightarrow (-a)^n = a^n$ only for even values of n.

c. There is no value of n for which $x - a$ is a factor of $x^n + a^n$. The possible rational zeros of $x^n + a^n$ are $\{\pm a, \pm a^2, \pm a^3, \ldots, \pm a^n\}$. If $x - a$ is a factor, then a is a root, and $a^n + a^n = 0 \Rightarrow a^n = -a^n$, which is not possible.

d. $x - a$ is a factor of $x^n - a^n$ for all positive integers n. The possible rational zeros of $x^n - a^n$ are $\{\pm a, \pm a^2, \pm a^3, \ldots, \pm a^n\}$. If $x - a$ is a factor, then a is a root, and $a^n - a^n = 0$, which is true for all values of n. However, if n is negative, then $x^n - a^n = \dfrac{1}{x^{-n}} - \dfrac{1}{a^{-n}}$ and $x - a$ is not a factor.

65. a. Divide the divisor and the dividend by 2 so that the leading coefficient of the divisor is 1:

$$\frac{2x^3 + 3x^2 + 6x - 2}{2x - 1} = \frac{2x^3 + 3x^2 + 6x - 2}{2\left(x - \frac{1}{2}\right)}$$

$$= \frac{x^3 + \frac{3}{2}x^2 + 3x - 1}{\left(x - \frac{1}{2}\right)}$$

$$
\begin{array}{r|rrrr}
\frac{1}{2} & 1 & \frac{3}{2} & 3 & -1 \\
 & & \frac{1}{2} & 1 & 2 \\
\hline
 & 1 & 2 & 4 & 1
\end{array}
$$

Because the original polynomials were divided by 2 to use synthetic division, multiply the remainder by 2 to find the remainder for the original division, 2.

b. Divide the divisor and the dividend by 2 so that the leading coefficient of the divisor is 1:

$$\frac{2x^3 - x^2 - 4x + 1}{2x + 3} = \frac{2x^3 - x^2 - 4x + 1}{2\left(x + \frac{3}{2}\right)}$$

$$= \frac{x^3 - \frac{1}{2}x^2 - 2x + \frac{1}{2}}{\left(x + \frac{3}{2}\right)}$$

$$
\begin{array}{r|rrrr}
-\frac{3}{2} & 1 & -\frac{1}{2} & -2 & \frac{1}{2} \\
 & & -\frac{3}{2} & 3 & -\frac{3}{2} \\
\hline
 & 1 & -2 & 1 & -1
\end{array}
$$

Because the original polynomials were divided by 2 to use synthetic division, multiply the remainder by 2 to find the remainder for the original division, −2.

67. $f(x) = ax^3 + bx^2 + cx + d, \ a \neq 0$

There are two possibilities for the end behavior of f:

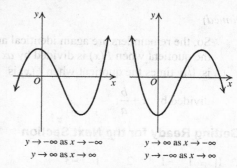

$y \to -\infty$ as $x \to -\infty$
$y \to \infty$ as $x \to \infty$

$y \to \infty$ as $x \to -\infty$
$y \to -\infty$ as $x \to \infty$

Therefore, there is one sign change from the left end of f to the right side of f or there are three sign changes. By the Intermediate Value Theorem, then, there is one zero (if there is one sign change) or there are three zeros (if there are three sign changes.)

69. $dp < 0$ means that d and p have opposite signs, and, thus, are on opposite sides of the x-axis. If $d > 0$ then $p < 0$. Since $kp < 0$, $k > 0$ also. The end behavior is $g(x) \to \infty$ as $x \to -\infty$ and $g(x) \to \infty$ as $x \to \infty$. However, f has end behavior shown in exercise 65. Therefore, the two graphs must intersect. Similar reasoning applies if $d < 0$.
Algebraic proof: Assume $d > 0$ and $p < 0$. Consider the fourth degree polynomial $g - f$. Then $(g - f)(0) = p - d < 0$. Since $k > 0$, $(g - f)(x)$ has upward end behavior on both ends, so $(g - f)(x)$ is eventually > 0. By the Intermediate Value Theorem, there is some c such that $(g - f)(c) = 0 \Rightarrow g(c) = f(c)$. Therefore, the graphs intersect.

71. Answers will vary. One example is
$$f(x) = x^4 + 2$$

3.3 Critical Thinking/Discussion/Writing

72. a. The quotient when $F(x)$ is divided by $2x - 6$ is one-half the quotient when $F(x)$ is divided by $x - 3$, but the remainders are identical.

b. If $\dfrac{F(x)}{x + \frac{b}{a}} = Q(x) + \dfrac{R(x)}{x + \frac{b}{a}}$, then

$$\frac{F(x)}{ax + b} = \frac{1}{a}\left[\frac{F(x)}{x + \frac{b}{a}}\right] = \frac{1}{a}\left[Q(x) + \frac{R(x)}{x + \frac{b}{a}}\right]$$

$$= \frac{1}{a}Q(x) + \frac{R(x)}{ax + b}$$

(continued on next page)

(continued)

So, the remainders are again identical and the quotient when $F(x)$ is divided by $ax + b$ is $1/a$ times the quotient when $F(x)$ is divided by $x + \dfrac{b}{a}$.

3.3 Getting Ready for the Next Section

73.
$$\begin{array}{r|rrrrr} 1| & 1 & -1 & 2 & 1 & -3 \\ & & 1 & 0 & 2 & 3 \\ \hline & 1 & 0 & 2 & 3 & 0 \end{array}$$
The quotient is $x^3 + 2x + 3$ and the remainder is 0.

75.
$$\begin{array}{r|rrrrrrr} 5| & -1 & 5 & 0 & 0 & 2 & -10 & 3 \\ & & -5 & 0 & 0 & 0 & 10 & 0 \\ \hline & -1 & 0 & 0 & 0 & 2 & 0 & 3 \end{array}$$
The quotient is $-x^5 + 2x$ and the remainder is 3.

77. The factors of 11 are $\pm 1,\ \pm 11$.

79. The factors of 51 are $\pm 1,\ \pm 3,\ \pm 17,\ \pm 51$.

81. $x^2 + 2x - 1$
Degree: 2
Leading coefficient: 1
Constant term: -1

83. $-7x^{10} + 3x^3 - 2x + 4$
Degree: 10
Leading coefficient: -7
Constant term: 4

85. $f(x) = (x-1)(x+2)^2$
The zeros are 1, with multiplicity 1, and -2, with multiplicity 2.

3.4 The Real Zeros of a Polynomial Function

3.4 Practice Problems

1. $F(x) = 2x^3 + 3x^2 - 6x - 8$
The factors of the constant term, -8, are $\{\pm 1, \pm 2, \pm 4, \pm 8\}$, and the factors of the leading coefficient, 2, are $\{\pm 1, \pm 2\}$. The possible rational zeros are
$$\left\{\pm \frac{1}{2}, \pm 1, \pm 2, \pm 4, \pm 8\right\}.$$

Use synthetic division to find one rational root:
$$\begin{array}{r|rrrr} -2| & 2 & 3 & -6 & -8 \\ & & -4 & 2 & 8 \\ \hline & 2 & -1 & -4 & 0 \end{array}$$
The remainder is 0, so -2 is a zero of the function.
$$2x^3 + 3x^2 - 6x - 8 = (x+2)\left(2x^2 - x - 4\right)$$
Now find the zeros of $2x^2 - x - 4$ using the quadratic formula:
$$x = \frac{1 \pm \sqrt{(-1)^2 - 4(2)(-4)}}{2(1)} = \frac{1 \pm \sqrt{33}}{2},\text{ which}$$
are not rational roots.
The only rational zero is $\{-2\}$.

2. $2x^3 - 9x^2 + 6x - 1 = 0$
The factors of the constant term, -1, are $\{\pm 1\}$, and the factors of the leading coefficient, 2, are $\{\pm 1, \pm 2\}$. The possible rational zeros are $\left\{\pm \frac{1}{2}, \pm 1\right\}$. Use synthetic division to find one rational zero:
$$\begin{array}{r|rrrr} \frac{1}{2}| & 2 & -9 & 6 & -1 \\ & & 1 & -4 & 1 \\ \hline & 2 & -8 & 2 & 0 \end{array}$$
Thus, $2x^3 - 9x^2 + 6x - 1 = 0 \Rightarrow$
$$\left(x - \frac{1}{2}\right)\left(2x^2 - 8x + 2\right) = 0 \Rightarrow$$
$$x - \frac{1}{2} = 0 \Rightarrow x = \frac{1}{2} \text{ or}$$
$$2x^2 - 8x + 2 = 0$$
$$x = \frac{-(-8) \pm \sqrt{(-8)^2 - 4(2)(2)}}{2(2)} = \frac{8 \pm \sqrt{48}}{4}$$
$$= 2 \pm \sqrt{3}$$
Solution set: $\left\{\frac{1}{2},\ 2 - \sqrt{3},\ 2 + \sqrt{3}\right\}$

3. $f(x) = 2x^5 + 3x^2 + 5x - 1$
There is one sign change in $f(x)$, so there is one positive zero.
$$f(-x) = 2(-x)^5 + 3(-x)^2 + 5(-x) - 1$$
$$= -2x^5 + 3x^2 - 5x - 1$$
There are two sign changes in $f(-x)$, so there are either 2 or 0 negative zeros.

4. $f(x) = 2x^3 + 5x^2 + x - 2$

The possible rational zeros are $\left\{\pm\dfrac{1}{2}, \pm 1, \pm 2\right\}$.

There is one sign change in $f(x)$, so there is one positive zero.

$f(-x) = 2(-x)^3 + 5(-x)^2 + (-x) - 2$
$ = -2x^3 + 5x^2 - x - 2$

There are two sign changes in $f(-x)$, so there are either 2 or 0 negative zeros. Try synthetic division by $x - k$ with $k = 1, 2, 3, \ldots$. The first integer that makes each number in the last row a 0 or a positive number is an upper bound on the zeros of $F(x)$. Then use synthetic division by $x - k$ with $k = -1, -2, -3, \ldots$.
The first negative integer for which the numbers in the last row alternate in sign is a lower bound on the zeros of $F(x)$. In this case, 1 is an upper bound and -3 is a lower bound.

1⌋	2	5	1	−2
		2	7	8
	2	7	8	6

−3⌋	2	5	1	−2
		−6	3	−12
	2	−1	4	−14

5. $f(x) = 3x^4 - 11x^3 + 22x - 12$

Step 1: f has at most 4 real zeros.

Step 2: $f(x) = 3x^4 - 11x^3 + 0x^2 + 22x - 12$

There are three sign changes in f, so f has 1 or 3 positive zeros.

$f(-x) = 3(-x)^4 - 11(-x)^3 + 0(-x)^2$
$ + 22(-x) - 12$
$ = 3x^4 + 11x^3 + 0x^2 - 22x - 12$

There is one sign change in $f(-x)$, so f has 1 negative zero.

Step 3: The factors of the constant term, -12, are $\{\pm 1, \pm 2, \pm 3, \pm 4, \pm 6, \pm 12\}$, and the factors of the leading coefficient, 3, are $\{\pm 1, \pm 3\}$. The possible rational zeros are

$\left\{\pm 1, \pm 2, \pm 3, \pm 4, \pm 6, \pm 12, \pm\dfrac{1}{3}, \pm\dfrac{2}{3}, \pm\dfrac{4}{3}\right\}$.

Steps 4−7: Test for zeros until a zero or an upper bound is found. Try a positive possibility:

3⌋	3	−11	0	22	−12
		9	−6	−18	12
	3	−2	−6	4	0

Since 3 is a zero, we have

$f(x) = (x - 3)(3x^3 - 2x^2 - 6x + 4)$.

Now find the a zero of
$Q_1(x) = 3x^3 - 2x^2 - 6x + 4$. The factors of the constant term, 4, are $\{\pm 1, \pm 2, \pm 4\}$, and the factors of the leading coefficient, 3, are $\{\pm 1, \pm 3\}$. The possible rational zeros are

$\left\{\pm 1, \pm 2, \pm 4, \pm\dfrac{1}{3}, \pm\dfrac{2}{3}, \pm\dfrac{4}{3}\right\}$. Try the positive possibilities first.

$\frac{2}{3}$⌋	3	−2	−6	4
		2	0	−4
	3	0	−6	0

Thus, $f(x) = (x - 3)\left(x - \dfrac{2}{3}\right)(3x^2 - 6)$.

Now solve $3x^2 - 6 = 0$.

$3x^2 - 6 \Rightarrow x^2 = 2 \Rightarrow x = \pm\sqrt{2}$

Solution set: $\left\{\dfrac{2}{3}, 3, \sqrt{2}, -\sqrt{2}\right\}$

6. $f(x) = 3x^3 - x^2 - 9x + 3$

Step 1: Since the degree, 3, is odd, and the leading coefficient, 3, is positive, the end behavior is similar to that of $y = x^3$.

Step 2: Solve $3x^3 - x^2 - 9x + 3 = 0$ to find the real zeros. There are two sign changes in f, so f has either 2 or 0 positive zeros.

$f(-x) = 3(-x)^3 - (-x)^2 - 9(-x) + 3$
$ = -3x^3 - x^2 + x + 3$

There is one sign change in $f(-x)$, so there is one negative zero. By the Rational Root Theorem, the possible rational roots are $\left\{\pm 1, \pm 3, \pm\dfrac{1}{3}\right\}$. Trying each value, we find that $\dfrac{1}{3}$ is a rational zero.

$\frac{1}{3}$⌋	3	−1	−9	3
		1	0	−3
	3	0	−9	0

(continued on next page)

(continued)

Solve the depressed equation $3x^2 - 9 = 0$ to find the remaining zeros.

$3x^2 - 9 = 0 \Rightarrow 3x^2 = 9 \Rightarrow x^2 = 3 \Rightarrow x = \pm\sqrt{3}$

Step 3: The three zeros divide the x-axis into four intervals, $\left(-\infty, -\sqrt{3}\right)$, $\left(-\sqrt{3}, \frac{1}{3}\right)$, $\left(\frac{1}{3}, \sqrt{3}\right)$, and $\left(\sqrt{3}, \infty\right)$.

Step 4: Determine the sign of a test value in each interval

Interval	Test point	Value of $f(x)$	Above/below x-axis
$\left(-\infty, -\sqrt{3}\right)$	-2	-7	below
$\left(-\sqrt{3}, \frac{1}{3}\right)$	0	3	above
$\left(\frac{1}{3}, \sqrt{3}\right)$	1	-4	below
$\left(\sqrt{3}, \infty\right)$	2	5	above

Plot the zeros, y-intercepts, and test points, and then join the points with a smooth curve.

$y = 3x^3 - x^2 - 9x + 3$

3.4 Concepts and Vocabulary

1. Let $P(x) = a_n x^n + \cdots + a_0$ with integer coefficients. If $\frac{p}{q}$ is a rational zero of $P(x)$, then $\frac{p}{q} = \dfrac{\text{possible factors of } a_0}{\text{possible factors of } a_n}$.

3. The number of positive zeros of a polynomial function $P(x)$, is equal to the number of <u>variations</u> of sign of $P(x)$ or less than that number by <u>an even integer</u>.

5. False. Possible rational zeros of $P(x) = 9x^3 - 9x^2 - x + 1$ are $\left\{\pm\frac{1}{9}, \pm\frac{1}{3}, \pm 1\right\}$.

7. True

3.4 Building Skills

9. The factors of the constant term, 5, are $\{\pm 1, \pm 5\}$, and the factors of the leading coefficient, 3, are $\{\pm 1, \pm 3\}$. The possible rational zeros are $\left\{\pm\frac{1}{3}, \pm 1, \pm\frac{5}{3}, \pm 5\right\}$.

11. The factors of the constant term, 6, are $\{\pm 1, \pm 2, \pm 3, \pm 6\}$, and the factors of the leading coefficient, 4, are $\{\pm 1, \pm 2, \pm 4\}$. The possible rational zeros are $\left\{\pm\frac{1}{4}, \pm\frac{1}{2}, \pm\frac{3}{4}, \pm 1, \pm\frac{3}{2}, \pm 2, \pm 3, \pm 6\right\}$.

13. The factors of the constant term, 4, are $\{\pm 1, \pm 2, \pm 4\}$, and the factors of the leading coefficient, 1, are $\{\pm 1\}$. The possible rational zeros are $\{\pm 1, \pm 2, \pm 4\}$. Use synthetic division to find one rational root:

$$
\begin{array}{r|rrrr}
1 & 1 & -1 & -4 & 4 \\
  &   & 1 & 0 & -4 \\
\hline
  & 1 & 0 & -4 & 0 \\
\end{array}
$$

So,
$$x^3 - x^2 - 4x + 4 = (x-1)(x^2 - 4)$$
$$= (x-1)(x-2)(x+2) \Rightarrow$$
the rational zeros are $\{-2, 1, 2\}$.

15. The factors of the constant term, 2, are $\{\pm 1, \pm 2\}$, and the factors of the leading coefficient, 1, are $\{\pm 1\}$. The possible rational zeros are $\{\pm 1, \pm 2\}$. Use synthetic division to find one rational root:

$$
\begin{array}{r|rrrr}
-1 & 1 & 1 & 2 & 2 \\
   &   & -1 & 0 & -2 \\
\hline
   & 1 & 0 & 2 & 0 \\
\end{array}
$$

So, $x^3 + x^2 + 2x + 2 = (x+1)(x^2 + 2) \Rightarrow$ the rational zero is $\{-1\}$.

17. The factors of the constant term, 6, are $\{\pm 1, \pm 2, \pm 3, \pm 6\}$, and the factors of the leading coefficient, 2, are $\{\pm 1, \pm 2\}$. The possible rational zeros are $\left\{\pm\dfrac{1}{2}, \pm 1, \pm\dfrac{3}{2}, \pm 2, \pm 3, \pm 6\right\}$. Use synthetic division to find one rational root:

$$
\begin{array}{r|rrrr}
2 & 2 & 1 & -13 & 6 \\
  &   & 4 & 10 & -6 \\
\hline
  & 2 & 5 & -3 & 0
\end{array}
$$

$$2x^3 + x^2 - 13x + 6 = (x-2)(2x^2 + 5x - 3)$$
$$= (x-2)(x+3)(2x-1) \Rightarrow$$

the rational zeros are $\left\{-3, \dfrac{1}{2}, 2\right\}$.

19. The factors of the constant term, −2, are $\{\pm 1, \pm 2\}$, and the factors of the leading coefficient, 3, are $\{\pm 1, \pm 3\}$. The possible rational zeros are $\left\{\pm\dfrac{1}{3}, \pm\dfrac{2}{3}, \pm 1, \pm 2\right\}$. Use synthetic division to find one rational root:

$$
\begin{array}{r|rrrr}
\frac{2}{3} & 3 & -2 & 3 & -2 \\
  &   & 2 & 0 & 2 \\
\hline
  & 3 & 0 & 3 & 0
\end{array}
$$

$$3x^3 - 2x^2 + 3x - 2 = \left(x - \dfrac{2}{3}\right)(3x^2 + 3) \Rightarrow \text{ the}$$

rational zero is $\left\{\dfrac{2}{3}\right\}$.

21. The factors of the constant term, 2, are $\{\pm 1, \pm 2\}$, and the factors of the leading coefficient, 3, are $\{\pm 1, \pm 3\}$. The possible rational zeros are $\left\{\pm\dfrac{1}{3}, \pm\dfrac{2}{3}, \pm 1, \pm 2\right\}$. Use synthetic division to find one rational root:

$$
\begin{array}{r|rrrr}
-\frac{1}{3} & 3 & 7 & 8 & 2 \\
  &   & -1 & -2 & -2 \\
\hline
  & 3 & 6 & 6 & 0
\end{array}
$$

$$3x^3 + 7x^2 + 8x + 2 = \left(x + \dfrac{1}{3}\right)(3x^2 + 6x + 6) \Rightarrow$$

the rational zero is $\left\{-\dfrac{1}{3}\right\}$.

23. The factors of the constant term, −2, are $\{\pm 1, \pm 2\}$, and the factors of the leading coefficient, 1, are $\{\pm 1\}$. The possible rational zeros are $\{\pm 1, \pm 2\}$. Use synthetic division to find one rational root:

$$
\begin{array}{r|rrrrr}
-1 & 1 & -1 & -1 & -1 & -2 \\
  &   & -1 & 2 & -1 & 2 \\
\hline
  & 1 & -2 & 1 & -2 & 0
\end{array}
$$

So, −1 is a rational zero. Use synthetic division again to find another rational zero:

$$x^4 - x^3 - x^2 - x - 2 = (x+1)(x^3 - 2x^2 + x - 2)$$

$$
\begin{array}{r|rrrr}
2 & 1 & -2 & 1 & -2 \\
  &   & 2 & 0 & 2 \\
\hline
  & 1 & 0 & 1 & 0
\end{array}
$$

$$x^4 - x^3 - x^2 - x - 2 = (x+1)(x-2)(x^2+1) \Rightarrow$$
the rational zeros are $\{-1, 2\}$.

25. The factors of the constant term, 12, are $\{\pm 1, \pm 2, \pm 3, \pm 4, \pm 6, \pm 12\}$, and the factors of the leading coefficient, 1, are $\{\pm 1\}$. The possible rational zeros are $\{\pm 1, \pm 2, \pm 3, \pm 4, \pm 6, \pm 12\}$. Use synthetic division to find one rational root:

$$
\begin{array}{r|rrrrr}
-3 & 1 & -1 & -13 & 1 & 12 \\
  &   & -3 & 12 & 3 & -12 \\
\hline
  & 1 & -4 & -1 & 4 & 0
\end{array}
$$

So, −3 is a rational zero. Use synthetic division again to find another rational zero:

$$x^4 - x^3 - 13x^2 + x + 12$$
$$= (x+3)(x^3 - 4x^2 - x + 4)$$

$$
\begin{array}{r|rrrr}
-1 & 1 & -4 & -1 & 4 \\
  &   & -1 & 5 & -4 \\
\hline
  & 1 & -5 & 4 & 0
\end{array}
$$

So, −1 is also a rational zero.

$$x^4 - x^3 - 13x^2 + x + 12$$
$$= (x+3)(x+1)(x^2 - 5x + 4)$$
$$= (x+3)(x+1)(x-4)(x-1) \Rightarrow$$
the rational zeros are $\{-3, -1, 1, 4\}$.

27. The factors of the constant term, 1, are $\{\pm 1\}$, and the factors of the leading coefficient, 1, are $\{\pm 1\}$. The possible rational zeros are $\{\pm 1\}$. Use synthetic division to find one rational root:

$$
\begin{array}{r|rrrrr}
1 & 1 & -2 & 10 & -1 & 1 \\
  &   & 1 & -1 & 9 & 8 \\
\hline
  & 1 & -1 & 9 & 8 & 9
\end{array}
$$

(continued on next page)

(*continued*)

The remainder is 9 so, 1 is not a rational zero.
Try −1:

$$\begin{array}{r|rrrrr} -1 & 1 & -2 & 10 & -1 & 1 \\ & & -1 & 3 & -13 & 14 \\ \hline & 1 & -3 & 13 & -14 & 15 \end{array}$$

The remainder is 15 so, −1 is not a rational zero. Therefore, there are no rational zeros.

29. $f(x) = 6x^3 + 13x^2 + x - 2$

Zeros: $\left\{-2, -\dfrac{1}{2}, \dfrac{1}{3}\right\}$

31. $f(x) = 2x^3 - 3x^2 - 14x + 21$

Zeros: $\left\{\pm\sqrt{7}, \dfrac{3}{2}\right\}$

33. $f(x) = 4x^4 + 4x^3 - 3x^2 - 2x + 1$

Zeros: $\left\{-1, \dfrac{1}{2}\right\}$

35. $f(x) = 5x^3 - 2x^2 - 3x + 4;$
$f(-x) = 5(-x)^3 - 2(-x)^2 - 3(-x) + 4$
$\qquad = -5x^3 - 2x^2 + 3x + 4$

There are two sign changes in $f(x)$, so there are either 2 or 0 positive zeros. There is one sign change in $f(-x)$, so there is one negative zero.

37. $f(x) = 2x^3 + 5x^2 - x + 2;$
$f(-x) = 2(-x)^3 + 5(-x)^2 - (-x) + 2$
$\qquad = -2x^3 + 5x^2 + x + 2$

There are two sign changes in $f(x)$, so there are either 2 or 0 positive zeros. There is one sign change in $f(-x)$, so there is one negative zero.

39. $h(x) = 2x^5 - 5x^3 + 3x^2 + 2x - 1$
$h(-x) = 2(-x)^5 - 5(-x)^3 + 3(-x)^2 + 2(-x) - 1$
$\qquad = -2x^5 + 5x^3 + 3x^2 - 2x - 1$

There are three sign changes in $h(x)$, so there are either 1 or 3 positive zeros. There are two sign changes in $h(-x)$, so there are either 2 or 0 negative zeros.

41. $G(x) = -3x^4 - 4x^3 + 5x^2 - 3x + 7;$
$G(-x) = -3(-x)^4 - 4(-x)^3 + 5(-x)^2$
$\qquad\qquad\qquad - 3(-x) + 7$
$\qquad = -3x^4 + 4x^3 + 5x^2 + 3x + 7$

There are three sign changes in $G(x)$, so there are either 3 or 1 positive zeros. There is one sign change in $G(-x)$, so there is 1 negative zero.

43. $f(x) = x^4 + 2x^2 + 4$
$f(-x) = (-x)^4 + 2(-x)^2 + 4 = x^4 + 2x^2 + 4$

There are no sign changes in $f(x)$, nor are there sign changes in $f(-x)$. Therefore, there are no positive zeros and no negative zeros.

45. $g(x) = 2x^5 + x^3 + 3x$
$g(-x) = 2(-x)^5 + (-x)^3 + 3(-x)$
$\qquad = -2x^5 - x^3 - 3x$

There are no sign changes in $g(x)$, nor are there sign changes in $g(-x)$. Therefore, there are no positive zeros and no negative zeros.

47. $h(x) = -x^5 - 2x^3 + 4$

$h(-x) = -(-x)^5 - 2(-x)^3 + 4$

$\quad\quad = x^5 + x^3 + 4$

There is one sign change in $h(x)$, and no sign changes in $g(-x)$. Therefore, there is one positive zero and there are no negative zeros.

49. The possible rational zeros are $\left\{\pm\dfrac{1}{3}, \pm 1, \pm 3\right\}$.

There are three sign changes in $f(x)$, so there are either 3 or 1 positive zeros. There are no sign changes in $f(-x)$, so there are no negative zeros. Using synthetic division with $k = 1, 2, 3, \ldots$ and $k = -1, -2, -3, \ldots$ gives an upper bound of 1 and a lower bound of -1.

1⌋	3	−1	9	−3
		3	2	11
	3	2	11	8

−1⌋	3	−1	9	−3
		−3	4	−13
	3	−4	13	−16

51. The possible rational zeros are $\left\{\pm\dfrac{1}{3}, \pm 1, \pm\dfrac{7}{3}, \pm 7\right\}$. There are no sign changes in $F(x)$, so there are no positive zeros. There are three sign changes in $F(-x)$, so there are either 3 or 1 negative zeros. Using synthetic division with $k = 1, 2, 3, \ldots$ and $k = -1, -2, -3, \ldots$ gives an upper bound of 1 and a lower bound of -3.

1⌋	3	2	5	7
		3	5	10
	3	5	10	17

−3⌋	3	2	5	7
		−9	21	−75
	3	−7	25	−68

53. The possible rational zeros are $\{\pm 1, \pm 31\}$.

There are two sign changes in $h(x)$, so there are either 2 or 0 positive zeros. There are two sign changes in $h(-x)$, so there are either 2 or 0 negative zeros. Using synthetic division with $k = 1, 2, 3, \ldots$ and $k = -1, -2, -3, \ldots$ gives an upper bound of 31 and a lower bound of -31.

31⌋	1	3	−15	−9	31
		31	1054	32,209	998,200
	1	34	1039	32,200	998,231

−31⌋	1	3	−15	−9	31
		−31	868	−26,443	820,012
	1	−28	853	−26,452	820,043

55. The possible rational zeros are
$\left\{\pm\dfrac{1}{6}, \pm\dfrac{1}{3}, \pm\dfrac{1}{2}, \pm 1, \pm\dfrac{7}{6}, \pm\dfrac{7}{3}, \pm\dfrac{7}{2}, \pm 7\right\}$.

There are two sign changes in $f(x)$, so there are either 2 or 0 positive zeros. There are two sign changes in $f(-x)$, so there are either 2 or 0 negative zeros. Using synthetic division with $k = 1, 2, 3, \ldots$ and $k = -1, -2, -3, \ldots$ gives an upper bound of 4 and a lower bound of -4.

4⌋	6	1	−43	−7	7
		24	100	228	884
	6	25	57	221	891

−4⌋	6	1	−43	−7	7
		−24	92	−196	812
	6	−23	49	−203	819

In exercises 57–76, first check to see the number of possible positive zeros and the number of possible negative zeros.

57. $f(x) = x^3 + 5x^2 - 8x + 2$

The function has degree 3, so there are three zeros, zero or two possible positive zeros; one possible negative zero.

The possible rational zeros are $\{\pm 1, \pm 2\}$.

1⌋	1	5	−8	2
		1	6	−2
	1	6	−2	0

So, 1 is a zero. Now solve the depressed equation $x^2 + 6x - 2 = 0$.

$$x = \frac{-6 \pm \sqrt{36 - 4(1)(-2)}}{2(1)} = \frac{-6 \pm \sqrt{44}}{2}$$

$$= -3 \pm \sqrt{11}$$

The solution set is $\left\{1, -3 \pm \sqrt{11}\right\}$.

59. $f(x) = 2x^3 - x^2 - 6x + 3$

The function has degree 3, so there are three zeros, zero or two possible positive zeros; one possible negative zero. The possible rational zeros are $\left\{\pm 1, \pm 3, \pm\dfrac{1}{2}, \pm\dfrac{3}{2}\right\}$.

$\frac{1}{2}$⌋	2	−1	−6	3
		1	0	−3
	2	0	−6	0

So, 1/2 is a zero.

(continued on next page)

(continued)

Now solve the depressed equation
$2x^2 - 6 = 0$.

$2x^2 - 6 = 0 \Rightarrow x^2 = 3 \Rightarrow x = \pm\sqrt{3}$

Solution set: $\left\{-\sqrt{3}, \frac{1}{2}, \sqrt{3}\right\}$

61. $f(x) = 2x^3 - 9x^2 + 6x - 1$

The function has degree 3, so there are three zeros, one or three possible positive zeros; no possible negative zeros.

The possible rational zeros are $\left\{\pm 1, \pm\frac{1}{2}\right\}$.

$$\begin{array}{r|rrrr} \frac{1}{2} & 2 & -9 & 6 & -1 \\ & & 1 & -4 & 1 \\ \hline & 2 & -8 & 2 & 0 \end{array}$$

So, 1/2 is a zero. Now solve the depressed equation $2x^2 - 8x + 2 = 0$

$x = \dfrac{-(-8) \pm \sqrt{(-8)^2 - 4(2)(2)}}{2(2)} = \dfrac{8 \pm \sqrt{48}}{4}$

$= 2 \pm \sqrt{3}$

Solution set: $\left\{2 - \sqrt{3}, \frac{1}{2}, 2 + \sqrt{3}\right\}$.

63. $f(x) = x^4 + x^3 - 5x^2 - 3x + 6$

The function has degree 4, so there are four zeros, zero or two possible positive zeros; zero or two possible negative zeros.
The possible rational zeros are
$\{\pm 1, \pm 2, \pm 3, \pm 6\}$.

$$\begin{array}{r|rrrrr} 1 & 1 & 1 & -5 & -3 & 6 \\ & & 1 & 2 & -3 & -6 \\ \hline & 1 & 2 & -3 & -6 & 0 \end{array}$$

So, 1 is a zero. Now find a zero of the depressed equation $x^3 + 2x^2 - 3x - 6 = 0$.

$$\begin{array}{r|rrrr} -2 & 1 & 2 & -3 & -6 \\ & & -2 & 0 & 6 \\ \hline & 1 & 0 & -3 & 0 \end{array}$$

So, −2 is a zero. Now solve the depressed equation $x^2 - 3 = 0$.

$x^2 - 3 = 0 \Rightarrow x^2 = 3 \Rightarrow x = \pm\sqrt{3}$

Solution set: $\left\{-2, -\sqrt{3}, 1, \sqrt{3}\right\}$

65. $f(x) = x^4 - 3x^3 + 3x - 1$

The function has degree 4, so there are four zeros, one or three possible positive real zeros; one possible negative real zero. The possible rational zeros are $\{\pm 1\}$.

$$\begin{array}{r|rrrrr} 1 & 1 & -3 & 0 & 3 & -1 \\ & & 1 & -2 & -2 & 1 \\ \hline & 1 & -2 & -2 & 1 & 0 \end{array}$$

So, 1 is a zero. Now find a zero of the depressed equation $g(x) = x^3 - 2x^2 - 2x + 1$.

$$\begin{array}{r|rrrr} -1 & 1 & -2 & -2 & 1 \\ & & -1 & 3 & -1 \\ \hline & 1 & -3 & 1 & 0 \end{array}$$

So, −1 is a zero. Now solve the depressed equation $x^2 - 3x + 1 = 0$

$x = \dfrac{3 \pm \sqrt{9 - 4(1)(1)}}{2(1)} = \dfrac{3 \pm \sqrt{5}}{2}$

The solution set is $\left\{\pm 1, \dfrac{3 \pm \sqrt{5}}{2}\right\}$.

67. $f(x) = 2x^4 - 5x^3 - 4x^2 + 15x - 6$

The function has degree 4, so there are four zeros, one or three possible positive real zeros; one possible negative real zero. The possible rational zeros are
$$\left\{\pm 1, \pm 2, \pm 3, \pm 6, \pm\frac{1}{2}, \pm\frac{3}{2}\right\}.$$

$$\begin{array}{r|rrrrr} 2 & 2 & -5 & -4 & 15 & -6 \\ & & 4 & -2 & -12 & 6 \\ \hline & 2 & -1 & -6 & 3 & 0 \end{array}$$

So, 2 is a zero. Now find a zero of the depressed equation $g(x) = 2x^3 - x^2 - 6x + 3$.

$$\begin{array}{r|rrrr} \frac{1}{2} & 2 & -1 & -6 & 3 \\ & & 1 & 0 & -3 \\ \hline & 2 & 0 & -6 & 0 \end{array}$$

So, 1/2 is a zero. Now solve the depressed equation $2x^2 - 6 = 0$.

$2x^2 - 6 = 0 \Rightarrow x^2 = 3 \Rightarrow x = \pm\sqrt{3}$

The solution set is $\left\{-\sqrt{3}, \frac{1}{2}, \sqrt{3}, 2\right\}$.

69. $f(x) = 6x^4 - x^3 - 13x^2 + 2x + 2$

The function has degree 4, so there are four zeros, zero or two possible positive real zeros; zero or two possible negative real zeros. The possible rational zeros are

$$\left\{ \pm 1, \pm 2, \pm \frac{1}{2}, \pm \frac{2}{3}, \pm \frac{1}{6}, \pm \frac{1}{3} \right\}.$$

$$\begin{array}{r|rrrrr} \frac{1}{2} & 6 & -1 & -13 & 2 & 2 \\ & & 3 & 1 & -6 & -2 \\ \hline & 6 & 2 & -12 & -4 & 0 \end{array}$$

So, 1/2 is a zero. Now find a zero of the depressed equation

$$g(x) = 6x^3 + 2x^2 - 12x - 4.$$

$$\begin{array}{r|rrrr} -\frac{1}{3} & 6 & 2 & -12 & -4 \\ & & -2 & 0 & 4 \\ \hline & 6 & 0 & -12 & 0 \end{array}$$

So, $-1/3$ is a zero. Now solve the depressed equation $6x^2 - 12 = 0$.

$6x^2 - 12 = 0 \Rightarrow x^2 = 2 \Rightarrow x = \pm\sqrt{2}$

The solution set is $\left\{ -\sqrt{2}, -\frac{1}{3}, \frac{1}{2}, \sqrt{2} \right\}$.

71. $f(x) = x^5 - 2x^4 - 4x^3 + 8x^2 + 3x - 6$

The function has degree 5, so there are five zeros. There are three sign changes, so there are either 1, 3, or 5 positive real zeros. There are two sign changes in $f(-x)$, so there are zero or two negative real zeros. The possible rational zeros are $\{\pm 1, \pm 3, \pm 6\}$. Using synthetic division to test the positive values, we find that one zero is 1:

$$\begin{array}{r|rrrrrr} 1 & 1 & -2 & -4 & 8 & 3 & -6 \\ & & 1 & -1 & -5 & 3 & 6 \\ \hline & 1 & -1 & -5 & 3 & 6 & 0 \end{array}$$

The zeros of the depressed function $x^4 - x^3 - 5x^2 + 3x + 6$ are also zeros of f. Use synthetic division again to find the next zero, 2:

$$\begin{array}{r|rrrrr} 2 & 1 & -1 & -5 & 3 & 6 \\ & & 2 & 2 & -6 & -6 \\ \hline & 1 & 1 & -3 & -3 & 0 \end{array}$$

$x^5 - 2x^4 - 4x^3 + 8x^2 + 3x - 6$
$= (x-1)(x-2)\left(x^3 + x^2 - 3x - 3\right)$

Now find a zero of the depressed function $x^3 + x^2 - 3x - 3$. Use synthetic division again to find the next zero, -1:

$$\begin{array}{r|rrrr} -1 & 1 & 1 & -3 & -3 \\ & & -1 & 0 & 3 \\ \hline & 1 & 0 & -3 & 0 \end{array}$$

$x^5 - 2x^4 - 4x^3 + 8x^2 + 3x - 6$
$= (x-1)(x-2)(x+1)\left(x^2 - 3\right)$

Now solve the depressed equation: $x^2 - 3 = 0$.

$x^2 - 3 = 0 \Rightarrow x^2 = 3 \Rightarrow x = \pm\sqrt{3}$

Solution set: $\left\{ -\sqrt{3}, -1, 1, \sqrt{3}, 2 \right\}$

73. $f(x) = 2x^5 + x^4 - 11x^3 - x^2 + 15x - 6$

The function has degree 5, so there are five zeros. There are three sign changes, so there are either 1 or 3 positive real zeros. There are two sign changes in $f(-x)$, so there are zero or two negative real zeros. The possible rational zeros are $\left\{\pm 1, \pm 2, \pm 3, \pm 6, \pm \frac{1}{2}, \pm \frac{3}{2}\right\}$. Using synthetic division to test the positive values, we find that one zero is 1:

$$\begin{array}{r|rrrrrr} 1 & 2 & 1 & -11 & -1 & 15 & -6 \\ & & 2 & 3 & -8 & -9 & 6 \\ \hline & 2 & 3 & -8 & -9 & 6 & 0 \end{array}$$

The zeros of the depressed function $2x^4 + 3x^3 - 8x^2 - 9x + 6$ are also zeros of f. Use synthetic division again to find the next zero, -2:

$$\begin{array}{r|rrrrr} -2 & 2 & 3 & -8 & -9 & 6 \\ & & -4 & 2 & 12 & -6 \\ \hline & 2 & -1 & -6 & 3 & 0 \end{array}$$

$2x^5 + x^4 - 11x^3 - x^2 + 15x - 6$
$= (x-1)(x+2)\left(2x^3 - x^2 - 6x + 3\right)$

Now find a zero of the depressed function $2x^3 - x^2 - 6x + 3$. Use synthetic division again to find the next zero, 1/2:

$$\begin{array}{r|rrrr} \frac{1}{2} & 2 & -1 & -6 & 3 \\ & & 1 & 0 & -3 \\ \hline & 2 & 0 & -6 & 0 \end{array}$$

$2x^5 + x^4 - 11x^3 - x^2 + 15x - 6$
$= (x-1)(x+2)\left(x - \frac{1}{2}\right)\left(2x^2 - 6\right)$

(continued on next page)

(*continued*)

Now solve the depressed equation:

$2x^2 - 6 = 0.$

$2x^2 - 6 = 0 \Rightarrow x^2 = 3 \Rightarrow x = \pm\sqrt{3}$

Solution set: $\left\{-2, -\sqrt{3}, \dfrac{1}{2}, 1, \sqrt{3}\right\}$

75. $f(x) = 2x^5 - 13x^4 + 27x^3 - 17x^2 - 5x + 6$

The function has degree 5, so there are five zeros. There are four sign changes, so there are either 0, 2, or 4 positive real zeros. There is one sign change in $f(-x)$, so there is one possible negative real zero. The possible rational zeros are

$$\left\{\pm 1, \pm 2, \pm 3, \pm 6, \pm\dfrac{1}{2}, \pm\dfrac{3}{2}\right\}.$$

Using synthetic division to test the positive values, we find that one zero is 1:

```
1| 2  -13   27  -17   -5    6
       2  -11   16   -1   -6
   2  -11   16   -1   -6    0
```

The zeros of the depressed function $2x^4 - 11x^3 + 16x^2 - x - 6$ are also zeros of f. Use synthetic division again to find the next zero, 1:

```
1| 2  -11   16   -1   -6
       2   -9    7    6
   2   -9    7    6    0
```

$2x^5 - 13x^4 + 27x^3 - 17x^2 - 5x + 6$
$= (x-1)(x-1)(2x^3 - 9x^2 + 7x + 6)$

The zeros of the depressed function $2x^3 - 9x^2 + 7x + 6$ are also zeros of P. Use synthetic division again to find the next zero, 2.

```
2| 2   -9    7    6
       4  -10   -6
   2   -5   -3    0
```

$2x^5 - 13x^4 + 27x^3 - 17x^2 - 5x + 6$
$= (x-1)(x-1)(x-2)(2x^2 - 5x - 3)$

Now solve the depressed equation:

$2x^2 - 5x - 3 = 0.$

$2x^2 - 5x - 3 = 0 \Rightarrow (2x+1)(x-3) = 0 \Rightarrow$

$x = -\dfrac{1}{2}, 3$

Solution set: $\left\{-\dfrac{1}{2}, 1 \text{ (multiplicity 2)}, 2, 3\right\}$

3.4 Applying the Concepts

77. The length of the rectangle is $x^2 - 2x + 3$ and its width is $x - 2$. Its area is 306 square units, so we have $\left(x^2 - 2x + 3\right)(x - 2) = 306 \Rightarrow$

$x^3 - 4x^2 + 7x - 6 = 306 \Rightarrow$

$x^3 - 4x^2 + 7x - 312 = 0.$

There are 3 sign changes in $f(x)$ and no sign changes in $f(-x)$, so there are 1 or 3 possible positive real zeros and no possible negative real zeros. The possible rational zeros are

$\{\pm 1, \pm 2, \pm 3, \pm 4, \pm 6, \pm 8, \pm 12, \pm 13, \pm 24,$
$\pm 26, \pm 39, \pm 52, \pm 78, \pm 104, \pm 156, \pm 312\}$

Using synthetic division, we find that 8 is a zero:

```
8| 1  -4    7  -312
       8   32   312
   1   4   39    0
```

Note that the discriminant of the depressed equation $x^2 + 4x + 39 = 0$, $4^2 - 4(1)(39) < 0$, so there are 2 complex solutions to the depressed equation. Therefore, $x = 8$. The width of the rectangle is $8 - 2 = 6$ units and the length of the rectangle is $8^2 - 2 \cdot 8 + 3 = 51$ units.

79. Use synthetic division to solve the equation $628 = 3x^3 - 6x^2 + 108x + 100 \Rightarrow$

$3x^3 - 6x^2 + 108x - 528 = 0.$ The factors of the constant term, -528, are
$\{\pm 1, \pm 2, \pm 3, \pm 4, \pm 6, \pm 8, \pm 11, \pm 48, \pm 66, \pm 88,$
$\pm 132, \pm 176, \pm 264, \pm 528\}$. The factors of the leading coefficient, 3, are $\{\pm 1, \pm 3\}$. Only the positive, whole number possibilities make sense for the problem, so the possible rational zeros are $\{1, 2, 3, 4, 6, 8, 11, 16, 22, 44, 48, 66,$ $88, 132, 176, 264, 528\}$.

```
4| 3   -6   108  -528
       12   24   528
   3    6  132    0
```

Thus, $x = 4$.

81. The cost function gives the result as a number of thousands, so set it equal to 125:

$x^3 - 15x^2 + 5x + 50 = 125 \Rightarrow$

$x^3 - 15x^2 + 5x - 75 = 0$. The factors of the constant term, -75, are $\{\pm 1, \pm 3, \pm 5, \pm 15, \pm 25, \pm 75\}$. The factors of the leading coefficient, 1, are $\{\pm 1\}$. Only the positive solutions make sense for the problem, so the possible rational zeros are $\{1, 3, 5, 15, 25, 75\}$. Use synthetic division to find the zero:

$$\begin{array}{r|rrrr} 15 & 1 & -15 & 5 & -75 \\ & & 15 & 0 & 75 \\ \hline & 1 & 0 & 5 & 0 \end{array}$$

The total monthly cost is $125,000 when 1500 units are produced.

83. a. The country exported oil for 5 years, from 2010 to 2015.

b. There are three zeros, so the minimum degree of the polynomial is 3. The zeros are 0, 3, and 8, so the polynomial is of the form $p(x) = a(x - 0)(x - 3)(x - 8)$. The graph goes through the point $(9, -3)$, so we have $-3 = a(9 - 0)(9 - 3)(9 - 8) \Rightarrow -3 = 54a \Rightarrow$

$a = -\dfrac{1}{18}$

Thus, the equation is

$p(x) = y = -\dfrac{1}{18} x(x - 3)(x - 8)$.

c. The year 2012 is represented by $x = 5$.

$p(5) = -\dfrac{1}{18}(5)(5 - 3)(5 - 8) \approx 1.7$

In 2012, the country exported about 1.7 million barrels of oil.

85. a. There are three zeros, so the minimum degree of the polynomial is 3. The zeros are 0, 6, and 8, so the polynomial is of the form $p(x) = a(x - 0)(x - 6)(x - 8)$. The graph goes through the point $(9, 0.5)$, so we have $0.5 = a(9 - 0)(9 - 6)(9 - 8) \Rightarrow$

$0.5 = 27a \Rightarrow a = \dfrac{1}{54}$

Thus, the equation is

$p(x) = y = \dfrac{1}{54} x(x - 6)(x - 8)$.

b. $p(4) = \dfrac{1}{54}(4)(4 - 6)(4 - 8) \approx 0.59259$

The profit in 2008 was about $592.59.

c. Set $Y_1 = \dfrac{1}{54} x(x - 6)(x - 8)$ and $Y_2 = 0.6$.

Find the intersection of the two graphs.

[0, 9, 1] by [−0.25, 1, 0.25]

Ms. Sharpy realized a profit of $600 in 2004 and 2007.

d.

The maximum profit was approximately $898.46 in 2006.

3.4 Beyond the Basics

87. Let $x = \sqrt{3} \Rightarrow x^2 = 3 \Rightarrow x^2 - 3 = 0$. The only possible rational zeros of this equation are ± 1 and ± 3. Because $\sqrt{3}$ is neither of these, it must be irrational.

89. $x = 3 - \sqrt{2} \Rightarrow 3 - x = \sqrt{2} \Rightarrow 9 - 6x + x^2 = 2 \Rightarrow$

$x^2 - 6x + 7 = 0$. The possible rational zeros of this equation are $\{\pm 1, \pm 7\}$. Since $3 - \sqrt{2}$ is not in the set, it must be irrational.

91. $y = (x - 1)(x + 1)(x - 0) = x^3 - x$

93. $y = \left(x + \dfrac{1}{2}\right)(x - 2)\left(x - \dfrac{7}{3}\right) \Rightarrow$

$y = (2x + 1)(x - 2)(3x - 7)$

$= 6x^3 - 23x^2 + 15x + 14$

95. $y = \left(x - \left(1 + \sqrt{2}\right)\right)\left(x - \left(1 - \sqrt{2}\right)\right)(x - 3)$

$= \left(x^2 - 2x - 1\right)(x - 3) = x^3 - 5x^2 + 5x + 3$

97. $y = \left(x - \left(1 + \sqrt{3}\right)\right)\left(x - \left(1 - \sqrt{3}\right)\right) \cdot \left(x - \left(3 - \sqrt{2}\right)\right)\left(x - \left(3 + \sqrt{2}\right)\right) = \left(x^2 - 2x - 2\right)\left(x^2 - 6x + 7\right)$
$= x^4 - 8x^3 + 17x^2 - 2x - 14$

99. $y = \left(x - \left(\sqrt{2} + \sqrt{3}\right)\right)\left(x - \left(\sqrt{2} - \sqrt{3}\right)\right) = x^2 - 2x\sqrt{2} - 1$

Since we must have integer coefficients, there must be additional zeros. Try $-\left(\sqrt{2} + \sqrt{3}\right)$:

$y = \left(x - \left(\sqrt{2} + \sqrt{3}\right)\right)\left(x - \left(\sqrt{2} - \sqrt{3}\right)\right) \cdot \left(x + \left(\sqrt{2} + \sqrt{3}\right)\right)\left(x + \left(\sqrt{2} - \sqrt{3}\right)\right) = x^4 - 10x^2 + 1$

101. $x = 3 - \sqrt{2} + \sqrt{5} \Rightarrow x - 3 = -\sqrt{2} + \sqrt{5} \Rightarrow \left(x - 3\right)^2 = \left(-\sqrt{2} + \sqrt{5}\right)^2 \Rightarrow x^2 - 6x + 9 = 7 - 2\sqrt{10} \Rightarrow$
$x^2 - 6x + 2 = -2\sqrt{10} \Rightarrow \left(x^2 - 6x + 2\right)^2 = \left(-2\sqrt{10}\right)^2 \Rightarrow x^4 - 12x^3 + 40x^2 - 24x + 4 = 40 \Rightarrow$
$y = x^4 - 12x^3 + 40x^2 - 24x - 36$

103. **i.** Simplifying the fraction if necessary, we can assume that $\dfrac{p}{q}$ is in lowest terms.

Since $\dfrac{p}{q}$ is a zero of F, we have

$$F\left(\frac{p}{q}\right) = 0.$$

ii. Substitute $\dfrac{p}{q}$ for x in the equation

$F(x) = 0.$

iii. Multiply the equation in (ii) by q^n.

iv. Subtract $a_0 q^n$ from both sides of the equation.

v. The left side of the equation in (iv) is
$a_n p^n + a_{n-1}p^{n-1}q + \cdots + a_1 pq^{n-1} = p\left(a_n p^{n-1} + a_{n-1}p^{n-2}q + \cdots + a_1 q^{n-1}\right)$.
Therefore p is a factor.

vi. $a = b \Leftrightarrow \dfrac{a}{p} = \dfrac{b}{p}$.

vii. Since p and q have no common prime factors, p must be a factor of a_0.

viii. Rearrange the terms of the equation in (iii).

ix. The left side of the equation in (viii) is
$a_{n-1}p^{n-1}q + \cdots + a_1 pq^{n-1} + a_0 q^n = q\left(a_{n-1}p^{n-1} + \cdots + a_1 pq^{n-2} + a_0 q^{n-1}\right)$.
Therefore q is a factor.

x. $a = b \Leftrightarrow \dfrac{a}{q} = \dfrac{b}{q}$.

xi. Since p and q have no common prime factors, q must be a factor of a_n.

3.4 Critical Thinking/Discussion/Writing

105. a. False. The factors of the constant term, 3, are $\{\pm 1, \pm 3\}$ and the factors of the leading coefficient, 1, are $\{\pm 1\}$. The possible rational zeros are $\{\pm 1, \pm 3\}$.

b. False. The factors of the constant term, 25, are $\{\pm 1, \pm 5, \pm 25\}$ and the factors of the leading coefficient, 2, are $\{\pm 1, \pm 2\}$. The possible rational zeros are
$\left\{\pm \dfrac{1}{2}, \pm 1, \pm \dfrac{5}{2}, \pm 5, \pm \dfrac{25}{2}, \pm 25\right\}$.

106. Since $f(x)$ has integer coefficients, if $x = \sqrt{2}$ is a zero, then so is $x = -\sqrt{2}$. Then, two of the factors of $f(x)$ are $\left(x + \sqrt{2}\right)$ and $\left(x - \sqrt{2}\right)$.

$\left(x + \sqrt{2}\right)\left(x - \sqrt{2}\right) = x^2 - 2$, so divide $f(x)$ by $x^2 - 2$ to find another factor.

$$
\begin{array}{r}
x^2 - 2x - 2 \\
x^2 + 0x - 2 \overline{)x^4 - 2x^3 - 4x^2 + 4x + 4} \\
\underline{x^4 + 0x^3 - 2x^2} \\
-2x^3 - 2x^2 + 4x \\
\underline{-2x^3 - 0x^2 + 4x} \\
-2x^2 + 0x + 4 \\
\underline{-2x^2 + 0x + 4}
\end{array}
$$

(continued on next page)

(*continued*)

Use the quadratic formula to solve $x^2 - 2x - 2 = 0$.

$$x = \frac{-(-2) \pm \sqrt{(-2)^2 - 4(1)(-2)}}{2(1)} = \frac{2 \pm \sqrt{4+8}}{2} = \frac{2 \pm 2\sqrt{3}}{2} = 1 \pm \sqrt{3}$$

The zeros of the equation are $x = \pm\sqrt{2}$ and $x = 1 \pm \sqrt{3}$.

3.4 Getting Ready for the Next Section

107. $(-2i)^5 = (-2)^5 (i)^5 = (-2)^5 (i)^4 i = -32i$

109. $(2x+i)(-2x+i) = -4x^2 + i^2 = -4x^2 - 1$

111. $(5x+2i)(5x-2i) = 25x^2 - 4i^2 = 25x^2 + 4$

113. $(x-1+2i)(x-1-2i) = ((x-1)+2i)((x-1)-2i) = (x-1)^2 - 4i^2 = x^2 - 2x + 1 + 4 = x^2 - 2x + 5$

115. $[x-(2+3i)][x-(2-3i)] = x^2 - (2-3i)x - (2+3i)x + (2+3i)(2-3i)$
$$= x^2 - 2x + 3ix - 2x - 3ix + 4 - 9i^2 = x^2 - 4x + 4 + 9 = x^2 - 4x + 13$$

3.5   The Complex Zeros of a Polynomial Function

3.5 Practice Problems

1. a. $P(x) = 3(x+2)(x-1)[x-(1+i)][x-(1-i)] = 3(x+2)(x-1)(x-1-i)(x-1+i)$

  b. $P(x) = 3(x+2)(x-1)[x-(1+i)][x-(1-i)] = 3(x+2)(x-1)(x^2 - 2x + 2)$
  $$= 3(x+2)(x^3 - 3x^2 + 4x - 2) = 3(x^4 - x^3 - 2x^2 + 6x - 4) = 3x^4 - 3x^3 - 6x^2 + 18x - 12$$

2. Since $2 - 3i$ is a zero of multiplicity 2, so is $2 + 3i$. Since i is a zero, so is $-i$. The eight zeros are $-3, -3, 2 - 3i, 2 - 3i, 2 + 3i, 2 + 3i, i,$ and $-i$.

3. The function has degree four, so there are four zeros. Since one zero is $2i$, another zero is $-2i$. So $(x - 2i)(x + 2i) = x^2 + 4$ is a factor of $P(x)$. Now divide to find the other factors.

$$
\begin{array}{r}
x^2 - 3x + 2 \\
x^2+4 \overline{\smash{\big)}\, x^4 - 3x^3 + 6x^2 - 12x + 8} \\
\underline{x^4 \qquad\;\; + 4x^2} \\
-3x^3 + 2x^2 - 12x \\
\underline{-3x^3 \qquad\;\; - 12x} \\
2x^2 \qquad + 8 \\
\underline{2x^2 \qquad + 8} \\
0
\end{array}
$$

So,
$$P(x) = x^4 - 3x^3 + 6x^2 - 12x + 8$$
$$= (x - 2i)(x + 2i)(x^2 - 3x + 2)$$
$$= (x - 2i)(x + 2i)(x - 2)(x - 1)$$
The zeros of $P(x)$ are $1, 2, 2i,$ and $-2i$.

4. $f(x) = x^4 - 8x^3 + 22x^2 - 28x + 16$
The function has degree 4, so there are four zeros. There are four sign changes, so there are either 0, 2, or 4 positive zeros.
$$f(-x) = (-x)^4 - 8(-x)^3 + 22(-x)^2 - 28(-x) + 16$$
$$= x^4 + 8x^3 + 22x^2 + 28x + 16$$
There are no sign changes in $f(-x)$, so there are no negative zeros. The possible rational zeros are $\{\pm 1, \pm 2, \pm 4, \pm 8, \pm 16\}$. Using synthetic division to test the positive values, we find that one zero is 2:

$$
\begin{array}{r|rrrrr}
2 & 1 & -8 & 22 & -28 & 16 \\
  &   & 2 & -12 & 20 & -16 \\
\hline
  & 1 & -6 & 10 & -8 & 0
\end{array}
$$

(*continued on next page*)

(continued)

The zeros of the depressed function
$x^3 - 6x^2 + 10x - 8$ are also zeros of P.
The possible rational zeros of the depressed
function are $\{\pm 1, \pm 2, \pm 4, \pm 8\}$. Examine only
the positive possibilities and find that 4 is a
zero:

$$
\begin{array}{r|rrrr}
4 & 1 & -6 & 10 & -8 \\
  &   & 4 & -8 & 8 \\
\hline
  & 1 & -2 & 2 & 0
\end{array}
$$

So,
$$x^4 - 8x^3 + 22x^2 - 28x + 16$$
$$= (x-2)(x-4)(x^2 - 2x + 2)$$

Now find the zeros of $x^2 - 2x + 2$ using the
quadratic formula:

$$x = \frac{2 \pm \sqrt{(-2)^2 - 4(1)(2)}}{2(1)} = \frac{2 \pm \sqrt{-4}}{2} = 1 \pm i$$

The zeros are 2, 4, 1 + i, and 1 − i.

3.5 Concepts and Vocabulary

1. The Fundamental Theorem of Algebra states
   that a polynomial function of degree $n \geq 1$ has
   at least one <u>complex</u> zero.

3. If P is a polynomial function with real
   coefficients and if $z = a + bi$ is a zero of P,
   then $\overline{z} = a - bi$ is also a zero of $P(x)$.

5. False. A polynomial function of degree $n \geq 1$
   has at least one complex zero.

7. True

3.5 Building Skills

9. $x^2 + 25 = 0 \Rightarrow x^2 = -25 \Rightarrow x = \pm 5i$

11. $x^2 + 4x + 4 = -9 \Rightarrow (x+2)^2 = -9 \Rightarrow$
    $x + 2 = \pm 3i \Rightarrow x = -2 \pm 3i$

13. $(x-2)(x-3i)(x+3i) = 0 \Rightarrow x = 2$ or $x = \pm 3i$

15. The remaining zero is $3 - i$.

17. The remaining zero is $-5 - i$.

19. The remaining zeros are $-i$ and $-3i$.

21. $P(x) = 2(x - (5-i))(x - (5+i))(x-3i)(x+3i)$
    $= 2x^4 - 20x^3 + 70x^2 - 180x + 468$

23. $P(x) = 7(x-5)^2(x-1)(x-(3-i))(x-(3+i))$
    $= 7x^5 - 119x^4 + 777x^3 - 2415x^2$
    $\qquad\qquad + 3500x - 1750$

25. The function has degree four, so there are four
    zeros. Since one zero is $3i$, another zero is $-3i$.
    So $(x - 3i)(x + 3i) = x^2 + 9$ is a factor of
    $P(x)$. Now divide to find the other factor:

    $$
    \begin{array}{r}
    x^2 + x \\
    x^2 + 9 \overline{\smash{\big)}\, x^4 + x^3 + 9x^2 + 9x} \\
    \underline{x^4 \qquad + 9x^2} \\
    x^3 \qquad\quad + 9x \\
    \underline{x^3 \qquad\quad + 9x} \\
    0
    \end{array}
    $$

    So,
    $$P(x) = x^4 + x^3 + 9x^2 + 9x$$
    $$= (x^2 + 9)(x^2 + x) = x(x+1)(x^2 + 9)$$

27. The function has degree five, so there are five
    zeros. Since one zero is $3 - i$, another zero is
    $3 + i$. So
    $(x - (3-i))(x - (3+i)) = x^2 - 6x + 10$ is a
    factor of $P(x)$. Now divide to find the other
    factor:

    $$
    \begin{array}{r}
    x^3 + x^2 - 2x \\
    x^2 - 6x + 10 \overline{\smash{\big)}\, x^5 - 5x^4 + 2x^3 + 22x^2 - 20x} \\
    \underline{x^5 - 6x^4 + 10x^3} \\
    x^4 - 8x^3 + 22x^2 \\
    \underline{x^4 - 6x^3 + 10x^2} \\
    -2x^3 + 12x^2 - 20x \\
    \underline{-2x^3 + 12x^2 - 20x} \\
    0
    \end{array}
    $$

    $$P(x) = x^5 - 5x^4 + 2x^3 + 22x^2 - 20x$$
    $$= (x^2 - 6x + 10)(x^3 + x^2 - 2x)$$
    $$= (x^2 - 6x + 10)(x^2 + x - 2)x$$
    $$= x(x+2)(x-1)(x^2 - 6x + 10)$$

29. The function has degree 3, so there are three zeros. There are three sign changes, so there are 1 or 3 positive zeros. The possible rational zeros are $\{\pm 1, \pm 17\}$. Using synthetic division we find that one zero is 1:

$$\begin{array}{r|rrrr} 1 & 1 & -9 & 25 & -17 \\ & & 1 & -8 & 17 \\ \hline & 1 & -8 & 17 & 0 \end{array}$$

$x^3 - 9x^2 + 25x - 17 = (x-1)(x^2 - 8x + 17).$
Now solve the depressed equation
$x^2 - 8x + 17 = 0 \Rightarrow$
$x = \dfrac{8 \pm \sqrt{(-8)^2 - 4(1)(17)}}{2(1)} \Rightarrow x = \dfrac{8 \pm \sqrt{-4}}{2} \Rightarrow$
$x = 4 \pm i.$ The zeros are $1, 4 \pm i.$

31. The function has degree 3, so there are three zeros. There are two sign changes, so there are either 2 or 0 positive zeros. There is one sign change in $f(-x)$, so there is one negative zero. The possible rational zeros are
$\left\{ \pm\dfrac{1}{3}, \pm\dfrac{2}{3}, \pm 1, \pm\dfrac{4}{3}, \pm\dfrac{5}{3}, \pm 2, \pm\dfrac{8}{3}, \pm\dfrac{10}{3}, \pm 4, \right.$
$\left. \pm 5, \pm\dfrac{20}{3}, \pm 8, \pm 10, \pm\dfrac{40}{3}, \pm 20, \pm 40 \right\}.$ Using synthetic division to test the negative values we find that one zero is $-4/3$.

$$\begin{array}{r|rrrr} -\frac{4}{3} & 3 & -2 & 22 & 40 \\ & & -4 & 8 & -40 \\ \hline & 3 & -6 & 30 & 0 \end{array}$$

$3x^3 - 2x^2 + 22x + 40 = \left(x + \dfrac{4}{3}\right)(3x^2 - 6x + 30)$
$= 3\left(x + \dfrac{4}{3}\right)(x^2 - 2x + 10)$
Now solve the depressed equation
$x^2 - 2x + 10 = 0 \Rightarrow$
$x = \dfrac{2 \pm \sqrt{(-2)^2 - 4(1)(10)}}{2(1)} \Rightarrow x = \dfrac{2 \pm \sqrt{-36}}{2} \Rightarrow$
$x = 1 \pm 3i.$ The zeros are $-\dfrac{4}{3}, 1 \pm 3i.$

33. The function has degree 4, so there are four zeros. There are four sign changes, so there are 4, 2, or 0 positive zeros. There are no sign changes in $f(-x)$, so there are no negative zeros. The possible rational zeros are
$\left\{ \pm\dfrac{1}{2}, \pm 1, \pm\dfrac{3}{2}, \pm 3, \pm\dfrac{9}{2}, \pm 9 \right\}.$

Using synthetic division to test the positive values we find that one zero is 1:

$$\begin{array}{r|rrrrr} 1 & 2 & -10 & 23 & -24 & 9 \\ & & 2 & -8 & 15 & -9 \\ \hline & 2 & -8 & 15 & -9 & 0 \end{array}$$

The zeros of the depressed function
$2x^3 - 8x^2 + 15x - 9$ are also zeros of P.
The possible rational zeros are
$\left\{ \pm\dfrac{1}{2}, \pm 1, \pm\dfrac{3}{2}, \pm 3, \pm\dfrac{9}{2}, \pm 9 \right\}.$

Using synthetic division to test the positive values we find that one zero is 1:

$$\begin{array}{r|rrrr} 1 & 2 & -8 & 15 & -9 \\ & & 2 & -6 & 9 \\ \hline & 2 & -6 & 9 & 0 \end{array}$$

Thus,
$2x^4 - 10x^3 + 23x^2 - 24x + 9$
$= (x-1)^2\left(2x^2 - 6x + 9\right).$

Now solve the depressed equation
$2x^2 - 6x + 9 = 0.$
$x = \dfrac{6 \pm \sqrt{(-6)^2 - 4(2)(9)}}{2(2)} \Rightarrow x = \dfrac{6 \pm \sqrt{-36}}{4} \Rightarrow$
$x = \dfrac{6 \pm 6i}{4} = \dfrac{3}{2} \pm \dfrac{3i}{2}.$

The zeros of P are 1 (multiplicity 2), $\dfrac{3}{2} \pm \dfrac{3i}{2}.$

35. The function has degree 4, so there are four zeros. There are three sign changes, so there are either 3 or 1 positive zeros. There is one sign change in $f(-x)$, so there is one negative zero. The possible rational zeros are
$\{\pm 1, \pm 2, \pm 3, \pm 5, \pm 6, \pm 10, \pm 15, \pm 30\}.$ Using synthetic division to test the negative values, we find that one zero is -3:

$$\begin{array}{r|rrrrr} -3 & 1 & -4 & -5 & 38 & -30 \\ & & -3 & 21 & -48 & 30 \\ \hline & 1 & -7 & 16 & -10 & 0 \end{array}$$

The zeros of the depressed function
$x^3 - 7x^2 + 16x - 10$ are also zeros of P.
The possible rational zeros of the depressed function are $\{\pm 1, \pm 2, \pm 5, \pm 10\}.$ Since we have already found the negative zero, we examine only the positive possibilities and find that 1 is a zero:

(continued on next page)

(continued)

$$\begin{array}{r|rrrr} 1] & 1 & -7 & 16 & -10 \\ & & 1 & -6 & 10 \\ \hline & 1 & -6 & 10 & 0 \end{array}$$

$$x^4 - 4x^3 - 5x^2 + 38x - 30$$
$$= (x+3)(x-1)(x^2 - 6x + 10)$$

Now solve the depressed equation:

$$x^2 - 6x + 10 = 0 \Rightarrow x = \frac{6 \pm \sqrt{(-6)^2 - 4(1)(10)}}{2(1)} \Rightarrow$$

$$x = \frac{6 \pm \sqrt{-4}}{2} \Rightarrow x = 3 \pm i.$$

The zeros are -3, 1, $3 \pm i$.

37. The function has degree 5, so there are five zeros. There are five sign changes, so there are either 5, 3, or 1 positive zeros. There are no sign changes in $f(-x)$, so there are no negative zeros. The possible rational zeros are $\left\{\frac{1}{2}, 1, \frac{3}{2}, 2, 3, 6\right\}$. Using synthetic division to test the positive values, we find that one zero is 1/2.

$$\begin{array}{r|rrrrrr} \frac{1}{2} & 2 & -11 & 19 & -17 & 17 & -6 \\ & & 1 & -5 & 7 & -5 & 6 \\ \hline & 2 & -10 & 14 & -10 & 12 & 0 \end{array}$$

The zeros of the depressed function $2x^4 - 10x^3 + 14x^2 - 10x + 12$ are also zeros of P. Use synthetic division again to find the next zero, 2:

$$\begin{array}{r|rrrrr} 2] & 2 & -10 & 14 & -10 & 12 \\ & & 4 & -12 & 4 & -12 \\ \hline & 2 & -6 & 2 & -6 & 0 \end{array}$$

$$2x^5 - 11x^4 + 19x^3 - 17x^2 + 17x - 6$$
$$= \left(x - \frac{1}{2}\right)(x-2)(2x^3 - 6x^2 + 2x - 6)$$
$$= \left(x - \frac{1}{2}\right)(x-2)(2x^3 - 6x^2 + 2x - 6)$$

Use factoring by grouping to factor $x^3 - 3x^2 + x - 3$:

$$x^3 - 3x^2 + x - 3 = x^2(x-3) + 1(x-3)$$
$$= (x^2 + 1)(x-3). \text{ So the remaining zeros are 3}$$
and $\pm i$. The zeros are $\frac{1}{2}, 2, 3, \pm i$.

39. Since one zero is $3i$, $-3i$ is another zero. There are two zeros, so the degree of the equation is at least 2. Thus, the equation is of the form
$$f(x) = a(x - 3i)(x + 3i) = a(x^2 + 9).$$
The graph passes through $(0, 3)$, so we have
$$3 = a(0^2 + 9) \Rightarrow a = \frac{1}{3}. \text{ Thus, the equation of}$$
the function is $f(x) = \frac{1}{3}(x^2 + 9)$.

41. Since i and $2i$ are zeros, so are $-i$, and $-2i$. From the graph, we see that 2 is also a zero. There are five zeros, so the degree of the equation is at least 5. Thus, the equation is of the form
$$f(x) = a(x - i)(x + i)(x - 2i)(x + 2i)(x - 2)$$
$$= a(x^2 + 1)(x^2 + 4)(x - 2)$$
The graph passes through $(0, 4)$, so we have
$$4 = a(0^2 + 1)(0^2 + 4)(0 - 2) \Rightarrow a = -\frac{1}{2}.$$
Thus, the equation is
$$f(x) = -\frac{1}{2}(x^2 + 1)(x^2 + 4)(x - 2) \text{ or}$$
$$f(x) = \frac{1}{2}(x^2 + 1)(x^2 + 4)(2 - x).$$

3.5 Beyond the Basics

43. There are three cube roots. We know that one root is 1. Using synthetic division, we find
$$0 = x^3 - 1 = (x - 1)(x^2 + x + 1):$$

$$\begin{array}{r|rrrr} 1] & 1 & 0 & 0 & -1 \\ & & 1 & 1 & 1 \\ \hline & 1 & 1 & 1 & 0 \end{array}$$

Solve the depressed equation
$$x^2 + x + 1 = 0 \Rightarrow x = \frac{-1 \pm \sqrt{1^2 - 4(1)(1)}}{2(1)}$$
$$= \frac{-1 \pm \sqrt{-3}}{2} = -\frac{1}{2} \pm \frac{i\sqrt{3}}{2}.$$

So, the cube roots of 1 are 1 and $-\frac{1}{2} \pm \frac{i\sqrt{3}}{2}$.

45. $P(x) = x^2 + (i - 2)x - 2i, \ x = -i$

$$\begin{array}{r|rrr} -i & 1 & i-2 & -2i \\ & & -i & 2i \\ \hline & 1 & -2 & 0 \end{array}$$

$$x^2 + (i - 2)x - 2i = (x + i)(x - 2)$$

47. $P(x) = x^3 - (3+i)x^2 - (4-3i)x + 4i, \ x = i$

$$\begin{array}{r|cccc} i & 1 & -(3+i) & -(4-3i) & 4i \\ & & i & -3i & -4i \\ \hline & 1 & -3 & -4 & 0 \end{array}$$

$x^3 - (3+i)x^2 - (4-3i)x + 4i$
$\quad = (x^2 - 3x - 4)(x - i) = (x-4)(x+1)(x-i)$

49. Since $1 + 2i$ is a zero, so is $1 - 2i$. The equation is of the form

$f(x) = a(x-2)\big(x-(1+2i)\big)\big(x-(1-2i)\big)$
$\quad = a(x-2)(x^2 - 2x + 5)$

The y-intercept is 40, so we have

$40 = a(0-2)(0^2 - 2(0) + 5) \Rightarrow a = -4$

Thus, the equation is

$f(x) = -4(x-2)(x^2 - 2x + 5)$.

Because $a < 0$, $y \to \infty$ as $x \to -\infty$ and $y \to -\infty$ as $x \to \infty$.

51. Since $3 + i$ is a zero, so is $3 - i$. The equation is of the form

$f(x) = a(x-1)(x+1)\big(x-(3+i)\big)\big(x+(3+i)\big)$
$\quad = a(x^2 - 1)(x^2 - 6x + 10)$

The y-intercept is 20, so we have

$20 = a(0^2 - 1)(0^2 - 6(0) + 10) \Rightarrow a = -2$

Thus, the equation is

$f(x) = -2(x^2 - 1)(x^2 - 6x + 10)$.

Because $a < 0$, $y \to -\infty$ as $x \to -\infty$ and $y \to -\infty$ as $x \to \infty$.

3.5 Critical Thinking/Discussion/Writing

53. Factoring the polynomial we have

$a_n x^n + a_{n-1} x^{n-1} + \cdots + a_1 x + a_0$
$\quad = a_n (x - r_1)(x - r_2) \cdots (x - r_n)$.

Expanding the right side, we find that the coefficient of x^{n-1} is $-a_n r_1 - a_n r_2 - \cdots - a_n r_n$ and the constant term is $(-1)^n a_n r_1 r_2 \cdots r_n$. Comparing the coefficients with those on the left side, we obtain

$a_{n-1} = -a_n(r_1 + r_2 + \cdots + r_n)$.

Because $a_n \neq 0$, $-\dfrac{a_{n-1}}{a_n} = r_1 + r_2 + \cdots + r_n$

and $(-1)^n \dfrac{a_0}{a_n} = r_1 r_2 \cdots r_n$.

54. a. $x^3 + 6x = 20 \Rightarrow x^3 + 6x - 20 = 0$. There is one sign change, so there is one positive root. There are no sign changes in $f(-x)$, so there are no negative roots. Therefore, there is only one real solution.

b. Substituting $v - u$ for x and remembering that $v^3 - u^3 = 20$ and $uv = 2$, we have

$(v-u)^3 + 6(v-u)$
$= v^3 - 3v^2 u + 3vu^2 - u^3 + 6(v-u)$
$= (v^3 - u^3) - (3v^2 u - 3vu^2) + 6(v-u)$
$= (v^3 - u^3) - 3vu(v-u) + 6(v-u)$
$= (v^3 - u^3) - (6 - 3uv)(v-u)$
$= 20 - (6 - 3(2))(v-u) = 20$. Therefore, $x = v - u$ is the solution.

c. To solve the system $v^3 - u^3 = 20, vu = 2$, solve the second equation for v, and substitute that value into the first equation, keeping in mind that u cannot be zero, so division by u is permitted:

$v^3 - u^3 = 20 \Rightarrow \left(\dfrac{2}{u}\right)^3 - u^3$

$\quad = \dfrac{8}{u^3} - u^3 = 20 \Rightarrow$

$-u^3 + \dfrac{8}{u^3} - 20 = 0$.

Let $-u^3 = a$, so

$a - \dfrac{8}{a} - 20 = 0 \Rightarrow a^2 - 20a - 8 = 0 \Rightarrow$

$a = \dfrac{20 \pm \sqrt{400 + 32}}{2} = 10 \pm 6\sqrt{3} = -u^3 \Rightarrow$

$\sqrt[3]{-10 \pm 6\sqrt{3}} = u \Rightarrow v = \dfrac{2}{\sqrt[3]{-10 \pm 6\sqrt{3}}}$

$= \dfrac{2}{\sqrt[3]{-10 \pm 6\sqrt{3}}} \cdot \dfrac{\sqrt[3]{10 \pm 6\sqrt{3}}}{\sqrt[3]{10 \pm 6\sqrt{3}}} = \sqrt[3]{10 \pm 6\sqrt{3}}$

d. If q is positive, then, according to Descartes's Rule of Signs, the polynomial $x^3 + px - q$ has one positive zero and no negative zeros. If q is negative, then it has no positive zeros and one negative zero. Either way, there is exactly one real solution. From (b), we have $v^3 - u^3 = q$, $vu = \dfrac{p}{3}$, and $x = v - u$. Substituting, we have

$$x^3 + px = (v-u)^3 + p(v-u)$$
$$= v^3 - 3v^2u + 3vu^2 - u^3 + p(v-u)$$
$$= (v^3 - u^3) - (3v^2u - 3vu^2) + p(v-u)$$
$$= (v^3 - u^3) - 3vu(v-u) + p(v-u)$$
$$= (v^3 - u^3) + (p - 3uv)(v-u)$$
$$= q + \left(p - 3 \cdot \frac{p}{3}\right)x = q \Rightarrow x = v-u \text{ is the}$$

solution.

Solving the system $v^3 - u^3 = q, \ vu = \dfrac{p}{3}$,

we obtain

$$v^3 - u^3 = q \Rightarrow \left(\frac{p}{3u}\right)^3 - u^3 = \frac{p^3}{3^3 u^3} - u^3 = q$$

$$\Rightarrow -u^3 + \frac{p^3}{3^3 u^3} - q = 0. \text{ Let } -u^3 = a, \text{ so}$$

we have $a - \dfrac{p}{3^3 a} - q = 0 \Rightarrow$

$$a^2 - qa - \frac{p^3}{3^3} = 0 \Rightarrow$$

$$a = \frac{q \pm \sqrt{q^2 + 4\left(\dfrac{p^3}{3^3}\right)}}{2} = \frac{q}{2} \pm \sqrt{\frac{q^2 + 4\left(\dfrac{p^3}{3^3}\right)}{4}}$$

$$= \frac{q}{2} \pm \sqrt{\left(\frac{q}{2}\right)^2 + \left(\frac{p}{3}\right)^3} = -u^3 \Rightarrow$$

$$u = \sqrt[3]{-\frac{q}{2} + \sqrt{\left(\frac{q}{2}\right)^2 + \left(\frac{p}{3}\right)^3}} \text{ and}$$

$$v = \sqrt[3]{\frac{q}{2} + \sqrt{\left(\frac{q}{2}\right)^2 + \left(\frac{p}{3}\right)^3}} \text{ or}$$

$$u = \sqrt[3]{-\frac{q}{2} - \sqrt{\left(\frac{q}{2}\right)^2 + \left(\frac{p}{3}\right)^3}} \text{ and}$$

$$v = \sqrt[3]{\frac{q}{2} - \sqrt{\left(\frac{q}{2}\right)^2 + \left(\frac{p}{3}\right)^3}}.$$

The difference $v - u$ is the same in both cases, so
$$x = v - u$$
$$= \sqrt[3]{\frac{q}{2} + \sqrt{\left(\frac{q}{2}\right)^2 + \left(\frac{p}{3}\right)^3}}$$
$$- \sqrt[3]{-\frac{q}{2} + \sqrt{\left(\frac{q}{2}\right)^2 + \left(\frac{p}{3}\right)^3}}.$$

e. Substituting $x = y - \dfrac{a}{3}$, we have

$$x^3 + ax^2 + bx + c = 0 \Rightarrow$$
$$\left(y - \frac{a}{3}\right)^3 + a\left(y - \frac{a}{3}\right)^2 + b\left(y - \frac{a}{3}\right) + c = 0 \Rightarrow$$
$$\left(y^3 - ay^2 + \frac{a^2}{3}y - \frac{a^3}{27}\right) + \left(ay^2 - \frac{2a^2y}{3} + \frac{a^3}{9}\right)$$
$$+ by - \frac{ab}{3} + c = 0 \Rightarrow$$

$$y^3 + py = q, \text{ where } p = b - \frac{a^2}{3} \text{ and}$$

$$q = -\frac{2a^3}{27} + \frac{ab}{3} - c$$

$$y^3 - \frac{a^2y}{3} + by + \frac{2a^3}{27} - \frac{ab}{3} + c = 0 \Rightarrow$$

$$y^3 + \left(b - \frac{a^2}{3}\right)y = -\frac{2a^3}{27} + \frac{ab}{3} - c$$

55. $x^3 + 6x^2 + 10x + 8 = 0 \Rightarrow a = 6, b = 10, c = 8$.

Substituting $x = y - \dfrac{6}{3} = y - 2$ as in (1e), we have

$$(y-2)^3 + 6(y-2)^2 + 10(y-2) + 8 = 0 \Rightarrow$$
$$(y^3 - 6y^2 + 12y - 8) + (6y^2 - 24y + 24)$$
$$+ 10y - 20 + 8 = 0 \Rightarrow$$
$$y^3 - 2y + 4 = 0 \Rightarrow y^3 - 2y = -4.$$

Then, using the results of (1d), we have

$$y = \sqrt[3]{\frac{-4}{2} + \sqrt{\left(\frac{-4}{2}\right)^2 + \left(\frac{-2}{3}\right)^3}}$$
$$- \sqrt[3]{-\frac{-4}{2} + \sqrt{\left(\frac{-4}{2}\right)^2 + \left(\frac{-2}{3}\right)^3}}.$$

Using a calculator, we find that $y = -2$. So $x = -2 - 2 = -4$. Now use synthetic division to find the depressed equation:

$$\begin{array}{r|rrrr} -4 & 1 & 6 & 10 & 8 \\ & & -4 & -8 & -8 \\ \hline & 1 & 2 & 2 & 0 \end{array}$$

$$x^3 + 6x^2 + 10x + 8 = (x+4)(x^2 + 2x + 2) = 0.$$
$$x^2 + 2x + 2 = 0 \Rightarrow$$

$$x = \frac{-2 \pm \sqrt{2^2 - 4(1)(2)}}{2(1)} = \frac{-2 \pm \sqrt{-4}}{2}$$

$$= \frac{-2 \pm 2i}{2} = -1 \pm i.$$

The solution set is $\{-4, -1 \pm i\}$.

3.5 Getting Ready for the Next Section

57. $2x^2 + x - 3 = 0 \Rightarrow (2x+3)(x-1) = 0 \Rightarrow$

$x = -\dfrac{3}{2},\ x = 1$

Solution set: $\left\{-\dfrac{3}{2},\ 1\right\}$

59. $x^4 - x^3 - 12x^2 = 0 \Rightarrow x^2\left(x^2 - x - 12\right) = 0 \Rightarrow$

$x^2(x-4)(x+3) = 0 \Rightarrow x = 0,\ x = 4,\ x = -3$

Solution set: $\{-3, 0, 4\}$

61. $\dfrac{x^2 - x - 4}{x - 2}$

$$
\begin{array}{r|rrr}
2| & 1 & -1 & -4 \\
 & & 2 & 2 \\
\hline
 & 1 & 1 & -2
\end{array}
$$

$Q(x) = x + 1;\ R(x) = -2$

63.
$$
\begin{array}{r}
4x + 3 \\
2x^3 + 0x^2 + x\overline{)8x^4 + 6x^3 - 0x^2 + 0x - 5} \\
\underline{8x^4 + 0x^3 + 4x^2} \\
6x^3 - 4x^2 + 0x \\
\underline{6x^3 + 0x^2 + 3x} \\
-4x^2 - 3x - 5
\end{array}
$$

$Q(x) = 4x + 3;\ R(x) = -4x^2 - 3x - 5$

65. $\dfrac{7 - (-1)}{2(-1)^2 + 3(-1)} = \dfrac{8}{-1} = -8$

67. $\dfrac{(2)^2 + 4(2) - 1}{9 - (2)^3} = 11$

3.6 Rational Functions

3.6 Practice Problems

1. $f(x) = \dfrac{x - 3}{x^2 - 4x - 5}$

The domain of f consists of all real numbers for which $x^2 - 4x - 5 \neq 0$.

$x^2 - 4x - 5 = 0 \Rightarrow (x-5)(x+1) = 0 \Rightarrow$

$x = 5$ or $x = -1$

Thus, the domain is

$(-\infty, -1) \cup (-1, 5) \cup (5, \infty)$.

2. a. $g(x) = \dfrac{3}{x - 2}$

Let $f(x) = \dfrac{1}{x}$. Then

$g(x) = \dfrac{3}{x-2} = 3\left(\dfrac{1}{x-2}\right) = 3f(x-2).$

The graph of $y = f(x-2)$ is the graph of $y = f(x)$ shifted two units to the right. This moves the vertical asymptote two units to the right. The graph of $y = 3f(x-2)$ is the graph of $y = f(x-2)$ stretched vertically three units.

The domain of g is $(-\infty, 2) \cup (2, \infty)$.

The range of g is $(-\infty, 0) \cup (0, \infty)$.

The vertical asymptote is $x = 2$.

The horizontal asymptote is $y = 0$.

$g(x) = \dfrac{3}{x - 2}$

b. $h(x) = \dfrac{2x + 5}{x + 1}$

$$
\begin{array}{r}
2 \\
x + 1\overline{)2x + 5} \\
\underline{2x + 2} \\
3
\end{array}
\qquad
h(x) = \dfrac{2x+5}{x+1} = 2 + \dfrac{3}{x+1}
$$

Let $f(x) = \dfrac{1}{x}$. Then

$h(x) = \dfrac{2x+5}{x+1} = 2 + \dfrac{3}{x+1} = 2 + 3f(x+1).$

The graph of $y = h(x)$ is the graph of $y = f(x)$ shifted one units to the left and then stretched vertically three units. The graph is then shifted two units up. This moves the vertical asymptote one unit to the left. The horizontal asymptote is shifted two units up. The domain of h is $(-\infty, -1) \cup (-1, \infty)$.

(continued on next page)

(continued)

The range of h is $(-\infty, 2) \cup (2, \infty)$.
The vertical asymptote is $x = -1$.
The horizontal asymptote is $y = 2$.

3. $f(x) = \dfrac{x+1}{x^2 + 3x - 10}$

The vertical asymptotes are located at the zeros of the denominator.
$x^2 + 3x - 10 = 0 \Rightarrow (x+5)(x-2) = 0 \Rightarrow$
$x = -5$ or $x = 2$
The vertical asymptotes are $x = -5$ and $x = 2$.

4. $f(x) = \dfrac{3-x}{x^2-9} = \dfrac{-(x-3)}{(x-3)(x+3)} = -\dfrac{1}{x+3}$

$x + 3 = 0 \Rightarrow x = -3$
The vertical asymptote is $x = -3$.

5. a. $f(x) = \dfrac{2x-5}{3x+4}$

Since the numerator and denominator both have degree 1, the horizontal asymptote is
$y = \dfrac{2}{3}$.

b. $g(x) = \dfrac{x^2 + 3}{x-1}$

Since the degree of the numerator is greater than the degree of the denominator, there are no horizontal asymptotes.

c. $h(x) = \dfrac{100x + 57}{0.01x^3 + 8x - 9}$

Since the degree of the numerator is less than the degree of the denominator, the horizontal asymptote is the line $y = 0$.

6. $f(x) = \dfrac{2x}{x^2 - 1}$

There are no common factors of the form $x - a$ between $2x$ and $x^2 - 1$.
First find the intercepts:
$\dfrac{2x}{x^2 - 1} = 0 \Rightarrow x = 0$
The graph passes through the origin.
Find the vertical asymptotes:
$x^2 - 1 = 0 \Rightarrow (x-1)(x+1) = 0 \Rightarrow$
$x = 1$ or $x = -1$
Find the horizontal asymptote: The degree of the numerator is less than the degree of the denominator, so the horizontal asymptote is the x-axis.

By long division, $f(x) = \dfrac{2x}{x^2 - 1} = 0 + \dfrac{2x}{x^2 - 1}$.

$R(x) = 2x$ has zero 0 and $D(x) = x^2 - 1$ has zeros -1 and 1. These zeros divide the x-axis into four intervals, $(-\infty, -1)$, $(-1, 0)$, $(0, 1)$, and $(1, \infty)$. Use test values to determine where the graph of f is above and below the x-axis.

Interval	Test point	Value of $f(x)$	Above/below x-axis
$(-\infty, -1)$	-3	$-\dfrac{3}{4}$	below
$(-1, 0)$	$-\dfrac{1}{2}$	$\dfrac{4}{3}$	above
$(0, 1)$	$\dfrac{1}{2}$	$-\dfrac{4}{3}$	below
$(1, \infty)$	2	$\dfrac{4}{3}$	above

7. $f(x) = \dfrac{2x^2 - 1}{2x^2 + x - 3}$

$f(x)$ is in lowest terms.

Find the intercepts:

$\dfrac{2x^2 - 1}{2x^2 + x - 3} = 0 \Rightarrow x = \pm\dfrac{\sqrt{2}}{2}$; $f(0) = \dfrac{1}{3}$

Find the vertical asymptotes:

$2x^2 + x - 3 = 0 \Rightarrow (2x + 3)(x - 1) = 0 \Rightarrow$

$x = -\dfrac{3}{2}$ or $x = 1$

Find the horizontal asymptote: The degree of the numerator is the same as the degree of the denominator, so the horizontal asymptote is

$y = \dfrac{2}{2} = 1$.

$f(x) = \dfrac{2x^2 - 1}{2x^2 + x - 3} = 1 + \dfrac{2 - x}{2x^2 + x - 3}$

The zero of $R(x) = 2 - x$ is 2 and the zeros of

$D(x) = 2x^2 + x - 3 = (2x + 3)(x - 1)$ are

$x = -\dfrac{3}{2}$ and $x = 1$.

Use test values to determine where the graph of f is above and below the horizontal asymptote

$y = 1$.

Interval	Test point	Value of $f(x)$	Above/below $y = 1$
$\left(-\infty, -\dfrac{3}{2}\right)$	-2	$\dfrac{7}{3}$	above
$\left(-\dfrac{3}{2}, 1\right)$	0	$\dfrac{1}{3}$	below
$(1, 2)$	$\dfrac{3}{2}$	$\dfrac{7}{6}$	above
$(2, \infty)$	3	$\dfrac{17}{18}$	above

Notice that the graph crosses the horizontal asymptote at $(2, 1)$.

8. $f(x) = \dfrac{x^2 + 1}{x^2 + 2}$

$f(x)$ is in lowest terms.

Find the intercepts:

$\dfrac{x^2 + 1}{x^2 + 2} = 0 \Rightarrow x = \pm i \Rightarrow$ there is no

x-intercept.

$f(0) = \dfrac{1}{2} \Rightarrow \left(0, \dfrac{1}{2}\right)$ is the y-intercept.

Find the vertical asymptotes:

$x^2 + 2 = 0 \Rightarrow x = i\sqrt{2} \Rightarrow$ there is no vertical asymptote.

Find the horizontal asymptote: The degree of the numerator is the same as the degree of the denominator, so the horizontal asymptote is

$y = \dfrac{1}{1} = 1$.

$f(x) = \dfrac{x^2 + 1}{x^2 + 2} = 1 + \dfrac{-1}{x^2 + 2}$

Neither $R(x) = -1$ nor $D(x) = x^2 + 2$ have

real zeros. Since $\dfrac{-1}{x^2 + 2}$ is negative for all

values of x, the graph of $f(x)$ is always below the line

$y = 1$.

9. $f(x) = \dfrac{x^2 + 2}{x - 1}$

$f(x)$ is in lowest terms.

Find the intercepts:

$\dfrac{x^2 + 2}{x - 1} = 0 \Rightarrow x = \pm i\sqrt{2} \Rightarrow$ there is no

x-intercept. $f(0) = -2 \Rightarrow (0, -2)$ is the

y-intercept.

Find the vertical asymptotes:

$x - 1 = 0 \Rightarrow x = 1$

(continued on next page)

(*continued*)

Find the horizontal asymptote: The degree of the numerator is greater than the degree of the denominator, so there is no horizontal asymptote. However, there is an oblique asymptote.

$$\frac{x^2 + 2}{x - 1} = x + 1 + \frac{3}{x - 1} \Rightarrow y = x + 1 \text{ is the}$$

oblique asymptote.

The graph is above the line $y = x + 1$ on $(1, \infty)$ and below the line on $(-\infty, 1)$.

The intervals determined by the zeros of the numerator and of the denominator of

$$f(x) - (x + 1) = \frac{x^2 + 2}{x - 1} - (x + 1) = \frac{3}{x - 1} \text{ divide}$$

the x-axis into two intervals, $(-\infty, 1)$ and $(1, \infty)$. Use test numbers to determine where the graph of f is above and below the x-axis.

Interval	Test point	Value of $f(x)$	Above/below $y = x + 1$
$(-\infty, 1)$	-1	$-\frac{3}{2}$	below
$(1, \infty)$	2	6	above

10. $R(x) = \dfrac{x(100 - x)}{x + 20}$

a. $R(10) = \dfrac{10(100 - 10)}{10 + 20} = 30$ billion dollars.

This means that if income is taxed at a rate of 10%, then the total revenue for the government will be 30 billion dollars. Similarly, $R(20) = \$40$ billion, $R(30) = \$42$ billion, $R(40) = \$40$ billion, $R(50) \approx \$35.7$ billion, $R(60) = \$30$ billion.

b.

c.

From the graphing calculator screen, we see that a tax rate of about 29% generates the maximum tax revenue of about $42.02 billion.

3.6 Concepts and Vocabulary

1. A rational function can be expressed in the form $\dfrac{N(x)}{D(x)}$, where *N(x)* and *D(x)* are polynomials and *D(x)* is not the zero polynomial.

3. The line $y = k$ is a horizontal asymptote of f if $f(x) \to k$ as $x \to \infty$ or as $x \to \underline{-\infty}$.

5. False. A rational function has a vertical asymptote only if the denominator has a real zero and the numerator and denominator have no common factors.

7. False. The graph of a rational function may cross its *horizontal* asymptote.

3.6 Building Skills

9. $x + 4 = 0 \Rightarrow x = -4$
The domain of the function is $(-\infty, -4) \cup (-4, \infty)$

11. $(-\infty, \infty)$

13. $x^2 - x - 6 = 0 \Rightarrow (x - 3)(x + 2) = 0 \Rightarrow x = -2, 3$
The domain of the function is $(-\infty, -2) \cup (-2, 3) \cup (3, \infty)$.

15. $x^2 - 6x + 8 = 0 \Rightarrow (x - 4)(x - 2) = 0 \Rightarrow x = 2, 4$
The domain of the function is $(-\infty, 2) \cup (2, 4) \cup (4, \infty)$.

17. As $x \to 1^+$, $f(x) \to \underline{\infty}$.

19. As $x \to -2^+$, $f(x) \to \underline{\infty}$.

21. As $x \to \infty$, $f(x) \to \underline{1}$.

23. The domain of f is $\underline{(-\infty, -2) \cup (-2, 1) \cup (1, \infty)}$.

25. The equations of the vertical asymptotes of the graph are $\underline{x = -2}$ and $\underline{x = 1}$.

27.

Domain: $(-\infty, 4) \cup (4, \infty)$

Range: $(-\infty, 0) \cup (0, \infty)$

Vertical asymptote: $x = 4$
Horizontal asymptote: $y = 0$

29.

Domain: $\left(-\infty, -\frac{1}{3}\right) \cup \left(-\frac{1}{3}, \infty\right)$

Range: $\left(-\infty, -\frac{1}{3}\right) \cup \left(-\frac{1}{3}, \infty\right)$

Vertical asymptote: $x = -\frac{1}{3}$

Horizontal asymptote: $y = -\frac{1}{3}$

31.

Domain: $(-\infty, -2) \cup (-2, \infty)$

Range: $(-\infty, -3) \cup (-3, \infty)$

Vertical asymptote: $x = -2$
Horizontal asymptote: $y = -3$

33.

Domain: $(-\infty, 4) \cup (4, \infty)$

Range: $(-\infty, 5) \cup (5, \infty)$

Vertical asymptote: $x = 4$
Horizontal asymptote: $y = 5$

In exercises 35−44, to find the vertical asymptotes, first eliminate any common factors in the numerator and denominator, and then set the denominator equal to zero and solve for x.

35. $x = 1$

37. $x = -4, x = 3$

39. $h(x) = \dfrac{x^2 - 1}{x^2 + x - 6} = \dfrac{(x-1)(x+1)}{(x-2)(x+3)}$.

The equations of the vertical asymptotes are $x = -3$ and $x = 2$.

41. $f(x) = \dfrac{x^2 - 6x + 8}{x^2 - x - 12} = \dfrac{(x-4)(x-2)}{(x-4)(x+3)}$.

Disregard the common factor. The vertical asymptote is $x = -3$.

43. There is no vertical asymptote.

For exercises 45–52, locate the horizontal asymptote as follows:

- If the degree of the numerator of a rational function is less than the degree of the denominator, then the x-axis ($y = 0$) if the horizontal asymptote.

- If the degree of the numerator of a rational function equals the degree of the denominator, the horizontal asymptote is the line with the equation $y = \dfrac{a_n}{b_m}$, where a_n is the coefficient of the leading term of the numerator and b_m is the coefficient of the leading term of the denominator.

- If the degree of the numerator of a rational function is greater than the degree of the denominator, then there is no horizontal asymptote.

45. $y = 0$

47. $y = \dfrac{2}{3}$

49. There is no horizontal asymptote.

51. $y = 0$

53. d   **55.** e   **57.** a

59. $0 = \dfrac{2x}{x-3} \Rightarrow x = 0$ is the x-intercept.

$\dfrac{2(0)}{0-3} = 0 \Rightarrow y = 0$ is the y-intercept. The vertical asymptote is $x = 3$. The horizontal asymptote is $y = 2$. The intervals to be tested are $(-\infty, 3)$ and $(3, \infty)$. The graph is above the horizontal asymptote on $(3, \infty)$ and below the horizontal asymptote on $(-\infty, 3)$.

61. $0 = \dfrac{x}{x^2 - 4} \Rightarrow x = 0$ is the x-intercept.

$\dfrac{0}{0^2 - 4} = 0 \Rightarrow y = 0$ is the y-intercept. The vertical asymptotes are $x = -2$ and $x = 2$. The horizontal asymptote is the x-axis. The intervals to be tested are $(-\infty, -2), (-2, 0), (0, 2)$ and $(2, \infty)$. The graph is above the x-axis on $(-2, 0) \cup (2, \infty)$ and below the x-axis on $(-\infty, -2) \cup (0, 2)$.

63. $0 = \dfrac{-2x^2}{x^2 - 9} \Rightarrow x = 0$ is the x-intercept.

$\dfrac{0^2}{0^2 - 4} = 0 \Rightarrow y = 0$ is the y-intercept. The vertical asymptotes are $x = -3$ and $x = 3$. The horizontal asymptote is $y = -2$. The intervals to be tested are $(-\infty, -3), (-3, 3),$ and $(3, \infty)$. The graph is above the horizontal asymptote on $(-3, 3)$ and below the horizontal asymptote on $(-\infty, -3) \cup (3, \infty)$.

65. $0 = \dfrac{2}{x^2 - 2} \Rightarrow$ there is no x-intercept.

$\dfrac{2}{0^2 - 2} = 0 \Rightarrow y = -1$ is the y-intercept. The vertical asymptotes are $x = \pm\sqrt{2}$. The horizontal asymptote is the x-axis.

(continued on next page)

(*continued*)

The intervals to be tested are

$(-\infty, -\sqrt{2}), (-\sqrt{2}, \sqrt{2})$ and $(\sqrt{2}, \infty)$.

he graph is above the x-axis on

$\left(-\infty, -\sqrt{2}\right) \cup \left(\sqrt{2}, \infty\right)$ and below the x-axis on

$\left(-\sqrt{2}, \sqrt{2}\right)$.

67. $0 = \dfrac{x+1}{(x-2)(x+3)} \Rightarrow x = -1$ is the x-intercept.

$\dfrac{0+1}{(0-2)(0+3)} = -\dfrac{1}{6} \Rightarrow y = -\dfrac{1}{6}$ is the

y-intercept. The vertical asymptotes are $x = -3$ and $x = 2$. The horizontal asymptote is the x-axis. The intervals to be tested are $(-\infty, -3), (-3, -1), (-1, 2)$, and $(2, \infty)$. The graph is above the x-axis on $(-3, -1) \cup (2, \infty)$ and below the x-axis on $(-\infty, -3) \cup (-1, 2)$.

69. $0 = \dfrac{x^2}{x^2+1} \Rightarrow x = 0$ is the x-intercept.

$\dfrac{0^2}{0^2+1} = 0 \Rightarrow y = 0$ is the y-intercept. There is no vertical asymptote. The horizontal asymptote is $y = 1$. The intervals to be tested are $(-\infty, 0)$ and $(0, \infty)$. The graph is above the x-axis on $(-\infty, 0) \cup (0, \infty)$ and below the horizontal asymptote on $(-\infty, \infty)$.

71. $f(x) = \dfrac{x^3 - 4x}{x^3 - 9x} = \dfrac{x(x-2)(x+2)}{x(x-3)(x+3)}$

$= \dfrac{(x-2)(x+2)}{(x-3)(x+3)}$

$f(x) = 0 \Rightarrow x = \pm 2$ are the x-intercepts.

$\dfrac{x^3 - 4x}{x^3 - 9x} = \dfrac{x(x^2 - 4)}{x(x^2 - 9)} = \dfrac{x^2 - 4}{x^2 - 9} \Rightarrow$

$\dfrac{0^2 - 4}{0^2 - 9} = \dfrac{4}{9} \Rightarrow \dfrac{4}{9}$ is the y-intercept. However,

there is a hole at $\left(0, \dfrac{4}{9}\right)$ since $0^3 - 9(0) = 0$

$x^2 - 9 = 0 \Rightarrow x(x+3)(x-3) = 0 \Rightarrow x = -3$ and $x = 3$ are the vertical asymptotes. The degree of the numerator is the same as the degree of the denominator, so the horizontal asymptote is $y = 1$. The intervals to be tested are $(-\infty, -3), (-3, 0), (0, 3)$, and $(3, \infty)$. The graph is above the horizontal asymptote on $(-\infty, -3) \cup (3, \infty)$ and below the horizontal asymptote on $(-3, 0) \cup (0, 3)$.

73. $0 = \dfrac{(x-2)^2}{x-2} \Rightarrow$ there is no x-intercept. There

is a hole at $(2, 0)$. $\dfrac{(0-2)^2}{0-2} = -2 \Rightarrow y = -2$ is

the y-intercept. There are no vertical asymptotes. There are no horizontal asymptotes. The intervals to be tested are $(-\infty, 2)$ and $(2, \infty)$.

The graph is above the x-axis on $(2, \infty)$ and below the x-axis on $(-\infty, 2)$.

75. The x-intercept is 1 and the vertical asymptote is $x = 2$. The horizontal asymptote is $y = -2$, so the degree of the numerator equals the degree of the denominator, and the ratio of the leading terms of the numerator and the denominator is -2. Thus, the equation is of the

form $y = a\left(\dfrac{x-1}{x-2}\right)$. The y-intercept is -1, so

we have $-1 = a\left(\dfrac{0-1}{0-2}\right) \Rightarrow a = -2$.

Thus, the equation is $f(x) = \dfrac{-2(x-1)}{x-2}$.

77. The x-intercepts are 1 and 3, and the vertical asymptotes are $x = 0$ and $x = 2$. The horizontal asymptote is $y = 1$, so the degree of the numerator equals the degree of the denominator, and the ratio of the leading terms of the numerator and the denominator is 1. Thus, the equation is of the form

$y = a\dfrac{(x-1)(x-3)}{x(x-2)}$. There is no y-intercept,

so the equation is $f(x) = \dfrac{(x-1)(x-3)}{x(x-2)}$.

79. $\dfrac{2x^2 + 1}{x} = 2x + \dfrac{1}{x}$.

The oblique asymptote is $y = 2x$.

81. $\dfrac{x^3 - 1}{x^2} = x - \dfrac{1}{x^2}$.

The oblique asymptote is $y = x$.

83.
$$\begin{array}{r} x-2 \\ x+1\overline{)x^2 - x + 1} \\ \underline{x^2 + x} \\ -2x + 1 \\ \underline{-2x - 2} \\ 3 \end{array}$$
The oblique asymptote is $y = x - 2$.

85.
$$\begin{array}{r} x-2 \\ x^2-1\overline{)x^3 - 2x^2 + 0x + 1} \\ \underline{x^3 \quad\quad - x} \\ -2x^2 + x + 1 \\ \underline{-2x^2 \quad\quad + 2} \\ x - 1 \end{array}$$

The oblique asymptote is $y = x - 2$. Note that there is a hole in the graph at $x = 1$.

(*continued on next page*)

(continued)

3.6 Applying the Concepts

87. a. $C(x) = 0.5x + 2000$

b. $\overline{C}(x) = \dfrac{C(x)}{x} = \dfrac{0.5x + 2000}{x} = 0.5 + \dfrac{2000}{x}$

c. $\overline{C}(100) = 0.5 + \dfrac{2000}{100} = 20.5$

$\overline{C}(500) = 0.5 + \dfrac{2000}{500} = 4.5$

$\overline{C}(1000) = 0.5 + \dfrac{2000}{1000} = 2.5$

These show the average cost of producing 100, 500, and 1000 trinkets, respectively.

d. The horizontal asymptote of $\overline{C}(x)$ is $y = 0.5$. It means that the average cost approaches the daily fixed cost of producing each trinket as the number of trinkets approaches ∞.

89. a.

b. $f(50) = \dfrac{4(50)+1}{100-50} \approx 4$ min

$f(75) = \dfrac{4(75)+1}{100-75} \approx 12$ min

$f(95) = \dfrac{4(95)+1}{100-95} \approx 76$ min

$f(99) = \dfrac{4(99)+1}{100-99} \approx 397$ min

c. (i) As $x \to 100^-$, $f(x) \to \infty$.

(ii) The statement is not applicable because the domain is $x < 100$.

d. No, the bird doesn't ever collect all the seed from the field.

91. a. $C(50) = \dfrac{3(50^2)+50}{50(100-50)} = \$3.02$ billion

b.

c. $30 = \dfrac{3x^2+50}{x(100-x)} = \dfrac{3x^2+50}{100x-x^2} \Rightarrow$

$3000x - 30x^2 = 3x^2 + 50 \Rightarrow$

$-33x^2 + 3000x - 50 = 0 \Rightarrow$

$x = \dfrac{-3000 \pm \sqrt{3000^2 - 4(-33)(-50)}}{2(-33)}$

$= \dfrac{-3000 \pm \sqrt{8,993,400}}{-66} \approx \dfrac{-3000 \pm 2998.9}{-66}$

≈ 90.89 or 0.017.

From the graph, we see that approximately 90.89% of the impurities can be removed at a cost of \$30 billion.

93. a. $P(0) = \dfrac{8(0)+16}{2(0)+1} = 16$ thousand $= 16,000$

b. The horizontal asymptote is $y = 4$. This means that the population will stabilize at 4000.

95. a. $f(x) = \dfrac{10x+200,000}{x-2500}$

b. $C(10,000) = \dfrac{10(10,000)+200,000}{10,000-2500} = \$40$

c. $20 > \dfrac{10x+200,000}{x-2500} \Rightarrow$

$20x - 50,000 > 10x + 200,000 \Rightarrow$

$10x > 250,000 \Rightarrow x > 25,000$

More than 25,000 books must be sold to bring the average cost under \$20.

d. The vertical asymptote is $x = 2500$. This represents the number of free samples. The horizontal asymptote is $y = 10$. This represents the cost of printing and binding one book.

3.6 Beyond the Basics

97. Stretch the graph of $y = \dfrac{1}{x}$ vertically by a factor of 2, and then reflect the graph about the x-axis.

99. Shift the graph of $y = \dfrac{1}{x^2}$ two units right.

101. Shift the graph of $y = \dfrac{1}{x^2}$ one unit right and two units down.

103. Shift the graph of $y = \dfrac{1}{x^2}$ six units left.

105. $x^2 - 2x + 1 \overline{\smash{\big)}\ x^2 - 2x + 2} \Rightarrow$

$$\dfrac{x^2 - 2x + 1}{1}$$

$$\dfrac{x^2 - 2x + 2}{x^2 - 2x + 1} = \dfrac{(x^2 - 2x + 1) + 1}{x^2 - 2x + 1} =$$

$$\dfrac{x^2 - 2x + 1}{x^2 - 2x + 1} + \dfrac{1}{x^2 - 2x + 1} = 1 + \dfrac{1}{(x-1)^2}$$

Shift the graph of $y = \dfrac{1}{x^2}$ one unit right and one unit up.

107. a. $f(x) \to 0$ as $x \to -\infty$; $f(x) \to 0$ as $x \to \infty$; $f(x) \to -\infty$ as $x \to 0^{-}$; $f(x) \to \infty$ as $x \to 0^{+}$. There are no x- or y-intercepts. The horizontal asymptote is the x-axis. The vertical asymptote is the y-axis. The graph is above the x-axis on $(0, \infty)$ and below the x-axis on $(-\infty, 0)$.

b. $f(x) \to 0$ as $x \to -\infty$; $f(x) \to 0$ as $x \to \infty$; $f(x) \to \infty$ as $x \to 0^-$; $f(x) \to -\infty$ as $x \to 0^+$. There are no *x*- or *y*-intercepts. The horizontal asymptote is the *x*-axis. The vertical asymptote is the *y*-axis. The graph is above the *x*-axis on $(-\infty, 0)$ and below the *x*-axis on $(0, \infty)$.

c. $f(x) \to 0$ as $x \to -\infty$; $f(x) \to 0$ as $x \to \infty$; $f(x) \to \infty$ as $x \to 0^-$; $f(x) \to \infty$ as $x \to 0^+$. There are no *x*- or *y*-intercepts. The horizontal asymptote is the *x*-axis. The vertical asymptote is the *y*-axis. The graph is above the *x*-axis on $(-\infty, 0) \cup (0, \infty)$.

d. $f(x) \to 0$ as $x \to -\infty$; $f(x) \to 0$ as $x \to \infty$; $f(x) \to -\infty$ as $x \to 0^-$; $f(x) \to -\infty$ as $x \to 0^+$. There are no *x*- or *y*-intercepts. The horizontal asymptote is the *x*-axis. The vertical asymptote is the *y*-axis. The graph is below the *x*-axis on $(-\infty, 0) \cup (0, \infty)$.

109.

111. $f(x) = 2x + 3 \Rightarrow [f(x)]^{-1} = \dfrac{1}{2x+3}$ and
$y = 2x + 3$ becomes $x = 2y + 3 \Rightarrow$
$\dfrac{x-3}{2} = \dfrac{x}{2} - \dfrac{3}{2} = y = f^{-1}(x)$.
Thus, the functions are different.

113. $g(x) = \dfrac{2x^3 + 3x^2 + 2x - 4}{x^2 - 1} = 2x + 3 + \dfrac{-1}{x^2 - 1}$
$g(x)$ has the oblique asymptote $y = 2x + 3$.
$g(x) \to -\infty$ as $x \to -\infty$, and $g(x) \to \infty$ as $x \to \infty$.

115. The horizontal asymptote is $y = 1$.
$\dfrac{x^2 + x - 2}{x^2 - 2x - 3} = 1 \Rightarrow x^2 + x - 2 = x^2 - 2x - 3 \Rightarrow$
$3x = -1 \Rightarrow x = -\dfrac{1}{3}$.
The point of intersection is $\left(-\dfrac{1}{3}, 1\right)$.

(*continued on next page*)

(*continued*)

117. Vertical asymptote at $x = 3 \Rightarrow$ the denominator is $x - 3$. Horizontal asymptote at $y = -1 \Rightarrow$ the ratio of the leading coefficients of the numerator and denominator $= -1$. The x–intercept $= 2 \Rightarrow$ the numerator is $x - 2$. So the equation is of the form

$$f(x) = -\frac{x-2}{x-3} \text{ or } \frac{2-x}{x-3}.$$

119. Vertical asymptote at $x = 0 \Rightarrow$ the denominator is x. The slant asymptote $y = -x \Rightarrow$ the degree of the numerator is one more than the degree of the denominator and the quotient of the numerator and the denominator is $-x$. The x-intercepts -1 and $1 \Rightarrow$ the numerator is $(x-1)(x+1)$. So the equation is of the form

$$f(x) = a\frac{(x-1)(x+1)}{x} = a\left(\frac{x^2-1}{x}\right).$$

$$\begin{array}{r} x \\ x\overline{)x^2-1} \\ \underline{x^2} \\ -1 \end{array}$$

However, the slant asymptote is $y = -x$, so $a = -1$ and the equation is

$$f(x) = -\frac{x^2-1}{x} = \frac{-x^2+1}{x}.$$

121. Vertical asymptote at $x = 2 \Rightarrow$ the denominator is $x - 2$. Horizontal asymptote at $y = 1 \Rightarrow$ the degrees of the numerator and denominator are the same and the leading coefficients of the numerator and denominator are the same. So the numerator is $x + a$, where a is chosen so that $f(0) = -2 \Rightarrow \frac{x+4}{x-2} = f(x)$.

123. $f(x) \to -\infty$ as $x \to 1^-$ and $f(x) \to \infty$ as $x \to 1^+ \Rightarrow$ the denominator is zero if $x = 1$. Since $x \to 1$ from both directions, the denominator is $(x-1)^2$. $f(x) \to 4$ as $x \to \pm\infty \Rightarrow$ the horizontal asymptote is 4.

So the leading coefficient of the numerator is 4 and the degree of the numerator is the same as the degree of the denominator. The numerator is $4x^2 + a$, where a is chosen so that

$$f\left(\frac{1}{2}\right) = 0 \Rightarrow \frac{4(1/2)^2 + a}{(1/2 - 1)^2} \Rightarrow a = -1. \text{ So,}$$

$$f(x) = \frac{4x^2 - 1}{(x-1)^2}.$$

125. Vertical asymptote at $x = 1 \Rightarrow$ the denominator could be $x - 1$. Because the oblique asymptote is $y = 3x + 2$, the numerator is $(3x + 2)(x - 1) + a$, where a can be any number (no x-intercepts or function values are given). Let $a = 1$. So.

$$f(x) = \frac{(3x+2)(x-1)+1}{x-1} = \frac{3x^2 - x + 1}{x-1}.$$

3.6 Critical Thinking/Discussion/Writing

127. Answers may vary. Sample answers are given:

a. $f(x) = \frac{1}{x}$

b. $f(x) = \frac{x^2 + 2}{x^2 + x - 2}$

c. $f(x) = \frac{x^3 + 2x^2 - 7x - 1}{x^3 + x^2 - 6x + 5}$

128. Answers may vary. Sample answers are given:

a. $f(x) = \frac{x^2 + 1}{x}$

b. $f(x) = \frac{x^3 + x - 2}{x^2}$

c. $f(x) = \frac{x^5 + x^4 + x^2 - 4}{x^4}$

129. Since $y = x + 1$ is the oblique asymptote, we know that $R(x) = x + 1 + \frac{r(x)}{D(x)}$. We are told that the graph of $R(x)$ intersects the asymptote at the points $(2, R(2))$ and $(5, R(5))$, so $R(x) = x + 1$ when $\frac{r(x)}{D(x)} = 0$. Thus, $r(x) = (x - 2)(x - 5)$.

(*continued on next page*)

(continued)

Recall that the numerator of a rational function must have degree greater than that of the denominator in order for there to be an oblique asymptote, so the denominator, $D(x)$, can be any polynomial whose degree is greater than or equal to 3. Thus, a possible rational function is

$$R(x) = x + 1 + \frac{(x-2)(x-5)}{x^5}$$
$$= \frac{(x+1)x^5 + (x-2)(x-5)}{x^5}.$$

130. See exercise 129 for explanation.

$$R(x) = (ax+b) + \frac{K(x-c_1)(x-c_2)\cdots(x-c_n)}{D(x)}$$
$$= \frac{(ax+b) + D(x) + K(x-c_1)(x-c_2)\cdots(x-c_n)}{D(x)},$$

where $K \neq 0$, $D(x)$ is a polynomial of degree $n+1$ and none of its zeros are at $c_1, c_2, \ldots, c_n$.

3.6 Getting Ready for the Next Section

131. a. $y - 3 = -\frac{2}{3}(x - (-5)) \Rightarrow y - 3 = -\frac{2}{3}(x+5) \Rightarrow$

$y = -\frac{2}{3}x - \frac{2}{3}(5) + 3 \Rightarrow y = -\frac{2}{3}x - \frac{1}{3}$

   b. $y = -\frac{2}{3}(1) - \frac{1}{3} = -1$

133. $3(1) + 2c = 11 \Rightarrow 2c = 8 \Rightarrow c = 4$

135. $12 = \frac{k}{\left(\frac{1}{2}\right)^2} \Rightarrow k = 3$

$y = \frac{3}{2^2} \Rightarrow y = \frac{3}{4}$

3.7 Variation

3.7 Practice Problems

1. $y = kx \Rightarrow 6 = 30k \Rightarrow \frac{1}{5} = k$

$y = \left(\frac{1}{5}\right) 120 = 24$

2. $I = kV \Rightarrow 60 = 220k \Rightarrow \frac{3}{11} = k$

$75 = \left(\frac{3}{11}\right) V \Rightarrow 275 = V$

A battery of 275 volts is needed to produce 60 amperes of current.

3. $y = kx^2 \Rightarrow 48 = k(2)^2 \Rightarrow 12 = k$

$y = 12(5)^2 = 300$

4. $A = \frac{k}{B} \Rightarrow 12 = \frac{k}{5} \Rightarrow 60 = k$

$A = \frac{60}{3} = 20$

5. $y = \frac{k}{\sqrt{x}} \Rightarrow \frac{3}{4} = \frac{k}{\sqrt{16}} \Rightarrow k = 3$

$2 = \frac{3}{\sqrt{x}} \Rightarrow \sqrt{x} = \frac{3}{2} \Rightarrow x = \frac{9}{4}$

6. $F = G \cdot \frac{m_1 m_2}{r^2} \Rightarrow m \cdot g = G \cdot \frac{mM_{\text{Mars}}}{R_{\text{Mars}}^2} \Rightarrow$

$g = G \cdot \frac{M_{\text{Mars}}}{R_{\text{Mars}}^2}$

Note that the radius of Mars is given in kilometers, which must be converted to meters.

$$g = \frac{\left(6.67 \times 10^{-11} \text{ m}^3 / \text{kg} / \text{sec}^2\right)\left(6.42 \times 10^{23} \text{ kg}\right)}{(3397 \text{ km})^2}$$

$$= \frac{\left(6.67 \times 10^{-11} \text{ m}^3 / \text{kg} / \text{sec}^2\right)\left(6.42 \times 10^{23} \text{ kg}\right)}{\left(3.397 \times 10^6 \text{ m}\right)^2}$$

$$\approx 3.7 \text{ m/sec}^2$$

3.7 Concepts and Vocabulary

1. y varies directly as x if $\underline{y = kx}$.

3. y varies directly as the nth power of x if $\underline{y = kx^n}$.

5. True

7. False.

3.7 Building Skills

9. $x = ky; 15 = 30k \Rightarrow k = \frac{1}{2}; x = \frac{1}{2}(28) = 14$

11. $s = kt^2; 64 = 2^2 k \Rightarrow k = 16; s = 5^2(16) = 400$

13. $r = \frac{k}{u}; 3 = \frac{k}{11} \Rightarrow k = 33; r = \frac{33}{1/3} = 99$

15. $B = \dfrac{k}{A^3}; 1 = \dfrac{k}{2^3} \Rightarrow k = 8; B = \dfrac{8}{4^3} = \dfrac{1}{8}$

17. $z = kxy; 42 = (2)(3)k \Rightarrow k = 7;\ 56 = 2(7)y \Rightarrow$
$y = 4$

19. $z = kx^2; 32 = 4^2 k \Rightarrow k = 2; z = 2(5^2) = 50$

21. $P = kTQ^2; 36 = (17)(6^2)k \Rightarrow k = \dfrac{1}{17};$
$P = \dfrac{1}{17}(4)(9^2) = \dfrac{324}{17}$

23. $z = \dfrac{k\sqrt{x}}{y^2}; 24 = \dfrac{\sqrt{16}k}{3^2} \Rightarrow k = 54$
$27 = \dfrac{54\sqrt{x}}{2^2} \Rightarrow 2 = \sqrt{x} \Rightarrow x = 4$

25. $\dfrac{16}{12} = \dfrac{8}{y} \Rightarrow y = 6$

27. $\dfrac{100}{x_0} = \dfrac{y}{2x_0} \Rightarrow y = 200$

3.7 Applying the Concepts

29. $y = Hx$, where y is the speed of the galaxies and x is the distance between them.

31. a. $y = 30.5x$, where y is the length in centimeters and x is the length in feet.

   b. (i) $y = 30.5(8) = 244$ cm

   (ii) $y = 30.5\left(5\dfrac{1}{3}\right) \approx 162.67$ cm

   c. (i) $57 = 30.5x \Rightarrow x \approx 1.87$ ft
   (ii) $124 = 30.5x \Rightarrow x \approx 4.07$ ft

33. $P = kQ; 7 = 20k \Rightarrow \dfrac{7}{20} = k; P = \dfrac{7}{20}(100) = 35$ g

35. $d = kt^2; 64 = 2^2 k \Rightarrow k = 16; 9 = 16t^2 \Rightarrow$
$t = \dfrac{3}{4}$ sec

37. $P = \dfrac{k}{V}; 20 = \dfrac{k}{300} \Rightarrow k = 6000;$
$P = \dfrac{6000}{100} = 60$ lb/in.^{2}

39. a. The astronaut is $6000 + 3960$ miles from the Earth's center.
$W = \dfrac{k}{d^2}; 120 = \dfrac{k}{3960^2} \Rightarrow k = 120(3960^2)$
$W = \dfrac{120(3960^2)}{(6000+3960)^2} \approx 18.97$ lb

   b. $200 = \dfrac{k}{3960^2} \Rightarrow k = 200(3960^2)$
$W = \dfrac{200(3960^2)}{3950^2} \approx 201.01$ lb

41. 1740 km $= 1{,}740{,}000$ m $= 1.740 \times 10^6$ m
$g = G \cdot \dfrac{M}{R^2} = 6.67 \times 10^{-11} \times \dfrac{7.4 \times 10^{22}}{\left(1.740 \times 10^6\right)^2}$
$= \dfrac{(6.67)(7.4)(10^{11})}{1.740^2 \times 10^{12}} \approx 1.63$ m/sec^2

43. a. $I = \dfrac{k}{d^2}; 320 = \dfrac{k}{10^2} \Rightarrow k = 32{,}000$
$I = \dfrac{32{,}000}{5^2} = 1280$ candlepower

   b. $400 = \dfrac{32{,}000}{d^2} \Rightarrow d^2 = 80 \Rightarrow d \approx 8.94$ ft
from the source.

45. $p = k\sqrt{l} \Rightarrow k = \dfrac{p}{\sqrt{l}} = \dfrac{2p}{2\sqrt{l}} = \dfrac{2p}{\sqrt{4l}} \Rightarrow$ the
length is multiplied by 4 if the period is doubled.

47. a. $H = kR^2 N$, where k is a constant

   b. $k(2R)^2 N = 4kR^2 N = 4H \Rightarrow$ the horsepower is multiplied by 4 if the radius is doubled.

   c. $kR^2(2N) = 2kR^2 N = 2H \Rightarrow$ the horsepower is doubled if the number of pistons is doubled.

   d. $k\left(\dfrac{R}{2}\right)^2 (2N) = \dfrac{2kR^2 N}{4} = \dfrac{kR^2 N}{2} = \dfrac{H}{2} \Rightarrow$
the horsepower is halved if the radius is halved and the number of pistons is doubled.

3.7 Beyond the Basics

49. a. $E = kl^2 v^3$

b. $1920 = k(10^2)(8^3) \Rightarrow k = \dfrac{3}{80} = 0.0375$

c. $E = 0.0375(8^2)(25^3) = 37,500$ watts

d. $kl^2(2v)^3 = 8kl^2v^3 = 8E$

e. $k(2l)^2v^3 = 4kl^2v^3 = 4E$

f. $k(2l)^2(2v)^3 = 4(8)kl^2v^3 = 32E$

51. a. $m = kw^{3/4}; 75 = k(75^{3/4}) \Rightarrow$
$k = \sqrt[4]{3}\sqrt{5} \approx 2.94$

b. $m = 2.94(450^{3/4}) \approx 287.25$ watts

c. $k(4w)^{3/4} = 4^{3/4}kw^{3/4} \approx 2.83kw^{3/4} \approx 2.83\,m$

d. $250 \approx 2.94w^{3/4} \Rightarrow w^{3/4} \approx 85.034 \Rightarrow$
$w \approx 373.93$ kg

53. a. $T^2 = \dfrac{4\pi^2 r^3}{G(M_1 + M_2)}$

b. Because gravity is in terms of cubic meters per kilogram per second squared, convert the distance from kilometers to meters:

$1.5 \times 10^8\,\text{km} = 1.5 \times 10^{11}\,\text{m}$.

$(3.15 \times 10^7)^2 \approx \dfrac{4\pi^2 (1.5 \times 10^{11})^3}{6.67 \times 10^{-11} M_{\text{sun}}} \Rightarrow$

$9.9225 \times 10^{14} \approx \dfrac{4\pi^2 (3.375) \times 10^{33}}{6.67 \times 10^{-11} M_{\text{sun}}} \Rightarrow$

$M_{\text{sun}} \approx \dfrac{133.24 \times 10^{24}}{6.67 \times 10^{-11} \times 9.9225 \times 10^{14}}$
$\approx 2.01 \times 10^{30}$ kg

55. a. $R = kN(P - N)$, where k is the constant of proportionality.

b. $45 = 1000(9000)k \Rightarrow k = 5 \times 10^{-6}$

c. $R = 5 \times 10^{-6}(5000)(5000) = 125$ people per day.

d. $100 = 5 \times 10^{-6} N(10,000 - N) \Rightarrow$
$20,000,000 = 10,000N - N^2 \Rightarrow$

$N^2 - 10,000N + 20,000,000 = 0 \Rightarrow$

$N = \dfrac{10,000 \pm \sqrt{10,000^2 - 4(1)(2 \times 10^7)}}{2(1)}$

$= \dfrac{10,000 \pm \sqrt{20,000,000}}{2} \approx 2764$ or 7236

3.7 Critical Thinking/Discussion/Writing

56. $I = \dfrac{kV}{R} \Rightarrow IR = kV$

$1.3I = \dfrac{kV}{1.2R} \Rightarrow 1.56IR = kV \Rightarrow$ the voltage

must increase by 56%.

57. a. $v = kw^2$. The diamond is cut into two

pieces whose weights are $\dfrac{2w}{5}$ and $\dfrac{3w}{5}$.

The value of the first piece is

$\left(\dfrac{4}{25}\right)(1000) = \$160$, and the value of the

second piece is $\left(\dfrac{9}{25}\right)(1000) = \$360$.

The two pieces together are valued at \$520, a loss of \$480.

b. The stone is broken into three pieces whose

weights are $\dfrac{5w}{25} = \dfrac{w}{5}, \dfrac{9w}{25}$, and $\dfrac{11w}{25}$. So,

the values of the three pieces are

$\left(\dfrac{1}{5}\right)^2 (25,000) = \$1000,$

$\left(\dfrac{9}{25}\right)^2 (25,000) = \$3240,$ and

$\left(\dfrac{11}{25}\right)^2 (25,000) = \$4840,$ respectively.

The total value is \$9,080, a loss of \$15,920.

c. The weights of the pieces are

$\dfrac{w}{15}, \dfrac{2w}{15}, \dfrac{3w}{15} = \dfrac{w}{5}, \dfrac{4w}{15}$, and $\dfrac{5w}{15} = \dfrac{w}{3}$,

respectively. If the original value is x, then

$x - 85,000 = \left(\dfrac{1}{15}\right)^2 x + \left(\dfrac{2}{15}\right)^2 x + \left(\dfrac{1}{5}\right)^2 x$

$+ \left(\dfrac{4}{15}\right)^2 x + \left(\dfrac{1}{3}\right)^2 x \Rightarrow$

$x - 85,000 = \dfrac{11}{45}x \Rightarrow x = \$112,500 =$ the

original value of the diamond.

(continued on next page)

(continued)

A diamond whose weight is twice that of the original diamond is worth 4 times the value of the original diamond = \$450,000.

58. $p = kd(n - f)$

$80 = k(40)(30 - f)$ and
$180 = k(60)(35 - f)$. Solving the first

equation for k we have $k = -\dfrac{2}{f - 30}$.

Substitute that value into the second equation and solve for f:

$180 = \left(\dfrac{2}{30 - f}\right)(60)(35 - f) \Rightarrow$

$180 = \dfrac{4200 - 120f}{30 - f} \Rightarrow$

$5400 - 180f = 4200 - 120f \Rightarrow$

$1200 = 60f \Rightarrow f = 20$

59. $w_{\text{solid}} = kr_o{}^3; w_{\text{hollow}} = kr_o{}^3 - kr_i{}^3 = \dfrac{7}{8}kr_o{}^3$

$kr_o{}^3 - kr_i{}^3 = \dfrac{7}{8}kr_o{}^3 \Rightarrow r_o{}^3 - r_i{}^3 = \dfrac{7}{8}r_o{}^3 \Rightarrow$

$\dfrac{1}{8}r_o{}^3 = r_i{}^3 \Rightarrow \dfrac{1}{8} = \dfrac{r_i{}^3}{r_o{}^3} \Rightarrow \dfrac{1}{2} = \dfrac{r_i}{r_o}$

60. $s = 24 - k\sqrt{w}; 20 = 24 - k\sqrt{4} \Rightarrow k = 2$

$0 \le 24 - 2\sqrt{w} \Rightarrow w \le 144$

The greatest number of wagons the engine can move is 144.

3.7 Getting Ready for the Next Sections

61. $5^0 = 1$

63. $3^{-2} = \dfrac{1}{3^2} = \dfrac{1}{9}$

65. $\left(\dfrac{1}{2}\right)^{-4} = 2^4 = 16$

67. $5^{2x-3} \cdot 5^{3-x} = 5^{(2x-3)+(3-x)} = 5^x$

69. $\left(2^{x-1}\right)^x = 2^{(x-1)(x)} = 2^{x^2 - x}$

71. $y = mx + b \Rightarrow y - b = mx \Rightarrow m = \dfrac{y - b}{x}$

73. $A = B(1 + C) \Rightarrow \dfrac{A}{B} = 1 + C \Rightarrow \dfrac{A}{B} - 1 = C$

75. $A = B \cdot 10^{-n} \Rightarrow \dfrac{A}{10^{-n}} = B \Rightarrow A \cdot 10^n = B$

Chapter 3 Review Exercises

Building Skills

1. **(i)**   Opens up
   (ii)   Vertex: $(1, 2)$
   (iii)   Axis: $x = 1$
   (iv) $0 = (x - 1)^2 + 2 \Rightarrow -2 = (x - 1)^2 \Rightarrow$
         there are no x-intercepts.
   (v) $y = (0 - 1)^2 + 2 \Rightarrow y = 3$ is the
         y-intercept.
   (vi)   The function is decreasing on $(-\infty, 1)$
         and increasing on $(1, \infty)$.

3. **(i)**   Opens down
   (ii)   vertex: $(3, 4)$
   (iii)   Axis: $x = 3$
   (iv) $0 = -2(x - 3)^2 + 4 \Rightarrow 2 = (x - 3)^2 \Rightarrow$
         $x = 3 \pm \sqrt{2}$ are the x-intercepts.
   (v) $y = -2(0 - 3)^2 + 4 \Rightarrow y = -14$ is the
         y-intercept.
   (vi)   The function is increasing on $(-\infty, 3)$
         and decreasing on $(3, \infty)$.

(continued on next page)

(continued)

5. **(i)**   Opens down

(ii)   Vertex: $(0, 3)$

(iii)   Axis: $x = 0$ (y-axis)

(iv) $0 = -2x^2 + 3 \Rightarrow \dfrac{3}{2} = x^2 \Rightarrow$

   $x = \pm \dfrac{\sqrt{6}}{2}$ are the x-intercepts.

(v) $y = -2(0)^2 + 3 \Rightarrow y = 3$ is the
   y-intercept.

(vi)   The function is increasing on $(-\infty, 0)$
   and decreasing on $(0, \infty)$.

7. **(i)**   Opens up

(ii)   To find the vertex, write the equation in
   standard form by completing the square:
   $$y = 2x^2 - 4x + 3$$
   $$y - 3 + 2 = 2(x^2 - 2x + 1)$$
   $$y = 2(x - 1)^2 + 1$$
   The vertex is $(1, 1)$.

(iii)   Axis: $x = 1$

(iv) $0 = 2x^2 - 4x + 3 \Rightarrow$

   $x = \dfrac{4 \pm \sqrt{(-4)^2 - 4(2)(3)}}{2(2)} = \dfrac{4 \pm \sqrt{-8}}{4} \Rightarrow$

   there are no x-intercepts.

(v) $y = 2(0)^2 - 4(0) + 3 \Rightarrow y = 3$ is the
   y-intercept.

(vi)   The function is decreasing on $(-\infty, 1)$
   and increasing on $(1, \infty)$.

9. **(i)**   Opens up

(ii)   To find the vertex, write the equation in
   standard form by completing the square.
   $$y = 3x^2 - 2x + 1$$
   $$y - 1 + \frac{1}{3} = 3\left(x^2 - \frac{2}{3}x + \frac{1}{9}\right)$$
   $$y = 3\left(x - \frac{1}{3}\right)^2 + \frac{2}{3}$$
   The vertex is $\left(\dfrac{1}{3}, \dfrac{2}{3}\right)$.

(iii)   Axis: $x = \dfrac{1}{3}$

(iv) $0 = 3x^2 - 2x + 1 \Rightarrow$

   $x = \dfrac{2 \pm \sqrt{(-2)^2 - 4(3)(1)}}{2(3)} = \dfrac{2 \pm \sqrt{-8}}{6} \Rightarrow$

   there are no x-intercepts.

(v) $y = 3(0)^2 - 3(0) + 1 \Rightarrow y = 1$ is the
   y-intercept.

(vi)   The function is decreasing on $\left(-\infty, \dfrac{1}{3}\right)$

   and increasing on $\left(\dfrac{1}{3}, \infty\right)$.

In exercises 11–14, find the vertex using the formula $\left(-\dfrac{b}{2a}, f\left(-\dfrac{b}{2a}\right)\right).$

11. $a > 0 \Rightarrow$ the graph opens up, so f has a minimum value at its vertex. The vertex is

$$\left(-\frac{-4}{2(1)}, f\left(-\frac{-4}{2(1)}\right)\right) = (2, -1).$$

The minimum value is -1.

13. $a < 0 \Rightarrow$ the graph opens down, so f has a maximum value at its vertex. The vertex is

$$\left(-\frac{-3}{2(-2)}, f\left(-\frac{-3}{2(-2)}\right)\right) = \left(-\frac{3}{4}, \frac{25}{8}\right).$$

The maximum value is $\frac{25}{8}$.

15. Shift the graph of $y = x^3$ one unit left and two units down.

17. Shift the graph of $y = x^3$ one unit right, reflect the resulting graph in the y-axis, and shift it one unit up.

19. (i) $f(x) \to -\infty$ as $x \to -\infty$
$f(x) \to \infty$ as $x \to \infty$

(ii) Zeros: $x = -2$, multiplicity 1, crosses the x-axis; $x = 0$, multiplicity 1, crosses the x-axis; $x = 1$, multiplicity 1, crosses the x-axis.

(iii) x-intercepts: $-2, 0, 1$;
$y = 0(0 - 1)(0 + 2) \Rightarrow y = 0$ is the y-intercept.

(iv) The intervals to be tested are $(-\infty, -2)$, $(-2, 0)$, $(0, 1)$, and $(1, \infty)$.
The graph is above the x-axis on $(-2, 0) \cup (1, \infty)$ and below the x-axis on $(-\infty, -2) \cup (0, 1)$.

(v) $f(-x) = -x(-x - 1)(-x + 2) \neq f(x) \Rightarrow$ f is not even.
$-f(x) = -(x(x - 1)(x + 2)) \neq f(-x) \Rightarrow$ f is not odd. There are no symmetries.

(vi)

21. (i) $f(x) \to -\infty$ as $x \to -\infty$
$f(x) \to -\infty$ as $x \to \infty$

(ii) Zeros: $x = 0$, multiplicity 2, touches but does not cross the x- axis; $x = 1$, multiplicity 2, touches but does not cross the x- axis.

(iii) x-intercepts: 0, 1;
$y = -0^2(0 - 1)^2 \Rightarrow y = 0$ is the y-intercept.

(iv) The intervals to be tested are $(-\infty, 0)$, $(0, 1)$, and $(1, \infty)$. The graph is below the x-axis on $(-\infty, 0) \cup (0, 1) \cup (1, \infty)$.

(v) $f(-x) = -(-x)^2(-x - 1)$
$= -x^2(-x - 1) \neq -f(x) \text{ or } f(x) \Rightarrow$
$= -f(x) \Rightarrow f$ is neither even nor odd.
There are no symmetries.

(vi)

23. **(i)** $f(x) \to -\infty$ as $x \to -\infty$
$f(x) \to -\infty$ as $x \to \infty$

(ii) Zeros: $x = -1$, multiplicity 1, crosses the x-axis; $x = 0$, multiplicity 2, touches but does not cross; $x = 1$, multiplicity 1, crosses the x-axis.

(iii) x-intercepts: $-1, 0, 1$;
$y = -0^2(0^2 - 1) \Rightarrow y = 0$ is the y-intercept.

(iv) The intervals to be tested are $(-\infty, -1)$, $(-1, 0), (0, 1)$, and $(1, \infty)$. The graph is above the x-axis on $(-1, 0) \cup (0, 1)$ and below the x-axis on $(-\infty, -1) \cup (1, \infty)$.

(v) $f(-x) = -(-x)^2((-x)^2 - 1)$
$= -x^2(x^2 - 1) = f(x) \Rightarrow f$ is even, and f is symmetric with respect to the y-axis.

(vi)

25.
$$\begin{array}{r} 2x+3 \\ 3x-2 \overline{\smash{)}6x^2+5x-13} \\ \underline{6x^2-4x} \\ 9x-13 \\ \underline{9x-6} \\ -7 \end{array}$$

27.
$$\begin{array}{r} 8x^3-12x^2+14x-21 \\ x+1 \overline{\smash{)}8x^4-4x^3+2x^2-7x+165} \\ \underline{8x^4+8x^3} \\ -12x^3+2x^2 \\ \underline{-12x^3-12x^2} \\ 14x^2-7x \\ \underline{14x^2+14x} \\ -21x+165 \\ \underline{-21x-21} \\ 186 \end{array}$$

29.
$$\begin{array}{r|rrrr} 3 & 1 & 0 & -12 & 3 \\ & & 3 & 9 & -9 \\ \hline & 1 & 3 & -3 & -6 \end{array}$$
Quotient: $x^2 + 3x - 3$ remainder -6.

31.
$$\begin{array}{r|rrrrr} -1 & 2 & -3 & 5 & -7 & 165 \\ & & -2 & 5 & -10 & 17 \\ \hline & 2 & -5 & 10 & -17 & 182 \end{array}$$
Quotient: $2x^3 - 5x^2 + 10x - 17$ remainder 182.

33. **(i)** $f(2) = 2^3 - 3(2^2) + 11(2) - 29 = -11$

(ii)
$$\begin{array}{r|rrrr} 2 & 1 & -3 & 11 & -29 \\ & & 2 & -2 & 18 \\ \hline & 1 & -1 & 9 & -11 \end{array}$$
The remainder is -11, so $f(2) = -11$.

35. **(i)** $f(2) = (-3)^4 - 2(-3)^2 - 5(-3) + 10 = 88$

(ii)
$$\begin{array}{r|rrrrr} -3 & 1 & 0 & -2 & -5 & 10 \\ & & -3 & 9 & -21 & 78 \\ \hline & 1 & -3 & 7 & -26 & 88 \end{array}$$
The remainder is 88, so $f(-3) = 88$.

37.
$$\begin{array}{r|rrrr} 2 & 1 & -7 & 14 & -8 \\ & & 2 & -10 & 8 \\ \hline & 1 & -5 & 4 & 0 \end{array}$$
The remainder is 0, so 2 is a zero. Now find the zeros of the depressed function
$x^2 - 5x + 4 = (x - 4)(x - 1) \Rightarrow$ 4 and 1 are
also zeros. So the zeros of $x^3 - 7x^2 + 14x - 8$ are 1, 2, and 4.

39.
$$\begin{array}{r|rrrr} \frac{1}{3} & 3 & 14 & 13 & -6 \\ & & 1 & 5 & 6 \\ \hline & 3 & 15 & 18 & 0 \end{array}$$
The remainder is 0, so 1/3 is a zero. Now find the zeros of the depressed function
$3x^2 + 15x + 18 = 3(x + 2)(x + 3) \Rightarrow -2$ and -3 are also zeros. So the zeros of
$3x^3 + 14x^2 + 13x - 6$ are -3, -2, and 1/3.

41. The factors of the constant term are $\{\pm 1, \pm 2, \pm 3, \pm 6\}$, and the factors of the leading coefficient are $\{\pm 1\}$. So the possible rational zeros are $\{\pm 1, \pm 2, \pm 3, \pm 6\}$.

43. $f(x) = 5x^3 + 11x^2 + 2x;$
$$f(-x) = 5(-x)^3 + 11(-x)^2 + 2(-x)$$
$$= -5x^3 + 11x^2 - 2x$$

There are no sign changes in $f(x)$, so there are no positive zeros. There are two sign changes in $f(-x)$, so there are 2 or 0 negative zeros.

$$5x^3 + 11x^2 + 2x = x(5x^2 + 11x + 2)$$
$$= x(5x+1)(x+2) \Rightarrow \text{ the zeros are } -2, -\frac{1}{5}, 0.$$

45. $f(x) = x^3 + 3x^2 - 4x - 12$
$$f(-x) = (-x)^3 + 3(-x)^2 - 4(-x) - 12$$
$$= -x^3 + 3x^2 + 4x - 12$$

There is one sign change in $f(x)$, so there is one positive zero. There are two sign changes in $f(-x)$, so there are 2 or 0 negative zeros.

The possible rational zeros are $\{\pm 1, \pm 2, \pm 3, \pm 4, \pm 6, \pm 12\}$. Use synthetic division to find one zero:

$$\begin{array}{r|rrrr} -3 & 1 & 3 & -4 & -12 \\ & & -3 & 0 & 12 \\ \hline & 1 & 0 & -4 & 0 \end{array}$$

The remainder is 0, so −3 is a zero. Now find the zeros of the depressed function

$x^2 - 4 = (x-2)(x+2) \Rightarrow -2$ and 2 are also zeros. So the zeros of $x^3 + 3x^2 - 4x - 12$ are −3, −2, and 2.

47. $f(x) = x^3 - 4x^2 - 5x + 14$
$$f(-x) = (-x)^3 - 4(-x)^2 - 5(-x) + 14$$
$$= -x^3 - 4x^2 + 5x + 14$$

There are two sign changes in $f(x)$, so there are 0 or 2 real positive zeros. There is one sign change in $f(-x)$, so there is one real negative zero. The possible rational zeros are $\{\pm 1, \pm 2, \pm 7, \pm 14\}$. Use synthetic division to find one zero:

$$\begin{array}{r|rrrr} -2 & 1 & -4 & -5 & 14 \\ & & -2 & 12 & -14 \\ \hline & 1 & -6 & 7 & 0 \end{array}$$

The remainder is 0, so −2 is a zero. Now find the zeros of the depressed function

$x^2 - 6x + 7$.

$$x = \frac{-(-6) \pm \sqrt{(-6)^2 - 4(1)(7)}}{2(1)}$$
$$= \frac{6 \pm \sqrt{8}}{2} = \frac{6 \pm 2\sqrt{2}}{2} = 3 \pm \sqrt{2}$$

So, the zeros of $x^3 - 4x^2 - 5x + 14$ are −2, $3 - \sqrt{2}$, and $3 + \sqrt{2}$.

49. $f(x) = 2x^3 - 5x^2 - 2x + 2$
$$f(-x) = 2(-x)^3 - 5(-x)^2 - 2(-x) + 2$$
$$= -2x^3 - 5x^2 + 2x + 2$$

There are two sign changes in $f(x)$, so there are 0 or 2 real positive zeros. There is one sign change in $f(-x)$, so there is one real negative zero. The possible rational zeros are $\left\{\pm 1, \pm 2, \pm \frac{1}{2}\right\}$. Use synthetic division to find one zero:

$$\begin{array}{r|rrrr} \frac{1}{2} & 2 & -5 & -2 & 2 \\ & & 1 & -2 & -2 \\ \hline & 2 & -4 & -4 & 0 \end{array}$$

The remainder is 0, so 1/2 is a zero. Now find the zeros of the depressed function

$2x^2 - 4x - 4$.

$$x = \frac{-(-4) \pm \sqrt{(-4)^2 - 4(2)(-4)}}{2(2)}$$
$$= \frac{4 \pm \sqrt{48}}{4} = \frac{4 \pm 4\sqrt{3}}{4} = 1 \pm \sqrt{3}$$

So, the zeros of $2x^3 - 4x^2 - 2x + 2$ are 1/2, $1 - \sqrt{3}$, and $1 + \sqrt{3}$.

51. The function has degree three, so there are three zeros. Use synthetic division to find the depressed function:

$$\begin{array}{r|rrrr} 2 & 1 & 0 & -7 & 6 \\ & & 2 & 4 & -6 \\ \hline & 1 & 2 & -3 & 0 \end{array}$$

Now find the zeros of the depressed function

$x^2 + 2x - 3 = (x+3)(x-1) \Rightarrow -3$ and 1 are zeros. The zeros of the original function are −3, 1, 2.

53. The function has degree four, so there are four zeros. Use synthetic division twice to find the depressed function:

$$\underline{-1|}\ 1\ -2\ \ 6\ -18\ -27 \qquad \underline{3|}\ 1\ -3\ \ 9\ -27$$
$$\quad\ \ \underline{\ -1\ \ \ 3\ -9\ \ \ 27\ } \qquad\qquad \underline{\ \ \ \ \ 3\ \ \ 0\ \ \ 27\ }$$
$$\quad\ \ 1\ -3\ \ \ 9\ -27\ \ \ \ 0 \qquad\qquad 1\ \ \ 0\ \ \ 9\ \ \ \ 0$$

Alternatively, divide

$x^4 - 2x^3 + 6x^2 - 18x - 27$ by

$(x+1)(x-3) = x^2 - 2x - 3.$

Now find the zeros of the depressed function

$x^2 + 9 \Rightarrow \pm 3i$ are zeros. The zeros of the original function are -1, 3, and $\pm 3i$.

55. The function has degree four, so there are four zeros. Since one zero is $-1+2i$, another zero is $-1-2i$. Divide $x^4 + 2x^3 + 9x^2 + 8x + 20$ by $(x-(-1+2i))(x-(-1-2i)) = x^2 + 2x + 5$ to find the depressed function:

$$
\begin{array}{r}
x^2 + 4 \\
x^2+2x+5\overline{)x^4 + 2x^3 + 9x^2 + 8x + 20}\\
\underline{x^4 + 2x^3 + 5x^2}\\
4x^2 + 8x + 20\\
\underline{4x^2 + 8x + 20}\\
0
\end{array}
$$

Now find the zeros of the depressed function:
$x^2 + 4 = 0 \Rightarrow x = \pm 2i$. The zeros of the original function are $\pm 2i$ and $-1 \pm 2i$.

57. The function has degree three, so there are three zeros. The possible rational zeros are $\{\pm 1, \pm 2, \pm 4\}$. Using synthetic division, we find that one zero is 1:

$$\underline{1|}\ 1\ -1\ -4\ \ 4$$
$$\quad\ \ \underline{\ \ \ \ \ 1\ \ \ 0\ -4\ }$$
$$\quad\ \ 1\ \ \ 0\ -4\ \ \ 0$$

Now find the zeros of the depressed function:
$x^2 - 4 = 0 \Rightarrow x = \pm 2$. The solution set is $\{-2, 1, 2\}$.

59. The function has degree three, so there are three zeros. The possible rational zeros are $\left\{\pm\frac{1}{4}, \pm\frac{1}{2}, \pm\frac{3}{4}, \pm 1, \pm\frac{3}{2}, \pm 3\right\}$. Using synthetic division, we find that one zero is -1:

$$\underline{-1|}\ 4\ \ \ 0\ -7\ -3$$
$$\quad\ \ \underline{\ -4\ \ \ 4\ \ \ 3\ }$$
$$\quad\ \ 4\ -4\ -3\ \ \ 0$$

Now find the zeros of the depressed function:
$4x^2 - 4x - 3 = 0 \Rightarrow (2x-3)(2x+1) = 0 \Rightarrow$
$x = \dfrac{3}{2}$ or $x = -\dfrac{1}{2}$.
The solution set is $\{-1/2, -1, 3/2\}$.

61. $x^3 - 8x^2 + 23x - 22 = 0$
The function has degree three, so there are three zeros. The possible rational zeros are $\{\pm 1, \pm 2, \pm 11, \pm 22\}$. Using synthetic division, we find that one zero is 2:

$$\underline{2|}\ 1\ -8\ \ 23\ -22$$
$$\quad\ \ \underline{\ \ \ 2\ -12\ \ \ 22\ }$$
$$\quad\ \ 1\ -6\ \ \ 11\ \ \ \ 0$$

Now find the zeros of the depressed function:
$x^2 - 6x + 11 = 0.$

$$x = \frac{-(-6) \pm \sqrt{(-6)^2 - 4(1)(11)}}{2(1)}$$
$$= \frac{6 \pm \sqrt{-8}}{2} = \frac{6 \pm 2i\sqrt{2}}{2} = 3 \pm i\sqrt{2}$$

Solution set: $\left\{2, 3 \pm i\sqrt{2}\right\}$

63. $3x^3 - 5x^2 + 16x + 6 = 0$
The function has degree three, so there are three zeros. The possible rational zeros are $\left\{\pm 1, \pm 2, \pm 3, \pm 6, \pm\frac{1}{3}, \pm\frac{2}{3}\right\}$. Using synthetic division, we find that one zero is $-1/3$:

$$\underline{-\frac{1}{3}|}\ 3\ -5\ \ \ 16\ \ \ \ 6$$
$$\qquad\ \ \underline{\ \ -1\ \ \ 2\ -6\ }$$
$$\qquad\ \ 3\ -6\ \ \ 18\ \ \ \ 0$$

Now find the zeros of the depressed function:
$3x^2 - 6x + 18 = 0 \Rightarrow 3(x^2 - 2x + 6) = 0.$

$$x = \frac{-(-2) \pm \sqrt{(-2)^2 - 4(1)(6)}}{2(1)}$$
$$= \frac{2 \pm \sqrt{-20}}{2} = \frac{2 \pm 2i\sqrt{5}}{2} = 1 \pm i\sqrt{5}$$

Solution set: $\left\{-1/3, 1 \pm i\sqrt{5}\right\}$

65. $x^4 - x^3 - x^2 - x - 2 = 0$

The function has degree four, so there are four zeros. The possible rational zeros are $\{\pm 1, \pm 2\}$. Using synthetic division, we find that one zero is 2:

$$\underline{2|}\ \begin{array}{rrrrr} 1 & -1 & -1 & -1 & -2 \\ & 2 & 2 & 2 & 2 \\ \hline 1 & 1 & 1 & 1 & 0 \end{array}$$

Now find the zeros of the depressed function: $x^3 + x^2 + x + 1 = 0$. We can use synthetic division or factor to find another zero:

$$x^3 + x^2 + x + 1 = 0$$
$$x^2(x+1) + (x+1) = 0$$
$$(x^2 + 1)(x + 1) = 0$$
$$x^2 + 1 = 0 \Rightarrow x^2 = -1 \Rightarrow x = \pm i \ \text{ or}$$
$$x + 1 = 0 \Rightarrow x = -1$$

Solution set: $\{-1, 2, \pm i\}$

67. $2x^4 - x^3 - 2x^2 + 13x - 6 = 0$

The function has degree four, so there are four zeros. The possible rational zeros are $\left\{ \pm 1, \pm 2, \pm 3, \pm 6, \pm \dfrac{1}{2}, \pm \dfrac{3}{2} \right\}$. Using synthetic division, we find that one zero is -2:

$$\underline{-2|}\ \begin{array}{rrrrr} 2 & -1 & -2 & 13 & -6 \\ & -4 & 10 & -16 & 6 \\ \hline 2 & -5 & 8 & -3 & 0 \end{array}$$

Now find the zeros of the depressed function: $2x^3 - 5x^2 + 8x - 3 = 0$. Again using synthetic division, we find that one zero is 1/2 :

$$\underline{\tfrac{1}{2}|}\ \begin{array}{rrrr} 2 & -5 & 8 & -3 \\ & 1 & -2 & 3 \\ \hline 2 & -4 & 6 & 0 \end{array}$$

Now find the zeros of the depressed function: $2x^2 - 4x + 6 = 0 \Rightarrow 2(x^2 - 2x + 3) = 0$.

$$x = \frac{-(-2) \pm \sqrt{(-2)^2 - 4(1)(3)}}{2(1)}$$
$$= \frac{2 \pm \sqrt{-8}}{2} = \frac{2 \pm 2i\sqrt{2}}{2} = 1 \pm i\sqrt{2}$$

Solution set: $\left\{ -2, \dfrac{1}{2}, 1 \pm i\sqrt{2} \right\}$.

69. The only possible rational roots are $\{\pm 1, \pm 2\}$. None of these satisfies the equation.

71. $f(1) = 1^3 + 6(1)^2 - 28 = -21;$

$f(2) = 2^3 + 6(2)^2 - 28 = 4$. Because the sign changes, there is a real zero between 1 and 2. The zero is approximately 1.88.

73. $1 + \dfrac{1}{x} = 0 \Rightarrow x = -1$ is the x-intercept. There is no y-intercept. The vertical asymptote is the y-axis ($x = 0$). The horizontal asymptote is $y = 1$. Testing the intervals $(-\infty, -1)$, $(-1, 0)$, and $(0, \infty)$, we find that the graph is above the x-axis on $(-\infty, -1) \cup (0, \infty)$ and below the x-axis on $(-1, 0)$.

75. $\dfrac{x}{x^2 - 1} = 0 \Rightarrow x = 0$ is the x-intercept.

$\dfrac{0}{0^2 - 1} = 0 \Rightarrow y = 0$ is the y-intercept. The vertical asymptotes are $x = 1$ and $x = -1$. The horizontal asymptote is the x-axis. Testing the intervals $(-\infty, -1), (-1, 0), (0, 1),$ and $(1, \infty)$, we find that the graph is above the x-axis on $(-\infty, -1) \cup (0, \infty)$ and below the x-axis on $(-1, 0) \cup (0, 1)$.

77. $\dfrac{x^3}{x^2-9}=0 \Rightarrow x=0$ is the x-intercept.

$\dfrac{0^3}{0^2-9}=0 \Rightarrow y=0$ is the y-intercept. The

vertical asymptotes are $x=3$ and $x=-3$.
There is no horizontal asymptote. The oblique
asymptote is $y=x$. Testing the intervals
$(-\infty,-3),(-3,0),(0,3),$ and $(3,\infty),$ we find
that the graph is above the x-axis on
$(-3,0)\bigcup(3,\infty),$ and below the x-axis on
$(-\infty,-3)\bigcup(0,3).$

79. $\dfrac{x^4}{x^2-4}=0 \Rightarrow x=0$ is the x-intercept.

$\dfrac{0^4}{0^2-4}=0 \Rightarrow y=0$ is the y-intercept. The

vertical asymptotes are $x=2$ and $x=-2$.
There is no horizontal asymptote.
Testing the intervals
$(-\infty,-2),\ (-2,0),(0,2),$ and $(2,\infty),$ we find
that the graph is above the x-axis on
$(-\infty,-2)\bigcup(2,\infty)$ and below the x-axis on
$(-2,0)\bigcup(0,2).$

Applying the Concepts

81. $y=kx; 12=4k \Rightarrow k=3; y=3(5)=15$

83. $s=kt^2; 20=2^2k \Rightarrow 5=k; s=3^2(5)=45$

85. The maximum height occurs at the vertex,

$$\left(-\dfrac{20}{2(-1/10)},\ f\left(-\dfrac{20}{2(-1/10)}\right)\right)=(100,1000).$$

The maximum height is 1000. To find where
the missile hits the ground, solve

$-\dfrac{1}{10}x^2+20x=0 \Rightarrow x=0$ or $x=200.$ The

missile hits the ground at $x=200.$

87. The area of each section is 400 square feet.

Since the width is $x,\ y=\dfrac{400}{x}.$ The total

amount of fencing needed is $4x+\dfrac{2400}{x}.$

Using a graphing calculator, we find that this
is a minimum at $x\approx24.5$ feet. So

$y\approx\dfrac{400}{24.5}\approx16.3$ feet.

The dimensions of each pen should be
approximately 24.5 ft by 16.3 ft.

89. a.

b.

The maximum occurs at $x=50.$

91. a. The revenue is $24x$. Profit = revenue – cost, so

$$P(x) = 24x - \left(150 + 3.9x + \frac{3}{1000}x^2\right)$$

$$= -\frac{3}{1000}x^2 + 20.1x - 150$$

b. The maximum occurs at the x-coordinate of the vertex: $-\dfrac{20.1}{2\left(-\dfrac{3}{1000}\right)} = 3350$

c. $\bar{C}(x) = \dfrac{C(x)}{x} = \dfrac{\dfrac{3}{1000}x^2 + 3.9x + 150}{x}$

$$= \frac{3}{1000}x + 3.9 + \frac{150}{x}$$

93. $280 = 40k \Rightarrow k = 7; s = 7(35) = \$245$

95. $I = \dfrac{ki}{d^2}$, where I = illumination, i = intensity, and d = distance from the source. The illumination 6 inches from the source is

$I_6 = \dfrac{300k}{6^2}$, while the illumination x inches

from the source is $I_x = \dfrac{300k}{x^2}$. $I_6 = 2I_x \Rightarrow$

$\dfrac{300k}{36} = 2\left(\dfrac{300k}{x^2}\right) \Rightarrow x^2 = 72 \Rightarrow x = 6\sqrt{2}$ in.

97. $I = \dfrac{k}{R}; 30 = \dfrac{k}{300} \Rightarrow k = 9000$

a. $I = \dfrac{9000}{250} = 36$ amp

b. $60 = \dfrac{9000}{R} \Rightarrow R = 150$ ohms

99. $R = ki(p - i)$

a. $255 = k(0.15)(20,000)(0.85)(20,000)$

$$= \frac{1}{200,000}$$

b. $R = \dfrac{1}{200,000}(10,000)(10,000) = 500$ people per day.

c. $95 = \dfrac{1}{200,000}x(20,000 - x) \Rightarrow$

$x^2 - 20,000x + 19,000,000 = 0 \Rightarrow$

$(x - 1000)(x - 19,000) = 0 \Rightarrow x = 1000$ or $x = 19,000$

Chapter 3 Practice Test A

1. $x^2 - 6x + 2 = 0 \Rightarrow x = \dfrac{6 \pm \sqrt{(-6)^2 - 4(1)(2)}}{2(1)} \Rightarrow$

$x = \dfrac{6 \pm \sqrt{28}}{2} = 3 \pm \sqrt{7}$ are the x-intercepts

2.

3. The vertex is at $\left(-\dfrac{b}{2a}, f\left(-\dfrac{b}{2a}\right)\right)$

$= \left(-\dfrac{14}{2(-7)}, f\left(-\dfrac{14}{2(-7)}\right)\right) = (1, 10)$.

4. The denominator is 0 when $x = -4$ or $x = 1$. The domain is $(-\infty, -4) \cup (-4, 1) \cup (1, \infty)$.

5. Using either long division or synthetic division, we find that the quotient is $x^2 - 4x + 3$ and the remainder is 0.

$$\begin{array}{r|rrrr} -2 & 1 & -2 & -5 & 6 \\ & & -2 & 8 & -6 \\ \hline & 1 & -4 & 3 & 0 \end{array}$$

6.

7. The function has degree three, so there are three zeros. Since 2 is a zero, use synthetic division to find the depressed function:

$$\begin{array}{r|rrrr} 2 & 2 & -2 & -8 & 8 \\ & & 4 & 4 & -8 \\ \hline & 2 & 2 & -4 & 0 \end{array}$$

Now find the zeros of $2x^2 + 2x - 4$:

$2x^2 + 2x - 4 = 2(x^2 + x - 2) = 2(x + 2)(x - 1) \Rightarrow$ the zeros are $x = -2$ and $x = 1$. The zeros of the original function are $-2, 1, 2$.

8.
$$\begin{array}{r} -3x^2 + 5x + 1 \\ 2x+3 \overline{\smash{\big)}\ -6x^3 + x^2 + 17x + 3} \\ \underline{-6x^3 - 9x^2} \\ 10x^2 + 17x \\ \underline{10x^2 + 15x} \\ 2x + 3 \\ \underline{2x + 3} \\ 0 \end{array}$$

9. Using synthetic division to find the remainder, we have $P(-2) = -53$.

$$\begin{array}{r|rrrrr} -2 & 1 & 5 & -7 & 9 & 17 \\ & & -2 & -6 & 26 & -70 \\ \hline & 1 & 3 & -13 & 35 & -53 \end{array}$$

10. The function has degree three, so there are three zeros. There are two sign changes in $f(x)$, so there are either 2 or 0 positive zeros. There is one sign change in $f(-x)$, so there is one negative zero. The possible rational zeros are $\{\pm 1, \pm 2, \pm 4, \pm 5, \pm 10, \pm 20\}$.

Using synthetic division, we find that -2 is a zero:

$$\begin{array}{r|rrrr} -2 & 1 & -5 & -4 & 20 \\ & & -2 & 14 & -20 \\ \hline & 1 & -7 & 10 & 0 \end{array}$$

Now find the zeros of the depressed function $x^2 - 7x + 10 : x^2 - 7x + 10 = (x - 5)(x - 2) \Rightarrow$ the zeros are $x = 2$ and $x = 5$. The zeros of the original function are $-2, 2, 5$.

11. The function has degree four, so there are four zeros. Factoring, we have $x^4 + x^3 - 15x^2$

$= x^2(x^2 + x - 15) \Rightarrow 0$ is a zero of multiplicity

2. Now find the zeros of $x^2 + x - 15$:

$$x = \frac{-1 \pm \sqrt{1^2 - 4(1)(-15)}}{2(1)} = \frac{-1 \pm \sqrt{61}}{2}.$$

The zeros of the original function are 0 and $\frac{-1 \pm \sqrt{61}}{2}$.

12. The factors of the constant term, 9, are $\{\pm 1, \pm 3, \pm 9\}$, while the factors of the leading coefficient are $\{\pm 1, \pm 2\}$. The possible rational zeros are $\left\{\pm \dfrac{1}{2}, \pm 1, \pm \dfrac{3}{2}, \pm 3, \pm \dfrac{9}{2}, \pm 9\right\}$.

13. $f(x) \to -\infty$ as $x \to -\infty$; $f(x) \to \infty$ as $x \to \infty$.

14. $f(x) = (x^2 - 4)(x + 2)^2$
$= (x - 2)(x + 2)(x + 2)^2 \Rightarrow$
the zeros are -2 (multiplicity 3) and 2 (multiplicity 1).

15. There are two sign changes in $f(x)$, so there are either two or zero positive zeros. There is one sign change in $f(-x)$, so there is one negative zero.

16. The zeros of the denominator are $x = 5$ and $x = -4$, so those are the vertical asymptotes. The degree of the numerator equals the degree of the denominator, so the horizontal asymptote is $y = 2$.

17. $y = \dfrac{kx}{t^2}$

18. $6 = \dfrac{8k}{2^2} \Rightarrow k = 3; y = \dfrac{3(12)}{3^2} = 4$

19. The minimum occurs at the x-coordinate of the vertex: $-\dfrac{-30}{2(1)} = 15$ (which represents 15 thousand units.)

20. $V = x(17 - 2x)(8 - 2x)$

Chapter 3 Practice Test B

1. $x^2 + 5x + 3 = 0 \Rightarrow x = \dfrac{-5 \pm \sqrt{5^2 - 4(1)(3)}}{2(1)}$

   $= \dfrac{-5 \pm \sqrt{13}}{2}$. The answer is B.

2. The graph of $f(x) = 4 - (x-2)^2$ is the graph of $f(x) = x^2$ shifted two units right, reflected across the x-axis, and then shifted 4 units up. The answer is D.

3. $\left(-\dfrac{b}{2a}, f\left(-\dfrac{b}{2a}\right)\right) = \left(-\dfrac{12}{2(6)}, f\left(-\dfrac{12}{2(6)}\right)\right)$

   $= (-1, -11)$. The answer is A.

4. The denominator is 0 when $x = -3$ or $x = 2$. The answer is B.

5. $\begin{array}{r|rrrr} -3 & 1 & 0 & -8 & 6 \\ & & -3 & 9 & -3 \\ \hline & 1 & -3 & 1 & 3 \end{array}$

   The answer is D.

6. $P(x) = x^4 + 2x^3 = x^3(x+2)$. So the zeros are 0 (multiplicity 3) and -2 (multiplicity 1). The only graph with those zeros is C.

7. $\begin{array}{r|rrrr} 3 & 3 & -26 & 61 & -30 \\ & & 9 & -51 & 30 \\ \hline & 3 & -17 & 10 & 0 \end{array}$

   The zeros of the depressed function $3x^2 - 17x + 10$ are $x = \dfrac{2}{3}$ and $x = 5$.

   The answer is C.

8. $\begin{array}{r} -5x^2 + 3x - 4 \\ 2x - 3 \overline{)\, -10x^3 + 21x^2 - 17x + 12} \\ \underline{-10x^3 + 15x^2} \\ 6x^2 - 17x \\ \underline{6x^2 - 9x} \\ -8x + 12 \\ \underline{-8x + 12} \\ 0 \end{array}$

   The answer is B.

9. $\begin{array}{r|rrrrr} -3 & 1 & 4 & 7 & 10 & 15 \\ & & -3 & -3 & -12 & 6 \\ \hline & 1 & 1 & 4 & -2 & 21 \end{array}$

   $P(-3) = 21$. The answer is C.

10. The polynomial has degree three, so there are three zeros. The possible rational zeros are $\{\pm 1, \pm 2, \pm 3, \pm 4, \pm 6, \pm 12\}$. Using synthetic division, we find that one zero is -3:

    $\begin{array}{r|rrrr} -3 & -1 & 1 & 8 & -12 \\ & & 3 & -12 & 12 \\ \hline & -1 & 4 & -4 & 0 \end{array}$

    The zeros of the depressed function $-x^2 + 4x - 4$ are $x = 2$ (multiplicity 2). The answer is A.

11. The polynomial has degree three, so there are three zeros. Factoring, we have $x^3 + x^2 - 30x = x(x^2 + x - 30) = x(x+6)(x-5)$. The answer is A.

12. The factors of the constant term, 60, are $\{\pm 1, \pm 2, \pm 3, \pm 4, \pm 5, \pm 6, \pm 10, \pm 12, \pm 15, \pm 20,$ $\pm 30, \pm 60\}$. Since the leading coefficient is 1, these are also the possible rational zeros. The answer is D.

13. The answer is C.

14. $(x^2 - 1)(x+1)^2 = (x-1)(x+1)(x+1)^2$. The answer is B.

15. There are three sign changes in $P(x)$, so there are 3 or 1 positive zeros. There are two sign changes in $P(-x)$, so there are 2 or 0 negative zeros. The answer is C.

16. The zeros of the denominator are $x = 3$ and $x = -4$, so those are the vertical asymptotes. The degree of the numerator equals the degree of the denominator, so the horizontal asymptote is $y = 1$. The answer is C.

17. The answer is D.

18. $27 = \dfrac{3^2 k}{1^3} \Rightarrow k = 3; S = \dfrac{3(6^2)}{3^3} = 4$. The answer is B.

19. The minimum occurs at the x-coordinate of the vertex: $-\dfrac{-24}{2(1)} = 12$. The answer is B.

20. $V = x(10 - 2x)(12 - 2x)$. The answer is D.

Cumulative Review Exercises
Chapters P–3

1. $d = \sqrt{(x_2 - x_1)^2 + (y_2 - y_1)^2}$
$= \sqrt{(-1-2)^2 + (3-5)^2} = \sqrt{13}$

2. $M = \left(\dfrac{x_1 + x_2}{2}, \dfrac{y_1 + y_2}{2} \right)$
$= \left(\dfrac{2 + (-8)}{2}, \dfrac{-5 + (-3)}{2} \right) = (-3, -4)$

3. $0 = x^2 - 2x - 8 \Rightarrow 0 = (x - 4)(x + 2) \Rightarrow$
$x - 4 = 0 \Rightarrow x = 4$ or $x + 2 = 0 \Rightarrow x = -2$
$f(0) = 0^2 - 2(0) - 8 = -8$
The x-intercepts are $(-2, 0)$ and $(4, 0)$, and the y-intercept is $(0, -8)$.

4. Write the equation in slope-intercept form:

$x + 3y - 6 = 0 \Rightarrow 3y = -x + 6 \Rightarrow y = -\dfrac{1}{3}x + 2$

The slope is $-\dfrac{1}{3}$, and the y-intercept is $(0, 2)$.

$x + 3(0) - 6 = 0 \Rightarrow x - 6 = 0 \Rightarrow x = 6$, so the x-intercept is $(6, 0)$.

5. Write the equation in slope-intercept form:

$x = 2y - 6 \Rightarrow x + 6 = 2y \Rightarrow \dfrac{1}{2}x + 3 = y$

The slope is $\dfrac{1}{2}$, and the y-intercept is $(0, 3)$. The x-intercept is $x = 2(0) - 6 = -6$.

6. $(x - 2)^2 + (y + 3)^2 = 16$

7. $x^2 + y^2 + 2x - 4y - 4 = 0 \Rightarrow$
$(x^2 + 2x) + (y^2 - 4y) = 4 \Rightarrow$
$(x^2 + 2x + 1) + (y^2 - 4y + 4) = 4 + 1 + 4 \Rightarrow$
$(x + 1)^2 + (y - 2)^2 = 9$. The center is $(-1, 2)$.
The radius is 3.

8. $y + 2 = 3(x - 1) \Rightarrow y = 3x - 5$

9. $2x + 3y = 5 \Rightarrow y = -\dfrac{2}{3}x + \dfrac{5}{3} \Rightarrow$ the slope is
$-\dfrac{2}{3}$.
$y - 3 = -\dfrac{2}{3}(x - 1) \Rightarrow y = -\dfrac{2}{3}x + \dfrac{11}{3}$.

10. $2x + 3 = 0 \Rightarrow x = -\dfrac{3}{2}$.

The domain is $\left(-\infty, -\dfrac{3}{2} \right) \cup \left(-\dfrac{3}{2}, \infty \right)$.

11. $4 - 2x > 0 \Rightarrow x < 2$. The domain is $(-\infty, 2)$.

12. $f(-2) = (-2)^2 - 2(-2) + 3 = 11$;
$f(3) = 3^2 - 2(3) + 3 = 6$;
$f(x + h) = (x + h)^2 - 2(x + h) + 3$
$= x^2 + 2xh + h^2 - 2x - 2h + 3$
$= x^2 + 2(h - 1)x + h^2 - 2h + 3$

(continued on next page)

(continued)

$$\frac{f(x+h)-f(x)}{h}$$

$$=\frac{(x^2+2(h-1)x+h^2-2h+3)-(x^2-2x+3)}{h}$$

$$=\frac{2(h-1)x+2x+h^2-2h}{h}$$

$$=\frac{2hx-2x+2x+h^2-2h}{h}=\frac{h^2+2hx-2h}{h}$$

$$=h+2x-2$$

13. a. $f(g(x))=\sqrt{x^2+1}$

    b. $g(f(x))=\left(\sqrt{x}\right)^2+1=x+1$

    c. $f(f(x))=\sqrt{\sqrt{x}}=\sqrt[4]{x}$

    d. $g(g(x))=\left(x^2+1\right)^2+1=x^4+2x^2+2$

14. a. $f(1)=3(1)+2=5;\ f(3)=4(3)-1=11;$
        $f(4)=6$

    b.

15. $y=2x-3.$ Interchange x and y, and then
      solve for y.

$$x=2y-3\Rightarrow y=\frac{x+3}{2}=\frac{1}{2}x+\frac{3}{2}=f^{-1}(x).$$

16. a.   Shift the graph of $y=\sqrt{x}$ two units left.

    b.   Shift the graph of $y=\sqrt{x}$ one unit left,
        stretch vertically by a factor of two, reflect
        about the x-axis, and then shift up three
        units.

17.   The factors of the constant term, -6, are
      $\{\pm1,\pm2,\pm3,\pm6\}$, and the factors of the leading
      coefficient, 2, are $\{\pm1,\pm2\}$. The possible
      rational zeros are $\left\{\pm\dfrac{1}{2},\pm1,\pm\dfrac{3}{2},\pm2,\pm3,\pm6\right\}$.

18.

19.

20.

21.

22. Since one zero is $1 + i$, another zero is $1 - i$.

So $(x - (1 - i))(x - (1 + i)) = x^2 - 2x + 2$ is a

factor of $f(x)$. Now divide to find the other

factor:

$$
\begin{array}{r}
x^2 - x - 2 \\
x^2 - 2x + 2 \overline{\smash{\big)}\, x^4 - 3x^3 + 2x^2 + 2x - 4} \\
\underline{x^4 - 2x^3 + 2x^2} \\
-x^3 + 0x^2 + 2x \\
\underline{-x^3 + 2x^2 - 2x} \\
-2x^2 + 4x - 4 \\
\underline{-2x^2 + 4x - 4} \\
0
\end{array}
$$

$f(x) = x^4 - 3x^3 + 2x^2 + 2x - 4$

$= (x - (1 - i))(x - (1 + i))(x^2 - x - 2)$

$= (x - (1 - i))(x - (1 + i))(x - 2)(x + 1) \Rightarrow$

the zeros of $f(x)$ are $-1, 2, 1 - i, 1 + i$.

23. $y = k\sqrt{x}; 6 = k\sqrt{4} \Rightarrow k = 3; y = 3\sqrt{9} = 9$

24. Profit = revenue – cost

$150x - (0.02x^2 + 100x + 3000)$

$\qquad = -0.02x^2 + 50x - 3000.$

The maximum occurs at the vertex:

$$
\left(-\frac{50}{2(-0.02)}, f\left(-\frac{50}{2(-0.02)} \right) \right)
$$

$$
= (1250, \$28,250).
$$

25.
$$
\begin{array}{r|rrrr}
10 & 0.02 & 48.8 & -2990 & 25{,}000 \\
   &   & 0.2 & 490   & -25{,}000 \\
\hline
   & 0.02 & 49   & -2500 & 0
\end{array}
$$

Now solve the depressed equation

$0.02x^2 + 49x - 2500 = 0.$

$x = \dfrac{-49 \pm \sqrt{49^2 - 4(0.02)(-2500)}}{2(0.02)}$

$= \dfrac{-49 \pm \sqrt{2601}}{0.04} = \dfrac{-49 \pm 51}{0.04} \Rightarrow x = 50$ or

$x = -2500$. There cannot be a negative amount

of units sold, so another break-even point is 50.

Chapter 4 Exponential and Logarithmic Functions

4.1 Exponential Functions

4.1 Practice Problems

1. $f(x) = \left(\dfrac{1}{4}\right)^x$

$f(2) = \left(\dfrac{1}{4}\right)^2 = \dfrac{1}{16}$

$f(0) = \left(\dfrac{1}{4}\right)^0 = 1$

$f(-1) = \left(\dfrac{1}{4}\right)^{-1} = 4$

$f\left(\dfrac{5}{2}\right) = \left(\dfrac{1}{4}\right)^{5/2} = \left(\sqrt{\dfrac{1}{4}}\right)^5 = \left(\dfrac{1}{2}\right)^5 = \dfrac{1}{32}$

$f\left(-\dfrac{3}{2}\right) = \left(\dfrac{1}{4}\right)^{-3/2} = 4^{3/2} = \left(\sqrt{4}\right)^3 = 2^3 = 8$

2. a. $3^{\sqrt{8}} \cdot 3^{\sqrt{2}} = 3^{\sqrt{8}+\sqrt{2}} = 3^{2\sqrt{2}+\sqrt{2}} = 3^{3\sqrt{2}} = 27^{\sqrt{2}}$

   b. $\left(a^{\sqrt{8}}\right)^{\sqrt{2}} = a^{\sqrt{8}\cdot\sqrt{2}} = a^{\sqrt{16}} = a^4$

3. a.

x	2^x
-3	$1/8$
-2	$1/4$
-1	$1/2$
0	1
1	2
2	4
3	8

   b.

x	$(1/3)^x$
-3	27
-2	9
-1	3
0	1
1	$1/3$
2	$1/9$
3	$1/27$

4. First reflect the graph of $f(x) = 2^x$ about the x-axis to obtain $f(x) = -2^x$. Then, shift the graph up three units.

5. $f(x) = 2^{x-1} + 3$

Shift the graph of $y = 2^x$ one unit to the right, then three units up.

Domain: $(-\infty, \infty)$; range: $(3, \infty)$; horizontal asymptote: $y = 3$

6. $(-2, 16)$ and $\left(3, \dfrac{1}{2}\right)$

$16 = ca^{-2}$

$\dfrac{1}{2} = ca^3$

$\dfrac{16}{1/2} = \dfrac{ca^{-2}}{ca^3}$

$32 = a^{-5} \Rightarrow a = \dfrac{1}{2}$

$16 = c\left(\dfrac{1}{2}\right)^{-2} \Rightarrow 16 = 4c \Rightarrow 4 = c$

So, $f(x) = 4\left(\dfrac{1}{2}\right)^x$.

7. $A = P + I = P + Prt$
   $= 10,000 + 10,000(0.075)(2)$
   $= 10,000 + 1500 = 11,500$
   There will be \$11,500 in the account.

8. a. $A = P(1+r)^t = 8000(1+0.075)^5$
$\approx 11,485.03$

There will be \$11,485.03 in the account.

b. Interest $= A - P = 11,485.03 - 8000$
$= 3485.03$

She will receive \$3485.03

9. The formula for compound interest is

$A = P\left(1+\dfrac{r}{n}\right)^{nt}$.

For each of the following, $t = 1$.

(i) Annual compounding ($n = 1$)

$A = \$5000\left(1+\dfrac{0.065}{1}\right)^1 = \$5325.00$

(ii) Semiannual compounding ($n = 2$)

$A = \$5000\left(1+\dfrac{0.065}{2}\right)^2 \approx \$5330.28$

(iii) Quarterly compounding ($n = 4$)

$A = \$5000\left(1+\dfrac{0.065}{4}\right)^4 \approx \$5333.01$

(iv) Monthly compounding ($n = 12$)

$A = \$5000\left(1+\dfrac{0.065}{12}\right)^{12} \approx \$5334.86$

(v) Daily compounding ($n = 365$)

$A = \$5000\left(1+\dfrac{0.065}{365}\right)^{365} \approx \$5335.76$

10. $20,000 = 9000\left(1+\dfrac{r}{12}\right)^{12\cdot8}$

$\left(1+\dfrac{r}{12}\right)^{96} = \dfrac{20,000}{9000} = \dfrac{20}{9}$

$1+\dfrac{r}{12} = \left(\dfrac{20}{9}\right)^{1/96}$

$\dfrac{r}{12} = \left(\dfrac{20}{9}\right)^{1/96} - 1$

$r = 12\left[\left(\dfrac{20}{9}\right)^{1/96} - 1\right] \approx 0.10023$

Carmen needs an interest rate of about 10.023%.

11. $A = Pe^{rt} = \$9000e^{(0.06)(8.25)} \approx \$14,764.48$

12. Use the exponential growth/decay formula
$A(t) = A_0 e^{kt}$

a. $A(30) = 6.08e^{0.016(30)} \approx 9.8257$

If the rate of growth is 1.6% per year, there will be about 9.83 billion people in the world in the year 2030.

b. $A(-10) = 6.08e^{0.016(-10)} \approx 5.1810$

If the rate of growth is 1.6% per year, there were about 5.18 billion people in the world in the year 1990.

13. $A(t) = A_0 e^{-kt}$

$A(6) = 22,000e^{(-0.18)(6)} \approx \$7471.10$

4.1 Concepts and Vocabulary

1. For the exponential function
$f(x) = ca^x, a > 0, \quad a \neq 1$, the domain is
$(-\infty, \infty),$ and for $c > 0$, the range is $\underline{(0, \infty)}$.

3. The horizontal asymptote of the graph of
$y = \left(\dfrac{1}{3}\right)^x$ is the *x*-axis.

5. False. The graphs are symmetric with respect to the *y*-axis.

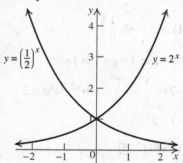

7. True

4.1 Building Skills

9. Not an exponential function. The base is not a constant.

11. Exponential function. The base is a constant, 1/2.

13. Not an exponential function. The base is not a constant.

15. Not an exponential function. The base is not a positive constant.

17. $f(0) = 5^{0-1} = 5^{-1} = \dfrac{1}{5}$

19. $g(3.2) = 3^{1-3.2} = 3^{-2.2} = 0.0892$
    $g(-1.2) = 3^{1-(-1.2)} = 3^{2.2} = 11.2116$

21. $h(1.5) = \left(\dfrac{2}{3}\right)^{2(1.5)-1} = \left(\dfrac{2}{3}\right)^2 = \dfrac{4}{9}$
    $h(-2.5) = \left(\dfrac{2}{3}\right)^{2(-2.5)-1} = \left(\dfrac{2}{3}\right)^{-6} = \dfrac{729}{64}$

23. $3^{\sqrt{2}} \cdot 3^{\sqrt{2}} = 3^{\sqrt{2}+\sqrt{2}} = 3^{2\sqrt{2}}$

25. $8^\pi \div 4^\pi = \left(2^3\right)^\pi \div \left(2^2\right)^\pi = 2^{3\pi} \div 2^{2\pi}$
    $\qquad = 2^{3\pi-2\pi} = 2^\pi$

27. $\left(3^{\sqrt{2}}\right)^{\sqrt{3}} = 3^{\sqrt{2}\cdot\sqrt{3}} = 3^{\sqrt{6}}$

29. $\left(a^{\sqrt{3}}\right)^{\sqrt{12}} = a^{\sqrt{3}\cdot\sqrt{12}} = a^{\sqrt{36}} = a^6$

31. a. (0, 1) and (2, 16)
    $f(0) = 1 \Rightarrow 1 = ca^0 \Rightarrow 1 = c$
    $f(2) = 16 \Rightarrow 16 = 1 \cdot a^2 \Rightarrow a = 4$
    $f(x) = 4^x$

    b. (0, 1) and $\left(-2, \dfrac{1}{9}\right)$
    $f(0) = 1 \Rightarrow 1 = ca^0 \Rightarrow 1 = c$
    $f(-2) = \dfrac{1}{9} \Rightarrow \dfrac{1}{9} = 1 \cdot a^{-2} \Rightarrow a = 3$
    $f(x) = 3^x$

33. a. (1, 1) and (2, 5)
    $f(1) = 1 \Rightarrow 1 = ca^1$
    $f(2) = 5 \Rightarrow 5 = ca^2$
    $\dfrac{1}{5} = \dfrac{ca^1}{ca^2} \Rightarrow \dfrac{1}{5} = a^{-1} \Rightarrow 5 = a$
    $f(1) = 1 \Rightarrow 1 = c \cdot 5^1 \Rightarrow c = \dfrac{1}{5}$
    $f(x) = \dfrac{1}{5} \cdot (5)^x$

b. (1, 1) and $\left(2, \dfrac{1}{5}\right)$
    $f(1) = 1 \Rightarrow 1 = ca^1$
    $f(2) = \dfrac{1}{5} \Rightarrow \dfrac{1}{5} = ca^2$
    $\dfrac{1}{1/5} = \dfrac{ca^1}{ca^2} \Rightarrow 5 = a^{-1} \Rightarrow \dfrac{1}{5} = a$
    $f(1) = 1 \Rightarrow 1 = c\left(\dfrac{1}{5}\right)^1 \Rightarrow c = 5$
    $f(x) = 5 \cdot \left(\dfrac{1}{5}\right)^x$

35.

x	4^x
-2	1/16
-1	1/4
0	1
1	4
2	16

37.

x	$(3/2)^{-x}$
-4	81/16
-2	9/4
-1	!!
0	1
1	2/3
2	4/9
4	16/81

39.

x	$(1/4)^x$
-2	16
-1	4
0	1
1	1/4
2	1/16

41.

x	1.3^{-x}
-4	≈ 2.86
-3	≈ 2.2
0	1
2	≈ 0.59
4	≈ 0.35

43. c   **45.** a

47. $g(x) = 2^x + 1$

Shift the graph of $f(x) = 2^x$ one unit up.

Domain: $(-\infty, \infty)$; Range: $(1, \infty)$

Horizontal asymptote: $y = 1$

49. $g(x) = 3^{x-1}$

Shift the graph of $f(x) = 3^x$ one unit right.

Domain: $(-\infty, \infty)$; Range: $(0, \infty)$

Horizontal asymptote: $y = 0$

51. $g(x) = 4^{-x}$

Reflect the graph of $f(x) = 4^x$ about the
y-axis.

Domain: $(-\infty, \infty)$; Range: $(0, \infty)$

Horizontal asymptote: $y = 0$

53. $g(x) = 2^{x-1} + 3$

Shift the graph of $f(x) = 2^x$ one unit right,
then up three units.

Domain: $(-\infty, \infty)$; Range: $(3, \infty)$

Horizontal asymptote: $y = 3$

55. $g(x) = -3^{x-2} + 1$

Shift the graph of $f(x) = 3^x$ two units right,
reflect across the x-axis, then shift the graph
up one unit.

Domain: $(-\infty, \infty)$; Range: $(-\infty, 1)$

Horizontal asymptote: $y = 1$

57. $g(x) = \left(\dfrac{1}{2}\right)^{x+1} - 1$

Shift the graph of $f(x) = \left(\frac{1}{2}\right)^x$ one unit left,
then shift the graph down one unit.

Domain: $(-\infty, \infty)$; Range: $(-1, \infty)$

Horizontal asymptote: $y = -1$

(*continued on next page*)

(*continued*)

59. $g(x) = -2 \cdot 5^{x-1} + 4$

Shift the graph of $f(x) = 5^x$ one unit to the right, reflect it about the x-axis, stretch vertically by a factor of 2, then shift four units up. Domain: $(-\infty, \infty)$; Range: $(-\infty, 4)$. Horizontal asymptote: $y = 4$.

61. The graph passes through the points $(0, 3)$ and $(1, 3.5)$.

$f(0) = 3 \Rightarrow 3 = a^0 + b \Rightarrow 3 = 1 + b \Rightarrow b = 2$
$f(1) = 3.5 \Rightarrow 3.5 = a^1 + 2 \Rightarrow 1.5 = a$
$f(x) = 1.5^x + 2$
$f(2) = 1.5^2 + 2 = 4.25$

63. The graph passes through $(-2, 7)$ and $(-1, 1)$.

$f(-2) = 7 \Rightarrow 7 = a^{-2} + b$ (1)
$f(-1) = 1 \Rightarrow 1 = a^{-1} + b$ (2)
Subtract (2) from (1).

$6 = a^{-2} - a^{-1} \Rightarrow 6 = \dfrac{1}{a^2} - \dfrac{1}{a} \Rightarrow$
$6a^2 = 1 - a \Rightarrow 6a^2 + a - 1 = 0 \Rightarrow$
$(2a+1)(3a-1) = 0 \Rightarrow a = -\dfrac{1}{2}, \dfrac{1}{3}$

If $a = -\dfrac{1}{2}$, then

$1 = \left(-\dfrac{1}{2}\right)^{-1} + b \Rightarrow 1 = -2 + b \Rightarrow b = 3.$

If $a = \dfrac{1}{3}$, then $1 = \left(\dfrac{1}{3}\right)^{-1} + b \Rightarrow 1 = 3 + b \Rightarrow$

$b = -2$. The horizontal asymptote of the given graph is $y = -2$, so the equation of the graph is

$f(x) = \left(\dfrac{1}{3}\right)^x - 2.$

$f(2) = \left(\dfrac{1}{3}\right)^2 - 2 = \dfrac{1}{9} - 2 = -\dfrac{17}{9}$

65. $y = 2^{x+2} + 5$

67. $y = 2\left(\dfrac{1}{2}\right)^x - 5$

69. The graph of $f(x) = 2^x$ was reflected about the x-axis, and then shifted up two units.

71. $I = Prt = \$5000 \cdot 0.1 \cdot 5 = \$2500$

73. $I = Prt = \$7800 \cdot 0.06875 \cdot 10.75 = \$5764.69$

75. a. $A = 3500\left(1 + \dfrac{0.065}{1}\right)^{13} = \$7936.21$

   b. interest $= \$7936.21 - \$3500 = \$4436.21$

77. a. $A = 7500e^{0.05(10)} = \$12,365.41$

   b. interest $= \$12,365.41 - \$7500 = \$4865.41$

79. $10,000 = P\left(1 + \dfrac{0.08}{1}\right)^{10} = P(1.08)^{10} \Rightarrow$
$P = \$4631.93$

81. $10,000 = P\left(1 + \dfrac{0.08}{365}\right)^{365(10)} \Rightarrow P = \$4493.68$

83. Reflect the graph of $y = e^x$ about the y-axis.

Horizontal asymptote: $y = 0$

85. Shift the graph of $y = e^x$ two units right.

Horizontal asymptote: $y = 0$

87. Shift the graph of $y = e^x$ one unit up.

Horizontal asymptote: $y = 1$

89. Shift the graph of $y = e^x$ two units right, reflect the graph about the *x*-axis, and then shift it three units up.

Horizontal asymptote: $y = 3$

4.1 Applying the Concepts

91. a. Exponential growth

   b. Linear growth

   c. Exponential growth

   d. Linear growth

93. $f(t) = 7.89e^{-0.42t}$

   a. The function represents exponential decay.

   b. $f(3) = 7.89e^{-0.42(3)} \approx 2.24$

   The average hard drive cost per gigabyte in 2003 was about \$2.24.

c. The price decreases quickly in the beginning, but then decreasing the price becomes more difficult as time goes on.

95. a. (i) $T = 200 \cdot 4^{-0.1(2)} + 25 = 176.6°C$

   (ii) $T = 200 \cdot 4^{-0.1(3.5)} + 25 = 148.1°C$

   b. $125 = 200 \cdot 4^{-0.1t} + 25 \Rightarrow 100 = 200 \cdot 4^{-0.1t} \Rightarrow$
   $\dfrac{1}{2} = 4^{-0.1t} \Rightarrow 2^{-1} = 2^{-0.2t} \Rightarrow -1 = -0.2t \Rightarrow$
   $t = 5$ hours

   c. As $t \to \infty$, $T \to 25$. Verify graphically.

   [0, 50, 10] by [0, 200, 25]

97. $A = 190,000\left(1 + \dfrac{0.03}{1}\right)^{1(5)} \approx \$220,262$

99. $A = 80,000\left(1 - \dfrac{0.15}{1}\right)^{1(5)} \approx \$35,496.43$

101. Assume that interest is compounded annually.
$22,000,000 = 5000\left(1 + \dfrac{r}{1}\right)^{1(54)} \Rightarrow$
$4400 = (1 + r)^{54} \Rightarrow 4400^{1/54} = 1 + r \Rightarrow$
$1.1681 \approx 1 + r \Rightarrow r \approx 0.1681 \approx 16.81\%$

103. $A = 10e^{-0.43(10)} \approx 0.1357 \text{ mm}^2$

105. $95,600 = 60,000\left(1 + \dfrac{r}{4}\right)^{4(12)} \Rightarrow$
$\dfrac{239}{150} = \left(1 + \dfrac{r}{4}\right)^{48} \Rightarrow \sqrt[48]{\dfrac{239}{150}} = 1 + \dfrac{r}{4} \Rightarrow$
$\sqrt[48]{\dfrac{239}{150}} - 1 = \dfrac{r}{4} \Rightarrow 4\left(\sqrt[48]{\dfrac{239}{150}} - 1\right) \approx 0.039 = r$

The interest rate on the bond was about 3.9%.

107. The amount Javier pays interest on is the principal plus the initial transfer fee

Bank A:

Transfer fee = $0.02(1000) = 20$

$$I = (1000 + 20)\left(1 + \frac{0.08}{12}\right)^{12(1)} \approx 1104.66$$

Bank B:

Transfer fee = $0.06(1000) = 60$

$$I = (1000 + 60)\left(1 + \frac{0.06}{12}\right)^{12(1)} \approx 1125.38$$

Javier will save $1125.38 - 1104.66 = \$20.72$ by choosing Bank A over Bank B.

109. Because the jar is full in 60 minutes, and the number of bacteria doubles every minute, it will be half full one minute earlier, or in 59 minutes.

4.1 Beyond the Basics

111. a. $f(x) + g(x) = \left(3^x + 3^{-x}\right) + \left(3^x - 3^{-x}\right)$
$$= 2 \cdot 3^x$$

b. $f(x) - g(x) = \left(3^x + 3^{-x}\right) - \left(3^x - 3^{-x}\right)$
$$= 2 \cdot 3^{-x}$$

c. $[f(x)]^2 - [g(x)]^2$
$$= \left[\left(3^x + 3^{-x}\right)\right]^2 - \left[\left(3^x - 3^{-x}\right)\right]^2$$
$$= \left(3^{2x} + 3^{-2x} + 2\right) - \left(3^{2x} + 3^{-2x} - 2\right) = 4$$

d. $[f(x)]^2 + [g(x)]^2$
$$= \left[\left(3^x + 3^{-x}\right)\right]^2 + \left[\left(3^x - 3^{-x}\right)\right]^2$$
$$= \left(3^{2x} + 3^{-2x} + 2\right) + \left(3^{2x} + 3^{-2x} - 2\right)$$
$$= 2\left(3^{2x} + 3^{-2x}\right)$$

113. $A = P\left(1 + \frac{r}{m}\right)^{m(1)} = P(1 + y) \Rightarrow$

$$\left(1 + \frac{r}{m}\right)^m = 1 + y \Rightarrow y = \left(1 + \frac{r}{m}\right)^m - 1$$

115. $y = e^x \to y = e^{x-1} \to y = e^{2x-1} \to y = 3e^{2x-1}$

117. $y = e^x \to y = e^{2+x} \to y = e^{2+3x} \to$
$$y = 5e^{2+3x} \to y = 5e^{2-3x} \to y = 5e^{2-3x} + 4$$

4.1 Critical Thinking/Discussion/Writing

119. The base a cannot be 1 because this makes the function a constant, $f(x) = 1$. Similarly, the base a cannot be 0 because this becomes $f(x) = 0^x = 0$, a constant. We rule out negative bases so that the domain can include all real numbers. For example, a cannot be -3 because $f\left(\frac{1}{2}\right) = (-3)^{1/2} = \sqrt{-3}$ is not a real number.

120. $g(x)$ is defined for the set of integers. For example, $g(-2) = (-3)^{-2} = \frac{1}{9}$ and

$$g(3) = (-3)^3 = -27.$$

121. If $0 < a < 1$, then, for $x < 0$, the denominator becomes very large, so $y = \dfrac{b}{1 + ca^x}$ approaches 0. For $x \geq 0$, ca^x approaches 0, so the denominator $1 + ca^x$ approaches 1 and $y = \dfrac{b}{1 + ca^x}$ approaches b.

If $a > 1$, then, for $x < 0$, ca^x approaches 0, so the denominator $1 + ca^x$ approaches 1 and $y = \dfrac{b}{1 + ca^x}$ approaches b. For $x \geq 0$, the denominator becomes very large, so $y = \dfrac{b}{1 + ca^x}$ approaches 0.

122. There are four possibilities, $0 < a < 1$ with $c < 0$, $0 < a < 1$ with $c > 0$, $a > 1$ with $c < 0$, and $a > 1$ with $c > 0$. They are illustrated below

$y = ca^x, 0 < a < 1, c > 0$

$y = ca^x, 0 < a < 1, c < 0$

$y = ca^x, a > 1, c > 0$

$y = ca^x, a > 1, c < 0$

123. The function $y = 2^x$ is an increasing function, so it is one-to-one. The horizontal line $y = k$ for $k > 0$ intersects the graph of $y = 2^x$ in exactly one point, so there is exactly one solution for each value of k.

$k = 16$
$k = 10$
$k = 6$
$y = 2^x$

124. The graphs of $y = 2^x$ and $y = x$ intersect in exactly two points: (1, 2) and (2, 4). So
$$2^x = 2x \Rightarrow x = 1 \text{ or } 2.$$

$y = 2^x$
$y = 2x$
(2, 4)
(1, 2)

4.1 Getting Ready for the Next Section

125. $10^0 = 1$

127. $(-8)^{1/3} = \sqrt[3]{-8} = -2$

129. $\left(\dfrac{1}{7}\right)^{-2} = 7^2 = 49$

131. $18^{1/2} = \sqrt{18} = \sqrt{9 \cdot 2} = 3\sqrt{2}$

133. $f(x) = y = 3x + 4$
Interchange x and y, then solve for y.
$$x = 3y + 4 \Rightarrow x - 4 = 3y \Rightarrow$$
$$\frac{x-4}{3} = y = f^{-1}(x)$$
The domain and range of f^{-1} are $(-\infty, \infty)$.

135. $f(x) = y = \sqrt{x}, \ x \geq 0$
Interchange x and y, then solve for y.
$$x = \sqrt{y} \Rightarrow x^2 = y = f^{-1}(x), \ x \geq 0$$
The domain and range of f^{-1} are $[0, \infty)$.

137. $f(x) = y = \dfrac{1}{x-1}, \ x \neq 1$
Interchange x and y, then solve for y.
$$x = \frac{1}{y-1} \Rightarrow x(y-1) = 1 \Rightarrow y - 1 = \frac{1}{x} \Rightarrow$$
$$y = f^{-1}(x) = 1 + \frac{1}{x}, \ x \neq 0$$
The domain of f^{-1} is $(-\infty, 0) \cup (0, \infty)$. The range of f^{-1} is $(-\infty, 1) \cup (1, \infty)$.

4.2 Logarithmic Functions

4.2 Practice Problems

1. a. $2^{10} = 1024$ is equivalent to $\log_2 1024 = 10$.

b. $9^{-1/2} = \dfrac{1}{3}$ is equivalent to $\log_9\left(\dfrac{1}{3}\right) = -\dfrac{1}{2}$.

c. $p = a^q$ is equivalent to $\log_a p = q$.

2. a. $\log_2 64 = 6$ is equivalent to $2^6 = 64$.

b. $\log_v u = w$ is equivalent to $v^w = u$.

3. a. $\log_3 9 = y \Rightarrow 3^y = 9 \Rightarrow 3^y = 3^2 \Rightarrow y = 2$
Thus, $\log_3 9 = 2$.

b. $\log_9 \dfrac{1}{3} = y \Rightarrow 9^y = \dfrac{1}{3} \Rightarrow 3^{2y} = 3^{-1} \Rightarrow y = -\dfrac{1}{2}$
Thus, $\log_9 \dfrac{1}{3} = -\dfrac{1}{2}$.

c. $\log_{1/2} 32 = y \Rightarrow \left(\dfrac{1}{2}\right)^y = 32 \Rightarrow 2^{-y} = 2^5 \Rightarrow$
$y = -5$
Thus, $\log_{1/2} 32 = -5$.

4. a. $\log_5 1 = 0$ **b.** $\log_3 3^5 = 5$

c. $7^{\log_7 5} = 5$

5. Since the domain of the logarithmic function is $(0, \infty)$, the expression $\sqrt{1-x}$ must be positive. $\sqrt{1-x}$ is defined for $(-\infty, 1)$, so the domain of $\log_{10}\sqrt{1-x}$ is $(-\infty, 1)$.

6. Create a table of values to find ordered pairs on the graph of $y = \log_2 x$.

x	$y = \log_2 x$	(x, y)
$\dfrac{1}{8}$	$2^{-3} = \dfrac{1}{8} \Rightarrow y = \log_2 \dfrac{1}{8} = -3$	$\left(\dfrac{1}{8}, -3\right)$
$\dfrac{1}{4}$	$2^{-2} = \dfrac{1}{4} \Rightarrow y = \log_2 \dfrac{1}{4} = -2$	$\left(\dfrac{1}{4}, -2\right)$
$\dfrac{1}{2}$	$2^{-1} = \dfrac{1}{2} \Rightarrow y = \log_2 \dfrac{1}{2} = -1$	$\left(\dfrac{1}{2}, -1\right)$
1	$2^0 = 1 \Rightarrow y = \log_2 1 = 0$	$(1, 0)$

x	$y = \log_2 x$	(x, y)
2	$2^1 = 2 \Rightarrow y = \log_2 2 = 1$	$(2, 1)$
4	$2^2 = 4 \Rightarrow y = \log_2 4 = 2$	$(4, 2)$
8	$2^3 = 8 \Rightarrow y = \log_2 8 = 3$	$(8, 3)$

Plot the ordered pairs and connect them with a smooth curve.

7. Shift the graph of $y = \log_2 x$ three units right, then reflect the resulting graph about the x-axis.

8. To sketch the graph of $f(x) = 3 - \log_2 x$, reflect the graph of $y = \log_2 x$ about the x-axis, and then shift three units up.

9. $P_2 = \log 3 - \log 2 \approx 0.176$
This means that about 17.6% of the data is expected to have 2 as the first digit.

10. a. $y = \ln \dfrac{1}{e} \Rightarrow e^y = \dfrac{1}{e} \Rightarrow e^y = e^{-1} \Rightarrow y = -1$
Thus, $\ln \dfrac{1}{e} = -1$.

b. Using a calculator, $\ln 2 \approx 0.693$.

11. a. If P dollars are invested, then the amount $A = 3P$.

$$A = Pe^{rt} \Rightarrow 3P = Pe^{0.065t} \Rightarrow 3 = e^{0.065t} \Rightarrow$$

$$\ln 3 = 0.065t \Rightarrow t = \frac{\ln 3}{0.065} \approx 16.9$$

It will take approximately 17 years to triple your money.

b. $A = Pe^{rt} \Rightarrow 3P = Pe^{5r} \Rightarrow 3 = e^{5r} \Rightarrow$

$$\ln 3 = 5r \Rightarrow t = \frac{\ln 3}{5} \approx 0.2197$$

The investment will triple in 5 years at the rate of 21.97%.

12. From Example 12, we have $k = -\ln\left(\frac{93}{108}\right)$.

With this value of k, find t when $T = 120°$.

$$\ln\left(\frac{T(t) - T_s}{T_0 - T_s}\right) = -kt$$

$$\ln\left(\frac{120 - 72}{180 - 72}\right) = -\left(-\ln\left(\frac{93}{108}\right)\right) \cdot t$$

$$\ln\left(\frac{48}{108}\right) = \ln\left(\frac{93}{108}\right)t$$

$$\left[\frac{1}{\ln\left(\frac{93}{108}\right)}\right]\ln\left(\frac{48}{108}\right) = t \Rightarrow t \approx 5.423$$

The employees should wait about 5.4 minutes (realistically 5.5 minutes) to deliver the coffee at 120°F.

4.2 Concepts and Vocabulary

1. The domain of the function $y = \log_a x$ is $(0, \infty)$ and it range is $(-\infty, \infty)$.

3. The logarithm with base 10 is called the <u>common</u> logarithm, and the logarithm with base e is called the <u>natural</u> logarithm.

5. False

7. False. The domain of $f(x) = \log_a x$, $a > 0$ and $a \neq 1$ is $(0, \infty)$.

4.2 Building Skills

9. $\log_5 25 = 2$

11. $\log_{1/16} 4 = -\frac{1}{2}$

13. $\log_{10} 1 = 0$

15. $\log_{10} 0.1 = -1$

17. $a^2 + 2 = 7 \Rightarrow a^2 = 5 \Rightarrow \log_a 5 = 2$

19. $2a^3 - 3 = 10 \Rightarrow 2a^3 = 13 \Rightarrow a^3 = \frac{13}{2} \Rightarrow$

$$\log_a\left(\frac{13}{2}\right) = 3$$

21. $2^5 = 32$

23. $10^2 = 100$

25. $10^0 = 1$

27. $10^{-2} = 0.01$

29. $3\log_8 2 = 1 \Rightarrow \log_8 2 = \frac{1}{3} \Rightarrow 8^{1/3} = 2$

31. $e^x = 2$

33. $\log_5 125 = 3$ because $5^3 = 125$

35. $\log 10,000 = 4$ because $10^4 = 10,000$

37. $\log_2 \frac{1}{8} = -3$ because $2^{-3} = \frac{1}{8}$

39. $\log_3 \sqrt{27} = \frac{3}{2}$ because $3^{3/2} = \sqrt{27}$

41. $\log_{16} 2 = \frac{1}{4}$ because $16^{1/4} = 2$

43. $\log_3 1 = 0$

45. $\log_7 7 = 1$

47. $\log_6 6^7 = 7$

49. $3^{\log_3 5} = 5$

51. $2^{\log_2 7} + \log_5 5^{-3} = 7 + (-3) = 4$

53. $4^{\log_4 6} - \log_4 4^{-2} = 6 - (-2) = 8$

55. Since the domain of the logarithmic function is $(0, \infty)$, the expression $x + 1$ must be positive. This occurs in the interval $(-1, \infty)$, so the domain of $\log_2(x + 1)$ is $(-1, \infty)$.

57. Since the domain of the logarithmic function is $(0, \infty)$, the expression $\sqrt{x - 1}$ must be positive. This occurs in the interval $(1, \infty)$, so the domain of $\log_3 \sqrt{x - 1}$ is $(1, \infty)$.

59. Since the domain of the logarithmic function is $(0, \infty)$, the expressions $(x - 2)$ and $(2x - 1)$ must be positive. This occurs in the interval $(2, \infty)$ for $x - 2$ and in the interval $\left(\frac{1}{2}, \infty\right)$ for $(2x - 1)$. The intersection of the two intervals is $(2, \infty)$, so the domain of f is $(2, \infty)$.

61. Since the domain of the logarithmic function is $(0, \infty)$, the expressions $(x - 1)$ and $(2 - x)$ must be positive. This occurs in the interval $(1, \infty)$ for $x - 1$ and in the interval $(-\infty, 2)$ for $(2 - x)$. The intersection of the two intervals is $(1, 2)$, so the domain of h is $(1, 2)$.

63. $f(x) = \ln|x|$

The domain of the logarithmic function is $(0, \infty)$, and the domain of $g(x) = |x|$ is $(-\infty, \infty)$. Thus, the domain of $f(x)$ is $(-\infty, 0) \cup (0, \infty)$.

65. $f(x) = \log_2\left(\dfrac{x}{3} - 4\right)$

The domain of the logarithmic function is $(0, \infty)$, so $\dfrac{x}{3} - 4 > 0 \Rightarrow \dfrac{x}{3} > 4 \Rightarrow x > 12$.

The domain of $f(x)$ is $(12, \infty)$.

67. $f(x) = \ln\left(\dfrac{x}{x + 1}\right)$

The domain of the logarithmic function is $(0, \infty)$, so $\dfrac{x}{x + 1} > 0$

Using the test interval method from section 2.5, we have

numerator: $x = 0$
denominator: $x = -1$

The boundary points are -1 and 0. The intervals are $(-\infty, -1)$, $(-1, 0)$, and $(0, \infty)$.

Interval	Test point	Value of $R(x) = \dfrac{x}{x + 1}$	Result
$(-\infty, -1)$	-2	2	$+$
$(-1, 0)$	$-\frac{1}{2}$	-1	$-$
$(0, \infty)$	1	$\frac{1}{2}$	$+$

The domain of $f(x)$ is $(-\infty, -1) \cup (0, \infty)$

69. $f(x) = \log_3\left(\dfrac{x - 2}{x + 1}\right)$

The domain of the logarithmic function is $(0, \infty)$, so $\dfrac{x - 2}{x + 1} > 0$

Using the test interval method from section 2.5, we have

numerator: $x = 2$
denominator: $x = -1$

The boundary points are -1 and 2. The intervals are $(-\infty, -1)$, $(-1, 2)$, and $(2, \infty)$.

Interval	Test point	Value of $R(x) = \dfrac{x - 2}{x + 1}$	Result
$(-\infty, -1)$	-2	4	$+$
$(-1, 2)$	0	-2	$-$
$(2, \infty)$	3	$\frac{1}{4}$	$+$

The domain of $f(x)$ is $(-\infty, -1) \cup (2, \infty)$.

71. a. F   **b.** A

    c. D   **d.** B

    e. E   **f.** C

73. $f(x) = \log_4(x + 3)$

Shift the graph of $y = \log_4 x$ three units left.

Domain: $(-3, \infty)$; range: $(-\infty, \infty)$; asymptote: $x = -3$

75. $f(x) = \log_{1/2}(x-1)$

Shift the graph of $y = \log_{1/2} x$ one unit right.

Domain: $(1, \infty)$; range: $(-\infty, \infty)$; asymptote: $x = 1$

77. $f(x) = -\log_5 x$

Reflect the graph of $y = \log_5 x$ about the x-axis. Domain: $(0, \infty)$; range: $(-\infty, \infty)$; asymptote: $x = 0$

79. $f(x) = 2 - \log_4(x-1)$

Shift the graph of $y = \log_4 x$ one unit right, reflect across the x-axis, then shift two units up. Domain: $(1, \infty)$; range: $(-\infty, \infty)$; asymptote: $x = 1$

81. $f(x) = 1 + \log_{1/5}(-x)$

Reflect the graph of $y = \log_{1/5} x$ in the y-axis, and then shift it up one unit. Domain: $(-\infty, 0)$; range: $(-\infty, \infty)$; asymptote: $x = 0$

83. $f(x) = |\log_3 x|$

On $(0, 1)$, reflect the graph of $y = \log_3 x$ about the x-axis. Domain: $(0, \infty)$; range: $[0, \infty)$; asymptote: $x = 0$ (the y-axis)

85. $f(x) = \log_2(x-1)$

Shift the graph of $y = \log_2 x$ one unit right. Domain: $(1, \infty)$; range: $(-\infty, \infty)$; asymptote: $x = 1$

87. $f(x) = \log_2(3-x)$

Shift the graph of $y = \log_2 x$ three units right and then reflect it about the y-axis. Domain: $(-\infty, 3)$; range: $(-\infty, \infty)$; asymptote: $x = 3$

89. $f(x) = -\log_2 x - 2$

Reflect the graph of $y = \log_2 x$ about the
x-axis and then shift it down two units.
Domain: $(0, \infty)$; range: $(-\infty, \infty)$;
asymptote: $x = 0$ (the y-axis)

91. $f(x) = 2 + \log_2(3 - x)$

Shift the graph of $y = \log_2 x$ three units right,
then reflect it about the y-axis, and then shift it
2 units up. Domain: $(-\infty, 3)$; range: $(-\infty, \infty)$;
asymptote: $x = 3$

93. $\log_4(\log_3 81) = \log_4 4 = 1$ because $3^4 = 81$ and
$4^1 = 4$

95. $\log_{\sqrt{2}} 2 = 2$ because $\sqrt{2}^2 = 2$.

97. $\log_{\sqrt{2}} 4 = 4$ because $(\sqrt{2})^4 = 4$.

99. $f(x) = \ln(x + 2)$

Shift the graph of $y = \ln x$ two units left.

101. $f(x) = -\ln(x - 1)$

Shift the graph of $y = \ln x$ graph one unit
right, then reflect the graph about the x-axis.

103. $f(x) = 2 + \ln(-x)$

Reflect the graph of $y = \ln x$ graph about the
y-axis and then shift it up two units.

105. $f(x) = 3 - 2\ln x$

Stretch the graph of $y = \ln x$ vertically by a
factor of 2, reflect the resulting graph about
the x-axis, and shift it three units up.

107. There was a horizontal shift of one unit left
and then a reflection about the x-axis.

4.2 Applying the Concepts

109. $f(t) = 221.69 - 66.9 \ln t$

a. Because $\ln 0$ is not defined, the year 2003
is represented by $t = 1$ and the year 2008 is
represented by $t = 6$.
$f(6) = 221.69 - 66.9 \ln 6 \approx 102$

In 2008, the consumer price index for
personal computers was about 102.

b. $221.69 - 66.9 \ln t = 30$

$-66.9 \ln t = -191.69$

$\ln t = \dfrac{191.69}{66.9}$

$t = e^{191.69/66.9} \approx 17.6$

The consumer price index will reach 30 sometime during $2003 + 17.6 = 2020$.

c. The price decreases faster in the beginning, but then the decreases in price become slower as time goes on.

111. $2P = Pe^{0.08t} \Rightarrow 2 = e^{0.08t} \Rightarrow \ln 2 = 0.08t \Rightarrow$

$t \approx 8.66$ years

113. $2P = Pe^{6k} \Rightarrow 2 = e^{6k} \Rightarrow \ln 2 = 6k \Rightarrow$

$k \approx 0.1155 = 11.55\%$

115. a. The population in 2010 was 109.7% of the population in 2000.

$308 = 1.097 A_0 \Rightarrow A_0 \approx 280.7657247$

The population in 2000 was about 280.8 million.

b. $308 = 280.7657247 e^{10r} \Rightarrow$

$\dfrac{308}{280.7657247} = e^{10r} \Rightarrow$

$\ln\left(\dfrac{308}{280.7657247}\right) = 10r \Rightarrow$

$r = \dfrac{\ln\left(\dfrac{308}{280.7657247}\right)}{10} \approx 0.009257 \approx 0.93\%$

c. $400 = 308 e^{0.0092t} \Rightarrow \dfrac{400}{308} = e^{0.0092t} \Rightarrow$

$\ln\left(\dfrac{400}{308}\right) = 0.0092t \Rightarrow t = \dfrac{\ln\left(\dfrac{400}{308}\right)}{0.0092} \approx 28.4$

years after 2010, or in the year 2039.

117. Find k using $T = 50, T_0 = 75, T_s = 20,$ and $t = 1$ minute.

$50 = 20 + (75 - 20)e^{-k(1)} \Rightarrow \dfrac{30}{55} = e^{-k} \Rightarrow$

$\ln\dfrac{30}{55} = -k \Rightarrow 0.6061 \approx k$

a. **(i)** $T = 20 + (75 - 20)e^{-0.6061(5)} \Rightarrow$

$T \approx 22.66°\text{F}$

(ii) $T = 20 + (75 - 20)e^{-0.6061(10)} \Rightarrow$

$T \approx 20.13°\text{F}$

(iii) $T = 20 + (75 - 20)e^{-0.6061(60)} \Rightarrow$

$T \approx 20°\text{F}$

b. $22 = 20 + (75 - 20)e^{-0.6061t} \Rightarrow$

$\dfrac{2}{55} = e^{-0.6061t} \Rightarrow \ln\dfrac{2}{55} = -0.6061t \Rightarrow$

$t \approx 5.5$ minutes

119. Find k using $T = 160, T_0 = 50, T_s = 400,$ and $t = 10$ minutes.

$160 = 400 + (50 - 400)e^{-10k} \Rightarrow \dfrac{240}{350} = e^{-10k} \Rightarrow$

$\ln\dfrac{24}{35} = -10k \Rightarrow k \approx 0.03773$

Use this value of k to find the time given $T = 220, T_0 = 50,$ and $T_s = 400$.

$220 = 400 + (50 - 400)e^{-0.03773t} \Rightarrow$

$\dfrac{18}{35} = e^{-0.03773t} \Rightarrow \ln\dfrac{18}{35} = -0.03773t \Rightarrow$

$t \approx 17.6$ minutes

121. a. $A(1) = 0.8(0.04) = 0.032$

Using the model $A(t) = A_0 e^{rt}$, we have

$0.032 = 0.04 e^r \Rightarrow 0.8 = e^r \Rightarrow r = \ln 0.8$

So, in one year (12 months), the concentration of the contaminant will be

$A(12) = 0.04 e^{12 \ln 0.8} = 0.0027 = 0.27\%$

b. $0.0001 \geq 0.04 e^{\ln 0.8 t} \Rightarrow 0.0025 \geq e^{\ln 0.8 t} \Rightarrow$

$\ln 0.0025 \geq \ln 0.8 t \Rightarrow t \geq 26.85$ months

123. a. $3000 = 1500 e^{3r} \Rightarrow \ln 2 = 3r \Rightarrow$

$r = \dfrac{\ln 2}{3} \approx 0.231$

Thus, the function is $P \approx 1500 e^{0.231t}$.

Alternatively, since the population doubles every three years, we have the geometric progression given by $P = 1500 \cdot 2^{t/3}$.

b. $P = 1500 e^{0.231(7)} \approx 7557$

There will be about 7557 sheep in the herd seven years from now.

c. $15,000 = 1500 e^{0.231t} \Rightarrow 10 = e^{0.231t} \Rightarrow$

$\ln 10 = 0.231t \Rightarrow t = \dfrac{\ln 10}{0.231} \approx 10$

The herd will have 15,000 sheep about 10 years from now.

125. a. $P_3 = \log 4 - \log 3 = \log\left(\dfrac{4}{3}\right) \approx 0.125$

About 12.5% of the data can be expected to have 3 as the first digit.

b. $P_1 = \log 2 - \log 1 \approx 0.3010$
$P_2 = \log 3 - \log 2 \approx 0.1761$
$P_3 = \log 4 - \log 3 \approx 0.1249$
$P_4 = \log 5 - \log 4 \approx 0.0969$
$P_5 = \log 6 - \log 5 \approx 0.0792$
$P_6 = \log 7 - \log 6 \approx 0.0669$
$P_7 = \log 8 - \log 7 \approx 0.0580$
$P_8 = \log 9 - \log 8 \approx 0.0512$
$P_9 = \log 10 - \log 9 \approx 0.0458$
$P_1 + P_2 + P_3 + \cdots + P_8 + P_9 = 1$

This means that one of the digits $1, \ldots, 9$ will appear as the first digit.

4.2 Beyond the Basics

127. a. $h(x) = \log_3 x$ and $g(x) = \log_2 x$. So, $f(x) = g(h(x))$. The domain of $h(x)$ is $(0, \infty) \Rightarrow$ the domain of $f(x)$ is $(1, \infty)$.

b. $h(x) = \ln(x-1)$ and $g(x) = \log x$. So, $f(x) = g(h(x))$. The domain of $h(x)$ is $(1, \infty) \Rightarrow$ the domain of $f(x)$ is $(2, \infty)$.

c. $h(x) = \log(x-1)$ and $g(x) = \ln x$. So, $f(x) = g(h(x))$. The domain of $h(x)$ is $(1, \infty) \Rightarrow$ the domain of $f(x)$ is $(2, \infty)$.

d. $h(x) = \log(x-1)$ and $g(x) = \log x$. So, $f(x) = g(g(h(x)))$. The domain of $h(x)$ is $(1, \infty)$. The domain of $g(h(x))$ is $(2, \infty)$. (Note that $\log(x-1) = 0 \Rightarrow x = 2$.) So, the domain of $g(g(h(x)))$ is $(11, \infty)$.

129. $y = \log x \rightarrow y = \log(x-2) \rightarrow$
$y = \log\left(\dfrac{1}{2}x - 2\right) \rightarrow y = 3\log\left(\dfrac{1}{2}x - 2\right) \rightarrow$
$y = -3\log\left(\dfrac{1}{2}x - 2\right) \rightarrow y = -3\log\left(\dfrac{1}{2}x - 2\right) + 4$

131. a. $P = 100,000e^{-0.07(20)} \approx \$24,659.70$

b. $50,000 = 75,000e^{-10r} \Rightarrow \dfrac{2}{3} = e^{-10r} \Rightarrow$
$\ln(2/3) = -10r \Rightarrow r \approx 0.0405 = 4.05\%$

4.2 Critical Thinking/Discussion/Writing

133. $2^{\log_2 3} - 3^{\log_3 2} = 3 - 2 = 1$

134. $\log_3 4 = \log_3\left(2^2\right) = 2\log_3 2$
$\log_2 9 = \log_2\left(3^2\right) = 2\log_2 3$
Let $x = \log_3 2$ and let $y = \log_2 3$.
$\left(\log_3 4 + \log_2 9\right)^2 - \left(\log_3 4 - \log_2 9\right)^2$
$= (2x + 2y)^2 - (2x - 2y)^2$
$= \left[(2x+2y) + (2x-2y)\right] \cdot$
$\qquad \left[(2x+2y) - (2x-2y)\right]$
$= (4x)(4y) = 16xy$
Note that
$x = \log_3 2 \Rightarrow 3^x = 2$ and $y = \log_2 3 \Rightarrow 2^y = 3$
$3^{xy} = \left(3^x\right)^y = 2^y = 3 \Rightarrow 3^{xy} = 3 \Rightarrow xy = 1$
Alternatively,
$xy = \log_3 2 \cdot \log_2 3 = \dfrac{\log 2}{\log 3} \cdot \dfrac{\log 3}{\log 2} = 1$.
Therefore, $16xy = 16 \cdot 1 = 16$, so
$\left(\log_3 4 + \log_2 9\right)^2 - \left(\log_3 4 - \log_2 9\right)^2 = 16$.

135. $\log_3\left[\log_4\left(\log_2 x\right)\right] = 0 \Rightarrow \log_4\left(\log_2 x\right) = 3^0 \Rightarrow$
$\log_4\left(\log_2 x\right) = 1 \Rightarrow \log_2 x = 4 \Rightarrow x = 2^4 = 16$

136. a. $f(x) = |\log x| = \begin{cases} -\log x & \text{if } 0 < x < 1 \\ \log x & \text{if } x \geq 1 \end{cases}$

137. a. Yes, the statement is always true.

b. The increasing property is used.

138. There are four possibilities, $0 < a < 1$ with $c < 0$, $0 < a < 1$ with $c > 0$, $a > 1$ with $c < 0$, and $a > 1$ with $c > 0$. They are illustrated below.

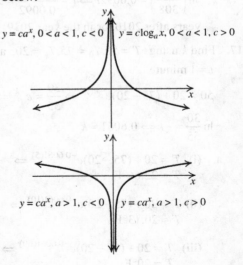

$y = ca^x, 0 < a < 1, c < 0$ \qquad $y = c\log_a x, 0 < a < 1, c > 0$

$y = ca^x, a > 1, c < 0$ \qquad $y = ca^x, a > 1, c > 0$

4.2 Getting Ready for the Next Sections

139. $a^2 \cdot a^7 = a^9$

141. $\sqrt{a^8} = \left(a^8\right)^{1/2} = a^{8\cdot(1/2)} = a^4$

143. $\left(\dfrac{243}{32}\right)^{4/5} = \left[\left(\dfrac{243}{32}\right)^{1/5}\right]^4 = \left(\dfrac{3}{2}\right)^4$

145. $\left(4.7 \times 10^7\right)\left(8.1 \times 10^5\right) = \left(4.7 \times 8.1\right)\left(10^7 \times 10^5\right)$
$$= 38.07 \times 10^{12}$$
$$= 3.807 \times 10^{13}$$

147. $\log_3 81 = 4$ because $3^4 = 81$.

$\log_3 3 + \log_3 27 = 1 + 3 = 4$ because $\log_3 3 = 1$

and $3^3 = 27$. Therefore,
$\log_3 81 = \log_3 3 + \log_3 27$.

149. $\log_2 16 = 4$ because $2^4 = 16$.

$2\log_2 4 = 2 \cdot 2 = 4$ because $2^2 = 4$.
Therefore, $\log_2 16 = 2\log_2 4$.

4.3 Rules of Logarithms

4.3 Practice Problems

1. Given $\log_5 z = 3$ and $\log_5 y = 2$

a. $\log_5 \left(y/z\right) = \log_5 y - \log_5 z = 2 - 3 = -1$

b. $\log_5\left(y^2 z^3\right) = \log_5\left(y^2\right) + \log_5\left(z^3\right)$
$$= 2\log_5 y + 3\log_5 z$$
$$= 2 \cdot 2 + 3 \cdot 3 = 13$$

2. a. $\ln \dfrac{2x-1}{x+4} = \ln(2x-1) - \ln(x+4)$

b. $\log\sqrt{\dfrac{4xy}{z}} = \log\left(\dfrac{4xy}{z}\right)^{1/2} = \dfrac{1}{2}\log\left(\dfrac{4xy}{z}\right)$
$$= \dfrac{1}{2}\left(\log 4xy - \log z\right)$$
$$= \dfrac{1}{2}\left(\log 4 + \log x + \log y - \log z\right)$$
$$= \dfrac{1}{2}\left(\log\left(2^2\right) + \log x + \log y - \log z\right)$$
$$= \log 2 + \dfrac{1}{2}\log x + \dfrac{1}{2}\log y - \dfrac{1}{2}\log z$$

3. $\dfrac{1}{2}\Big[\log\left(x+1\right) + \log\left(x-1\right)\Big]$
$$= \dfrac{1}{2}\Big[\log\left((x+1)(x-1)\right)\Big]$$
$$= \dfrac{1}{2}\log\left(x^2-1\right) = \log\left(x^2-1\right)^{1/2}$$
$$= \log\sqrt{x^2-1}$$

4. $\qquad K = 234^{567}$

$\log K = \log\left(234^{567}\right)$

$\log K = 567\log 234 \approx 1343.345391$

Since $\log K$ lies between the integers 1343 and 1344, the number K requires 1344 digits to the left of the decimal point. By definition of the common logarithm, we have

$K = 10^{1343.345391} = 10^{0.345391} \times 10^{1343}$
$$\approx 2.215088 \times 10^{1343}$$

5. $\log_3 15 = \dfrac{\log 15}{\log 3} \approx 2.46497$

6. a., b., Substitute $(3, 3)$ and $(9, 1)$ in the equation $y = c + b\log x$ to obtain

$3 = c + b\log 3$ (1)
$1 = c + b\log 9$ (2)

Subtract equation (1) from equation (2) and solve the resulting equation for b.

$-2 = b\log 9 - b\log 3 = b\left(\log 9 - \log 3\right)$
$$= b\log\dfrac{9}{3} = b\log 3$$

$b = -\dfrac{2}{\log 3}$

Substitute the value for b into equation (1) and solve for c.

$3 = c + \left(-\dfrac{2}{\log 3}\right)\log 3 \Rightarrow 3 = c - 2 \Rightarrow c = 5$

Substituting the values for b and c into
$y = c + b\log x$ gives

$y = 5 - \dfrac{2}{\log 3}\left(\log x\right) = 5 - 2\left(\dfrac{\log x}{\log 3}\right)$
$$= 5 - 2\log_3 x$$

7. $\ln\left(\dfrac{A(t)}{A_0}\right) = kt \Rightarrow \ln\left(\dfrac{66}{100}\right) = 15k \Rightarrow$

$\ln(0.66) = 15k \Rightarrow k = \dfrac{\ln(0.66)}{15}$

To find the half-life, we use the formula

$h = -\dfrac{\ln 2}{k} = -\dfrac{\ln 2}{\ln(0.66)/15} \approx 25.$

The half-life of strontium-90 is about 25 years.

8. King Tut died in 1346 B.C., so the object was made in 1540 B.C., so the time elapsed between when the object was made and 1960 is $1540 + 1960 = 3500$ years. The decay function for carbon-14 in exponential form is

$A(t) = A_0 e^{-0.0001216t}$. (See example 8 in the text.) Let $x =$ the percent of the original amount of carbon-14 in the object remaining after t years. Then

$xA_0 = A_0 e^{-0.0001216t} \Rightarrow x = e^{-0.0001216t}$.
$t = 3500$, so
$x = e^{-0.0001216(3500)} \approx 0.6534 \approx 65.34\%$.

4.3 Concepts and Vocabulary

1. $\log_a MN = \log_a M + \log_a N$.

3. $\log_a M^r = r \log_a M$.

5. False. There is no rule for the logarithm of a sum. $\log_a u + \log_a v = \log_a (uv)$.

7. False. $\ln\left(2\sqrt{x}\right) = \ln 2 + \ln\sqrt{x} = \ln 2 + \dfrac{1}{2}\ln x$

4.3 Building Skills

9. $\log 6 = \log(2 \cdot 3) = \log 2 + \log 3$
$= 0.3 + 0.48 = 0.78$

11. $\log 5 = \log\left(\dfrac{10}{2}\right) = \log 10 - \log 2 = 1 - 0.3 = 0.7$

13. $\log\left(\dfrac{2}{x}\right) = \log 2 - \log x = 0.3 - 2 = -1.7$

15. $\log(2x^2 y) = \log 2 + 2\log x + \log y$
$= 0.3 + 2(2) + 3 = 7.3$

17. $\log\sqrt[3]{x^2 y^4} = \log(x^{2/3} y^{4/3}) = \dfrac{2}{3}\log x + \dfrac{4}{3}\log y$
$= \dfrac{2}{3}(2) + \dfrac{4}{3}(3) = \dfrac{16}{3}$

19. $\log\sqrt[3]{48} = \log(2^4 \cdot 3)^{1/3} = \dfrac{4}{3}\log 2 + \dfrac{1}{3}\log 3$
$= \dfrac{4}{3}(0.3) + \dfrac{1}{3}(0.48) = 0.56$

21. $\ln[x(x-1)] = \ln x + \ln(x-1)$

23. $\ln\dfrac{x+1}{(x-2)^2} = \ln(x+1) - \ln(x-2)^2$
$= \ln(x+1) - 2\ln(x-2)$

25. $\log_a \sqrt{x}y^3 = \log_a \sqrt{x} + \log_a y^3$
$= \log_a x^{1/2} + \log_a y^3$
$= \dfrac{1}{2}\log_a x + 3\log_a y$

27. $\log_a \sqrt[3]{\dfrac{x}{y}} = \log_a\left(\dfrac{x}{y}\right)^{1/3} = \dfrac{1}{3}\log_a\left(\dfrac{x}{y}\right)$
$= \dfrac{1}{3}\log_a x - \dfrac{1}{3}\log_a y$

29. $\log_2 \sqrt[4]{\dfrac{xy^2}{8}} = \log_2\left(\dfrac{xy^2}{8}\right)^{1/4} = \dfrac{1}{4}\log_2\left(\dfrac{xy^2}{8}\right)$
$= \dfrac{1}{4}\log_2\left(xy^2\right) - \dfrac{1}{4}\log_2 8$
$= \dfrac{1}{4}\log_2 x + \dfrac{1}{4}\log_2 y^2 - \dfrac{1}{4}(3)$
$= \dfrac{1}{4}\log_2 x + 2\cdot\dfrac{1}{4}\log_2 y - \dfrac{3}{4}$
$= \dfrac{1}{4}\log_2 x + \dfrac{1}{2}\log_2 y - \dfrac{3}{4}$

31. $\log\dfrac{\sqrt{x^2+1}}{x+3} = \log\sqrt{x^2+1} - \log(x+3)$
$= \dfrac{1}{2}\log(x^2+1) - \log(x+3)$

33. $\log_3 \dfrac{(x-1)(x+1)}{x^2-4} = \log_3\left[(x-1)(x+1)\right] - \log_3\left(x^2-4\right)$

$\qquad = \log_3(x-1) + \log_3(x+1) - \log_3\left[(x-2)(x+2)\right]$

$\qquad = \log_3(x-1) + \log_3(x+1) - \left(\log_3(x-2) + \log_3(x+2)\right)$

$\qquad = \log_3(x-1) + \log_3(x+1) - \log_3(x-2) - \log_3(x+2)$

35. $\log_b x^2 y^3 z = 2\log_b x + 3\log_b y + \log_b z$

37. $\ln\left(\dfrac{x\sqrt{x-1}}{x^2+2}\right) = \ln\left(x\sqrt{x-1}\right) - \ln\left(x^2+2\right) = \ln x + \ln\left((x-1)^{1/2}\right) - \ln\left(x^2+2\right) = \ln x + \dfrac{1}{2}\ln(x-1) - \ln\left(x^2+2\right)$

39. $\ln\left(\dfrac{(x+1)^2}{(x-3)\sqrt{x+4}}\right) = \ln\left((x+1)^2\right) - \ln\left((x-3)\sqrt{x+4}\right) = 2\ln(x+1) - \left(\ln(x-3) + \ln(x+4)^{1/2}\right)$

$\qquad = 2\ln(x+1) - \ln(x-3) - \dfrac{1}{2}\ln(x+4)$

41. $\ln\left((x+1)\sqrt{\dfrac{x^2+2}{x^2+5}}\right) = \ln(x+1) + \ln\left(\dfrac{x^2+2}{x^2+5}\right)^{1/2} = \ln(x+1) + \dfrac{1}{2}\left(\ln\left(x^2+2\right) - \ln\left(x^2+5\right)\right)$

$\qquad = \ln(x+1) + \dfrac{1}{2}\ln\left(x^2+2\right) - \dfrac{1}{2}\ln\left(x^2+5\right)$

43. $\ln\left(\dfrac{x^3(3x+1)^4}{\sqrt{x^2+1}\,(x+2)^{-5}(x-3)^2}\right) = \ln\left(\dfrac{x^3(3x+1)^4(x+2)^5}{\left(x^2+1\right)^{1/2}(x-3)^2}\right)$

$\qquad = \ln\left[x^3(3x+1)^4(x+2)^5\right] - \ln\left[\left(x^2+1\right)^{1/2}(x-3)^2\right]$

$\qquad = 3\ln x + 4\ln(3x+1) + 5\ln(x+2) - \dfrac{1}{2}\ln\left(x^2+1\right) - 2\ln(x-3)$

45. $\log_2 x + \log_2 7 = \log_2(7x)$

47. $\log(3x+2) - \log x = \log\dfrac{3x+2}{x}$

49. $\dfrac{1}{2}\ln x + 2\ln y = \ln x^{1/2} + \ln y^2 = \ln\sqrt{x} + \ln y^2 = \ln y^2\sqrt{x}$

51. $\dfrac{1}{2}\log x - \log y + \log z = \log\left(\dfrac{z\sqrt{x}}{y}\right)$

53. $\dfrac{1}{5}(\log_2 z + 2\log_2 y) = \log_2\left(y^2 z\right)^{1/5} = \log_2\sqrt[5]{y^2 z}$

55. $\ln x + 2\ln y + 3\ln z = \ln(xy^2 z^3)$

57. $3\log x - \log y + \dfrac{1}{2}\log\left(x^2+4\right) = \log x^3 - \log y + \log\left(x^2+4\right)^{1/2} = \log x^3 - \log y + \log\sqrt{x^2+4}$

$\qquad = \log\dfrac{x^3\sqrt{x^2+4}}{y}$

59. $2\ln x - \dfrac{1}{2}\ln(x^2+1) = \ln x^2 - \ln\sqrt{x^2+1}$

$$= \ln\left(\dfrac{x^2}{\sqrt{x^2+1}}\right)$$

61. $K = e^{500}$

$\log K = \log\left(e^{500}\right) = 500\log e = 217.147241 \Rightarrow$

$K = 10^{217.147241} = 10^{0.147241}\times 10^{217}$

$\approx 1.4036\times 10^{217}$

63. $K = 324^{756}$

$\log K = \log\left(324^{756}\right) = 756\log 324$

$= 1897.972028 \Rightarrow$

$K = 10^{1897.972028} = 10^{0.972028}\times 10^{1897}$

$\approx 9.3762\times 10^{1897}$

65. $K = 234^{567}$

$\log K = \log\left(234^{567}\right) = 567\log 234$

$= 1343.345$

$M = 567^{234}$

$\log M = \log\left(567^{234}\right) = 234\log 567$

$= 664.338$

Since $\log K > \log M$, $K > M$. Thus, $234^{567} > 567^{234}$.

67. $K = 17^{200}\cdot 53^{67}$

$\log K = \log\left(17^{200}\cdot 53^{67}\right)$

$= \log\left(17^{200}\right) + \log\left(53^{67}\right)$

$= 200\log 17 + 67\log 53$

$= 361.6163$

There are 362 digits in the given product.

69. $\log_2 5 = \dfrac{\log 5}{\log 2}\approx 2.322$

71. $\log_{1/2} 3 = \dfrac{\log 3}{\log\frac{1}{2}}\approx -1.585$

73. $\log_{\sqrt{5}}\sqrt{17} = \dfrac{\log\sqrt{17}}{\log\sqrt{5}}\approx 1.760$

75. $\log_2 7 + \log_4 3 = \dfrac{\log 7}{\log 2} + \dfrac{\log 3}{\log 4}\approx 3.6$

77. $\log_3\sqrt{3} = \log_3 3^{1/2} = \dfrac{1}{2}$

79. $\log_3(\log_2 8) = \log_3(\log_2 2^3) = \log_3 3 = 1$

81. $5^{2\log_5 3 + \log_5 2} = 5^{\log_5 3^2 + \log_5 2} = 5^{\log_5 (9\cdot 2)}$

$= 5^{\log_5 18} = 18$

83. $\log 4 + 2\log 5 = \log(4\cdot 5^2) = \log 100$

$= \log 10^2 = 2$

85. Substitute $(10, 1)$ and $(1, 2)$ in the equation $y = c + b\log x$ to obtain

$1 = c + b\log 10$ (1)

$2 = c + b\log 1$ (2)

Subtract equation (1) from equation (2) and solve the resulting equation for b.

$1 = b\log 1 - b\log 10 = b\cdot 0 - b\cdot 1 = -b \Rightarrow$

$b = -1$

Substitute the value for b into equation (1) and solve for c.

$1 = c - \log 10 \Rightarrow 1 = c - 1 \Rightarrow c = 2$

Substituting the values for b and c into $y = c + b\log x$ gives $y = 2 - \log x$.

87. Substitute $(e, 1)$ and $(1, 2)$ in the equation $y = c + b\log x$ to obtain

$1 = c + b\log e$ (1)

$2 = c + b\log 1$ (2)

Since $\log 1 = 0$, equation (2) becomes $c = 2$. Substitute the value for c into equation (1) and solve for b.

$1 = 2 + b\log e \Rightarrow -1 = b\log e \Rightarrow b = -\dfrac{1}{\log e}$

Substituting the values for b and c into $y = c + b\log x$ gives

$y = 2 - \left(\dfrac{1}{\log e}\right)\log x = 2 - \left(\dfrac{\log x}{\log e}\right) = 2 - \ln x.$

89. Substitute $(5, 4)$ and $(25, 7)$ in the equation $y = c + b\log x$ to obtain

$4 = c + b\log 5$ (1)

$7 = c + b\log 25$ (2)

Subtract equation (1) from equation (2) and solve the resulting equation for b.

$3 = b\log 25 - b\log 5 = b(\log 25 - \log 5)$

$= b\log\left(\dfrac{25}{5}\right) = b\log 5 \Rightarrow$

$b = \dfrac{3}{\log 5}$

Substitute the value for b into equation (1) and solve for c.

$4 = c + \left(\dfrac{3}{\log 5}\right)\log 5 \Rightarrow 4 = c + 3 \Rightarrow c = 1$

(continued on next page)

(*continued*)

Substituting the values for b and c into
$y = c + b \log x$ gives

$$y = 1 + \left(\frac{3}{\log 5} \right) \log x = 1 + 3 \left(\frac{\log x}{\log 5} \right)$$

$$= 1 + 3 \log_5 x.$$

91. Substitute $(2, 4)$ and $(4, 9)$ in the equation
$y = c + b \log x$ to obtain

$$4 = c + b \log 2 \quad (1)$$
$$9 = c + b \log 4 \quad (2)$$

Subtract equation (1) from equation (2) and
solve the resulting equation for b.

$$5 = b \log 4 - b \log 2 = b \left(\log 4 - \log 2 \right)$$

$$= b \log \left(\frac{4}{2} \right) \Rightarrow 5 = b \log 2 \Rightarrow b = \frac{5}{\log 2}$$

Substitute the value for b into equation (1) and
solve for c.

$$4 = c + \left(\frac{5}{\log 2} \right) \log 2 = c + 5 \Rightarrow -1 = c$$

Substituting the values for b and c into
$y = c + b \log x$ gives

$$y = -1 + \left(\frac{5}{\log 2} \right) \log x = -1 + 5 \left(\frac{\log x}{\log 2} \right)$$

$$= -1 + 5 \log_2 x.$$

93. $A(t) = A_0 e^{kt} \Rightarrow 23 = 50 e^{12k} \Rightarrow$
$0.46 = e^{12k} \Rightarrow \ln(0.46) = 12k \Rightarrow$

$$k = \frac{\ln(0.46)}{12}$$

To find the half-life, we use the formula

$$h = -\frac{\ln 2}{k} = -\frac{\ln 2}{\ln(0.46)/12} \approx 10.7.$$

The half-life is about 10.7 years.

95. $A(t) = A_0 e^{kt} \Rightarrow 3.8 = 10.3 e^{15k} \Rightarrow$

$$\frac{3.8}{10.3} = e^{15k} \Rightarrow \ln \left(\frac{3.8}{10.3} \right) = 15k \Rightarrow$$

$$k = \frac{\ln \left(\frac{3.8}{10.3} \right)}{15}$$

To find the half-life, we use the formula

$$h = -\frac{\ln 2}{k} = -\frac{\ln 2}{\ln(3.8/10.3)/15} \approx 10.4.$$

The half-life is about 10.4 hours.

4.3 Applying the Concepts

97. $7.09 = 7 e^{1k} \Rightarrow$

$$\ln \frac{7.09}{7} = k \approx 0.012775 = 1.2775\%$$

99. $7 e^{100k} < 20 \Rightarrow 100k < \ln \left(\frac{20}{7} \right) \Rightarrow$
$0.010498 = 1.0498\% < k$
The maximum rate of growth is 1.0498%.

101. $3500 = 1000 e^{0.1t} \Rightarrow \ln 3.5 = 0.1t \Rightarrow$
$t \approx 12.53$ years

103. Because the half-life is 8 days, there will be
10 grams left after 8 days. Use this to find k:

$$10 = 20 e^{8k} \Rightarrow \frac{1}{2} = e^{8k} \Rightarrow \ln \left(\frac{1}{2} \right) = 8k \Rightarrow$$

$$k \approx -0.08664$$

$$A = 20 e^{-0.08664(5)} \approx 12.969 \text{ grams}$$

105. $\frac{1}{2} = e^{-0.055t} \Rightarrow \ln \left(\frac{1}{2} \right) = -0.055t \Rightarrow t \approx 12.6$ yr.

107. $1.5 = 5 e^{-10k} \Rightarrow 0.3 = e^{-10k} \Rightarrow \ln 0.3 = -10k \Rightarrow$
$k \approx 0.1204$

109. Find k using $A = 8, A_0 = 16,$ and $t = 36$:

$$8 = 16 e^{-36k} \Rightarrow \frac{1}{2} = e^{-36k} \Rightarrow$$

$$\ln \left(\frac{1}{2} \right) = 36k \Rightarrow k \approx 0.0193$$

$$A = 16 e^{-0.0193(8)} \Rightarrow A \approx 13.7 \text{ grams.}$$

In exercises 111−114, we use the formula

$$P = \frac{r \cdot M}{1 - \left(1 + \frac{r}{n} \right)^{-nt}} \div n,$$

where P = the payment, r = the annual interest rate,
M = the mortgage amount, t = the number of years,
and n = the number of payments per year.

111. $P = \dfrac{0.06 \cdot 120,000}{1 - \left(1 + \frac{0.06}{12} \right)^{-12 \cdot 20}} \div 12 = 859.72$

The monthly payment is \$859.72.
There are 240 payments so the total amount
paid is $240 \cdot \$859.72 = \$206,332.80.$
The amount of interest paid is
$206,332.80 - 120,000 = \$86,332.80.$

113. $850 = \dfrac{0.085 \cdot M}{1 - \left(1 + \frac{0.085}{12} \right)^{-12 \cdot 30}} \div 12$

$$10,200 = \frac{0.085 \cdot M}{1 - \left(1 + \frac{0.085}{12} \right)^{-12 \cdot 30}}$$

(*continued on next page*)

(*continued*)

$$M = \frac{10,200\left[1-\left(1+\frac{0.085}{12}\right)^{-12\cdot30}\right]}{0.085}$$

$$\approx 110,545.60$$

Andy can afford a mortgage of about $110,545.60.

4.3 Beyond the Basics

115. $\log_b\left(\sqrt{x^2+1}-x\right)+\log_b\left(\sqrt{x^2+1}+x\right)$

$$= \log_b\left(\left(\sqrt{x^2+1}-x\right)\left(\sqrt{x^2+1}+x\right)\right)$$

$$= \log_b\left(x^2+1-x^2\right)=\log_b 1 = 0$$

117. $(\log_b a)(\log_a b) = \dfrac{\log a}{\log b}\cdot\dfrac{\log b}{\log a} = 1$

119.

[0, 10, 1] by [−4, 4,1]

121. a. $\log\left(\dfrac{a}{b}\right)+\log\left(\dfrac{b}{a}\right)+\log\left(\dfrac{c}{a}\right)+\log\left(\dfrac{a}{c}\right)=0$

(See exercise 118.)

b. $\log\left(\dfrac{a^2}{bc}\right)+\log\left(\dfrac{b^2}{ca}\right)+\log\left(\dfrac{c^2}{ab}\right)$

$$= 2\log a - (\log b + \log c)$$

$$+ 2\log b - (\log c + \log a)$$

$$+ 2\log c - (\log a + \log b) = 0$$

c. $\log_2 3 \cdot \log_3 4 = \dfrac{\log 3}{\log 2}\cdot\dfrac{\log 2^2}{\log 3}$

$$= \dfrac{\log 3}{\log 2}\cdot\dfrac{2\log 2}{\log 3} = 2$$

d. $\log_a b \cdot \log_b c \cdot \log_c a$

$$= \dfrac{\log b}{\log a}\cdot\dfrac{\log c}{\log b}\cdot\dfrac{\log a}{\log c} = 1$$

123. $f(x) = \log_4\left(\log_5\left(\log_3\left(18x-x^2-77\right)\right)\right)$

$$18x-x^2-77 = -\left(x^2-18x+77\right)$$

$$= -(x-11)(x-7)$$

$\log_3\left(18x-x^2-77\right)$ is defined only for those values of x which make $18x-x^2-77>0$. Thus, the domain of $\log_3\left(18x-x^2-77\right)$ is $(7, 11)$.

$\log_5\left[\log_3\left(18x-x^2-77\right)\right]$ is defined only for those values of x in the interval $(7, 11)$ which make $\log_3\left(18x-x^2-77\right)>0$. We use a graphing calculator to solve this. Note that we use the change of base formula to define the function.

$$Y_1 = \log_3\left(18x-x^2-77\right) = \frac{\log\left(18x-x^2-77\right)}{\log 3}.$$

[6, 12] by [−1 ,2]

Thus, the domain of $\log_5\left[\log_3\left(18x-x^2-77\right)\right]$ is approximately $(7.27, 10.73)$.

$\log_4\left(\log_5\left(\log_3\left(18x-x^2-77\right)\right)\right)$ is defined only for those values of x in the interval $(7.27, 10.73)$ which make

$$\log_5\left[\log_3\left(18x-x^2-77\right)\right]>0.$$

We use a graphing calculator to solve this. Note that we use the change of base formula to define the function.

$$Y_2 = \log_5\left(\log_3\left(18x-x^2-77\right)\right)$$

$$= \log_5\left(Y_1\right) = \frac{Y_1}{\log 5}$$

[6, 12] by [−1 ,1]

Thus, the domain of $\log_4\left(\log_5\left(\log_3\left(18x-x^2-77\right)\right)\right)$ is $(8, 10)$.

125. a. False **b.** True
 c. False **d.** False
 e. True **f.** True
 g. False **h.** True
 i. False **j.** True

127.
$$\log\left(\frac{a+b}{3}\right) = \frac{1}{2}(\log a + \log b)$$
$$2\log\left(\frac{a+b}{3}\right) = \log(ab)$$
$$2\log(a+b) - 2\log 3 = \log(ab)$$
$$\log(a+b)^2 - \log 9 = \log(ab)$$
$$\log(a+b)^2 = \log(ab) + \log 9$$
$$\log(a+b)^2 = \log(9ab)$$
$$(a+b)^2 = 9ab$$
$$a^2 + 2ab + b^2 = 9ab \Rightarrow a^2 + b^2 - 7ab = 0$$

4.3 Critical Thinking/Discussion/Writing

129. In step 2, $\log\left(\frac{1}{2}\right)$ is negative, so $3 < 4 \Rightarrow$
$$3\log\left(\frac{1}{2}\right) > 4\log\left(\frac{1}{2}\right).$$

130. The domain of $2\log x$ is $(0, \infty)$, while the domain of $\log(x^2)$ is $(-\infty, 0) \cup (0, \infty)$.

131. If $p = 2^m - 1$ is a prime number, then the only way p and 2^m have a different number of digits is if $2^m = 10^k$ or $2^{n-k} = 5^k$. However, this is impossible because 2^{n-k} is even and 5^k is odd. Thus, p and 2^m have the same number of digits.

$$K = 2^{43112609}$$
$$\log K = \log\left(2^{43112609}\right) = 43112609\log 2$$
$$\approx 12978188.5$$

There are 12,978,189 digits in this prime number.

132. $\frac{1}{\log(1-x)}$ is defined only for those values of $\log(1-x) \neq 0$ and for those values of x such that $\log(1-x)$ is defined.
$$\log(1-x) = 0 \Rightarrow 1 - x = 1 \Rightarrow x = 0$$
$\log(1-x)$ is defined for
$$1 - x > 0 \Rightarrow 1 > x \text{ or } x < 1.$$
Thus, the domain of $\frac{1}{\log(1-x)}$ is
$(-\infty, 0) \cup (0, 1)$.

$\frac{1}{\ln(x+2)}$ is defined only for those values of $\ln(x+2) \neq 0$ and for those values of x such that $\ln(x+2)$ is defined.
$$\ln(x+2) = 0 \Rightarrow x + 2 = 1 \Rightarrow x = -1$$
$\ln(x+2)$ is defined for
$$x + 2 > 0 \Rightarrow x > -2 \text{ or } -2 < x.$$
Thus, the domain of $\frac{1}{\ln(x+2)}$ is
$(-2, -1) \cup (-1, \infty)$.
The intersection of the two domains is the domain of $\frac{1}{\log(1-x)} + \frac{1}{\ln(x+2)}$. Thus, the domain is $(-2, -1) \cup (-1, 0) \cup (0, 1)$.

4.3 Getting Ready for the Next Section

133. $11 \cdot 3^0 = 11 \cdot 1 = 11$

135. $4^x \cdot 2^{-2x+1} = \left(2^2\right)^x \cdot 2^{-2x+1} = 2^{2x} \cdot 2^{-2x+1}$
$$= 2^{2x+(-2x+1)} = 2^1 = 2$$

For exercises 137–140, let $t = 5^x$. Then $5^{2x} = t^2$.

137. $5^{2x} - 5^x = -1 \Rightarrow t^2 - t = -1 \Rightarrow t^2 - t + 1 = 0$

139. $\frac{5^x + 3 \cdot 5^{-x}}{5^x} = \frac{1}{4} \Rightarrow \frac{t + \frac{3}{t}}{t} = \frac{1}{4} \Rightarrow \frac{\frac{t^2+3}{t}}{t} = \frac{1}{4} \Rightarrow$
$$\frac{t^2+3}{t^2} = \frac{1}{4} \Rightarrow 4t^2 + 12 = t^2 \Rightarrow 3t^2 + 12 = 0 \Rightarrow$$
$$t^2 + 4 = 0$$

141. $2x - (11 + x) = 8x + (7 + 2x)$
$$x - 11 = 10x + 7$$
$$-18 = 9x \Rightarrow x = -2$$
Solution set: $\{-2\}$

143.
$$x^2 + 3x - 1 = 3$$
$$x^2 + 3x - 4 = 0$$
$$(x-1)(x+4) = 0$$
$$x - 1 = 0 \mid x + 4 = 0$$
$$x = 1 \mid x = -4$$
Solution set: $\{-4, 1\}$

145. $2x < 7 + x \Rightarrow x < 7$
Solution set: $(-\infty, 7)$

147.
$$12x > 30 - 3x$$
$$15x > 30 \Rightarrow x > 2$$
Solution set: $(2, \infty)$

4.4 Exponential and Logarithmic Equations and Inequalities

4.4 Practice Problems

1. a. $3^x = 243 \Rightarrow 3^x = 3^5 \Rightarrow x = 5$

b. $8^x = 4 \Rightarrow \left(2^3\right)^x = 2^2 \Rightarrow 2^{3x} = 2^2 \Rightarrow$
$$3x = 2 \Rightarrow x = \frac{2}{3}$$

2. $7 \cdot 3^{x+1} = 11 \Rightarrow 3^{x+1} = \frac{11}{7} \Rightarrow$
$$\ln\left(3^{x+1}\right) = \ln\left(\frac{11}{7}\right) \Rightarrow (x+1)\ln 3 = \ln\left(\frac{11}{7}\right) \Rightarrow$$
$$x + 1 = \frac{\ln(11/7)}{\ln 3} \Rightarrow x = \frac{\ln(11/7)}{\ln 3} - 1 \approx -0.589$$

3.
$$3^{x+1} = 2^{2x}$$
$$\ln\left(3^{x+1}\right) = \ln\left(2^{2x}\right)$$
$$(x+1)\ln 3 = 2x \ln 2$$
$$x \ln 3 + \ln 3 = 2x \ln 2$$
$$x \ln 3 - 2x \ln 2 = -\ln 3$$
$$x(\ln 3 - 2 \ln 2) = -\ln 3$$
$$x = -\frac{\ln 3}{\ln 3 - 2 \ln 2} \approx 3.819$$

4.
$$e^{2x} - 4e^x - 5 = 0$$
$$\left(e^x - 5\right)\left(e^x + 1\right) = 0$$
$$e^x - 5 = 0 \text{ or } e^x + 1 = 0$$
$$e^x - 5 = 0 \Rightarrow e^x = 5 \Rightarrow \ln\left(e^x\right) = \ln 5 \Rightarrow$$
$$x \ln e = \ln 5 \Rightarrow x = \ln 5 \approx 1.609$$
$e^x + 1 = 0 \Rightarrow e^x = -1$, which is not possible.

5. The given model is $P(t) = P_0 \left(1 + r\right)^t$.

a. In 2020, ten years after the base year, the population of the United States will be
$$P(10) = 308\left(1 + 0.011\right)^{10} \approx 343.61 \text{ million.}$$
The population of Pakistan will be
$$P(10) = 185\left(1 + 0.033\right)^{10} \approx 255.96 \text{ million.}$$

b. To find when the population of the U.S. will be 350 million, solve for t:
$$350 = 308\left(1 + 0.011\right)^t \Rightarrow \frac{350}{308} = 1.011^t \Rightarrow$$
$$\ln\left(\frac{350}{308}\right) = t \ln 1.011 \Rightarrow$$
$$t = \frac{\ln(350/308)}{\ln 1.011} \approx 11.69$$
The population of the U.S. will be 350 million approximately 11.69 years after 2010, sometime in the year 2022.

c. To find when the population of the two countries will be the same, solve for t:
$$308\left(1 + .011\right)^t = 185\left(1 + .033\right)^t$$
$$308(1.011)^t = 185(1.033)^t$$
$$\frac{308}{185} = \frac{1.033^t}{1.011^t} = \left(\frac{1.033}{1.011}\right)^t$$
$$\ln\left(\frac{308}{185}\right) = t \ln\left(\frac{1.033}{1.011}\right)$$
$$\frac{\ln(308/185)}{\ln(1.033/1.011)} \approx 23.68 = t$$

The populations of the two countries will be the same about 23.68 years after 2010, sometime in the year 2033.

6. $1 + 2\ln x = 4 \Rightarrow 2\ln x = 3 \Rightarrow \ln x = \frac{3}{2} \Rightarrow x = e^{3/2}$

7.
$$\log_3\left(x - 8\right) + \log_3 x = 2$$
$$\log_3\left[x(x-8)\right] = 2$$
$$x^2 - 8x = 3^2$$
$$x^2 - 8x - 9 = 0$$
$$(x-9)(x+1) = 0$$
$$x - 9 = 0 \text{ or } x + 1 = 0$$
$$x = 9 \qquad x = -1$$
Now check each possible solution in the original equation.
$$\log_3\left(9 - 8\right) + \log_3 9 \overset{?}{=} 2$$
$$\log_3 1 + \log_3 9 \overset{?}{=} 2$$
$$0 + 2 = 2 \checkmark$$

(continued on next page)

(continued)

$$\log_3(-1-8) + \log_3(-1) \overset{?}{=} 2$$

This is not possible because logarithms are not defined for negative values.

The solution set is $\{9\}$.

8. $\ln(x+5) + \ln(x+1) = \ln(x-1)$
    $\ln[(x+5)(x+1)] = \ln(x-1)$
    $\ln(x^2 + 6x + 5) = \ln(x-1)$
    $x^2 + 6x + 5 = x - 1$
    $x^2 + 5x + 6 = 0$
    $(x+2)(x+3) = 0$
    $x + 2 = 0 \Rightarrow x = -2$ or $x + 3 = 0 \Rightarrow x = -3$

    Now check each possible solution in the original equation.

    $$\ln(-2+5) + \ln(-2+1) \overset{?}{=} \ln(-2-1)$$
    $$\ln 3 + \ln(-1) = \ln(-3)$$

    This is not possible because logarithms are not defined for negative values.

    $$\ln(-3+5) + \ln(-3+1) \overset{?}{=} \ln(-3-1)$$
    $$\ln 2 + \ln(-2) = \ln(-4)$$

    This is not possible because logarithms are not defined for negative values.

    The solution set is $\varnothing$.

9. The year 1987 represents $t = 0$, so the year 2005 represents $t = 18$. Using equation (19) in example 9 in the text, we have

    $$6.5 = \frac{35}{1 + 6e^{-18k}} \Rightarrow 6.5 + 39e^{-18k} = 35 \Rightarrow$$

    $$e^{-18k} = \frac{28.5}{39} \Rightarrow -18k = \ln\left(\frac{28.5}{39}\right) \Rightarrow$$

    $$k = -\frac{\ln(28.5/39)}{18} \approx 0.0174 = 1.74\%$$

    The growth rate was about 1.74%.

10. $3(0.5)^x + 7 > 19 \Rightarrow 3(0.5)^x > 12 \Rightarrow$
    $(0.5)^x > 4 \Rightarrow x \ln 0.5 > \ln 4 \Rightarrow x < \dfrac{\ln 4}{\ln 0.5} \Rightarrow$
    $x < -2$

11. Since $1 - 3x$ must be positive, $x < \dfrac{1}{3}$.

    $\ln(1-3x) > 2 \Rightarrow 1 - 3x > e^2 \Rightarrow$
    $-3x > e^2 - 1 \Rightarrow x < \dfrac{e^2 - 1}{-3} \Rightarrow$
    $x < \dfrac{1 - e^2}{3}$ or $x < -2.13$

4.4 Concepts and Vocabulary

1. An equation that contains terms of the form a^x is called a(n) <u>exponential</u> equation.

3. The equation $y = \dfrac{m}{1 + ae^{-bx}}$ represents a(n) <u>logistic</u> model.

5. True

7. True.
    $$8^{2x} = 4^{3x} \Rightarrow (2^3)^{2x} = (2^2)^{3x} \Rightarrow 2^{6x} = 2^{6x}$$

4.4 Building Skills

9. $2^x = 16 \Rightarrow 2^x = 2^4 \Rightarrow x = 4$

11. $8^x = 32 \Rightarrow 2^{3x} = 2^5 \Rightarrow 3x = 5 \Rightarrow x = \dfrac{5}{3}$

13. $4^{|x|} = 128 \Rightarrow 2^{2|x|} = 2^7 \Rightarrow 2|x| = 7 \Rightarrow x = \pm\dfrac{7}{2}$

15. $5^{-|x|} = 625 \Rightarrow 5^{-|x|} = 5^4 \Rightarrow -|x| = 4 \Rightarrow$ there is no solution.

17. $\ln x = 0 \Rightarrow x = 1$

19. $\log_2 x = -1 \Rightarrow 2^{-1} = x \Rightarrow x = \dfrac{1}{2}$

21. $\log_3 |x| = 2 \Rightarrow 3^2 = |x| \Rightarrow x = \pm 9$

23. $\dfrac{1}{2}\log x - 2 = 0 \Rightarrow \dfrac{1}{2}\log x = 2 \Rightarrow \log x = 4 \Rightarrow$
    $10^4 = 10,000 = x$

In exercises 25–60, the equations can be solved using either the common logarithm or the natural logarithm.

25. $2^x = 3 \Rightarrow x \ln 2 = \ln 3 \Rightarrow x = \dfrac{\ln 3}{\ln 2} \approx 1.585$

27. $2^{2x+3} = 15 \Rightarrow (2x+3)\ln 2 = \ln 15 \Rightarrow$
    $2x + 3 = \dfrac{\ln 15}{\ln 2} \Rightarrow$
    $x = \dfrac{\dfrac{\ln 15}{\ln 2} - 3}{2} = \dfrac{\ln 15 - 3\ln 2}{2\ln 2} \approx 0.453$

29. $e^{x+1} = 3 \Rightarrow (x+1)\ln e = \ln 3 \Rightarrow x + 1 = \ln 3 \Rightarrow$
    $x = \ln 3 - 1 \approx 0.099$

31. $5 \cdot 2^x - 7 = 10 \Rightarrow 2^x = \dfrac{17}{5} \Rightarrow$

$x \ln 2 = \ln\left(\dfrac{17}{5}\right) \Rightarrow x = \dfrac{\ln 17 - \ln 5}{\ln 2} \approx 1.766$

33. $3 \cdot 4^{2x-1} + 4 = 14 \Rightarrow 4^{2x-1} = \dfrac{10}{3} \Rightarrow$

$(2x - 1) \ln 4 = \ln\left(\dfrac{10}{3}\right) \Rightarrow$

$x = \dfrac{\dfrac{\ln 10 - \ln 3}{\ln 4} + 1}{2} = \dfrac{\ln 10 - \ln 3 + \ln 4}{2 \ln 4} \approx 0.934$

35. $2e^{x-2} + 3 = 7 \Rightarrow 2e^{x-2} = 4 \Rightarrow e^{x-2} = 2 \Rightarrow$

$(x - 2) \ln e = \ln 2 \Rightarrow x - 2 = \ln 2 \Rightarrow$

$x = \ln 2 + 2 \approx 2.693$

37. $5^{1-x} = 2^x \Rightarrow (1 - x) \ln 5 = x \ln 2 \Rightarrow$

$\ln 5 - x \ln 5 = x \ln 2 \Rightarrow \ln 5 = x(\ln 5 + \ln 2) \Rightarrow$

$x = \dfrac{\ln 5}{\ln 5 + \ln 2} \approx 0.699$

39. $2^{1-x} = 3^{4x+6} \Rightarrow (1 - x) \ln 2 = (4x + 6) \ln 3 \Rightarrow$

$\ln 2 - x \ln 2 = 4x \ln 3 + 6 \ln 3 \Rightarrow$

$\ln 2 - 6 \ln 3 = 4x \ln 3 + x \ln 2 \Rightarrow$

$\ln 2 - 6 \ln 3 = x(4 \ln 3 + \ln 2) \Rightarrow$

$\dfrac{\ln 2 - 6 \ln 3}{4 \ln 3 + \ln 2} = x \Rightarrow x \approx -1.159$

41. $2 \cdot 3^{x-1} = 5^{x+1}$

$\ln\left(2 \cdot 3^{x-1}\right) = \ln\left(5^{x+1}\right)$

$\ln 2 + (x - 1) \ln 3 = (x + 1) \ln 5$

$\ln 2 + x \ln 3 - \ln 3 = x \ln 5 + \ln 5$

$x \ln 3 - x \ln 5 = \ln 5 - \ln 2 + \ln 3$

$x(\ln 3 - \ln 5) = \ln 5 - \ln 2 + \ln 3$

$x = \dfrac{\ln 5 - \ln 2 + \ln 3}{\ln 3 - \ln 5}$

$ = \dfrac{\ln 2 - \ln 3 - \ln 5}{\ln 5 - \ln 3} \approx -3.944$

43. $1.065^t = 2 \Rightarrow t \ln 1.065 = \ln 2 \Rightarrow$

$t = \dfrac{\ln 2}{\ln 1.065} \approx 11.007$

45. Let $y = 2^x$. Then $2^{2x} - 4 \cdot 2^x = 21 \Rightarrow$

$y^2 - 4y - 21 = 0 \Rightarrow (y - 7)(y + 3) = 0 \Rightarrow$

$y = 7$ or $y = -3$. Reject the negative solution.

$2^x = 7 \Rightarrow x \ln 2 = \ln 7 \Rightarrow x = \dfrac{\ln 7}{\ln 2} \approx 2.807$

47. $9^x - 6 \cdot 3^x + 8 = 0 \Rightarrow 3^{2x} - 6 \cdot 3^x + 8 = 0.$

Let $y = 3^x$. Then $3^{2x} - 6 \cdot 3^x + 8 = 0 \Rightarrow$

$y^2 - 6y + 8 = 0 \Rightarrow (y - 4)(y - 2) = 0 \Rightarrow y = 2$

or $y = 4$. Substituting, we have $3^x = 2 \Rightarrow$

$x \ln 3 = \ln 2 \Rightarrow x = \dfrac{\ln 2}{\ln 3} \approx 0.631$ or $3^x = 4 \Rightarrow$

$x \ln 3 = \ln 4 \Rightarrow x = \dfrac{\ln 4}{\ln 3} \approx 1.262.$

The solution set is $\{0.631, 1.262\}$.

49. $3^{3x} - 4 \cdot 3^{2x} + 2 \cdot 3^x = 8$

$3^{3x} - 4 \cdot 3^{2x} + 2 \cdot 3^x - 8 = 0$

$3^{2x}\left(3^x - 4\right) + 2\left(3^x - 4\right) = 0$

$\left(3^{2x} + 2\right)\left(3^x - 4\right) = 0$

Now check each possible solution in the

$3^{2x} + 2 = 0 \Rightarrow 3^{2x} = -2 \Rightarrow$ there is no

solution.

$3^x - 4 = 0 \Rightarrow 3^x = 4 \Rightarrow x \ln 3 = \ln 4 \Rightarrow$

$x = \dfrac{\ln 4}{\ln 3} \approx 1.262$

Solution set: $\left\{\dfrac{\ln 4}{\ln 3} \approx 1.262\right\}$

51. $e^{2x} - 2e^x - 3 = 0$

$\left(e^x - 3\right)\left(e^x + 1\right) = 0$

$e^x - 3 = 0 \Rightarrow e^x = 3 \Rightarrow x \ln e = \ln 3 \Rightarrow x = \ln 3$

$e^x + 1 = 0 \Rightarrow e^x = -1 \Rightarrow$ there is no solution.

Solution set: $\{\ln 3 \approx 1.099\}$

53. $\dfrac{3^x - 3^{-x}}{3^x + 3^{-x}} = \dfrac{1}{4} \Rightarrow 4 \cdot 3^x - 4 \cdot 3^{-x} = 3^x + 3^{-x} \Rightarrow$

$3 \cdot 3^x - 5 \cdot 3^{-x} = 0 \Rightarrow 3^x\left(3 \cdot 3^x - 5 \cdot 3^{-x} = 0\right) \Rightarrow$

$3 \cdot 3^{2x} - 5 \cdot 3^0 = 0 \Rightarrow 3 \cdot 3^{2x} = 5 \Rightarrow 3^{2x} = \dfrac{5}{3} \Rightarrow$

$2x \ln 3 = \ln\left(\dfrac{5}{3}\right) \Rightarrow$

$x = \dfrac{\ln(5/3)}{2 \ln 3} = \dfrac{\ln 5 - \ln 3}{2 \ln 3} \approx 0.232$

55. $\dfrac{4}{2 + 3^x} = 1 \Rightarrow 4 = 2 + 3^x \Rightarrow 2 = 3^x \Rightarrow$

$\ln 2 = x \ln 3 \Rightarrow x = \dfrac{\ln 2}{\ln 3} \approx 0.631$

57. $\dfrac{17}{5 - 3^x} = 7 \Rightarrow 17 = 35 - 7 \cdot 3^x \Rightarrow$

$7 \cdot 3^x = 18 \Rightarrow \ln 7 + x \ln 3 = \ln 18 \Rightarrow$

$x = \dfrac{\ln 18 - \ln 7}{\ln 3} \approx 0.860$

59. $\dfrac{5}{2+3^x} = 4 \Rightarrow 5 = 4(2+3^x) \Rightarrow$

$5 = 8 + 4 \cdot 3^x \Rightarrow -\dfrac{3}{4} = 3^x \Rightarrow$ there is no solution.

61. $3 + \log(2x+5) = 2 \Rightarrow \log(2x+5) = -1 \Rightarrow$

$10^{-1} = 2x+5 \Rightarrow -\dfrac{49}{10} = 2x \Rightarrow x = -\dfrac{49}{20}$

63. $\log\left(x^2 - x - 5\right) = 0 \Rightarrow x^2 - x - 5 = 10^0 \Rightarrow$

$x^2 - x - 6 = 0 \Rightarrow (x-3)(x+2) = 0 \Rightarrow x = -2$
or $x = 3$

65. $\log_4\left(x^2 - 7x + 14\right) = 1 \Rightarrow x^2 - 7x + 14 = 4^1 \Rightarrow$

$x^2 - 7x + 10 = (x-2)(x-5) \Rightarrow x = 2$ or $x = 5$

67. $\ln(2x-3) - \ln(x+5) = 0 \Rightarrow$

$\ln(2x-3) = \ln(x+5) \Rightarrow 2x-3 = x+5 \Rightarrow x = 8$

69. $\log x + \log(x+9) = 1 \Rightarrow \log\left(x(x+9)\right) = 1 \Rightarrow$

$x^2 + 9x = 10 \Rightarrow x^2 + 9x - 10 = 0 \Rightarrow$
$(x+10)(x-1) = 0 \Rightarrow x = -10$ or $x = 1$.
Reject the negative solution because the
logarithm of a negative number is undefined.
The solution is $\{1\}$.

71. $\log_a(5x-2) - \log_a(3x+4) = 0 \Rightarrow$

$\log_a \dfrac{5x-2}{3x+4} = 0 \Rightarrow \dfrac{5x-2}{3x+4} = a^0 = 1 \Rightarrow$
$5x - 2 = 3x + 4 \Rightarrow x = 3$

73. $\log_6(x+2) + \log_6(x-3) = 1 \Rightarrow$

$\log_6\left((x+2)(x-3)\right) = 1 \Rightarrow x^2 - x - 6 = 6 \Rightarrow$
$x^2 - x - 12 = 0 \Rightarrow (x-4)(x+3) = 0 \Rightarrow$
$x = 4$ or $x = -3$.
Reject the negative solution because the
logarithm of a negative number is undefined.
The solution is $\{4\}$.

75. $\log_3(2x-7) - \log_3(4x-1) = 2 \Rightarrow$

$\log_3 \dfrac{2x-7}{4x-1} = 2 \Rightarrow \dfrac{2x-7}{4x-1} = 3^2 = 9 \Rightarrow$

$2x - 7 = 36x - 9 \Rightarrow 2 = 34x \Rightarrow x = \dfrac{1}{17}$.

$2\left(\dfrac{1}{17}\right) - 7 = -\dfrac{117}{17}$, so there is no solution.

77. $\log_7 3x + \log_7(2x-1) = \log_7(16x-10) \Rightarrow$

$\log_7\left(3x(2x-1)\right) = \log_7(16x-10) \Rightarrow$
$6x^2 - 3x = 16x - 10 \Rightarrow 6x^2 - 19x + 10 = 0 \Rightarrow$
$(2x-5)(3x-2) = 0 \Rightarrow x = \dfrac{5}{2}$ or $x = \dfrac{2}{3}$

The solution is $\{1\}$.

79. $f(x) = 20 + a \cdot 2^{kx};\ f(0) = 50;\ f(1) = 140$

$50 = 20 + a \cdot 2^{k \cdot 0} \Rightarrow 50 = 20 + a \Rightarrow a = 30$
$140 = 20 + 30 \cdot 2^{k \cdot 1} \Rightarrow 4 = 2^k \Rightarrow k = 2$
$f(2) = 20 + 30 \cdot 2^{2 \cdot 2} = 20 + 480 = 500$

81. $f(x) = 16 + a \cdot 3^{kx};\ f(0) = 21;\ f(4) = 61$

$21 = 16 + a \cdot 3^{k \cdot 0} \Rightarrow a = 5$
$61 = 16 + 5 \cdot 3^{k \cdot 4} \Rightarrow 9 = 3^{4k} \Rightarrow 3^2 = 3^{4k} \Rightarrow$
$2 = 4k \Rightarrow k = \dfrac{1}{2}$
$f(2) = 16 + 5 \cdot 3^{\frac{1}{2} \cdot 2} = 16 + 15 = 31$

83. $f(x) = \dfrac{10}{3 + ae^{kx}};\ f(0) = 2;\ f(1) = \dfrac{1}{2}$

$2 = \dfrac{10}{3 + ae^{k \cdot 0}} \Rightarrow 2 = \dfrac{10}{3 + a} \Rightarrow 6 + 2a = 10 \Rightarrow$
$2a = 4 \Rightarrow a = 2$
$\dfrac{1}{2} = \dfrac{10}{3 + 2e^{k \cdot 1}} \Rightarrow \dfrac{1}{2} = \dfrac{10}{3 + 2e^k} \Rightarrow$
$3 + 2e^k = 20 \Rightarrow e^k = \dfrac{17}{2} \Rightarrow k = \ln\dfrac{17}{2}$
$f(2) = \dfrac{10}{3 + 2e^{2\left(\ln\frac{17}{2}\right)}} \approx 0.068 = \dfrac{4}{59}$

85. $f(x) = \dfrac{4}{a + 4e^{kx}};\ f(0) = 2;\ f(1) = 9$

$2 = \dfrac{4}{a + 4e^{k \cdot 0}} = \dfrac{4}{a + 4} \Rightarrow 2a + 8 = 4 \Rightarrow a = -2$
$9 = \dfrac{4}{-2 + 4e^{k \cdot 1}} = \dfrac{4}{-2 + 4e^k} \Rightarrow$
$-18 + 36e^k = 4 \Rightarrow e^k = \dfrac{22}{36} = \dfrac{11}{18} \Rightarrow k = \ln\dfrac{11}{18}$
$f(2) = \dfrac{4}{-2 + 4e^{2\ln(11/18)}} \approx -7.902$

87. $5(0.3)^x + 1 \le 11 \Rightarrow (0.3)^x \le 2 \Rightarrow$

$x \ln 0.3 \le \ln 2 \Rightarrow x \ge \dfrac{\ln 2}{\ln 0.3}$ (Note that $\ln 0.3 < 0$)

89. $-3(1.2)^x + 11 \ge 8 \Rightarrow -3(1.2)^x \ge -3 \Rightarrow$

$1.2^x \le 1 \Rightarrow x \ln 1.2 \le \ln 1 \Rightarrow x \le 0$

91. Note that the domain of $\log(5x+15)$ is $(-3, \infty)$

since $5x + 15$ must be greater than 0.
$\log(5x+15) < 2 \Rightarrow 5x + 15 < 10^2 \Rightarrow$
$5x < 85 \Rightarrow x < 17$
Solution set: $(-3, 17)$

93. Note that the domain of $\ln(x-5)$ is $(5, \infty)$ since $x-5$ must be greater than 0.
$$\ln(x-5) \geq 1 \Rightarrow x-5 \geq e \Rightarrow x \geq e+5$$
Solution set: $[e+5, \infty)$

95. Note that the domain of $\log_2(3x-7)$ is $\left(\frac{7}{3}, \infty\right)$ since $3x-7$ must be greater than 0.
$$\log_2(3x-7) < 3 \Rightarrow 3x-7 < 2^3 \Rightarrow 3x < 15 \Rightarrow x < 5$$
Solution set: $\left(\frac{7}{3}, 5\right)$

97. $1+4e^{2x} > 9 \Rightarrow 4e^{2x} > 8 \Rightarrow e^{2x} > 2 \Rightarrow$
$$2x\ln e > \ln 2 \Rightarrow 2x > \ln 2 \Rightarrow x > \frac{\ln 2}{2}$$

99. $\dfrac{50}{1+2e^{-x}} \geq 10 \Rightarrow 50 \geq 10\left(1+2e^{-x}\right) \Rightarrow$
$$50 \geq 10 + 20e^{-x} \Rightarrow 40 \geq 20e^{-x} \Rightarrow$$
$$2 \geq e^{-x} \Rightarrow \ln 2 \geq -x\ln e \Rightarrow \ln 2 \geq -x \Rightarrow$$
$$x \geq -\ln 2$$

4.4 Applying the Concepts

101. $f(x) = 0.885 - 0.14\ln x$

   a. $f(1) = 0.885 - 0.14\ln 1 \approx 0.885 \approx 89\%$

   The relative risk of mortality associated with exercising one hour per week is about 89%, so the reduced risk is 100% − 89% = 11%.

   b. If the risk of mortality is reduced by 20%, then it is 80% or 0.80.
$$0.885 - 0.14\ln x = 0.80$$
$$-0.14\ln x = -0.085$$
$$\ln x = \frac{0.085}{0.14}$$
$$x = e^{0.085/0.14} \approx 1.835$$
   To reduce the risk of mortality by 20%, a person should exercise about 1.835 hours or 1 hour, 50 minutes per week.

103. a. $18,000 = 10,000\left(1+\dfrac{0.06}{1}\right)^{(1)t} \Rightarrow$
$$1.8 = 1.06^t \Rightarrow t = \frac{\ln 1.8}{\ln 1.06} \approx 10.087 \text{ years}$$

   b. $18,000 = 10,000\left(1+\dfrac{0.06}{4}\right)^{4t} \Rightarrow$
$$1.8 = 1.015^{4t} \Rightarrow 4t = \frac{\ln 1.8}{\ln 1.015} \Rightarrow t \approx 9.870 \text{ yr}$$

   c. $18,000 = 10,000\left(1+\dfrac{0.06}{12}\right)^{12t} \Rightarrow$
$$1.8 = 1.005^{12t} \Rightarrow 12t = \frac{\ln 1.8}{\ln 1.005} \Rightarrow$$
$$t \approx 9.821 \text{ years}$$

   d. $18,000 = 10,000\left(1+\dfrac{0.06}{365}\right)^{365t} \Rightarrow$
$$1.8 = \left(1+\frac{0.06}{365}\right)^{365t} \Rightarrow$$
$$365t = \frac{\ln 1.8}{\ln\left(1+\dfrac{0.06}{365}\right)} \Rightarrow t \approx 9.797 \text{ years}$$

   e. $18,000 = 10,000e^{0.06t} \Rightarrow 1.8 = e^{0.06t} \Rightarrow$
$$\ln 1.8 = 0.06t \Rightarrow t \approx 9.796 \text{ years}$$

105. $40,000 = 20,000e^{8r} \Rightarrow 2 = e^{8r} \Rightarrow$
$$\ln 2 = 8r \Rightarrow r = \frac{\ln 2}{8} \approx 0.0866 = 8.66\%$$
$$c_5 = 20,000e^{5(0.0866)} \approx \$30,837.52$$
The car will cost \$30,837.52 in five years. The annual rate of inflation is about 8.66%.

107. a. $\log\left(\dfrac{I}{12}\right) = -0.025(30) \Rightarrow$
$$\frac{I}{12} = 10^{-0.025(30)} \Rightarrow I \approx 2.134 \text{ lumens}$$

   b. $\log\left(\dfrac{4}{12}\right) = -0.025x \Rightarrow x \approx 19.08 \text{ feet}$

109. a. First, find a by substituting $t = 0$ and
$$f(0) = 1000: 1000 = \frac{20,000}{1+ae^{-k(0)}} \Rightarrow$$
$$1000(1+a) = 20,000 \Rightarrow 1+a = 20 \Rightarrow a = 19.$$
Now find k using $t = 4$ and $f(t) = 8999$.
$$8999 = \frac{20,000}{1+19e^{-k(4)}} \Rightarrow$$
$$8999 + 170,981e^{-4k} = 20,000 \Rightarrow$$
$$e^{-4k} = \frac{11,001}{170,981} \Rightarrow -4k = \ln\left(\frac{11,001}{170,981}\right) \Rightarrow$$
$$k \approx 0.68589$$
Use this value of k to find the number of people infected:
$$f(8) = \frac{20,000}{1+19e^{-0.68589(8)}} \approx 18,542$$

b. $12,400 = \dfrac{20,000}{1+19e^{-0.68589t}} \Rightarrow$

$12,400 + 12,400 \cdot 19e^{-0.68589t} = 20,000 \Rightarrow$

$12,400 \cdot 19e^{-0.68589t} = 7600 \Rightarrow$

$e^{-0.68589t} = \dfrac{7600}{12,400 \cdot 19} \Rightarrow$

$-0.68589t = \ln\left(\dfrac{7600}{12,400 \cdot 19}\right) \Rightarrow$

$t = \dfrac{\ln\left(\dfrac{7600}{12,400 \cdot 19}\right)}{-0.68589} \approx 5$

12,400 people will be infected after 5 weeks.

111. a. $\dfrac{4490}{1+e^{5.4094-1.0255(0)}} \approx 20$

b. The carrying capacity is the numerator, 4490.

c.

113. $14.7e^{-0.21h} \le 8 \Rightarrow e^{-0.21h} \le \dfrac{8}{14.7} \Rightarrow$

$-0.21h \ln e \le \ln\left(\dfrac{8}{14.7}\right) \Rightarrow$

$-0.21h \le \ln 8 - \ln 14.7 \Rightarrow h \ge \dfrac{\ln 8 - \ln 14.7}{-0.21} \Rightarrow$

$h \ge 2.9$

P is at most 8 psi for $h \ge 2.9$ miles.

4.4 Beyond the Basics

115. $P = \dfrac{M}{1+e^{-kt}} \Rightarrow P\left(1+e^{-kt}\right) = M \Rightarrow$

$P + Pe^{-kt} = M \Rightarrow \dfrac{M-P}{P} = e^{-kt} \Rightarrow$

$\dfrac{M-P}{P} = \dfrac{1}{e^{kt}} \Rightarrow \dfrac{P}{M-P} = e^{kt} \Rightarrow$

$\ln\left(\dfrac{P}{M-P}\right) = kt \Rightarrow t = \dfrac{1}{k}\ln\left(\dfrac{P}{M-P}\right)$

117. a. $\dfrac{\log x}{2} = \dfrac{\log y}{3} = \dfrac{\log z}{5} = k \Rightarrow$

$\log x = 2k \Rightarrow x = 10^{2k}.$

$\log y = 3k \Rightarrow y = 10^{3k}.$

$\log z = 5k \Rightarrow z = 10^{5k}.$

So, $xy = 10^{2k} \cdot 10^{3k} = 10^{5k} = z.$

b. $\dfrac{\log x}{2} = \dfrac{\log y}{3} = \dfrac{\log z}{5} \Rightarrow$

$3\log x = 2\log y \Rightarrow \log x^3 = \log y^2 \Rightarrow$

$x^3 = y^2.$

$5\log x = 2\log z \Rightarrow \log x^5 = \log z^2 \Rightarrow$

$x^5 = z^2.$

So, $y^2 z^2 = x^3 x^5 = x^8$

119. $(\log x)^2 = \log x \Rightarrow (\log x)^2 - \log x = 0 \Rightarrow$

$\log x(\log x - 1) = 0 \Rightarrow \log x = 0 \Rightarrow x = 1$ or

$\log x - 1 = 0 \Rightarrow \log x = 1 \Rightarrow x = 10$

121. $(\log_3 x)(\log_3 3x) = 2 \Rightarrow$

$(\log_3 x)(\log_3 3 + \log_3 x) = 2 \Rightarrow$

$(\log_3 x)(1 + \log_3 x) = 2$

Let $u = \log_3 x.$ Then we have

$u(1+u) = 2 \Rightarrow u^2 + u - 2 = 0 \Rightarrow$

$(u+2)(u-1) = 0 \Rightarrow u = -2, 1 \Rightarrow$

$\log_3 x = -2 \Rightarrow x = 3^{-2} = \dfrac{1}{9}$ or

$\log_3 x = 1 \Rightarrow x = 3^1 = 3$

123. $\log_4 x^2(x-1)^2 - \log_2(x-1) = 1 \Rightarrow$

$\dfrac{\log_2 x^2(x-1)^2}{\log_2 4} - \log_2(x-1) = 1 \Rightarrow$

$\dfrac{\log_2 [x(x-1)]^2}{2} - \log_2(x-1) = 1 \Rightarrow$

$\dfrac{2\log_2 [x(x-1)]}{2} - \log_2(x-1) = 1 \Rightarrow$

$\log_2 [x(x-1)] - \log_2(x-1) = 1 \Rightarrow$

$\log_2 \dfrac{x(x-1)}{x-1} = 1 \Rightarrow \log_2 x = 1 \Rightarrow x = 2.$

125. $\dfrac{\log(3x-5)}{2} = \log x \Rightarrow \log(3x-5) = 2\log x \Rightarrow \log(3x-5) = \log x^2 \Rightarrow 3x-5 = x^2 \Rightarrow$

$x^2 - 3x + 5 = 0$

$x = \dfrac{-(-3) \pm \sqrt{(-3)^2 - 4(1)(5)}}{2(1)} = \dfrac{3 \pm \sqrt{9-20}}{2} = \dfrac{3 \pm \sqrt{-11}}{2}$

Since logarithms are not defined for complex numbers, there is no solution. Solution set: $\varnothing$

127. $f(x) = y = 3^x + 5$

Interchange x and y, then solve for y.

$x = 3^y + 5 \Rightarrow x - 5 = 3^y \Rightarrow \log_3(x-5) = y = f^{-1}(x)$

129. $f(x) = y = 3 \cdot 4^x + 7$

Interchange x and y, then solve for y.

$x = 3 \cdot 4^y + 7 \Rightarrow \dfrac{x-7}{3} = 4^y \Rightarrow \log_4\left(\dfrac{x-7}{3}\right) = y = f^{-1}(x)$

131. $f(x) = y = 1 + \log_2(x-1)$

Interchange x and y, then solve for y.

$x = 1 + \log_2(y-1) \Rightarrow x - 1 = \log_2(y-1) \Rightarrow 2^{x-1} = y - 1 \Rightarrow 2^{x-1} + 1 = y = f^{-1}(x)$

133. $f(x) = y = \dfrac{1}{2}\ln\left(\dfrac{x-1}{x+1}\right)$

Interchange x and y, then solve for y.

$x = \dfrac{1}{2}\ln\left(\dfrac{y-1}{y+1}\right) \Rightarrow 2x = \ln\left(\dfrac{y-1}{y+1}\right) \Rightarrow e^{2x} = \dfrac{y-1}{y+1} \Rightarrow (y+1)e^{2x} = y - 1 \Rightarrow$

$ye^{2x} + e^{2x} = y - 1 \Rightarrow e^{2x} + 1 = y - ye^{2x} \Rightarrow e^{2x} + 1 = y\left(1 - e^{2x}\right) \Rightarrow \dfrac{1 + e^{2x}}{1 - e^{2x}} = y = f^{-1}(x)$

135. $7^n > 43^{67} \Rightarrow n\ln 7 > 67\ln 43 \Rightarrow n > \dfrac{67\ln 43}{\ln 7} \Rightarrow n > 129.5$

Thus, the smallest integer for which $7^n > 43^{67}$ is 130.

137. $8^{1/n} < 1.01 \Rightarrow \dfrac{1}{n}\ln 8 < \ln 1.01 \Rightarrow \dfrac{\ln 8}{\ln 1.01} < n \Rightarrow 208.98 < n$

Thus, the smallest integer for which $8^{1/n} < 1.01$ is 209.

139. If 31^n has 567 digits, then $566 \le \log 31^n < 567$. (See Section 4.3, Example 4.)

$566 \le \log 31^n < 567 \Rightarrow 566 \le n\log 31 < 567 \Rightarrow \dfrac{566}{\log 31} \le n < \dfrac{567}{\log 31} \Rightarrow 379.5 \le n < 380.2$

Thus, $n = 380$.

141. $\dfrac{x^{\log y}}{x^{\log z}} \cdot \dfrac{y^{\log z}}{y^{\log x}} \cdot \dfrac{z^{\log x}}{z^{\log y}} = \dfrac{\left(10^{\log x}\right)^{\log y}}{\left(10^{\log x}\right)^{\log z}} \cdot \dfrac{\left(10^{\log y}\right)^{\log z}}{\left(10^{\log y}\right)^{\log x}} \cdot \dfrac{\left(10^{\log z}\right)^{\log x}}{\left(10^{\log z}\right)^{\log y}} = \dfrac{10^{\log x \log y}}{10^{\log x \log z}} \cdot \dfrac{10^{\log y \log z}}{10^{\log x \log y}} \cdot \dfrac{10^{\log x \log z}}{10^{\log y \log z}}$

$= \dfrac{10^{\log x \log y + \log y \log z + \log x \log z}}{10^{\log x \log z + \log x \log y + \log y \log z}} = 1$

4.4 Critical Thinking/Discussion/Writing

143. $\dfrac{P}{2} = \dfrac{P}{1+ae^{-kt}} \Rightarrow \dfrac{1}{2} = \dfrac{1}{1+ae^{-kt}} \Rightarrow$

$1+ae^{-kt} = 2 \Rightarrow ae^{-kt} = 1 \Rightarrow e^{-kt} = \dfrac{1}{a} = a^{-1} \Rightarrow$

$-kt = \ln\left(a^{-1}\right) = -\ln a \Rightarrow t = \dfrac{\ln a}{k}$

144. a. $\log_4 (x-1)^2 = 3 \Rightarrow (x-1)^2 = 4^3 = 64 \Rightarrow$

$x^2 - 2x - 63 = 0 \Rightarrow (x+7)(x-9) = 0 \Rightarrow$

$x = -7,\, 9$

b. $2\log_4(x-1) = 3 \Rightarrow \log_4(x-1) = \dfrac{3}{2} \Rightarrow$

$x - 1 = 4^{3/2} = 8 \Rightarrow x = 9$

c. $2\log_4|x-1| = 3 \Rightarrow \log_4|x-1| = \dfrac{3}{2} \Rightarrow$

$|x-1| = 4^{3/2} = 8 \Rightarrow x - 1 = 8 \Rightarrow x = 9$ or
$x - 1 = -8 \Rightarrow x = -7$

The three equations do not have identical solutions because the equation in (a) has a quadratic term, the equation in (b) has a linear term, and the equation in (c) has an absolute value.

4.4 Getting Ready for the Next Section

145. $9^2 = 81 \Rightarrow \log_9 81 = 2$

147. $2 \cdot 10^x + 1 = 7 \Rightarrow 2 \cdot 10^x = 6 \Rightarrow 10^x = 3 \Rightarrow$
$\log 3 = x$

149. $\log_2 64 = 6 \Rightarrow 2^6 = 64$

151. $\log\left(\dfrac{A}{2}\right) = 3 \Rightarrow 10^3 = \dfrac{A}{2} \Rightarrow A = 2 \cdot 10^3$

153. $\log_5 x = -3 \Rightarrow x = 5^{-3} = \dfrac{1}{125}$

155. $\log_x 1000 = 3 \Rightarrow 1000 = x^3 \Rightarrow 10^3 = x^3 \Rightarrow$
$x = 10$

157. $2^{x+1} = 32 \Rightarrow 2^{x+1} = 2^5 \Rightarrow x+1 = 5 \Rightarrow x = 4$

159. $3^{x+1} = 5^{2x-3}$
$(x+1)\ln 3 = (2x-3)\ln 5$
$x\ln 3 + \ln 3 = 2x\ln 5 - 3\ln 5$
$\ln 3 + 3\ln 5 = 2x\ln 5 - x\ln 3$
$\ln 3 + 3\ln 5 = x(2\ln 5 - \ln 3)$
$x = \dfrac{\ln 3 + 3\ln 5}{2\ln 5 - \ln 3}$

4.5 Logarithmic Scales; Modeling

4.5 Practice Problems

1. a. $\text{pH} = -\log\left[H^+\right] = -\log\left(2.68 \times 10^{-6}\right)$
$= -\left(\log 2.68 + \log 10^{-6}\right)$
$= -\log 2.68 + 6 \approx 5.57$

b. $8.47 = -\log\left[H^+\right]$
$-8.47 = \log\left[H^+\right]$
$\left[H^+\right] = 10^{-8.47} = 10^{0.53} \times 10^{-9} \approx 3.39 \times 10^{-9}$

2. $2.8 = \text{pH}_{\text{acid rain}} = -\log\left[H^+\right]_{\text{acid rain}} \Rightarrow$
$\left[H^+\right]_{\text{acid rain}} = 10^{-2.8}$
$6.2 = \text{pH}_{\text{ordinary rain}} = -\log\left[H^+\right]_{\text{ordinary rain}} \Rightarrow$
$\left[H^+\right]_{\text{ordinary rain}} = 10^{-6.2}$

$\dfrac{\left[H^+\right]_{\text{acid rain}}}{\left[H^+\right]_{\text{ordinary rain}}} = \dfrac{10^{-2.8}}{10^{-6.2}} = 10^{-2.8-(-6.2)}$

$= 10^{3.4} \approx 2512$

This acid rain is about 2512 times more acidic than the ordinary rain.

3. $M = \log\left(\dfrac{I}{I_0}\right)$

$6.5 = \log\left(\dfrac{I}{I_0}\right) \Rightarrow \left(\dfrac{I}{I_0}\right) = 10^{6.5} \Rightarrow$

$I = 10^{6.5} I_0 \approx 3{,}162{,}278 I_0$

4. Let I_M denote the intensity of the Mozambique earthquake and let I_C denote the intensity of the southern California earthquake. Then we have

$7.0 = \log\left(\dfrac{I_M}{I_0}\right) \Rightarrow \dfrac{I_M}{I_0} = 10^{7.0} \Rightarrow I_M = 10^{7.0} I_0$

$5.2 = \log\left(\dfrac{I_C}{I_0}\right) \Rightarrow \dfrac{I_C}{I_0} = 10^{5.2} \Rightarrow I_C = 10^{5.2} I_0$

$\dfrac{I_M}{I_C} = \dfrac{10^{7.0} I_0}{10^{5.2} I_0} = \dfrac{10^{7.0}}{10^{5.2}} = 10^{7.0-5.2} = 10^{1.8} \approx 63.1$

The intensity of the Mozambique earthquake was about 63 times that of the southern California earthquake.

5. Let M_{2011} and E_{2011} represent the magnitude and energy of the 2011 Japan earthquake. Let M_{1997} and E_{1997} represent the magnitude and energy of the 1977 Iran earthquake. From Table 4.7, we have $M_{2011} = 9.0$ and $M_{1997} = 7.5$.

$$\frac{E_{2011}}{E_{1997}} = 10^{1.5(M_{2011}-M_{1997})} = 10^{1.5(9.0-7.5)} = 10^{2.25}$$
$$\approx 177.8$$

The energy released by the 2011 Japan earthquake was about 178 times that released by the 1997 Iran earthquake.

6. $I = 200 \times 10^{-7} \text{ W/m}^2$ and $I_0 = 10^{-12} \text{ W/m}^2$.

$$L = 10\log\left(\frac{I}{I_0}\right) = 10\log\left(\frac{200 \times 10^{-7}}{10^{-12}}\right)$$
$$= 10\log\left(200 \times 10^5\right) = 10\left[\log 200 + \log 10^5\right]$$
$$= 10\left[\log 200 + 5\log 10\right] = 10\left[\log 200 + 5\right]$$
$$= 10\log 200 + 50 \approx 73.01$$

The decibel level is approximately 73 dB.

7. $I = I_0 \times 10^{L/10}$
$$I_{75} = I_0 \times 10^{75/10}$$
$$I_{55} = I_0 \times 10^{55/10}$$
$$\frac{I_{75}}{I_{55}} = \frac{I_0 \times 10^{75/10}}{I_0 \times 10^{55/10}} = 10^{20/10} = 10^2 = 100$$

A 75 dB sound is 100 times more intense than a 55 dB sound.

8. $I = I_0 \times 10^{L/10}$
$$I_{48} = 10^{-12} \times 10^{48/10} = 10^{-12} \times 10^{4.8} = 10^{-7.2}$$
$$= 10^{0.8} \times 10^{-8} \approx 6.3 \times 10^{-8} \text{ W/m}^2$$

The intensity of a 48 dB sound is about $6.3 \times 10^{-8} \text{ W/m}^2$.

9. B is two semitones above A, so $P(f) = 200$.
$$f_0 = 440 \text{ Hz}$$
$$P(f) = 1200\log_2\frac{f}{f_0}$$
$$200 = 1200\log_2\frac{f}{440} \Rightarrow \frac{1}{6} = \log_2\frac{f}{440} \Rightarrow$$
$$2^{1/6} = \frac{f}{440} \Rightarrow f = 2^{1/6} \cdot 440 \approx 493.9$$

The frequency of B is about 494 Hz.

10. If the reference frequency is f_0 and it increases by 25%, then
$$f = f_0 + 0.25 f_0 = 1.25 f_0. \text{ Then,}$$
$$P(f) = 1200\log_2\frac{1.25 f_0}{f_0}$$
$$= 1200\log_2 1.25 = 1200\frac{\log 1.25}{\log 2}$$
$$\approx 386.3$$

Since an equal tempered major third is 400 cents, the difference is about $400 - 386.3 = 13.7$ cents, or a little less than 14 cents.

11. a. $m_2 - m_1 = 2.5\log\left(\frac{b_1}{b_2}\right)$
$$2 - 0 = \frac{5}{2}\log\left(\frac{b_1}{b_2}\right) \Rightarrow \frac{4}{5} = \log\left(\frac{b_1}{b_2}\right) \Rightarrow$$
$$\frac{b_1}{b_2} = 10^{4/5} \approx 6.3$$

A magnitude 0 star is approximately 6.3 times brighter than a magnitude 2 star.

b. Let $m_1 = 4.6$ and $b_2 = 150 b_1$.
$$m_2 - m_1 = 2.5\log\left(\frac{b_1}{b_2}\right)$$
$$m_2 - 4.6 = 2.5\log\left(\frac{b_1}{1.50 b_1}\right) = 2.5\log\left(\frac{1}{1.50}\right)$$
$$= 2.5\log 1.50^{-1} = -2.5\log 1.50$$
$$m_2 = -2.5\log 1.50 + 4.6 \approx 4.1598$$

The magnitude of the star that is 50% brighter than a star of magnitude 4.6 is about 4.1598.

12. Set x as the elapsed time and y as the temperature inside the car. Using a graphing calculator, the logarithmic regression function is $y = 72.88 + 11.66\ln x$.

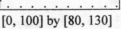

[0, 100] by [80, 130]   [0, ln 100] by [80, 130]

The graph on the left shows a scatterplot of the data along with the logarithmic regression function. The graph on the right shows a scatter-plot of the transformed data along with the linear regression, $Y = 72.88 + 11.66X$ for the transformed data $Y = y$ and $X = \ln x$. From the scatterplot of the transformed data, we see that the logarithmic model is a good choice.

13. Set x as the body weight and y as the metabolism rate. Using a graphing calculator, the power regression function is
$$y = 2.00x^{0.75}.$$

[0, 50] by [0, 50]   [0, ln 50] by [0, ln 50]

The graph on the left shows a scatterplot of the data along with the power regression function. The graph on the right shows a scatter-plot of the transformed data along with the linear regression, $Y = 0.69 + 0.75X$ for the transformed data $Y = \ln y$ and $X = \ln x$. From the scatterplot of the transformed data, we see that the power model is a good choice. Convert the linear function to the power function: $y = e^{0.69}x^{0.75} \approx 1.99x^{0.75}$, which is close to the power regression function determined with the calculator.

4.5 Concepts and Vocabulary

1. On the Richter scale, the magnitude of an earthquake $M = \underline{\log \frac{I}{I_0}}$. The energy E released by an earthquake of magnitude M is given by $\log E = \underline{4.4 + 1.5M}$.

3. If f_0 is a reference frequency and f is the frequency of a note, then the change in pitch $P(f) = \underline{1200\log_2\left(\frac{f}{f_0}\right)}$.

5. False. If the pH value of a solution is more than 7, the solution is alkaline.

7. True

4.5 Building Skills

9. $\text{pH} = -\log\left[\text{H}^+\right] = -\log\left(10^{-8}\right) = 8\log 10 = 8$
The substance is a base.

11. $\text{pH} = -\log\left[\text{H}^+\right] = -\log\left(2.3 \times 10^{-5}\right)$
$= -\left(\log 2.3 + \log 10^{-5}\right)$
$= -\left(\log 2.3 - 5\log 10\right) = -\left(\log 2.3 - 5\right)$
$= -\log 2.3 + 5 \approx 4.64$
The substance is an acid.

13. $\text{pH} = 6 = -\log\left[\text{H}^+\right] \Rightarrow \left[\text{H}^+\right] = 10^{-6}$
$\left[\text{H}^+\right]\left[\text{OH}^-\right] = 10^{-14}$
$10^{-6}\left[\text{OH}^-\right] = 10^{-14}$
$\left[\text{OH}^-\right] = \frac{10^{-14}}{10^{-6}} = 10^{-8}$

15. $\text{pH} = 9.5 = -\log\left[\text{H}^+\right] \Rightarrow \left[\text{H}^+\right] = 10^{-9.5} \Rightarrow$
$\left[\text{H}^+\right] = 10^{0.5} \times 10^{-10} \approx 3.16 \times 10^{-10}$
$\left[\text{H}^+\right]\left[\text{OH}^-\right] = 10^{-14}$
$\left(3.16 \times 10^{-10}\right)\left[\text{OH}^-\right] = 10^{-14}$
$\left[\text{OH}^-\right] = \frac{10^{-14}}{3.16 \times 10^{-10}} = \frac{1}{3.16} \times 10^{-4}$
$\approx 0.316 \times 10^{-4} = 3.16 \times 10^{-5}$

17. a. $M = \log\left(\frac{I}{I_0}\right)$
$5 = \log\left(\frac{I}{I_0}\right) \Rightarrow \left(\frac{I}{I_0}\right) = 10^5 \Rightarrow I = 10^5 I_0$

 b. $\log E = 4.4 + 1.5M$
$\log E = 4.4 + 1.5(5) = 11.9 \Rightarrow$
$E = 10^{11.9} = 10^{0.9} \times 10^{11}$
$\approx 7.94 \times 10^{11}$ joules

19. a. $M = \log\left(\frac{I}{I_0}\right)$
$7.8 = \log\left(\frac{I}{I_0}\right) \Rightarrow \left(\frac{I}{I_0}\right) = 10^{7.8} \Rightarrow$
$I = 10^{0.8} \times 10^7 I_0 \approx 6.3 \times 10^7 I_0$

 b. $\log E = 4.4 + 1.5M$
$\log E = 4.4 + 1.5(7.8) = 16.1 \Rightarrow$
$E = 10^{16.1} = 10^{0.1} \times 10^{16}$
$\approx 1.26 \times 10^{16}$ joules

21. a. $\log E = 4.4 + 1.5M$
$\log 10^{13.4} = 4.4 + 1.5M \Rightarrow 13.4 = 4.4 + 1.5M$
$M = \frac{13.4 - 4.4}{1.5} = \frac{9}{1.5} = 6$

 b. $M = \log\left(\frac{I}{I_0}\right)$
$6 = \log\left(\frac{I}{I_0}\right) \Rightarrow \left(\frac{I}{I_0}\right) = 10^6 \Rightarrow I = 10^6 I_0$

23. a.
$$\log E = 4.4 + 1.5M$$
$$\log 10^{12} = 4.4 + 1.5M \Rightarrow 12 = 4.4 + 1.5M$$
$$M = \frac{12 - 4.4}{1.5} = \frac{7.6}{1.5} \approx 5.1$$

b.
$$M = \log\left(\frac{I}{I_0}\right)$$
$$5.1 = \log\left(\frac{I}{I_0}\right) \Rightarrow \left(\frac{I}{I_0}\right) = 10^{5.1}$$
$$I = 10^{5.1} I_0 = 10^{0.1} \times 10^5 I_0 \approx 1.26 \times 10^5 I_0$$

25.
$$L = 10\log\left(\frac{I}{I_0}\right) = 10\log\left(\frac{10^{-8}}{10^{-12}}\right) = 10\log 10^4$$
$$= 40$$

27.
$$L = 10\log\left(\frac{I}{I_0}\right) = 10\log\left(\frac{3.5 \times 10^{-7}}{10^{-12}}\right)$$
$$= 10\log\left(3.5 \times 10^5\right) = 10\left(\log 3.5 + \log 10^5\right)$$
$$= 10\log 3.5 + 50 \approx 55.4$$

29.
$$I = I_0 \times 10^{L/10} = 10^{-12} \times 10^{80/10}$$
$$= 10^{-12} \times 10^8 = 10^{-4}$$

31.
$$I = I_0 \times 10^{L/10} = 10^{-12} \times 10^{64.7/10}$$
$$= 10^{-12} \times 10^{6.47} = 10^{-12} \times 10^6 \times 10^{0.47}$$
$$\approx 2.95 \times 10^{-6}$$

33. A# is one semitone above A, so $P(f) = 100$.

$f_0 = 440$ Hz
$$P(f) = 1200\log_2 \frac{f}{f_0}$$
$$100 = 1200\log_2 \frac{f}{440} \Rightarrow \frac{1}{12} = \log_2 \frac{f}{440} \Rightarrow$$
$$2^{1/12} = \frac{f}{440} \Rightarrow f = 2^{1/12} \cdot 440 \approx 466 \text{ Hz}$$

The frequency of A# is about 466 Hz.

C is three semitones above A, so $P(f) = 300$.

$f_0 = 440$ Hz
$$P(f) = 1200\log_2 \frac{f}{f_0}$$
$$300 = 1200\log_2 \frac{f}{440} \Rightarrow \frac{1}{4} = \log_2 \frac{f}{440} \Rightarrow$$
$$2^{1/4} = \frac{f}{440} \Rightarrow f = 2^{1/4} \cdot 440 \approx 523 \text{ Hz}$$

The frequency of C is about 523 Hz.

35.
$$P(f) = 1200\log_2 \frac{f}{f_0} = 1200\log_2\left(\frac{10}{9}\right)$$
$$= 1200\left(\frac{\log\left(\frac{10}{9}\right)}{\log 2}\right) \approx 182$$

The difference, $200 - 182 = 18$ cents, is noticeable.

37. Let the magnitude and brightness of star A be m_1 and b_1, and let the magnitude and brightness of star B be m_2 and b_2.
$$m_2 - m_1 = 2.5\log\left(\frac{b_1}{b_2}\right)$$
$$20 - 4 = 2.5\log\left(\frac{b_1}{b_2}\right) \Rightarrow 16 = 2.5\log\left(\frac{b_1}{b_2}\right) \Rightarrow$$
$$6.4 = \log\left(\frac{b_1}{b_2}\right) \Rightarrow \frac{b_1}{b_2} = 10^{6.4} \Rightarrow$$
$$b_1 = 10^{0.4} \times 10^6 b_2 \approx 2.5 \times 10^6 b_2$$

Star A is 2.5×10^6 times as bright as star B.

39. Let the magnitude and brightness of star A be m_1 and b_1, and let the magnitude of star B be $m_2 = m_1 - 2$ and b_2.
$$m_2 - m_1 = 2.5\log\left(\frac{b_1}{b_2}\right)$$
$$(m_1 - 2) - m_1 = 2.5\log\left(\frac{b_1}{b_2}\right) \Rightarrow$$
$$-2 = 2.5\log\left(\frac{b_1}{b_2}\right) \Rightarrow -0.8 = \log\left(\frac{b_1}{b_2}\right) \Rightarrow$$
$$\frac{b_1}{b_2} = 10^{-0.8} \approx \frac{1}{6.3} \Rightarrow 6.3b_1 = b_2$$

Star B is about 6.3 times brighter than star A.

4.5 Applying the Concepts

41.
$$\text{pH} = -\log\left[H^+\right] = -\log\left(3.98 \times 10^{-8}\right)$$
$$= -\left(\log 3.98 + \log 10^{-8}\right) = -\left(\log 3.98 - 8\right)$$
$$= -\log 3.98 + 8 \approx 7.4$$

The pH of human blood is about 7.4; it is basic.

43. a.
$$\text{pH} = 3.15 = -\log\left[H^+\right] \Rightarrow \left[H^+\right] = 10^{-3.15} \Rightarrow$$
$$\left[H^+\right] = 10^{0.85} \times 10^{-4} \approx 7.1 \times 10^{-4}$$

b.
$$\text{pH} = 7.2 = -\log\left[H^+\right] \Rightarrow \left[H^+\right] = 10^{-7.2} \Rightarrow$$
$$\left[H^+\right] = 10^{0.8} \times 10^{-8} \approx 6.3 \times 10^{-8}$$

c. $\text{pH} = 7.78 = -\log\left[H^+\right] \Rightarrow \left[H^+\right] = 10^{-7.78} \Rightarrow$
$\left[H^+\right] = 10^{0.22} \times 10^{-8} \approx 1.7 \times 10^{-8}$

d. $\text{pH} = 3 = -\log\left[H^+\right] \Rightarrow \left[H^+\right] = 10^{-3} \Rightarrow$
$\left[H^+\right] = 10^{-3}$

45. $\text{pH}_{\text{acid rain}} = 3.8 = -\log\left[H^+\right] \Rightarrow$
$\left[H^+\right] = 10^{-3.8} \Rightarrow$
$\left[H^+\right] = 10^{0.2} \times 10^{-4} \approx 1.6 \times 10^{-4}$

The average concentration of hydrogen ions in the acid rain is 1.6×10^{-4} moles per liter.

$\text{pH}_{\text{ordinary rain}} = 6 = -\log\left[H^+\right] \Rightarrow$
$\left[H^+\right] = 10^{-6}$

$\dfrac{\left[H^+\right]_{\text{acid rain}}}{\left[H^+\right]_{\text{ordinary rain}}} = \dfrac{1.6 \times 10^{-4}}{10^{-6}} = 1.6 \times 10^2 = 160$

The acid rain is 160 times more acidic than ordinary rain.

47. $\left[H^+_A\right] = 100\left[H^+_B\right]$
$\text{pH}_A = -\log\left[H^+_A\right] = -\log\left(100\left[H^+_B\right]\right)$
$= -\left(\log 100 + \log\left[H^+_B\right]\right)$
$= -\left(2 + \log\left[H^+_B\right]\right) = -\log\left[H^+_B\right] - 2$
$= \text{pH}_B - 2$

49. $\left[H^+_{\text{new}}\right] = 50\left[H^+_{\text{original}}\right]$
$\text{pH}_{\text{new}} = -\log\left(50\left[H^+_{\text{original}}\right]\right)$
$= -\log 50 - \log\left[H^+_{\text{original}}\right]$
$= -\log 50 - \text{pH}_{\text{original}}$

The pH decreases by $\log 50 \approx 1.7$.
The solution becomes more acidic.

51. a. $M = \log\left(\dfrac{I}{I_0}\right)$

$7.8 = \log\left(\dfrac{I}{I_0}\right) \Rightarrow \left(\dfrac{I}{I_0}\right) = 10^{7.8} \Rightarrow$
$I = 10^{7.8} I_0 = 10^{0.8} \times 10^7 I_0 \approx 6.31 \times 10^7 I_0$

b. $\log E = 4.4 + 1.5(7.8) = 16.1 \Rightarrow$
$E = 10^{16.1} = 10^{0.1} \times 10^{16} \approx 1.259 \times 10^{16} \text{ j}$

53. a. $M_A = M_B + 1$. Then we have
$M_A = M_B + 1 = \log\left(\dfrac{I_A}{I_0}\right) \Rightarrow$

$10^{(M_B+1)} = \dfrac{I_A}{I_0} \Rightarrow 10^{M_B} \times 10 = \dfrac{I_A}{I_0} \Rightarrow$

$10^{M_B} \times 10 I_0 = I_A \quad (1)$

$M_B = \log\left(\dfrac{I_B}{I_0}\right) \Rightarrow 10^{M_B} = \dfrac{I_B}{I_0} \Rightarrow$

$10^{M_B} \times I_0 = I_B$

Using equation (1), we have
$10^{M_B} \times 10 I_0 = I_A \Rightarrow 10^{M_B} \times I_0 \times 10 = I_A \Rightarrow$
$I_B \times 10 = I_A$

The intensity of the earthquake A was 10 times that of earthquake B.

b. $\dfrac{E_A}{E_B} = 10^{1.5(M_A - M_B)} = 10^{1.5(M_B + 1 - M_B)}$

$\qquad = 10^{1.5} \approx 31.6 \Rightarrow E_A \approx 31.6 E_B$

The energy released by the earthquake A was about 31.6 times that released by earthquake B.

55. $M = \log\left(\dfrac{I}{I_0}\right)$

$I_A = 150 I_B \Rightarrow$

$\log\left(\dfrac{I_A}{I_0}\right) = \log\left(\dfrac{150 I_B}{I_0}\right) = \log\left(150 \cdot \dfrac{I_B}{I_0}\right)$

$\qquad = \log 150 + \log\left(\dfrac{I_B}{I_0}\right) = \log 150 + M_B$

The difference in Richter scale readings is $\log 150 \approx 2.18$.

57. $L = 10\log\left(\dfrac{I}{I_0}\right) = 10\log\left(\dfrac{5.2 \times 10^{-5}}{10^{-12}}\right)$
$= 10\log\left(5.2 \times 10^7\right) = 10\left[\log 5.2 + \log 10^7\right]$
$= 10\left[\log 5.2 + 7\right] = 10\log 5.2 + 70 \approx 77.2 \text{ dB}$

59. $I = I_0 \times 10^{L/10}$
$I_{130} = I_0 \times 10^{130/10}$
$I_{65} = I_0 \times 10^{65/10}$
$\dfrac{I_{130}}{I_{65}} = \dfrac{I_0 \times 10^{130/10}}{I_0 \times 10^{65/10}} = 10^{65/10} = 10^{6.5}$
$\qquad = 10^{0.5} \times 10^6 \approx 3.16 \times 10^6$

The 130 dB sound is about 3.16×10^6 times as intense as the 65 dB sound.

61. A sound at the threshold of pain has intensity $10\,\text{W/m}^2$. A sound that is 1000 times as intense has intensity $10^4\,\text{W/m}^2$.

$$L = 10\log\left(\frac{I}{I_0}\right) = 10\log\left(\frac{10^4}{10^{-12}}\right) = 10\log 10^{16}$$
$$= 160\text{ dB}$$

63. $L_2 = L_1 + 1$

$I_{L_2} = I_0 \times 10^{L_2/10} = I_0 \times 10^{0.1(L_1+1)}$

$I_{L_1} = I_0 \times 10^{L_1/10} = I_0 \times 10^{0.1L_1}$

$\dfrac{I_{L_2}}{I_{L_1}} = \dfrac{I_0 \times 10^{0.1(L_1+1)}}{I_0 \times 10^{0.1L_1}} = 10^{0.1(L_1+1-L_1)}$ The

$= 10^{0.1} \approx 1.26 \Rightarrow I_{L_2} \approx 1.26 \times I_{L_1}$

intensity of the louder sound is about 1.26 times the intensity of the softer sound.

65. $P(f) = 1200\log_2\dfrac{f}{f_0}$

$P(441) = 1200\log_2\left(\dfrac{441}{440}\right) = 1200\left(\dfrac{\log\left(\frac{441}{440}\right)}{\log 2}\right)$

$\approx 3.93 \approx 4$

The difference in pitch is about 4 cents higher.

67. Let m_1 represent the magnitude of the sun and let m_2 represent the magnitude of the full Moon.

$m_2 - m_1 = 2.5\log\left(\dfrac{b_1}{b_2}\right)$

$-13 - (-27) = 2.5\log\left(\dfrac{b_1}{b_2}\right) \Rightarrow$

$\dfrac{14}{2.5} = 5.6 = \log\left(\dfrac{b_1}{b_2}\right) \Rightarrow$

$\dfrac{b_1}{b_2} = 10^{5.6} = 10^{0.6} \times 10^5 \approx 3.98 \times 10^5 \Rightarrow$

$b_1 = 3.98 \times 10^5 b_2$

The sun is about 3.98×10^5 times brighter than the full Moon.

69. Let m_1 represent the magnitude of the Sun and let m_2 represent the magnitude of Venus.

$m_2 - m_1 = 2.5\log\left(\dfrac{b_1}{b_2}\right)$

$-4 - (-27) = 2.5\log\left(\dfrac{b_1}{b_2}\right)$

$\dfrac{23}{2.5} = 9.2 = \log\left(\dfrac{b_1}{b_2}\right)$

$\dfrac{b_1}{b_2} = 10^{9.2} = 10^{0.2} \times 10^9 \approx 1.58 \times 10^9 \Rightarrow$

$b_1 = 1.58 \times 10^9 b_2$

The Sun is about 1.58×10^9 times brighter than Venus.

71. Let m_1 and b_1 represent the magnitude and brightness of the unknown star and let m_2 and b_2 represent the magnitude and brightness of Saturn.

$m_2 - m_1 = 2.5\log\left(\dfrac{b_1}{b_2}\right)$; $b_1 = 560b_2$

$1.47 - m_1 = 2.5\log\left(\dfrac{560b_2}{b_2}\right) \Rightarrow$

$1.47 - m_1 = 2.5\log 560 \Rightarrow$

$1.47 - 2.5\log 560 = m_1 \Rightarrow -5.4 \approx m_1$

The magnitude of the unknown star is about -5.4.

73. $y = 25.72 \cdot 1.6667^x = 25.72e^{x\ln 1.6667}$
$= 25.72e^{0.511x}$

75. $y = 66.654 - 10.367\ln x$

4.5 Beyond the Basics

77. $L = 10\log\left(\dfrac{4 \times 10^4}{r^2}\right)$

  a. $r = 10$ ft

$L = 10\log\left(\dfrac{4 \times 10^4}{10^2}\right) \approx 26$ dB

$r = 25$ ft

$L = 10\log\left(\dfrac{4 \times 10^4}{25^2}\right) \approx 18$ dB

$r = 50$ ft

$L = 10\log\left(\dfrac{4 \times 10^4}{50^2}\right) \approx 12$ dB

$r = 100$ ft

$L = 10\log\left(\dfrac{4 \times 10^4}{100^2}\right) \approx 6$ dB

b. $0 = 10\log\left(\dfrac{4\times10^4}{r^2}\right) \Rightarrow 0 = \log\left(\dfrac{4\times10^4}{r^2}\right) \Rightarrow$

$10^0 = 1 = \dfrac{4\times10^4}{r^2} \Rightarrow r^2 = 40,000 \Rightarrow$

$r = 200$ ft

The decibel level drops to 0 when a listener is 200 ft away.

c. $L = 10\log\left(\dfrac{4\times10^4}{r^2}\right)$

$= 10\left(\log 4 + \log(10^4) - \log(r^2)\right)$

$= 10\left((4 + \log 4) - 2\log r\right)$

$= 40 + 10\log 4 - 20\log r$

$a = 40 + 10\log 4 = 10(4 + \log 4)$

$b = -20$

d. $L = 10\log\left(\dfrac{4\times10^4}{r^2}\right) = 10\log\left(\dfrac{2\times10^2}{r}\right)^2 \Rightarrow$

$L = 20\log\left(\dfrac{200}{r}\right) \Rightarrow \dfrac{L}{20} = \log\left(\dfrac{200}{r}\right) \Rightarrow$

$10^{L/20} = \dfrac{200}{r} \Rightarrow r = \dfrac{200}{10^{L/20}}$

79. The energy released by the earthquake is given by

$E = (2.5\times10^4)\times10^{1.5(8)} = 2.5\times10^{16}$ joules.

$\dfrac{E_{\text{earthquake}}}{E_{\text{bomb}}} = \dfrac{2.5\times10^{16}}{5\times10^{15}} = 0.5\times10 = 5$

The earthquake released five times as much energy as the nuclear bomb did.

81. $10 = 1200\log_2\dfrac{f}{880} \Rightarrow \dfrac{1}{120} = \log_2\dfrac{f}{880} \Rightarrow$

$2^{1/120} = \dfrac{f}{880} \Rightarrow f = 880\times2^{1/120} \approx 885$

$-10 = 1200\log_2\dfrac{f}{880} \Rightarrow -\dfrac{1}{120} = \log_2\dfrac{f}{880} \Rightarrow$

$2^{-1/120} = \dfrac{f}{880} \Rightarrow f = 880\times2^{-1/120} \approx 875$

The frequency range is [875, 885].

83. Let m_1 and b_1 represent the magnitude and brightness of the unknown star and let m_2 and b_2 represent the magnitude and brightness of the given star.

$m_2 - m_1 = 2.5\log\left(\dfrac{b_1}{b_2}\right);\ b_2 = 176b_1$

$m_2 - 3.42 = 2.5\log\left(\dfrac{b_1}{176b_1}\right) \Rightarrow$

$m_2 - 3.42 = 2.5\log\left(\dfrac{1}{176}\right) \Rightarrow$

$m_2 = 2.5\log\left(\dfrac{1}{176}\right) + 3.42 \approx -2.2$

The magnitude of the unknown star is about −2.2.

4.5 Getting Ready for the Next Section

85. $3x + 2y = 7,\ y = \dfrac{1}{2}$

Substituting, we have

$3x + 2\left(\dfrac{1}{2}\right) = 7 \Rightarrow 3x + 1 = 7 \Rightarrow 3x = 6 \Rightarrow$

$x = 2$

87. $-4x + 7y = 7,\ x = 0$

Substituting, we have

$-4(0) + 7y = 7 \Rightarrow 7y = 7 \Rightarrow y = 1$

89. Rearrange the equation to put it in slope-intercept form.

$10x - 2y = 28 \Rightarrow -2y = -10x + 28 \Rightarrow$

$y = 5x - 14$

The slope is 5.

91. Parallel lines have the same slope. Rearrange the equation to put it in slope-intercept form.

$6x - 3y = 7 \Rightarrow -3y = -6x + 7 \Rightarrow$

$y = 3x - \dfrac{7}{3}$

The slope is 2. Now use the point-slope form to write the equation of the parallel line.

$y - (-1) = 2(x - 5) \Rightarrow y + 1 = 2(x - 5) \Rightarrow$

$y + 1 = 2x - 10 \Rightarrow 2x - y = 11$

93. Two equations have the same graph if the coefficients of one of the equations are multiples are the coefficients of the second equation. So any equation that has the same graph as the equation $2x - 4y = 12$ will have the form $2nx - 4ny = 12n$. We are given that $12n = 3$, so $n = \dfrac{1}{4}$. Therefore, $a = 2\cdot\dfrac{1}{4} = \dfrac{1}{2}$

and $b = -4\cdot\dfrac{1}{4} = -1$.

Chapter 4 Review Exercises

Concepts and Vocabulary

1. False. $f(x) = a^x$ is an exponential function if $a > 0$ and $a \neq 1$.

3. False. The domain of $f(x) = \log(2 - x)$ is $(-\infty, 2)$.

5. True

7. False. $\ln M + \ln N = \ln(MN)$

9. True

Building Skills

11. h   **13.** f

15. d   **17.** a

19. Domain: $(-\infty, \infty)$
range: $(0, \infty)$
asymptote: $y = 0$

21. Domain: $(-\infty, \infty)$
range: $(3, \infty)$
asymptote: $y = 3$

23. Domain: $(-\infty, \infty)$
range: $(0, 1]$
asymptote: $y = 0$

25. Domain: $(-\infty, 0)$
range: $(-\infty, \infty)$
asymptote: $x = 0$

27. Domain: $(1, \infty)$
range: $(-\infty, \infty)$
asymptote: $x = 1$

29. Domain: $(-\infty, 0)$
range: $(-\infty, \infty)$
asymptote: $x = 0$

31. a. y-intercept: $y = 3 - 2e^0 = 1$

x-intercept: $0 = 3 - 2e^{-x} \Rightarrow 2e^{-x} = 3 \Rightarrow$

$e^{-x} = \dfrac{3}{2} \Rightarrow e^x = \dfrac{2}{3} \Rightarrow x = \ln\left(\dfrac{2}{3}\right)$

b. As $x \to \infty$, $f(x) \to 3$.
As $x \to -\infty$, $f(x) \to -\infty$.

33. a. y-intercept: $y = e^{0^2} = 1$

x-intercept: $0 = e^{-x^2} \Rightarrow \ln 0 = -x^2 \Rightarrow$
there is no x-intercept.

b. As $x \to \infty$, $f(x) \to 0$.

As $x \to -\infty$, $f(x) \to 0$.

35. $10 = a(2^{0k}) \Rightarrow a = 10$

$640 = 10(2^{3k}) \Rightarrow 64 = 8^k \Rightarrow k = 2$

$f(2) = 10(2^{2(2)}) = 160$

37. $1 = \dfrac{4}{1 + ae^{-k(0)}} = \dfrac{4}{1+a} \Rightarrow 1 + a = 4 \Rightarrow a = 3$

$\dfrac{1}{2} = \dfrac{4}{1 + 3e^{-3k}} \Rightarrow 1 + 3e^{-3k} = 8 \Rightarrow e^{-3k} = \dfrac{7}{3} \Rightarrow$

$-3k = \ln\left(\dfrac{7}{3}\right) \Rightarrow k \approx -0.2824$

$f(4) = \dfrac{4}{1 + 3e^{-(-0.2824)(4)}} \approx 0.38898$

39. The graph passes through $(0, 3)$, so we have $3 = ca^0 \Rightarrow c = 3$. The graph passes through $(2, 12)$, so we have $12 = 3a^2 \Rightarrow 4 = a^2 \Rightarrow$ $a = 2$. The equation is $f(x) = 3 \cdot 2^x$.

41. The graph has been shifted right one unit, so the equation is of the form $y = \log_a(x-1)$.

The graph passes through the point $(6, 1)$, so we have

$1 = \log_a(6-1) \Rightarrow 1 = \log_a 5 \Rightarrow a = 5$.

Thus, the equation is $y = \log_5(x-1)$.

43. a. $y = -\left(2^{x-1} + 3\right) = -2^{x-1} - 3$

b. $y = -2^{x-1} + 3$

45. $\ln(xy^2z^3) = \ln x + 2\ln y + 3\ln z$

47. $\ln\left[\dfrac{x\sqrt{x^2+1}}{(x^2+3)^2}\right] = \ln x + \dfrac{1}{2}\ln(x^2+1) - 2\ln(x^2+3)$

49. $\ln y = \ln x + \ln 3 = \ln(3x) \Rightarrow y = 3x$

51. $\ln y = \ln(x-3) - \ln(y+2) \Rightarrow$

$\ln y + \ln(y+2) = \ln(x-3) \Rightarrow$

$\ln(y(y+2)) = \ln(x-3) \Rightarrow$

$\ln(y^2 + 2y) = \ln(x-3) \Rightarrow y^2 + 2y = x - 3 \Rightarrow$

$y^2 + 2y + 1 = x - 3 + 1 \Rightarrow (y+1)^2 = x - 2 \Rightarrow$

$y + 1 = \pm\sqrt{x-2} \Rightarrow y = -1 \pm \sqrt{x-2}$

We disregard the negative solution, so

$y = -1 + \sqrt{x-2}$.

53. $\ln y = \dfrac{1}{2}\ln(x-1) + \dfrac{1}{2}\ln(x+1) - \ln(x^2+1) \Rightarrow$

$\ln y = \ln\left(\dfrac{\sqrt{x-1}\sqrt{x+1}}{x^2+1}\right) = \ln\left(\dfrac{\sqrt{x^2-1}}{x^2+1}\right) \Rightarrow$

$y = \dfrac{\sqrt{x^2-1}}{x^2+1}$

55. $3^x = 81 \Rightarrow 3^x = 3^4 \Rightarrow x = 4$

57. $2^{x^2+2x} = 16 \Rightarrow 2^{x^2+2x} = 2^4 \Rightarrow x^2 + 2x = 4 \Rightarrow$

$x^2 + 2x - 4 = 0 \Rightarrow x = \dfrac{-2 \pm \sqrt{4+16}}{2} = -1 \pm \sqrt{5}$

59. $3^x = 23 \Rightarrow x\ln 3 = \ln 23 \Rightarrow x = \dfrac{\ln 23}{\ln 3} \approx 2.854$

61. $273^x = 19 \Rightarrow x\ln 273 = \ln 19 \Rightarrow$

$x = \dfrac{\ln 19}{\ln 273} \approx 0.525$

63. $3^{2x} = 7^x \Rightarrow 2x\ln 3 = x\ln 7 \Rightarrow$

$2x\ln 3 - x\ln 7 = 0 \Rightarrow x(2\ln 3 - \ln 7) = 0 \Rightarrow$

$x = 0$

65. $1.7^{3x} = 3^{2x-1} \Rightarrow 3x\ln 1.7 = (2x-1)\ln 3 \Rightarrow$

$3x\ln 1.7 = 2x\ln 3 - \ln 3 \Rightarrow$

$3x\ln 1.7 - 2x\ln 3 = -\ln 3 \Rightarrow$

$x(3\ln 1.7 - 2\ln 3) = -\ln 3 \Rightarrow$

$x = \dfrac{-\ln 3}{3\ln 1.7 - 2\ln 3} \approx 1.815$

67. $\log_3(x+2) - \log_3(x-1) = 1 \Rightarrow$

$\log_3\left(\dfrac{x+2}{x-1}\right) = 1 \Rightarrow \dfrac{x+2}{x-1} = 3 \Rightarrow$

$x + 2 = 3x - 3 \Rightarrow 2x = 5 \Rightarrow x = \dfrac{5}{2}$

69. $\log_2(x+2) + \log_2(x+4) = 3 \Rightarrow$

$\log_2[(x+2)(x+4)] = 3 \Rightarrow$

$x^2 + 6x + 8 = 2^3 \Rightarrow x^2 + 6x = 0 \Rightarrow$

$x(x+6) = 0 \Rightarrow x = -6, 0$

Reject -6 since $\log_2(-6+2) = \log_2(-4)$ is not defined. Solution set: $\{0\}$

71. $\log_5\left(x^2-5x+6\right)-\log_5\left(x-2\right)=1\Rightarrow$

$\log_5\left(\dfrac{x^2-5x+6}{x-2}\right)=1\Rightarrow\dfrac{x^2-5x+6}{x-2}=5\Rightarrow$

$x^2-5x+6=5x-10\Rightarrow x^2-10x+16=0\Rightarrow$

$(x-2)(x-8)=0\Rightarrow x=2,8$

If $x=2$, then $\log_5\left(x-2\right)=\log_5 0$, which is undefined, so reject $x=2$. Solution set: $\{8\}$

73. $\log_6\left(x-2\right)+\log_6\left(x+1\right)$

$\qquad=\log_6\left(x+4\right)+\log_6\left(x-3\right)\Rightarrow$

$\log_6\left[(x-2)(x+1)\right]=\log_6\left[(x+4)(x-3)\right]\Rightarrow$

$x^2-x-2=x^2+x-12\Rightarrow-2x=-10\Rightarrow x=5$

75. $2\ln 3x=3\ln x\Rightarrow\ln(3x)^2=\ln x^3\Rightarrow$

$9x^2=x^3\Rightarrow x=9$

77. Multipling both sides of $2^x-8\cdot 2^{-x}-7=0$ by 2^x we have $2^{2x}-8\cdot 2^{-x+x}-7\cdot 2^x=0\Rightarrow$

$2^{2x}-7\cdot 2^x-8=0$. Letting $u=2^x$, we have

$u^2-7u-8=0\Rightarrow(u-8)(u+1)=0\Rightarrow u=8$

or $u=-1$. (Reject this solution) $2^x=8\Rightarrow x=3$

79. $3\left(0.2^x\right)+5\le 20\Rightarrow 3\left(0.2^x\right)\le 15\Rightarrow$

$\left(0.2^x\right)\le 5\Rightarrow x\ln 0.2\le\ln 5\Rightarrow x\ge\dfrac{\ln 5}{\ln 0.2}\Rightarrow$

$x\ge-1$

81. Note that the domain of $\log\left(2x+7\right)$ is

$\left(-\dfrac{7}{2},\infty\right)$ since $2x+7$ must be greater than 0.

$\log\left(2x+7\right)<2\Rightarrow 2x+7<10^2\Rightarrow 2x<93\Rightarrow$

$x<\dfrac{93}{2}$

Solution set: $\left(-\dfrac{7}{2},\dfrac{93}{2}\right)$

Applying the Concepts

83. $2P=P\left(1+0.0625\right)^t\Rightarrow 2=1.0625^t\Rightarrow$

$\ln 2=t\ln 1.0625\Rightarrow t=\dfrac{\ln 2}{\ln 1.0625}\approx 11.4$

It will take about 11.4 years to double the investment.

85. Find k: $\dfrac{1}{2}=e^{20k}\Rightarrow\ln\left(\dfrac{1}{2}\right)=20k\Rightarrow$

$k=\dfrac{\ln\left(1/2\right)}{20}$

$0.25=e^{\frac{\ln(1/2)}{20}t}\Rightarrow\ln 0.25=\dfrac{\ln\left(1/2\right)}{20}t\Rightarrow$

$t=\dfrac{\ln 0.25}{\dfrac{\ln\left(1/2\right)}{20}}=40$

It will take 40 hours to decay to 25% of its original value.

87. At 5% compounded yearly,

$A=7000\left(1+\dfrac{0.05}{1}\right)^{7(1)}=9849.70$.

At 4.75% compounded monthly,

$A=7000\left(1+\dfrac{0.0475}{12}\right)^{7(12)}=9754.74$.

The 5% investment provides the greater return.

89. a. $t=10;P(10)=33e^{0.003(10)}\approx 34$ million

b. $60=33e^{0.003t}\Rightarrow\dfrac{60}{33}=e^{0.003t}\Rightarrow$

$\ln\left(\dfrac{60}{33}\right)=0.003t\Rightarrow t=\dfrac{\ln\left(\frac{60}{33}\right)}{0.003}\approx 199.3$ yr

The population will be 60 million sometime during the year 2206 (after 199.3 years).

91. a.

b. $0.029=0.3e^{-0.47t}\Rightarrow\dfrac{0.029}{0.3}=e^{-0.47t}\Rightarrow$

$\ln\left(\dfrac{0.029}{0.3}\right)=-0.47t\Rightarrow t=\dfrac{\ln\left(\frac{0.029}{0.3}\right)}{-0.47}\Rightarrow$

$t\approx 4.97\approx 5$ hours

93. First find a and k:

$f(0)=212=75+ae^{-k(0)}\Rightarrow 137=a$

$f(1)=200=75+137e^{-k(1)}\Rightarrow\dfrac{125}{137}=e^{-k}\Rightarrow$

$\ln(125/137)=-k\Rightarrow k\approx 0.0917$

$150=75+137e^{-0.0917t}\Rightarrow\dfrac{75}{137}=e^{-0.0917t}\Rightarrow$

$t=\dfrac{\ln(75/137)}{-0.0917}\approx 6.57$ minutes

95. a. $D(0) = 5e^{0.08(0)} \approx 5$ thousand $= 5000$ people per square mile

b. $D(5) = 5e^{0.08(5)} \approx 7.459$ thousand $= 7459$ people per square mile

c. $15 = 5e^{0.08t} \Rightarrow 3 = e^{0.08t} \Rightarrow$
$t = \dfrac{\ln 3}{0.08} \approx 13.73$ miles

97. $I = I_0 e^{-0.73(2)} \approx 0.2322 I_0$

99. a. $s(470,000) = 0.04 + 0.86\ln(470,000)$
≈ 11.27 feet per second

b. $s(450) = 0.04 + 0.86\ln(450)$
≈ 5.29 feet per second

c. $4.6 = 0.04 + 0.86\ln p \Rightarrow \dfrac{4.56}{0.86} = \ln p \Rightarrow$
$p = e^{4.56/0.86} \approx 200.8 \approx 201$ people

101. a. $m(0) = \dfrac{6}{1 + 5e^{-0.7(0)}} = \dfrac{6}{1+5} = 1$ gram

b. The mass approaches 6 grams

c. $5 = \dfrac{6}{1 + 5e^{-0.7t}} \Rightarrow 5 + 25e^{-0.7t} = 6 \Rightarrow$
$e^{-0.7t} = \dfrac{1}{25} \Rightarrow t = \dfrac{\ln\left(\dfrac{1}{25}\right)}{-0.7} \approx 4.6$ days

103. $P(f) = 1200\log_2 \dfrac{f}{f_0}$
$P(f) = 1200\log_2 \left(\dfrac{1.2 f_0}{f_0}\right) = 1200\log_2(1.2)$
$= 1200\left(\dfrac{\log 1.2}{\log 2}\right) \approx 315.6 \approx 316$

The difference, $316 - 300 = 16$ cents, is noticeable.

105. $pH_{\text{very acid rain}} = 2.4 = -\log\left[H^+\right] \Rightarrow$
$\left[H^+\right] = 10^{-2.4} \Rightarrow$
$\left[H^+\right] = 10^{0.6} \times 10^{-3} \approx 4.0 \times 10^{-3}$

The average concentration of hydrogen ions in the very acid rain is 4.0×10^{-3} moles per liter.
$pH_{\text{acid rain}} = 5.6 = -\log\left[H^+\right] \Rightarrow$
$\left[H^+\right] = 10^{-5.6} = 10^{0.4} \times 10^{-6} \approx 2.5 \times 10^{-6}$

$\dfrac{\left[H^+\right]_{\text{very acid rain}}}{\left[H^+\right]_{\text{acid rain}}} = \dfrac{10^{-2.4}}{10^{-5.6}} = 10^{3.2} = 10^{0.2} \times 10^3$
$\approx 1.585 \times 10^3 = 1585$

The very acid rain is 1585 times more acidic than acid rain.

107. $I = I_0 \times 10^{L/10}$
$I_{115} = I_0 \times 10^{115/10}$
$I_{95} = I_0 \times 10^{95/10}$
$\dfrac{I_{115}}{I_{95}} = \dfrac{I_0 \times 10^{115/10}}{I_0 \times 10^{95/10}} = 10^{20/10} = 10^2 = 100$

The 115 dB sound is 100 times as intense as the 95 dB sound.

109. Let m_1 represent the magnitude of Sirius and let m_2 represent the magnitude of Saturn.
$m_2 - m_1 = 2.5\log\left(\dfrac{b_1}{b_2}\right)$
$1.47 - (-1) = 2.5\log\left(\dfrac{b_1}{b_2}\right) \Rightarrow$
$\dfrac{2.47}{2.5} = 0.988 = \log\left(\dfrac{b_1}{b_2}\right) \Rightarrow$
$\dfrac{b_1}{b_2} = 10^{0.988} \approx 9.7 \Rightarrow b_1 = 9.7 b_2$

Sirius is about 9.7 times brighter than Saturn.

Chapter 4 Practice Test A

1. $5^{-x} = 125 \Rightarrow 5^{-x} = 5^3 \Rightarrow x = -3$

2. $\log_2 x = 5 \Rightarrow x = 2^5 = 32$

3. Range: $(-\infty, 1)$; asymptote: $y = 1$

4. $\log_2 \dfrac{1}{8} \Rightarrow 2^x = \dfrac{1}{8} = 2^{-3} \Rightarrow x = -3$

5. $\left(\dfrac{1}{4}\right)^{2-x} = 4 \Rightarrow 4^{x-2} = 4 \Rightarrow x - 2 = 1 \Rightarrow x = 3$

6. $\log 0.001 \Rightarrow 10^x = 0.001 = 10^{-3} \Rightarrow x = -3$

7. $\ln 3 + 5\ln x = \ln 3 + \ln x^5 = \ln(3x^5)$

8. $2^{x+1} = 5 \Rightarrow (x+1)\ln 2 = \ln 5 \Rightarrow$
$x\ln 2 + \ln 2 = \ln 5 \Rightarrow x\ln 2 = \ln 5 - \ln 2 \Rightarrow$
$x\ln 2 = \ln\left(\dfrac{5}{2}\right) \Rightarrow x = \dfrac{\ln(5/2)}{\ln 2}$

9. $e^{2x} + e^x - 6 = 0 \Rightarrow (e^x + 3)(e^x - 2) = 0 \Rightarrow$
$e^x = -3$ (reject this) or $e^x - 2 = 0 \Rightarrow$
$e^x = 2 \Rightarrow x = \ln 2$

10. $\ln \dfrac{2x^3}{(x+1)^5} = \ln(2x^3) - \ln(x+1)^5$
$= \ln 2 + 3 \ln x - 5 \ln(x+1)$

11. $\ln e^{-5} = -5$

12. $y = \ln(x-1) + 3$

13.

14. Domain: $(-\infty, 0)$

15. $3 \ln x + \ln(x^3 + 2) - \dfrac{1}{2} \ln(3x^2 + 2)$
$= \ln x^3 + \ln(x^3 + 2) - \ln \sqrt{3x^2 + 2}$
$= \ln \dfrac{x^3(x^3 + 2)}{\sqrt{3x^2 + 2}}$

16. $\log x = \log 6 - \log(x-1) \Rightarrow \log x = \log \dfrac{6}{x-1} \Rightarrow$
$x = \dfrac{6}{x-1} \Rightarrow x^2 - x = 6 \Rightarrow x^2 - x - 6 = 0 \Rightarrow$
$(x-3)(x+2) = 0 \Rightarrow x = -2$ (reject this) or $x = 3$

17. $\log_x 9 = 2 \Rightarrow x^2 = 9 \Rightarrow x = 3$

18. $A = 15{,}000\left(1 + \dfrac{0.07}{4}\right)^{4t} = 15{,}000(1.0175)^{4t}$

19. $1{,}500{,}000 = 15{,}000 e^{0.2t} \Rightarrow 100 = e^{0.2t} \Rightarrow$
$\ln 100 = 0.2t \Rightarrow t = \dfrac{\ln 100}{0.2} \approx 23.03$

The Hispanic population reached 1.5 million in 1983, 23.03 years after 1960.

20. Let I_{1960} denote the intensity of the 1960 Morocco earthquake and let I_{1972} denote the intensity of the 1972 Nicaragua earthquake. Then,

$5.8 = \log\left(\dfrac{I_{1960}}{I_0}\right) \Rightarrow \dfrac{I_{1960}}{I_0} = 10^{5.8} \Rightarrow$
$I_{1960} = 10^{5.8} I_0$
$6.2 = \log\left(\dfrac{I_{1972}}{I_0}\right) \Rightarrow \dfrac{I_{1972}}{I_0} = 10^{6.2} \Rightarrow$
$I_{1972} = 10^{6.2} I_0$
$\dfrac{I_{1972}}{I_{1960}} = \dfrac{10^{6.2} I_0}{10^{5.8} I_0} = 10^{0.4} \approx 2.5 \Rightarrow I_{1972} \approx 2.5 I_{1960}$

The intensity of the 1972 Nicaragua earthquake was about 2.5 times that of the 1960 Morocco earthquake.

Chapter 4 Practice Test B

1. $3^{-x} = 9 \Rightarrow 3^{-x} = 3^2 \Rightarrow x = -2$.
The answer is B.

2. $\log_5 x = 2 \Rightarrow 5^2 = 25 = x$. The answer is B.

3. The answer is B.

4. $\log_4 64 \Rightarrow 4^x = 64 = 4^3 \Rightarrow x = 3$.
The answer is D.

5. $\left(\dfrac{1}{3}\right)^{1-x} = 3 \Rightarrow 3^{x-1} = 3 \Rightarrow x - 1 = 1 \Rightarrow x = 2$.
The answer is D.

6. $\log 0.01 \Rightarrow 10^x = 10^{-2} \Rightarrow x = -2$.
The answer is B.

7. $\ln 7 + 2 \ln x = \ln(7x^2)$. The answer is B.

8. $2^{x^2} = 3^x \Rightarrow x^2 \ln 2 = x \ln 3 \Rightarrow$
$x^2 \ln 2 - x \ln 3 = 0 \Rightarrow x(x \ln 2 - \ln 3) = 0 \Rightarrow$
$x = 0$ or $x \ln 2 - \ln 3 = 0 \Rightarrow x = \dfrac{\ln 3}{\ln 2}$
The answer is C.

9. $e^{2x} - e^x - 6 = 0 \Rightarrow (e^x - 3)(e^x + 2) = 0 \Rightarrow$
$e^x = -2$ (reject this) or $e^x - 3 = 0 \Rightarrow$
$e^x = 3 \Rightarrow x = \ln 3$.
The answer is D.

10. $\ln \dfrac{3x^2}{(x+1)^{10}} = \ln(3x^2) - \ln(x+1)^{10}$
$= \ln 3 + 2 \ln x - 10 \ln(x+1)$.
The answer is D.

11. $\ln e^{3x} = 3x$. The answer is B.

12. The answer is D.

13. The answer is B.

14. The answer is A.

15. $\ln x - 2\ln(x^2+1) + \dfrac{1}{2}\ln(x^4+1)$

$\qquad = \ln x - \ln(x^2+1)^2 + \ln\sqrt{x^4+1}$

$\qquad = \ln \dfrac{x\sqrt{x^4+1}}{(x^2+1)^2}.$

The answer is A.

16. $\log x = \log 12 - \log(x+1) \Rightarrow \log x = \log\dfrac{12}{x+1} \Rightarrow$

$x = \dfrac{12}{x+1} \Rightarrow x^2 + x = 12 \Rightarrow x^2 + x - 12 = 0 \Rightarrow$

$(x+4)(x-3) = 0 \Rightarrow x = -4$ (reject this) or

$x = 3$

The answer is D.

17. $\log_x 16 = 4 \Rightarrow x^4 = 16 = 2^4 \Rightarrow x = 2$

The answer is B.

18. $A = 12,000\left(1 + \dfrac{0.105}{12}\right)^{12t} = 12,000(1.00875)^{12t}$

The answer is D.

19. $t = 2020 - 2000 = 20$

$P(20) = 10,000\log_5(20+5) = 10,000\log_5 25$

$\qquad = 20,000.$ The answer is B.

20. Let I_{1994} denote the intensity of the 1994 Northridge, CA earthquake and let I_{1988} denote the intensity of the 1988 Armenia earthquake. Then,

$6.7 = \log\left(\dfrac{I_{1994}}{I_0}\right) \Rightarrow \dfrac{I_{1994}}{I_0} = 10^{6.7} \Rightarrow$

$I_{1994} = 10^{6.7}I_0$

$7.0 = \log\left(\dfrac{I_{1988}}{I_0}\right) \Rightarrow \dfrac{I_{1988}}{I_0} = 10^{7.0} \Rightarrow$

$I_{1988} = 10^{7.0}I_0$

$\dfrac{I_{1988}}{I_{1994}} = \dfrac{10^{7.0}I_0}{10^{6.7}I_0} = 10^{0.3} \approx 2.0 \Rightarrow$

$I_{1988} \approx 2.0 I_{1994}$

The intensity of the 1988 Armenia earthquake was about twice that of the 1994 Northridge, CA earthquake. The answer is B.

Cumulative Review Exercises (Chapters P–4)

1. $x^2 + (0-1)^2 = 1 \Rightarrow x^2 = 0 \Rightarrow x = 0.$

$0^2 + (y-1)^2 = 1 \Rightarrow y^2 - 2y + 1 = 1 \Rightarrow$

$y^2 - 2y = 0 \Rightarrow y(y-2) = 0 \Rightarrow y = 0$ or $y = 2.$

The x-intercept is 0.

The y-intercepts are 0 and 2.

$x^2 + (y-1)^2 = 1 \Rightarrow (y-1)^2 = 1 - x^2 \Rightarrow$

$y - 1 = \pm\sqrt{1-x^2} \Rightarrow y = 1 \pm \sqrt{1-x^2} = f(x)$

$f(-x) = 1 \pm \sqrt{1-(-x)^2} = 1 \pm \sqrt{1-x^2} = f(x) \Rightarrow$

$f(x)$ is even $\Rightarrow f$ is symmetric about the y-axis.

2.

3. $x^2 + 2x + y^2 - 4y - 20 = 0$

Group the terms and complete the squares.

$\left(x^2 + 2x + 1\right) + \left(y^2 - 4y + 4\right) = 20 + 1 + 4$

$\qquad\qquad (x+1)^2 + (y-2)^2 = 25$

The center of the circle is $(-1, 2)$ and its radius is 5.

4. $2x + 3y = 17 \Rightarrow y = -\dfrac{2}{3}x + \dfrac{17}{3} \Rightarrow$ the slope of the perpendicular line is $\dfrac{3}{2}$.

$4 = \dfrac{3}{2}(-2) + b \Rightarrow 7 = b.$

The equation is $y = \dfrac{3}{2}x + 7.$

5. Logarithms are positive, so the domain is $(-\infty, -3) \cup (2, \infty)$.

6. a. $f(-2) = 2(-2) - 3 = -7$

    b. $f(0) = 2(0) - 3 = -3$

    c. $f(3) = 2(3^2) + 1 = 19$

7. Shift the graph of $f(x) = \sqrt{x}$ one unit left, stretch vertically by a factor of 3, then shift the graph two units down.

8. $(f \circ g)(x) = f\left(-\dfrac{1}{x}\right) = 3\sqrt{-\dfrac{1}{x}} \Rightarrow$ the domain is $(-\infty, 0)$.

9. a. The graph of $f(x)$ passes the horizontal line test, so there is an inverse. $y = 2x^3 + 1$

    becomes $x = 2y^3 + 1 \Rightarrow \dfrac{x-1}{2} = y^3 \Rightarrow$

    $y = \sqrt[3]{\dfrac{x-1}{2}} = f^{-1}(x)$.

    b. The graph of $f(x)$ does not pass the horizontal line test, so there is no inverse.

    c. The graph of $f(x)$ passes the horizontal line test, so there is an inverse. $y = \ln x$

    becomes $x = \ln y \Rightarrow e^x = y = f^{-1}(x)$.

10. a. $x^2 + 1 \overline{)2x^3 + 0x^2 + 3x + 1}$
    $\underline{2x^3 + 0x^2 + 2x}$
    $x + 1$

    The quotient is $2x + \dfrac{x+1}{x^2+1}$.

    b. $\begin{array}{r|rrrrr} -3 & 2 & 7 & 0 & -2 & 3 \\ & & -6 & -3 & 9 & -21 \\ \hline & 2 & 1 & -3 & 7 & -18 \end{array}$

$\dfrac{2x^4 + 7x^3 - 2x + 3}{x+3}$
$= 2x^3 + x^2 - 3x + 7 - \dfrac{18}{x+3}$

11. a. $(x-1)(x-2)(x+3) = x^3 - 7x + 6$

    b. $(x-1)(x+1)(x-i)(x+i) = x^4 - 1$

12. a. $\log_2 5x^3 = \log_2 5 + 3\log_2 x$

    b. $\log_a \sqrt[3]{\dfrac{xy^2}{z}} = \dfrac{1}{3}\log_a \dfrac{xy^2}{z}$
    $= \dfrac{1}{3}(\log_a x + 2\log_a y - \log_a z)$

    c. $\ln \dfrac{3\sqrt{x}}{5y} = \ln 3\sqrt{x} - \ln 5y$
    $= \ln 3 + \dfrac{1}{2}\ln x - (\ln 5 + \ln y)$
    $= \ln 3 + \dfrac{1}{2}\ln x - \ln 5 - \ln y$

13. a. $\dfrac{1}{2}(\log x + \log y) = \dfrac{1}{2}\log(xy) = \log\sqrt{xy}$

    b. $3\ln x - 2\ln y = \ln x^3 - \ln y^2 = \ln \dfrac{x^3}{y^2}$

14. a. $\log_9 17 = \dfrac{\log 17}{\log 9} \approx 1.289$

    b. $\log_3 25 = \dfrac{\log 25}{\log 3} \approx 2.930$

    c. $\log_{1/2} 0.3 = \dfrac{\log 0.3}{\log(1/2)} \approx 1.737$

15. a. $\log_6(x+3) + \log_6(x-2) = 1 \Rightarrow$
    $\log_6((x+3)(x-2)) = 1 \Rightarrow$
    $(x+3)(x-2) = 6 \Rightarrow x^2 + x - 6 = 6 \Rightarrow$
    $x^2 + x - 12 = 0 \Rightarrow (x+4)(x-3) = 0 \Rightarrow$
    $x = -4$ (reject this) or $x = 3$

    b. $5^{x^2 - 4x + 5} = 25 = 5^2 \Rightarrow x^2 - 4x + 5 = 2 \Rightarrow$
    $x^2 - 4x + 3 = 0 \Rightarrow (x-3)(x-1) = 0 \Rightarrow$
    $x = 1$ or $x = 3$

    c. $3.1^{x-1} = 23 \Rightarrow (x-1)\ln 3.1 = \ln 23 \Rightarrow$
    $x - 1 = \dfrac{\ln 23}{\ln 3.1} \Rightarrow x = 1 + \dfrac{\ln 23}{\ln 3.1} \approx 3.771$

16. The possible rational zeros are $\{\pm 1, \pm 2, \pm 4\}$.

Using synthetic division, we find that one zero is 1:

```
1|  1  -1  -2  -2   4
         1   0  -2  -4
    _____
    1   0  -2  -4   0
```

Using synthetic division again, we find that 2 is a root of the depressed equation

$f(x) = x^3 - 2x - 4$:

```
2|  1   0  -2  -4
        2   4   4
    _____
    1   2   2   0
```

There are no rational zeros of the depressed equation $f(x) = x^2 + 2x + 2$. So the only rational zeros are $\{1, 2\}$.

Find the upper bound for the zeros by testing 1, 2, 3,... using synthetic division. The smallest value that makes all the terms in the bottom row nonnegative is the upper bound. The upper bound is 3.

```
3|  1  -1  -2  -2   4
        3   6  12  30
    _____
    1   2   4  10  34
```

Find the lower bound for the zeros by testing –1, –2, –3,...using synthetic division. The largest value that makes the terms in the bottom row alternate signs is the lower bound. The lower bound is –1.

```
-1|  1  -1  -2  -2   4
         -1   2   0   2
     _____
     1  -2   0  -2   6
```

17. a. The zeros are 1 (multiplicity 3), –2 (multiplicity 2), and 3 (multiplicity 1).

b. At $x = 1$ and $x = 3$, the graph crosses the x-axis. At $x = -2$, the graph touches, but does not cross, the x-axis.

c. As $x \to -\infty$, $f(x) \to \infty$.

As $x \to \infty$, $f(x) \to \infty$.

d.

18. a. The vertical asymptotes are $x = 3$ and $x = -4$.

b. $f(x) = \dfrac{(x-1)(x+2)}{(x-3)(x+4)} = \dfrac{x^2 + x - 2}{x^2 + x - 12} \Rightarrow$ the horizontal asymptote is $y = 1$.

c. $f(x)$ lies above the horizontal asymptote on $(-\infty, -4) \cup (3, \infty)$. $f(x)$ lies below the horizontal asymptote on $(-4, 3)$.

d.

Chapter 5 Systems of Equations and Inequalities

5.1 Systems of Linear Equations in Two Variables

5.1 Practice Problems

1. a. Check $(2, 2)$.

Equation (1)	Equation (2)
$x + y = 4$	$3x - y = 0$
$\overset{?}{2 + 2 = 4}$	$\overset{?}{3(2) - 2 = 0}$
$4 = 4$ ✓	$4 = 0$ ✗

Because $(2, 2)$ does not satisfy both equations, it is not a solution of the system.

b. Check $(1, 3)$.

Equation (1)	Equation (2)
$x + y = 4$	$3x - y = 0$
$\overset{?}{1 + 3 = 4}$	$\overset{?}{3(1) - 3 = 0}$
$4 = 4$ ✓	$0 = 0$ ✓

Because $(1, 3)$ satisfies both equations, it is a solution of the system.

2. The solution is $\{(-1, 3)\}$.

$$\begin{cases} x + y = 2 \\ 4x + y = -1 \end{cases}$$

3. $\begin{cases} x - y = 5 \quad (1) \\ 2x + y = 7 \quad (2) \end{cases}$

Solve the equation (1) for x, then substitute that expression into equation (2) and solve for y.

$x - y = 5 \Rightarrow x = y + 5$

$2x + y = 7 \Rightarrow 2(y + 5) + y = 7 \Rightarrow$

$3y + 10 = 7 \Rightarrow 3y = -3 \Rightarrow y = -1$

Now substitute the value for y into equation (1) and solve for x.

$x - y = 5 \Rightarrow x - (-1) = 5 \Rightarrow x = 4$

The solution is $\{(4, -1)\}$.

4. $\begin{cases} x - 3y = 1 \quad (1) \\ -2x + 6y = 3 \quad (2) \end{cases}$

Solve the equation (1) for x, then substitute that expression into the equation (2) and solve for y.

$x - 3y = 1 \Rightarrow x = 1 + 3y$

$-2x + 6y = 3 \Rightarrow -2(1 + 3y) + 6y = 3 \Rightarrow$

$-2 - 6y + 6y = 3 \Rightarrow -2 = 3$

Since the equation $-2 = 3$ is false, the system is inconsistent. The solution set is $\varnothing$.

5. $\begin{cases} -2x + y = -3 \quad (1) \\ 4x - 2y = 6 \quad (2) \end{cases}$

Solve the equation (1) for y, then substitute that expression into equation (2) and solve for x.

$-2x + y = -3 \Rightarrow y = 2x - 3$

$4x - 2y = 6 \Rightarrow 4x - 2(2x - 3) = 6 \Rightarrow$

$4x - 4x + 6 = 6 \Rightarrow 6 = 6$

The equation $6 = 6$ is true for every value of x. Thus, any value of x can be used in the equation $y = 2x - 3$. The solutions of the system are of the form $\{(x, 2x - 3)\}$.

6. $\begin{cases} 3x + 2y = 3 \quad (1) \\ 9x - 4y = 4 \quad (2) \end{cases}$

Multiply the equation (1) by 2, then add the two equations.

$\begin{cases} 3x + 2y = 3 \\ 9x - 4y = 4 \end{cases} \Rightarrow \begin{cases} 6x + 4y = 6 \\ 9x - 4y = 4 \end{cases} \Rightarrow 15x = 10 \Rightarrow$

$x = \dfrac{2}{3}$

Now substitute $x = \dfrac{2}{3}$ into the equation (2) and solve for y:

$9x - 4y = 4 \Rightarrow 9\left(\dfrac{2}{3}\right) - 4y = 4 \Rightarrow 6 - 4y = 4 \Rightarrow$

$-4y = -2 \Rightarrow y = \dfrac{1}{2}$

The solution is $\left\{\left(\dfrac{2}{3}, \dfrac{1}{2}\right)\right\}$.

7. $\begin{cases} \dfrac{4}{x} + \dfrac{3}{y} = 1 \\ \dfrac{2}{x} - \dfrac{6}{y} = 3 \end{cases}$

Let $u = \dfrac{1}{x}$ and $v = \dfrac{1}{y}$. Then the equations

become $\begin{cases} 4u + 3v = 1 \\ 2u - 6v = 3 \end{cases}$. Multiply the first equation

by 2, then add the equations:

$\begin{cases} 4u + 3v = 1 \\ 2u - 6v = 3 \end{cases} \Rightarrow \begin{cases} 8u + 6v = 2 \\ 2u - 6v = 3 \end{cases} \Rightarrow 10u = 5 \Rightarrow u = \dfrac{1}{2}$

Substitute this value into the first equation to solve for v:

$4\left(\dfrac{1}{2}\right) + 3v = 1 \Rightarrow 3v = -1 \Rightarrow v = -\dfrac{1}{3}$.

Now substitute these values into the expressions for u and v to solve for x and y.

$u = \dfrac{1}{x} \Rightarrow \dfrac{1}{2} = \dfrac{1}{x} \Rightarrow x = 2.$

$v = \dfrac{1}{y} \Rightarrow -\dfrac{1}{3} = \dfrac{1}{y} \Rightarrow y = -3$

The solution is $\{(2, -3)\}$.

8. $\begin{cases} p = 20 + 0.002x & (1) \\ p = 77 - 0.008x & (2) \end{cases}$

Solve by substitution.

$20 + 0.002x = 77 - 0.008x \Rightarrow 0.010x = 57 \Rightarrow$
$x = 5700$

Substitute this value into equation (1) and solve for p.

$p = 20 + 0.002x \Rightarrow p = 20 + 0.002(5700) \Rightarrow$
$p = 31.4$

The equilibrium point is $(5700, 31.4)$.

9. Let $x =$ the amount invested at 12%.
Let $y =$ the amount invested at 8%.
Then $0.12x =$ the income from the 12% investment and $0.08y =$ the income from the 8% investment. The system of equations is

$\begin{cases} x + y = 150,000 & (1) \\ 0.12x + 0.08y = 15,400 & (2) \end{cases}$

We will use the elimination method to solve the system.

$\begin{array}{ll} -12x - 12y = -1,800,000 & \text{Multiply by } -12. \\ \underline{12x + 8y = 1,540,000} & \text{Multiply by 100.} \\ -4y = -260,000 & \text{Add.} \end{array}$

$y = \dfrac{-260,000}{-4} = 65,000$

Solve for y.

Back-substitute $y = 65,000$ into equation (1) and solve for x.

$x + 65,000 = 150,000$
$x = 85,000$

Check:
12% of 85,000 = 10,200
8% of 65,000 = 5200
85,000 + 65,000 = 150,000 and
10,200 + 5200 = 15,400, as given.
Solution: \$85,000 was invested at 12% and \$65,000 was invested at 8%.

5.1 Concepts and Vocabulary

1. The ordered pair (a, b) is a <u>solution</u> of a system of equations in x and y provided that when x is replaced by a and y is replace by b, the resulting equations are true.

3. If, in the process of solving a system of equations, you get an equation of the form $0 = k$, where k is not zero, then the system is <u>inconsistent</u>.

5. False. A system consisting of two identical equations has an infinite number of solutions.

7. True

5.1 Building Skills

9. Substituting each ordered pair into the system
$\begin{cases} 2x + 3y = 3 \\ 3x - 4y = 13 \end{cases}$,
we find that $(3, -1)$ is a solution.
$\begin{cases} 2(3) + 3(-1) = 6 - 3 = 3 \\ 3(3) - 4(-1) = 9 + 4 = 13 \end{cases}$

11. Substituting each ordered pair into the system
$\begin{cases} 5x - 2y = 7 \\ -10x + 4y = 11 \end{cases}$,
we find that none of the ordered pairs are solutions.

13. Substituting each ordered pair into the system
$\begin{cases} x + y = 1 \\ \dfrac{1}{2}x + \dfrac{1}{3}y = 2 \end{cases}$,
we find that $(10, -9)$ is a solution.
$\begin{cases} 10 - 9 = 1 \\ \dfrac{1}{2}(10) + \dfrac{1}{3}(-9) = 5 - 3 = 2 \end{cases}$

15. The solution is $\{(2, 1)\}$.

$$\begin{cases} 2+1=3 \\ 2-1=1 \end{cases}$$

17. The solution is $\{(2, 2)\}$.

$$\begin{cases} 2+2(2)=2+4=6 \\ 2(2)+2=4+2=6 \end{cases}$$

19. The system is inconsistent.

21. The solution is $\left\{ \left(\dfrac{7}{3}, \dfrac{14}{3} \right) \right\}$.

$$\begin{cases} \dfrac{7}{3} + \dfrac{14}{3} = \dfrac{21}{3} = 7 \\ \dfrac{14}{3} = 2\left(\dfrac{7}{3} \right) \end{cases}$$

23. The system is dependent. The general solution is $\{(x, 12-3x)\}$.

25. The slopes of the two lines are different, so the system is independent.

27. The slopes of the two lines are different, so the system is independent.

29. The slopes and y-intercepts of the two lines are the same, so they coincide, and the system is dependent.

31. The slopes of the two lines are different, so the system is independent.

33. The slopes of the two lines are the same while the y-intercepts are different, so the system is inconsistent.

35. The slopes of the two lines are the same while the y-intercepts are different, so the system is inconsistent.

37. The slopes of the two lines are different, so the system is independent.

In exercises 39–48, check your answers by substituting the values of x and y into both of the original equations.

39. Substitute $y = 2x+1$ into

$5x+2y=9 \Rightarrow 5x+2(2x+1)=9 \Rightarrow$
$5x+4x+2=9 \Rightarrow 9x=7 \Rightarrow x=7/9$

Use this value of x to find y: $y = 2\left(\dfrac{7}{9} \right)+1 \Rightarrow$

$y = \dfrac{23}{9}$. The solution is $\left\{ \left(\dfrac{7}{9}, \dfrac{23}{9} \right) \right\}$.

41. Solve the second equation for y, and then substitute $y = 7 - x$ into the first equation:
$3x - (7 - x) = 5 \Rightarrow 3x - 7 + x = 5 \Rightarrow 4x = 12 \Rightarrow$
$x = 3$. Substitute this value into the second equation to find y: $3 + y = 7 \Rightarrow y = 4$
The solution is $\{(3,4)\}$.

43. Solve the first equation for y, and then substitute $y = 2x - 5$ into the second equation:
$-4x + 2(2x - 5) = 7 \Rightarrow -4x + 4x - 10 = 7 \Rightarrow$
$-10 \neq 7 \Rightarrow$ there is no solution.
Solution set: $\varnothing$

45. Solve the first equation for y, and then substitute $y = 3 - \dfrac{2}{3}x$ into the second equation:
$3x + 2\left(3 - \dfrac{2}{3}x\right) = 1 \Rightarrow 3x + 6 - \dfrac{4}{3}x = 1 \Rightarrow$
$\dfrac{5}{3}x = -5 \Rightarrow x = -3$. Substitute this value into the first equation to find y:
$\dfrac{2}{3}(-3) + y = 3 \Rightarrow y = 5$.
The solution is $\{(-3, 5)\}$.

47. Solve the first equation for y, and then substitute $y = \dfrac{1}{2}(x - 5)$ into the second equation:
$-3x + 6\left(\dfrac{1}{2}(x - 5)\right) = -15 \Rightarrow -3x + 3x - 15 = -15 \Rightarrow$
$-15 = -15 \Rightarrow$ the system is dependent.
The general form of the solution is
$\left\{\left(x, \dfrac{1}{2}(x - 5)\right)\right\}$.

49. Add the two equations and solve for x:
$\begin{cases} x - y = 1 \\ x + y = 5 \end{cases} \Rightarrow 2x = 6 \Rightarrow x = 3$. Now substitute
$x = 3$ into the second equation and solve for y:
$3 + y = 5 \Rightarrow y = 2$. The solution is $\{(3, 2)\}$.

51. Multiply the first equation by –2:
$\begin{cases} x + y = 0 \\ 2x + 3y = 3 \end{cases} \Rightarrow \begin{cases} -2x - 2y = 0 \\ 2x + 3y = 3 \end{cases}$
Add the equations and solve for y:
$\begin{cases} -2x - 2y = 0 \\ 2x + 3y = 3 \end{cases} \Rightarrow y = 3$.
Substitute this value into the first equation and solve for x: $x + 3 = 0 \Rightarrow x = -3$.
The solution is $\{(-3, 3)\}$.

53. Multiply the first equation by 2:
$\begin{cases} 5x - y = 5 \\ 3x + 2y = -10 \end{cases} \Rightarrow \begin{cases} 10x - 2y = 10 \\ 3x + 2y = -10 \end{cases}$
Add the equations and solve for x:
$\begin{cases} 10x - 2y = 10 \\ 3x + 2y = -10 \end{cases} \Rightarrow 13x = 0 \Rightarrow x = 0$.
Substitute this value into the first equation and solve for y: $5(0) - y = 5 \Rightarrow y = -5$. The solution is $\{(0, -5)\}$.

55. Multiply the first equation by 2:
$\begin{cases} x - y = 2 \\ -2x + 2y = 5 \end{cases} \Rightarrow \begin{cases} 2x - 2y = 4 \\ -2x + 2y = 5 \end{cases}$. Adding the
equations gives $0 = 9 \Rightarrow$ the equations are inconsistent, and there is no solution.
Solution set: $\varnothing$

57. Multiply the second equation by –2:
$\begin{cases} 4x + 6y = 12 \\ 2x + 3y = 6 \end{cases} \Rightarrow \begin{cases} 4x + 6y = 12 \\ -4x - 6y = -12 \end{cases}$. Adding
the equations gives $0 = 0 \Rightarrow$ the equations are dependent. Solving the second equation for y,
we have $2x + 3y = 6 \Rightarrow y = -\dfrac{2x}{3} + 2$.
The general form of the solution is
$\left\{\left(x, -\dfrac{2x}{3} + 2\right)\right\}$.

59. Solve the first equation for y, and then substitute this value into the second equation to solve for x: $2x + y = 9 \Rightarrow y = -2x + 9$.
$2x - 3(-2x + 9) = 5 \Rightarrow 8x - 27 = 5 \Rightarrow x = 4$.
Substitute this value into the first equation and solve for y: $2(4) + y = 9 \Rightarrow y = 1$.
The solution is $\{(4, 1)\}$.

61. Multiply the second equation by –2 and then add the equations:
$\begin{cases} 2x + 5y = 2 \\ x + 3y = 2 \end{cases} \Rightarrow \begin{cases} 2x + 5y = 2 \\ -2x - 6y = -4 \end{cases} \Rightarrow -y = -2 \Rightarrow$
$y = 2$. Substitute this value into the second equation and solve for x: $x + 3(2) = 2 \Rightarrow x = -4$.
The solution is $\{(-4, 2)\}$.

63. Multiply the second equation by –3 and then add the equations:
$\begin{cases} 2x + 3y = 7 \\ 3x + y = 7 \end{cases} \Rightarrow \begin{cases} 2x + 3y = 7 \\ -9x - 3y = -21 \end{cases} \Rightarrow$
$-7x = -14 \Rightarrow x = 2$. Substitute this value into the second equation and solve for y:
$3(2) + y = 7 \Rightarrow y = 1$.
The solution is $\{(2, 1)\}$.

65. Multiply the first equation by −2 and the second equation by 3, then add the equations to solve for x:

$$\begin{cases} 2x + 3y = 9 \\ 3x + 2y = 11 \end{cases} \Rightarrow \begin{cases} -4x - 6y = -18 \\ 9x + 6y = 33 \end{cases} \Rightarrow$$

$5x = 15 \Rightarrow x = 3.$

Substitute this value into the first equation to solve for y: $2(3) + 3y = 9 \Rightarrow 3y = 3 \Rightarrow y = 1.$

The solution is $\{(3, 1)\}$.

67. First, simplify both equations:

$$\begin{cases} \dfrac{x}{4} + \dfrac{y}{6} = 1 \\ x + 2(x - y) = 7 \end{cases} \Rightarrow \begin{cases} 6x + 4y = 24 \\ 3x - 2y = 7 \end{cases}.$$

Now multiply the second equation by 2, and then add the equations:

$$\begin{cases} 6x + 4y = 24 \\ 3x - 2y = 7 \end{cases} \Rightarrow \begin{cases} 6x + 4y = 24 \\ 6x - 4y = 14 \end{cases} \Rightarrow$$

$12x = 38 \Rightarrow x = \dfrac{19}{6}.$

Substitute this value into the second equation and solve for y:

$$\dfrac{19}{6} + 2\left(\dfrac{19}{6} - y\right) = 7 \Rightarrow \dfrac{57}{6} - 2y = 7 \Rightarrow$$

$$-2y = -\dfrac{5}{2} \Rightarrow y = \dfrac{5}{4}.$$

The solution is $\left\{\left(\dfrac{19}{6}, \dfrac{5}{4}\right)\right\}$.

69. Simplify the first equation:

$3x = 2(x + y) \Rightarrow 3x = 2x + 2y \Rightarrow x = 2y.$

Substitute this expression for x into the second equation: $3(2y) - 5y = 2 \Rightarrow y = 2.$ Substitute this value into the first equation to solve for x:

$3x = 2(x + 2) \Rightarrow 3x = 2x + 4 \Rightarrow x = 4.$

The solution is $\{(4, 2)\}$.

71. Multiply the first equation by −2, then add the equations:

$$\begin{cases} 0.2x + 0.7y = 1.5 \\ 0.4x - 0.3y = 1.3 \end{cases} \Rightarrow \begin{cases} -0.4x - 1.4y = -3 \\ 0.4x - 0.3y = 1.3 \end{cases} \Rightarrow$$

$-1.7y = -1.7 \Rightarrow y = 1.$ Substitute this value into the first equation to solve for x:

$0.2x + 0.7(1) = 1.5 \Rightarrow 0.2x = 0.8 \Rightarrow x = 4.$

The solution is $\{(4, 1)\}$.

73. $\begin{cases} \dfrac{x}{2} + \dfrac{y}{3} = 1 & (1) \\ \dfrac{3x}{4} + \dfrac{y}{2} = 1 & (2) \end{cases}$

Clear the fractions by multiplying equation (1) by 6 and equation (2) by 8.

$$\begin{cases} 3x + 2y = 6 & (1) \\ 6x + 4y = 8 & (2) \end{cases}$$

Now multiply equation (1) by −2, then add the two equations.

$$\begin{array}{r} -6x - 4y = -24 \\ \underline{6x + 4y = 8} \\ 0 = -16 \quad \text{False} \end{array}$$

The system is inconsistent.
The solution set is $\varnothing$.

75. $\begin{cases} \dfrac{x}{3} - \dfrac{y}{2} = 1 & (1) \\ \dfrac{3y}{8} - \dfrac{x}{4} = -\dfrac{3}{4} & (2) \end{cases}$

Clear the fractions by multiplying equation (1) by 6 and equation (2) by 8.

$$\begin{cases} 2x - 3y = 6 & (1) \\ 3y - 2x = -6 & (2) \end{cases}$$

Now add the two equations.

$$\begin{array}{r} 2x - 3y = 6 \\ \underline{-2x + 3y = -6} \\ 0 = 0 \quad \checkmark \end{array}$$

The system is dependent.
Solve equation (1) for y in terms of x.

$$\dfrac{x}{3} - \dfrac{y}{2} = 1 \Rightarrow 2x - 3y = 6 \Rightarrow -3y = -2x + 6 \Rightarrow$$

$$3y = 2(x - 3) \Rightarrow y = \dfrac{2}{3}(x - 3)$$

The solution set is $\left\{\left(x, \dfrac{2}{3}(x - 3)\right)\right\}$.

77. $\begin{cases} \dfrac{2}{x} + \dfrac{5}{y} = -5 & (1) \\ \dfrac{3}{x} - \dfrac{2}{y} = -17 & (2) \end{cases}$

Let $u = \dfrac{1}{x}$ and $v = \dfrac{1}{y}$. Then the system

becomes $\begin{cases} 2u + 5v = -5 \\ 3u - 2v = -17 \end{cases}$.

(*continued on next page*)

(continued)

Multiply the first equation by 2 and the second equation by 5, then add the equations:

$$\begin{cases} 2u + 5v = -5 \\ 3u - 2v = -17 \end{cases} \Rightarrow$$

$$\begin{cases} 4u + 10v = -10 \\ 15u - 10v = -85 \end{cases} \Rightarrow 19u = -95 \Rightarrow u = -5.$$

Substitute this value into the first equation to solve for v: $2(-5) + 5v = -5 \Rightarrow 5v = 5 \Rightarrow v = 1$.

$$u = \frac{1}{x} \Rightarrow -5 = \frac{1}{x} \Rightarrow x = -\frac{1}{5}.$$

$$v = \frac{1}{y} \Rightarrow 1 = \frac{1}{y} \Rightarrow y = 1.$$

The solution is $\left\{ \left(-\dfrac{1}{5}, 1 \right) \right\}$.

79. $\begin{cases} \dfrac{3}{x} + \dfrac{1}{y} = 4 & (1) \\[2mm] \dfrac{6}{x} - \dfrac{1}{y} = 2 & (2) \end{cases}$

Let $u = \dfrac{1}{x}$ and $v = \dfrac{1}{y}$. Then the system becomes $\begin{cases} 3u + v = 4 \\ 6u - v = 2 \end{cases}$. Add the equations and

solve for u: $9u = 6 \Rightarrow u = \dfrac{2}{3}$. Substitute this value into the first equation and solve for v:

$$3\left(\frac{2}{3} \right) + v = 4 \Rightarrow v = 2.$$

$$u = \frac{1}{x} \Rightarrow \frac{2}{3} = \frac{1}{x} \Rightarrow x = \frac{3}{2}.$$

$$v = \frac{1}{y} \Rightarrow 2 = \frac{1}{y} \Rightarrow y = \frac{1}{2}.$$

The solution is $\left\{ \left(\dfrac{3}{2}, \dfrac{1}{2} \right) \right\}$.

81. $\begin{cases} \dfrac{5}{x} + \dfrac{10}{y} = 3 & (1) \\[2mm] \dfrac{2}{x} - \dfrac{12}{y} = -2 & (2) \end{cases}$

Let $u = \dfrac{1}{x}$ and $v = \dfrac{1}{y}$. Then the system

becomes $\begin{cases} 5u + 10v = 3 \\ 2u - 12v = -2 \end{cases}$. Multiply the first

equation by 2 and the second equation by −5, then add the equation to solve for v:

$$\begin{cases} 5u + 10v = 3 \\ 2u - 12v = -2 \end{cases} \Rightarrow \begin{cases} 10u + 20v = 6 \\ -10u + 60v = 10 \end{cases} \Rightarrow$$

$80v = 16 \Rightarrow v = \dfrac{1}{5}$. Substitute this value into the first equation to solve for u:

$$5u + 10\left(\frac{1}{5} \right) = 3 \Rightarrow 5u = 1 \Rightarrow u = \frac{1}{5}.$$

$$u = \frac{1}{x} \Rightarrow \frac{1}{5} = \frac{1}{x} \Rightarrow x = 5.$$

$$v = \frac{1}{y} \Rightarrow \frac{1}{5} = \frac{1}{y} \Rightarrow y = 5.$$

The solution is $\{(5, 5)\}$.

83. $\begin{cases} \dfrac{2}{x} + \dfrac{1}{y} = 4 & (1) \\[2mm] x + 2y = 6xy & (2) \end{cases}$

Divide equation (2) by $6xy$.

$$\begin{cases} \dfrac{2}{x} + \dfrac{1}{y} = 4 \\[2mm] \dfrac{1}{6y} + \dfrac{1}{3x} = 1 \end{cases} \Rightarrow \begin{cases} \dfrac{2}{x} + \dfrac{1}{y} = 4 & (1) \\[2mm] \dfrac{1}{3x} + \dfrac{1}{6y} = 1 & (2) \end{cases}$$

Let $u = \dfrac{1}{x}$ and $v = \dfrac{1}{y}$. Then the system

becomes $\begin{cases} 2u + v = 4 \\[1mm] \dfrac{1}{3}u + \dfrac{1}{6}v = 1 \end{cases}$.

Clear the fractions in equation (2) by multiplying by 6.

$$\begin{cases} 2u + v = 4 \\[1mm] \dfrac{1}{3}u + \dfrac{1}{6}v = 1 \end{cases} \Rightarrow \begin{cases} 2u + v = 4 \\ 2u + v = 6 \end{cases}$$

Multiply equation (2) by −1, then add the equations.

$$\begin{aligned} 2u + v &= 4 \\ \underline{-2u - v} &= \underline{-6} \\ 0 &= -2 \quad \text{False} \end{aligned}$$

The system is inconsistent.
The solution set is $\varnothing$.

5.1 Applying the Concepts

In exercises 85–88, the equilibrium point is the point that satisfies both the demand equation and the supply equation.

85. Add the equations to solve for p:

$$\begin{cases} 2p + x = 140 \\ 12p - x = 280 \end{cases} \Rightarrow 14p = 420 \Rightarrow p = 30.$$

Substitute this value into the first equation to solve for x: $2(30) + x = 140 \Rightarrow x = 80$.

The equilibrium point is (80, 30).

87. Multiply the second equation by −1, then add the equations to solve for x:

$$\begin{cases} 2p + x = 25 \\ x - p = 13 \end{cases} \Rightarrow \begin{cases} 2p + x = 25 \\ p - x = -13 \end{cases} \Rightarrow 3p = 12 \Rightarrow$$

$p = 4$. Substitute this value into the second equation to solve for x: $x - 4 = 13 \Rightarrow x = 17$. The equilibrium point is $(17, 4)$.

89. Let x = the diameter of the largest pizza, and let y = the diameter of the smallest pizza. Then

$$\begin{cases} x + y = 29 \\ x - y = 13 \end{cases} \Rightarrow 2x = 42 \Rightarrow x = 21.$$

$21 + y = 29 \Rightarrow y = 8$.
The largest pizza has diameter 21 inches, and the smallest pizza has diameter 8 inches.

91. Let x = the percentage of paper trash, and let y = the percentage of plastic trash. If the total amount of trash is t, then

$$\begin{cases} x + y = 48 \\ x = 5y \end{cases}.$$

Substitute the expression for x from the second equation into the first equation to solve for y: $5y + y = 48 \Rightarrow y = 8$. Substitute this value into the first equation to solve for x: $x + 8 = 48 \Rightarrow x = 40$. So, 8% of the trash is plastic and 40% is paper.

93. Let x = the number of beads, and let y = the number of doubloons . Then

$$\begin{cases} 0.4x + 0.3y = 265 \\ x + y = 770 \end{cases}.$$ Multiply the second

equation by −0.3 and then add the equations to solve for x: $\begin{cases} 0.4x + 0.3y = 265 \\ x + y = 770 \end{cases} \Rightarrow$

$$\begin{cases} 0.4x + 0.3y = 265 \\ -0.3x - 0.3y = -231 \end{cases} \Rightarrow 0.1x = 34 \Rightarrow x = 340.$$

Substitute this value into the second equation to solve for y: $340 + y = 770 \Rightarrow y = 430$. Levon bought 340 beads and 430 doubloons.

95. Let x = the number of Egg McMuffins, and let y = the number of Breakfast Burritos. Then

$$\begin{cases} 27x + 21y = 123 \\ 17x + 13y = 77 \end{cases}.$$ Multiply the first equation

by 13 and the second equation by −21, then add the equations to solve for x:

$$\begin{cases} 27x + 21y = 123 \\ 17x + 13y = 77 \end{cases} \Rightarrow \begin{cases} 351x + 273y = 1599 \\ -357x - 273y = -1617 \end{cases} \Rightarrow$$

$-6x = -18 \Rightarrow x = 3$.

Substitute this value into the first equation to solve for y:

$27(3) + 21y = 123 \Rightarrow 21y = 42 \Rightarrow y = 2$. You will need to eat three Egg McMuffins and two Breakfast Burritos.

97. Let x = the amount invested at 7.5%, and let y = the amount invested at 12%. Then

$$\begin{cases} x + y = 50,000 \\ 0.075x + 0.12y = 5190 \end{cases}.$$ Solve the first

equation for x and substitute this value into the second equation to solve for y:
$x + y = 50,000 \Rightarrow x = 50,000 - y$.

$0.075(50,000 - y) + 0.12y = 5190 \Rightarrow$
$3750 - 0.075y + 0.12y = 5190 \Rightarrow$
$0.045y = 1440 \Rightarrow y = \$32,000$.

Substitute this value into the first equation to solve for x:
$x + 32,000 = 50,000 \Rightarrow x = 18,000$.
Mrs. García invested \$18,000 at 7.5% and \$32,000 at 12%.

99. Let x = the amount earned tutoring, and let y = the amount earned working at McDougal's.

Then $\begin{cases} x = 2y \\ \dfrac{x + y}{2} = 11.25 \end{cases}$.

Substitute the expression for x from the first equation into the second equation and solve for

y. $\dfrac{2y + y}{2} = 11.25 \Rightarrow 3y = 22.50 \Rightarrow y = 7.50$.

Substitute this value into the first equation to solve for x: $x = 2(7.50) = 15$. She earned \$15 tutoring and \$7.50 working at McDougal's.

101. Let x = the number of pounds of the herb, and let y = the number of pounds of tea. Then

$$\begin{cases} x + y = 100 \\ 5.50x + 3.20y = 3.66(100) \end{cases}.$$ Solve the first

equation for x, and substitute this expression into the second equation to solve for y:
$x + y = 100 \Rightarrow x = 100 - y$.

$5.5(100 - y) + 3.20y = 3.66(100) \Rightarrow$
$550 - 5.5y + 3.2y = 366 \Rightarrow -2.3y = -184 \Rightarrow$
$y = 80$. Substitute this value into the first equation to solve for x: $x + 80 = 100 \Rightarrow x = 20$. There are 20 pounds of the herb and 80 pounds of tea in the mixture.

103. Let x = the speed of the plane in still air, and let y = the wind speed. So, the speed of the plane with the wind is $x + y$, and the speed of the plane against the wind is $x - y$. Then, using the fact that rate × time = distance, we have

$$\begin{cases} 5(x + y) = 3000 \\ 6(x - y) = 3000 \end{cases} \Rightarrow \begin{cases} x + y = 600 \\ x - y = 500 \end{cases}.$$ Add the

equations to solve for x: $2x = 1100 \Rightarrow x = 550$. Substitute this value into the first equation to solve for y: $5(550 + y) = 3000 \Rightarrow$ $550 + y = 600 \Rightarrow y = 50$. The speed of the plane is 550 kph, and the wind speed is 50 kph.

105. a. $C(x) = y = 2x + 30,000$; $R(x) = y = 3.50x$

b.

c. $\begin{cases} y = 2x + 30,000 \\ y = 3.5x \end{cases}$.

Substitute the expression for y from the second equation into the first equation to solve for x:
$2x + 30,000 = 3.5x \Rightarrow 1.5x = 30,000 \Rightarrow$ $x = 20,000$ magazines

107. Let x = Shanaysha's weekly sales. Then her salary at store B is $150 + 0.04x$.
$150 + 0.04x > 400 \Rightarrow 0.04x > 250 \Rightarrow x > 6250$. Her weekly sales should be more than $6250.

5.1 Beyond the Basics

109. $\begin{cases} 2\log_3 x + 3\log_3 y = 8 & (1) \\ 3\log_3 x - \log_3 y = 1 & (2) \end{cases}$

Multiply equation (2) by 3. Then add the two equations.
$\begin{aligned} 2\log_3 x + 3\log_3 y &= 8 \\ 9\log_3 x - 3\log_3 y &= 3 \\ \hline 11\log_3 x &= 11 \end{aligned}$
Solve for x.
$11\log_3 x = 11 \Rightarrow \log_3 x = 1 \Rightarrow x = 3$

Substitute $x = 3$ into equation (2) and solve for y:
$3\log_3 3 - \log_3 y = 1 \Rightarrow 3(1) - \log_3 y = 1 \Rightarrow$
$-\log_3 y = -2 \Rightarrow \log_3 y = 2 \Rightarrow y = 3^2 = 9$
Be sure to check your answer in the original system of equations. The solution is $\{(3, 9)\}$.

111. $\begin{cases} 3e^x - 4e^y = 4 & (1) \\ 2e^x + 5e^y = 18 & (2) \end{cases}$

Multiply equation (1) by 5 and equation (2) by 4. Then add the equations.
$\begin{aligned} 15e^x - 20e^y &= 20 \\ 8e^x + 20e^y &= 72 \\ \hline 23e^x &= 92 \end{aligned}$
Now solve for x.
$23e^x = 92 \Rightarrow e^x = \dfrac{92}{23} \Rightarrow x = \ln 4.$
Substitute this value in equation (1) and solve for y.
$3e^{\ln 4} - 4e^y = 4 \Rightarrow 3(4) - 4e^y = 4 \Rightarrow$
$-4e^y = -8 \Rightarrow e^y = 2 \Rightarrow y = \ln 2$
The solution is $(\ln 4, \ln 2)$.

113. First solve the system $\begin{cases} x + 2y = 7 \\ 3x + 5y = 11 \end{cases}$.

Multiply the first equation by –3, then add the equations to solve for y:
$\begin{cases} x + 2y = 7 \\ 3x + 5y = 11 \end{cases} \Rightarrow \begin{cases} -3x - 6y = -21 \\ 3x + 5y = 11 \end{cases} \Rightarrow -y = -10 \Rightarrow$
$y = 10$. Substitute this value into the first equation to solve for x: $x + 2(10) = 7 \Rightarrow$
$x = -13$. Now substitute $(-13, 10)$ into the third equation to solve for c:
$-13c + 3(10) = 4 \Rightarrow -13c = -26 \Rightarrow c = 2$.

115. $\begin{cases} 2x + cy = 11 \\ 5x - 7y = 5 \end{cases}$

The system is inconsistent if the lines are parallel, i.e., if the coefficients of one line are multiples of the coefficients of the other line, but the constant in the first line is not the same multiple of the constant in the second line.
$2x = \dfrac{2}{5}(5x)$, so $c = \dfrac{2}{5}(-7) = -\dfrac{14}{5}$

117. $\ell_1 = a_1 x + b_1 y + c_1 = 0$

$\ell_2 = a_2 x + b_2 y + c_2 = 0 \Rightarrow \ell_1 + k\ell_2 = 0$ (1)

If $P(h, k)$ is the point of intersection of ℓ_1 and ℓ_2, then $a_1 h + b_1 k + c_1 = 0$ and $a_2 h + b_2 k + c_2 = 0$. Since $\ell_1 + k\ell_2 = 0$, P lies on this line also. Therefore,

$$\ell_1 + k\ell_2 = a_1 x + b_1 y + c_1 + k\left(a_2 x + b_2 y + c_2\right)$$
$$= 0$$

is an equation of the line passing through P.

119. To find the intersection of the two lines, solve the system $\begin{cases} 5x + 2y = 7 \\ 6x - 5y = 38 \end{cases}$. Multiply the first equation by 5 and the second equation by 2, then add the equations to solve for x:

$\begin{cases} 5x + 2y = 7 \\ 6x - 5y = 38 \end{cases} \Rightarrow \begin{cases} 25x + 10y = 35 \\ 12x - 10y = 76 \end{cases} \Rightarrow 37x = 111 \Rightarrow$

$x = 3$. Substitute this value into the first equation to solve for y: $5(3) + 2y = 7 \Rightarrow$

$2y = -8 \Rightarrow y = -4$. So the line passes through the point $(3, -4)$. Because the line also passes through the point $(1, 3)$, the slope of the line is

$\dfrac{3 - (-4)}{1 - 3} = -\dfrac{7}{2}$. The equation of the line is

$y - 3 = -\dfrac{7}{2}(x - 1) \Rightarrow y = -\dfrac{7}{2} x + \dfrac{13}{2}$. Verify graphically:

5.1 Critical Thinking/Discussion/Writing

121. a. There is only one solution if the equations are independent. Solve for x by multiplying the first equation by b_2 and multiplying the second equation by $-b_1$, then adding the equations:

$\begin{cases} a_1 x + b_1 y = c_1 \\ a_2 x + b_2 y = c_2 \end{cases} \Rightarrow$

$\begin{cases} b_2 a_1 x + b_2 b_1 y = b_2 c_1 \\ -b_1 a_2 x - b_1 b_2 y = -b_1 c_2 \end{cases} \Rightarrow$

$b_2 a_1 x - b_1 a_2 x = b_2 c_1 - b_1 c_2 \Rightarrow$

$x(b_2 a_1 - b_1 a_2) = b_2 c_1 - b_1 c_2 \Rightarrow$

$x = \dfrac{b_2 c_1 - b_1 c_2}{a_1 b_2 - a_2 b_1}$

Solve for y by multiplying the first equation by a_2 and multiplying the second equation by $-a_1$, then adding the equations.

$\begin{cases} a_1 x + b_1 y = c_1 \\ a_2 x + b_2 y = c_2 \end{cases} \Rightarrow$

$\begin{cases} a_1 a_2 x + a_2 b_1 y = a_2 c_1 \\ -a_1 a_2 x - a_1 b_2 y = -a_1 c_2 \end{cases} \Rightarrow$

$a_2 b_1 y - a_1 b_2 y = a_2 c_1 - a_1 c_2 \Rightarrow$

$y(a_2 b_1 - a_1 b_2) = a_2 c_1 - a_1 c_2 \Rightarrow$

$y = \dfrac{a_2 c_1 - a_1 c_2}{a_2 b_1 - a_1 b_2} = \dfrac{a_1 c_2 - a_2 c_1}{a_1 b_2 - a_2 b_1}$.

Note that the denominator of x and y, $a_1 b_2 - a_2 b_1$, cannot equal zero.

b. There is no solution if the equations are inconsistent. They are inconsistent if they are parallel. The slopes of the lines are

equal and $\dfrac{b_1}{a_1} = \dfrac{b_2}{a_2} \Rightarrow \dfrac{a_1}{a_2} = \dfrac{b_1}{b_2}$. The intercepts are different. If $c_1 = c_2$, then

$\dfrac{c_1}{a_1} \neq \dfrac{c_2}{a_2}$.

c. There are infinitely many solutions if the equations are dependent. From part (b), we

have $\dfrac{a_1}{a_2} = \dfrac{b_1}{b_2}$. The intercepts of the lines

must be the same, so $\dfrac{a_1}{a_2} = \dfrac{b_1}{b_2} = \dfrac{c_1}{c_2}$.

122. a. The slope of ℓ_1 is $-\dfrac{3}{4}$, so the slope of the

perpendicular is $\dfrac{4}{3}$. The equation of ℓ_2 is

$y - 8 = \dfrac{4}{3}(x - 5) \Rightarrow 3y - 24 = 4x - 20 \Rightarrow$

$3y - 4x = 4$.

b. To find Q, solve the system $\begin{cases} 3x + 4y = 12 \\ -4x + 3y = 4 \end{cases}$.

Multiply the first equation by 4 and the second equation by 3, then add the equations to solve

for x: $\begin{cases} 3x + 4y = 12 \\ -4x + 3y = 4 \end{cases} \Rightarrow$

$\begin{cases} 12x + 16y = 48 \\ -12x + 9y = 12 \end{cases} \Rightarrow 25y = 60 \Rightarrow y = \dfrac{12}{5}$.

Substitute this value into the first equation to

solve for x: $3x + 4\left(\dfrac{12}{5}\right) = 12 \Rightarrow 3x = \dfrac{12}{5} \Rightarrow$

$x = \dfrac{4}{5}$. The coordinates of Q are $\left(\dfrac{4}{5}, \dfrac{12}{5}\right)$.

c. $d(P, Q) = \sqrt{\left(5 - \dfrac{4}{5}\right)^2 + \left(8 - \dfrac{12}{5}\right)^2} = \sqrt{49} = 7$

123. a. The slope of ℓ_1 is $-\dfrac{a}{b}$, so the slope of the

perpendicular is $\dfrac{b}{a}$. The equation of ℓ_2 is

$y - y_1 = \dfrac{b}{a}(x - x_1) \Rightarrow$

$ay - ay_1 = bx - bx_1 \Rightarrow ay - bx = ay_1 - bx_1$ or

$b(x - x_1) - a(y - y_1) = 0$.

b. The intersection of ℓ_1 and ℓ_2 is the solution of the system

$\begin{cases} ax + by = -c \\ -bx + ay = ay_1 - bx_1 \end{cases} \Rightarrow$

$\begin{cases} abx + b^2 y = -bc \\ -abx + a^2 y = a^2 y_1 - abx_1 \end{cases} \Rightarrow$

$y(a^2 + b^2) = -bc + a^2 y_1 - abx_1 \Rightarrow$

$y = \dfrac{-bc + a^2 y_1 - abx_1}{a^2 + b^2}$

$\begin{cases} ax + by = -c \\ -bx + ay = ay_1 - bx_1 \end{cases} \Rightarrow$

$\begin{cases} -a^2 x - aby = ac \\ -b^2 x + aby = aby_1 - b^2 x_1 \end{cases} \Rightarrow$

$-x(a^2 + b^2) = ac + aby_1 - b^2 x_1 \Rightarrow$

$x = \dfrac{-ac - aby_1 + b^2 x_1}{a^2 + b^2}$

$\quad = -\dfrac{ac + aby_1 - b^2 x_1}{a^2 + b^2}$

The intersection of the two lines is

$\left(-\dfrac{ac + aby_1 - b^2 x_1}{a^2 + b^2}, \dfrac{-abx_1 + a^2 y_1 - bc}{a^2 + b^2}\right)$.

c. Now find the distance between (x_1, y_1) and

$\left(-\dfrac{ac + aby_1 - b^2 x_1}{a^2 + b^2}, \dfrac{-bc + a^2 y_1 - abx_1}{a^2 + b^2}\right)$:

$\sqrt{\left(x_1 + \dfrac{ac + aby_1 - b^2 x_1}{a^2 + b^2}\right)^2 + \left(y_1 - \dfrac{-bc + a^2 y_1 - abx_1}{a^2 + b^2}\right)^2}$

$= \sqrt{\left(\dfrac{a(ax_1 + by_1 + c)}{a^2 + b^2}\right)^2 + \left(\dfrac{b(ax_1 + by_1 + c)}{a^2 + b^2}\right)^2}$

$= \sqrt{\dfrac{a^2 (ax_1 + by_1 + c)^2}{\left(a^2 + b^2\right)^2} + \dfrac{b^2 (ax_1 + by_1 + c)^2}{\left(a^2 + b^2\right)^2}}$

$= \sqrt{\dfrac{(a^2 + b^2)(ax_1 + by_1 + c)^2}{\left(a^2 + b^2\right)^2}}$

$= \sqrt{\dfrac{(ax_1 + by_1 + c)^2}{a^2 + b^2}} = \dfrac{|ax_1 + by_1 + c|}{\sqrt{a^2 + b^2}}$.

124. a. $a = 1, b = 1, c = -7, x_1 = 2, y_1 = 3$

$\dfrac{|ax_1 + by_1 + c|}{\sqrt{a^2 + b^2}} = \dfrac{|1(2) + 1(3) - 7|}{\sqrt{1^2 + 1^2}} = \dfrac{2}{\sqrt{2}} = \sqrt{2}$

b. $a = 2, b = -1, c = 3, x_1 = -2, y_1 = 5$

$\dfrac{|ax_1 + by_1 + c|}{\sqrt{a^2 + b^2}} = \dfrac{|2(-2) - 1(5) + 3|}{\sqrt{2^2 + (-1)^2}}$

$\quad = \dfrac{6}{\sqrt{5}} = \dfrac{6\sqrt{5}}{5}$

c. $a = 5, b = -2, c = -7, x_1 = 3, y_1 = 4$

$\dfrac{|ax_1 + by_1 + c|}{\sqrt{a^2 + b^2}} = \dfrac{|5(3) - 2(4) - 7|}{\sqrt{3^2 + 4^2}} = 0$

d. $a = a, b = b, c = c, x_1 = 0, y_1 = 0$

$\dfrac{|ax_1 + by_1 + c|}{\sqrt{a^2 + b^2}} = \dfrac{|a(0) + b(0) + c|}{\sqrt{a^2 + b^2}}$

$\quad = \dfrac{|c|}{\sqrt{a^2 + b^2}}$

125.

Let $Q(a, b)$ be the reflection point. Using the hint, we have two equations, one developed from the fact that the midpoint of $\overline{PQ}$ lies on line ℓ, and the other developed from the fact that ℓ is perpendicular to the line containing the segment $\overline{PQ}$.

Equation (1): The midpoint of $\overline{PQ}$ is

$M\left(\dfrac{a+2}{2}, \dfrac{b+1}{2}\right)$. Since M lies on line ℓ, we

have $2\left(\dfrac{a+2}{2}\right)+3\left(\dfrac{b+1}{2}\right)=20$. We can

simplify this to

$(a+2)+\dfrac{3}{2}b+\dfrac{3}{2}=20 \Rightarrow a+\dfrac{3}{2}b=\dfrac{33}{2} \Rightarrow$
$2a+3b=33$

Equation (2): The slope of ℓ is $-\dfrac{2}{3}$, so the

slope of the line perpendicular to line ℓ is $\dfrac{3}{2}$.

Using the formula for the slope of $\overline{PQ}$, we have

$\dfrac{3}{2}=\dfrac{b-1}{a-2} \Rightarrow 3a-6=2b-2 \Rightarrow 3a-2b=4.$

So, the system is

$\begin{cases} 2a+3b=33 & (1) \\ 3a-2b=4 & (2) \end{cases}$

Multiply the first equation by 2 and the second equation by 3. Then add the two equations, then solve for a.

$4a+6b=66$
$\underline{9a-6b=12}$
$\quad 13a=78 \Rightarrow a=6$

Substitute $a=6$ into equation (2) and solve for b.

$3(6)-2b=4 \Rightarrow -2b=-14 \Rightarrow b=7$

Be sure to check your answer.
The reflection point is $Q(6, 7)$.

126.

Let $Q(a, b)$ be the reflection point. Using the hint from exercise 122, we have two equations, one developed from the fact that the midpoint of $\overline{PQ}$ lies on line ℓ, and the other developed from the fact that ℓ is perpendicular to the line containing the segment $\overline{PQ}$.

Equation (1): The midpoint of $\overline{PQ}$ is

$M\left(\dfrac{a+2}{2}, \dfrac{b+1}{2}\right)$. Since M lies on line ℓ, we

have $3\left(\dfrac{a+2}{2}\right)+\left(\dfrac{b+1}{2}\right)=12$. We can

simplify this to

$3(a+2)+(b+1)=24 \Rightarrow$
$3a+6+b+1=24 \Rightarrow 3a+b=17$

Equation (2): The slope of ℓ is -3, so the

slope of the line perpendicular to line ℓ is $\dfrac{1}{3}$.

Using the formula for the slope of $\overline{PQ}$, we have

$\dfrac{1}{3}=\dfrac{b-1}{a-2} \Rightarrow 3b-3=a-2 \Rightarrow a-3b=-1.$

So, the system is

$\begin{cases} 3a+b=17 & (1) \\ a-3b=-1 & (2) \end{cases}$

Multiply the first equation by 3. Then add the two equations, then solve for a.

$9a+3b=51$
$\underline{a-3b=-1}$
$\quad 10a=50 \Rightarrow a=5$

Substitute $a=5$ into equation (1) and solve for b.

$3(5)+b=17 \Rightarrow b=2$

Be sure to check your answer.
The reflection point is $Q(5, 2)$.

5.1 Getting Ready for the Next Section

127. $y+2z=7; \ z=2$
$y+2(2)=7 \Rightarrow y+4=7 \Rightarrow y=3$

129. $2x + 3y + 5z = 21; \; y = 1; \; z = 2$

$2x + 3(1) + 5(2) = 21 \Rightarrow 2x + 13 = 21 \Rightarrow$

$2x = 8 \Rightarrow x = 4$

131. $2x + 3y - 2z = 5; \; y - z = 4, \; z = -3$

$y - (-3) = 4 \Rightarrow y + 3 = 4 \Rightarrow y = 1$

$2x + 3(1) - 2(-3) = 5 \Rightarrow 2x + 9 = 5 \Rightarrow$

$2x = -4 \Rightarrow x = -2$

5.2 Systems of Linear Equations in Three Variables

5.2 Practice Problems

1. Substituting the ordered triple into the system

$$\begin{cases} x + y + z = 1 \\ 3x + 4y + z = -4 \\ 2x + y + 2z = 5 \end{cases},$$

we find that $(2, -3, 2)$ is a solution.

$$\begin{cases} 2 + (-3) + 2 = 1 \\ 3(2) + 4(-3) + 2 = -4 \\ 2(2) + (-3) + 2(2) = 5 \end{cases}$$

2. $$\begin{cases} 2x + 5y = 1 & (1) \\ x - 3y + 2z = 1 & (2) \\ -x + 2y + z = 7 & (3) \end{cases}$$

Interchange equations (1) and (2).

$$\begin{cases} x - 3y + 2z = 1 & (2) \\ 2x + 5y = 1 & (1) \\ -x + 2y + z = 7 & (3) \end{cases}$$

Multiply equation (2) by -2, then add the resulting equation and equation (1) to eliminate x, then replace equation (1) with the resulting equation.

$$\begin{array}{r} -2x + 6y - 4z = -2 \quad (2) \\ 2x + 5y = 1 \quad (1) \\ \hline 11y - 4z = -1 \quad (4) \end{array}$$

$$\begin{cases} x - 3y + 2z = 1 & (2) \\ 11y - 4z = -1 & (4) \\ -x + 2y + z = 7 & (3) \end{cases}$$

Now add equations (2) and (3), then replace equation (3) with the resulting equation.

$$\begin{array}{r} x - 3y + 2z = 1 \quad (2) \\ -x + 2y + z = 7 \quad (3) \\ \hline -y + 3z = 8 \quad (5) \end{array}$$

$$\begin{cases} x - 3y + 2z = 1 & (2) \\ 11y - 4z = -1 & (4) \\ -y + 3z = 8 & (5) \end{cases}$$

Now multiply equation (5) by 11, add the resulting equation to equation (4), then solve for z.

$$\begin{array}{r} 11y - 4z = -1 \quad (4) \\ -11y + 33z = 88 \quad (5) \\ \hline 29z = 87 \\ z = 3 \quad (6) \end{array}$$

The system becomes

$$\begin{cases} x - 3y + 2z = 1 & (2) \\ 11y - 4z = -1 & (4). \\ z = 3 & (6) \end{cases}$$

Now substitute $z = 3$ into equation (4) and solve for y.

$11y - 4(3) = -1 \Rightarrow 11y = 11 \Rightarrow y = 1$

Substitute the values of y and z into equation (2) and solve for x.

$x - 3(1) + 2(3) = 1 \Rightarrow x + 3 = 1 \Rightarrow x = -2$

So, we have $x = -2$, $y = 1$, and $z = 3$. Now check in the original system.

$$\begin{cases} 2(-2) + 5(1) = 1 \quad \checkmark \\ -2 - 3(1) + 2(3) = 1 \quad \checkmark \\ -(-2) + 2(1) + 3 = 7 \quad \checkmark \end{cases}$$

The solution is $\{(-2, 1, 3)\}$.

3. $$\begin{cases} 2x + 2y + 2z = 12 & (1) \\ -3x + y - 11z = -6 & (2) \\ 2x + y + 4z = -8 & (3) \end{cases}$$

Divide equation (1) by 2 and replace equation (1) with the resulting equation.

$$\begin{cases} x + y + z = 6 & (4) \\ -3x + y - 11z = -6 & (2) \\ 2x + y + 4z = -8 & (3) \end{cases}$$

Multiply equation (4) by 3, add the resulting equation to equation (2), and replace equation (2) with the result. Similarly, multiply equation (4) by -2, add the resulting equation to equation (3), and replace equation (3) with the result.

$$\begin{cases} x + y + z = 6 & (4) \\ 4y - 8z = 12 & (5) \\ -y + 2z = -20 & (6) \end{cases}$$

Multiply equation (6) by 4, add the resulting equation to equation (5), and replace equation (6) with the result.

$$\begin{array}{r} 4y - 8z = 12 \\ -4y + 8z = -80 \\ \hline 0 = -68 \end{array}$$

$$\begin{cases} x + y + z = 6 & (4) \\ 4y - 8z = 12 & (5) \\ 0 = -68 & (7) \end{cases}$$

Equation (7) is false, so the solution set of the system is $\varnothing$.

4.
$$\begin{cases} x + y + z = 5 & (1) \\ -4x - y - 8z = -29 & (2) \\ 2x + 5y - 2z = 1 & (3) \end{cases}$$

Multiply equation (1) by 4, add the resulting equation to equation (2), and replace equation (2) with the result. Similarly, multiply equation (1) by −2, add the resulting equation to equation (3), and replace equation (3) with the resulting equation.

$$\begin{cases} x + y + z = 5 & (1) \\ 3y - 4z = -9 & (4) \\ 3y - 4z = -9 & (5) \end{cases}$$

Now multiply equation (4) by −1, add the result to equation (5), and replace equation (5) with the resulting equation.

$$\begin{cases} x + y + z = 5 & (1) \\ 3y - 4z = -9 & (4) \\ 0 = 0 & (6) \end{cases}$$

Solve equation (4) for y.

$$3y - 4z = -9 \Rightarrow y = \frac{-9 + 4z}{3} = -3 + \frac{4}{3}z$$

Substitute this expression into equation (1) and solve for x.

$$x + y + z = 5 \Rightarrow x + \left(-3 + \frac{4}{3}z\right) + z = 5 \Rightarrow$$

$$x + \frac{7}{3}z - 3 = 5 \Rightarrow x = -\frac{7}{3}z + 8$$

The solution set is $\left\{\left(8 - \frac{7}{3}z, -3 + \frac{4}{3}z, z\right)\right\}$.

5.
$$\begin{cases} x + 3y + 2z = 4 & (1) \\ 2x + 7y - z = 5 & (2) \end{cases}$$

Eliminate x.

$$\begin{array}{r} -2x - 6y - 4z = -8 \\ 2x + 7y - z = 5 \\ \hline y - 5z = -3 \end{array}$$

$$\begin{cases} x + 3y + 2z = 4 & (1) \\ y - 5z = -3 & (3) \end{cases}$$

Solve equation (3) for y in terms of z.
$$y - 5z = -3 \Rightarrow y = 5z - 3$$

Substitute this expression into equation (1) and then solve for x in terms of z.
$$x + 3(5z - 3) + 2z = 4 \Rightarrow$$
$$x + 15z - 9 + 2z = 4 \Rightarrow x = -17z + 13$$
The solution set is $\{(-17z + 13, 5z - 3, z)\}$.

6. The system is
$$\begin{cases} x + y = 0.65 & (\text{beam 1}) \\ x + z = 0.70 & (\text{beam 2}) \\ y + z = 0.55 & (\text{beam 3}) \end{cases}.$$

Multiply the first equation by −1, add the result to the second equation, and replace the second equation with the new equation:
$$-1(x + y = 0.65) \Rightarrow -x - y = -0.65$$
$$\begin{cases} -x - y = -0.65 \\ x + z = 0.70 \end{cases} \Rightarrow -y + z = 0.05$$

$$\begin{cases} x + y = 0.65 \\ x + z = 0.70 \\ y + z = 0.55 \end{cases} \Rightarrow \begin{cases} x + y = 0.65 \\ -y + z = 0.05 \\ y + z = 0.55 \end{cases}$$

Add the second and third equations and solve for z: $2z = 0.60 \Rightarrow z = 0.30$. Substituting this value into the original second equation, we have $x + 0.30 = 0.70 \Rightarrow x = 0.40$. Substituting this value into the original first equation, we have $0.40 + y = 0.65 \Rightarrow y = 0.25$. Referring to table 8.1, we see that cell A contains bone (since $x = 0.40$), cell B contains healthy tissue (since $y = 0.25$), and cell C contains tumorous tissue (since $z = 0.30$).

5.2 Concepts and Vocabulary

1. Systems of equations that have the same solution sets are called <u>equivalent</u> systems.

3. If any of the equations in a system has no solution, then the system is <u>inconsistent</u>.

5. True

7. False. The system will have no solution.

5.2 Building Skills

9. Substituting the ordered triple into the system
$$\begin{cases} 2x - 2y - 3z = 1 \\ 3y + 2z = -1 \\ y + z = 0 \end{cases}$$
we find that $(1, -1, 1)$ *is* a solution.
$$\begin{cases} 2(1) - 2(-1) - 3(1) = 1 \\ 3(-1) + 2(1) = -1 \\ -1 + 1 = 0 \end{cases}$$

11. Substituting the ordered triple into the system
$$\begin{cases} x + 3y - 2z = 0 \\ 2x - y + 4z = 5 \\ x - 11y + 14z = 0 \end{cases}$$
we find that $(-10, 8, 7)$ *is not* a solution.
$$\begin{cases} -10 + 3(8) - 2(7) = 0 \\ 2(-10) - 8 + 4(7) = 0 \neq 5 \\ (-10) - 11(8) + 14(7) = 0 \end{cases}$$

13. To solve the system $\begin{cases} x+y+z=4 \\ y-2z=4 \\ z=-1 \end{cases}$,

substitute $z=-1$ into the second equation, and solve for y: $y-2(-1)=4 \Rightarrow y=2$.
Now substitute the values for y and z into the first equation, and solve for x:
$x+2-1=4 \Rightarrow x=3$.
The solution is $\{(3, 2, -1)\}$.

15. To solve the system $\begin{cases} x-5y+3z=-1 \\ y-2z=-6 \\ z=4 \end{cases}$,

substitute $z=4$ into the second equation, and solve for y: $y-2(4)=-6 \Rightarrow y=2$.
Now substitute the values for y and z into the first equation, and solve for x:
$x-5(2)+3(4)=-1 \Rightarrow x=-3$. The solution is $\{(-3, 2, 4)\}$.

17. $\begin{cases} 2x-2y-3z=1 \\ 3y+2z=-1 \\ y+z=0 \end{cases} \Rightarrow \begin{cases} 2x-2y-3z=1 \\ y+z=0 \\ 3y+2z=-1 \end{cases}$.

Multiply the first equation by $1/2$:
$\frac{1}{2}(2x-2y-3z=1) \Rightarrow x-y-\frac{3}{2}z=\frac{1}{2}$
Now multiply the second equation by -3, add the result to the third equation, and replace the third equation with the new equation:
$-3(y+z=0) \Rightarrow -3y-3z=0$.

$\begin{cases} -3y-3z=0 \\ 3y+2z=-1 \end{cases} \Rightarrow -z=-1 \Rightarrow z=1$.

So the system becomes
$\begin{cases} x-y-\frac{3}{2}z=\frac{1}{2} \\ y+\phantom{\frac{3}{2}}z=0 \\ \phantom{x-y-\frac{3}{2}}z=1 \end{cases}$.

19. Multiply the first equation by -2, add the result to the second equation, and replace the second equation with the new equation:
$-2(x+3y-2z=0) \Rightarrow -2x-6y+4z=0$

$\begin{cases} -2x-6y+4z=0 \\ 2x-y+4z=5 \end{cases} \Rightarrow -7y+8z=5$.

The system becomes:
$\begin{cases} x+3y-2z=0 \\ 2x-y+4z=5 \\ x-11y+14z=0 \end{cases} \Rightarrow \begin{cases} x+3y-2z=0 \\ -7y+8z=5 \\ x-11y+14z=0 \end{cases}$

Multiply the first equation by -1, add the result to the third equation, and replace the third equation with the new equation:
$-1(x+3y-2z=0) \Rightarrow -x-3y+2z=0$

$\begin{cases} -x-3y+2z=0 \\ x-11y+14z=5 \end{cases} \Rightarrow -14y+16z=5$.

The system becomes:
$\begin{cases} x+3y-2z=0 \\ -7y+8z=5 \\ x-11y+14z=0 \end{cases} \Rightarrow \begin{cases} x+3y-2z=0 \\ -7y+8z=5 \\ -14y+16z=5 \end{cases}$

Multiply the second equation by -2, add the result to the third equation, and replace the third equation with the new equation:
$-2(-7y+8z=5) \Rightarrow 14y-16z=-10$

$\begin{cases} 14y-16z=-10 \\ -14y+16z=5 \end{cases} \Rightarrow 0=-5$.

The system becomes:
$\begin{cases} x+3y-2z=0 \\ -7y+8z=5 \\ x-11y+14z=0 \end{cases} \Rightarrow \begin{cases} x+3y-2z=0 \\ y-\frac{8}{7}z=\frac{5}{7} \\ 0=-5 \end{cases}$

21. $\begin{cases} x-y+z=6 \\ 2y+3z=5 \\ 2z=6 \end{cases} \Rightarrow \begin{cases} x-y+z=6 \\ 2y+3z=5 \\ z=3 \end{cases}$.

Substitute $z=3$ into the second equation to solve for y: $2y+3(3)=5 \Rightarrow y=-2$. Now substitute the values for y and z into the first equation to solve for x: $x-(-2)+3=6 \Rightarrow x=1$. The solution is $\{(1, -2, 3)\}$.

23. Multiply the first equation by $-3/4$ add the result to the second equation, then replace the second equation with the new equation:

$-\frac{3}{4}(4x+4y+4z=7) \Rightarrow$

$-3x-3y-3z=-\frac{21}{4}$

$\begin{cases} -3x-3y-3z=-\frac{21}{4} \\ 3x-8y=14 \end{cases} \Rightarrow -11y-3z=\frac{35}{4} \Rightarrow$

$y+\frac{3}{11}z=-\frac{35}{44}$

$\begin{cases} 4x+4y+4z=7 \\ 3x-8y=14 \\ 4z=-1 \end{cases} \Rightarrow \begin{cases} x+y+z=\frac{7}{4} \\ y+\frac{3}{11}z=-\frac{35}{44} \\ z=-\frac{1}{4} \end{cases}$

(continued on next page)

(*continued*)

Substitute $z = -1/4$ into the second equation to solve for y:

$$y + \frac{3}{11}\left(-\frac{1}{4}\right) = -\frac{35}{44} \Rightarrow y = -\frac{8}{11}.$$

Now substitute the values for y and z into the first equation to solve for x:

$$4x + 4\left(-\frac{8}{11}\right) + 4\left(-\frac{1}{4}\right) = 7 \Rightarrow x = \frac{30}{11}.$$

The solution is $\left\{\left(\frac{30}{11}, -\frac{8}{11}, -\frac{1}{4}\right)\right\}$.

In exercises 25–46, be sure to check the answers by substituting the values into the original system of equations.

25. Multiply the first equation by -1, add the result to the second equation, and replace the second equation with the new equation:

$$-1(x + y + z = 6) \Rightarrow -x - y - z = -6$$

$$\begin{cases} -x - y - z = -6 \\ x - y + z = 2 \end{cases} \Rightarrow -2y = -4$$

$$\begin{cases} x + y + z = 6 \\ x - y + z = 2 \\ 2x + y - z = 1 \end{cases} \Rightarrow \begin{cases} x + y + z = 6 \\ -2y = -4 \\ 2x + y - z = 1 \end{cases}$$

Simplify the new second equation, and then multiply the first equation by -2, add the result to the third equation, and replace the third equation with the new equation:

$$-2(x + y + z = 6) \Rightarrow -2x - 2y - 2z = -12$$

$$\begin{cases} -2x - 2y - 2z = -12 \\ 2x + y - z = 1 \end{cases} \Rightarrow -y - 3z = -11$$

$$\begin{cases} x + y + z = 6 \\ -2y = -4 \\ 2x + y - z = 1 \end{cases} \Rightarrow \begin{cases} x + y + z = 6 \\ y = 2 \\ -y - 3z = -11 \end{cases}$$

Add the second and third equations, replacing the third equation with the new equation:

$$\begin{cases} x + y + z = 6 \\ y = 2 \\ -y - 3z = -11 \end{cases} \Rightarrow \begin{cases} x + y + z = 6 \\ y = 2 \\ -3z = -9 \end{cases} \Rightarrow$$

$$\begin{cases} x + y + z = 6 \\ y = 2 \\ z = 3 \end{cases}$$

Substitute the values for y and z into the original first equation to solve for x:

$$x + 2 + 3 = 6 \Rightarrow x = 1.$$

The solution is $\{(1, 2, 3)\}$.

27. Switch the first and second equations.

$$\begin{cases} 2x + 3y + z = 9 \\ x + 2y + 3z = 6 \\ 3x + y + 2z = 8 \end{cases} \Rightarrow \begin{cases} x + 2y + 3z = 6 \\ 2x + 3y + z = 9 \\ 3x + y + 2z = 8 \end{cases}$$

Multiply the first equation by -2, add the result to the second equation, then replace the second equation with the new equation:

$$-2(x + 2y + 3z = 6) = -2x - 4y - 6z = -12$$

$$\begin{cases} -2x - 4y - 6z = -12 \\ 2x + 3y + z = 9 \end{cases} \Rightarrow -y - 5z = -3 \Rightarrow$$

$$y + 5z = 3$$

$$\begin{cases} x + 2y + 3z = 6 \\ 2x + 3y + z = 9 \\ 3x + y + 2z = 8 \end{cases} \Rightarrow \begin{cases} x + 2y + 3z = 6 \\ y + 5z = 3 \\ 3x + y + 2z = 8 \end{cases}$$

Multiply the first equation by -3, add the result to the third equation, then replace the third equation with the new equation:

$$-3(x + 2y + 3z = 6) = -3x - 6y - 9z = -18$$

$$\begin{cases} -3x - 6y - 9z = -18 \\ 3x + y + 2z = 8 \end{cases} \Rightarrow -5y - 7z = -10 \Rightarrow$$

$$5y + 7z = 10$$

$$\begin{cases} x + 2y + 3z = 6 \\ y + 5z = 3 \\ 3x + y + 2z = 8 \end{cases} \Rightarrow \begin{cases} x + 2y + 3z = 6 \\ y + 5z = 3 \\ 5y + 7z = 10 \end{cases}$$

Multiply the second equation by -5, and add the result to the third equation, then replace the third equation with the new equation:

$$-5(y + 5z = 3) \Rightarrow -5y - 25z = -15$$

$$\begin{cases} -5y - 25z = -15 \\ 5y + 7z = 10 \end{cases} \Rightarrow -18z = -5 \Rightarrow z = \frac{5}{18}$$

$$\begin{cases} x + 2y + 3z = 6 \\ y + 5z = 3 \\ 5y + 7z = 10 \end{cases} \Rightarrow \begin{cases} x + 2y + 3z = 6 \\ y + 5z = 3 \\ z = \frac{5}{18} \end{cases}$$

Substitute the value of z into the second equation to solve for y:

$$y + 5\left(\frac{5}{18}\right) = 3 \Rightarrow y = \frac{29}{18}.$$

Substitute the values for y and z into the first equation to solve for x:

$$x + 2\left(\frac{29}{18}\right) + 3\left(\frac{5}{18}\right) = 6 \Rightarrow x = \frac{35}{18}.$$

The solution is $\left\{\left(\frac{35}{18}, \frac{29}{18}, \frac{5}{18}\right)\right\}$.

29. $\begin{cases} x - 4y + 7z = 14 \\ 3x + 8y - 2z = 13 \\ 7x - 8y + 26z = 5 \end{cases}$

Multiply the first equation by –3, add the result to the second equation, then replace the second equation with the new equation:

$-3(x - 4y + 7z = 14) \Rightarrow -3x + 12y - 21z = -42$

$\begin{cases} -3x + 12y - 21z = -42 \\ 3x + 8y - 2z = 3 \end{cases} \Rightarrow 20y - 23z = -39$

$\begin{cases} x - 4y + 7z = 14 \\ 3x + 8y - 2z = 13 \\ 7x - 8y + 26z = 5 \end{cases} \Rightarrow \begin{cases} x - 4y + 7z = 14 \\ 20y - 23z = -39 \\ 7x - 8y + 26z = 5 \end{cases}$

Multiply the first equation by –7, add the result to the third equation, then replace the third equation:

$-7(x - 4y + 7z = 14) \Rightarrow -7x + 28y - 49z = -98$

$\begin{cases} -7x + 28y - 49z = -98 \\ 7x - 8y + 26z = 5 \end{cases} \Rightarrow 20y - 23z = -93$

$\begin{cases} x - 4y + 7z = 14 \\ 20y - 23z = -39 \\ 7x - 8y + 26z = 5 \end{cases} \Rightarrow \begin{cases} x - 4y + 7z = 14 \\ 20y - 23z = -39 \\ 20y - 23z = -93 \end{cases}$

Multiplying the second equation by –1, and adding the result to the third equation gives:

$-1(20y - 23z = -39) \Rightarrow -20y + 23z = 39$

$\begin{cases} -20y + 23z = 39 \\ 20y - 23z = -93 \end{cases} \Rightarrow 0 = -54$ False

Thus, the system is inconsistent and the solution set is $\varnothing$.

31. Multiply the second equation by –2, add the result to the first equation, then replace the second equation with the new equation:

$-2(x + 3y - z = -2) \Rightarrow -2x - 6y + 2z = 4$

$\begin{cases} -2x - 6y + 2z = 4 \\ 2x + 3y + 2z = 7 \end{cases} \Rightarrow -3y + 4z = 11$

$\begin{cases} 2x + 3y + 2z = 7 \\ x + 3y - z = -2 \\ x - y + 2z = 8 \end{cases} \Rightarrow \begin{cases} 2x + 3y + 2z = 7 \\ -3y + 4z = 11 \\ x - y + 2z = 8 \end{cases}$

Multiply the third equation by –2, add the result to the first equation, then replace the third equation with the new equation:

$-2(x - y + 2z = 8) \Rightarrow -2x + 2y - 4z = -16$

$\begin{cases} -2x + 2y - 4z = -16 \\ 2x + 3y + 2z = 7 \end{cases} \Rightarrow 5y - 2z = -9$

$\begin{cases} 2x + 3y + 2z = 7 \\ x + 3y - z = -2 \\ x - y + 2z = 8 \end{cases} \Rightarrow \begin{cases} 2x + 3y + 2z = 7 \\ -3y + 4z = 11 \\ 5y - 2z = -9 \end{cases}$

Multiply the third equation by 2, and add the result to the second equation to solve for y:

$2(5y - 2z = -9) \Rightarrow 10y - 4z = -18$

$\begin{cases} 10y - 4z = -18 \\ -3y + 4z = 11 \end{cases} \Rightarrow 7y = -7 \Rightarrow y = -1$

$\begin{cases} 2x + 3y + 2z = 7 \\ -3y + 4z = 11 \\ 5y - 2z = -9 \end{cases} \Rightarrow \begin{cases} 2x + 3y + 2z = 7 \\ y = -1 \\ 5y - 2z = -9 \end{cases}$

Multiply the second equation by –5, then add the result to the third equation and replace the third equation with the new equation:

$\begin{cases} -5y = 5 \\ 5y - 2z = -9 \end{cases} \Rightarrow -2z = -4 \Rightarrow z = 2$

$\begin{cases} 2x + 3y + 2z = 7 \\ y = -1 \\ 5y - 2z = -9 \end{cases} \Rightarrow \begin{cases} 2x + 3y + 2z = 7 \\ y = -1 \\ z = 2 \end{cases}$

Substitute the values for y and z into the first equation to solve for x:

$2x + 3(-1) + 2(2) = 7 \Rightarrow x = 3$.

The solution is $\{(3, -1, 2)\}$.

33. Switch the first and third equations:

$\begin{cases} 4x - 2y + z = 5 \\ 2x + y - 2z = 4 \\ x + 3y - 2z = 6 \end{cases} \Rightarrow \begin{cases} x + 3y - 2z = 6 \\ 2x + y - 2z = 4 \\ 4x - 2y + z = 5 \end{cases}$

Multiply the first equation by –2, add the result to the second equation, and replace the second equation with the new equation:

$-2(x + 3y - 2z = 6) \Rightarrow -2x - 6y + 4z = -12$

$\begin{cases} -2x - 6y + 4z = -12 \\ 2x + y - 2z = 4 \end{cases} \Rightarrow -5y + 2z = -8$

$\begin{cases} x + 3y - 2z = 6 \\ 2x + y - 2z = 4 \\ 4x - 2y + z = 5 \end{cases} \Rightarrow \begin{cases} x + 3y - 2z = 6 \\ -5y + 2z = -8 \\ 4x - 2y + z = 5 \end{cases}$

Multiply the first equation by –4, add the result to the third equation, and replace the third equation with the new equation:

$-4(x + 3y - 2z = 6) \Rightarrow -4x - 12y + 8z = -24$

$\begin{cases} -4x - 12y + 8z = -24 \\ 4x - 2y + z = 5 \end{cases} \Rightarrow -14y + 9z = -19$

$\begin{cases} x + 3y - 2z = 6 \\ -5y + 2z = -8 \\ 4x - 2y + z = 5 \end{cases} \Rightarrow \begin{cases} x + 3y - 2z = 6 \\ -5y + 2z = -8 \\ -14y + 9z = -19 \end{cases}$

Multiply the second equation by $-\frac{14}{5}$ and add the result to the third equation and replace the third equation with the new equation:

(continued on next page)

(continued)

$$-\frac{14}{5}(-5y+2z=-8) \Rightarrow 14y - \frac{28}{5}z = \frac{112}{5}$$

$$\begin{cases} 14y - \frac{28}{5}z = \frac{112}{5} \\ -14y + 9z = -19 \end{cases} \Rightarrow \frac{17}{5}z = \frac{17}{5} \Rightarrow z = 1$$

$$\begin{cases} x+3y-2z=6 \\ -5y+2z=-8 \\ -14y+9z=-19 \end{cases} \Rightarrow \begin{cases} x+3y-2z=6 \\ -5y+2z=-8 \\ z=1 \end{cases}$$

Substitute $z = 1$ into the second equation to obtain $-5y + 2(1) = -8 \Rightarrow y = 2$. Substitute the values for y and z into the first equation to solve for x: $x + 3(2) - 2(1) = 6 \Rightarrow x = 2$. The solution is $\{(2, 2, 1)\}$.

35. Multiply the second equation by -1 and then add the result to the third equation:

$$-(x+y-z=1) \Rightarrow -x-y+z=-1$$

$$\begin{cases} -x-y+z=-1 \\ x+y+2z=4 \end{cases} \Rightarrow 3z=3 \Rightarrow z=1$$

Multiply the second equation by -2, add the result to the first equation, and replace the second equation with the new equation:

$$-2(x+y-z=1) \Rightarrow -2x-2y+2z=-2$$

$$\begin{cases} 2x+y+z=6 \\ -2x-2y+2z=-2 \end{cases} \Rightarrow -y+3z=4$$

$$\begin{cases} 2x+y+z=6 \\ x+y-z=1 \\ x+y+2z=4 \end{cases} \Rightarrow \begin{cases} 2x+y+z=6 \\ -y+3z=4 \\ z=1 \end{cases}$$

Substitute $z = 1$ into the second equation to solve for y: $-y + 3(1) = 4 \Rightarrow y = -1$. Substitute the values for y and z into the first equation to solve for x: $2x - 1 + 1 = 6 \Rightarrow x = 3$. The solution is $(3, -1, 1)$.

37. $$\begin{cases} x-y-z=3 \\ x+9y+z=3 \\ 2x+3y-z=6 \end{cases}$$

Multiply the first equation by -1, add the result to the second equation, and replace the second equation with the new equation:

$$-1(x-y-z=3) \Rightarrow -x+y+z=-3$$

$$\begin{cases} -x+y+z=-3 \\ x+9y+z=3 \end{cases} \Rightarrow 10y+2z=0$$

$$\begin{cases} x-y-z=3 \\ x+9y+z=3 \\ 2x+3y-z=6 \end{cases} \Rightarrow \begin{cases} x-y-z=3 \\ 10y+2z=0 \\ 2x+3y-z=6 \end{cases}$$

Multiply the first equation by -2, add the result to the third equation, and replace the third equation with the new equation:

$$-2(x-y-z=3) \Rightarrow -2x+2y+2z=-6$$

$$\begin{cases} -2x+2y+2z=-6 \\ 2x+3y-z=6 \end{cases} \Rightarrow 5y+z=0$$

$$\begin{cases} x-y-z=3 \\ 10y+2z=0 \\ 2x+3y-z=6 \end{cases} \Rightarrow \begin{cases} x-y-z=3 \\ 10y+2z=0 \\ 5y+z=0 \end{cases}$$

Multiply the third equation by -2, then add the result to the second equation:

$$-2(5y+z=0) \Rightarrow -10y-2z=0$$

$$\begin{cases} 10y+2z=0 \\ -10y-2z=0 \end{cases} \Rightarrow 0=0$$

The equation $0 = 0$ is equivalent to $0z = 0$, which is true for every value of z. Solving the second equation for y, we have

$$10y+2z=0 \Rightarrow y=-\frac{1}{5}z.$$ Substituting this into the first equation, we have

$$x-y-z=3 \Rightarrow$$

$$x-\left(-\frac{1}{5}z\right)-z=3 \Rightarrow x=3+\frac{4}{5}z$$

Thus, the solution is $\left\{\left(3+\frac{4}{5}z, -\frac{1}{5}z, z\right)\right\}$.

39. Rearrange the equations as shown:

$$\begin{cases} x+y=0 \\ y+2z=-4 \\ y+z=4-x \end{cases} \Rightarrow \begin{cases} x+y=0 \\ y+2z=-4 \\ x+y+z=4 \end{cases} \Rightarrow$$

$$\begin{cases} x+y+z=4 \\ y+2z=-4 \\ x+y=0 \end{cases}$$

Subtract the third equation from the first equation, and replace the third equation:

$$\begin{cases} x+y+z=4 \\ y+2z=-4 \\ x+y=0 \end{cases} \Rightarrow \begin{cases} x+y+z=4 \\ y+2z=-4 \\ z=4 \end{cases}$$

Substitute $z = 4$ into the second equation to solve for y: $y + 2(4) = -4 \Rightarrow y = -12$.

Substitute the values for x and y into the first equation to solve for x:
$x - 12 + 4 = 4 \Rightarrow x = 12$.
The solution is $\{(12, -12, 4)\}$.

41. Add the first and second equations, and replace the second equation:

$$\begin{cases} 2y - z = -4 \\ x + z = 3 \end{cases} \Rightarrow x + 2y = -1.$$

$$\begin{cases} 2y - z = -4 \\ x + z = 3 \\ 2x + 3y = -1 \end{cases} \Rightarrow \begin{cases} 2y - z = -4 \\ x + 2y = -1 \\ 2x + 3y = -1 \end{cases}$$

Switch the first and third equations, multiply the second equation by −2, add the result to the first equation, and replace the second equation with the new equation:

$$-2(x + 2y = -1) \Rightarrow -2x - 4y = 2$$

$$\begin{cases} -2x - 4y = 2 \\ 2x + 3y = -1 \end{cases} \Rightarrow -y = 1 \Rightarrow y = -1$$

$$\begin{cases} 2x + 3y = -1 \\ y = -1 \\ 2y - z = -4 \end{cases}$$

Multiply the second equation by −2, and add the result to the third equation.

$$\begin{cases} -2y = 2 \\ 2y - z = -4 \end{cases} \Rightarrow -z = -2 \Rightarrow z = 2$$

$$\begin{cases} 2x + 3y = -1 \\ y = -1 \\ z = 2 \end{cases}$$

Substitute the value for y into the first equation to solve for x:

$$2x + 3(-1) = -1 \Rightarrow x = 1.$$

The solution is $\{(1, -1, 2)\}$.

43. Multiply the second equation by −2, add the result to the third equation, and replace the second equation with the result:

$$-2(2x + y = 8) \Rightarrow -4x - 2y = -16$$

$$\begin{cases} 2y + 3z = 1 \\ -4x - 2y = -16 \end{cases} \Rightarrow -4x + 3z = -15$$

$$\begin{cases} 3x - 2z = 11 \\ 2x + y = 8 \\ 2y + 3z = 1 \end{cases} \Rightarrow \begin{cases} 3x - 2z = 11 \\ -4x + 3z = -15 \\ 2y + 3z = 1 \end{cases}$$

Multiply the first equation by 4 and the second equation by 3, add the two result and replace the second equation:

$$4(3x - 2z = 11) \Rightarrow 12x - 8z = 44$$

$$3(-4x + 3z = -15) \Rightarrow -12x + 9x = -45$$

$$\begin{cases} 12x - 8z = 44 \\ -12x + 9z = -45 \end{cases} \Rightarrow z = -1$$

$$\begin{cases} 3x - 2z = 11 \\ -4x + 3z = -15 \\ 2y + 3z = 1 \end{cases} \Rightarrow$$

$$\begin{cases} 3x - 2z = 11 \\ z = -1 \\ 2y + 3z = 1 \end{cases} \Rightarrow \begin{cases} 3x - 2z = 11 \\ 2y + 3z = 1 \\ z = -1 \end{cases}$$

Substitute $z = -1$ into the first and second equations to solve for x and y:

$$2y + 3(-1) = 1 \Rightarrow y = 2.$$

$$3x - 2(-1) = 11 \Rightarrow x = 3.$$

The solution is $\{(3, 2, -1)\}$.

45. Multiplying the second equation by −1, and then adding the two equations, we have:

$$\begin{cases} 2x + 6y + 11 = 0 \\ -6y + 18z - 1 = 0 \end{cases} \Rightarrow 2x + 18z + 10 = 0 \Rightarrow$$

$$2x + 18z = -10 \Rightarrow x = -5 - 9z$$

$$6y - 18z + 1 = 0 \Rightarrow y = -\frac{1}{6} + 3z.$$

The solution is $\left\{ \left(-5 - 9z, -\frac{1}{6} + 3z, z \right) \right\}$.

The system is dependent.

5.2 Applying the Concepts

47. Let x = the amount invested at 4%, y = the amount invested at 5%, and z = the amount invested at 6%. Then, we have the system:

$$\begin{cases} x + y + z = 20,000 \\ 0.04x + 0.05y + 0.06z = 1060 \\ 0.06z = 2(0.05y) \end{cases} \Rightarrow$$

$$\begin{cases} x + y + z = 20,000 \\ 4x + 5y + 6z = 106,000 \\ 6z = 10y \end{cases} \Rightarrow$$

$$\begin{cases} x + y + z = 20,000 \\ 4x + 5y + 6z = 106,000 \\ -10y + 6z = 0 \end{cases}$$

Multiply the first equation by −4, add the result to the second equation, and replace the second equation with the new equation:

$$-4(x + y + z = 20,000) \Rightarrow$$

$$-4x - 4y - 4z = -80,000$$

$$\begin{cases} -4x - 4y - 4z = -80,000 \\ 4x + 5y + 6z = 106,000 \end{cases} \Rightarrow y + 2z = 26,000$$

$$\begin{cases} x + y + z = 20,000 \\ y + 2z = 26,000 \\ -10y + 6z = 0 \end{cases}$$

(continued on next page)

(continued)

Multiply the second equation by 10, add the result to the third equation, and solve for z:

$$10(y + 2z = 26,000) \Rightarrow 10y + 20z = 260,000$$

$$\begin{cases} 10y + 20z = 260,000 \\ -10y + 6z = 0 \end{cases} \Rightarrow 26z = 260,000 \Rightarrow$$

$$z = 10,000$$

$$\begin{cases} x + y + z = 20,000 \\ y + 2z = 26,000 \Rightarrow \\ -10y + 6z = 0 \end{cases}$$

$$\begin{cases} x + y + z = 20,000 \\ y + 2z = 26,000 \\ z = 10,000 \end{cases}$$

Substitute $z = 10,000$ into the second equation to solve for y:

$$y + 2(10,000) = 26,000 \Rightarrow y = 6000.$$

Substitute the values for y and z into the first equation to solve for x:

$$x + 6000 + 10,000 = 20,000 \Rightarrow x = 4000.$$

Miguel invested \$4000 at 4%, \$6000 at 5%, and \$10,000 at 6%.

49. Let a = the number of hours Alex worked, b = the number of hours Becky worked, and c = the number of hours Courtney worked. Then, we have the system:

$$\begin{cases} a + b + c = 6 \\ 124a + 118b + 132c = 741 \Rightarrow \\ b + c = 2a \end{cases}$$

$$\begin{cases} a + b + c = 6 \\ 124a + 118b + 132c = 741 \\ -2a + b + c = 0 \end{cases}$$

Subtract the third equation from the first, and replace the first equation with the result:

$$\begin{cases} a + b + c = 6 \\ 124a + 118b + 132c = 741 \Rightarrow \\ -2a + b + c = 0 \end{cases}$$

$$\begin{cases} 3a = 6 \\ 124a + 118b + 132c = 741 \Rightarrow \\ -2a + b + c = 0 \end{cases}$$

$$\begin{cases} a = 2 \\ 124a + 118b + 132c = 741 \Rightarrow \\ -2a + b + c = 0 \end{cases}$$

Substitute $a = 2$ into the second and third equations and simplify:

$$\begin{cases} a = 2 \\ 124(2) + 118b + 132c = 741 \Rightarrow \\ -2(2) + b + c = 0 \end{cases}$$

$$\begin{cases} a = 2 \\ 118b + 132c = 493 \\ b + c = 4 \end{cases}$$

Multiply the third equation by -118, and add the result to the second equation to solve for c:

$$-118(b + c = 4) \Rightarrow -118b - 118c = -472$$

$$\begin{cases} 118b + 132c = 493 \\ -118b - 118c = -472 \end{cases} \Rightarrow 14c = 21 \Rightarrow c = 1.5$$

Substitute $c = 1.5$ into the third equation to solve for b: $b + 1.5 = 4 \Rightarrow b = 2.5$.

Alex worked 2 hours, Becky worked 2.5 hours, and Courtney worked 1.5 hours.

51. Let n = the number of nickels, d = the number of dimes, and q = the number of quarters. Then, we have the system

$$\begin{cases} n + d + q = 300 \\ d = 3(n + q) \\ 0.05n + 0.1d + 0.25q = 30.05 \end{cases} \Rightarrow$$

$$\begin{cases} n + d + q = 300 \\ -3n + d - 3q = 0 \\ 5n + 10d + 25q = 3005 \end{cases}$$

Multiply the first equation by 3, add the result to the second equation, and replace the second equation with the new equation:

$$3(n + d + q = 300) \Rightarrow 3n + 3d + 3q = 900$$

$$\begin{cases} 3n + 3d + 3q = 900 \\ -3n + d - 3q = 0 \end{cases} \Rightarrow 4d = 900 \Rightarrow d = 225$$

$$\begin{cases} n + d + q = 300 \\ d = 225 \\ 5n + 10d + 25q = 3005 \end{cases}$$

Multiply the first equation by -5, add the result to the third equation, and replace the third equation with the new equation:

$$-5(n + d + q = 300) \Rightarrow -5n - 5d - 5q = -1500$$

$$\begin{cases} -5n - 5d - 5q = -1500 \\ 5n + 10d + 25q = 3005 \end{cases} \Rightarrow 5d + 20q = 1505$$

$$\begin{cases} n + d + q = 300 \\ d = 225 \Rightarrow \\ 5n + 10d + 25q = 3005 \end{cases}$$

$$\begin{cases} n + d + q = 300 \\ d = 225 \\ 5d + 20q = 1505 \end{cases}$$

Substitute $d = 225$ into the third equation to solve for q, then substitute the values for d and q into the first equation to solve for n:

$$5(225) + 20q = 1505 \Rightarrow q = 19.$$

$$n + 225 + 19 = 300 \Rightarrow n = 56.$$

There are 56 nickels, 225 dimes, and 19 quarters.

53. Let x = the number of daytime hours Amy worked, y = the number of night hours Amy worked, and z = the number of holiday hours Amy worked. Then, we have the system

$$\begin{cases} x + y + z = 53 \\ 7.40x + 9.20y + 11.75z = 452.20 \\ x = y + z + 9 \end{cases} \Rightarrow$$

$$\begin{cases} x + y + z = 53 \\ 7.40x + 9.20y + 11.75z = 452.20 \\ x - y - z = 9 \end{cases}$$

Add the first and third equations, and replace the first equation with the result:

$$\begin{cases} x + y + z = 53 \\ 7.40x + 9.20y + 11.75z = 452.20 \\ x - y - z = 9 \end{cases} \Rightarrow$$

$$\begin{cases} 2x = 62 \\ 7.40x + 9.20y + 11.75z = 452.20 \\ x - y - z = 9 \end{cases} \Rightarrow$$

$$\begin{cases} x = 31 \\ 7.40x + 9.20y + 11.75z = 452.20 \\ x - y - z = 9 \end{cases}$$

Substitute $x = 31$ into the second and third equations and simplify:

$$\begin{cases} x = 31 \\ 7.40x + 9.20y + 11.75z = 452.20 \\ x - y - z = 9 \end{cases} \Rightarrow$$

$$\begin{cases} x = 31 \\ 7.40(31) + 9.20y + 11.75z = 452.20 \\ 31 - y - z = 9 \end{cases} \Rightarrow$$

$$\begin{cases} x = 31 \\ 9.20y + 11.75z = 222.80 \\ -y - z = -22 \end{cases}$$

Multiply the third equation by 9.2, add the result to the second equation, and solve for z:
$9.2(-y - z = -22) \Rightarrow -9.2y - 9.2z = 202.4$

$$\begin{cases} 9.20y + 11.75z = 222.80 \\ -9.20y - 9.20z = -202.40 \end{cases} \Rightarrow$$

$2.55z = 20.40 \Rightarrow z = 8$

$$\begin{cases} x = 31 \\ 9.20y + 11.75z = 222.80 \\ -y - z = -22 \end{cases} \Rightarrow$$

$$\begin{cases} x = 31 \\ 9.20y + 11.75z = 222.80 \\ z = 8 \end{cases}$$

$9.20y + 11.75(8) = 222.80 \Rightarrow y = 14$

Amy worked 31 daytime hours, 14 night hours, and 8 holiday hours.

For exercises 55–58, refer to Table 8.1 in the text.

55. The system is

$$\begin{cases} x + y = 0.54 \text{ (beam 1)} \\ x + z = 0.40 \text{ (beam 2)} \\ y + z = 0.52 \text{ (beam 3)} \end{cases}$$

Multiply the first equation by -1, add the result to the second equation, and replace the second equation with the new equation:
$-1(x + y = 0.54) \Rightarrow -x - y = -0.54$

$$\begin{cases} -x - y = -0.54 \\ x + z = 0.40 \end{cases} \Rightarrow -y + z = -0.14$$

$$\begin{cases} x + y = 0.54 \\ x + z = 0.40 \\ y + z = 0.52 \end{cases} \Rightarrow \begin{cases} x + y = 0.54 \\ -y + z = -0.14 \\ y + z = 0.52 \end{cases}$$

Add the second and third equations and solve for z: $2z = 0.38 \Rightarrow z = 0.19$. Substituting this value into the original second equation, we have $x + 0.19 = 0.40 \Rightarrow x = 0.21$. Substituting this value into the original first equation, we have $0.21 + y = 0.54 \Rightarrow y = 0.33$.

Cell A contains healthy tissue (because $x = 0.21$), cell B contains tumorous tissue (because $y = 0.33$), and cell C contains healthy tissue (because $z = 0.19$).

57. The system is

$$\begin{cases} x + y = 0.51 \text{ (beam 1)} \\ x + z = 0.49 \text{ (beam 2)} \\ y + z = 0.44 \text{ (beam 3)} \end{cases}$$

Multiply the first equation by -1, add the result to the second equation, and replace the second equation with the new equation:
$-1(x + y = 0.51) \Rightarrow -x - y = -0.51$

$$\begin{cases} -x - y = -0.51 \\ x + z = 0.49 \end{cases} \Rightarrow -y + z = -0.02$$

$$\begin{cases} x + y = 0.51 \\ x + z = 0.49 \\ y + z = 0.44 \end{cases} \Rightarrow \begin{cases} x + y = 0.51 \\ -y + z = -0.02 \\ y + z = 0.44 \end{cases}$$

Add the second and third equations and solve for z: $2z = 0.42 \Rightarrow z = 0.21$. Substituting this value into the original second equation, we have $x + 0.21 = 0.49 \Rightarrow x = 0.28$. Substituting this value into the original first equation, we have $0.28 + y = 0.51 \Rightarrow y = 0.23$.

Cell A contains healthy tissue (because $x = 0.28$), cell B contains healthy tissue (because $y = 0.23$), and cell C contains healthy tissue (because $z = 0.21$).

5.2 Beyond the Basics

59. Because each of the ordered triples satisfies the given equation, we can find the values of the coefficients by solving the system

$$\begin{cases} 1 + 0b + 0c = d \\ 0 + 1b + 0c = d \\ 0 + 0b + 1c = d \end{cases} \Rightarrow \begin{cases} 1 = d \\ b = d \\ c = d \end{cases} \Rightarrow b = c = d = 1.$$

The equation is $x + y + z = 1$.

61. Because each of the ordered triples satisfies the given equation, we can find the values of the coefficients by solving the system

$$\begin{cases} 3 - 4b + 0c = d \\ 0 + \dfrac{1}{4}b + \dfrac{1}{2}c = d \\ 1 + 1b - 4c = d \end{cases} \Rightarrow \begin{cases} -4b - \quad d = -3 \\ b + 2c - 4d = 0 \\ b - 4c - \ d = -1 \end{cases}$$

Multiplying the second equation by 2, adding it to the third equation, and replacing the second equation with the new equation, we have

$$2(b + 2c - 4d = 0) \Rightarrow 2b + 4c - 8d = 0$$

$$\begin{cases} 2b + 4c - 8d = 0 \\ b - 4c - \ d = -1 \end{cases} \Rightarrow 3b - 9d = -1$$

$$\begin{cases} -4b - \quad d = -3 \\ b + 2c - 4d = 0 \\ b - 4c - \ d = -1 \end{cases} \Rightarrow \begin{cases} -4b - \quad d = -3 \\ 3b - \quad 9d = -1 \\ b - 4c - \ d = 1 \end{cases}$$

From the first equation, we have $d = -3 - 4b$. Substituting this into the second equation, we have

$$3b - 9(-3 - 4b) = -1 \Rightarrow 39b = 26 \Rightarrow b = \frac{2}{3}.$$

Then $d = 3 - 4\left(\dfrac{2}{3}\right) - 3 = \dfrac{1}{3}$.

$$\frac{1}{4}\left(\frac{2}{3}\right) + \frac{1}{2}c = \frac{1}{3} \Rightarrow \frac{1}{2}c = \frac{1}{6} \Rightarrow c = \frac{1}{3}.$$

The equation is $x + \dfrac{2}{3}y + \dfrac{1}{3}z = \dfrac{1}{3}$.

63. Because each of the ordered triples satisfies the given equation, we can find the values of the coefficients by solving the system

$$\begin{cases} a(0)^2 + \ b(0) + \ c = 1 \\ a(-1)^2 + b(-1) + c = 0 \\ a(1)^2 + \ b(1) + \ c = 4 \end{cases} \Rightarrow \begin{cases} c = 1 \\ a - b + c = 0 \\ a + b + c = 4 \end{cases}$$

Substituting $c = 1$ into the second and third equations and then solving for b and c, we have $\begin{cases} a - b + 1 = 0 \\ a + b + 1 = 4 \end{cases} \Rightarrow 2a = 2 \Rightarrow a = 1$

$1 - b + 1 = 0 \Rightarrow b = 2$

The equation is $y = x^2 + 2x + 1$.

65. Because each of the ordered triples satisfies the given equation, we can find the values of the coefficients by solving the system

$$\begin{cases} a(1)^2 + \ b(1) + \ c = 2 \\ a(-1)^2 + b(-1) + c = 4 \\ a(2)^2 + \ b(2) + \ c = 4 \end{cases} \Rightarrow \begin{cases} c = 2 \\ a - b + c = 4 \\ 4a + 2b + c = 4 \end{cases}$$

Substituting $c = 2$ into the second and third equations and then solving for b and c, we have

$$\begin{cases} a - b + 2 = 4 \\ 4a + 2b + 2 = 4 \end{cases} \Rightarrow \begin{cases} 2a - 2b + 4 = 8 \\ 4a + 2b + 2 = 4 \end{cases} \Rightarrow$$

$6a = 6 \Rightarrow a = 1$

$1 - b + 2 = 4 \Rightarrow b = -1$

The equation is $y = x^2 - x + 2$.

67. Because each of the ordered triples satisfies the given equation, we can find the values of the coefficients by solving the system

$$\begin{cases} 0^2 + 4^2 + a(0) + b(4) + c = 0 \\ \left(2\sqrt{2}\right)^2 + \left(2\sqrt{2}\right)^2 + a\left(2\sqrt{2}\right) + b\left(2\sqrt{2}\right) + c = 0 \\ (-4)^2 + 0^2 + a(-4) + b(0) + c = 0 \end{cases} \Rightarrow$$

$$\begin{cases} 4b + c = -16 \\ \left(2\sqrt{2}\right)a + \left(2\sqrt{2}\right)b + c = -16 \\ -4a + c = -16 \end{cases}$$

From the third equation, we have $c = 4a - 16$. Substituting this expression into the first and second equations, we have

$$\begin{cases} 4b + 4a - 16 = -16 \\ \left(2\sqrt{2}\right)a + \left(2\sqrt{2}\right)b + 4a - 16 = -16 \end{cases} \Rightarrow$$

$$\begin{cases} 4a + 4b = 0 \\ \left(2\sqrt{2} + 4\right)a + \left(2\sqrt{2}\right)b = 0 \end{cases} \Rightarrow$$

$$\begin{cases} a + b = 0 \\ \left(2\sqrt{2} + 4\right)a + \left(2\sqrt{2}\right)b = 0 \end{cases} \Rightarrow$$

$$\begin{cases} a = -b \\ \left(2\sqrt{2} + 4\right)a + \left(2\sqrt{2}\right)b = 0 \end{cases} \Rightarrow$$

$$\left(2\sqrt{2} + 4\right)(-b) + \left(2\sqrt{2}\right)b = 0 \Rightarrow$$

$-4b = 0 \Rightarrow b = 0$. Substituting this value into the first equation, we have $4(0) + c = -16 \Rightarrow$ $c = -16$. Substituting into the third equation, we have $-4a - 16 = -16 \Rightarrow a = 0$. The equation is $x^2 + y^2 - 16 = 0$.

69. Because each of the ordered triples satisfies the given equation, we can find the values of the coefficients by solving the system

$$\begin{cases} 1^2 + 2^2 + a(1) + b(2) + c = 0 \\ 6^2 + (-3)^2 + a(6) + b(-3) + c = 0 \\ (4)^2 + 1^2 + a(4) + b(1) + c = 0 \end{cases} \Rightarrow$$

$$\begin{cases} a + 2b + c = -5 \\ 6a - 3b + c = -45 \\ 4a + b + c = -17 \end{cases}$$

Multiply the first equation by -6, add the result to the second equation, then replace the second equation with the new equation. Similarly, multiply the first equation by -4, add that result to the third equation, then replace the third equation with the new equation.

$$-6(a + 2b + c = -5) \Rightarrow -6a - 12b - 6c = 30$$

$$\begin{cases} -6a - 12b - 6c = 30 \\ 6a - 3b + c = -45 \end{cases} \Rightarrow -15b - 5c = -15 \Rightarrow$$

$$3b + c = 3$$

$$-4(a + 2b + c = -5) \Rightarrow -4a - 8b - 4c = 20$$

$$\begin{cases} -4a - 8b - 4c = 20 \\ 4a + b + c = -17 \end{cases} \Rightarrow -7b - 3c = 3$$

$$\begin{cases} a + 2b + c = -5 \\ 6a - 3b + c = -45 \\ 4a + b + c = -17 \end{cases} \Rightarrow \begin{cases} a + 2b + c = -5 \\ 3b + c = 3 \\ -7b - 3c = 3 \end{cases}$$

From the second equation, we have $c = -3b + 3$. Substitute this expression into the third equation, and solve for b.

$$-7b - 3(-3b + 3) = 3 \Rightarrow 2b = 12 \Rightarrow b = 6.$$

So, $c = -3(6) + 3 \Rightarrow c = -15$. Substituting into the original first equation, we have $a + 2(6) - 15 = -5 \Rightarrow a = -2$.

The equation is $x^2 + y^2 - 2x + 6y - 15 = 0$.

71. Letting $u = 1/x, v = 1/y, w = 1/z$, we have

$$\begin{cases} \dfrac{1}{x} + \dfrac{3}{y} - \dfrac{1}{z} = 5 \\ \dfrac{2}{x} + \dfrac{4}{y} + \dfrac{6}{z} = 4 \\ \dfrac{2}{x} + \dfrac{3}{y} + \dfrac{1}{z} = 3 \end{cases} \Rightarrow \begin{cases} u + 3v - w = 5 \\ 2u + 4v + 6w = 4 \\ 2u + 3v + w = 3 \end{cases}$$

Multiplying the first equation by -2, adding the result to the second and third equations, and replacing those equations with the new equations, we have

$$-2(u + 3v - w = 5) \Rightarrow -2u - 6v + 2w = -10$$

$$\begin{cases} -2u - 6v + 2w = -10 \\ 2u + 4v + 6w = 4 \end{cases} \Rightarrow -2v + 8w = -6 \Rightarrow$$

$$v - 4w = 3$$

$$\begin{cases} -2u - 6v + 2w = -10 \\ 2u + 3v + w = 3 \end{cases} \Rightarrow -3v + 3w = -7$$

$$\begin{cases} u + 3v - w = 5 \\ 2u + 4v + 6w = 4 \\ 2u + 3v + w = 3 \end{cases} \Rightarrow \begin{cases} u + 3v - w = 5 \\ v - 4w = 3 \\ -3v + 3w = -7 \end{cases}$$

Multiplying the second equation by 3 and adding the result to the third equation, we have

$$3(v - 4w = 3) \Rightarrow 3v - 12w = 9$$

$$\begin{cases} 3v - 12w = 9 \\ -3v + 3w = -7 \end{cases} \Rightarrow -9w = 2 \Rightarrow w = -\dfrac{2}{9}.$$

Substituting, we have

$$v - 4\left(-\dfrac{2}{9}\right) = 3 \Rightarrow v = \dfrac{19}{9} \text{ and}$$

$$u + 3\left(\dfrac{19}{9}\right) - \left(-\dfrac{2}{9}\right) = 5 \Rightarrow u = -\dfrac{14}{9}. \text{ So,}$$

$$x = -\dfrac{9}{14}, y = \dfrac{9}{19}, z = -\dfrac{9}{2}.$$

73. Solve the first three equations in the system for x, y, and z:

$$-3(x + 2y - 5z = 9) \Rightarrow -3x - 6y + 15z = -27$$

$$\begin{cases} -3x - 6y + 15z = -27 \\ 3x - y + 2z = 14 \end{cases} \Rightarrow -7y + 17z = -13$$

$$-2(x + 2y - 5z = 9) \Rightarrow -2x - 4y + 10z = -18$$

$$\begin{cases} -2x - 4y + 10z = -18 \\ 2x + 3y - z = 3 \end{cases} \Rightarrow -y + 9z = -15$$

$$\begin{cases} x + 2y - 5z = 9 \\ 3x - y + 2z = 14 \\ 2x + 3y - z = 3 \end{cases} \Rightarrow \begin{cases} x + 2y - 5z = 9 \\ -7y + 17z = -13 \\ -y + 9z = -15 \end{cases}$$

$$-7(-y + 9z = -15) \Rightarrow 7y - 63z = 105$$

$$\begin{cases} -7y + 17z = -13 \\ 7y - 63z = 105 \end{cases} \Rightarrow -46z = 92 \Rightarrow z = -2$$

$$-7y + 17(-2) = -13 \Rightarrow -7y = 21 \Rightarrow y = -3$$

$$x + 2(-3) - 5(-2) = 9 \Rightarrow x = 5$$

Substituting $x = 5$, $y = -3$, and $z = -2$ into the fourth equation of the original system, we have

$$5c - 5(-3) + (-2) + 3 = 0 \Rightarrow$$

$$5c = -16 \Rightarrow c = -\dfrac{16}{5}.$$

75. Each of the ordered triples satisfies the given equation, so find the values of the coefficients by solving the system

$$\begin{cases} a(-1)^2 + b(-1) + c = -1 \\ a(0)^2 + b(0) + c = 5 \\ a(2)^2 + b(2) + c = 5 \end{cases} \Rightarrow \begin{cases} a - b + c = -1 \\ c = 5 \\ 4a + 2b + c = 5 \end{cases}$$

Substitute $c = 5$ into the first and third equations and then solve those equations for a and b

$$\begin{cases} a - b + 5 = -1 \\ 4a + 2b + 5 = 5 \end{cases} \Rightarrow \begin{cases} a - b = -6 \\ 2a + b = 0 \end{cases} \Rightarrow$$

$3a = -6 \Rightarrow a = -2$

$-2 - b + 5 = -1 \Rightarrow b = -4.$

The equation is $y = -2x^2 + 4x + 5$.

5.2 Critical Thinking/Discussion/Writing

Answers may vary for exercises 77 and 78.

77. $\begin{cases} 1 + (-1) - 2 = -2 \\ 2(1) - (-1) + 3(2) = 9 \\ 1 + (-1) + 2 = 2 \end{cases} \Rightarrow \begin{cases} x + y - z = -2 \\ 2x - y + 3z = 9 \\ x + y + z = 2 \end{cases}$

78. a. $\begin{cases} x + 3y + 3z = 15 \\ x + 2y + z = 1 \\ 2y + 4z = 11 \end{cases}$ **b.** $\begin{cases} x + 2y - z = 4 \\ x + 3y + 2z = 5 \\ y + 3z = 1 \end{cases}$

8.2 Getting Ready for the Next Section

79. $\dfrac{1}{2 \cdot 3} = \dfrac{1}{2} + \dfrac{B}{3} \Rightarrow 1 = 3 + 2B \Rightarrow -2 = 2B \Rightarrow$

$B = -1$

81. $\dfrac{1}{n \cdot (n+1)} = \dfrac{A}{n} - \dfrac{1}{n+1} \Rightarrow 1 = A(n+1) - n \Rightarrow$

$1 + n = A(n+1) \Rightarrow 1 = A$

83. $2x + 3 = A(x + 3) + B(x - 1)$

$2x + 3 = Ax + 3A + Bx - B = (A + B)x + 3A - B$

Equating the coefficients gives the system

$\begin{cases} A + B = 2 \\ 3A - B = 3 \end{cases} \Rightarrow 4A = 5 \Rightarrow A = \dfrac{5}{4}$

$\dfrac{5}{4} + B = 2 \Rightarrow B = \dfrac{3}{4}$

85. $x^2 + 5x + 6 = (x + 2)(x + 3)$

87. $2x^2 + 5x - 3 = (2x - 1)(x + 3)$

89. $4x^2 - 9 = (2x - 3)(2x + 3)$

5.3 Partial-Fraction Decomposition

5.3 Practice Problems

1. $\dfrac{2x - 7}{(x+1)(x-2)} = \dfrac{A}{x+1} + \dfrac{B}{x-2} \Rightarrow$

$2x - 7 = A(x - 2) + B(x + 1) \Rightarrow$

$2x - 7 = (A + B)x + (-2A + B) \Rightarrow$

$\begin{cases} A + B = 2 \\ -2A + B = -7 \end{cases} \Rightarrow \begin{cases} -A - B = -2 \\ -2A + B = -7 \end{cases} \Rightarrow$

$-3A = -9 \Rightarrow A = 3$

$3 + B = 2 \Rightarrow B = -1$

$\dfrac{2x - 7}{(x+1)(x-2)} = \dfrac{3}{x+1} - \dfrac{1}{x-2}$

2. $\dfrac{3x^2 + 4x + 3}{x^3 - x} = \dfrac{3x^2 + 4x + 3}{x(x-1)(x+1)}$

$= \dfrac{A}{x} + \dfrac{B}{x-1} + \dfrac{C}{x+1} \Rightarrow$

$3x^2 + 4x + 3$

$= A(x+1)(x-1) + Bx(x+1) + Cx(x-1)$

$= Ax^2 - A + Bx^2 + Bx + Cx^2 - Cx$

$= (A + B + C)x^2 + (B - C)x - A$

$\begin{cases} A + B + C = 3 \\ B - C = 4 \\ -A = 3 \end{cases} \Rightarrow A = -3$

$\begin{cases} -3 + B + C = 3 \\ B - C = 4 \end{cases} \Rightarrow -3 + 2B = 7 \Rightarrow B = 5$

$5 - C = 4 \Rightarrow C = 1$

$\dfrac{3x^2 + 4x + 3}{x^3 - x} = -\dfrac{3}{x} + \dfrac{5}{x-1} + \dfrac{1}{x+1}$

3. $\dfrac{x + 5}{x(x-1)^2} = \dfrac{A}{x} + \dfrac{B}{(x-1)} + \dfrac{C}{(x-1)^2} \Rightarrow$

$x + 5 = A(x-1)^2 + Bx(x-1) + Cx \Rightarrow$

$x + 5 = (A + B)x^2 + (-2A - B + C)x + A$

$\begin{cases} A + B = 0 \\ -2A - B + C = 1 \\ A = 5 \end{cases} \Rightarrow A = 5, \ B = -5$

$-2A - B + C = 1 \Rightarrow -2(5) + 5 + C = 1 \Rightarrow C = 6$

$\dfrac{x + 5}{x(x-1)^2} = \dfrac{5}{x} - \dfrac{5}{(x-1)} + \dfrac{6}{(x-1)^2}$

4. $\dfrac{2x+1}{(x+3)^2} = \dfrac{A}{(x+3)} + \dfrac{B}{(x+3)^2} \Rightarrow$

$2x+1 = A(x+3) + B \Rightarrow 2x+1 = Ax + 3A + B$

$\begin{cases} A = 2 \\ 3A + B = 1 \end{cases} \Rightarrow A = 2$

$3A + B = 1 \Rightarrow 3(2) + B = 1 \Rightarrow B = -5$

$\dfrac{2x+1}{(x+3)^2} = \dfrac{2}{x+3} - \dfrac{5}{(x+3)^2}$

5. $\dfrac{3x^2 + 5x - 2}{x(x^2+2)} = \dfrac{A}{x} + \dfrac{Bx+C}{x^2+2} \Rightarrow$

$3x^2 + 5x - 2 = A(x^2+2) + (Bx+C)x \Rightarrow$

$3x^2 + 5x - 2 = (A+B)x^2 + Cx + 2A \Rightarrow$

$\begin{cases} A + B = 3 \\ \quad C = 5 \\ 2A = -2 \end{cases} \Rightarrow A = -1, \; C = 5$

$A + B = 3 \Rightarrow -1 + B = 3 \Rightarrow B = 4$

$\dfrac{3x^2 + 5x - 2}{x(x^2+2)} = -\dfrac{1}{x} + \dfrac{4x+5}{x^2+2}$

6. $\dfrac{x^2 + 3x + 1}{(x^2+1)^2} = \dfrac{Ax+B}{x^2+1} + \dfrac{Cx+D}{(x^2+1)^2} \Rightarrow$

$x^2 + 3x + 1 = (Ax+B)(x^2+1) + Cx + D \Rightarrow$

$x^2 + 3x + 1 = Ax^3 + Bx^2 + (A+C)x + (B+D) \Rightarrow$

$\begin{cases} A = 0 \\ \quad B = 1 \\ A + \quad C = 3 \\ \quad B + D = 1 \end{cases} \Rightarrow A = 0, B = 1, C = 3, \; D = 0$

$\dfrac{x^2 + 3x + 1}{(x^2+1)^2} = \dfrac{1}{x^2+1} + \dfrac{3x}{(x^2+1)^2}$

7. $\dfrac{1}{4 \cdot 5} + \dfrac{1}{5 \cdot 6} + \dfrac{1}{6 \cdot 7} + \cdots + \dfrac{1}{3111 \cdot 3112}$

$= \left(\dfrac{1}{4} - \dfrac{1}{5}\right) + \left(\dfrac{1}{5} - \dfrac{1}{6}\right) + \left(\dfrac{1}{6} - \dfrac{1}{7}\right) + \cdots$

$\qquad + \left(\dfrac{1}{3111} - \dfrac{1}{3112}\right)$

$= \dfrac{1}{4} - \dfrac{1}{3112} = \dfrac{778 - 1}{3112} = \dfrac{777}{3112}$

8. $R = \dfrac{(x+1)(x+2)}{4x+7} \Rightarrow \dfrac{1}{R} = \dfrac{4x+7}{(x+1)(x+2)}$

$\dfrac{4x+7}{(x+1)(x+2)} = \dfrac{A}{x+1} + \dfrac{B}{x+2} \Rightarrow$

$4x+7 = A(x+2) + B(x+1) \Rightarrow$

$4x+7 = (A+B)x + (2A+B) \Rightarrow$

$\begin{cases} A + B = 4 \\ 2A + B = 7 \end{cases} \Rightarrow A = 3, B = 1$

$\dfrac{4x+7}{(x+1)(x+2)} = \dfrac{3}{x+1} + \dfrac{1}{x+2}$

$\qquad\qquad = \dfrac{1}{\dfrac{x+1}{3}} + \dfrac{1}{x+2} = \dfrac{1}{R}$

This means that if two resistances $R_1 = \dfrac{x+1}{3}$

and $R_2 = x+2$ are connected in parallel, they will produce a total resistance given by $\dfrac{(x+1)(x+2)}{4x+7}$.

5.3 Concepts and Vocabulary

1. In a rational expression, if the degree of the numerator, $P(x)$, is less than the degree of the denominator, $Q(x)$, then the expression is a <u>proper</u> fraction.

3. In a rational expression, if $(x - 8)$ is a linear factor that is repeated three times in the denominator, then the portion of the expression's partial-fraction decomposition that corresponds to $(x-8)^3$ has <u>three</u> terms.

5. False. The factors are repeated linear factors.

7. True

5.3 Building Skills

9. $\dfrac{1}{(x-1)(x+2)} = \dfrac{A}{x-1} + \dfrac{B}{x+2}$

11. $\dfrac{1}{x^2 + 7x + 6} = \dfrac{1}{(x+6)(x+1)} = \dfrac{A}{x+6} + \dfrac{B}{x+1}$

13. $\dfrac{2}{x^3 - x^2} = \dfrac{2}{x^2(x-1)} = \dfrac{A}{x^2} + \dfrac{B}{x} + \dfrac{C}{x-1}$

15. $\dfrac{x^2 - 3x + 3}{(x+1)(x^2 - x + 1)} = \dfrac{A}{x+1} + \dfrac{Bx+C}{x^2 - x + 1}$

17. $\dfrac{3x - 4}{(x^2+1)^2} = \dfrac{Ax+B}{x^2+1} + \dfrac{Cx+D}{(x^2+1)^2}$

19. $\dfrac{2x+1}{(x+1)(x+2)} = \dfrac{A}{x+1} + \dfrac{B}{x+2} \Rightarrow 2x+1 = A(x+2) + B(x+1) \Rightarrow 2x+1 = (A+B)x + (2A+B) \Rightarrow$

$\begin{cases} A+B=2 \\ 2A+B=1 \end{cases} \Rightarrow \begin{cases} -A-B=-2 \\ 2A+B=1 \end{cases} \Rightarrow A=-1, B=3$

$\dfrac{2x+1}{(x+1)(x+2)} = -\dfrac{1}{x+1} + \dfrac{3}{x+2} = \dfrac{3}{x+2} - \dfrac{1}{x+1}$

21. $\dfrac{1}{x^2+4x+3} = \dfrac{1}{(x+3)(x+1)} = \dfrac{A}{x+3} + \dfrac{B}{x+1} \Rightarrow 1 = A(x+1) + B(x+3) = (A+B)x + (A+3B) \Rightarrow$

$\begin{cases} A+B=0 \\ A+3B=1 \end{cases} \Rightarrow \begin{cases} -A-B=0 \\ A+3B=1 \end{cases} \Rightarrow 2B=1 \Rightarrow B=\dfrac{1}{2}, A=-\dfrac{1}{2}$

$\dfrac{1}{x^2+4x+3} = -\dfrac{1}{2(x+3)} + \dfrac{1}{2(x+1)}$

23. $\dfrac{2}{x^2+2x} = \dfrac{2}{x(x+2)} = \dfrac{A}{x} + \dfrac{B}{x+2} \Rightarrow 2 = A(x+2) + Bx = (A+B)x + 2A \Rightarrow \begin{cases} A+B=0 \\ 2A=2 \end{cases} \Rightarrow A=1, B=-1 \Rightarrow$

$\dfrac{2}{x^2+2x} = \dfrac{1}{x} - \dfrac{1}{x+2}$

25. $\dfrac{x}{(x+1)(x+2)(x+3)} = \dfrac{A}{x+1} + \dfrac{B}{x+2} + \dfrac{C}{x+3} \Rightarrow x = A(x+2)(x+3) + B(x+1)(x+3) + C(x+1)(x+2) \Rightarrow$

$x = A(x^2+5x+6) + B(x^2+4x+3) + C(x^2+3x+2) = (A+B+C)x^2 + (5A+4B+3C)x + (6A+3B+2C)$

$\begin{cases} A+B+C=0 \\ 5A+4B+3C=1 \\ 6A+3B+2C=0 \end{cases}$

$-5(A+B+C=0) = -5A-5B-5C=0$

$\begin{cases} -5A-5B-5C=0 \\ 5A+4B+3C=1 \end{cases} \Rightarrow -B-2C=1$

$-6(A+B+C=0) = -6A-6B-6C=0$

$\begin{cases} -6A-6B-6C=0 \\ 6A+3B+2C=0 \end{cases} \Rightarrow -3B-4C=0$

$\begin{cases} A+B+C=0 \\ 5A+3B+3C=1 \\ 6A+3B+2C=0 \end{cases} \Rightarrow \begin{cases} A+B+C=0 \\ -B-2C=1 \\ -3B-4C=0 \end{cases}$

$-3(-B-2C=1) = 3B+6C=-3$

$\begin{cases} 3B+6C=-3 \\ -3B-4C=0 \end{cases} \Rightarrow 2C=-3 \Rightarrow C=-\dfrac{3}{2}$

$-B-2\left(-\dfrac{3}{2}\right) = 1 \Rightarrow -B=-4 \Rightarrow B=2$

$A+2-\dfrac{3}{2}=0 \Rightarrow A=-\dfrac{1}{2}$

$\dfrac{x}{(x+1)(x+2)(x+3)} = -\dfrac{1}{2(x+1)} + \dfrac{2}{x+2} - \dfrac{3}{2(x+3)}$

27. $\dfrac{x^2}{(x-1)(x^2+5x+4)} = \dfrac{x^2}{(x-1)(x+1)(x+4)} = \dfrac{A}{x-1} + \dfrac{B}{x+1} + \dfrac{C}{x+4} \Rightarrow$

$x^2 = A(x+1)(x+4) + B(x-1)(x+4) + C(x-1)(x+1) \Rightarrow$

$x^2 = A(x^2+5x+4) + B(x^2+3x-4) + C(x^2-1) \Rightarrow$

$x^2 = (A+B+C)x^2 + (5A+3B)x + (4A-4B-C) \Rightarrow$

$\begin{cases} A+\ B+C=1 \\ 5A+3B\ \ \ \ =0 \\ 4A-4B-C=0 \end{cases} \Rightarrow \begin{cases} A+\ B+C=1 \\ 4A-4B-C=0 \end{cases} \Rightarrow 5A-3B=1$

$\begin{cases} A+\ B+C=1 \\ 5A+3B\ \ \ \ =0 \\ 4A-4B-C=0 \end{cases} \Rightarrow \begin{cases} A+\ B+C=1 \\ 5A+3B\ \ \ \ =0 \\ 5A-3B\ \ \ \ =1 \end{cases} \Rightarrow 10A=1 \Rightarrow A=\dfrac{1}{10}$

$5\left(\dfrac{1}{10}\right) + 3B = 0 \Rightarrow 3B = -\dfrac{1}{2} \Rightarrow B = -\dfrac{1}{6}$

$\dfrac{1}{10} - \dfrac{1}{6} + C = 1 \Rightarrow C = \dfrac{16}{15}$

$\dfrac{x^2}{(x-1)(x^2+5x+4)} = \dfrac{1}{10(x-1)} - \dfrac{1}{6(x+1)} + \dfrac{16}{15(x+4)}$

29. $\dfrac{3x+1}{(x-5)(x+3)^2} = \dfrac{A}{(x+3)^2} + \dfrac{B}{x+3} + \dfrac{C}{x-5}$

$3x+1 = A(x-5) + B(x+3)(x-5) + C(x+3)^2$

$3x+1 = A(x-5) + B(x^2-2x-15) + C(x^2+6x+9)$

$3x+1 = x^2(B+C) + x(A-2B+6C) + (-5A-15B+9C) \Rightarrow$

$\begin{cases} B+\ C=0 \\ A-\ 2B+6C=3 \\ -5A-15B+9C=1 \end{cases} \Rightarrow \begin{cases} A-\ 2B+6C=3 \\ -5A-15B+9C=1 \end{cases} \Rightarrow \begin{cases} 5A-10B+30C=15 \\ -5A-15B+\ 9C=1 \end{cases} \Rightarrow -25B+39C=16$

$\begin{cases} B+\ \ C=0 \\ -25B+39C=16 \end{cases} \Rightarrow \begin{cases} 25B+25C=0 \\ -25B+39C=16 \end{cases} \Rightarrow 64C=16 \Rightarrow C=\dfrac{1}{4}$

$B+\dfrac{1}{4}=0 \Rightarrow B=-\dfrac{1}{4}$

$A-2\left(-\dfrac{1}{4}\right) + 6\left(\dfrac{1}{4}\right) = 3 \Rightarrow A+\dfrac{1}{2}+\dfrac{3}{2}=3 \Rightarrow A=1$

$\dfrac{3x+1}{(x-5)(x+3)^2} = \dfrac{1}{(x+3)^2} - \dfrac{1}{4(x+3)} + \dfrac{1}{4(x-5)}$

31. $\dfrac{8x+1}{(x+2)^2(x-3)} = \dfrac{A}{(x+2)^2} + \dfrac{B}{x+2} + \dfrac{C}{x-3}$

$8x+1 = A(x-3) + B(x+2)(x-3) + C(x+2)^2$

$8x+1 = A(x-3) + B(x^2-x-6) + C(x^2+4x+4)$

$8x+1 = x^2(B+C) + x(A-B+4C) + (-3A-6B+4C) \Rightarrow$

$\begin{cases} B+\ C=0 \\ A-\ B+4C=8 \\ -3A-6B+4C=1 \end{cases} \Rightarrow \begin{cases} A-\ B+4C=8 \\ -3A-6B+4C=1 \end{cases} \Rightarrow \begin{cases} 3A-3B+12C=24 \\ -3A-6B+\ 4C=1 \end{cases} \Rightarrow -9B+16C=25$

$\begin{cases} B+\ \ C=0 \\ -9B+16C=25 \end{cases} \Rightarrow \begin{cases} 9B+\ 9C=0 \\ -9B+16C=25 \end{cases} \Rightarrow 25C=25 \Rightarrow C=1$

$B+1=0 \Rightarrow B=-1$

(continued on next page)

(continued)

$$A - (-1) + 4(1) = 8 \Rightarrow A = 3$$

$$\frac{8x+1}{(x+2)^2(x-3)} = \frac{3}{(x+2)^2} - \frac{1}{x+2} + \frac{1}{x-3}$$

33. $\dfrac{x-1}{(x+1)^2} = \dfrac{A}{(x+1)^2} + \dfrac{B}{(x+1)} \Rightarrow x - 1 = A + B(x+1) = Bx + (A+B) \Rightarrow \begin{cases} B = 1 \\ A + B = -1 \end{cases} \Rightarrow B = 1, A = -2$

$$\frac{x-1}{(x+1)^2} = -\frac{2}{(x+1)^2} + \frac{1}{(x+1)}$$

35. $\dfrac{x-1}{(2x-3)^2} = \dfrac{A}{2x-3} + \dfrac{B}{(2x-3)^2} \Rightarrow x - 1 = A(2x-3) + B \Rightarrow x - 1 = 2Ax + (-3A + B)$

$$\begin{cases} 2A = 1 \\ -3A + B = -1 \end{cases} \Rightarrow A = \frac{1}{2}$$

$$-3\left(\frac{1}{2}\right) + B = -1 \Rightarrow B = \frac{1}{2}$$

$$\frac{x-1}{(2x-3)^2} = \frac{1}{2(2x-3)} + \frac{1}{2(2x-3)^2}$$

37. $\dfrac{2x^2 + x}{(x+1)^3} = \dfrac{A}{(x+1)^3} + \dfrac{B}{(x+1)^2} + \dfrac{C}{x+1} \Rightarrow$

$$2x^2 + x = A + B(x+1) + C(x+1)^2 = A + Bx + B + C(x^2 + 2x + 1) = Cx^2 + (B + 2C)x + (A + B + C) \Rightarrow$$

$$\begin{cases} A + B + C = 0 \\ B + 2C = 1 \Rightarrow C = 2, B = -3, A = 1 \\ C = 2 \end{cases}$$

$$\frac{2x^2 + x}{(x+1)^3} = \frac{2}{x+1} - \frac{3}{(x+1)^2} + \frac{1}{(x+1)^3}$$

39. $\dfrac{2x+3}{(x+2)^3} = \dfrac{A}{x+2} + \dfrac{B}{(x+2)^2} + \dfrac{C}{(x+2)^3} \Rightarrow$

$$2x + 3 = A(x+2)^2 + B(x+2) + C = A(x^2 + 4x + 4) + Bx + 2B + C = Ax^2 + x(4A + B) + 2B + C$$

$$\begin{cases} A = 0 \\ 4A + B = 2 \Rightarrow 4(0) + B = 2 \Rightarrow B = 2 \\ 2B + C = 3 \end{cases}$$

$$2(2) + C = 3 \Rightarrow C = -1$$

$$\frac{2x+3}{(x+2)^3} = \frac{2}{(x+2)^2} - \frac{1}{(x+2)^3}$$

41. $\dfrac{-x^2 + 3x + 1}{x^3 + 2x^2 + x} = \dfrac{-x^2 + 3x + 1}{x(x+1)^2} = \dfrac{A}{(x+1)^2} + \dfrac{B}{x+1} + \dfrac{C}{x} \Rightarrow$

$$-x^2 + 3x + 1 = Ax + Bx(x+1) + C(x+1)^2 = Ax + Bx^2 + Bx + Cx^2 + 2Cx + C = (B+C)x^2 + (A+B+2C)x + C$$

$$\begin{cases} B + C = -1 \\ A + B + 2C = 3 \Rightarrow C = 1, B = -2 \\ C = 1 \end{cases}$$

$$A + 2(-2) + 3(1) = 2 \Rightarrow A = 3$$

$$\frac{-x^2 + 3x + 1}{x^3 + 2x^2 + x} = \frac{1}{x} - \frac{2}{x+1} + \frac{3}{(x+1)^2}$$

43. $\dfrac{3x+1}{(x+1)(x^2-1)} = \dfrac{3x+1}{(x+1)(x+1)(x-1)} = \dfrac{A}{x-1} + \dfrac{B}{x+1} + \dfrac{C}{(x+1)^2} \Rightarrow$

$3x+1 = A(x+1)^2 + B(x+1)(x-1) + C(x-1) = A(x^2+2x+1) + B(x^2-1) + C(x-1)$

$3x+1 = x^2(A+B) + x(2A+C) + (A-B-C) \Rightarrow$

$\begin{cases} A+B \quad\quad = 0 \\ 2A+ \quad C = 3 \\ A-B-C = 1 \end{cases} \Rightarrow \begin{cases} A+B \quad\quad = 0 \\ A-B-C = 1 \end{cases} \Rightarrow 2A - C = 1$

$\begin{cases} 2A+C = 3 \\ 2A-C = 1 \end{cases} \Rightarrow A = 1, \, C = 1$

$1 + B = 0 \Rightarrow B = -1$

$\dfrac{3x+1}{(x+1)(x^2-1)} = \dfrac{1}{x-1} - \dfrac{1}{x+1} + \dfrac{1}{(x+1)^2}$

45. $\dfrac{1}{x^2(x+1)^2} = \dfrac{A}{x} + \dfrac{B}{x^2} + \dfrac{C}{x+1} + \dfrac{D}{(x+1)^2} \Rightarrow 1 = Ax(x+1)^2 + B(x+1)^2 + Cx(x+1) + Dx^2$

Letting $x = -1$, we have $D = 1$. Letting $x = 0$, we have $B = 1$. Substitute the values for B and D, expand the equation, and simplify:

$1 = Ax(x+1)^2 + 1(x+1)^2 + Cx^2(x+1) + 1x^2 = (A+C)x^3 + (2+2A+C)x^2 + (2+A)x + 1 \Rightarrow$

$\begin{cases} A+C = 0 \\ 2A+C = -2 \\ A = -2 \end{cases} \Rightarrow A = -2, \, C = 2$

$\dfrac{1}{x^2(x+1)^2} = -\dfrac{2}{x} + \dfrac{1}{x^2} + \dfrac{2}{x+1} + \dfrac{1}{(x+1)^2}$

47. $\dfrac{1}{(x^2-1)^2} = \dfrac{1}{((x-1)(x+1))^2} = \dfrac{A}{x-1} + \dfrac{B}{(x-1)^2} + \dfrac{C}{x+1} + \dfrac{D}{(x+1)^2} \Rightarrow$

$1 = A(x-1)(x+1)^2 + B(x+1)^2 + C(x-1)^2(x+1) + D(x-1)^2$

Letting $x = 1$, we have $1 = B(2)^2 \Rightarrow B = 1/4$. Letting $x = -1$, we have $1 = D(-2)^2 \Rightarrow D = 1/4$. Substitute the values for B and D, expand the equation, and simplify:

$1 = A(x-1)(x+1)^2 + \dfrac{1}{4}(x+1)^2 + C(x-1)^2(x+1) + \dfrac{1}{4}(x-1)^2$

$1 = (A+C)x^3 + \left(\dfrac{1}{2} + A - C\right)x^2 + (-A-C)x + \left(\dfrac{1}{2} - A + C\right) \Rightarrow$

$\begin{cases} A+C = 0 \\ A-C = -\dfrac{1}{2} \\ -A-C = 0 \\ -A+C = \dfrac{1}{2} \end{cases} \Rightarrow A = -\dfrac{1}{4}, \, C = \dfrac{1}{4}$

$\dfrac{1}{(x^2-1)^2} = -\dfrac{1}{4(x-1)} + \dfrac{1}{4(x-1)^2} + \dfrac{1}{4(x+1)} + \dfrac{1}{4(x+1)^2}$

49. $\dfrac{x^2-5}{(x-2)\left(x^2-2x+3\right)} = \dfrac{A}{x-2} + \dfrac{Bx+C}{x^2-2x+3}$

$x^2-5 = A\left(x^2-2x+3\right) + (Bx+C)(x-2) = A\left(x^2-2x+3\right) + \left(Bx^2-2Bx+Cx-2C\right)$

$\qquad = A\left(x^2-2x+3\right) + \left(Bx^2 + (-2B+C)x - 2C\right) = x^2(A+B) + x(-2A-2B+C) + (3A-2C)$

$\begin{cases} A+\ B\ \ \ \ \ \ =1 \\ -2A-2B+C=0 \\ 3A-\ \ \ \ \ 2C=-5 \end{cases} \Rightarrow \begin{cases} A+\ B\ \ \ \ \ \ =1 \\ -2A-2B+C=0 \end{cases} \Rightarrow \begin{cases} 2A+2B\ \ \ \ \ \ =2 \\ -2A-2B+C=0 \end{cases} \Rightarrow C=2$

$3A-2(2)=-5 \Rightarrow 3A=-1 \Rightarrow A=-\dfrac{1}{3}$

$-\dfrac{1}{3}+B=1 \Rightarrow B=\dfrac{4}{3}$

$\dfrac{x^2-5}{(x-2)\left(x^2-2x+3\right)} = -\dfrac{1}{3(x-2)} + \dfrac{\frac{4}{3}x+2}{x^2-2x+3} = -\dfrac{1}{3(x-2)} + \dfrac{4x+6}{3\left(x^2-2x+3\right)} = -\dfrac{1}{3(x-2)} + \dfrac{2(2x+3)}{3\left(x^2-2x+3\right)}$

51. $\dfrac{x-1}{(x+1)^2\left(x^2+2x+2\right)} = \dfrac{A}{x+1} + \dfrac{B}{(x+1)^2} + \dfrac{Cx+D}{x^2+2x+2}$

$x-1 = A(x+1)\left(x^2+2x+2\right) + B\left(x^2+2x+2\right) + (Cx+D)(x+1)^2$

Substitute $x=-1$ into the equation to obtain

$-2 = B\left((-1)^2 + 2(-1) + 2\right) \Rightarrow B=-2$

Substitute the value for B, expand the equation, and simplify:

$x-1 = A\left(x^3+3x^2+4x+2\right) - 2\left(x^2+2x+2\right) + \left(Cx^3+2Cx^2+Cx+Dx^2+2Dx+D\right)$

$\qquad = x^3(A+C) + x^2(3A+2C+D-2) + x(4A+C+2D-4) + (2A+D-4)$

$\begin{cases} A+C\ \ \ \ \ \ =0 \\ 3A+2C+D-2=0 \\ 4A+C+2D-4=1 \\ 2A+\ \ \ \ \ D-4=-1 \end{cases} \Rightarrow \begin{cases} A+\ C\ \ \ \ \ \ =0 \\ 3A+2C\ +D=2 \\ 4A+\ C+2D=5 \\ 2A+\ \ \ \ \ \ \ D=3 \end{cases} \Rightarrow \begin{cases} 3A+2C+\ D=2 \\ 4A+\ C+2D=5 \end{cases} \Rightarrow \begin{cases} 3A+2C+\ D=2 \\ -8A-2C-4D=-10 \end{cases} \Rightarrow$

$-5A-3D=-8$

$\begin{cases} -5A-3D=-8 \\ 2A+\ D=3 \end{cases} \Rightarrow \begin{cases} -5A-3D=-8 \\ 6A+3D=9 \end{cases} \Rightarrow A=1,\ D=1$

$A+C=0 \Rightarrow C=-1$

$\dfrac{x-1}{(x+1)^2\left(x^2+2x+2\right)} = \dfrac{1}{x+1} - \dfrac{2}{(x+1)^2} + \dfrac{-x+1}{x^2+2x+2}$

53. $\dfrac{x^2+1}{(x-2)\left(x^2-3x+2\right)} = \dfrac{x^2+1}{(x-2)(x-2)(x-1)} = \dfrac{A}{x-2} + \dfrac{B}{(x-2)^2} + \dfrac{C}{x-1}$

$x^2+1 = A(x-2)(x-1) + B(x-1) + C(x-2)^2$

$\qquad = A\left(x^2-3x+2\right) + B(x-1) + C\left(x^2-4x+4\right) = x^2(A+C) + x(-3A+B-4C) + (2A-B+4C)$

$\begin{cases} A+\ \ \ \ \ \ C=1 \\ -3A+B-4C=0 \\ 2A-B+4C=1 \end{cases} \Rightarrow \begin{cases} -3A+B-4C=0 \\ 2A-B+4C=1 \end{cases} \Rightarrow -A=1 \Rightarrow A=-1,\ C=2$

$3+B-8=0 \Rightarrow B=5$

$\dfrac{x^2+1}{(x-2)\left(x^2-3x+2\right)} = -\dfrac{1}{x-2} + \dfrac{5}{(x-2)^2} + \dfrac{2}{x-1}$

55. $\dfrac{6x+7}{4x^2+12x+9} = \dfrac{6x+7}{(2x+3)^2} = \dfrac{A}{2x+3} + \dfrac{B}{(2x+3)^2}$

$6x+7 = A(2x+3) + B = 2Ax + (3A+B)$

$\begin{cases} 2A = 6 \\ 3A + B = 7 \end{cases} \Rightarrow A = 3, B = -2$

$\dfrac{6x+7}{4x^2+12x+9} = \dfrac{3}{2x+3} - \dfrac{2}{(2x+3)^2}$

57. $\dfrac{x-3}{x^3+x^2} = \dfrac{x-3}{x^2(x+1)} = \dfrac{A}{x} + \dfrac{B}{x^2} + \dfrac{C}{x+1}$

$x-3 = Ax(x+1) + B(x+1) + Cx^2 = (A+C)x^2 + (A+B)x + B$

$\begin{cases} A + C = 0 \\ A + B = 1 \\ B = -3 \end{cases} \Rightarrow B = -3, A = 4, C = -4$

$\dfrac{x-3}{x^3+x^2} = \dfrac{4}{x} - \dfrac{3}{x^2} - \dfrac{4}{x+1}$

59. $\dfrac{x^2+2x+4}{x^3+x^2} = \dfrac{x^2+2x+4}{x^2(x+1)} = \dfrac{A}{x} + \dfrac{B}{x^2} + \dfrac{C}{x+1}$

$x^2+2x+4 = Ax(x+1) + B(x+1) + Cx^2 = (A+C)x^2 + (A+B)x + B$

$\begin{cases} A + C = 1 \\ A + B = 2 \\ B = 4 \end{cases} \Rightarrow B = 4, A = -2, C = 3$

$\dfrac{x^2+2x+4}{x^3+x^2} = -\dfrac{2}{x} + \dfrac{4}{x^2} + \dfrac{3}{x+1}$

61. $\dfrac{x}{x^4-1} = \dfrac{x}{(x-1)(x+1)(x^2+1)} = \dfrac{A}{x-1} + \dfrac{B}{x+1} + \dfrac{Cx+D}{x^2+1}$

$x = A(x+1)(x^2+1) + B(x-1)(x^2+1) + (Cx+D)(x^2-1)$

$x = (A+B+C)x^3 + (A-B+D)x^2 + (A+B-C)x + (A-B-D)$

$\begin{cases} A + B + C = 0 \\ A - B + D = 0 \\ A + B - C = 1 \\ A - B - D = 0 \end{cases} \Rightarrow A = \dfrac{1}{4}, B = \dfrac{1}{4}, C = -\dfrac{1}{2}, D = 0$

$\dfrac{x}{x^4-1} = \dfrac{1}{4(x-1)} + \dfrac{1}{4(x+1)} - \dfrac{x}{2(x^2+1)}$

63. $\dfrac{1}{x(x^2+1)^2} = \dfrac{A}{x} + \dfrac{Bx+C}{x^2+1} + \dfrac{Dx+E}{(x^2+1)^2} \Rightarrow$

$1 = A(x^2+1)^2 + (Bx+C)x(x^2+1) + (Dx+E)x = (A+B)x^4 + Cx^3 + (2A+B+D)x^2 + (C+E)x + A$

$\begin{cases} A + B = 0 \\ C = 0 \\ 2A + B + D = 0 \\ C + E = 0 \\ A = 1 \end{cases} \Rightarrow A = 1, B = -1, C = 0, D = -1, E = 0$

$\dfrac{1}{x(x^2+1)^2} = \dfrac{1}{x} - \dfrac{x}{x^2+1} - \dfrac{x}{(x^2+1)^2}$

65. $\dfrac{2x^2 + 3x}{(x^2 + 1)(x^2 + 2)} = \dfrac{Ax + B}{x^2 + 1} + \dfrac{Cx + D}{x^2 + 2}$

$2x^2 + 3x = (Ax + B)(x^2 + 2) + (Cx + D)(x^2 + 1) = (A + C)x^3 + (B + D)x^2 + (2A + C)x + (2B + D)$

$\begin{cases} A + C = 0 \\ B + D = 2 \\ 2A + C = 3 \\ 2B + D = 0 \end{cases} \Rightarrow A = 3, B = -2, C = -3, D = 4$

$\dfrac{2x^2 + 3x}{(x^2 + 1)(x^2 + 2)} = \dfrac{3x - 2}{x^2 + 1} + \dfrac{-3x + 4}{x^2 + 2}$

5.3 Applying the Concepts

67. $\dfrac{1}{1 \cdot 2} + \dfrac{1}{2 \cdot 3} + \dfrac{1}{3 \cdot 4} + \cdots + \dfrac{1}{n(n+1)} = \left(1 - \dfrac{1}{2}\right) + \left(\dfrac{1}{2} - \dfrac{1}{3}\right) + \left(\dfrac{1}{3} - \dfrac{1}{4}\right) + \cdots + \left(\dfrac{1}{n} - \dfrac{1}{n+1}\right) = 1 - \dfrac{1}{n+1} = \dfrac{n}{n+1}$

69. Each term is of the form $\dfrac{2}{(2n-1)(2n+1)}$. Decomposing $\dfrac{2}{(2n-1)(2n+1)}$ we have

$\dfrac{2}{(2n-1)(2n+1)} = \dfrac{A}{2n-1} + \dfrac{B}{2n+1} \Rightarrow$

$2 = A(2n+1) + B(2n-1) = (2A + 2B)n + (A - B) \Rightarrow$

$\begin{cases} 2A + 2B = 0 \\ A - B = 2 \end{cases} \Rightarrow A = 1, B = -1$

$\dfrac{2}{(2n-1)(2n+1)} = \dfrac{1}{2n-1} - \dfrac{1}{2n+1}.$

So,

$\dfrac{2}{1 \cdot 3} + \dfrac{2}{3 \cdot 5} + \dfrac{2}{5 \cdot 7} + \cdots + \dfrac{2}{(2n-1)(2n+1)} = \left(\dfrac{1}{1} - \dfrac{1}{3}\right) + \left(\dfrac{1}{3} - \dfrac{1}{5}\right) + \left(\dfrac{1}{5} - \dfrac{1}{7}\right) + \cdots + \dfrac{1}{2n-1} - \dfrac{1}{2n+1}$

$= 1 - \dfrac{1}{2n+1} = \dfrac{2n}{2n+1}$

71. $\dfrac{1}{R} = \dfrac{2x + 4}{(x+1)(x+3)}$

$\dfrac{2x + 4}{(x+1)(x+3)} = \dfrac{A}{x+1} + \dfrac{B}{x+3} \Rightarrow 2x + 4 = A(x+3) + B(x+1) = (A + B)x + (3A + B)$

$\begin{cases} A + B = 2 \\ 3A + B = 4 \end{cases} \Rightarrow A = 1, B = 1$

$\dfrac{2x + 4}{(x+1)(x+3)} = \dfrac{1}{x+1} + \dfrac{1}{x+3} = \dfrac{1}{R}.$

This means that if two resistances $R_1 = x + 1$ and $R_2 = x + 3$ are connected in parallel, they will produce a

total resistance given by $\dfrac{(x+1)(x+3)}{2x + 4}$.

73. $\dfrac{1}{R} = \dfrac{R_1R_2 + R_2R_3 + R_3R_1}{R_1R_2R_3} = \dfrac{A}{R_1} + \dfrac{B}{R_2} + \dfrac{C}{R_3}$

$R_1R_2 + R_2R_3 + R_3R_1 = AR_2R_3 + BR_1R_3 + CR_1R_2$

$\begin{cases} A = 1 \\ B = 1 \\ C = 1 \end{cases}$

$\dfrac{R_1R_2 + R_2R_3 + R_3R_1}{R_1R_2R_3} = \dfrac{1}{R_1} + \dfrac{1}{R_2} + \dfrac{1}{R_3} = \dfrac{1}{R}$

This means that if three resistances $R_1, R_2,$ and R_3 are connected in parallel, they will produce a total

resistance given by $\dfrac{R_1R_2R_3}{R_1R_2 + R_2R_3 + R_3R_1}$.

5.3 Beyond the Basics

75. $\begin{cases} A + B + C = 0 \\ \quad\ 2B - 2C = 10 \\ -4A \qquad\quad = -4 \end{cases} \Rightarrow \begin{cases} A + B + C = 0 \\ \quad\ B - C = 5 \\ A \qquad\quad = 1 \end{cases}$

Subtract the third equation from the first equation, and replace the first equation with the new equation:

$\begin{cases} A + B + C = 0 \\ \quad\ B - C = 5 \\ A \qquad\quad = 1 \end{cases} \Rightarrow \begin{cases} \quad\ B + C = -1 \\ \quad\ B - C = 5 \\ A \qquad\quad = 1 \end{cases}$

Add the first and second equations to solve for B:

$\begin{cases} B + C = -1 \\ B - C = 5 \\ A \qquad = 1 \end{cases} \Rightarrow \begin{cases} B + C = -1 \\ 2B \qquad\quad = 4 \\ A \qquad\quad = 1 \end{cases} \Rightarrow \begin{cases} B + C = -1 \\ B \qquad = 2 \\ A \qquad = 1 \end{cases}$

Substitute the value for B into the first equation to solve for C:

$\begin{cases} B + C = -1 \\ B \qquad = 2 \\ A \qquad = 1 \end{cases} \Rightarrow \begin{cases} 2 + C = -1 \\ B \qquad = 2 \\ A \qquad = 1 \end{cases} \Rightarrow \begin{cases} C = -3 \\ B = 2 \\ A = 1 \end{cases}$

77. $\dfrac{4x}{\left(x^2 - 1\right)^2} = \dfrac{4x}{(x-1)^2(x+1)^2} = \dfrac{A}{x-1} + \dfrac{B}{(x-1)^2} + \dfrac{C}{x+1} + \dfrac{D}{(x+1)^2}$

$4x = A(x-1)(x+1)^2 + B(x+1)^2 + C(x-1)^2(x+1) + D(x-1)^2$

$\quad = A\left(x^3 + x^2 - x - 1\right) + B\left(x^2 + 2x + 1\right) + C\left(x^3 - x^2 - x + 1\right) + D\left(x^2 - 2x + 1\right)$

$4x = (A + C)x^3 + (A + B - C + D)x^2 + (-A + 2B - C - 2D)x + (-A + B + C + D)$

$\begin{cases} A + \qquad C \qquad\quad = 0 \quad (1) \\ A + B - C + D = 0 \quad (2) \\ -A + 2B - C - 2D = 4 \quad (3) \\ -A + B + C + D = 0 \quad (4) \end{cases}$

Add equations (1) and (3), and (2) and (4), and replace equations (3) and (4):

$\begin{cases} A + \quad C \qquad\quad = 0 \\ A + B - C + D = 0 \\ -A + 2B - C - 2D = 4 \\ -A + B + C + D = 0 \end{cases} \Rightarrow \begin{cases} A + \quad C \qquad\quad = 0 \quad (1) \\ A + B - C + D = 0 \quad (2) \\ \quad 2B - 2D = 4 \quad (3) \\ \quad 2B + 2D = 0 \quad (4) \end{cases} \Rightarrow \begin{cases} A + \quad C \qquad\quad = 0 \\ A + B - C + D = 0 \\ \quad 4B \qquad\quad = 4 \\ \quad 2B + 2D = 0 \end{cases} \Rightarrow$

$B = 1, D = -1, A = 0, C = 0$

$\dfrac{4x}{\left(x^2 - 1\right)^2} = \dfrac{1}{(x-1)^2} - \dfrac{1}{(x+1)^2}$

79. $\dfrac{x+1}{(x^2+1)(x-1)^2} = \dfrac{A}{x-1} + \dfrac{B}{(x-1)^2} + \dfrac{Cx+D}{x^2+1}$

$x+1 = A(x-1)(x^2+1) + B(x^2+1) + (Cx+D)(x-1)^2$

$\quad = (A+C)x^3 + (-A+B-2C+D)x^2 + (A+B+C-2D)x + (-A+B+D)$

$\begin{cases} A+\quad C \quad = 0 \\ -A+B-2C+\ D = 0 \\ A+B+\ C-2D = 1 \\ -A+B+\quad\ D = 1 \end{cases}$

From the first equation, we have $C = -A$. Substitute this into the last three equations and simplify:

$\begin{cases} A+\ C\ =0 \\ -A+B-2C+D=0 \\ A+\ C-2D=1 \\ -A+B+\ D=1 \end{cases} \Rightarrow \begin{cases} C=-A \\ -A+B-2(-A)+D=0 \\ A+\ (-A)-2D=1 \\ -A+B+\ D=1 \end{cases} \Rightarrow \begin{cases} C=-A \\ A+B+\ D=0 \\ \quad -2D=1 \\ -A+B+\ D=1 \end{cases} \Rightarrow \begin{cases} C=-A \\ A+B+D=0 \\ D=-\dfrac{1}{2} \\ -A+B+D=1 \end{cases}$

Substitute the value for D into the second and fourth equations, and simplify:

$\begin{cases} C=-A \\ A+B=\dfrac{1}{2} \\ \quad D=-\dfrac{1}{2} \\ -A+B=\dfrac{3}{2} \end{cases} \Rightarrow B=1, A=-\dfrac{1}{2}, C=\dfrac{1}{2}$

$\dfrac{x+1}{(x^2+1)(x-1)^2} = -\dfrac{1}{2(x-1)} + \dfrac{1}{(x-1)^2} + \dfrac{x-1}{2(x^2+1)}$

81. $\dfrac{x^3}{(x+1)^2(x+2)^2} = \dfrac{A}{x+1} + \dfrac{B}{(x+1)^2} + \dfrac{C}{x+2} + \dfrac{D}{(x+2)^2}$

$x^3 = A(x+1)(x+2)^2 + B(x+2)^2 + C(x+2)(x+1)^2 + D(x+1)^2$

Letting $x=-2$, we have $(-2)^3 = D(-1)^2 \Rightarrow D=-8$. Letting $x=-1$, we have $(-1)^3 = B(1)^2 \Rightarrow B=-1$.

Expand the equation and substitute the values for B and D:

$x^3 = (A+C)x^3 + (-9+5A+4C)x^2 + (-20+8A+5C)x + (-12+4A+2C)$

$\begin{cases} A+\ C=1 \\ 5A+4C=9 \\ 8A+5C=20 \\ 4A+2C=12 \end{cases} \Rightarrow A=5, C=-4$

$\dfrac{x^3}{(x+1)^2(x+2)^2} = \dfrac{5}{x+1} - \dfrac{1}{(x+1)^2} - \dfrac{4}{x+2} - \dfrac{8}{(x+2)^2}$

83.

$\begin{array}{r} 1 \\ x^2+3x+2 \overline{)\, x^2+4x+5} \\ \underline{x^2+3x+2} \\ x+3 \end{array}$

$\dfrac{x^2+4x+5}{x^2+3x+2} = 1 + \dfrac{x+3}{x^2+3x+2}$

Decompose $\dfrac{x+3}{x^2+3x+2}$:

$\dfrac{x+3}{x^2+3x+2} = \dfrac{x+3}{(x+1)(x+2)} = \dfrac{A}{x+1} + \dfrac{B}{x+2} \Rightarrow$

$x+3 = A(x+2) + B(x+1) = (A+B)x + (2A+B)$

$\begin{cases} A+B=1 \\ 2A+B=3 \end{cases} \Rightarrow A=2, B=-1$

$\dfrac{x^2+4x+5}{x^2+3x+2} = 1 + \dfrac{2}{x+1} - \dfrac{1}{x+2}$

85. Each term is of the form

$$\frac{1}{\sqrt{2n+1}+\sqrt{2n+3}} \cdot \frac{\sqrt{2n+1}-\sqrt{2n+3}}{\sqrt{2n+1}-\sqrt{2n+3}} = \frac{\sqrt{2n+1}-\sqrt{2n+3}}{2n+1-(2n+3)} = \frac{\sqrt{2n+1}-\sqrt{2n+3}}{-2} = \frac{\sqrt{2n+3}-\sqrt{2n+1}}{2}.$$

So,

$$\frac{1}{\sqrt{3}+\sqrt{5}} + \frac{1}{\sqrt{5}+\sqrt{7}} + \frac{1}{\sqrt{7}+\sqrt{9}} + \cdots + \frac{1}{\sqrt{2n-1}+\sqrt{2n+1}} + \frac{1}{\sqrt{2n+1}+\sqrt{2n+3}}$$

$$= \frac{\sqrt{5}-\sqrt{3}}{2} + \frac{\sqrt{7}-\sqrt{5}}{2} + \frac{\sqrt{9}-\sqrt{7}}{2} + \cdots + \frac{\sqrt{2n+1}-\sqrt{2n-1}}{2} + \frac{\sqrt{2n+3}-\sqrt{2n+1}}{2}$$

$$= \frac{\sqrt{2n+3}-\sqrt{3}}{2}$$

5.3 Critical Thinking/Discussion/Writing

87. Two equal polynomials have equal corresponding coefficients.

88. $f(x) = \dfrac{x+8}{x^2+x-2} = \dfrac{x+8}{(x+2)(x-1)}$

$$= \frac{A}{x+2} + \frac{B}{x-1} \Rightarrow$$

$$x+8 = A(x-1) + B(x+2)$$

$$= (A+B)x + (-A+2B)$$

$$\begin{cases} A+B=1 \\ -A+2B=8 \end{cases} \Rightarrow 3B=9 \Rightarrow B=3, \ A=-2$$

$$\frac{x+8}{x^2+x-2} = -\frac{2}{x+2} + \frac{3}{x-1}$$

$$g(x) = -\frac{2}{x+2}, \quad h(x) = \frac{3}{x-1}$$

a.

b. The graphs of *f* and *g* have the same vertical asymptote, $x = -2$, and the same horizontal asymptote, $y = 0$, or the *x*-axis. Also *g* is an asymptote of *f* as *x* approaches 1 from the right or from the left.

c. The graphs of *f* and *h* have the same vertical asymptote, $x = 1$, and the same horizontal asymptote, $y = 0$, or the *x*-axis. Also *h* is an asymptote of *f* as *x* approaches −2 from the right or from the left.

5.3 Getting Ready for the Next Section

89. $x^2 + 10x + 21 = 0 \Rightarrow (x+7)(x+3) = 0 \Rightarrow$
$x = -7, \ x = -3$

91. $2x^2 + x - 10 = 0 \Rightarrow (2x+5)(x-2) = 0 \Rightarrow$
$x = -\dfrac{5}{2}, \ x = 2$

In exercises 93−100, be sure to check the solution in both equations.

93. $\begin{cases} x + y = 3 & (1) \\ 3x + 2y = 7 & (2) \end{cases}$

From equation (1), we have $y = 3 - x$.
Substituting in equation (2), we have
$$3x + 2(3-x) = 7 \Rightarrow x + 6 = 7 \Rightarrow x = 1$$
Substituting $x = 1$ in equation (1) gives
$1 + y = 3 \Rightarrow y = 2$.
Solution set: $\{(1, 2)\}$

95. $\begin{cases} x - 2y = 1 & (1) \\ 2x + 3y = 16 & (2) \end{cases}$

From equation (1), we have $x = 2y + 1$.
Substituting in equation (2), we have
$$2(2y+1) + 3y = 16 \Rightarrow 7y + 2 = 16 \Rightarrow$$
$$7y = 14 \Rightarrow y = 2$$
Substituting $y = 2$ in equation (1) gives
$x - 2(2) = 1 \Rightarrow x - 4 = 1 \Rightarrow x = 5$
Solution set: $\{(5, 2)\}$

97. $\begin{cases} 2x - y = 4 & (1) \\ 3x + 2y = 13 & (2) \end{cases}$

Multiply equation (1) by 2, then add the two equations and solve for x.

$4x - 2y = 8$
$3x + 2y = 13$
$\overline{7x = 21} \Rightarrow x = 3$

Substitute $x = 3$ in equation (1), then solve for y.

$2(3) - y = 4 \Rightarrow 6 - y = 4 \Rightarrow y = 2$

Solution set: $\{(3, 2)\}$

99. $\begin{cases} 2x + 5y = 1 & (1) \\ 3x - 2y = -8 & (2) \end{cases}$

Multiply equation (1) by 2 and equation (2) by 5 to eliminate y. Add the resulting equations and solve for x.

$4x + 10y = 2$
$15x - 10y = -40$
$\overline{19x = -38} \Rightarrow x = -2$

Substitute $x = -2$ in equation (1), then solve for y.

$2(-2) + 5y = 1 \Rightarrow -4 + 5y = 1 \Rightarrow 5y = 5 \Rightarrow y = 1$

Solution set: $\{(-2, 1)\}$

5.4 Systems of Nonlinear Equations

5.4 Practice Problems

1. $\begin{cases} x^2 + y = 2 \\ 2x + y = 3 \end{cases} \Rightarrow \begin{cases} y = 2 - x^2 \\ 2x + y = 3 \end{cases} \Rightarrow$

$2x + (2 - x^2) = 3 \Rightarrow -x^2 + 2x - 1 = 0 \Rightarrow$
$x^2 - 2x + 1 = 0 \Rightarrow (x - 1)^2 = 0 \Rightarrow x = 1$
$2(1) + y = 3 \Rightarrow y = 1$

The solution is $\{(1, 1)\}$.

2. $\begin{cases} x^2 + 2y^2 = 34 \\ x^2 - y^2 = 7 \end{cases} \Rightarrow 3y^2 = 27 \Rightarrow y^2 = 9 \Rightarrow$

$y = \pm 3$
$x^2 - (-3)^2 = 7 \Rightarrow x^2 = 16 \Rightarrow x = \pm 4$
$x^2 - (3)^2 = 7 \Rightarrow x^2 = 16 \Rightarrow x = \pm 4$

The solution is $\{(-4, -3), (-4, 3), (4, -3), (4, 3)\}$.

3. Let $x =$ the number of shares received as dividends, and let $p =$ the selling price per share. Then $xp = 1950$. The number of shares she sold at p dollars per share is $240 + x$. Thus, revenue $= (240 + x)p$ and the cost of her stock was $(240)(40) + 100 = 9700$.

Since revenue $-$ cost $=$ profit, we have

$(240 + x)p - 9700 = 7850 \Rightarrow$
$240p + xp = 17,550$

Thus, the system of equations is

$\begin{cases} xp = 1950 \\ 240p + xp = 17,550 \end{cases}$

$\begin{cases} xp = 1950 \\ 240p + xp = 17,550 \end{cases} \Rightarrow$

$240p + 1950 = 17,550 \Rightarrow 240p = 15,600 \Rightarrow$
$p = 65$
$65x = 1950 \Rightarrow x = 30$

Danielle received 30 shares as dividends and sold her stock at $65 per share.

5.4 Concepts and Vocabulary

1. In a system of nonlinear equations, <u>at least one</u> equation must be nonlinear.

3. The solutions of the system

$\begin{cases} ax + by = c & (1) \\ (x - h)^2 + (y - k)^2 = r^2 & (2) \end{cases}$

represent the points of <u>intersection</u> of the graphs of equations (1) and (2).

5. True

7. True

5.4 Building Skills

9. Substituting each ordered pair into the system

$\begin{cases} 2x + 3y = 3 \\ x - y^2 = 2 \end{cases}$

we find that $(3, -1)$ is a solution.

$\begin{cases} 2(3) + 3(-1) = 6 - 3 = 3 \\ 3 - (-1)^2 = 3 - 1 = 2 \end{cases}$

11. Substituting each ordered pair into the system

$\begin{cases} 5x - 2y = 7 \\ x^2 + y^2 = 2 \end{cases}$,

we find that $(1, -1)$ is a solution:

$\begin{cases} 5(1) - 2(-1) = 5 + 2 = 7 \\ (1)^2 + (-1)^2 = 1 + 1 = 2 \end{cases}$

13. Substituting each ordered pair into the system

$$\begin{cases} 4x^2 + 5y^2 = 180 \\ x^2 - y^2 = 9 \end{cases}$$

we find that (5, 4), (−5, 4), and (−5, −4) are solutions.

$$\begin{cases} 4(5)^2 + 5(4)^2 = 100 + 80 = 180 \\ 5^2 - 4^2 = 25 - 16 = 9 \end{cases}$$

$$\begin{cases} 4(-5)^2 + 5(4)^2 = 100 + 80 = 180 \\ (-5)^2 - (4)^2 = 25 - 16 = 9 \end{cases}$$

$$\begin{cases} 4(-5)^2 + 5(-4)^2 = 100 + 80 = 180 \\ (-5)^2 - (-4)^2 = 25 - 16 = 9 \end{cases}$$

15. Substituting each ordered pair into the system

$$\begin{cases} y = e^{x-1} \\ y = 2x - 1 \end{cases}$$

we find that (1, 1) is a solution:

$$\begin{cases} y = e^{1-1} = e^0 = 1 \\ y = 2(1) - 1 = 1 \end{cases}$$

17. $\begin{cases} y = x^2 \\ y = x + 2 \end{cases} \Rightarrow x^2 = x + 2 \Rightarrow x^2 - x - 2 = 0 \Rightarrow$

$(x-2)(x+1) = 0 \Rightarrow x = 2$ or $x = -1$

$y = 2^2 = 4$ or $y = (-1)^2 = 1$

The solution is {(2, 4), (−1, 1)}.

19. $\begin{cases} x^2 - y = 6 \\ x - y = 0 \end{cases} \Rightarrow \begin{cases} x^2 - y = 6 \\ x = y \end{cases} \Rightarrow x^2 - x = 6 \Rightarrow$

$x^2 - x - 6 = 0 \Rightarrow (x-3)(x+2) = 0 \Rightarrow$

$x = 3$ or $x = -2$

$3 - y = 0 \Rightarrow y = 3$ or $-2 - y = 0 \Rightarrow y = -2$

The solution is {(3, 3), (−2, −2)}.

21. $\begin{cases} x^2 + y^2 = 9 \\ x = 3 \end{cases} \Rightarrow 3^2 + y^2 = 9 \Rightarrow y = 0$

The solution is {(3, 0)}.

23. $\begin{cases} x^2 + y^2 = 5 \\ x - y = -3 \end{cases} \Rightarrow \begin{cases} x^2 + y^2 = 5 \\ x = y - 3 \end{cases} \Rightarrow$

$(y-3)^2 + y^2 = 5 \Rightarrow 2y^2 - 6y + 4 = 0 \Rightarrow$

$y^2 - 3y + 2 = 0 \Rightarrow (y-2)(y-1) = 0 \Rightarrow$

$y = 2$ or $y = 1$

$x - 2 = -3 \Rightarrow x = -1$

$x - 1 = -3 \Rightarrow x = -2$

The solution is {(−2, 1), (−1, 2)}.

25. $\begin{cases} x^2 - 4x + y^2 = -2 \\ x - y = 2 \end{cases} \Rightarrow \begin{cases} x^2 - 4x + y^2 = -2 \\ x = y + 2 \end{cases} \Rightarrow$

$(y+2)^2 - 4(y+2) + y^2 = -2 \Rightarrow 2y^2 = 2 \Rightarrow$

$y = \pm 1$

$x - (-1) = 2 \Rightarrow x = 1$

$x - 1 = 2 \Rightarrow x = 3$

The solution is {(1, −1), (3, 1)}.

27. $\begin{cases} x - y = -2 \\ xy = 3 \end{cases} \Rightarrow \begin{cases} x = y - 2 \\ xy = 3 \end{cases} \Rightarrow y(y-2) = 3 \Rightarrow$

$y^2 - 2y - 3 = 0 \Rightarrow (y-3)(y+1) = 0 \Rightarrow$

$y = 3$ or $y = -1$

$x - 3 = -2 \Rightarrow x = 1$

$x - (-1) = -2 \Rightarrow x = -3$

The solution is {(1, 3), (−3, −1)}.

29. $\begin{cases} 4x^2 + y^2 = 25 \\ x + y = 5 \end{cases} \Rightarrow \begin{cases} 4x^2 + y^2 = 25 \\ x = 5 - y \end{cases} \Rightarrow$

$4(5-y)^2 + y^2 = 25 \Rightarrow 5y^2 - 40y + 75 = 0 \Rightarrow$

$y^2 - 8y + 15 = 0 \Rightarrow (y-5)(y-3) = 0 \Rightarrow$

$y = 5$ or $y = 3$

$x + 5 = 5 \Rightarrow x = 0$

$x + 3 = 5 \Rightarrow x = 2$

The solution is {(0, 5), (2, 3)}.

31. $\begin{cases} x^2 - y^2 = 24 \\ 5x - 7y = 0 \end{cases} \Rightarrow \begin{cases} x^2 - y^2 = 24 \\ x = \frac{7}{5}y \end{cases} \Rightarrow$

$\left(\frac{7}{5}y\right)^2 - y^2 = 24 \Rightarrow \frac{24}{25}y^2 = 24 \Rightarrow y^2 = 25 \Rightarrow$

$y = \pm 5$

$5x - 7(-5) = 0 \Rightarrow x = -7$

$5x - 7(5) = 0 \Rightarrow x = 7$

The solution is {(7, 5), (−7, −5)}.

33. $\begin{cases} x^2 + y^2 = 20 \\ x^2 - y^2 = 12 \end{cases} \Rightarrow 2x^2 = 32 \Rightarrow x = \pm 4$

$(-4)^2 + y^2 = 20 \Rightarrow y^2 = 4 \Rightarrow y = \pm 2$

$(4)^2 + y^2 = 20 \Rightarrow y^2 = 4 \Rightarrow y = \pm 2$

The solution is {(−4, −2), (−4, 2), (4, −2), (4, 2)}.

35. $\begin{cases} x^2 + 2y^2 = 12 \\ 7y^2 - 5x^2 = 8 \end{cases} \Rightarrow \begin{cases} 5x^2 + 10y^2 = 60 \\ 7y^2 - 5x^2 = 8 \end{cases} \Rightarrow$

$17y^2 = 68 \Rightarrow y^2 = 4 \Rightarrow y = \pm 2$

$x^2 + 2(-2)^2 = 12 \Rightarrow x^2 = 4 \Rightarrow x = \pm 2$

$x^2 + 2(2)^2 = 12 \Rightarrow x^2 = 4 \Rightarrow x = \pm 2$

The solution is {(−2, −2), (−2, 2), (2, −2), (2, 2)}.

37. $\begin{cases} x^2 - y = 2 \\ 2x - y = 4 \end{cases} \Rightarrow \begin{cases} x^2 - y = 2 \\ -2x + y = -4 \end{cases} \Rightarrow$

$x^2 - 2x = -2 \Rightarrow x^2 - 2x + 2 = 0 \Rightarrow$

$x = \dfrac{2 \pm \sqrt{4 - 8}}{2} = 1 \pm i \Rightarrow$ there are no real

solutions. Solution set: $\varnothing$

39. $\begin{cases} x^2 + y^2 = 5 \\ 3x^2 - 2y^2 = -5 \end{cases} \Rightarrow \begin{cases} 2x^2 + 2y^2 = 10 \\ 3x^2 - 2y^2 = -5 \end{cases} \Rightarrow$

$5x^2 = 5 \Rightarrow x = \pm 1$

$(1)^2 + y^2 = 5 \Rightarrow y = \pm 2$

$(-1)^2 + y^2 = 5 \Rightarrow y = \pm 2$

The solution is $\{(-1, -2), (-1, 2), (1, -2),$
$(1, 2)\}$.

41. $\begin{cases} x^2 + y^2 + 2x = 9 \\ x^2 + 4y^2 + 3x = 14 \end{cases} \Rightarrow$

$\begin{cases} 4x^2 + 4y^2 + 8x = 36 \\ x^2 + 4y^2 + 3x = 14 \end{cases} \Rightarrow 3x^2 + 5x - 22 = 0 \Rightarrow$

$(x - 2)(3x + 11) = 0 \Rightarrow x = 2 \text{ or } x = -\dfrac{11}{3}$

$(2)^2 + y^2 + 2(2) = 9 \Rightarrow y = \pm 1$

$\left(-\dfrac{11}{3}\right)^2 + y^2 + 2\left(-\dfrac{11}{3}\right) = 9 \Rightarrow y^2 + \dfrac{55}{9} = 9 \Rightarrow$

$y = \pm \dfrac{\sqrt{26}}{3}$

The solution is $\left\{(2, -1), (2, 1), \left(-\dfrac{11}{3}, -\dfrac{\sqrt{26}}{3}\right),\right.$

$\left.\left(-\dfrac{11}{3}, \dfrac{\sqrt{26}}{3}\right)\right\}$.

43. Using substitution, we have

$\begin{cases} x + y = 8 \\ xy = 15 \end{cases} \Rightarrow \begin{cases} x = 8 - y \\ xy = 15 \end{cases} \Rightarrow$

$y(8 - y) = 15 \Rightarrow -y^2 + 8y - 15 = 0 \Rightarrow$

$y^2 - 8y + 15 = 0 \Rightarrow (y - 3)(y - 5) = 0 \Rightarrow$

$y = 3 \text{ or } y = 5$

$x + 3 = 8 \Rightarrow x = 5$

$x + 5 = 8 \Rightarrow x = 3$

The solution is $\{(3, 5), (5, 3)\}$.

45. Using elimination, we have

$\begin{cases} x^2 + y^2 = 2 \\ 3x^2 + 3y^2 = 9 \end{cases} \Rightarrow \begin{cases} -3x^2 - 3y^2 = -6 \\ 3x^2 + 3y^2 = 9 \end{cases} \Rightarrow$

$0 = 3 \Rightarrow$ there is no solution.

Solution set: $\varnothing$

47. Using substitution, we have

$\begin{cases} y^2 = 4x + 4 \\ y = 2x - 2 \end{cases} \Rightarrow (2x - 2)^2 = 4x + 4 \Rightarrow$

$4x^2 - 12x = 0 \Rightarrow 4x(x - 3) = 0 \Rightarrow x = 0 \text{ or } x = 3$

$y = 2(0) - 2 = -2$

$y = 2(3) - 2 = 4$

The solution is $\{(0, -2), (3, 4)\}$.

49. Using substitution, we have

$\begin{cases} x^2 + 4y^2 = 25 \\ x - 2y + 1 = 0 \end{cases} \Rightarrow \begin{cases} x^2 + 4y^2 = 25 \\ x = 2y - 1 \end{cases} \Rightarrow$

$(2y - 1)^2 + 4y^2 = 25 \Rightarrow 8y^2 - 4y - 24 = 0 \Rightarrow$

$4(y - 2)(2y + 3) = 0 \Rightarrow y = 2 \text{ or } y = -\dfrac{3}{2}$

$x - 2(2) + 1 = 0 \Rightarrow x = 3$

$x - 2\left(-\dfrac{3}{2}\right) + 1 = 0 \Rightarrow x = -4$

The solution is $\left\{(3, 2), \left(-4, -\dfrac{3}{2}\right)\right\}$.

51. Using elimination, we have

$\begin{cases} x^2 - 3y^2 = 1 \\ x^2 + 4y^2 = 8 \end{cases} \Rightarrow -7y^2 = -7 \Rightarrow y = \pm 1$

$x^2 - 3(-1)^2 = 1 \Rightarrow x = \pm 2$

$x^2 - 3(1)^2 = 1 \Rightarrow x = \pm 2$

The solution is $\{(-2, -1), (-2, 1), (2, -1),$
$(2, 1)\}$.

53. Using elimination, we have

$\begin{cases} x^2 - xy + 5x = 4 \\ 2x^2 - 3xy + 10x = -2 \end{cases} \Rightarrow$

$\begin{cases} -3x^2 + 3xy - 15x = -12 \\ 2x^2 - 3xy + 10x = -2 \end{cases} \Rightarrow -x^2 - 5x = -14 \Rightarrow$

$x^2 + 5x - 14 = 0 \Rightarrow (x + 7)(x - 2) = 0 \Rightarrow$

$x = -7 \text{ or } x = 2$

$(-7)^2 - (-7)y + 5(-7) = 4 \Rightarrow y = -\dfrac{10}{7}$

$(2)^2 - 2y + 5(2) = 4 \Rightarrow y = 5$

The solution is $\left\{\left(-7, -\dfrac{10}{7}\right), (2, 5)\right\}$.

55. Using elimination, we have

$$\begin{cases} x^2 + y^2 - 8x = -8 \\ x^2 - 4y^2 + 6x = 0 \end{cases} \Rightarrow$$

$$\begin{cases} 4x^2 + 4y^2 - 32x = -32 \\ x^2 - 4y^2 + 6x = 0 \end{cases} \Rightarrow 5x^2 - 26x = -32 \Rightarrow$$

$$5x^2 - 26x + 32 = 0 \Rightarrow (x - 2)(5x - 16) = 0 \Rightarrow$$

$$x = 2 \text{ or } x = \frac{16}{5}$$

$$2^2 + y^2 - 8(2) = -8 \Rightarrow y = \pm 2$$

$$\left(\frac{16}{5}\right)^2 + y^2 - 8\left(\frac{16}{5}\right) = -8 \Rightarrow y^2 = \frac{184}{25} \Rightarrow$$

$$y = \pm \frac{2\sqrt{46}}{5}$$

The solution is

$$\left\{ (2, -2), (2, 2), \left(\frac{16}{5}, -\frac{2\sqrt{46}}{5}\right), \left(\frac{16}{5}, \frac{2\sqrt{46}}{5}\right) \right\}.$$

5.4 Applying the Concepts

57. Using elimination, we have

$$\begin{cases} p + 2x^2 = 96 \\ p - 13x = 39 \end{cases} \Rightarrow 2x^2 + 13x = 57 \Rightarrow$$

$$2x^2 + 13x - 57 = 0 \Rightarrow (x - 3)(2x + 19) = 0 \Rightarrow$$

$$x = 3 \text{ or } x = -\frac{19}{2} \text{ (reject this)}$$

$$p - 13(3) = 39 \Rightarrow p = 78$$

Market equilibrium occurs when 3 (hundred) units are sold and the price is $78/unit.

59. Let x = the first positive number and let y = the second positive number. Then

$$\begin{cases} x + y = 24 \\ xy = 143 \end{cases} \Rightarrow \begin{cases} y = 24 - x \\ xy = 143 \end{cases} \Rightarrow$$

$$x(24 - x) = 143 \Rightarrow -x^2 + 24x - 143 = 0 \Rightarrow$$

$$-(x - 13)(x - 11) = 0 \Rightarrow x = 11 \text{ or } x = 13$$

$$11 + y = 24 \Rightarrow y = 13$$

$$13 + y = 24 \Rightarrow y = 11$$

The numbers are 11 and 13.

61. a. Let x = the length of the two equal sides and let y = the length of the third side.

Using the Pythagorean theorem, we have $x^2 + (y - x)^2 = 100^2$. So,

$$\begin{cases} 2x + y + 100 = 360 \\ x^2 + (y - x)^2 = 100^2 \end{cases} \Rightarrow$$

$$\begin{cases} y = 260 - 2x \\ x^2 + (y - x)^2 = 100^2 \end{cases} \Rightarrow$$

$$x^2 + ((260 - 2x) - x)^2 = 10,000 \Rightarrow$$

$$10x^2 - 1560x + 57,600 = 0 \Rightarrow$$

$$10(x - 60)(x - 96) = 0 \Rightarrow x = 60 \text{ or } x = 96$$

$$2(60) + y + 100 = 360 \Rightarrow y = 140$$

$$2(96) + y + 100 = 360 \Rightarrow y = 68$$

Because $y > x$ (from the diagram), we reject $x = 96$ and $y = 68$.

So, $x = 60$ m and $y = 140$ m

b. The area is $60^2 + \frac{1}{2}(60)(80) = 6000 \text{ m}^2$.

63. Let x = the original number of students in the group and let y = the original cost per student. Then

$$\begin{cases} xy = 960 \\ (x + 8)(y - 6) = 960 \end{cases} \Rightarrow$$

$$\begin{cases} y = \frac{960}{x} \\ (x + 8)(y - 6) = 960 \end{cases} \Rightarrow$$

$$(x + 8)\left(\frac{960}{x} - 6\right) = 960 \Rightarrow$$

$$-6x + \frac{7680}{x} + 912 = 960 \Rightarrow$$

$$-6x + \frac{7680}{x} - 48 = 0 \Rightarrow$$

$$-6x^2 - 48x + 7680 = 0 \Rightarrow$$

$$-6(x - 32)(x + 40) = 0 \Rightarrow x = 32 \text{ or}$$

$$x = -40 \text{ (reject this)}$$

$$32y = 960 \Rightarrow y = 30$$

There were originally 32 students at a cost of $30 each.

65. Let x = the number of shares of stock she bought and let y = the original price per share. Then

$$\begin{cases} xy + 100 = 10,000 \\ (x+30)(y+3) = 11,900 + 100 \end{cases} \Rightarrow$$

$$\begin{cases} y = \dfrac{9900}{x} \\ (x+30)(y+3) = 12,000 \end{cases} \Rightarrow$$

$$(x+30)\left(\dfrac{9900}{x} + 3\right) = 12,000 \Rightarrow$$

$$3x + \dfrac{297,000}{x} - 2010 = 0 \Rightarrow$$

$$3x^2 - 2010x + 297,000 = 0 \Rightarrow$$

$$3(x-450)(x-220) = 0 \Rightarrow x = 450 \text{ or } x = 220$$

The problem says that she bought more than 400 shares, so reject $x = 220$.

$$450y + 100 = 10,000 \Rightarrow y = 22$$

She bought 450 shares at $22 per share.

5.4 Beyond the Basics

67. $\begin{cases} x^2 + y^2 - 7x + 5y + 6 = 0 \\ y = 0 \end{cases} \Rightarrow$

$$x^2 - 7x + 6 = 0 \Rightarrow (x-6)(x-1) = 0 \Rightarrow$$

$$x = 6 \text{ or } x = 1$$

The circle intersects the x-axis at $A(1, 0)$ and

$B(6, 0)$. $d(A, B) = \sqrt{(6-1)^2 + (0-0)^2} = 5.$

69. $\begin{cases} x^2 + y^2 + 2x - 4y - 5 = 0 \\ x - y + 1 = 0 \end{cases} \Rightarrow$

$$\begin{cases} x^2 + y^2 + 2x - 4y - 5 = 0 \\ x = y - 1 \end{cases} \Rightarrow$$

$$(y-1)^2 + y^2 + 2(y-1) - 4y - 5 = 0 \Rightarrow$$

$$2y^2 - 4y - 6 = 0 \Rightarrow 2(y-3)(y+1) = 0 \Rightarrow$$

$$y = 3 \text{ or } y = -1$$

$$x - 3 + 1 = 0 \Rightarrow x = 2$$

$$x - (-1) + 1 \Rightarrow x = -2$$

The circle and the line intersect at $A(-2, -1)$ and $B(2, 3)$.

$$d(A, B) = \sqrt{(-2-2)^2 + (-1-3)^2} =$$

$$\sqrt{16+16} = \sqrt{32} = 4\sqrt{2}.$$

71. $\begin{cases} x^2 + y^2 - 2x + 2y - 3 = 0 \\ x + 2y + 6 = 0 \end{cases} \Rightarrow$

$$\begin{cases} x^2 + y^2 - 2x + 2y - 3 = 0 \\ x = -2y - 6 \end{cases} \Rightarrow$$

$$(-2y - 6)^2 + y^2 - 2(-2y - 6) + 2y - 3 = 0 \Rightarrow$$

$$5y^2 + 30y + 45 = 0 \Rightarrow 5(y+3)^2 = 0 \Rightarrow y = -3$$

$$x + 2(-3) + 6 = 0 \Rightarrow x = 0$$

The line and the circle intersect at only one point $(0, -3)$, so the line is tangent to the circle.

73. $\begin{cases} x + y = 2 \quad (1) \\ x^2 + y^2 = c^2 \quad (2) \end{cases}$

From equation (1), we have $y = 2 - x$. Substituting in equation (2) and then solving for x, we have

$$x^2 + (2-x)^2 = c^2 \Rightarrow x^2 + 4 - 4x + x^2 = c^2 \Rightarrow$$

$$2x^2 - 4x + 4 - c^2 = 0 \Rightarrow$$

$$x = \dfrac{-(-4) \pm \sqrt{(-4)^2 - 4(2)(4 - c^2)}}{2(2)}$$

Since the equations are tangent, there is only one solution. Recall that the value of the discriminant is 0 when there is exactly one real solution, so solve

$$(-4)^2 - 4(2)(4 - c^2) = 0 \Rightarrow$$

$$16 - 32 + 8c^2 = 0 \Rightarrow c^2 = 2 \Rightarrow c = \pm\sqrt{2}$$

75. $\begin{cases} 4x^2 + y^2 = 25 & (1) \\ 8x + 3y = c & (2) \end{cases}$

From equation (2), we have $y = \dfrac{c - 8x}{3}$.

Substituting in equation (1) and then solving for x, we have

$$4x^2 + \left(\frac{c - 8x}{3}\right)^2 = 25$$

$$4x^2 + \frac{c^2 - 16cx + 64x^2}{9} = 25$$

$$36x^2 + c^2 - 16cx + 64x^2 = 225$$

$$100x^2 - 16cx + c^2 - 225 = 0$$

$$x = \frac{-(-16c) \pm \sqrt{(-16c)^2 - 4(100)(c^2 - 225)}}{2(100)}$$

Since the equations are tangent, there is only one solution. Recall that the value of the discriminant is 0 when there is exactly one real solution, so solve

$$(-16c)^2 - 4(100)(c^2 - 225) = 0 \Rightarrow$$

$$256c^2 - 400c^2 + 90,000 = 0 \Rightarrow$$

$$90,000 - 144c^2 = 0 \Rightarrow$$

$$(300 - 12c)(300 + 12c) = 0 \Rightarrow c = \pm 25$$

In exercises 77–80, let $u = \dfrac{1}{x}$ and $v = \dfrac{1}{y}$.

77. $\begin{cases} \dfrac{1}{x} - \dfrac{1}{y} = 5 \\ \dfrac{1}{xy} - 24 = 0 \end{cases} \Rightarrow \begin{cases} u - v = 5 & (1) \\ uv - 24 = 0 & (2) \end{cases}$

From equation (1), we have $u = v + 5$. Substitute this into equation (2) and solve for v.

$$v(v + 5) - 24 = 0 \Rightarrow v^2 + 5v - 24 = 0 \Rightarrow$$

$$(v + 8)(v - 3) = 0 \Rightarrow v = -8, 3$$

$$v = -8 \Rightarrow u + 8 = 5 \Rightarrow u = -3 \Rightarrow$$

$$x = -\frac{1}{3}, \; y = -\frac{1}{8}$$

$$v = 3 \Rightarrow u - 3 = 5 \Rightarrow u = 8 \Rightarrow x = \frac{1}{8}, \; y = \frac{1}{3}$$

Solution set: $\left\{ \left(-\dfrac{1}{3}, -\dfrac{1}{8}\right), \left(\dfrac{1}{8}, \dfrac{1}{3}\right) \right\}$

79. $\begin{cases} \dfrac{5}{x^2} - \dfrac{3}{y^2} = 2 \\ \dfrac{6}{x^2} + \dfrac{1}{y^2} = 7 \end{cases} \Rightarrow \begin{cases} 5u^2 - 3v^2 = 2 & (1) \\ 6u^2 + v^2 = 7 & (2) \end{cases}$

Solve equation (2) for v^2, then substitute this expression in equation (1) and solve for u.

$$v^2 = 7 - 6u^2$$

$$5u^2 - 3(7 - 6u^2) = 2 \Rightarrow 23u^2 - 21 = 2 \Rightarrow$$

$$u^2 = 1 \Rightarrow u = \pm 1 \Rightarrow x = \pm 1$$

If $u = 1$, then

$$6(1^2) + v^2 = 7 \Rightarrow v^2 = 1 \Rightarrow v = \pm 1 \Rightarrow y = \pm 1$$

If $u = -1$, then

$$6(-1^2) + v^2 = 7 \Rightarrow v^2 = 1 \Rightarrow v = \pm 1 \Rightarrow y = \pm 1$$

Solution set: $\{(-1, -1), (-1, 1), (1, -1), (1, 1)\}$

81. $(a + bi)^2 = -5 + 12i \Rightarrow$

$$a^2 + 2abi - b^2 = -5 + 12i \Rightarrow$$

$$\begin{cases} a^2 - b^2 = -5 \\ 2ab = 12 \end{cases} \Rightarrow \begin{cases} a^2 - b^2 = -5 & (1) \\ ab = 6 & (2) \end{cases}$$

From equation (2), we have $b = \dfrac{6}{a}$. Substitute this into equation (1) and solve for a.

$$a^2 - \left(\frac{6}{a}\right)^2 = -5 \Rightarrow a^4 + 5a^2 - 36 = 0 \Rightarrow$$

$$(a^2 + 9)(a^2 - 4) = 0 \Rightarrow$$

$$(a^2 + 9)(a - 2)(a + 2) = 0 \Rightarrow a = \pm 2, \; \pm 3i$$

Since a is real, reject $\pm 3i$. Using equation (2), if $a = -2$, then $b = -3$. If $a = 2$, then $b = 3$. Thus, the square roots of $w = -5 + 12i$ are $-2 - 3i$ and $2 + 3i$.

83. $(a + bi)^2 = 7 - 24i \Rightarrow$

$$a^2 + 2abi - b^2 = 7 - 24i \Rightarrow$$

$$\begin{cases} a^2 - b^2 = 7 \\ 2ab = -24 \end{cases} \Rightarrow \begin{cases} a^2 - b^2 = 7 & (1) \\ ab = -12 & (2) \end{cases}$$

From equation (2), we have $b = -\dfrac{12}{a}$.

Substitute this into equation (1) and solve for a.

$$a^2 - \left(-\frac{12}{a}\right)^2 = 7 \Rightarrow a^4 - 7a^2 - 144 = 0 \Rightarrow$$

$$(a^2 - 16)(a^2 + 9) = 0 \Rightarrow$$

$$(a^2 + 9)(a - 4)(a + 4) = 0 \Rightarrow a = \pm 4, \; \pm 3i$$

Since a is real, reject $\pm 3i$.
Using equation (2), if $a = -4$, then $b = 3$.
If $a = 4$, then $b = -3$. Thus, the square roots of $w = 7 - 24i$ are $-4 + 3i$ and $4 - 3i$.

85. $(a+bi)^2 = 13+8\sqrt{3}i \Rightarrow$

$a^2 + 2abi - b^2 = 13 + 8\sqrt{3}i \Rightarrow$

$\begin{cases} a^2 - b^2 = 13 \\ 2ab = 8\sqrt{3} \end{cases} \Rightarrow \begin{cases} a^2 - b^2 = 13 & (1) \\ ab = 4\sqrt{3} & (2) \end{cases}$

From equation (2), we have $b = \dfrac{4\sqrt{3}}{a}$.

Substitute this into equation (1) and solve for a.

$a^2 - \left(\dfrac{4\sqrt{3}}{a}\right)^2 = 13 \Rightarrow a^4 - 13a^2 - 48 = 0 \Rightarrow$

$\left(a^2 - 16\right)\left(a^2 + 3\right) = 0 \Rightarrow$

$\left(a^2 + 3\right)(a - 4)(a + 4) = 0 \Rightarrow a = \pm 4,\ \pm\sqrt{3}i$

Since a is real, reject $\pm\sqrt{3}i$. Using equation (2), if $a = -4$, then $b = -\sqrt{3}$. If $a = 4$, then $b = \sqrt{3}$. Thus, the square roots of $w = 13 + 8\sqrt{3}i$ are $-4 - \sqrt{3}i$ and $4 + \sqrt{3}i$.

87. Using substitution, we have

$\begin{cases} y = 3^x + 4 \\ y = 3^{2x} - 2 \end{cases} \Rightarrow 3^{2x} - 2 = 3^x + 4 \Rightarrow$

$3^{2x} - 3^x - 6 = 0$

Let $u = 3^x$. Then $u^2 - u - 6 = 0 \Rightarrow$

$(u - 3)(u + 2) = 0 \Rightarrow u = 3$ or $u = -2$

(Reject the negative solution.)

$3 = 3^x \Rightarrow x = 1$

$y = 3^1 + 4 = 7$

The solution is $\{(1, 7)\}$.

89. Using substitution, we have

$\begin{cases} y = 2^x + 3 \\ y = 2^{2x} + 1 \end{cases} \Rightarrow 2^x + 3 = 2^{2x} + 1 \Rightarrow$

$2^{2x} - 2^x - 2 = 0.$

Let $u = 2^x$. Then $u^2 - u - 2 = 0 \Rightarrow$

$(u - 2)(u + 1) = 0 \Rightarrow u = 2$ or $u = -1$

(reject the negative solution.)

$2 = 2^x \Rightarrow x = 1;\quad y = 2^1 + 3 = 5$

The solution is $\{(1, 5)\}$.

91. $\begin{cases} x = 3^y \\ 3^{2y} = 3x - 2 \end{cases} \Rightarrow 3^{2y} = 3 \cdot 3^y - 2 \Rightarrow$

$3^{2y} - 3 \cdot 3^y + 2 = 0$. Let $u = 3^y$. Then

$u^2 - 3u + 2 = 0 \Rightarrow (u - 2)(u - 1) = 0 \Rightarrow$

$u = 2$ or $u = 1$

$2 = 3^y \Rightarrow \ln 2 = y \ln 3 \Rightarrow \dfrac{\ln 2}{\ln 3} = y;$

$x = 3^{\ln 2/\ln 3} = 2$ or $1 = 3^y \Rightarrow y = 0$

$x = 3^0 = 1$

The solution is $\left\{(1, 0), \left(2, \dfrac{\ln 2}{\ln 3}\right)\right\}$.

5.4 Critical Thinking/Discussion/Writing

93. a. Not possible   **b.** Possible

    c. Possible   **d.** Possible

    e. Not possible   **f.** Not possible.

94. a. Possible   **b.** Possible

    c. Possible   **d.** Possible

    e. Possible   **f.** Not possible.

5.4 Getting Ready for the Next Section

95.

$2x + y = 0$

97.

$3x + 4y = 12$

99.

$y - 2x^2 = 0$

101.

$y + x^2 - 2 = 0$

103.

Point	$5x + 3y = 0$?
$(0, 0)$	$5(0) + 3(0) \overset{?}{=} 0$ $0 = 0$ ✓
$(-3, 5)$	$5(-3) + 3(5) \overset{?}{=} 0$ $0 = 0$ ✓
$(-1, 1)$	$5(-1) + 3(1) \overset{?}{=} 0$ $-2 = 0$ ✗
$(1, -1)$	$5(1) + 3(-1) \overset{?}{=} 0$ $2 = 0$ ✗

$(0, 0)$ and $(-3, 5)$ lie on the graph of the equation $5x + 3y = 0$.

105.

Point	$y = x^2 + 3$?
$(0, 0)$	$0 \overset{?}{=} 0^2 + 3$ $0 = 3$ ✗
$(1, 4)$	$4 \overset{?}{=} 1^2 + 3$ $3 = 3$ ✓
$(-1, 4)$	$4 \overset{?}{=} (-1)^2 + 3$ $4 = 4$ ✓
$(2, 5)$	$5 \overset{?}{=} 2^2 + 3$ $5 = 7$ ✗

$(1, 4)$ and $(-1, 4)$ lie on the graph of the equation $y = x^2 + 3$.

5.5   Systems of Inequalities

5.5 Practice Problems

1.

$3x + y = 6$

2.

$y = -2x + 4$

3.

$x + y = 9$

4.

$x - y = 1$ $3x - y = 3$

5. Solve the systems $\begin{cases} 2x + 3y = 16 \\ 4x + 2y = 16 \end{cases}$,

$\begin{cases} 4x + 2y = 16 \\ 2x - \ y = 8 \end{cases}$, and $\begin{cases} 2x + 3y = 16 \\ 2x - \ y = 8 \end{cases}$ to find the

corner points: (4, 0), (5, 2), and (2, 4).

$2x - y = 8$ $2x + 3y = 16$ $4x + 2y = 16$

6.

$y = x^2 - 2$

7.

$y = x^2 + 1$ (0, 13)   (3, 10)   (−2, 5) $y = 4x + 13$ $y = -x + 13$

5.5 Concepts and Vocabulary

1. In the graph of $x - 3y > 1$, the corresponding equation $x - 3y = 1$ is graphed as a <u>dashed</u> line.

3. In a system of inequalities containing both $2x + y > 5$ and $2x - y \le 3$, the point of intersection of the lines $2x + y = 5$ and $2x - y = 3$ is <u>not a</u> solution of the system.

5. True

7. True

5.5 Building Skills

9.

$x \ge 0$

11.

$x > -1$

13.

$x \geq 2$

15.

$y - x < 0$

17.

$x + 2y < 6$

19.

$2x + 3y \geq 12$

21.

$x - 2y \leq 4$

23.

$3x + 5y < 15$

25. Substitute each ordered pair into the system:

$\begin{cases} 0 + 0 < 2 \\ 2(0) + 0 \geq 6 \end{cases} \quad \begin{cases} -4 - 1 < 2 \\ 2(-4) - 1 \geq 6 \end{cases}$

$\begin{cases} 3 + 0 < 2 \\ 2(3) + 0 \geq 6 \end{cases} \quad \begin{cases} 0 + 3 < 2 \\ 2(0) + 3 \geq 6 \end{cases}$

None of the ordered pairs are a solution.

27. Substitute each ordered pair into the system:

$\begin{cases} 0 < 0 + 2 \\ 0 + 0 \leq 4 \end{cases} \begin{cases} 0 < 1 + 2 \\ 1 + 0 \leq 4 \end{cases} \begin{cases} 1 < 0 + 2 \\ 0 + 1 \leq 4 \end{cases} \begin{cases} 1 < 1 + 2 \\ 1 + 1 \leq 4 \end{cases}$

All of the ordered pairs are solutions.

29. Substitute each ordered pair into the system:

$\begin{cases} 3(0) - 4(0) \leq 12 \\ 0 + 0 \leq 4 \\ 5(0) - 2(0) \geq 6 \end{cases} \quad \begin{cases} 3(2) - 4(0) \leq 12 \\ 2 + 0 \leq 4 \\ 5(2) - 2(0) \geq 6 \end{cases}$

$\begin{cases} 3(3) - 4(1) \leq 12 \\ 3 + 1 \leq 4 \\ 5(3) - 2(1) \geq 6 \end{cases} \quad \begin{cases} 3(2) - 4(2) \leq 12 \\ 2 + 2 \leq 4 \\ 5(2) - 2(2) \geq 6 \end{cases}$

The solutions are (2, 0), (3, 1) and (2, 2).

31.

33.

35.

The system is inconsistent. There are no vertices of the solution set.

37.

39.

41.

43.

45.

47. A   **49.** B   **51.** E

53.

55.

57.

59.

61.

63.

65.

67. D, K

69. E, L

71. E, B

73.

75.

77.

79.

5.5 Applying the Concepts

81. Let $x =$ the number of cases of Coke and $y =$ the number of cases of Sprite. Then we have

$$\begin{cases} x + y \leq 40 & (1) \\ \quad x \geq 10 & (2) \\ \quad y \geq 5 & (3) \end{cases}$$

To find the vertices, solve the systems

$$\begin{cases} x + y = 40 & (1) \\ \quad x = 10 & (2) \end{cases}, \begin{cases} x + y = 40 & (1) \\ \quad y = 5 & (3) \end{cases}, \text{ and}$$

$$\begin{cases} x = 10 & (2) \\ y = 5 & (3) \end{cases}.$$

The vertices are $(10, 30)$, $(35, 5)$, and $(10, 5)$.

83. Let $x =$ the number of two story houses and $y =$ the number of one story houses. Then we have

$$\begin{cases} 7x + 5y \leq 43 \\ 4x + 3y \leq 25 \\ \quad x \geq 0 \\ \quad y \geq 0 \end{cases}$$

To find the vertices, solve the systems

$$\begin{cases} 7x + 5y = 43 \\ 4x + 3y = 25 \end{cases}, \begin{cases} 7x + 5y = 43 \\ \quad y = 0 \end{cases}, \begin{cases} 4x + 3y = 25 \\ \quad x = 0 \end{cases},$$

and $\begin{cases} x = 0 \\ y = 0 \end{cases}$.

$$\begin{cases} 7x + 5y = 43 \\ 4x + 3y = 25 \end{cases} \Rightarrow \begin{cases} 28x + 20y = 172 \\ -28x - 21y = -175 \end{cases} \Rightarrow$$

$$-y = -3 \Rightarrow y = 3$$

$$7x + 15 = 43 \Rightarrow x = 4$$

$$\begin{cases} 7x + 5y = 43 \\ \quad y \geq 0 \end{cases} \Rightarrow x = \frac{43}{7}$$

$$\begin{cases} 4x + 3y = 25 \\ \quad x = 0 \end{cases} \Rightarrow y = \frac{25}{3}$$

The vertices are $(0, 0)$, $\left(0, \dfrac{25}{3}\right)$, $\left(\dfrac{43}{7}, 0\right)$,

and $(4, 3)$.

85. Let $x =$ number of 3 hp engines and $y =$ the number of 5 hp engines. Then we have

$$\begin{cases} 3x + \quad 4.5y \leq 360 & (1) \\ 2x + \quad\quad y \leq 200 & (2) \\ 0.5x + 0.75y \leq 60 & (3) \\ \quad\quad\quad x \geq 0 \\ \quad\quad\quad y \geq 0 \end{cases}$$

Equations (1) and (3) coincide. To find the vertices, solve the systems

$$\begin{cases} 3x + 4.5y = 360 \\ 2x + \quad y = 200 \end{cases}, \begin{cases} 3x + 4.5y = 360 \\ \quad\quad x = 0 \end{cases}, \text{ and}$$

$$\begin{cases} 2x + y = 200 \\ \quad\quad y = 0 \end{cases}.$$

$$\begin{cases} 3x + 4.5y = 360 \\ 2x + \quad y = 200 \end{cases} \Rightarrow$$

$$3x + 4.5(200 - 2x) = 360 \Rightarrow 900 - 6x = 360 \Rightarrow$$

$$x = 90$$

$$2(90) + y = 200 \Rightarrow y = 20$$

$$\begin{cases} 3x + 4.5y = 360 \\ \quad\quad x = 0 \end{cases} \Rightarrow 4.5y = 360 \Rightarrow y = 80$$

$$\begin{cases} 2x + y = 200 \\ \quad\quad y = 0 \end{cases} \Rightarrow 2x = 200 \Rightarrow x = 100$$

The vertices are $(0, 0)$ $(100, 0)$, $(0, 80)$, and $(90, 20)$.

5.5 Beyond the Basics

87. The equation of the line connecting $(0, 2)$ and $(3, 0)$ is $y = -\dfrac{2}{3}x + 2$. The system is

$$\begin{cases} y \le -\dfrac{2}{3}x + 2 \\ x \ge 0 \\ y \ge 0 \end{cases}.$$

89. $\begin{cases} x \ge -1 \\ x \le 3 \end{cases}.$

91. The equation of the line connecting $(0, 4)$ and $(2, 3)$ is $y = -\dfrac{1}{2}x + 4$. The equation of the line connecting $(2, 3)$ and $(4, 0)$ is $y = -\dfrac{3}{2}x + 6$. The system is

$$\begin{cases} x \ge 0 \\ y \ge 0 \\ y \le -\dfrac{1}{2}x + 4 \\ y \le -\dfrac{3}{2}x + 6 \end{cases}.$$

93. The equation of the line connecting $(0, 16)$ and $(10, 6)$ is $y = -x + 16$. The equation of the line connecting $(10, 6)$ and $(5, 1)$ is $y = x - 4$. The equation of the line connecting $(5, 1)$ and $(0, 6)$ is $y = -x + 6$. The system is

$$\begin{cases} x \ge 0 \\ y \le -x + 16 \\ y \ge x - 4 \\ y \ge -x + 6 \end{cases}.$$

95.

97.

99.

101.

103.

105.

5.5 Critical Thinking/Discussion/Writing

107.

108.

5.5 Getting Ready for the Next Section

109. To find the point of intersection, solve the system

$$\begin{cases} x + 2y = 40 & (1) \\ 3x + y = 30 & (2) \end{cases}$$

Solve equation (2) for y in terms of x, then substitute the expression in equation (1) and solve for x.

$3x + y = 30 \Rightarrow y = -3x + 30$

$x + 2(-3x + 30) = 40 \Rightarrow -5x + 60 = 40 \Rightarrow$

$-5x = -20 \Rightarrow x = 4$

Substitute $x = 4$ into equation (1) and solve for y.

$4 + 2y = 40 \Rightarrow 2y = 36 \Rightarrow y = 18$

The point of intersection is $(4, 18)$.

111. To find the point of intersection, solve the system

$$\begin{cases} x - 2y = 2 & (1) \\ 3x + 2y = 12 & (2) \end{cases}$$

Add the two equations, then solve for x.

$x - 2y = 2$
$\underline{3x + 2y = 12}$
$\qquad 4x = 14 \Rightarrow x = \dfrac{14}{4} = \dfrac{7}{2}$

Substitute $x = \dfrac{7}{2}$ into equation (1) and solve for y.

$\dfrac{7}{2} - 2y = 2 \Rightarrow -2y = -\dfrac{3}{2} \Rightarrow y = \dfrac{3}{4}$

The point of intersection is $\left(\dfrac{7}{2}, \dfrac{3}{4}\right)$.

113.

Solve each pair of equations to find the points of intersection.

I: $\begin{cases} 2y + x = 8 \\ y - x = 1 \end{cases}$

II: $\begin{cases} y - x = 1 \\ y + 2x = 13 \end{cases}$

III: $\begin{cases} y + 2x = 13 \\ 2y + x = 8 \end{cases}$

For system I, we will use substitution.

$y = x + 1$

$2(x + 1) + x = 8 \Rightarrow 3x + 2 = 8 \Rightarrow 3x = 6 \Rightarrow$
$x = 2$

$y - 2 = 1 \Rightarrow y = 3$

The point of intersection is $(2, 3)$.

For system II, we will use substitution.

$y = x + 1$

$(x + 1) + 2x = 13 \Rightarrow 3x + 1 = 13 \Rightarrow 3x = 12 \Rightarrow$
$x = 4; \quad y - 4 = 1 \Rightarrow y = 5$

The point of intersection is $(4, 5)$.

For system III, we will use substitution.

$y = -2x + 13$

$2(-2x + 13) + x = 8 \Rightarrow -3x + 26 = 8 \Rightarrow$
$-3x = -18 \Rightarrow x = 6$

$y + 2(6) = 13 \Rightarrow y + 12 = 13 \Rightarrow y = 1$

The point of intersection is $(6, 1)$.
The vertices of the triangle are $(2, 3)$, $(4, 5)$, and $(6, 1)$.

115. a.

b. Solve the systems
$$\begin{cases} x = 0 \\ y = 0 \end{cases}, \begin{cases} 3x + 4y = 12 \\ y = 0 \end{cases}, \begin{cases} 3x + 4y = 12 \\ 4x + 3y = 12 \end{cases},$$

and $\begin{cases} 4x + 3y = 12 \\ x = 0 \end{cases}$

to find the vertices $(0, 0)$, $(0, 3)$, $\left(\dfrac{12}{7}, \dfrac{12}{7}\right)$,

and $(3, 0)$.

5.6   Linear Programming

5.6 Practice Problems

1. Maximize $f = 4x + 5y$ subject to the constraints
$$\begin{cases} 5x + 7y \le 35 \\ x \ge 0 \\ y \ge 0 \end{cases}.$$

First, graph the solution set of the constraints.

Solve the systems $\begin{cases} x = 0 \\ y = 0 \end{cases}, \begin{cases} 5x + 7y = 35 \\ y = 0 \end{cases}$, and

$\begin{cases} 5x + 7y = 35 \\ x = 0 \end{cases}$ to find the vertices: $(0, 0)$,

$(7, 0)$ and $(0, 5)$. Now find the values of the objective function at each vertex:

Ordered pair	$f = 4x + 5y$
$(0, 0)$	0
$(7, 0)$	28
$(0, 5)$	25

The maximum is 28 at $(7, 0)$.

2. Minimize $f = \dfrac{1}{2}x + y$ subject to the constraints
$$\begin{cases} x + y \le 8 \\ x + 2y \ge 4 \\ 3x + 2y \ge 6 \\ x \ge 0 \\ y \ge 0 \end{cases}.$$

The graph and vertices of the set of feasible solutions are as given in Example 2. The vertices are: $(0, 8)$, $(0, 3)$, $(1, 1.5)$, $(4, 0)$, and $(8, 0)$. Now find the values of the objective function at each vertex:

Ordered pair	$f = \dfrac{1}{2}x + y$
$(0, 8)$	8
$(0, 3)$	3
$(1, 1.5)$	2
$(4, 0)$	2
$(8, 0)$	4

The minimum is 2 at $(1, 1.5)$ and $(4, 0)$. In fact, the unique minimum value occurs at every point on the line segment AE. So the linear programming problem has infinitely many solutions, each of which satisfies the equation $x + 2y = 4$.

3. Let $x =$ the number of ounces of soup and let $y =$ the number of ounces of salad. The number of calories in the two items is $f = 30x + 60y$. (This is the objective function.) The constraints are as given in Example 3.
$$\begin{cases} x + y \ge 10 \\ 3x + 2y \ge 24 \\ x \ge 0 \\ y \ge 0 \end{cases}.$$

The graph and vertices of the set of feasible solutions are as given in Example 3. The vertices are $(0, 12)$, $(4, 6)$, and $(10, 0)$.

(continued on next page)

(*continued*)

Now find the values of the objective function at each vertex:

Ordered pair	$f = 30x + 60y$
(0, 12)	720
(4, 6)	480
(10, 0)	300

The minimum is 300 at (10, 0). This means that the lunch menu for Peter Griffin should contain 10 ounces of soup and 0 ounces of salad.

5.6 Concepts and Vocabulary

1. The process of finding the maximum or minimum value of a quantity is called <u>optimization</u>.

3. The inequalities that determine the region S in a linear programming problem are called <u>constraints</u>, and S is called the set of <u>feasible</u> solutions.

5. True

7. True

5.6 Building Skills

9.

Ordered pair	$f = x + y$
(10, 0)	10
(3, 2)	5
(1, 4)	5
(0, 8)	8
(10, 8)	18

Maximum: 18; minimum: 5

11.

Ordered pair	$f = x + 2y$
(10, 0)	10
(3, 2)	7
(1, 4)	9
(0, 8)	16
(10, 8)	26

Maximum: 26; minimum: 7

13.

Ordered pair	$f = 5x + 2y$
(10, 0)	50
(3, 2)	19
(1, 4)	13
(0, 8)	16
(10, 8)	66

Maximum: 66; minimum: 13

15.

Solve the systems

$$\begin{cases} x = 0 \\ 5x + 8y = 40 \end{cases} \begin{cases} 5x + 8y = 40 \\ 3x + y = 12 \end{cases}, \begin{cases} 3x + y = 12 \\ y = 0 \end{cases},$$

and $\begin{cases} x = 0 \\ y = 0 \end{cases}$ to find the vertices: (0, 5),

$\left(\dfrac{56}{19}, \dfrac{60}{19}\right)$, (4, 0, and (0, 0). Now find the

values of the objective function at each vertex:

Ordered pair	$f = 9x + 13y$
(0, 5)	65
$\left(\dfrac{56}{19}, \dfrac{60}{19}\right)$	$\dfrac{1284}{19}$
(4, 0)	36
(0, 0)	0

The maximum is $\dfrac{1284}{19}$ at $\left(\dfrac{56}{19}, \dfrac{60}{19}\right)$.

17.

Solve the systems

$$\begin{cases} x = 0 \\ x + 2y = 100 \end{cases}, \begin{cases} x + 2y = 100 \\ x + y = 70 \end{cases}, \begin{cases} x + y = 70 \\ 2x + y = 120 \end{cases},$$

$$\begin{cases} 2x + y = 120 \\ y = 0 \end{cases}, \text{ and } \begin{cases} x = 0 \\ y = 0 \end{cases} \text{ to find the}$$

vertices: (0, 50), (40, 30), (50, 20), (60, 0), and (0, 0). Now find the values of the objective function at each vertex:

Ordered pair	$f = 5x + 7y$
(0, 50)	350
(40, 30)	410
(50, 20)	390
(60, 0)	300
(0, 0)	0

The maximum is 410 at (40, 30).

19.

Solve the systems

$$\begin{cases} x = 0 \\ 2x + y = 2 \end{cases}, \begin{cases} 2x + y = 2 \\ x + 3y = 3 \end{cases}, \text{and } \begin{cases} x + 3y = 3 \\ y = 0 \end{cases} \text{ to}$$

find the vertices: (0, 2), $\left(\dfrac{3}{5}, \dfrac{4}{5} \right)$, and (3, 0).

Now find the values of the objective function at each vertex:

Ordered pair	$f = x + 4y$
(0, 2)	8
$\left(\dfrac{3}{5}, \dfrac{4}{5} \right)$	$\dfrac{19}{5}$
(3, 0)	3

The minimum is 3 at (3, 0).

21.

Solve the systems

$$\begin{cases} x = 0 \\ 2x + y = 100 \end{cases}, \begin{cases} 2x + y = 100 \\ 3x + 4y = 360 \end{cases}, \text{ and}$$

$$\begin{cases} 3x + 4y = 360 \\ y = 0 \end{cases} \text{ to find the vertices: (0, 100),}$$

(8, 84), and (120, 0). Now find the values of the objective function at each vertex.

Ordered pair	$f = 13x + 15y$
(0, 100)	1500
(8, 84)	1364
(120, 0)	1560

The minimum is 1364 at (8, 84).

23.

Solve the systems $\begin{cases} x = 0 \\ y = 0 \end{cases}, \begin{cases} 5x + 3y = 18 \\ y = 0 \end{cases},$

$$\begin{cases} 2x + 2y = 8 \\ 5x + 3y = 18 \end{cases}, \begin{cases} 3x + 8y = 24 \\ 2x + 2y = 8 \end{cases}, \text{ and}$$

$$\begin{cases} 3x + 8y = 24 \\ x = 0 \end{cases} \text{ to find the vertices: (0, 0),}$$

(3.6, 0), (3, 1), (1.6, 2.4), and (0, 3). Now find the values of the objective function at each vertex:

Ordered pair	$f = 15x + 7y$
(0, 0)	0
(3.6, 0)	54
(3, 1)	52
(1.6, 2.4)	40.8
(0, 3)	21

The maximum is 54 at (3.6, 0).

25.

Solve the systems

$$\begin{cases} y = 0 \\ x + 2y = 50 \end{cases}, \quad \begin{cases} x + 2y = 50 \\ x + y = 35 \end{cases}, \quad \begin{cases} x + y = 35 \\ 2x + y = 40 \end{cases},$$

and $\begin{cases} 2x + y = 40 \\ x = 0 \end{cases}$ to find the vertices: (50, 0),

(20, 15), (30, 5), and (0, 40). Now find the values of the objective function at each vertex:

Ordered pair	$f = 8x + 16y$
(50, 0)	400
(20, 15)	400
(10, 20)	400
(5, 30)	520
(0, 40)	640

The minimum is 400 for all (x, y) on the line segments between (10, 20) and (20, 15), and (20, 15) and (50, 0).

5.6 Applying the Concepts

27. Let x = the number of corn acres, and let y = the number of soybean acres. Then, the profit $p = 50x + 40y$. The constraints are $x \ge 0$, $y \ge 0$, $x + y \le 240$ (number of acres) and $2x + y \le 320$ (number of labor hours).

Solve the systems

$$\begin{cases} x = 0 \\ x + y = 240 \end{cases}, \quad \begin{cases} x + y = 240 \\ 2x + y = 320 \end{cases}, \quad \text{and}$$

$$\begin{cases} 2x + y = 320 \\ y = 0 \end{cases} \text{ to find the vertices: (0, 240),}$$

(80, 160), and (160, 0). Note that (0, 0) is also a vertex.

Now find the values of the profit function at each vertex:

Ordered pair	$p = 50x + 40y$
(0, 240)	9600
(80, 160)	10,400
(160, 0)	8000

The maximum profit is \$10,400 when 80 acres of corn and 160 acres of soybeans are planted.

29. Let x = the number of hours machine I operates, and let y = the number hours machine II operates. Then, the cost is $c = 50x + 80y$. The constraints are $x \ge 0, y \ge 0, 20x + 30y \ge 1400$ (number of units of Grade A plywood) and $10x + 40y \ge 1200$ (number of units of Grade B plywood.) Solve the systems

$$\begin{cases} x = 0 \\ 20x + 30y = 1400 \end{cases}, \quad \begin{cases} 20x + 30y = 1400 \\ 10x + 40y = 1200 \end{cases}, \quad \text{and}$$

$$\begin{cases} 10x + 40y = 1200 \\ y = 0 \end{cases} \text{ to find the vertices}$$

$\left(0, \dfrac{140}{3}\right)$, (40, 20), and (120, 0).

Now find the values of the cost function at each vertex:

Ordered pair	$c = 50x + 80y$
$\left(0, \frac{140}{3}\right)$	≈ 3733.33
(40, 20)	3600
(120, 0)	6000

The minimum cost is \$3600 when machine I operates for 40 hours and machine II operates for 20 hours.

31. Let $x =$ the number of minutes of television time, and let $y =$ the number of pages of newspaper advertising. Then, the exposure is $f = 60{,}000x + 20{,}000y$. The constraints are $x \geq 1$, $y \geq 2$, and $1000x + 500y \leq 6000$ (budget.) Solve the systems

$$\begin{cases} x = 1 \\ 1000x + 500y = 6000 \end{cases}, \begin{cases} 1000x + 500y = 6000 \\ y = 2 \end{cases},$$

and $\begin{cases} x = 1 \\ y = 2 \end{cases}$ to find the vertices $(1, 10)$, $(5, 2)$, and $(1, 2)$.

Now find the values of the exposure function at each vertex:

Ordered pair	$f = 60{,}000x + 20{,}000y$
$(1, 10)$	$260{,}000$
$(5, 2)$	$340{,}000$
$(1, 2)$	$100{,}000$

There are a maximum of 340,000 viewers if the company buys 5 minutes of television time and 2 pages of newspaper advertising.

33. Let $x =$ the number of orange acres, and let $y =$ the number of grapefruit acres. Then, the profit is $p = 40x + 30y - 3000$. The constraints are $x \geq 0$, $y \geq 0$, $x + y \leq 480$ (number of acres) and $2x + y \leq 800$ (number of labor hours). Solve the systems

$$\begin{cases} x = 0 \\ x + y = 480 \end{cases}, \begin{cases} x + y = 480 \\ 2x + y = 800 \end{cases}, \text{ and}$$

$\begin{cases} 2x + y = 800 \\ y = 0 \end{cases}$ to find the vertices: $(0, 480)$,

$(320, 160)$, and $(400, 0)$. Note that $(0, 0)$ is also a vertex.

Now find the values of the profit function at each vertex:

Ordered pair	$40x + 30y - 3000$
$(0, 480)$	$11{,}400$
$(320, 160)$	$14{,}600$
$(400, 0)$	$13{,}000$

The maximum profit is $14,600 when 320 acres of oranges and 160 acres of grapefruits are planted.

35. Let $x =$ the number of rectangular tables, and let $y =$ the number of circular tables. The profit is $p = 3x + 4y$. The number of hours to assemble the rectangular tables is x and the number of hours to assemble the circular tables is y. The number of hours to finish the rectangular tables is x and the number of hours to finish the circular tables is $2y$. The 20 assemblers work a total of $(20)(40) = 800$ hours, and the 30 finishers work a total of $(30)(40) = 1200$ hours. So, the constraints are $x \geq 0$, $y \geq 0$, $x + y \leq 800$, and $x + 2y \leq 1200$.

Solve the systems

$$\begin{cases} x = 0 \\ x + 2y = 1200 \end{cases}, \begin{cases} x + 2y = 1200 \\ x + y = 800 \end{cases}, \text{ and}$$

$\begin{cases} x + y = 800 \\ y = 0 \end{cases}$ to find the vertices: $(0, 600)$,

$(400, 400)$, and $(800, 0)$. Note that $(0, 0)$ is also a vertex.

Now find the values of the profit function at each vertex:

Ordered pair	$p = 3x + 4y$.
$(0, 600)$	2400
$(400, 400)$	2800
$(800, 0)$	2400

The maximum profit is $2800 for 400 rectangular tables and 400 circular tables.

37. Let x = the number of terraced houses, and let y = the number of cottages. Then, the revenue is $r = 40,000x + 45,000y$. The number of units of concrete for the two house types is $x + y$; the number of units of wood for the two house types is $2x + y$; and the number of units of glass for the two house types is $2x + 5y$. The constraints are $x \geq 0$, $y \geq 0$, $x + y \leq 100$, $2x + y \leq 160$, and $2x + 5y \leq 400$. Solve the systems

$$\begin{cases} x = 0 \\ 2x + 5y = 400 \end{cases}, \begin{cases} 2x + 5y = 400 \\ x + y = 100 \end{cases},$$

$$\begin{cases} x + y = 100 \\ 2x + y = 160 \end{cases}, \text{ and } \begin{cases} 2x + y = 160 \\ y = 0 \end{cases} \text{ to find the}$$

vertices: $(0, 80)$, $\left(\dfrac{100}{3}, \dfrac{200}{3} \right)$, $(60, 40)$, and $(80, 0)$. Note that $(0, 0)$ is also a vertex.

Now find the values of the profit function at each vertex. Note that we cannot use $\left(\dfrac{100}{3}, \dfrac{200}{3} \right)$ because there cannot be a fraction of a house. This value becomes $(33, 66)$.

Ordered pair	$r = 40,000x + 45,000y$
$(0, 80)$	$3,600,000$
$(33, 66)$	$4,290,000$
$(60, 40)$	$4,200,000$
$(80, 0)$	$3,200,000$

The maximum profit is $4,290,000 when 33 terraced houses and 66 cottages are built.

39. Let x = the number of female guests, and let y = the number of male guests. Then, the number of guests who eat in the restaurant is $0.5x + 0.3y$, and the number of guests who do not eat in the restaurant is $0.5x + 0.7y$. The profit $p = 15(0.5x + 0.3y) - 2(0.5x + 0.7y)$ $= 6.5x + 3.1y$. The constraints are $x \geq 0$, $y \geq 0$, $x \leq 125$, $y \leq 220$, $x + y \geq 100$ and $x + y \leq 300$. Solve the systems $\begin{cases} x = 0, \\ y = 220 \end{cases}$,

$$\begin{cases} y = 220 \\ x + y = 300 \end{cases}, \begin{cases} x + y = 300 \\ x = 125 \end{cases}, \begin{cases} x = 125 \\ y = 0 \end{cases},$$

$$\begin{cases} y = 0 \\ x + y = 100 \end{cases}, \text{ and } \begin{cases} x + y = 100 \\ x = 0 \end{cases} \text{ to find the}$$

vertices: $(0, 220)$, $(80, 220)$, $(125, 175)$, $(125, 0)$, $(100, 0)$, and $(0, 100)$.

Now find the values of the profit function at each vertex:

Ordered pair	$p = 6.5x + 3.1y$
$(0, 220)$	682
$(80, 220)$	1202
$(125, 175)$	1355
$(125, 0)$	812.50
$(100, 0)$	650
$(0, 100)$	310

The maximum profit is $1355 when there are 125 female guests and 175 male guests.

41. Let x = the number of pounds of enchiladas, and let y = the number of pounds of vegetable loaf. The cost $c = 3.50x + 2.25y$. The number of units of carbohydrate is $5x + 2y$. The number of units of protein is $3x + 2y$. The number of units of fat is $5x + y$.

The constraints are
$x \geq 0$, $y \geq 0$, $5x + 2y \geq 60$, $3x + 2y \geq 40$, and $5x + y \geq 35$. Solve the systems

$$\begin{cases} x = 0, \\ 5x + y = 35 \end{cases} \begin{cases} 5x + y = 35 \\ 5x + 2y = 60 \end{cases} \begin{cases} 5x + 2y = 60 \\ 3x + 2y = 40 \end{cases}$$

and $\begin{cases} 3x + 2y = 40 \\ x = 0 \end{cases}$ to find the vertices:

$(0, 35)$, $(2, 25)$, $(10, 5)$, and $(40/3, 0)$.

Now find the values of the profit function at each vertex:

Ordered pair	$c = 3.50x + 2.25y$
(0, 35)	78.75
(2, 25)	63.25
(10, 5)	46.25
(40/3, 0).	46.67

The minimum cost is \$46.25 when Elisa buys 10 enchilada meals and 5 vegetable loafs.

5.6 Beyond the Basics

43.

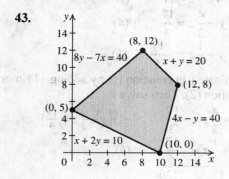

Solve the systems

$$\begin{cases} x + 2y = 10 \\ 4x - y = 40 \end{cases} \begin{cases} 4x - y = 40 \\ x + y = 20 \end{cases} \begin{cases} x + y = 20 \\ 8y - 7x = 40 \end{cases}$$

and $\begin{cases} 8y - 7x = 40 \\ x + 2y = 10 \end{cases}$ to find the vertices:

$(10, 0)$, $(12, 8)$, $(8, 12)$, and $(0, 5)$. Now find the values of the objective function at each vertex:

Ordered pair	$f = 8x + 7y$
(10, 0)	80
(12, 8)	152
(8, 12)	148
(0, 5)	35

The maximum is 152 at (12, 8).

45.

Solve the systems

$$\begin{cases} x = 0 \\ y = 0 \end{cases} \begin{cases} y = 0 \\ x = 3 \end{cases} \begin{cases} x = 3 \\ 7x + 2y = 28 \end{cases} \begin{cases} 7x + 2y = 28 \\ y = 6 \end{cases}$$

and $\begin{cases} y = 6 \\ y = x + 5 \end{cases}$ to find the vertices: $(0, 0)$,

$(3, 0)$, $\left(3, \frac{7}{2}\right)$, $\left(\frac{16}{7}, 6\right)$, $(1, 6)$, and $(0, 5)$.

Now find the values of the objective function at each vertex:

Ordered pair	$f = 5x + 6y$
(0, 0)	0
(3, 0)	15
$\left(3, \frac{7}{2}\right)$	36
$\left(\frac{16}{7}, 6\right)$	$\frac{332}{7}$
(1, 6)	41
(0, 5)	30

The minimum is 0 at (0, 0).

47. $ax_1 + by_1 = M = ax_2 + by_2 \Rightarrow$
$ax_1 - ax_2 = by_2 - by_1 \Rightarrow$
$a(x_1 - x_2) = b(y_2 - y_1) \Rightarrow$
$\dfrac{a}{b} = \dfrac{y_2 - y_1}{x_1 - x_2} = -\dfrac{y_2 - y_1}{x_2 - x_1} = -m$, where m is the

slope of $\overline{PQ}$.

Since the slope is the same between any two points on a given line segment, we can conclude that f has the same value M at every point of the line segment PQ.

5.6 Critical Thinking/Discussion/Writing

49.

a. S is unbounded because there is no upper limit on the values for x.

b.

Ordered pair	$f = 2x + 3y$
$(0, 0)$	0
$(0, 6)$	18
$(2, 8)$	28
$(x, 8)$	$2x + 24$
$(x, 0)$	$2x$

As x increases, the value of the objective function $f = 2x + 3y$ increases. Since S is unbounded, there is no solution.

50. Minimize $f = 3.2x + 2.5y + 1.3z$ subject to the constraints
$x + 2y + 5 \geq 1$
$10x + 5y + 4z \geq 50$
$6x + 20y + 7z \geq 10$
$x \geq 0,\ y \geq 0,\ z \geq 0$

5.6 Getting Ready for the Next Section

51. $\begin{cases} x - 2y = 4 & (1) \\ -3x + 5y = -7 & (2) \end{cases}$

The first equation already has a leading coefficient of 1. To eliminate x from equation (2), add 3 times equation (1) to equation (2). Then solve for y.

$3x - 6y = 12 \qquad 3(x - 2y = 12)$
$\underline{-3x + 5y = -7 \qquad (2)}$
${-y = 5}$
${y = -5} \qquad (3)$

The equivalent system in triangular form is
$\begin{cases} x - 2y = 4 & (1) \\ y = -5 & (3) \end{cases}$

Back-substitute the value of y into equation (1) and solve for x.

$x - 2(-5) = 4 \Rightarrow x + 10 = 4 \Rightarrow x = -6$

The solution set is $\{(-6, -5)\}$. Be sure to check the solution in each of the original equations.

53. $\begin{cases} x - 4y - z = 11 & (1) \\ 2x - 5y + 2z = 39 & (2) \\ -3x + 2y + z = 1 & (3) \end{cases}$

The first equation already has a leading coefficient of 1. To eliminate x from equation (2), add -2 times equation (1) to equation (2).

$-2x + 8y + 2z = -22 \qquad (-2)(x - 4y - z = 11)$
$\underline{2x - 5y + 2z = 39 \qquad (2)}$
$3y + 4z = 17 \qquad (4)$

Multiply equation (4) by $\frac{1}{3}$.
$\begin{cases} x - 4y - z = 11 & (1) \\ y + \dfrac{4}{3}z = \dfrac{17}{4} & (4) \\ -3x + 2y + z = 1 & (3) \end{cases}$

To eliminate x from equation (3), add 3 times equation (1) to equation (3).

$3x - 12y - 3z = 33 \qquad (3)(x - 4y - z = 11)$
$\underline{-3x + 2y + z = 1 \qquad (3)}$
${-10y - 2z = 34} \qquad (5)$

Multiply equation (5) by $-\frac{1}{10}$.
$\begin{cases} x - 4y - z = 11 & (1) \\ y + \dfrac{4}{3}z = \dfrac{17}{3} & (4) \\ y + \dfrac{1}{5}z = -\dfrac{17}{5} & (5) \end{cases}$

Eliminate y in equation (5) by adding -1 times equation (2). Then solve for z.

$-y - \dfrac{4}{3}z = -\dfrac{17}{3} \qquad (-1)\left(y + \frac{4}{3}z = \frac{17}{4}\right)$
$\underline{y + \dfrac{1}{5}z = -\dfrac{17}{5} \qquad (5)}$
${-\dfrac{17}{15}z = -\dfrac{136}{15}}$
$z = 8 \qquad (6)$

(continued on next page)

(continued)

The equivalent system in triangular form is

$$\begin{cases} x - 4y - z = 11 & (1) \\ y + \dfrac{4}{3}z = \dfrac{17}{3} & (4) \\ z = 8 & (6) \end{cases}$$

Back-substitute the value of z into equation (4) and solve for y.

$$y + \frac{4}{3}(8) = \frac{17}{3} \Rightarrow y + \frac{32}{3} = \frac{17}{3} \Rightarrow y = -5$$

Back-substitute $z = 8$ and $y = -5$ into equation (1) and solve for x.

$$x - 4(-5) - 8 = 11 \Rightarrow x + 12 = 11 \Rightarrow x = -1$$

The solution set is $\{(-1, -5, 8)\}$. Be sure to check the solution in each of the original equations.

For exercises 55–58, refer to the following array.

$$\begin{bmatrix} 2 & 4 & -6 & 10 \\ 3 & -2 & 4 & 12 \end{bmatrix} \begin{matrix} \leftarrow \text{row 1} \\ \leftarrow \text{row 2} \end{matrix}$$

55. Multiply each number in row 1 by $\frac{1}{2}$.

$$\begin{bmatrix} 1 & 2 & -3 & 5 \\ 3 & -2 & 4 & 12 \end{bmatrix}$$

57. Using the rectangular array in exercise 56, multiply each number in row 2 by $-\frac{1}{8}$.

$$\begin{bmatrix} 1 & 2 & -3 & 5 \\ 0 & 1 & -\dfrac{13}{8} & \dfrac{3}{8} \end{bmatrix}$$

Chapter 5 Review Exercises

Building Skills

1. Using elimination, we have

$$\begin{cases} 3x - y = -5 \\ x + 2y = 3 \end{cases} \Rightarrow \begin{cases} 6x - 2y = -10 \\ x + 2y = 3 \end{cases} \Rightarrow 7x = -7 \Rightarrow$$

$$x = -1$$
$$3(-1) - y = -5 \Rightarrow y = 2$$
The solution is $\{(-1, 2)\}$.

3. Using elimination, we have

$$\begin{cases} 2x + 4y = 3 \\ 3x + 6y = 10 \end{cases} \Rightarrow \begin{cases} -6x - 12y = -9 \\ 6x + 12y = 20 \end{cases} \Rightarrow 0 = 11 \Rightarrow$$

there is no solution. Solution set: $\varnothing$

5. Using elimination, we have

$$\begin{cases} 3x - y = 3 \\ \dfrac{1}{2}x + \dfrac{1}{3}y = 2 \end{cases} \Rightarrow \begin{cases} x - \dfrac{1}{3}y = 1 \\ \dfrac{1}{2}x + \dfrac{1}{3}y = 2 \end{cases} \Rightarrow \dfrac{3}{2}x = 3 \Rightarrow$$

$$x = 2$$
$$3(2) - y = 3 \Rightarrow y = 3$$
The solution is $\{(2, 3)\}$.

7. Multiply the first equation by -2, add the result to the second equation, and replace the second equation with the new equation:

$$\begin{cases} x + 3y + z = 0 \\ 2x - y + z = 5 \\ 3x - 3y + 2z = 10 \end{cases} \Rightarrow \begin{cases} x + 3y + z = 0 \\ -7y - z = 5 \\ 3x - 3y + 2z = 10 \end{cases}$$

Multiply the first equation by -3, add the result to the third equation, and replace the third equation with the new equation:

$$\begin{cases} x + 3y + z = 0 \\ -7y - z = 5 \\ 3x - 3y + 2z = 10 \end{cases} \Rightarrow \begin{cases} x + 3y + z = 0 \\ -7y - z = 5 \\ -12y - z = 10 \end{cases}$$

Subtract the third equation from the second equation, and replace the second equation with the result:

$$\begin{cases} x + 3y + z = 0 \\ -7y - z = 5 \\ -12y - z = 10 \end{cases} \Rightarrow \begin{cases} x + 3y + z = 0 \\ 5y = -5 \\ -12y - z = 10 \end{cases} \Rightarrow$$

$$\begin{cases} x + 3y + z = 0 \\ y = -1 \\ -12y - z = 10 \end{cases}$$

Substitute $y = -1$ into the third equation to solve for z and then substitute the values for z and y into the first equation to solve for x:

$$\begin{cases} x + 3y + z = 0 \\ y = -1 \\ -12y - z = 10 \end{cases} \Rightarrow \begin{cases} x + 3y + z = 0 \\ y = -1 \\ -12(-1) - z = 10 \end{cases} \Rightarrow$$

$$\begin{cases} x + 3y + z = 0 \\ y = -1 \\ z = 2 \end{cases} \Rightarrow \begin{cases} x + 3(-1) + 2 = 0 \\ y = -1 \\ z = 2 \end{cases} \Rightarrow$$

$$\begin{cases} x = 1 \\ y = -1 \\ z = 2 \end{cases}$$

The solution is $\{(1, -1, 2)\}$.

9. Switch the second and third equations, multiply the first equation by -2, add the result to the second equation, and replace the second equation with the result.

$$\begin{cases} x + y = 1 \\ 3y + 2z = 0 \\ 2x - 3z = 7 \end{cases} \Rightarrow \begin{cases} x + y = 1 \\ 2x - 3z = 7 \\ 3y + 2z = 0 \end{cases} \Rightarrow$$

(continued on next page)

(*continued*)

$$\begin{cases} x + y = 1 \\ -2y - 3z = 5 \\ 3y + 2z = 0 \end{cases}$$

Multiply the second equation by 3, multiply the third equation by 2, add the results, replace the third equation with the new equation, and solve for z:

$$\begin{cases} x + y = 1 \\ -2y - 3z = 5 \\ 3y + 2z = 0 \end{cases} \Rightarrow \begin{cases} x + y = 1 \\ -2y - 3z = 5 \\ -5z = 15 \end{cases}$$

$$\begin{cases} x + y = 1 \\ -2y - 3z = 5 \\ z = -3 \end{cases}$$

Substitute $z = -3$ into the second equation to solve for y, and then substitute the value for y into the first equation to solve for x:

$$\begin{cases} x + y = 1 \\ -2y - 3z = 5 \\ z = -3 \end{cases} \Rightarrow \begin{cases} x + y = 1 \\ -2y - 3(-3) = 5 \\ z = -3 \end{cases} \Rightarrow$$

$$\begin{cases} x + y = 1 \\ y = 2 \\ z = -3 \end{cases} \Rightarrow \begin{cases} x + 2 = 1 \\ y = 2 \\ z = -3 \end{cases} \Rightarrow \begin{cases} x = -1 \\ y = 2 \\ z = -3 \end{cases}$$

The solution is $\{(-1, 2, -3)\}$.

11. Multiply the first equation by -5, add the result to the second equation, and replace the second equation with the result, then switch the first and second equations:

$$\begin{cases} x + y + z = 1 \\ x + 5y + 5z = -1 \\ 3x - y - z = 4 \end{cases} \Rightarrow \begin{cases} x + y + z = 1 \\ -4x = -6 \\ 3x - y - z = 4 \end{cases} \Rightarrow$$

$$\begin{cases} x = \dfrac{3}{2} \\ x + y + z = 1 \\ 3x - y - z = 4 \end{cases}$$

Add the second and third equations, and replace the second equation with the result.

$$\begin{cases} x = \dfrac{3}{2} \\ x + y + z = 1 \\ 3x - y - z = 4 \end{cases} \Rightarrow \begin{cases} x = \dfrac{3}{2} \\ 4x = 5 \\ 3x - y - z = 4 \end{cases} \Rightarrow$$

$$\begin{cases} x = 3/2 \\ x = 5/4 \Rightarrow \text{There is no solution.} \\ 3x - y - z = 4 \end{cases}$$

Solution set: $\varnothing$

13. The system is a dependent system because there are two equations in three unknowns. Multiply the first equation by -5, and multiply the second equation by 4, then add the resulting equations and solve for z:

$$\begin{cases} x + 4y + 3z = 1 \\ 2x + 5y + 4z = 4 \end{cases} \Rightarrow \begin{cases} -5x - 20y - 15z = -5 \\ 8x + 20y + 16z = 16 \end{cases} \Rightarrow$$

$3x + z = 11 \Rightarrow z = -3x + 11$

Substitute the expression for z into the first equation and solve for y:

$x + 4y + 3(-3x + 11) = 1 \Rightarrow -8x + 4y + 33 = 1 \Rightarrow$

$y = 2x - 8$.

The solution is $\{(x, 2x - 8, -3x + 11)\}$.

15. Add the first and third equations, and replace the first equation with the result:

$$\begin{cases} 3x - y = 2 \\ x + 2y = 9 \\ 3x + y = 10 \end{cases} \Rightarrow \begin{cases} 6x = 12 \\ x + 2y = 9 \\ 3x + y = 10 \end{cases} \Rightarrow \begin{cases} x = 2 \\ x + 2y = 9 \\ 3x + y = 10 \end{cases}$$

Substitute $x = 2$ into the second equation and solve for y:

$$\begin{cases} x = 2 \\ x + 2y = 9 \\ 3x + y = 10 \end{cases} \Rightarrow \begin{cases} x = 2 \\ 2 + 2y = 9 \\ 3x + y = 10 \end{cases} \Rightarrow \begin{cases} x = 2 \\ y = \dfrac{7}{2} \\ 3x + y = 10 \end{cases}$$

$3(2) + \dfrac{7}{2} = \dfrac{19}{2} \neq 10 \Rightarrow$ there is no solution.

Solution set: $\varnothing$

17. $\begin{cases} x + y + z = 3 & (1) \\ 2x - y + z = 4 & (2) \\ x + 4y + 2z = 5 & (3) \end{cases}$

Multiply equation (1) by -2, then add the result to equation (2), and replace equation (2) with the new equation:

$$\begin{cases} x + y + z = 3 & (1) \\ 2x - y + z = 4 & (2) \\ x + 4y + 2z = 5 & (3) \end{cases} \Rightarrow$$

$$\begin{cases} x + y + z = 3 & (1) \\ -3y - z = -2 & (4) \\ x + 4y + 2z = 5 & (3) \end{cases}$$

Multiply equation (1) by -1, then add the result to equation (3), and replace equation (3) with the new equation:

$$\begin{cases} x + y + z = 3 & (1) \\ 2x - y + z = 4 & (2) \\ x + 4y + 2z = 5 & (3) \end{cases} \Rightarrow \begin{cases} x + y + z = 3 & (1) \\ -3y - z = -2 & (4) \\ 3y + z = 2 & (5) \end{cases}$$

(*continued on next page*)

(*continued*)

Add equation (4) to equation (5) to solve for z:

$$\begin{cases} x + y + z = 3 & (1) \\ -3y - z = -2 & (4) \\ 3y + z = 2 & (5) \end{cases} \Rightarrow 0 = 0$$

The equation $0 = 0$ is equivalent to $0z = 0$, which is true for every value of z. Solving the second equation for y, we have

$-3y - z = -2 \Rightarrow y = \frac{2}{3} - \frac{1}{3}z$. Substituting this

into the first equation, we have $x + y + z = 3 \Rightarrow$

$$x + \left(\frac{2}{3} - \frac{1}{3}z\right) - z = 3 \Rightarrow x = \frac{7}{3} - \frac{2}{3}z$$

Thus, the solution is $\left\{\left(\frac{7}{3} - \frac{2}{3}z, \frac{2}{3} - \frac{1}{3}z, z\right)\right\}$.

19.

21.

23.

25.

Solve the systems

$$\begin{cases} x = 0 \\ 3x + 7y = 21 \end{cases}, \begin{cases} 3x + 7y = 21 \\ y = 0 \end{cases}, \text{ and } \begin{cases} y = 0 \\ x = 0 \end{cases} \text{ to}$$

find the vertices: $(0, 3)$, $(7, 0)$, and $(0, 0)$. Now find the values of the objective function at each vertex:

Ordered pair	$z = 2x + 3y$
$(0, 3)$	9
$(7, 0)$	14
$(0, 0)$	0

The maximum is 14 at $(7, 0)$.

27.

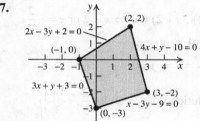

Solve the systems

$$\begin{cases} 2x - 3y = -2 \\ 4x + y = 10 \end{cases}, \begin{cases} 4x + y = 10 \\ x - 3y = 9 \end{cases}, \begin{cases} x - 3y = 9 \\ 3x + y = -3 \end{cases},$$

and $\begin{cases} 3x + y = -3 \\ 2x - 3y = -2 \end{cases}$ to find the vertices: $(2, 2)$,

$(3, -2)$, $(0, -3)$, and $(-1, 0)$. Now find the values of the objective function at each vertex:

Ordered pair	$z = x + 3y$
$(2, 2)$	8
$(3, -2)$	-3
$(0, -3)$	-9
$(-1, 0)$	-1

The minimum is -9 at $(0, -3)$.

29. Using substitution, we have

$$\begin{cases} x + 3y = 1 \\ x^2 - 3x = 7y + 3 \end{cases} \Rightarrow \begin{cases} x = 1 - 3y \\ x^2 - 3x = 7y + 3 \end{cases} \Rightarrow$$

$$(1 - 3y)^2 - 3(1 - 3y) = 7y + 3 \Rightarrow$$

$$9y^2 - 4y - 5 = 0 \Rightarrow (y - 1)(9y + 5) = 0 \Rightarrow$$

$$y = 1 \text{ or } y = -\frac{5}{9}$$

$$x + 3(1) = 1 \Rightarrow x = -2$$

$$x + 3\left(-\frac{5}{9}\right) = 1 \Rightarrow x = \frac{24}{9} = \frac{8}{3}$$

The solution is $\left\{ (-2, 1), \left(\frac{8}{3}, -\frac{5}{9}\right) \right\}$.

31. Using substitution, we have

$$\begin{cases} x - y = 4 \\ 5x^2 + y^2 = 24 \end{cases} \Rightarrow \begin{cases} y = x - 4 \\ 5x^2 + y^2 = 24 \end{cases} \Rightarrow$$

$$5x^2 + (x - 4)^2 = 24 \Rightarrow 6x^2 - 8x - 8 = 0 \Rightarrow$$

$$2(x - 2)(3x + 2) = 0 \Rightarrow x = 2 \text{ or } x = -\frac{2}{3}$$

$$2 - y = 4 \Rightarrow y = -2$$

$$-\frac{2}{3} - y = 4 \Rightarrow y = -\frac{14}{3}$$

The solution is $\left\{ (2, -2), \left(-\frac{2}{3}, -\frac{14}{3}\right) \right\}$.

33. Using substitution, we have

$$\begin{cases} xy = 2 \\ x^2 + 2y^2 = 9 \end{cases} \Rightarrow \begin{cases} y = \frac{2}{x} \\ x^2 + 2y^2 = 9 \end{cases} \Rightarrow$$

$$x^2 + 2\left(\frac{2}{x}\right)^2 = 9 \Rightarrow x^2 - 9 + \frac{8}{x^2} = 0 \Rightarrow$$

$$x^4 - 9x^2 + 8 = 0$$

Let $u = x^2$. Then $u^2 - 9u + 8 = 0 \Rightarrow$

$$(u - 8)(u - 1) = 0 \Rightarrow u = 8 \Rightarrow x^2 = 8 \Rightarrow$$

$$x = \pm 2\sqrt{2} \text{ or } u = 1 \Rightarrow x^2 = 1 \Rightarrow x^2 = \pm 1$$

$$\left(2\sqrt{2}\right)y = 2 \Rightarrow y = \frac{\sqrt{2}}{2}$$

$$\left(-2\sqrt{2}\right)y = 2 \Rightarrow y = -\frac{\sqrt{2}}{2}$$

$$(1)y = 2 \Rightarrow y = 2; (-1)y = 2 \Rightarrow y = -2$$

None of the values are extraneous, so the

solution is $\left\{ (1, 2), (-1, -2), \left(2\sqrt{2}, \frac{\sqrt{2}}{2}\right), \right.$

$$\left. \left(-2\sqrt{2}, -\frac{\sqrt{2}}{2}\right) \right\}.$$

35.

37.

39.

41. $\dfrac{x + 4}{x^2 + 5x + 6} = \dfrac{x + 4}{(x + 2)(x + 3)} = \dfrac{A}{x + 2} + \dfrac{B}{x + 3} \Rightarrow$

$$x + 4 = A(x + 3) + B(x + 2) \Rightarrow$$

$$x + 4 = (A + B)x + (3A + 2B) \Rightarrow$$

$$\begin{cases} A + B = 1 \\ 3A + 2B = 4 \end{cases} \Rightarrow \begin{cases} -2A - 2B = -2 \\ 3A + 2B = 4 \end{cases} \Rightarrow$$

$$A = 2, B = -1$$

$$\frac{x + 4}{x^2 + 5x + 6} = \frac{2}{x + 2} - \frac{1}{x + 3}$$

43. $\dfrac{3x^2 + x + 1}{x(x - 1)^2} = \dfrac{A}{x} + \dfrac{B}{x - 1} + \dfrac{C}{(x - 1)^2} \Rightarrow$

$$3x^2 + x + 1 = A(x - 1)^2 + Bx(x - 1) + Cx \Rightarrow$$

$$3x^2 + x + 1$$

$$= (A + B)x^2 + (-2A - B + C)x + A \Rightarrow$$

$$\begin{cases} A + B = 3 \\ -2A - B + C = 1 \Rightarrow A = 1, B = 2, C = 5 \\ A = 1 \end{cases}$$

$$\frac{3x^2 + x + 1}{x(x - 1)^2} = \frac{1}{x} + \frac{2}{x - 1} + \frac{5}{(x - 1)^2}$$

45. $\dfrac{x^2+2x+3}{(x^2+4)^2}=\dfrac{Ax+B}{x^2+4}+\dfrac{Cx+D}{(x^2+4)^2}\Rightarrow$

$x^2+2x+3=(Ax+B)(x^2+4)+Cx+D\Rightarrow$

x^2+2x+3
$\quad=Ax^3+Bx^2+(4A+C)x+(4B+D)\Rightarrow$

$\begin{cases}A=0\\B=1\\4A+C=2\\4B+D=3\end{cases}\Rightarrow A=0,B=1,C=2,D=-1$

$\dfrac{x^2+2x+3}{(x^2+4)^2}=\dfrac{1}{x^2+4}+\dfrac{2x-1}{(x^2+4)^2}$

Applying the Concepts

47. Let x = the amount invested at high-risk. Let y = the amount invested at 4% interest. Then

$\begin{cases}x+y=15,000\\0.12x+0.04y=1300\end{cases}\Rightarrow$

$\begin{cases}-0.04x-0.04y=-600\\0.12x+0.04y=1300\end{cases}\Rightarrow 0.08x=700\Rightarrow$

$x=8750$

$8750+y=15,000\Rightarrow y=6250$

The speculator invested $8750 at 12% and $6250 at 4%.

49. Let x = the length of the rectangle. Let y = the width of the rectangle. Then

$\begin{cases}xy=63\\2x+2y=33\end{cases}\Rightarrow\begin{cases}y=\dfrac{63}{x}\\2x+2y=33\end{cases}\Rightarrow$

$2x+2\left(\dfrac{63}{x}\right)=33\Rightarrow 2x-33+\dfrac{126}{x}=0\Rightarrow$

$2x^2-33x+126=0\Rightarrow(x-6)(2x-21)=0\Rightarrow$

$x=6$ or $x=\dfrac{21}{2}=10.5;\ 6y=63\Rightarrow y=\dfrac{21}{2}=10.5$

$\dfrac{21}{2}y=63\Rightarrow y=6$

The rectangle is 10.5 feet by 6 feet.

51. Let x = the length of one leg. Let y = the length of the other leg. Then

$\begin{cases}x^2+y^2=17^2\\(x+1)^2+(y+4)^2=20^2\end{cases}\Rightarrow$

$\begin{cases}y^2=289-x^2\\x^2+2x+y^2+8y-383=0\end{cases}\Rightarrow$

$x^2+2x+289-x^2+8\sqrt{289-x^2}-383=0\Rightarrow$

$8\sqrt{289-x^2}=-2x+94\Rightarrow$

$64(289-x^2)=(-2x+94)^2\Rightarrow$

$18,496-64x^2=4x^2-376x+8836\Rightarrow$

$68x^2-376x-9660=0\Rightarrow$

$4(x-15)(17x+161)=0\Rightarrow x=15$ or $x=-\dfrac{161}{7}$

(Reject the negative solution.)

$15^2+y^2=17^2\Rightarrow y=8$

The legs of the triangle are 15 and 8.

53. Let x = the width of the smaller pastures. Let y = the length of the pastures.

Then,

$\begin{cases}4xy=6400\\4x+2y=240\end{cases}\Rightarrow\begin{cases}y=\dfrac{1600}{x}\\4x+2y=240\end{cases}\Rightarrow$

$4x+2\left(\dfrac{1600}{x}\right)=240\Rightarrow 4x-240+\dfrac{3200}{x}=0\Rightarrow$

$4x^2-240x+3200=0\Rightarrow$

$4(x-40)(x-20)=0\Rightarrow x=40$ or $x=20$

$4(40)y=6400\Rightarrow y=40$

$4(20)y=6400\Rightarrow y=80$

The dimensions of the pasture are 40 meters by 160 meters.

55. a. $C(x)=12x+60,000;\ R(x)=20x$

b.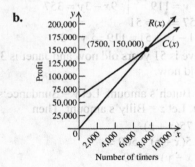

c. $C(x)=R(x)\Rightarrow 12x+60,000=20x\Rightarrow$
$x=7500$
$R(12)=20(7500)=150,000$
The breakeven point is (7500, 150,000).

d. $P(x)=R(x)-C(x)=0.15C(x)\Rightarrow$
$20x-(12x+60,000)$
$\quad=0.15(12x+60,000)\Rightarrow$
$8x-60,000=1.8x+9000\Rightarrow$
$6.2x=69,000\Rightarrow x\approx 11,129$
11,129 timers should be sold.

57. Let x = the double occupancy rate. Let y = the single occupancy rate. Each pays half of the double occupancy rate, so Alisha paid

$$6\left(\frac{x}{2}\right) + 3(x - y) = 6x - 3y \text{ for the nine}$$

months, and Sunita paid $6\left(\frac{x}{2}\right) + 3y = 3x + 3y$

for the nine months.

$$\begin{cases} 6x - 3y = 2250 \\ 3x + 3y = 3150 \end{cases} \Rightarrow 9x = 5400 \Rightarrow x = 600$$

$$3(600) + 3y = 3150 \Rightarrow y = 450$$

The double occupancy rate is \$600 per month, and the single occupancy rate is \$450 per month.

59. Let x = the number of students who passed the exam. Let y = the number of students who failed the exam. The total number of points scored for the class is $26 \times 72 = 1872$. Then

$$\begin{cases} x + y = 26 \\ 78x + 26y = 1872 \end{cases} \Rightarrow \begin{cases} -26x - 26y = -676 \\ 78x + 26y = 1872 \end{cases} \Rightarrow$$

$$52x = 1196 \Rightarrow x = 23, y = 3$$

Three students failed the exam.

61. Let x = Steve's age now. Let y = Janet's age now. Then $x - y$ = the difference in their ages.

$$\begin{cases} y = 2(y - (x - y)) \\ x + ((x - y) + x) = 119 \end{cases} \Rightarrow \begin{cases} y = 4y - 2x \\ 3x - y = 119 \end{cases} \Rightarrow$$

$$\begin{cases} -2x + 3y = 0 \\ 3x - y = 119 \end{cases} \Rightarrow \begin{cases} -2x + 3y = 0 \\ 9x - 3y = 357 \end{cases} \Rightarrow$$

$$7x = 357 \Rightarrow x = 51$$

$$51 + (51 - y) + 51 = 119 \Rightarrow y = 34$$

So, Steve is 51 years old now and Janet is 34 years old now.

63. Let x = Butch's amount. Let y = Sundance's amount. Let z = Billy's amount. Then

$$\begin{cases} x = 0.75(y + z) \\ y = 0.5(x + z) + 500 \\ z = 3(x - y) - 1000 \end{cases} \Rightarrow$$

$$\begin{cases} x - 0.75y - 0.75z = 0 \\ -0.5x + y - 0.5z = 500 \\ -3x + 3y + z = -1000 \end{cases}$$

Multiply the first equation by 0.5, add the result to the second equation, and replace the second equation with the new equation. Then, multiply the first equation by 3, add the result to the third equation, and replace the third equation with the new equation.

$$\begin{cases} x - 0.75y - 0.75z = 0 \\ -0.5x + y - 0.5z = 500 \\ -3x + 3y + z = -1000 \end{cases} \Rightarrow$$

$$\begin{cases} x - 0.75y - 0.75z = 0 \\ 0.625y - 0.875z = 500 \\ 0.75y - 1.25z = -1000 \end{cases}$$

Multiply the second equation by -1.2, add the result to the third equation, replace the third equation with the new equation, then solve for z:

$$\begin{cases} x - 0.75y - 0.75z = 0 \\ 0.625y - 0.875z = 500 \\ 0.75y - 1.25z = -1000 \end{cases} \Rightarrow$$

$$\begin{cases} x - 0.75y - 0.75z = 0 \\ 0.625y - 0.875z = 500 \\ -0.2z = -1600 \end{cases} \Rightarrow$$

$$\begin{cases} x - 0.75y - 0.75z = 0 \\ 0.625y - 0.875z = 500 \\ z = 8000 \end{cases}$$

Substitute $z = 8000$ into the second equation to solve for y, then substitute the values for y and z into the first equation to solve for x:

$$\begin{cases} x - 0.75y - 0.75z = 0 \\ 0.625y - 0.875z = 500 \\ z = 8000 \end{cases} \Rightarrow$$

$$\begin{cases} x - 0.75y - 0.75z = 0 \\ 0.625y - 0.875(8000) = 500 \\ z = 8000 \end{cases} \Rightarrow$$

$$\begin{cases} x - 0.75y - 0.75z = 0 \\ y = 12,000 \\ z = 8000 \end{cases} \Rightarrow$$

$$\begin{cases} x - 0.75(12,000) - 0.75(8000) = 0 \\ y = 12,000 \\ z = 8000 \end{cases} \Rightarrow$$

$$\begin{cases} x = 15,000 \\ y = 12,000 \\ z = 8000 \end{cases}$$

So, they stole \$35,000. Butch received \$15,000, Sundance received \$12,000, and Billy received \$8000.

65. $2^x = 8^y \Rightarrow 2^x = 2^{3y} \Rightarrow x = 3y$

$9^y = 3^{x-2} \Rightarrow 3^{2y} = 3^{x-2} \Rightarrow 2y = x - 2$

$$\begin{cases} x = 3y \\ 2y = x - 2 \end{cases} \Rightarrow 2y = 3y - 2 \Rightarrow y = 2, x = 6$$

67. Let x = the number of two-story houses. Let y = the number of one-story houses. The profit $p = 10,000x + 4000y.$ The two-story houses use $7x$ units of material, while the one-story houses use y units of material. The two-story houses use x units of labor, while the one-story houses use $2y$ units of labor. So, the constraints are $x \geq 0, y \geq 0, 7x + y \leq 42$, and $x + 2y \leq 32.$

Solve the systems $\begin{cases} x = 0 \\ x + 2y = 32 \end{cases}, \begin{cases} x + 2y = 32 \\ 7x + y = 42 \end{cases},$

and $\begin{cases} 7x + y = 42 \\ y = 0 \end{cases}$ to find the vertices: (0, 16), (4, 14), and (6, 0).

Now find the values of the profit function at each vertex:

Ordered pair	$p = 10,000x + 4000y.$
(0, 16)	64,000
(4, 14)	96,000
(6, 0)	60,000

The builder should build 4 two-story houses and 14 one-story houses for a maximum profit of $96,000.

69. Let x = the cost of the small lobster, y = the cost of the medium lobster, and z = the cost of the large lobster. Then we have

$\begin{cases} 4x + 2y + z = 344 & (1) \\ 3x + 2z = 255 & (2) \\ 5x + 2y + 2z = 449 & (3) \end{cases}$

Interchange equations (2) and (3).

$\begin{cases} 4x + 2y + z = 344 & (1) \\ 5x + 2y + 2z = 449 & (3) \\ 3x + 2z = 255 & (2) \end{cases}$

Multiply equation (1) by -1, add the result to equation (3), then replace equation (1) with the result (4).

$\begin{cases} 4x + 2y + z = 344 \\ 5x + 2y + 2z = 449 \\ 3x + 2z = 255 \end{cases} \Rightarrow$

$\begin{cases} -4x - 2y - z = -344 \\ 5x + 2y + 2z = 449 \\ 3x + 2z = 255 \end{cases} \Rightarrow$

$\begin{cases} x + z = 105 & (4) \\ 5x + 2y + 2z = 449 & (3) \\ 3x + 2z = 255 & (2) \end{cases}$

Multiply equation (4) by -3, add the result to equation (2), then replace equation (2) with the result (5).

$\begin{cases} x + z = 105 & (4) \\ 5x + 2y + 2z = 499 & (3) \\ 3x + 2z = 255 & (2) \end{cases} \Rightarrow$

$\begin{cases} -3x - 3z = -315 \\ 5x + 2y + 2z = 449 \\ 3x + 2z = 255 \end{cases} \Rightarrow$

$\begin{cases} x + z = 105 & (4) \\ 5x + 2y + 2z = 449 & (3) \\ -z = -60 \Rightarrow z = 60 & (5) \end{cases}$

Substitute $z = 60$ in equation (4) and solve for x: $x + 60 = 105 \Rightarrow x = 45$

Now substitute $x = 45$ and $z = 60$ into equation (3) and solve for y:
$5(45) + 2y + 2(60) = 449 \Rightarrow 2y = 104 \Rightarrow y = 52$

A small lobster costs $45, a medium lobster costs $52, and a large lobster costs $60.

Chapter 5 Practice Test A

1. Using elimination, we have
$\begin{cases} 2x - y = 4 \\ 2x + y = 4 \end{cases} \Rightarrow 4x = 8 \Rightarrow x = 2$
$2(2) + y = 4 \Rightarrow y = 0$
The solution is $\{(2, 0)\}.$

2. Using elimination, we have
$\begin{cases} x + 2y = 8 \\ 3x + 6y = 24 \end{cases} \Rightarrow \begin{cases} 3x + 6y = 24 \\ 3x + 6y = 24 \end{cases} \Rightarrow$ the system

is consistent. The solution is $\left\{ \left(x, 4 - \dfrac{x}{2} \right) \right\}.$

3. Using substitution, we have
$\begin{cases} -2x + y = 4 \\ 4x - 2y = 4 \end{cases} \Rightarrow \begin{cases} y = 2x + 4 \\ 4x - 2y = 4 \end{cases} \Rightarrow$
$4x - 2(2x + 4) = 4 \Rightarrow -8 = 4 \Rightarrow$ there is no solution.
Solution set: $\varnothing$

4. $\begin{cases} 3x+3y=-15 \\ 2x-2y=-10 \end{cases} \Rightarrow \begin{cases} x+y=-5 \\ x-y=-5 \end{cases} \Rightarrow 2x=-10 \Rightarrow$

$x=-5$

$3(-5)+3y=-15 \Rightarrow y=0$

The solution is $\{(-5, 0)\}$.

5. Using elimination, we have

$\begin{cases} \dfrac{5}{3}x+\dfrac{y}{2}=14 \\ \dfrac{2}{3}x-\dfrac{y}{8}=3 \end{cases} \Rightarrow \begin{cases} 10x+3y=84 \\ 16x-3y=72 \end{cases} \Rightarrow$

$26x=156 \Rightarrow x=6$

$\dfrac{5}{3}(6)+\dfrac{y}{2}=14 \Rightarrow y=8$

The solution is $\{(6, 8)\}$.

6. Using substitution, we have

$\begin{cases} y=x^2 \\ 3x-y+4=0 \end{cases} \Rightarrow 3x-x^2+4=0 \Rightarrow$

$x^2-3x-4=0 \Rightarrow (x-4)(x+1)=0 \Rightarrow$

$x=4$ or $x=-1$

$y=4^2=16$

$y=(-1)^2=1$

The solution is $\{(4, 16), (-1, 1)\}$.

7. Using elimination, we have

$\begin{cases} x-3y=-4 \\ 2x^2+3x-3y=8 \end{cases} \Rightarrow -2x^2-2x=-12 \Rightarrow$

$x^2+x-6=0 \Rightarrow (x+3)(x-2)=0 \Rightarrow$

$x=-3$ or $x=2$

$-3-3y=-4 \Rightarrow y=\dfrac{1}{3}$

$2-3y=-4 \Rightarrow y=2$

The solution is $\left\{\left(-3, \dfrac{1}{3}\right), (2,2)\right\}$.

8. a. Let x = the weight of one bar. Let y = the weight of the other bar. Then

$\begin{cases} x+y=485 \\ x-y=15 \end{cases}$

b. $\begin{cases} x+y=485 \\ x-y=15 \end{cases} \Rightarrow 2x=500 \Rightarrow x=250$

$250+y=485 \Rightarrow y=235$

One bar weight 250 pounds and the other weighs 235 pounds.

9. Multiply the first equation by -2, add the result to the second equation, and replace the second equation with the new equation. Then divide the second equation by 4.

$\begin{cases} 2x+y+2z=4 \\ 4x+6y+z=15 \\ 2x+2y+7z=-1 \end{cases} \Rightarrow \begin{cases} 2x+y+2z=4 \\ y-\dfrac{3}{4}z=\dfrac{7}{4} \\ 2x+2y+7z=-1 \end{cases}$

Subtract the third equation from the first equation and replace the third equation with the new equation:

$\begin{cases} 2x+y+2z=4 \\ y-\dfrac{3}{4}z=\dfrac{7}{4} \\ 2x+2y+7z=-1 \end{cases} \Rightarrow \begin{cases} 2x+y+2z=4 \\ y-\dfrac{3}{4}z=\dfrac{7}{4} \\ -y-5z=5 \end{cases}$

Add the second and third equations, replace the third equation with the new equation

$\begin{cases} 2x+y+2z=4 \\ y-\dfrac{3}{4}z=\dfrac{7}{4} \\ -y-5z=5 \end{cases} \Rightarrow \begin{cases} 2x+y+2z=4 \\ y-\dfrac{3}{4}z=\dfrac{7}{4} \\ -\dfrac{23}{4}z=\dfrac{27}{4} \end{cases}$

Solve for the third equation for z. Divide the first equation by 2:

$\begin{cases} x+\dfrac{1}{2}y+z=2 \\ y-\dfrac{3}{4}z=\dfrac{7}{4} \\ z=-\dfrac{27}{23} \end{cases}$

10. $\begin{cases} x-6y+3z=-2 \\ 9y-5z=2 \\ 2z=10 \end{cases} \Rightarrow \begin{cases} x-6y+3z=-2 \\ 9y-5z=2 \\ z=5 \end{cases}$

$9y-5(5)=2 \Rightarrow y=3$

$x-6(3)+3(5)=-2 \Rightarrow x=1$

The solution is $\{(1, 3, 5)\}$.

11. Subtract the third equation from the first equation, replace the third equation with the new equation, and solve for z.

$\begin{cases} x+y+z=8 \\ 2x-2y+2z=4 \\ x+y-z=12 \end{cases} \Rightarrow \begin{cases} x+y+z=8 \\ 2x-2y+2z=4 \\ z=-2 \end{cases}$

Multiply the first equation by -2, add the result to the second equation, and replace the second equation with the new equation. Solve for y:

$\begin{cases} x+y+z=8 \\ 2x-2y+2z=4 \\ z=-2 \end{cases} \Rightarrow \begin{cases} x+y+z=8 \\ y=3 \\ z=-2 \end{cases}$

(continued on next page)

(*continued*)

Substitute the values for y and z into the first equation to solve for x:

$$\begin{cases} x+3-2=8 \\ \quad y \quad =3 \\ \quad\quad z=-2 \end{cases} \Rightarrow \begin{cases} x=7 \\ y=3 \\ z=-2 \end{cases}$$

The solution is $\{(7, 3, -2)\}$.

12. Add the second and third equations, and replace the third equation with the result:

$$\begin{cases} x+\quad\quad z=-1 \\ \quad 3y+2z=5 \\ 3x-3y+\ z=-8 \end{cases} \Rightarrow \begin{cases} x+\quad\quad z=-1 \\ \quad 3y+2z=5 \\ 3x\quad +3z=-3 \end{cases} \Rightarrow$$

$$\begin{cases} x+\quad z=-1 \\ \ 3y+2z=5 \\ x\ +\ z=-1 \end{cases}$$

The first and third equations are the same, so we have two equations in three unknowns. Solve the first equation for z, then substitute that expression into the second equation to solve for y:

$$z=-x-1;\ 3y+2(-x-1)=5 \Rightarrow$$

$$3y=2x+7 \Rightarrow y=\frac{2}{3}x+\frac{7}{3}$$

The solution is $\left\{\left(x, \frac{2}{3}x+\frac{7}{3}, -x-1\right)\right\}$.

13. Add the first and second equations, replace the first equation with the solution, then solve for x:

$$\begin{cases} 2x-\ y+z=2 \\ x+\ y-z=-1 \\ x-5y+5z=7 \end{cases} \Rightarrow \begin{cases} x\quad\quad =\dfrac{1}{3} \\ x+\ y-z=-1 \\ x-5y+5z=7 \end{cases}$$

Substitute $x=1/3$ into the second and third equations, then solve the system

$$\begin{cases} \dfrac{1}{3}+\ y-\ z=-1 \\ \dfrac{1}{3}-5y+5z=7 \end{cases} \Rightarrow \begin{cases} 5y-5z=-\dfrac{20}{3} \\ -5y+5z=\dfrac{20}{3} \end{cases} \Rightarrow 0=0$$

Now, solve for y: $\dfrac{1}{3}+y-z=-1 \Rightarrow z=z-\dfrac{4}{3}$.

The solution is $\left\{\left(\dfrac{1}{3}, -\dfrac{4}{3}+z, z\right)\right\}$.

14. Let n = the number of nickels. Let d = the number of dimes. Let q = the number of quarters. Then

$$\begin{cases} n+d+q=300 \\ d=3(n+q) \\ 0.05n+0.1d+0.25q=30.65 \end{cases} \Rightarrow$$

$$\begin{cases} n+\ d+\ q=300 \\ -3n+\ d-\ 3q=0 \\ 5n+10d+25q=3065 \end{cases}$$

Multiply the first equation by 3, add the result to the second equation, and replace the second equation with the new equation

$$\begin{cases} n+\ d+\ q=300 \\ -3n+\ d-\ 3q=0 \\ 5n+10d+25q=3065 \end{cases} \Rightarrow$$

$$\begin{cases} n+\ d+\ q=300 \\ \quad 4d\quad =900 \\ 5n+10d+25q=3065 \end{cases} \Rightarrow$$

$$\begin{cases} n+\ d+\ q=300 \\ \quad d\quad =225 \\ 5n+10d+25q=3065 \end{cases}$$

Multiply the first equation by -5, add the result to the third equation, and replace the third equation with the new equation:

$$\begin{cases} n+\ d+\ q=300 \\ \quad d\quad =225 \\ 5n+10d+25q=3065 \end{cases} \Rightarrow$$

$$\begin{cases} n+d+\ q=300 \\ \quad d\quad =225 \\ \quad 5d+20q=1565 \end{cases}$$

Substitute $d=225$ into the third equation, solve for q, then substitute the values for d and q into the first equation to solve for n:

$$5(225)+20q=1565 \Rightarrow q=22$$

$$n+225+22=300 \Rightarrow n=53$$

There are 53 nickels, 225 dimes, and 22 quarters.

15. $$\frac{2x}{(x-5)(x+1)}=\frac{A}{x-5}+\frac{B}{x+1}$$

16. $$\frac{-5x^2+x-8}{(x-2)(x^2+1)^2}$$

$$=\frac{A}{x-2}+\frac{Bx+C}{x^2+1}+\frac{Dx+E}{(x^2+1)^2}$$

17. $\dfrac{x+3}{(x+4)^2(x-7)} = \dfrac{A}{x-7} + \dfrac{B}{x+4} + \dfrac{C}{(x+4)^2} \Rightarrow$

$x+3 = A(x+4)^2 + B(x-7)(x+4) + C(x-7)$

Letting $x = 7$, we have $10 = 121A \Rightarrow A = \dfrac{10}{121}$.

Letting $x = -4$, we have

$-1 = -11C \Rightarrow \dfrac{1}{11} = C$. Substitute the values

for A and C into the equation and expand.

$x+3 = \left(B + \dfrac{10}{121}\right)x^2 + \left(\dfrac{91}{121} - 3B\right)x$

$+ \left(\dfrac{83}{121} - 28B\right) \Rightarrow$

$B + \dfrac{10}{121} = 0 \Rightarrow B = -\dfrac{10}{121}$

$\dfrac{x+3}{(x+4)^2(x-7)}$

$= \dfrac{10}{121(x-7)} - \dfrac{10}{121(x+4)} + \dfrac{1}{11(x+4)^2}$

18.

19.

20. Solve the systems $\begin{cases} x = 0 \\ x+3y = 3 \end{cases}$, $\begin{cases} x+3y = 3 \\ y = 0 \end{cases}$,

and $\begin{cases} y = 0 \\ x = 0 \end{cases}$ to find the vertices: $(0, 1)$, $(3, 0)$,

and $(0, 0)$.

Now find the values of the objective function at each vertex:

Ordered pair	$z = 2x + y$
$(0, 1)$	1
$(3, 0)$	6
$(0, 0)$	0

The maximum value is 6 at $(3, 0)$.

Chapter 5 Practice Test B

1. Substitute each of the ordered pairs into the system.

$\begin{cases} 2(5) - 3(-2) = 16 \\ 5 - (-2) = 7 \end{cases}$ The answer is A.

2. Substitute each of the ordered pairs into the system.

$\begin{cases} 6\left(\dfrac{44}{3}\right) - 9(10) = -2 \\ 3\left(\dfrac{44}{3}\right) - 5(10) = -6 \end{cases}$

The answer is C.

3. Substitute each of the ordered pairs into the system.

$\begin{cases} 2\left(\dfrac{5}{2}y + \dfrac{9}{2}\right) - 5y = 9 \\ 4\left(\dfrac{5}{2}y + \dfrac{9}{2}\right) - 10y = 18 \end{cases}$

The answer is B.

4. Substituting each of the ordered pairs into the system, we find that none are solutions. The answer is C.

5. Substitute each of the ordered pairs into the system.

$\begin{cases} \dfrac{1}{5}(1) + \dfrac{2}{5}(2) = 1 \\ \dfrac{1}{4}(1) - \dfrac{1}{3}(2) = -\dfrac{5}{12} \end{cases}$

The answer is B.

6. Substitute each of the ordered pairs into the system.

$\begin{cases} \left(3y + \dfrac{1}{2}\right) - 3y = \dfrac{1}{2} \\ -2\left(3y + \dfrac{1}{2}\right) + 6y = -1 \end{cases}$

The answer is D.

7. Substitute each of the ordered pairs into the system.

$$\begin{cases} 3(4) + 4(0) = 12 \\ 3(4)^2 + 16(0)^2 = 48 \end{cases} \quad \begin{cases} 3(2) + 4\left(\dfrac{3}{2}\right) = 12 \\ 3(2)^2 + 16\left(\dfrac{3}{2}\right)^2 = 48 \end{cases}$$

The answer is D.

8. Let x = the number of student tickets and y = the number of nonstudent tickets. Then

$$\begin{cases} x + y = 300 \\ 2x + 5y = 975 \end{cases} \Rightarrow \begin{cases} -2x - 2y = -600 \\ 2x + 5y = 975 \end{cases} \Rightarrow$$

$$3y = 375 \Rightarrow y = 125$$
$$x + 125 = 300 \Rightarrow x = 175$$

The answer is D.

9. Multiply the first equation by 2, subtract the second equation from that result, then replace the second equation with the new equation:

$$\begin{cases} x + 3y + 3z = 4 \\ 2x + 5y + 4z = 5 \\ x + 2y + 2z = 6 \end{cases} \Rightarrow \begin{cases} x + 3y + 3z = 4 \\ y + 2z = 3 \\ x + 2y + 2z = 6 \end{cases}$$

Subtract the third equation from the first equation, then replace the third equation with the new equation:

$$\begin{cases} x + 3y + 3z = 4 \\ y + 2z = 3 \\ x + 2y + 2z = 6 \end{cases} \Rightarrow \begin{cases} x + 3y + 3z = 4 \\ y + 2z = 3 \\ y + z = -2 \end{cases}$$

Subtract the third equation from the second equation and replace the third equation with the new equation:

$$\begin{cases} x + 3y + 3z = 4 \\ y + 2z = 3 \\ y + z = -2 \end{cases} \Rightarrow \begin{cases} x + 3y + 3z = 4 \\ y + 2z = 3 \\ z = 5 \end{cases}$$

The answer is C.

10. From the third equation, $z = 3$.
$$10y - 2(3) = 4 \Rightarrow y = 1; 2x + 1 + 3 = 0 \Rightarrow$$
$$x = -2$$
The answer is A.

11. Substitute each of the ordered pairs into the system.

$$\begin{cases} 2(1) + 13(-1) + 6(2) = 1 \\ 3(1) + 10(-1) + 11(2) = 15 \\ 2(1) + 10(-1) + 8(2) = 8 \end{cases}$$

The answer is B.

12. Substituting each of the ordered pairs into the system, we find that none are solutions.
The answer is D.

13. Substitute each of the ordered pairs into the system.

$$\begin{cases} x - 2(5x + 1) + 3(3x + 2) = 4 \\ 2x - (5x + 1) + (3x + 2) = 1 \\ x + (5x + 1) - 2(3x + 2) = -3 \end{cases}$$

The answer is C.

14. Let x = the number of students going to France. Let y = the number of students going to Italy. Let z = the number of students going to Spain. Then

$$\begin{cases} x + y + z = 46 \\ x + y = 4 + z \\ x = z - 2 \end{cases} \Rightarrow \begin{cases} x + y + z = 46 \\ x + y - z = 4 \\ x - z = -2 \end{cases} \text{ Subtract}$$

the second equation from the first equation, replace the second equation with the new equation, and solve for z:

$$\begin{cases} x + y + z = 46 \\ x + y - z = 4 \\ x - z = -2 \end{cases} \Rightarrow \begin{cases} x + y + z = 46 \\ 2z = 42 \\ x - z = -2 \end{cases} \Rightarrow$$

$$\begin{cases} x + y + z = 46 \\ z = 21 \\ x - z = -2 \end{cases}$$

Substitute $z = 21$ into the third equation to solve for x. Then substitute the values for x and z into the first equation to solve for y:
$$x - 21 = -2 \Rightarrow x = 19; 19 + y + 21 = 46 \Rightarrow$$
$$y = 6$$
The answer is B.

15. C   **16.** A

17.
$$\dfrac{x^2 + 15x + 18}{x^3 - 9x} = \dfrac{x^2 + 15x + 18}{x(x-3)(x+3)}$$
$$= \dfrac{A}{x} + \dfrac{B}{x-3} + \dfrac{C}{x+3} \Rightarrow$$
$$x^2 + 15x + 18$$
$$= A(x^2 - 9) + Bx(x+3) + Cx(x-3)$$
Letting $x = -3$, we have
$$(-3)^2 + 15(-3) + 18 = -3C(-3-3) \Rightarrow$$
$$-18 = 18C \Rightarrow -1 = C$$
Letting $x = 3$, we have
$$(3)^2 + 15(3) + 18 = 3B(3+3) \Rightarrow$$
$$72 = 18B \Rightarrow 4 = B$$
Substituting the values for B and C and expanding the equation, we have
$$x^2 + 15x + 18$$
$$= A(x^2 - 9) + 4x(x+3) - x(x-3)$$
$$= (A+3)x^2 + 15x - 9A \Rightarrow$$
$$\begin{cases} A + 3 = 1 \\ -9A = 18 \end{cases} \Rightarrow A = -2$$

(*continued on next page*)

(continued)

$$\frac{x^2+15x+18}{x^3-9x}=-\frac{2}{x}+\frac{4}{x-3}-\frac{1}{x+3}$$

The answer is C.

18. B   19. A

20. Solve the systems

$$\begin{cases} x=0 \\ 2x+3y=16 \end{cases}, \begin{cases} 2x+3y=16 \\ 2x+y=8 \end{cases}, \begin{cases} 2x+y=8 \\ y=0 \end{cases} \text{ and }$$

$$\begin{cases} y=0 \\ x=0 \end{cases} \text{ to find the vertices: } \left(0,\frac{16}{3}\right),$$

(2, 4), (4, 0), and (0, 0).

Now find the values of the objective function at each vertex:

Ordered pair	$z=3x+21y$
$\left(0,\dfrac{16}{3}\right)$	112
(2, 4)	90
(4, 0)	12
(0, 0)	0

The maximum value is 112 at $\left(0,\frac{16}{3}\right)$.

The answer is B.

Cumulative Review Exercises (Chapters P–5)

1. $\dfrac{1}{x-1}+\dfrac{4}{x-4}=\dfrac{5}{x-5}\Rightarrow$
    $(x-4)(x-5)+4(x-1)(x-5)$
    $\qquad\qquad =5(x-1)(x-4)\Rightarrow$
    $5x^2-33x+40=5x^2-25x+20\Rightarrow$
    $-8x=-20\Rightarrow x=\dfrac{5}{2}$

2. Let $u=x+\dfrac{1}{x}$. Then

    $2u^2-7u+5=0\Rightarrow(u-1)(2u-5)=0\Rightarrow$
    $u=1$ or $u=5/2$

    $x+\dfrac{1}{x}=1\Rightarrow x^2-x+1=0\Rightarrow$

    $x=\dfrac{1\pm\sqrt{1-4}}{2}=\dfrac{1\pm i\sqrt{3}}{2}=\dfrac{1}{2}\pm\dfrac{i\sqrt{3}}{2}$

    $x+\dfrac{1}{x}=\dfrac{5}{2}\Rightarrow 2x^2-5x+2=0\Rightarrow$
    $(x-2)(2x-1)=0\Rightarrow x=2$ or $x=1/2$

    The solution is $\left\{\dfrac{1}{2}\pm\dfrac{i\sqrt{3}}{2},2,\dfrac{1}{2}\right\}$.

3. $\sqrt{3x-5}=x-3\Rightarrow 3x-5=(x-3)^2\Rightarrow$
    $x^2-9x+14=0\Rightarrow(x-7)(x-2)=0\Rightarrow$
    $x=7$ or $x=2$.
    Check each answer to see if either is extraneous:

    $\sqrt{3(7)-5}=\sqrt{16}=4=7-3$

    $\sqrt{3(2)-5}=\sqrt{1}=1\neq 2-3\Rightarrow 2$ is extraneous.
    The solution is {7}.

4. $\dfrac{x-1}{x+3}=0\Rightarrow x=1$ is the x-intercept. The
    vertical asymptote is $x=-3$. The intervals to
    be tested are $(-\infty,-3)$, $(-3,1]$, and $[1,\infty)$.

Interval	Test Point	Value of $\dfrac{x-1}{x+3}$	Result
$(-\infty,-3)$	-4	5	+
$(-3,1]$	0	$-\dfrac{1}{3}$	−
$[1,\infty)$	2	$\dfrac{1}{5}$	+

The solution is $(-3,1]$.

5. Solve the associated equation to find the test intervals:

$x^2 - 9x + 20 = 0 \Rightarrow (x-4)(x-5) = 0 \Rightarrow$

$x = 4$ or $x = 5$.

The intervals to be tested are $(-\infty, 4), (4, 5),$ and $(5, \infty)$.

Interval	Test Point	Value of $x^2 - 9x + 20$	Result
$(-\infty, 4)$	0	20	+
$(4, 5)$	4.5	−0.25	−
$(5, \infty)$	10	30	+

The solution is $(-\infty, 4) \cup (5, \infty)$.

6. $2^{x-1} = 5 \Rightarrow (x-1)\ln 2 = \ln 5 \Rightarrow$

$x \ln 2 - \ln 2 = \ln 5 \Rightarrow x = \dfrac{\ln 5 + \ln 2}{\ln 2}$

7. $\log_x 16 = 4 \Rightarrow x^4 = 16 \Rightarrow x^4 = 2^4 \Rightarrow x = 2$

8. $\log(x-3) + \log(x-1) = \log(2x-5) \Rightarrow$

$\log\big((x-3)(x-1)\big) = \log(2x-5) \Rightarrow$

$x^2 - 4x + 3 = 2x - 5 \Rightarrow x^2 - 6x + 8 = 0 \Rightarrow$

$(x-4)(x-2) = 0 \Rightarrow x = 4$ or $x = 2$

Reject $x = 2$ because it makes $\log(2-3) = \log(-1)$, which is not possible. The solution is $\{4\}$.

9. Shift the graph of $y = |x|$ one unit left and two units down.

10. Shift the graph of $y = x^2$ two units right, reflect it across the x-axis, then shift it three units up.

11. Shift the graph of $y = 2^x$ one unit right and three units down.

12. Shift the graph of $y = \sqrt{x}$ one unit right and three units up.

13. a. Switch the variables and solve for y:

$y = 2x - 2 \Rightarrow x = 2y - 2 \Rightarrow y = f^{-1} = \dfrac{x+2}{2}$

b.

14. a. The factors of the constant term are $\{\pm 1, \pm 2, \pm 3, \pm 6\}$. The factors of the leading coefficient are $\{\pm 1\}$. The possible rational zeros are $\{\pm 1, \pm 2, \pm 3, \pm 6\}$.

b.
$$\begin{array}{r|rrrr} 2 & 1 & -1 & 1 & -6 \\ & & 2 & 2 & 6 \\ \hline & 1 & 1 & 3 & 0 \end{array}$$
A real zero is 2.

c. $x^3 - x^2 + x - 6 = (x-2)(x^2 + x + 3)$
$$x^2 + x + 3 = 0 \Rightarrow x = \frac{-1 \pm \sqrt{1-12}}{2} \Rightarrow$$
$$x = \frac{-1 \pm i\sqrt{11}}{2}$$
The zeros are $\left\{2, -\dfrac{1}{2} \pm \dfrac{i\sqrt{11}}{2}\right\}$.

15. $\log_3(9x^4) = \log_3 9 + \log_3 x^4 = 2 + 4\log_3 x$

16. a. $0.5 = e^{-0.05t} \Rightarrow \ln 0.5 = -0.05t \Rightarrow$
$$t = -\frac{\ln 0.5}{0.05} \approx 13.86 \text{ years}$$

b. $0.5 = e^{-0.0002t} \Rightarrow \ln 0.5 = -0.0002t \Rightarrow$
$$t = -\frac{\ln 0.5}{0.0002} \approx 3465.74 \text{ years}$$

17. $f^{-1}(9) = 2$

18. $A = Pe^{rt} \Rightarrow 2P = Pe^{0.075t} \Rightarrow 2 = e^{0.075t} \Rightarrow$
$$\ln 2 = 0.075t \Rightarrow t = \frac{\ln 2}{0.075} \approx 9.24$$
It will take about 9.24 years to double the money.

19. $\begin{cases} 5x - 2y + 25 = 0 \\ 4y - 3x - 29 = 0 \end{cases} \Rightarrow \begin{cases} 10x - 4y = -50 \\ -3x + 4y = 29 \end{cases} \Rightarrow$
$7x = -21 \Rightarrow x = -3$
$4y - 3(-3) - 29 = 0 \Rightarrow y = 5$
The solution is $\{(-3, 5)\}$.

20. Multiply the second equation by -2, add the result to the first equation, and replace the second equation with the new equation:
$$\begin{cases} 2x - y + z = 3 \\ x + 3y - 2z = 11 \\ 3x - 2y + z = 4 \end{cases} \Rightarrow \begin{cases} 2x - y + z = 3 \\ -7y + 5z = -19 \\ 3x - 2y + z = 4 \end{cases}$$

Multiply the first equation by 3, multiply the third equation by -2, add the two and replace the third equation with the new equation:
$$\begin{cases} 2x - y + z = 3 \\ -7y + 5z = -19 \\ 3x - 2y + z = 4 \end{cases} \Rightarrow \begin{cases} 2x - y + z = 3 \\ -7y + 5z = -19 \\ y + z = 1 \end{cases}$$

Multiply the third equation by 7, add the result to the second equation, replace the third equation and solve for z:
$$\begin{cases} 2x - y + z = 3 \\ -7y + 5z = -19 \\ y + z = 1 \end{cases} \Rightarrow \begin{cases} 2x - y + z = 3 \\ -7y + 5z = -19 \\ 12z = -12 \end{cases} \Rightarrow$$
$z = -1$
$-7y + 5(-1) = -19 \Rightarrow y = 2$
$2x - 2 - 1 = 3 \Rightarrow x = 3$
The solution is $\{(3, 2, -1)\}$.

Chapter 6 Matrices and Determinants

6.1 Matrices and Systems of Equations

6.1 Practice Problems

1. a. 3×2 **b.** 1×2

2. $\begin{bmatrix} 0 & 3 & -1 & 8 \\ 1 & 4 & 0 & 14 \\ 0 & -2 & 9 & 0 \end{bmatrix}$

3. $\begin{bmatrix} 3 & 4 & 5 \\ 2 & 4 & 6 \end{bmatrix} \xrightarrow{R_1 \leftrightarrow R_2} \begin{bmatrix} 2 & 4 & 6 \\ 3 & 4 & 5 \end{bmatrix} \xrightarrow{\frac{1}{2}R_1}$

$\begin{bmatrix} 1 & 2 & 3 \\ 3 & 4 & 5 \end{bmatrix} \xrightarrow{-3R_1 + R_2 \to R_2} \begin{bmatrix} 1 & 2 & 3 \\ 0 & -2 & -4 \end{bmatrix}$

4. First write the system as an augmented matrix.

$\begin{cases} x - 6y + 3z = -2 \\ 3x + 3y - 2z = -2 \\ 2x - 3y + z = -2 \end{cases} \Rightarrow \begin{bmatrix} 1 & -6 & 3 & -2 \\ 3 & 3 & -2 & -2 \\ 2 & -3 & 1 & -2 \end{bmatrix}$

Now perform the row operations.

$\begin{bmatrix} 1 & -6 & 3 & -2 \\ 3 & 3 & -2 & -2 \\ 2 & -3 & 1 & -2 \end{bmatrix}$

$\xrightarrow[2R_1 - R_3 \to R_3]{3R_1 - R_2 \to R_2} \begin{bmatrix} 1 & -6 & 3 & -2 \\ 0 & -21 & 11 & -4 \\ 0 & -9 & 5 & -2 \end{bmatrix}$

$\xrightarrow{9R_2 - 21R_3 \to R_3} \begin{bmatrix} 1 & -6 & 3 & -2 \\ 0 & -21 & 11 & -4 \\ 0 & 0 & -6 & 6 \end{bmatrix}$

$\xrightarrow[-\frac{1}{6}R_3 \to R_3]{-\frac{1}{21}R_2 \to R_2} \begin{bmatrix} 1 & -6 & 3 & -2 \\ 0 & 1 & -\frac{11}{21} & \frac{4}{21} \\ 0 & 0 & 1 & -1 \end{bmatrix}$

Thus, $z = -1$.
Using back-substitution, we have

$y - \frac{11}{21}(-1) = \frac{4}{21} \Rightarrow y = -\frac{7}{21} = -\frac{1}{3}$. Then,

$x - 6\left(-\frac{1}{3}\right) + 3(-1) = -2 \Rightarrow x - 1 = -2 \Rightarrow$
$x = -1$

The solution is $\left\{ \left(-1, -\frac{1}{3}, -1 \right) \right\}$.

5. First write the system as an augmented matrix.

$\begin{cases} x - 2y = 1 \\ 2x + 3y = 16 \end{cases} \Rightarrow \begin{bmatrix} 1 & -2 & 1 \\ 2 & 3 & 16 \end{bmatrix}$

Now perform the row operations.

$\begin{bmatrix} 1 & -2 & 1 \\ 2 & 3 & 16 \end{bmatrix} \xrightarrow{-2R_1 + R_2 \to R_2} \begin{bmatrix} 1 & -2 & 1 \\ 0 & 7 & 14 \end{bmatrix}$

$\xrightarrow{\frac{1}{7}R_2 \to R_2} \begin{bmatrix} 1 & -2 & 1 \\ 0 & 1 & 2 \end{bmatrix}$

Thus, $y = 2$. Using back-substitution, we have
$x - 2(2) = 1 \Rightarrow x = 5$. The solution is $\{(5, 2)\}$.

6. First write the system as an augmented matrix.

$\begin{cases} 2x + y - z = 7 \\ x - 3y - 3z = 4 \\ 4x + y + z = 3 \end{cases} \Rightarrow \begin{bmatrix} 2 & 1 & -1 & 7 \\ 1 & -3 & -3 & 4 \\ 4 & 1 & 1 & 3 \end{bmatrix}$

Now perform the row operations.

$\begin{bmatrix} 2 & 1 & -1 & 7 \\ 1 & -3 & -3 & 4 \\ 4 & 1 & 1 & 3 \end{bmatrix}$

$\xrightarrow{R_1 \leftrightarrow R_2} \begin{bmatrix} 1 & -3 & -3 & 4 \\ 2 & 1 & -1 & 7 \\ 4 & 1 & 1 & 3 \end{bmatrix}$

$\xrightarrow[-4R_1 + R_3 \to R_3]{-2R_1 + R_2 \to R_2} \begin{bmatrix} 1 & -3 & -3 & 4 \\ 0 & 7 & 5 & -1 \\ 0 & 13 & 13 & -13 \end{bmatrix}$

$\xrightarrow[\frac{1}{13}R_3 \to R_3]{\frac{1}{7}R_2 \to R_2} \begin{bmatrix} 1 & -3 & -3 & 4 \\ 0 & 1 & \frac{5}{7} & -\frac{1}{7} \\ 0 & 1 & 1 & -1 \end{bmatrix}$

$\xrightarrow{\frac{7}{2}(-R_2 + R_3) \to R_3} \begin{bmatrix} 1 & -3 & -3 & 4 \\ 0 & 1 & \frac{5}{7} & -\frac{1}{7} \\ 0 & 0 & 1 & -3 \end{bmatrix}$

Thus, $z = -3$. Using back-substitution, we
have $y + \frac{5}{7}(-3) = -\frac{1}{7} \Rightarrow y = 2$ and
$x - 3(2) - 3(-3) = 4 \Rightarrow x = 1$.
The solution is $\{(1, 2, -3)\}$.

7. $\begin{cases} 6x+8y-14z=3 \\ 3x+4y-7z=12 \\ 6x+3y+z=0 \end{cases} \Rightarrow \begin{bmatrix} 6 & 8 & -14 & | & 3 \\ 3 & 4 & -7 & | & 12 \\ 6 & 3 & 1 & | & 0 \end{bmatrix}$

Now perform the row operations.

$\begin{bmatrix} 6 & 8 & -14 & | & 3 \\ 3 & 4 & -7 & | & 12 \\ 6 & 3 & 1 & | & 0 \end{bmatrix}$

$\xrightarrow{R_1 \leftrightarrow R_2} \begin{bmatrix} 3 & 4 & -7 & | & 12 \\ 6 & 8 & -14 & | & 3 \\ 6 & 3 & 1 & | & 0 \end{bmatrix}$

$\xrightarrow{\frac{1}{3}R_1 \to R_1} \begin{bmatrix} 1 & \frac{4}{3} & -\frac{7}{3} & | & 4 \\ 6 & 8 & -14 & | & 3 \\ 6 & 3 & 1 & | & 0 \end{bmatrix}$

$\xrightarrow{-6R_1+R_2 \to R_2} \begin{bmatrix} 1 & \frac{4}{3} & -\frac{7}{3} & | & 4 \\ 0 & 0 & 0 & | & -21 \\ 6 & 3 & 1 & | & 0 \end{bmatrix}$

Since the second row is equivalent to $0 = -21$, the system is inconsistent and the solution is $\varnothing$.

8. $\begin{cases} 2x-3y-2z=0 \\ x+y-2z=7 \\ 3x-5y-5z=3 \end{cases} \Rightarrow \begin{bmatrix} 2 & -3 & -2 & | & 0 \\ 1 & 1 & -2 & | & 7 \\ 3 & -5 & -5 & | & 3 \end{bmatrix}$

$\begin{bmatrix} 2 & -3 & -2 & | & 0 \\ 1 & 1 & -2 & | & 7 \\ 3 & -5 & -5 & | & 3 \end{bmatrix} \xrightarrow{R_1 \leftrightarrow R_2} \begin{bmatrix} 1 & 1 & -2 & | & 7 \\ 2 & -3 & -2 & | & 0 \\ 3 & -5 & -5 & | & 3 \end{bmatrix}$

$\xrightarrow[-3R_1+R_3 \to R_3]{-2R_1+R_2 \to R_2} \begin{bmatrix} 1 & 1 & -2 & | & 7 \\ 0 & -5 & 2 & | & -14 \\ 0 & -8 & 1 & | & -18 \end{bmatrix}$

$\xrightarrow{-\frac{1}{5}R_2 \to R_2} \begin{bmatrix} 1 & 1 & -2 & | & 7 \\ 0 & 1 & -\frac{2}{5} & | & \frac{14}{5} \\ 0 & -8 & 1 & | & -18 \end{bmatrix}$

$\xrightarrow[8R_2+R_3 \to R_3]{R_1-R_2 \to R_1} \begin{bmatrix} 1 & 0 & -\frac{8}{5} & | & \frac{21}{5} \\ 0 & 1 & -\frac{2}{5} & | & \frac{14}{5} \\ 0 & 0 & -\frac{11}{5} & | & \frac{22}{5} \end{bmatrix}$

$\xrightarrow{-\frac{5}{11}R_3 \to R_3} \begin{bmatrix} 1 & 0 & -\frac{8}{5} & | & \frac{21}{5} \\ 0 & 1 & -\frac{2}{5} & | & \frac{14}{5} \\ 0 & 0 & 1 & | & -2 \end{bmatrix}$

$\xrightarrow[R_2+\frac{2}{5}R_3 \to R_2]{R_1+\frac{8}{5}R_3 \to R_1} \begin{bmatrix} 1 & 0 & 0 & | & 1 \\ 0 & 1 & 0 & | & 2 \\ 0 & 0 & 1 & | & -2 \end{bmatrix}$

The solution is $\{(1, 2, -2)\}$.

9. $\begin{cases} x+z=-1 \\ 3y+2z=5 \\ 3x-3y+z=-8 \end{cases} \Rightarrow \begin{bmatrix} 1 & 0 & 1 & | & -1 \\ 0 & 3 & 2 & | & 5 \\ 3 & -3 & 1 & | & -8 \end{bmatrix}$

$\begin{bmatrix} 1 & 0 & 1 & | & -1 \\ 0 & 3 & 2 & | & 5 \\ 3 & -3 & 1 & | & -8 \end{bmatrix}$

$\xrightarrow{-3R_1+R_3 \to R_3} \begin{bmatrix} 1 & 0 & 1 & | & -1 \\ 0 & 3 & 2 & | & 5 \\ 0 & -3 & -2 & | & -5 \end{bmatrix}$

$\xrightarrow[R_2+R_3 \to R_3]{\frac{1}{3}R_2 \to R_2} \begin{bmatrix} 1 & 0 & 1 & | & -1 \\ 0 & 1 & \frac{2}{3} & | & \frac{5}{3} \\ 0 & 0 & 0 & | & 0 \end{bmatrix}$

The matrix is in row echelon form. The equivalent system is $\begin{cases} x+z=-1 \\ y+\frac{2}{3}z=\frac{5}{3} \end{cases}$. Solving for x and y in terms of z, we have $x = -z-1$ and $y=\frac{5}{3}-\frac{2}{3}z$.

The solution set is $\left\{ \left(-z-1, \frac{5}{3}-\frac{2}{3}z, z \right) \right\}$.

10. a. $\begin{cases} s=0.01s+0.1c+0.1t+12.46 \\ c=0.2s+0.02c+0.01t+3 \\ t=0.25s+0.3c+2.7 \end{cases}$ or

$\begin{cases} 0.99s-0.1c-0.1t=12.46 \\ -0.2s+0.98c-0.01t=3 \\ -0.25s-0.3c+t=2.7 \end{cases}$

b. $0.99(14)-0.1(6)-0.1(8)=12.46$ [a]
$-0.2(14)+0.98(6)-0.01(8)=3$ [a]
$-0.25(14)-0.3(6)+(8)=2.7$ [a]

6.1 Concepts and Vocabulary

1. A matrix is any rectangular array of <u>numbers</u>.

3. Two matrices are row equivalent if one can be obtained form the other by a sequence of <u>row operations</u>.

5. False. A 3×4 matrix has three rows with four entries in each row. In other words, the matrix has three rows and four columns.

7. False. The augmented matrix of the system is
$$\begin{bmatrix} 2 & 3 & 0 & | & 5 \\ 0 & 3 & -4 & | & -1 \\ 1 & 0 & 2 & | & 3 \end{bmatrix}.$$

6.1 Building Skills

9. 1×1 **11.** 2×4 **13.** 2×3

15. $a_{13} = 3, a_{31} = 9, a_{33} = 11, a_{34} = 12$

17. No, the array is not a matrix because it is not a rectangular array of numbers.

19. $\begin{bmatrix} 2 & 4 & | & 2 \\ 1 & -3 & | & 1 \end{bmatrix}$

21. $\begin{bmatrix} 5 & -7 & | & 11 \\ -13 & 17 & | & 19 \end{bmatrix}$

23. $\begin{bmatrix} -1 & 2 & 3 & | & 8 \\ 2 & -3 & 9 & | & 16 \\ 4 & -5 & -6 & | & 32 \end{bmatrix}$

25. $\begin{cases} x + 2y - 3z = 4 \\ -2x - 3y + z = 5 \\ 3x - 3y + 2z = 7 \end{cases}$

27. $\begin{cases} x - y + z = 2 \\ 2x + y - 3z = 6 \end{cases}$

29.(i) $\begin{bmatrix} 2 & 3 & 5 \\ 1 & 2 & 3 \end{bmatrix} \xrightarrow{R_1 \leftrightarrow R_2} \begin{bmatrix} 1 & 2 & 3 \\ 2 & 3 & 5 \end{bmatrix}$

(ii) $\begin{bmatrix} 1 & 2 & 3 \\ 2 & 3 & 5 \end{bmatrix} \xrightarrow{-2R_1 + R_2 \rightarrow R_2} \begin{bmatrix} 1 & 2 & 3 \\ 0 & -1 & -1 \end{bmatrix}$

(iii) $\begin{bmatrix} 1 & 2 & 3 \\ 0 & -1 & -1 \end{bmatrix} \xrightarrow{-R_2} \begin{bmatrix} 1 & 2 & 3 \\ 0 & 1 & 1 \end{bmatrix}$

31.(i) $\begin{bmatrix} 1 & 2 & 3 & 4 \\ 0 & 4 & -3 & 11 \\ 0 & 1 & 5 & -3 \end{bmatrix} \xrightarrow{R_2 \leftrightarrow R_3}$
$$\begin{bmatrix} 1 & 2 & 3 & 4 \\ 0 & 1 & 5 & -3 \\ 0 & 4 & -3 & 11 \end{bmatrix}$$

(ii) $\begin{bmatrix} 1 & 2 & 3 & 4 \\ 0 & 1 & 5 & -3 \\ 0 & 4 & -3 & 11 \end{bmatrix} \xrightarrow{-4R_2 + R_3 \rightarrow R_3}$
$$\begin{bmatrix} 1 & 2 & 3 & 4 \\ 0 & 1 & 5 & -3 \\ 0 & 0 & -23 & 23 \end{bmatrix}$$

(iii) $\begin{bmatrix} 1 & 2 & 3 & 4 \\ 0 & 1 & 5 & -3 \\ 0 & 0 & -23 & 23 \end{bmatrix} \xrightarrow{-\frac{1}{23}R_3}$
$$\begin{bmatrix} 1 & 2 & 3 & 4 \\ 0 & 1 & 5 & -3 \\ 0 & 0 & 1 & -1 \end{bmatrix}$$

33. $\begin{bmatrix} 4 & 5 & -7 \\ 5 & 4 & -2 \end{bmatrix} \xrightarrow{\frac{1}{4}R_1} \begin{bmatrix} 1 & \frac{5}{4} & -\frac{7}{4} \\ 5 & 4 & -2 \end{bmatrix}$

$\xrightarrow{-5R_1 + R_2 \rightarrow R_2} \begin{bmatrix} 1 & \frac{5}{4} & -\frac{7}{4} \\ 0 & -\frac{9}{4} & \frac{27}{4} \end{bmatrix}$

$\xrightarrow{-\frac{4}{9}R_2 \rightarrow R_2} \begin{bmatrix} 1 & \frac{5}{4} & -\frac{7}{4} \\ 0 & 1 & -3 \end{bmatrix}$

35. $\begin{bmatrix} 1 & 4 & 3 & 1 \\ 0 & -3 & -2 & 0 \\ 0 & 7 & 5 & -3 \end{bmatrix} \xrightarrow{-\frac{1}{3}R_2} \begin{bmatrix} 1 & 4 & 3 & 1 \\ 0 & 1 & \frac{2}{3} & 0 \\ 0 & 7 & 5 & -3 \end{bmatrix}$

$\xrightarrow{-7R_2 + R_3 \rightarrow R_3} \begin{bmatrix} 1 & 4 & 3 & 1 \\ 0 & 1 & \frac{2}{3} & 0 \\ 0 & 0 & \frac{1}{3} & -3 \end{bmatrix}$

$\xrightarrow{3R_3} \begin{bmatrix} 1 & 4 & 3 & 1 \\ 0 & 1 & \frac{2}{3} & 0 \\ 0 & 0 & 1 & -9 \end{bmatrix}$

37. No. The matrix does not have a step-like pattern that moves down and to the right. (Property 2)

39. Yes. The matrix is in reduced row-echelon form.

41. Yes. The matrix is in reduced row-echelon form.

43. Yes. The matrix is in reduced row-echelon form.

45. $\begin{cases} x + 2y = 1 \\ \quad\quad y = -2 \end{cases} \Rightarrow x + 2(-2) = 1 \Rightarrow x = 5$

The solution is $\{(5, -2)\}$.

47. $\begin{cases} x + 4y + 2z = 2 \\ \quad\quad\quad\quad z = 3 \end{cases} \Rightarrow x + 4y = -4 \Rightarrow x = -4y - 4$

The solution is $\left\{ (-4y - 4, y, 3) \right\}$.

49. $\begin{cases} x + 2y + 3z = 2 \\ \quad\quad y - 2z = 4 \Rightarrow y - 2(-1) = 4 \Rightarrow y = 2 \\ \quad\quad\quad\quad z = -1 \end{cases}$

$x + 2(2) + 3(-1) = 2 \Rightarrow x = 1$
The solution is $\{(1, 2, -1)\}$.

51. $\begin{cases} x = 2 \\ y = -5 \\ z + 2w = 3 \end{cases} \Rightarrow z = -2w + 3$

The solution is $\{(2, -5, -2w + 3, w)\}$.

53. $\begin{cases} x \quad\quad\quad = -5 \\ y \quad\quad\quad = 4 \\ z + 2w = 3 \\ \quad\quad w = 0 \end{cases} \Rightarrow z + 2(0) = 3 \Rightarrow z = 3$

The solution is $\{(-5, 4, 3, 0)\}$.

55. $\begin{bmatrix} 1 & -2 & | & 11 \\ 2 & -1 & | & 13 \end{bmatrix} \xrightarrow{-2R_1 + R_2 \to R_2} \begin{bmatrix} 1 & -2 & | & 11 \\ 0 & 3 & | & -9 \end{bmatrix}$

$\xrightarrow{\frac{1}{3}R_2 \to R_2} \begin{bmatrix} 1 & -2 & | & 11 \\ 0 & 1 & | & -3 \end{bmatrix} \Rightarrow$

$y = -3; x - 2(-3) = 11 \Rightarrow x = 5$
The solution is $\{(5, -3)\}$.

57. $\begin{bmatrix} 2 & -3 & | & 3 \\ 4 & -1 & | & 11 \end{bmatrix} \xrightarrow{\frac{1}{2}R_1 \to R_1} \begin{bmatrix} 1 & -\dfrac{3}{2} & | & \dfrac{3}{2} \\ 4 & -1 & | & 11 \end{bmatrix}$

$\xrightarrow{-4R_1 + R_2 \to R_2} \begin{bmatrix} 1 & -\dfrac{3}{2} & | & \dfrac{3}{2} \\ 0 & 5 & | & 5 \end{bmatrix}$

$\xrightarrow{\frac{1}{5}R_2 \to R_2} \begin{bmatrix} 1 & -\dfrac{3}{2} & | & \dfrac{3}{2} \\ 0 & 1 & | & 1 \end{bmatrix} \Rightarrow$

$y = 1; x - \dfrac{3}{2}(1) = \dfrac{3}{2} \Rightarrow x = 3$
The solution is $\{(3, 1)\}$.

59. $\begin{bmatrix} 3 & -5 & | & 4 \\ 4 & -15 & | & 13 \end{bmatrix} \xrightarrow{\frac{1}{3}R_1 \to R_1} \begin{bmatrix} 1 & -\dfrac{5}{3} & | & \dfrac{4}{3} \\ 4 & -15 & | & 13 \end{bmatrix}$

$\xrightarrow{-4R_1 + R_2 \to R_2} \begin{bmatrix} 1 & -\dfrac{5}{3} & | & \dfrac{4}{3} \\ 0 & -\dfrac{25}{3} & | & \dfrac{23}{3} \end{bmatrix}$

$\xrightarrow{-\frac{3}{25}R_2 \to R_2} \begin{bmatrix} 1 & -\dfrac{5}{3} & | & \dfrac{4}{3} \\ 0 & 1 & | & -\dfrac{23}{25} \end{bmatrix} \Rightarrow$

$y = -\dfrac{23}{25}; x - \left(\dfrac{5}{3}\right)\left(-\dfrac{23}{25}\right) = \dfrac{4}{3} \Rightarrow x = -\dfrac{1}{5}$

The solution is $\left\{ \left(-\dfrac{1}{5}, -\dfrac{23}{25} \right) \right\}$.

61. $\begin{bmatrix} 1 & -1 & | & 1 \\ 2 & 1 & | & 5 \\ 3 & -4 & | & 2 \end{bmatrix} \xrightarrow[-3R_1 + R_3 \to R_3]{-2R_1 + R_2 \to R_2} \begin{bmatrix} 1 & -1 & | & 1 \\ 0 & 3 & | & 3 \\ 0 & -1 & | & -1 \end{bmatrix}$

$\xrightarrow[-R_3 \to R_3]{\frac{1}{3}R_2 \to R_2} \begin{bmatrix} 1 & -1 & | & 1 \\ 0 & 1 & | & 1 \\ 0 & 1 & | & 1 \end{bmatrix} \Rightarrow$

$y = 1; x - 1 = 1 \Rightarrow x = 2$
The solution is $\{(2, 1)\}$.

63. $\begin{bmatrix} 1 & 1 & 1 & | & 6 \\ 1 & -1 & 1 & | & 2 \\ 2 & 1 & -1 & | & 1 \end{bmatrix}$

$\xrightarrow[-2R_1 + R_3 \to R_3]{R_1 - R_2 \to R_2} \begin{bmatrix} 1 & 1 & 1 & | & 6 \\ 0 & 2 & 0 & | & 4 \\ 0 & -1 & -3 & | & -11 \end{bmatrix}$

$\xrightarrow{\frac{1}{2}R_2 \to R_2} \begin{bmatrix} 1 & 1 & 1 & | & 6 \\ 0 & 1 & 0 & | & 2 \\ 0 & -1 & -3 & | & -11 \end{bmatrix}$

$\xrightarrow{R_2 + R_3 \to R_3} \begin{bmatrix} 1 & 1 & 1 & | & 6 \\ 0 & 1 & 0 & | & 2 \\ 0 & 0 & -3 & | & -9 \end{bmatrix}$

$\xrightarrow{-\frac{1}{3}R_3 \to R_3} \begin{bmatrix} 1 & 1 & 1 & | & 6 \\ 0 & 1 & 0 & | & 2 \\ 0 & 0 & 1 & | & 3 \end{bmatrix} \Rightarrow$

$z = 3; y = 2; x + 2 + 3 = 6 \Rightarrow x = 1$
The solution is $\{(1, 2, 3)\}$.

65. $\begin{bmatrix} 2 & 3 & -1 & | & 9 \\ 1 & 1 & 1 & | & 9 \\ 3 & -1 & -1 & | & -1 \end{bmatrix}$

$\xrightarrow{R_1 \leftrightarrow R_2} \begin{bmatrix} 1 & 1 & 1 & | & 9 \\ 2 & 3 & -1 & | & 9 \\ 3 & -1 & -1 & | & -1 \end{bmatrix}$

$\xrightarrow[-3R_1+R_3\to R_3]{-2R_1+R_2\to R_2} \begin{bmatrix} 1 & 1 & 1 & | & 9 \\ 0 & 1 & -3 & | & -9 \\ 0 & -4 & -4 & | & -28 \end{bmatrix}$

$\xrightarrow{4R_2+R_3\to R_3} \begin{bmatrix} 1 & 1 & 1 & | & 9 \\ 0 & 1 & -3 & | & -9 \\ 0 & 0 & -16 & | & -64 \end{bmatrix}$

$\xrightarrow{-\frac{1}{16}R_3\to R_3} \begin{bmatrix} 1 & 1 & 1 & | & 9 \\ 0 & 1 & -3 & | & -9 \\ 0 & 0 & 1 & | & 4 \end{bmatrix} \Rightarrow$

$z = 4; \; y - 3(4) = -9 \Rightarrow y = 3$

$x + 3 + 4 = 9 \Rightarrow x = 2$

The solution is $\{(2, 3, 4)\}$.

67. $\begin{bmatrix} 3 & 2 & 4 & | & 19 \\ 2 & -1 & 1 & | & 3 \\ 6 & 7 & -1 & | & 17 \end{bmatrix}$

$\xrightarrow{R_1 \leftrightarrow \frac{1}{2}R_2} \begin{bmatrix} 1 & -\frac{1}{2} & \frac{1}{2} & | & \frac{3}{2} \\ 3 & 2 & 4 & | & 19 \\ 6 & 7 & -1 & | & 17 \end{bmatrix}$

$\xrightarrow[-6R_1+R_3\to R_3]{-3R_1+R_2\to R_2} \begin{bmatrix} 1 & -\frac{1}{2} & \frac{1}{2} & | & \frac{3}{2} \\ 0 & \frac{7}{2} & \frac{5}{2} & | & \frac{29}{2} \\ 0 & 10 & -4 & | & 8 \end{bmatrix}$

$\xrightarrow{\frac{2}{7}R_2\to R_2} \begin{bmatrix} 1 & -\frac{1}{2} & \frac{1}{2} & | & \frac{3}{2} \\ 0 & 1 & \frac{5}{7} & | & \frac{29}{7} \\ 0 & 10 & -4 & | & 8 \end{bmatrix}$

$\xrightarrow{-10R_2+R_3\to R_3} \begin{bmatrix} 1 & -\frac{1}{2} & \frac{1}{2} & | & \frac{3}{2} \\ 0 & 1 & \frac{5}{7} & | & \frac{29}{7} \\ 0 & 0 & -\frac{78}{7} & | & -\frac{234}{7} \end{bmatrix}$

$\xrightarrow{-\frac{7}{78}R_3\to R_3} \begin{bmatrix} 1 & -\frac{1}{2} & \frac{1}{2} & | & \frac{3}{2} \\ 0 & 1 & \frac{5}{7} & | & \frac{29}{7} \\ 0 & 0 & 1 & | & 3 \end{bmatrix} \Rightarrow$

$z = 3$

$y + \left(\frac{5}{7}\right)(3) = \frac{29}{7} \Rightarrow y = 2$

$x - \frac{1}{2}(2) + \frac{1}{2}(3) = \frac{3}{2} \Rightarrow x = 1$

The solution is $\{(1, 2, 3)\}$.

69. $\begin{bmatrix} 1 & -1 & 0 & | & 1 \\ 1 & 0 & -1 & | & -1 \\ 2 & 1 & -1 & | & 3 \end{bmatrix}$

$\xrightarrow[-2R_1+R_3\to R_3]{R_1-R_2\to R_2} \begin{bmatrix} 1 & -1 & 0 & | & 1 \\ 0 & -1 & 1 & | & 2 \\ 0 & 3 & -1 & | & 1 \end{bmatrix}$

$\xrightarrow[3R_2+R_3\to R_3]{-R_2+R_1\to R_1} \begin{bmatrix} 1 & 0 & -1 & | & -1 \\ 0 & -1 & 1 & | & 2 \\ 0 & 0 & 2 & | & 7 \end{bmatrix}$

$\xrightarrow{\frac{1}{2}R_3\to R_3} \begin{bmatrix} 1 & 0 & -1 & | & -1 \\ 0 & -1 & 1 & | & 2 \\ 0 & 0 & 1 & | & \frac{7}{2} \end{bmatrix}$

$\xrightarrow[-R_2+R_3\to R_2]{R_3+R_1\to R_1} \begin{bmatrix} 1 & 0 & 0 & | & \frac{5}{2} \\ 0 & 1 & 0 & | & \frac{3}{2} \\ 0 & 0 & 1 & | & \frac{7}{2} \end{bmatrix} \Rightarrow$

$x = \frac{5}{2}, y = \frac{3}{2}, z = \frac{7}{2}$

The solution is $\left\{\left(\frac{5}{2}, \frac{3}{2}, \frac{7}{2}\right)\right\}$.

71. $\begin{bmatrix} 1 & 1 & -1 & | & 4 \\ 1 & 3 & 5 & | & 10 \\ 3 & 5 & 3 & | & 18 \end{bmatrix}$

$\xrightarrow[-3R_1+R_3\to R_3]{R_1-R_2\to R_2} \begin{bmatrix} 1 & 1 & -1 & | & 4 \\ 0 & -2 & -6 & | & -6 \\ 0 & 2 & 6 & | & 6 \end{bmatrix}$

$\xrightarrow{R_2+R_3\to R_3} \begin{bmatrix} 1 & 1 & -1 & | & 4 \\ 0 & -2 & -6 & | & -6 \\ 0 & 0 & 0 & | & 0 \end{bmatrix}$

(continued on next page)

(continued)

$$\xrightarrow{-\frac{1}{2}R_2 \to R_2}\begin{bmatrix} 1 & 1 & -1 & 4 \\ 0 & 1 & 3 & 3 \\ 0 & 0 & 0 & 0 \end{bmatrix}$$

$$\xrightarrow{R_1 - R_2 \to R_1}\begin{bmatrix} 1 & 0 & -4 & 1 \\ 0 & 1 & 3 & 3 \\ 0 & 0 & 0 & 0 \end{bmatrix} \Rightarrow$$

$x - 4z = 1 \Rightarrow x = 1 + 4z$
$y + 3z = 3 \Rightarrow y = 3 - 3z$
The solution is $\{(1+4z, 3-3z, z)\}$.

73. $\begin{bmatrix} 1 & 2 & -1 & 6 \\ 3 & 1 & 2 & 3 \\ 2 & 5 & 3 & 9 \end{bmatrix}$

$$\xrightarrow[-2R_1+R_3 \to R_3]{-3R_1+R_2 \to R_2}\begin{bmatrix} 1 & 2 & -1 & 6 \\ 0 & -5 & 5 & -15 \\ 0 & 1 & 5 & -3 \end{bmatrix}$$

$$\xrightarrow{-\frac{1}{5}R_2 \to R_2}\begin{bmatrix} 1 & 2 & -1 & 6 \\ 0 & 1 & -1 & 3 \\ 0 & 1 & 5 & -3 \end{bmatrix}$$

$$\xrightarrow[R_2-R_3 \to R_3]{-2R_2+R_1 \to R_1}\begin{bmatrix} 1 & 0 & 1 & 0 \\ 0 & 1 & -1 & 3 \\ 0 & 0 & -6 & 6 \end{bmatrix}$$

$$\xrightarrow{-\frac{1}{6}R_3 \to R_3}\begin{bmatrix} 1 & 0 & 1 & 0 \\ 0 & 1 & -1 & 3 \\ 0 & 0 & 1 & -1 \end{bmatrix}$$

$$\xrightarrow[R_2+R_3 \to R_2]{R_1-R_3 \to R_1}\begin{bmatrix} 1 & 0 & 0 & 1 \\ 0 & 1 & 0 & 2 \\ 0 & 0 & 1 & -1 \end{bmatrix} \Rightarrow$$

$x = 1, y = 2, z = -1$
The solution is $\{(1, 2, -1)\}$.

75. a. There is one solution: $\{(2, 3)\}$.

b. There are infinitely many solutions:
$\{(-2y+2, y, 2)\}$.

c. There is no solution.

77. False. Each row of matrix A has seven entries.

6.1 Applying the Concepts

79. a. $\begin{cases} a = 0.1b + 1000 \\ b = 0.2a + 780 \end{cases}$

b. Rewrite the system as
$\begin{cases} a - 0.1b = 1000 \\ -0.2a + b = 780 \end{cases} \Rightarrow \begin{bmatrix} 1 & -0.1 & 1000 \\ -0.2 & 1 & 780 \end{bmatrix}$

c. $\begin{bmatrix} 1 & -0.1 & 1000 \\ -0.2 & 1 & 780 \end{bmatrix}$

$$\xrightarrow{0.2R_1+R_2 \to R_2}\begin{bmatrix} 1 & -0.1 & 1000 \\ 0 & 0.98 & 980 \end{bmatrix}$$

$$\xrightarrow{\frac{100}{98}R_2 \to R_2}\begin{bmatrix} 1 & -0.1 & 1000 \\ 0 & 1 & 1000 \end{bmatrix}$$

$$\xrightarrow{0.1R_2+R_1 \to R_1}\begin{bmatrix} 1 & 0 & 1100 \\ 0 & 1 & 1000 \end{bmatrix} \Rightarrow$$

$a = 1100, b = 1000$

81. a. $\begin{cases} l = 0.4t + 0.2f + 10,000 \\ t = 0.5l + 0.3t + 20,000 \\ f = 0.5l + 0.05t + 0.35f + 10,000 \end{cases}$

b. Rewrite the system as
$\begin{cases} l - 0.4t - 0.2f = 10,000 \\ -0.5l + 0.7t - 0f = 20,000 \Rightarrow \\ -0.5l - 0.05t + 0.65f = 10,000 \end{cases}$

$\begin{bmatrix} 1 & -0.4 & -0.2 & 10,000 \\ -0.5 & 0.7 & 0 & 20,000 \\ -0.5 & -0.05 & 0.65 & 10,000 \end{bmatrix}$

c. $\begin{bmatrix} 1 & -0.4 & -0.2 & 10,000 \\ -0.5 & 0.7 & 0 & 20,000 \\ -0.5 & -0.05 & 0.65 & 10,000 \end{bmatrix}$

$$\xrightarrow[R_1+2R_3 \to R_3]{R_1+2R_2 \to R_2}\begin{bmatrix} 1 & -0.4 & -0.2 & 10,000 \\ 0 & 1 & -0.2 & 50,000 \\ 0 & -0.5 & 1.1 & 30,000 \end{bmatrix}$$

$$\xrightarrow{R_2+2R_3 \to R_3}\begin{bmatrix} 1 & -0.4 & -0.2 & 10,000 \\ 0 & 1 & -0.2 & 50,000 \\ 0 & 0 & 2 & 110,000 \end{bmatrix}$$

$$\xrightarrow[-\frac{1}{2}R_3 \to R_3]{R_1+0.4R_2 \to R_1}\begin{bmatrix} 1 & 0 & -0.28 & 30,000 \\ 0 & 1 & -0.2 & 50,000 \\ 0 & 0 & 1 & 55,000 \end{bmatrix}$$

$$\xrightarrow[R_2+0.2R_3 \to R_2]{R_1+0.28R_3 \to R_1}\begin{bmatrix} 1 & 0 & 0 & 45,400 \\ 0 & 1 & 0 & 61,000 \\ 0 & 0 & 1 & 55,000 \end{bmatrix}$$

$l = \$45,400, t = \$61,000, f = \$55,000$

83. $T_1 = \dfrac{0+0+300+T_2}{4} \Rightarrow 4T_1 - T_2 = 300$

$T_2 = \dfrac{T_1+T_3+190+200}{4} \Rightarrow -T_1+4T_2-T_3 = 390$

$T_3 = \dfrac{0+0+100+T_2}{4} \Rightarrow -T_2+4T_3 = 100$

(continued on next page)

(*continued*)

$$\begin{bmatrix} 4 & -1 & 0 & | & 300 \\ -1 & 4 & -1 & | & 390 \\ 0 & -1 & 4 & | & 100 \end{bmatrix}$$

$$\xrightarrow{R_1 \leftrightarrow -R_2} \begin{bmatrix} 1 & -4 & 1 & | & -390 \\ 4 & -1 & 0 & | & 300 \\ 0 & -1 & 4 & | & 100 \end{bmatrix}$$

$$\xrightarrow{-4R_1 + R_2 \to R_2} \begin{bmatrix} 1 & -4 & -1 & | & -390 \\ 0 & 15 & -4 & | & 1860 \\ 0 & -1 & 4 & | & 100 \end{bmatrix}$$

$$\xrightarrow{R_2 + R_3 \to R_2} \begin{bmatrix} 1 & -4 & 1 & | & -390 \\ 0 & 14 & 0 & | & 1960 \\ 0 & -1 & 4 & | & 100 \end{bmatrix}$$

$$\xrightarrow{\frac{1}{14}R_2 \to R_2} \begin{bmatrix} 1 & -4 & 1 & | & -390 \\ 0 & 1 & 0 & | & 140 \\ 0 & -1 & 4 & | & 100 \end{bmatrix}$$

$$\xrightarrow{R_2 + R_3 \to R_3} \begin{bmatrix} 1 & -4 & 1 & | & -390 \\ 0 & 1 & 0 & | & 140 \\ 0 & 0 & 4 & | & 240 \end{bmatrix}$$

$$\xrightarrow[\frac{1}{4}R_3 \to R_3]{R_1 + 4R_2 \to R_1} \begin{bmatrix} 1 & 0 & 1 & | & 170 \\ 0 & 1 & 0 & | & 140 \\ 0 & 0 & 1 & | & 60 \end{bmatrix}$$

$$\xrightarrow{R_1 - R_3 \to R_1} \begin{bmatrix} 1 & 0 & 0 & | & 110 \\ 0 & 1 & 0 & | & 140 \\ 0 & 0 & 1 & | & 60 \end{bmatrix} \Rightarrow$$

$$T_1 = 110, \; T_2 = 140, \; T_3 = 60$$

85. a. $\begin{cases} x - y & = 270 - 200 \\ -x + & z = 180 - 300 \\ y - z = 40 + 70 - 60 \end{cases}$

$$\Rightarrow \begin{cases} x - y & = 70 \\ -x + & z = -120 \\ y - z = 50 \end{cases}$$

b.
$$\begin{bmatrix} 1 & -1 & 0 & | & 70 \\ -1 & 0 & 1 & | & -120 \\ 0 & 1 & -1 & | & 50 \end{bmatrix}$$

$$\xrightarrow{R_1 + R_2 \to R_2} \begin{bmatrix} 1 & -1 & 0 & | & 70 \\ 0 & -1 & 1 & | & -50 \\ 0 & 1 & -1 & | & 50 \end{bmatrix}$$

$$\xrightarrow[R_2 + R_3 \to R_3]{R_1 - R_2 \to R_1} \begin{bmatrix} 1 & 0 & -1 & | & 120 \\ 0 & -1 & 1 & | & -50 \\ 0 & 0 & 0 & | & 0 \end{bmatrix} \Rightarrow$$

(*continued on next page*)

$x - z = 120 \Rightarrow x = z + 120$

$-y + z = -50 \Rightarrow y = z + 50$

The solution is $\{(z + 120, z + 50, z)\}$.

There are $300 + 200 + 60 = 560$ cars entering the system, so

$0 \le (z + 120) + (z + 50) + z \le 560$

$0 \le 3z + 170 \le 560$

$0 \le 3z \le 390$

$0 \le z \le 130$

Thus, the system has 131 solutions.

87. $9 = a(-1)^2 + b(-1) + c = a - b + c$

$3 = a(1)^2 + b(1) + c = a + b + c$

$6 = a(2)^2 + b(2) + c = 4a + 2b + c$

$$\begin{bmatrix} 1 & -1 & 1 & | & 9 \\ 1 & 1 & 1 & | & 3 \\ 4 & 2 & 1 & | & 6 \end{bmatrix}$$

$$\xrightarrow[-4R_1 + R_3 \to R_3]{-R_1 + R_2 \to R_2} \begin{bmatrix} 1 & -1 & 1 & | & 9 \\ 0 & 2 & 0 & | & -6 \\ 0 & 6 & -3 & | & -30 \end{bmatrix}$$

$$\xrightarrow[-\frac{1}{3}R_3 \to R_3]{\frac{1}{2}R_2 \to R_2} \begin{bmatrix} 1 & -1 & 1 & | & 9 \\ 0 & 1 & 0 & | & -3 \\ 0 & -2 & 1 & | & 10 \end{bmatrix}$$

$$\xrightarrow{2R_2 + R_3 \to R_3} \begin{bmatrix} 1 & -1 & 1 & | & 9 \\ 0 & 1 & 0 & | & -3 \\ 0 & 0 & 1 & | & 4 \end{bmatrix}$$

$$\xrightarrow{R_1 + R_2 \to R_1} \begin{bmatrix} 1 & 0 & 1 & | & 6 \\ 0 & 1 & 0 & | & -3 \\ 0 & 0 & 1 & | & 4 \end{bmatrix}$$

$$\xrightarrow{R_1 - R_3 \to R_1} \begin{bmatrix} 1 & 0 & 0 & | & 2 \\ 0 & 1 & 0 & | & -3 \\ 0 & 0 & 1 & | & 4 \end{bmatrix}$$

Thus, $a = 2$, $b = -3$, and $c = 4$. The equation is $y = 2x^2 - 3x + 4$.

89. $-3 = a + b\sin\left(\dfrac{\pi}{4}\right) + c\sin\left(2\left(\dfrac{\pi}{4}\right)\right) \Rightarrow$

$-3 = a + \dfrac{\sqrt{2}}{2}b + c \Rightarrow -6 = 2a + \sqrt{2}b + 2c$

$2 = a + b\sin\left(\dfrac{3\pi}{4}\right) + c\sin\left(2\left(\dfrac{3\pi}{4}\right)\right) \Rightarrow$

$2 = a + \dfrac{\sqrt{2}}{2}b - c \Rightarrow 4 = 2a + \sqrt{2}b - 2c$

(*continued on next page*)

(continued)

$$5 = a + b\sin\left(\frac{7\pi}{4}\right) + c\sin\left(2\left(\frac{7\pi}{4}\right)\right) \Rightarrow$$

$$5 = a - \frac{\sqrt{2}}{2}b - c \Rightarrow 10 = 2a - \sqrt{2}b - 2c$$

The system of equations is

$$\begin{cases} 2a + \sqrt{2}b + 2c = -6 \\ 2a + \sqrt{2}b - 2c = 4 \\ 2a - \sqrt{2}b - 2c = 10 \end{cases}.$$

$$\begin{bmatrix} 2 & \sqrt{2} & 2 & | & -6 \\ 2 & \sqrt{2} & -2 & | & 4 \\ 2 & -\sqrt{2} & -2 & | & 10 \end{bmatrix}$$

$$\xrightarrow[R_1 + R_2 \to R_2]{\frac{(R_1 + R_3)}{4} \to R_1} \begin{bmatrix} 1 & 0 & 0 & | & 1 \\ 4 & 2\sqrt{2} & 0 & | & -2 \\ 2 & -\sqrt{2} & -2 & | & 10 \end{bmatrix}$$

$$\xrightarrow[-2R_1 + R_3 \to R_3]{-4R_1 + R_2 \to R_2} \begin{bmatrix} 1 & 0 & 0 & | & 1 \\ 0 & 2\sqrt{2} & 0 & | & -6 \\ 0 & -\sqrt{2} & -2 & | & 8 \end{bmatrix}$$

$$\xrightarrow{\frac{R_2}{2\sqrt{2}} \to R_2} \begin{bmatrix} 1 & 0 & 0 & | & 1 \\ 0 & 1 & 0 & | & -\frac{3\sqrt{2}}{2} \\ 0 & -\sqrt{2} & -2 & | & 8 \end{bmatrix}$$

$$\xrightarrow{\frac{\sqrt{2}R_2 + R_3}{-2} \to R_3} \begin{bmatrix} 1 & 0 & 0 & | & 1 \\ 0 & 1 & 0 & | & -\frac{3\sqrt{2}}{2} \\ 0 & 0 & 1 & | & -\frac{5}{2} \end{bmatrix}$$

The equation is $f(x) = 1 - \frac{3\sqrt{2}}{2}\sin x - \frac{5}{2}\sin(2x)$.

91. $\begin{cases} a(1)^3 + b(1)^2 + c(1) + d = 5 \\ a(-1)^3 + b(-1)^2 + c(-1) + d = 1 \\ a(2)^3 + b(2)^2 + c(2) + d = 7 \\ a(-2)^3 + b(-2)^2 + c(-2) + d = 11 \end{cases} \Rightarrow$

$$\begin{cases} a + b + c + d = 5 \\ -a + b - c + d = 1 \\ 8a + 4b + 2c + d = 7 \\ -8a + 4b - 2c + d = 11 \end{cases}$$

Using a graphing calculator, we have

```
[A]
[[1   1   1   1   5]
 [-1  1  -1   1   1]
 [8   4   2   1   7]
 [-8  4  -2   1  11]]
■
```
```
rref([A])
[[1 0 0 0  -1]
 [0 1 0 0   2]
 [0 0 1 0   3]
 [0 0 0 1   1]]
■
```

The equation is $y = -x^3 + 2x^2 + 3x + 1$.

93. For $1 \le x \le 4$, the system to be solved is

$$\begin{cases} 4a + b = 4 \\ a + b = 2 \end{cases}.$$

$$\begin{bmatrix} 4 & 1 & | & 4 \\ 1 & 1 & | & 2 \end{bmatrix} \xrightarrow{\frac{R_1 - R_2}{3} \to R_1} \begin{bmatrix} 1 & 0 & | & \frac{2}{3} \\ 1 & 1 & | & 2 \end{bmatrix}$$

$$\xrightarrow{R_2 - R_1 \to R_2} \begin{bmatrix} 1 & 0 & | & \frac{2}{3} \\ 0 & 1 & | & \frac{4}{3} \end{bmatrix}$$

The equation is $f(x) = \frac{2}{3}x + \frac{4}{3}$.

For $4 \le x \le 6$, the system to be solved is

$$\begin{cases} 4a + b = 4 \\ 6a + b = 1 \end{cases}.$$

$$\begin{bmatrix} 4 & 1 & | & 4 \\ 6 & 1 & | & 1 \end{bmatrix} \xrightarrow{\frac{R_1 - R_2}{-2} \to R_1} \begin{bmatrix} 1 & 0 & | & -\frac{3}{2} \\ 6 & 1 & | & 1 \end{bmatrix}$$

$$\xrightarrow{R_2 - 6R_1 \to R_2} \begin{bmatrix} 1 & 0 & | & -\frac{3}{2} \\ 0 & 1 & | & 10 \end{bmatrix}$$

The equation is $f(x) = -\frac{3}{2}x + 10$.

The piecewise linear function is

$$f(x) = \begin{cases} \frac{2}{3}x + \frac{4}{3}, & 1 \le x \le 4 \\ -\frac{3}{2}x + 10, & 4 \le x \le 6 \end{cases}$$

6.1 Beyond the Basics

95. $\begin{bmatrix} 1 & 1 & 1 & 1 & | & 0 \\ 1 & 3 & 2 & 4 & | & 0 \\ 2 & 0 & 1 & -1 & | & 0 \end{bmatrix}$

$$\xrightarrow[-2R_1 + R_3 \to R_3]{R_1 - R_2 \to R_2} \begin{bmatrix} 1 & 1 & 1 & 1 & | & 0 \\ 0 & -2 & -1 & -3 & | & 0 \\ 0 & -2 & -1 & -3 & | & 0 \end{bmatrix}$$

$$\xrightarrow[R_2 - R_3 \to R_3]{\frac{1}{2}R_2 + R_1 \to R_1} \begin{bmatrix} 1 & 0 & \frac{1}{2} & -\frac{1}{2} & | & 0 \\ 0 & 1 & \frac{1}{2} & \frac{3}{2} & | & 0 \\ 0 & 0 & 0 & 0 & | & 0 \end{bmatrix}$$

$$\xrightarrow{-\frac{1}{2}R_2 \to R_2} \begin{bmatrix} 1 & 0 & \frac{1}{2} & -\frac{1}{2} & | & 0 \\ 0 & 1 & \frac{1}{2} & \frac{3}{2} & | & 0 \\ 0 & 0 & 0 & 0 & | & 0 \end{bmatrix} \Rightarrow$$

(continued on next page)

(*continued*)

$$y + \frac{1}{2}z + \frac{3}{2}w = 0 \Rightarrow z = -2y - 3w$$

$$x + \frac{1}{2}z - \frac{1}{2}w = 0 \Rightarrow z = w - 2x$$

$$-2y - 3w = w - 2x \Rightarrow x = 2w + y$$

The solution is $\{(y + 2w, \ y, \ -2y - 3w, \ w)\}$.

97. a. $\begin{bmatrix} 1 & 2 & -3 & 1 \\ 1 & 0 & -3 & 2 \\ 0 & 1 & 1 & 0 \\ 2 & 3 & 0 & -2 \end{bmatrix}$

$\xrightarrow{R_1 \leftrightarrow R_4} \begin{bmatrix} 2 & 3 & 0 & -2 \\ 1 & 0 & -3 & 2 \\ 0 & 1 & 1 & 0 \\ 1 & 2 & -3 & 1 \end{bmatrix}$

$\xrightarrow[R_1 - 2R_4 \to R_4]{R_2 - R_4 \to R_2} \begin{bmatrix} 2 & 3 & 0 & -2 \\ 0 & -2 & 0 & 1 \\ 0 & 1 & 1 & 0 \\ 0 & -1 & 6 & -4 \end{bmatrix}$

$\xrightarrow[R_2 + 2R_3 \to R_3]{R_2 - 2R_4 \to R_4} \begin{bmatrix} 2 & 3 & 0 & -2 \\ 0 & -2 & 0 & 1 \\ 0 & 0 & 2 & 1 \\ 0 & 0 & -12 & 9 \end{bmatrix}$

$\xrightarrow{6R_3 + R_4 \to R_4} \begin{bmatrix} 2 & 3 & 0 & -2 \\ 0 & -2 & 0 & 1 \\ 0 & 0 & 2 & 1 \\ 0 & 0 & 0 & 15 \end{bmatrix}$

$\xrightarrow[\substack{\frac{1}{2}R_1 \to R_1 \\ -\frac{1}{2}R_2 \to R_2 \\ \frac{1}{2}R_3 \to R_3 \\ \frac{1}{15}R_4 \to R_4}]{} \begin{bmatrix} 1 & \frac{3}{2} & 0 & -1 \\ 0 & 1 & 0 & -\frac{1}{2} \\ 0 & 0 & 1 & \frac{1}{2} \\ 0 & 0 & 0 & 1 \end{bmatrix} = B$

$\begin{bmatrix} 1 & 2 & -3 & 1 \\ 1 & 0 & -3 & 2 \\ 0 & 1 & 1 & 0 \\ 2 & 3 & 0 & -2 \end{bmatrix}$

$\xrightarrow[2R_1 - R_4 \to R_4]{R_1 - R_2 \to R_2} \begin{bmatrix} 1 & 2 & -3 & 1 \\ 0 & 2 & 0 & -1 \\ 0 & 1 & 1 & 0 \\ 0 & 1 & -6 & 4 \end{bmatrix}$

$\xrightarrow{R_2 - 2R_4 \to R_4} \begin{bmatrix} 1 & 2 & -3 & 1 \\ 0 & 2 & 0 & -1 \\ 0 & 1 & 1 & 0 \\ 0 & 0 & 12 & -9 \end{bmatrix}$

$\xrightarrow[\substack{\frac{1}{2}R_2 \to R_2 \\ \frac{1}{12}R_4 \to R_4}]{} \begin{bmatrix} 1 & 2 & -3 & 1 \\ 0 & 1 & 0 & -\frac{1}{2} \\ 0 & 1 & 1 & 0 \\ 0 & 0 & 1 & -\frac{3}{4} \end{bmatrix}$

$\xrightarrow{R_2 - R_3 \to R_3} \begin{bmatrix} 1 & 2 & -3 & 1 \\ 0 & 1 & 0 & -\frac{1}{2} \\ 0 & 0 & -1 & -\frac{1}{2} \\ 0 & 0 & 1 & -\frac{3}{4} \end{bmatrix}$

$\xrightarrow{R_3 + R_4 \to R_4} \begin{bmatrix} 1 & 2 & -3 & 1 \\ 0 & 1 & 0 & -\frac{1}{2} \\ 0 & 0 & -1 & -\frac{1}{2} \\ 0 & 0 & 0 & -\frac{5}{4} \end{bmatrix}$

$\xrightarrow[-\frac{4}{5}R_4 \to R_4]{-R_3 \to R_3} \begin{bmatrix} 1 & 2 & -3 & 1 \\ 0 & 1 & 0 & -\frac{1}{2} \\ 0 & 0 & 1 & \frac{1}{2} \\ 0 & 0 & 0 & 1 \end{bmatrix} = C$

b. Use a calculator to show that the reduced row-echelon form of all three matrices is

$$\begin{bmatrix} 1 & 0 & 0 & 0 \\ 0 & 1 & 0 & 0 \\ 0 & 0 & 1 & 0 \\ 0 & 0 & 0 & 1 \end{bmatrix}.$$

99. a.

$$\begin{bmatrix} a & b & m \\ c & d & n \end{bmatrix} \xrightarrow[\frac{1}{c}R_2 \to R_2]{\frac{1}{a}R_1 \to R_1} \begin{bmatrix} 1 & \dfrac{b}{a} & \dfrac{m}{a} \\ 1 & \dfrac{d}{c} & \dfrac{n}{c} \end{bmatrix}$$

$$\xrightarrow{R_1 - R_2 \to R_2} \begin{bmatrix} 1 & \dfrac{b}{a} & \dfrac{m}{a} \\ 0 & \dfrac{b}{a}-\dfrac{d}{c} & \dfrac{m}{a}-\dfrac{n}{c} \end{bmatrix}$$

$$= \begin{bmatrix} 1 & \dfrac{b}{a} & \dfrac{m}{a} \\ 0 & \dfrac{bc-ad}{ac} & \dfrac{cm-an}{ac} \end{bmatrix}$$

$$\xrightarrow{\frac{ac}{bc-da}R_2 \to R_2} \begin{bmatrix} 1 & \dfrac{b}{a} & \dfrac{m}{a} \\ 0 & 1 & \dfrac{cm-an}{bc-ad} \end{bmatrix}$$

$$\xrightarrow{-\frac{b}{a}R_2 + R_1 \to R_1} \begin{bmatrix} 1 & 0 & \dfrac{dm-bn}{ad-bc} \\ 0 & 1 & \dfrac{cm-an}{bc-ad} \end{bmatrix}$$

The solution is $\left(\dfrac{dm-bn}{ad-bc}, \dfrac{cm-an}{bc-ad} \right)$.

b. **(i)** There is a unique solution if $bc \neq ad$.

    (ii) There is no solution if $bc = ad$ and

      $\dfrac{m}{b} \neq \dfrac{n}{d}$.

    (iii) There are infinitely many solutions if

      $bc = ad$ and $\dfrac{m}{b} = \dfrac{n}{d}$.

101. Using the hint suggested, we have

$$\begin{bmatrix} 1 & 1 & 1 & 6 \\ 3 & -1 & 3 & 10 \\ 5 & 5 & -4 & 3 \end{bmatrix}$$

$$\xrightarrow[-5R_1 + R_3 \to R_3]{-3R_1 + R_2 \to R_2} \begin{bmatrix} 1 & 1 & 1 & 6 \\ 0 & -4 & 0 & -8 \\ 0 & 0 & -9 & -27 \end{bmatrix}$$

$$\xrightarrow[-\frac{1}{9}R_3 \to R_3]{-\frac{1}{4}R_2 \to R_2} \begin{bmatrix} 1 & 1 & 1 & 6 \\ 0 & 1 & 0 & 2 \\ 0 & 0 & 1 & 3 \end{bmatrix} \Rightarrow$$

$w = 3, v = 2, u + 2 + 3 = 6 \Rightarrow u = 1$

$\log x = 1 \Rightarrow x = 10;$

$\log y = 2 \Rightarrow y = 100$

$\log z = 3 \Rightarrow y = 1000$

The solution is $\{(10, 100, 1000)\}$.

103. Substitute the given points into the given system, subtract c from both sides of each equation, and then write and solve the augmented matrix.

$$\begin{bmatrix} 2 & 1 & -1 & 0 \\ \frac{1}{3} & -\frac{1}{9} & -1 & 0 \\ -3 & -\frac{7}{3} & -1 & 0 \end{bmatrix}$$

$$\xrightarrow{\frac{R_1}{2} \to R_1} \begin{bmatrix} 1 & \frac{1}{2} & -\frac{1}{2} & 0 \\ \frac{1}{3} & -\frac{1}{9} & -1 & 0 \\ -3 & -\frac{7}{3} & -1 & 0 \end{bmatrix}$$

$$\xrightarrow[3R_1 + R_3 \to R_3]{R_1 - 3R_2 \to R_2} \begin{bmatrix} 1 & \frac{1}{2} & -\frac{1}{2} & 0 \\ 0 & \frac{5}{6} & \frac{5}{2} & 0 \\ 0 & -\frac{5}{6} & -\frac{5}{2} & 0 \end{bmatrix}$$

$$\xrightarrow[\frac{5}{3}R_1 + R_3 \to R_3]{\substack{R_1 + \frac{1}{5}R_2 \to R_1 \\ R_2 + R_3 \to R_2}} \begin{bmatrix} 1 & \frac{2}{3} & 0 & 0 \\ 0 & 0 & 0 & 0 \\ \frac{5}{3} & 0 & -\frac{10}{3} & 0 \end{bmatrix}$$

$$\xrightarrow{\frac{3}{5}R_3 \to R_3} \begin{bmatrix} 1 & \frac{2}{3} & 0 & 0 \\ 0 & 0 & 0 & 0 \\ 1 & 0 & -2 & 0 \end{bmatrix}$$

This yields the system

$$\begin{cases} a + \frac{2}{3}b = 0 \\ a - 2c = 0 \end{cases} \Rightarrow \begin{cases} b = -\frac{3}{2}a \\ c = \frac{1}{2}a \end{cases}.$$

Because this holds for values other than $a = b = c = 0$, the points are collinear.

105. Substitute the given points into the given system, subtract d from both sides of each equation, and then write and solve the augmented matrix.

$$\begin{bmatrix} 2 & 1 & -\frac{5}{4} & -1 & 0 \\ -1 & 0 & -\frac{3}{2} & -1 & 0 \\ 4 & 3 & -\frac{7}{4} & -1 & 0 \\ 1 & -2 & 0 & -1 & 0 \end{bmatrix}$$

$$\xrightarrow[\frac{-3R_1 + R_3}{-2} \to R_3]{2R_1 + R_4 \to R_4} \begin{bmatrix} 2 & 1 & -\frac{5}{4} & -1 & 0 \\ -1 & 0 & -\frac{3}{2} & -1 & 0 \\ 1 & 0 & -1 & -1 & 0 \\ 5 & 0 & -\frac{5}{2} & -3 & 0 \end{bmatrix}$$

(continued on next page)

(continued)

$$\xrightarrow{2\left(-\frac{5}{2}R_3+R_4\right)\to R_4}\begin{bmatrix}2&1&-\frac54&-1&|&0\\-1&0&-\frac32&-1&|&0\\1&0&-1&-1&|&0\\5&0&0&-1&|&0\end{bmatrix}$$

$$\xrightarrow[\frac{-R_3+R_4}{4}\to R_3]{R_1-R_2\to R_2}\begin{bmatrix}2&1&-\frac54&-1&|&0\\3&1&\frac14&0&|&0\\1&0&\frac14&0&|&0\\5&0&0&-1&|&0\end{bmatrix}$$

$$\xrightarrow{R_2-R_3\to R_2}\begin{bmatrix}2&1&-\frac54&-1&|&0\\2&1&0&0&|&0\\1&0&\frac14&0&|&0\\5&0&0&-1&|&0\end{bmatrix}$$

$$\xrightarrow{R_1-R_2\to R_1}\begin{bmatrix}0&0&-\frac54&-1&|&0\\2&1&0&0&|&0\\1&0&\frac14&0&|&0\\5&0&0&-1&|&0\end{bmatrix}$$

$$\xrightarrow{R_1+5R_3\to R_1}\begin{bmatrix}5&0&0&-1&|&0\\2&1&0&0&|&0\\1&0&\frac14&0&|&0\\5&0&0&-1&|&0\end{bmatrix}$$

$$\xrightarrow{R_1-R_4\to R_1}\begin{bmatrix}0&0&0&0&|&0\\2&1&0&0&|&0\\1&0&\frac14&0&|&0\\5&0&0&-1&|&0\end{bmatrix}\Rightarrow$$

$2a+b=0\Rightarrow b=-2a$
$a+\frac14 c=0\Rightarrow c=-4a$
$5a-d=0\Rightarrow d=5a$
Because this holds for values other than
$a=b=c=d=0$, the points are collinear.

107. Three points that satisfy the function are $(1,1)$,
$(2,5)$, and $(3,12)$. Find the equation by solving the
system

$$\begin{cases}a(1)^2+b(1)+c=1\\a(2)^2+b(2)+c=5\\a(3)^2+b(3)+c=12\end{cases}\Rightarrow\begin{cases}a+b+c=1\\4a+2b+c=5\\9a+3b+c=12\end{cases}.$$

$$\begin{bmatrix}1&1&1&|&1\\4&2&1&|&5\\9&3&1&|&12\end{bmatrix}$$

$$\xrightarrow[-9R_1+R_3\to R_3]{-4R_1+R_2\to R_2}\begin{bmatrix}1&1&1&|&1\\0&-2&-3&|&1\\0&-6&-8&|&3\end{bmatrix}$$

$$\xrightarrow{-3R_2+R_3\to R_3}\begin{bmatrix}1&1&1&|&1\\0&-2&-3&|&1\\0&0&1&|&0\end{bmatrix}$$

$$\xrightarrow[\frac{R_2+3R_3}{-2}\to R_2]{R_1-R_3\to R_1}\begin{bmatrix}1&1&0&|&1\\0&1&0&|&-\frac12\\0&0&1&|&0\end{bmatrix}$$

$$\xrightarrow{R_1-R_2\to R_1}\begin{bmatrix}1&0&0&|&\frac32\\0&1&0&|&-\frac12\\0&0&1&|&0\end{bmatrix}$$

The equation is $P(n)=\frac32 n^2-\frac12 n$.

6.1 Critical Thinking/Discussion/Writing

109. If $a\ne 0$,
$$\begin{bmatrix}a\\b\end{bmatrix}\xrightarrow[\frac1b R_2\to R_2]{\frac1a R_1\to R_1}\begin{bmatrix}1\\1\end{bmatrix}\xrightarrow{R_1-R_2\to R_2}\begin{bmatrix}1\\0\end{bmatrix}.$$
If $a=0$, $b=0$, $\begin{bmatrix}a\\b\end{bmatrix}=\begin{bmatrix}0\\0\end{bmatrix}$.

110. If the matrix is $\begin{bmatrix}0&0\\0&0\end{bmatrix}$, then it is in reduced
row-echelon form already. If the matrix is
$\begin{bmatrix}1&k\\0&0\end{bmatrix}$ (for k any real number), then it is in
reduced row-echelon form already.

If the matrix is $\begin{bmatrix}0&1\\0&0\end{bmatrix}$ then it is in reduced
row-echelon form already. If the matrix is
$\begin{bmatrix}a&b\\c&d\end{bmatrix}$, then we have

$$\begin{bmatrix}a&b\\c&d\end{bmatrix}\xrightarrow[\frac1c R_2\to R_2]{\frac1a R_1\to R_1}\begin{bmatrix}1&\frac ba\\1&\frac dc\end{bmatrix}$$

(continued on next page)

(*continued*)

$$\xrightarrow{R_1 - R_2 \to R_2} \begin{bmatrix} 1 & \dfrac{b}{a} \\ 0 & \dfrac{b}{a} - \dfrac{d}{c} \end{bmatrix} = \begin{bmatrix} 1 & \dfrac{b}{a} \\ 0 & \dfrac{bc - ad}{ac} \end{bmatrix}$$

$$\xrightarrow{\frac{ac}{bc-ad} R_2 \to R_2} \begin{bmatrix} 1 & \dfrac{b}{a} \\ 0 & 1 \end{bmatrix}$$

$$\xrightarrow{-\frac{b}{a} R_2 + R_1 \to R_1} \begin{bmatrix} 1 & 0 \\ 0 & 1 \end{bmatrix}$$

111. Yes. Use the inverse of the operation used to transform A to B.

112. a. True. For example,

$$\begin{bmatrix} 1 & 0 & 0 & 6 \\ 0 & 1 & 0 & 3 \\ 0 & 0 & 1 & 2 \end{bmatrix} \to \begin{bmatrix} 1 & 0 & 0 & 6 \\ 0 & 0 & 1 & 2 \end{bmatrix}$$

b. False. For example,

$$\begin{bmatrix} 1 & 0 & 0 & 6 \\ 0 & 1 & 0 & 3 \\ 0 & 0 & 1 & 2 \end{bmatrix} \to \begin{bmatrix} 1 & 0 & 6 \\ 0 & 0 & 3 \\ 0 & 1 & 2 \end{bmatrix}$$

6.1 Maintaining Skills

113. $3(x-1) = 5 - x \Rightarrow 3x - 3 = 5 - x \Rightarrow$
$4x = 8 \Rightarrow x = 2$
Solution set: $\{2\}$

115. $-3(x+4) + 2 = 8 - x \Rightarrow -3x - 12 + 2 = 8 - x \Rightarrow$
$-3x - 10 = 8 - x \Rightarrow -2x = 18 \Rightarrow x = -9$
Solution set: $\{-9\}$

Be sure to check your solutions in the original equations in exercises 117–120.

117. $\begin{cases} 2x - y = 5 & (1) \\ x + 2y = 25 & (2) \end{cases}$

From equation (1) we have $y = 2x - 5$. Substituting this expression in equation (2) gives
$x + 2(2x - 5) = 25 \Rightarrow x + 4x - 10 = 25 \Rightarrow$
$5x = 35 \Rightarrow x = 7$
Substitute $x = 7$ in equation (1) and solve for y.
$2(7) - y = 5 \Rightarrow 14 - y = 5 \Rightarrow y = 9$
Solution set: $\{(7, 9)\}$

119. $\begin{cases} x + 3y = 6 & (1) \\ 2x + 6y = 8 & (2) \end{cases}$

Multiply equation (1) by -2, then add the two equations.

$$\begin{array}{r} -2x - 6y = -12 \\ 2x + 6y = 8 \\ \hline 0 = -4 \quad \text{False} \end{array}$$

Solution set: $\varnothing$

121. True

123. True. This is an example of the distributive property.

6.2 Matrix Algebra

6.2 Practice Problems

1. $\begin{bmatrix} 1 & 2x - y \\ x + y & 5 \end{bmatrix} = \begin{bmatrix} 1 & 1 \\ 2 & 5 \end{bmatrix} \Rightarrow$
$\begin{cases} 2x - y = 1 \\ x + y = 2 \end{cases} \Rightarrow x = 1, \ y = 1$

2. $\begin{bmatrix} 2 & -1 & 4 \\ 5 & 0 & 9 \end{bmatrix} + \begin{bmatrix} -8 & 2 & 9 \\ 7 & 3 & 6 \end{bmatrix} = \begin{bmatrix} -6 & 1 & 13 \\ 12 & 3 & 15 \end{bmatrix}$

3. $2A - 3B = 2\begin{bmatrix} 7 & -4 \\ 3 & 6 \\ 0 & -2 \end{bmatrix} - 3\begin{bmatrix} 1 & -3 \\ 2 & 2 \\ 5 & 8 \end{bmatrix}$

$= \begin{bmatrix} 14 & -8 \\ 6 & 12 \\ 0 & -4 \end{bmatrix} - \begin{bmatrix} 3 & -9 \\ 6 & 6 \\ 15 & 24 \end{bmatrix}$

$= \begin{bmatrix} 11 & 1 \\ 0 & 6 \\ -15 & -28 \end{bmatrix}$

4. $5\begin{bmatrix} 1 & -1 \\ 3 & 5 \end{bmatrix} + 3X = 2\begin{bmatrix} 2 & 7 \\ -3 & -5 \end{bmatrix} \Rightarrow$

$\begin{bmatrix} 5 & -5 \\ 15 & 25 \end{bmatrix} + 3X = \begin{bmatrix} 4 & 14 \\ -6 & -10 \end{bmatrix} \Rightarrow$

$3X = \begin{bmatrix} 4 & 14 \\ -6 & -10 \end{bmatrix} - \begin{bmatrix} 5 & -5 \\ 15 & 25 \end{bmatrix} = \begin{bmatrix} -1 & 19 \\ -21 & -35 \end{bmatrix} \Rightarrow$

$X = \begin{bmatrix} -\dfrac{1}{3} & \dfrac{19}{3} \\ -7 & -\dfrac{35}{3} \end{bmatrix}$

5. Yes, the product matrix AB is defined. A is a 2×3 matrix, while B is a 3×1 matrix. The product matrix has order 2×1.

6. $N = \begin{bmatrix} S & C & M \end{bmatrix} = \begin{bmatrix} 10 & 30 & 45 \end{bmatrix}$; $P = \begin{bmatrix} 41 \\ 26 \\ 19 \end{bmatrix}$

$$NP = \begin{bmatrix} 10 & 30 & 45 \end{bmatrix} \begin{bmatrix} 41 \\ 26 \\ 19 \end{bmatrix} = 10(41) + 30(26) + 45(19) = \$2045 \text{ thousand}$$

7. $\begin{bmatrix} 3 & -1 & 2 & 7 \end{bmatrix} \begin{bmatrix} -2 \\ 0 \\ 1 \\ 5 \end{bmatrix} = \begin{bmatrix} 3(-2) - 1(0) + 2(1) + 7(5) \end{bmatrix} = \begin{bmatrix} 31 \end{bmatrix}$

8. AB is not defined because A is a 2×2 matrix and B is a 3×2 matrix.

$$BA = \begin{bmatrix} 8 & 1 \\ -2 & 6 \\ 0 & 4 \end{bmatrix} \begin{bmatrix} 5 & 0 \\ 2 & -1 \end{bmatrix} = \begin{bmatrix} 8(5)+1(2) & 8(0)+1(-1) \\ -2(5)+6(2) & -2(0)+6(-1) \\ 0(5)+4(2) & 0(0)+4(-1) \end{bmatrix} = \begin{bmatrix} 42 & -1 \\ 2 & -6 \\ 8 & -4 \end{bmatrix}$$

9. $AB = \begin{bmatrix} 7 & 1 \\ 0 & 3 \end{bmatrix} \begin{bmatrix} 2 & -1 \\ 4 & 4 \end{bmatrix} = \begin{bmatrix} 7(2)+1(4) & 7(-1)+1(4) \\ 0(2)+3(4) & 0(-1)+3(4) \end{bmatrix} = \begin{bmatrix} 18 & -3 \\ 12 & 12 \end{bmatrix}$

$BA = \begin{bmatrix} 2 & -1 \\ 4 & 4 \end{bmatrix} \begin{bmatrix} 7 & 1 \\ 0 & 3 \end{bmatrix} = \begin{bmatrix} 2(7)-1(0) & 2(1)-1(3) \\ 4(7)+4(0) & 4(1)+4(3) \end{bmatrix} = \begin{bmatrix} 14 & -1 \\ 28 & 16 \end{bmatrix}$

10. $AD = \begin{bmatrix} 0 & 1 \\ 1 & 0.25 \end{bmatrix} \begin{bmatrix} 0 & 4 & 4 & 1 & 1 & 0 \\ 0 & 0 & 1 & 1 & 6 & 6 \end{bmatrix}$

$= \begin{bmatrix} 0(0)+1(0) & 0(4)+1(0) & 0(4)+1(1) & 0(1)+1(1) & 0(1)+1(6) & 0(0)+1(6) \\ 1(0)+0.25(0) & 1(4)+0.25(0) & 1(4)+0.25(1) & 1(1)+0.25(1) & 1(1)+0.25(6) & 1(0)+0.25(6) \end{bmatrix}$

$= \begin{bmatrix} 0 & 0 & 1 & 1 & 6 & 6 \\ 0 & 4 & 4.25 & 1.25 & 2.5 & 1.5 \end{bmatrix}$

6.2 Concepts and Vocabulary

1. Two $m \times n$ matrices $A = \begin{bmatrix} a_{ij} \end{bmatrix}$ and $B = \begin{bmatrix} b_{ij} \end{bmatrix}$ are equal if $\underline{a_{ij} = b_{ij}}$ for all i and j.

3. The product of a $1 \times n$ matrix A and $n \times 1$ matrix B is $\underline{1 \times 1}$ matrix.

5. False. It is possible that $AB = BA$, but not necessarily true.

7. False. For example, suppose $A = \begin{bmatrix} 0 & 1 \\ 0 & 0 \end{bmatrix}$. Then $A^2 = \begin{bmatrix} 0(0)+1(0) & 0(1)+1(0) \\ 0(0)+0(0) & 0(0)+0(0) \end{bmatrix} = \begin{bmatrix} 0 & 0 \\ 0 & 0 \end{bmatrix}$

6.2 Building Skills

9. $\begin{bmatrix} 2 \\ -3 \end{bmatrix} = \begin{bmatrix} x \\ u \end{bmatrix} \Rightarrow x = 2, u = -3$

11. $\begin{bmatrix} 2 & x \\ y & -3 \end{bmatrix} = \begin{bmatrix} 2 & -1 \\ 3 & -3 \end{bmatrix} \Rightarrow x = -1, y = 3$

13. $\begin{bmatrix} 2x-3y & -4 \\ 5 & 3x+y \end{bmatrix} = \begin{bmatrix} 1 & -4 \\ 5 & 7 \end{bmatrix} \Rightarrow$

$\begin{cases} 2x-3y=1 \\ 3x+y=7 \end{cases} \Rightarrow x=2, y=1$

15. $\begin{bmatrix} x-y & 1 & 2 \\ 4 & 3x-2y & 3 \\ 5 & 6 & 5x-10y \end{bmatrix} = \begin{bmatrix} -1 & 1 & 2 \\ 4 & -1 & 3 \\ 5 & 6 & 6 \end{bmatrix} \Rightarrow$

$\begin{cases} x-y=-1 \\ 3x-2y=-1 \\ 5x-10y=6 \end{cases} \Rightarrow \left[\begin{array}{rr|r} 1 & -1 & -1 \\ 3 & -2 & -1 \\ 5 & -10 & 6 \end{array}\right]$

$\xrightarrow[-5R_1+R_3 \to R_3]{-3R_1+R_2 \to R_2} \left[\begin{array}{rr|r} 1 & -1 & -1 \\ 0 & 1 & 2 \\ 0 & -5 & 11 \end{array}\right]$

$\xrightarrow{5R_2+R_3 \to R_3} \left[\begin{array}{rr|r} 1 & -1 & -1 \\ 0 & 1 & 2 \\ 0 & 0 & 21 \end{array}\right] \Rightarrow 0=21$

There is no solution.

17. a. $A+B = \begin{bmatrix} 1 & 2 \\ 3 & 4 \end{bmatrix} + \begin{bmatrix} -1 & 0 \\ 2 & -3 \end{bmatrix} = \begin{bmatrix} 0 & 2 \\ 5 & 1 \end{bmatrix}$

b. $A-B = \begin{bmatrix} 1 & 2 \\ 3 & 4 \end{bmatrix} - \begin{bmatrix} -1 & 0 \\ 2 & -3 \end{bmatrix} = \begin{bmatrix} 2 & 2 \\ 1 & 7 \end{bmatrix}$

c. $-3A = -3\begin{bmatrix} 1 & 2 \\ 3 & 4 \end{bmatrix} = \begin{bmatrix} -3 & -6 \\ -9 & -12 \end{bmatrix}$

d. $3A-2B = 3\begin{bmatrix} 1 & 2 \\ 3 & 4 \end{bmatrix} - 2\begin{bmatrix} -1 & 0 \\ 2 & -3 \end{bmatrix}$

$= \begin{bmatrix} 3 & 6 \\ 9 & 12 \end{bmatrix} - \begin{bmatrix} -2 & 0 \\ 4 & -6 \end{bmatrix} = \begin{bmatrix} 5 & 6 \\ 5 & 18 \end{bmatrix}$

e. $(A+B)^2 = \left(\begin{bmatrix} 1 & 2 \\ 3 & 4 \end{bmatrix} + \begin{bmatrix} -1 & 0 \\ 2 & -3 \end{bmatrix}\right)^2$

$= \left(\begin{bmatrix} 0 & 2 \\ 5 & 1 \end{bmatrix}\right)^2 = \begin{bmatrix} 0 & 2 \\ 5 & 1 \end{bmatrix}\begin{bmatrix} 0 & 2 \\ 5 & 1 \end{bmatrix}$

$= \begin{bmatrix} 0(0)+2(5) & 0(2)+2(1) \\ 5(0)+1(5) & 5(2)+1(1) \end{bmatrix}$

$= \begin{bmatrix} 10 & 2 \\ 5 & 11 \end{bmatrix}$

f. $A^2 = \begin{bmatrix} 1 & 2 \\ 3 & 4 \end{bmatrix}\begin{bmatrix} 1 & 2 \\ 3 & 4 \end{bmatrix}$

$= \begin{bmatrix} 1(1)+2(3) & 1(2)+2(4) \\ 3(1)+4(3) & 3(2)+4(4) \end{bmatrix} = \begin{bmatrix} 7 & 10 \\ 15 & 22 \end{bmatrix}$

$B^2 = \begin{bmatrix} -1 & 0 \\ 2 & -3 \end{bmatrix}\begin{bmatrix} -1 & 0 \\ 2 & -3 \end{bmatrix}$

$= \begin{bmatrix} -1(-1)+0(2) & -1(0)+0(-3) \\ 2(-1)-3(2) & 2(0)-3(-3) \end{bmatrix}$

$= \begin{bmatrix} 1 & 0 \\ -8 & 9 \end{bmatrix}$

$A^2-B^2 = \begin{bmatrix} 7 & 10 \\ 15 & 22 \end{bmatrix} - \begin{bmatrix} 1 & 0 \\ -8 & 9 \end{bmatrix} = \begin{bmatrix} 6 & 10 \\ 23 & 13 \end{bmatrix}$

19. a. $A+B$ is not defined.

b. $A-B$ is not defined.

c. $-3A = -3\begin{bmatrix} 2 & 3 \\ -4 & 5 \end{bmatrix} = \begin{bmatrix} -6 & -9 \\ 12 & -15 \end{bmatrix}$

d. $3A-2B$ is not defined

e. $(A+B)^2$ is not defined.

f. A^2-B^2 is not defined.

21. a. $A+B = \begin{bmatrix} 4 & 0 & -1 \\ -2 & 5 & 2 \\ 0 & 0 & 1 \end{bmatrix} + \begin{bmatrix} 3 & 1 & 0 \\ 1 & -4 & 2 \\ 2 & 1 & 3 \end{bmatrix}$

$= \begin{bmatrix} 7 & 1 & -1 \\ -1 & 1 & 4 \\ 2 & 1 & 4 \end{bmatrix}$

b. $A-B = \begin{bmatrix} 4 & 0 & -1 \\ -2 & 5 & 2 \\ 0 & 0 & 1 \end{bmatrix} - \begin{bmatrix} 3 & 1 & 0 \\ 1 & -4 & 2 \\ 2 & 1 & 3 \end{bmatrix}$

$= \begin{bmatrix} 1 & -1 & -1 \\ -3 & 9 & 0 \\ -2 & -1 & -2 \end{bmatrix}$

c. $-3A = -3\begin{bmatrix} 4 & 0 & -1 \\ -2 & 5 & 2 \\ 0 & 0 & 1 \end{bmatrix} = \begin{bmatrix} -12 & 0 & 3 \\ 6 & -15 & -6 \\ 0 & 0 & -3 \end{bmatrix}$

d. $3A - 2B = 3\begin{bmatrix} 4 & 0 & -1 \\ -2 & 5 & 2 \\ 0 & 0 & 1 \end{bmatrix} - 2\begin{bmatrix} 3 & 1 & 0 \\ 1 & -4 & 2 \\ 2 & 1 & 3 \end{bmatrix} = \begin{bmatrix} 12 & 0 & -3 \\ -6 & 15 & 6 \\ 0 & 0 & 3 \end{bmatrix} - \begin{bmatrix} 6 & 2 & 0 \\ 2 & -8 & 4 \\ 4 & 2 & 6 \end{bmatrix} = \begin{bmatrix} 6 & -2 & -3 \\ -8 & 23 & 2 \\ -4 & -2 & -3 \end{bmatrix}$

e. $(A+B)^2 = \begin{bmatrix} 7 & 1 & -1 \\ -1 & 1 & 4 \\ 2 & 1 & 4 \end{bmatrix}\begin{bmatrix} 7 & 1 & -1 \\ -1 & 1 & 4 \\ 2 & 1 & 4 \end{bmatrix} = \begin{bmatrix} 7(7)+1(-1)-1(2) & 7(1)+1(1)-1(1) & 7(-1)+1(4)-1(4) \\ -1(7)+1(-1)+4(2) & -1(1)+1(1)+4(1) & -1(-1)+1(4)+4(4) \\ 2(7)+1(-1)+4(2) & 2(1)+1(1)+4(1) & 2(-1)+1(4)+4(4) \end{bmatrix}$

$= \begin{bmatrix} 46 & 7 & -7 \\ 0 & 4 & 21 \\ 21 & 7 & 18 \end{bmatrix}$

f. $A^2 = \begin{bmatrix} 4 & 0 & -1 \\ -2 & 5 & 2 \\ 0 & 0 & 1 \end{bmatrix}\begin{bmatrix} 4 & 0 & -1 \\ -2 & 5 & 2 \\ 0 & 0 & 1 \end{bmatrix} = \begin{bmatrix} 4(4)+0(-2)-1(0) & 4(0)+0(5)-1(0) & 4(-1)+0(2)-1(1) \\ -2(4)+5(-2)+2(0) & -2(0)+5(5)+2(0) & -2(-1)+5(2)+2(1) \\ 0(4)+0(-2)+1(0) & 0(0)+0(5)+1(0) & 0(-1)+0(2)+1(1) \end{bmatrix}$

$= \begin{bmatrix} 16 & 0 & -5 \\ -18 & 25 & 14 \\ 0 & 0 & 1 \end{bmatrix}$

$B^2 = \begin{bmatrix} 3 & 1 & 0 \\ 1 & -4 & 2 \\ 2 & 1 & 3 \end{bmatrix}\begin{bmatrix} 3 & 1 & 0 \\ 1 & -4 & 2 \\ 2 & 1 & 3 \end{bmatrix} = \begin{bmatrix} 3(3)+1(1)+0(2) & 3(1)+1(-4)+0(1) & 3(0)+1(2)+0(3) \\ 1(3)-4(1)+2(2) & 1(1)-4(-4)+2(1) & 1(0)-4(2)+2(3) \\ 2(3)+1(1)+3(2) & 2(1)+1(-4)+3(1) & 2(0)+1(2)+3(3) \end{bmatrix} = \begin{bmatrix} 10 & -1 & 2 \\ 3 & 19 & -2 \\ 13 & 1 & 11 \end{bmatrix}$

$A^2 - B^2 = \begin{bmatrix} 16 & 0 & -5 \\ -18 & 25 & 14 \\ 0 & 0 & 1 \end{bmatrix} - \begin{bmatrix} 10 & -1 & 2 \\ 3 & 19 & -2 \\ 13 & 1 & 11 \end{bmatrix} = \begin{bmatrix} 6 & 1 & -7 \\ -21 & 6 & 16 \\ -13 & -1 & -10 \end{bmatrix}$

23. a. $A + B = \begin{bmatrix} 1 & 2 & -3 \\ 3 & 4 & 5 \\ 2 & -1 & 0 \end{bmatrix} + \begin{bmatrix} 3 & 1 & 0 \\ 1 & -4 & 2 \\ 2 & 1 & 3 \end{bmatrix} = \begin{bmatrix} 4 & 3 & -3 \\ 4 & 0 & 7 \\ 4 & 0 & 3 \end{bmatrix}$

b. $A - B = \begin{bmatrix} 1 & 2 & -3 \\ 3 & 4 & 5 \\ 2 & -1 & 0 \end{bmatrix} - \begin{bmatrix} 3 & 1 & 0 \\ 1 & -4 & 2 \\ 2 & 1 & 3 \end{bmatrix} = \begin{bmatrix} -2 & 1 & -3 \\ 2 & 8 & 3 \\ 0 & -2 & -3 \end{bmatrix}$

c. $-3A = -3\begin{bmatrix} 1 & 2 & -3 \\ 3 & 4 & 5 \\ 2 & -1 & 0 \end{bmatrix} = \begin{bmatrix} -3 & -6 & 9 \\ -9 & -12 & -15 \\ -6 & 3 & 0 \end{bmatrix}$

d. $3A - 2B = 3\begin{bmatrix} 1 & 2 & -3 \\ 3 & 4 & 5 \\ 2 & -1 & 0 \end{bmatrix} - 2\begin{bmatrix} 3 & 1 & 0 \\ 1 & -4 & 2 \\ 2 & 1 & 3 \end{bmatrix} = \begin{bmatrix} 3 & 6 & -9 \\ 9 & 12 & 15 \\ 6 & -3 & 0 \end{bmatrix} - \begin{bmatrix} 6 & 2 & 0 \\ 2 & -8 & 4 \\ 4 & 2 & 6 \end{bmatrix} = \begin{bmatrix} -3 & 4 & -9 \\ 7 & 20 & 11 \\ 2 & -5 & -6 \end{bmatrix}$

e. $(A+B)^2 = \begin{bmatrix} 4 & 3 & -3 \\ 4 & 0 & 7 \\ 4 & 0 & 3 \end{bmatrix}\begin{bmatrix} 4 & 3 & -3 \\ 4 & 0 & 7 \\ 4 & 0 & 3 \end{bmatrix} = \begin{bmatrix} 4(4)+3(4)-3(4) & 4(3)+3(0)-3(0) & 4(-3)+3(7)-3(3) \\ 4(4)+0(4)+7(4) & 4(3)+0(0)+7(0) & 4(-3)+0(7)+7(3) \\ 4(4)+0(4)+3(4) & 4(3)+0(0)+3(0) & 4(-3)+0(7)+3(3) \end{bmatrix}$

$= \begin{bmatrix} 16 & 12 & 0 \\ 44 & 12 & 9 \\ 28 & 12 & -3 \end{bmatrix}$

f. $A^2 = \begin{bmatrix} 1 & 2 & -3 \\ 3 & 4 & 5 \\ 2 & -1 & 0 \end{bmatrix}\begin{bmatrix} 1 & 2 & -3 \\ 3 & 4 & 5 \\ 2 & -1 & 0 \end{bmatrix} = \begin{bmatrix} 1(1)+2(3)-3(2) & 1(2)+2(4)-3(-1) & 1(-3)+2(5)-3(0) \\ 3(1)+4(3)+5(2) & 3(2)+4(4)+5(-1) & 3(-3)+4(5)+5(0) \\ 2(1)-1(3)+0(2) & 2(2)-1(4)+0(-1) & 2(-3)-1(5)+0(0) \end{bmatrix}$

$= \begin{bmatrix} 1 & 13 & 7 \\ 25 & 17 & 11 \\ -1 & 0 & -11 \end{bmatrix}$

$B^2 = \begin{bmatrix} 3 & 1 & 0 \\ 1 & -4 & 2 \\ 2 & 1 & 3 \end{bmatrix}\begin{bmatrix} 3 & 1 & 0 \\ 1 & -4 & 2 \\ 2 & 1 & 3 \end{bmatrix} = \begin{bmatrix} 3(3)+1(1)+0(2) & 3(1)+1(-4)+0(1) & 3(0)+1(2)+0(3) \\ 1(3)-4(1)+2(2) & 1(1)-4(-4)+2(1) & 1(0)-4(2)+2(3) \\ 2(3)+1(1)+3(2) & 2(1)+1(-4)+3(1) & 2(0)+1(2)+3(3) \end{bmatrix} = \begin{bmatrix} 10 & -1 & 2 \\ 3 & 19 & -2 \\ 13 & 1 & 11 \end{bmatrix}$

$A^2 - B^2 = \begin{bmatrix} 1 & 13 & 7 \\ 25 & 17 & 11 \\ -1 & 0 & -11 \end{bmatrix} - \begin{bmatrix} 10 & -1 & 2 \\ 3 & 19 & -2 \\ 13 & 1 & 11 \end{bmatrix} = \begin{bmatrix} -9 & 14 & 5 \\ 22 & -2 & 13 \\ -14 & -1 & -22 \end{bmatrix}$

25. $\begin{bmatrix} 2 & 3 & -1 \\ 1 & -2 & 4 \end{bmatrix} + X = \begin{bmatrix} -2 & 1 & 0 \\ 2 & 3 & 4 \end{bmatrix} \Rightarrow X = \begin{bmatrix} -2 & 1 & 0 \\ 2 & 3 & 4 \end{bmatrix} - \begin{bmatrix} 2 & 3 & -1 \\ 1 & -2 & 4 \end{bmatrix} = \begin{bmatrix} -4 & -2 & 1 \\ 1 & 5 & 0 \end{bmatrix}$

27. $2X - \begin{bmatrix} 2 & 3 & -1 \\ 1 & -2 & 4 \end{bmatrix} = \begin{bmatrix} -2 & 1 & 0 \\ 2 & 3 & 4 \end{bmatrix} \Rightarrow 2X = \begin{bmatrix} -2 & 1 & 0 \\ 2 & 3 & 4 \end{bmatrix} + \begin{bmatrix} 2 & 3 & -1 \\ 1 & -2 & 4 \end{bmatrix} \Rightarrow$

$2X = \begin{bmatrix} 0 & 4 & -1 \\ 3 & 1 & 8 \end{bmatrix} \Rightarrow X = \begin{bmatrix} 0 & 2 & -\frac{1}{2} \\ \frac{3}{2} & \frac{1}{2} & 4 \end{bmatrix}$

29. $2X + 3\begin{bmatrix} 2 & 3 & -1 \\ 1 & -2 & 4 \end{bmatrix} = \begin{bmatrix} -2 & 1 & 0 \\ 2 & 3 & 4 \end{bmatrix} \Rightarrow 2X + \begin{bmatrix} 6 & 9 & -3 \\ 3 & -6 & 12 \end{bmatrix} = \begin{bmatrix} -2 & 1 & 0 \\ 2 & 3 & 4 \end{bmatrix} \Rightarrow$

$2X = \begin{bmatrix} -2 & 1 & 0 \\ 2 & 3 & 4 \end{bmatrix} - \begin{bmatrix} 6 & 9 & -3 \\ 3 & -6 & 12 \end{bmatrix} \Rightarrow 2X = \begin{bmatrix} -8 & -8 & 3 \\ -1 & 9 & -8 \end{bmatrix} \Rightarrow X = \begin{bmatrix} -4 & -4 & \frac{3}{2} \\ -\frac{1}{2} & \frac{9}{2} & -4 \end{bmatrix}$

31. $2\begin{bmatrix} 2 & 3 & -1 \\ 1 & -2 & 4 \end{bmatrix} + 3\begin{bmatrix} -2 & 1 & 0 \\ 2 & 3 & 4 \end{bmatrix} + 4X = 0 \Rightarrow \begin{bmatrix} 4 & 6 & -2 \\ 2 & -4 & 8 \end{bmatrix} + \begin{bmatrix} -6 & 3 & 0 \\ 6 & 9 & 12 \end{bmatrix} = -4X \Rightarrow$

$\begin{bmatrix} -2 & 9 & -2 \\ 8 & 5 & 20 \end{bmatrix} = -4X \Rightarrow X = \begin{bmatrix} \frac{1}{2} & -\frac{9}{4} & \frac{1}{2} \\ -2 & -\frac{5}{4} & -5 \end{bmatrix}$

33. a. $\begin{bmatrix} 1 & 2 \\ 3 & 4 \end{bmatrix}\begin{bmatrix} -2 & 1 \\ 3 & 5 \end{bmatrix} = \begin{bmatrix} 1(-2)+2(3) & 1(1)+2(5) \\ 3(-2)+4(3) & 3(1)+4(5) \end{bmatrix} = \begin{bmatrix} 4 & 11 \\ 6 & 23 \end{bmatrix}$

b. $\begin{bmatrix} -2 & 1 \\ 3 & 5 \end{bmatrix}\begin{bmatrix} 1 & 2 \\ 3 & 4 \end{bmatrix} = \begin{bmatrix} -2(1)+1(3) & -2(2)+1(4) \\ 3(1)+5(3) & 3(2)+5(4) \end{bmatrix} = \begin{bmatrix} 1 & 0 \\ 18 & 26 \end{bmatrix}$

35. a. $\begin{bmatrix} 2 & -1 & 0 \\ -3 & 1 & 2 \end{bmatrix}\begin{bmatrix} 1 & 5 \\ -2 & 3 \\ 4 & 0 \end{bmatrix} = \begin{bmatrix} 2(1)-1(-2)+0(4) & 2(5)-1(3)+0(0) \\ -3(1)+1(-2)+2(4) & -3(5)+1(3)+2(0) \end{bmatrix} = \begin{bmatrix} 4 & 7 \\ 3 & -12 \end{bmatrix}$

b. $\begin{bmatrix} 1 & 5 \\ -2 & 3 \\ 4 & 0 \end{bmatrix}\begin{bmatrix} 2 & -1 & 0 \\ -3 & 1 & 2 \end{bmatrix} = \begin{bmatrix} 1(2)+5(-3) & 1(-1)+5(1) & 1(0)+5(2) \\ -2(2)+3(-3) & -2(-1)+3(1) & -2(0)+3(2) \\ 4(2)+0(-3) & 4(-1)+0(1) & 4(0)+0(2) \end{bmatrix} = \begin{bmatrix} -13 & 4 & 10 \\ -13 & 5 & 6 \\ 8 & -4 & 0 \end{bmatrix}$

37. a. $\begin{bmatrix} 2 & 3 & 5 \end{bmatrix}\begin{bmatrix} 1 \\ -2 \\ 4 \end{bmatrix} = \begin{bmatrix} 2(1)+3(-2)+5(4) \end{bmatrix} = \begin{bmatrix} 16 \end{bmatrix}$

b. $\begin{bmatrix} 1 \\ -2 \\ 4 \end{bmatrix}\begin{bmatrix} 2 & 3 & 5 \end{bmatrix} = \begin{bmatrix} 1(2) & 1(3) & 1(5) \\ -2(2) & -2(3) & -2(5) \\ 4(2) & 4(3) & 4(5) \end{bmatrix} = \begin{bmatrix} 2 & 3 & 5 \\ -4 & -6 & -10 \\ 8 & 12 & 20 \end{bmatrix}$

39. a. $\begin{bmatrix} 1 & 2 & 3 \end{bmatrix}\begin{bmatrix} 1 & 2 & -1 \\ 0 & 3 & 1 \\ 2 & 0 & -3 \end{bmatrix} = \begin{bmatrix} 1(1)+2(0)+3(2) & 1(2)+2(3)+3(0) & 1(-1)+2(1)+3(-3) \end{bmatrix} = \begin{bmatrix} 7 & 8 & -8 \end{bmatrix}$

b. The product BA is not defined.

41. a. $\begin{bmatrix} 2 & 0 & 1 \\ 1 & 4 & 2 \\ 3 & -1 & 0 \end{bmatrix}\begin{bmatrix} 3 & 1 & 0 \\ -1 & 2 & 0 \\ 4 & 5 & 2 \end{bmatrix} = \begin{bmatrix} 2(3)+0(-1)+1(4) & 2(1)+0(2)+1(5) & 2(0)+0(0)+1(2) \\ 1(3)+4(-1)+2(4) & 1(1)+4(2)+2(5) & 1(0)+4(0)+2(2) \\ 3(3)-1(-1)+0(4) & 3(1)-1(2)+0(5) & 3(0)-1(0)+0(2) \end{bmatrix} = \begin{bmatrix} 10 & 7 & 2 \\ 7 & 19 & 4 \\ 10 & 1 & 0 \end{bmatrix}$

b. $\begin{bmatrix} 3 & 1 & 0 \\ -1 & 2 & 0 \\ 4 & 5 & 2 \end{bmatrix}\begin{bmatrix} 2 & 0 & 1 \\ 1 & 4 & 2 \\ 3 & -1 & 0 \end{bmatrix} = \begin{bmatrix} 3(2)+1(1)+0(3) & 3(0)+1(4)+0(-1) & 3(1)+1(2)+0(0) \\ -1(2)+2(1)+0(3) & -1(0)+2(4)+0(-1) & -1(1)+2(2)+0(0) \\ 4(2)+5(1)+2(3) & 4(0)+5(4)+2(-1) & 4(1)+5(2)+2(0) \end{bmatrix} = \begin{bmatrix} 7 & 4 & 5 \\ 0 & 8 & 3 \\ 19 & 18 & 14 \end{bmatrix}$

43. $AB = \begin{bmatrix} 3 & 1 & 2 \\ 0 & 4 & 3 \\ 1 & -2 & 2 \end{bmatrix}\begin{bmatrix} 2 & 5 & 0 \\ 1 & 2 & -1 \\ 3 & 0 & 2 \end{bmatrix} = \begin{bmatrix} 3(2)+1(1)+2(3) & 3(5)+1(2)+2(0) & 3(0)+1(-1)+2(2) \\ 0(2)+4(1)+3(3) & 0(5)+4(2)+3(0) & 0(0)+4(-1)+3(2) \\ 1(2)-2(1)+2(3) & 1(5)-2(2)+2(0) & 1(0)-2(-1)+2(2) \end{bmatrix} = \begin{bmatrix} 13 & 17 & 3 \\ 13 & 8 & 2 \\ 6 & 1 & 6 \end{bmatrix}$

$BA = \begin{bmatrix} 2 & 5 & 0 \\ 1 & 2 & -1 \\ 3 & 0 & 2 \end{bmatrix}\begin{bmatrix} 3 & 1 & 2 \\ 0 & 4 & 3 \\ 1 & -2 & 2 \end{bmatrix} = \begin{bmatrix} 2(3)+5(0)+0(1) & 2(1)+5(4)+0(-2) & 2(2)+5(3)+0(2) \\ 1(3)+2(0)-1(1) & 1(1)+2(4)-1(-2) & 1(2)+2(3)-1(2) \\ 3(3)+0(0)+2(1) & 3(1)+0(4)+2(-2) & 3(2)+0(3)+2(2) \end{bmatrix} = \begin{bmatrix} 6 & 22 & 19 \\ 2 & 11 & 6 \\ 11 & -1 & 10 \end{bmatrix}$

So, $AB \neq BA$.

45. $(AB)C = \left(\begin{bmatrix} 1 & 2 \\ 3 & 4 \end{bmatrix}\begin{bmatrix} 2 & -3 \\ 3 & 5 \end{bmatrix}\right)\begin{bmatrix} 0 & 1 \\ 2 & 4 \end{bmatrix} = \begin{bmatrix} 1(2)+2(3) & 1(-3)+2(5) \\ 3(2)+4(3) & 3(-3)+4(5) \end{bmatrix}\begin{bmatrix} 0 & 1 \\ 2 & 4 \end{bmatrix} = \begin{bmatrix} 8 & 7 \\ 18 & 11 \end{bmatrix}\begin{bmatrix} 0 & 1 \\ 2 & 4 \end{bmatrix}$

$= \begin{bmatrix} 8(0)+7(2) & 8(1)+7(4) \\ 18(0)+11(2) & 18(1)+11(4) \end{bmatrix} = \begin{bmatrix} 14 & 36 \\ 22 & 62 \end{bmatrix}$

$A(BC) = \begin{bmatrix} 1 & 2 \\ 3 & 4 \end{bmatrix}\left(\begin{bmatrix} 2 & -3 \\ 3 & 5 \end{bmatrix}\begin{bmatrix} 0 & 1 \\ 2 & 4 \end{bmatrix}\right) = \begin{bmatrix} 1 & 2 \\ 3 & 4 \end{bmatrix}\begin{bmatrix} 2(0)-3(2) & 2(1)-3(4) \\ 3(0)+5(2) & 3(1)+5(4) \end{bmatrix} = \begin{bmatrix} 1 & 2 \\ 3 & 4 \end{bmatrix}\begin{bmatrix} -6 & -10 \\ 10 & 23 \end{bmatrix}$

$= \begin{bmatrix} 1(-6)+2(10) & 1(-10)+2(23) \\ 3(-6)+4(10) & 3(-10)+4(23) \end{bmatrix} = \begin{bmatrix} 14 & 36 \\ 22 & 62 \end{bmatrix} \Rightarrow (AB)C = A(BC)$

47. $(A+B)C = \left(\begin{bmatrix} 1 & 2 \\ 3 & 4 \end{bmatrix} + \begin{bmatrix} 2 & -3 \\ 3 & 5 \end{bmatrix}\right)\begin{bmatrix} 0 & 1 \\ 2 & 4 \end{bmatrix} = \begin{bmatrix} 3 & -1 \\ 6 & 9 \end{bmatrix}\begin{bmatrix} 0 & 1 \\ 2 & 4 \end{bmatrix} = \begin{bmatrix} 3(0)-1(2) & 3(1)-1(4) \\ 6(0)+9(2) & 6(1)+9(4) \end{bmatrix} = \begin{bmatrix} -2 & -1 \\ 18 & 42 \end{bmatrix}$

$AC + BC = \begin{bmatrix} 1 & 2 \\ 3 & 4 \end{bmatrix}\begin{bmatrix} 0 & 1 \\ 2 & 4 \end{bmatrix} + \begin{bmatrix} 2 & -3 \\ 3 & 5 \end{bmatrix}\begin{bmatrix} 0 & 1 \\ 2 & 4 \end{bmatrix} = \begin{bmatrix} 1(0)+2(2) & 1(1)+2(4) \\ 3(0)+4(2) & 3(1)+4(4) \end{bmatrix} + \begin{bmatrix} 2(0)-3(2) & 2(1)-3(4) \\ 3(0)+5(2) & 3(1)+5(4) \end{bmatrix}$

$= \begin{bmatrix} 4 & 9 \\ 8 & 19 \end{bmatrix} + \begin{bmatrix} -6 & -10 \\ 10 & 23 \end{bmatrix} = \begin{bmatrix} -2 & -1 \\ 18 & 42 \end{bmatrix} \Rightarrow (A+B)C = AC + BC$

6.2 Applying the Concepts

49.

	Steel	Glass	Wood	
$A+B = \begin{bmatrix} 7 & 3 & 18 \\ 4 & 1 & 3 \end{bmatrix} + \begin{bmatrix} 6 & 2 & 20 \\ 3 & 1 & 4 \end{bmatrix} =$	13	5	38	Material
	7	2	7	Transportation

51. $\begin{bmatrix} 100 & 300 & 400 \end{bmatrix}\begin{bmatrix} 60 \\ 38 \\ 17 \end{bmatrix} = \begin{bmatrix} 100(60)+300(38)+400(17) \end{bmatrix} = \begin{bmatrix} 24,200 \end{bmatrix}$

The total cost is \$24,200.

53. a.

	Chairman	President	Vice President
Salary	2,500,000	1,250,000	100,000
Bonus	1,500,000	750,000	150,000
Stock	50,000	25,000	5000

b. $\begin{bmatrix} 1 \\ 1 \\ 4 \end{bmatrix}$ Chairman, President, Vice President

c. $\begin{bmatrix} 2,500,000 & 1,250,000 & 100,000 \\ 1,500,000 & 750,000 & 150,000 \\ 50,000 & 25,000 & 5000 \end{bmatrix}\begin{bmatrix} 1 \\ 1 \\ 4 \end{bmatrix} = \begin{bmatrix} 2,500,000(1)+1,250,000(1)+100,000(4) \\ 1,500,000(1)+750,000(1)+150,000(4) \\ 50,000(1)+25,000(1)+5000(4) \end{bmatrix}$

	Totals	
$=$	4,150,000	Salary
	2,850,000	Bonus
	95,000	Stock

55. $AD = \begin{bmatrix} 1 & 0 \\ 0 & -1 \end{bmatrix}\begin{bmatrix} 0 & 4 & 4 & 1 & 1 & 0 \\ 0 & 0 & 1 & 1 & 6 & 6 \end{bmatrix}$

$= \begin{bmatrix} 1(0)+0(0) & 1(4)+0(0) & 1(4)+0(1) & 1(1)+0(1) & 1(1)+0(6) & 1(0)+0(6) \\ 0(0)-1(0) & 0(4)-1(0) & 0(4)-1(1) & 0(1)-1(1) & 0(1)-1(6) & 0(0)-1(6) \end{bmatrix}$

$= \begin{bmatrix} 0 & 4 & 4 & 1 & 1 & 0 \\ 0 & 0 & -1 & -1 & -6 & -6 \end{bmatrix}$

57. $AD = \begin{bmatrix} 1 & 0 \\ 0.25 & 1 \end{bmatrix}\begin{bmatrix} 0 & 4 & 4 & 1 & 1 & 0 \\ 0 & 0 & 1 & 1 & 6 & 6 \end{bmatrix}$

$= \begin{bmatrix} 1(0)+0(0) & 1(4)+0(0) & 1(4)+0(1) & 1(1)+0(1) & 1(1)+0(6) & 1(0)+0(6) \\ 0.25(0)+1(0) & 0.25(4)+1(0) & 0.25(4)+1(1) & 0.25(1)+1(1) & 0.25(1)+1(6) & 0.25(0)+1(6) \end{bmatrix}$

$= \begin{bmatrix} 0 & 4 & 4 & 1 & 1 & 0 \\ 0 & 1 & 2 & 1.25 & 6.25 & 6 \end{bmatrix}$

In exercises 59–62, the matrix representation of the quadrilateral $PQRS = D = \begin{bmatrix} 0 & 2 & 1 & 0 \\ 0 & 0 & 2 & 1 \end{bmatrix}$, where row 1 contains the x-coordinates of P, Q, R, S, and row 2 contains the y-coordinates.

59. $AD = \begin{bmatrix} 2 & 0 \\ 0 & 1 \end{bmatrix} \begin{bmatrix} 0 & 2 & 1 & 0 \\ 0 & 0 & 2 & 1 \end{bmatrix} = \begin{bmatrix} 0 & 4 & 2 & 0 \\ 0 & 0 & 2 & 1 \end{bmatrix}$

The coordinates of the vertices are $(0, 0)$, $(4, 0)$, $(2, 2)$, and $(0, 1)$.

61. $AD = \begin{bmatrix} 0.6 & 0.8 \\ -0.8 & 0.6 \end{bmatrix} \begin{bmatrix} 0 & 2 & 1 & 0 \\ 0 & 0 & 2 & 1 \end{bmatrix} = \begin{bmatrix} 0 & 1.2 & 2.2 & 0.8 \\ 0 & -1.6 & 0.4 & 0.6 \end{bmatrix}$

The coordinates of the vertices are $(0, 0)$, $(1.2, -1.6)$, $(2.2, 0.4)$, and $(0.8, 0.6)$.

In exercises 63–66, the matrix representation of the quadrilateral $PQRS = D = \begin{bmatrix} 0 & 1 & 2 & 1 \\ 0 & 0 & 1 & 2 \end{bmatrix}$, where row 1 contains the x-coordinates of P, Q, R, S, and row 2 contains the y-coordinates.

63.

$AD = \begin{bmatrix} 0 & 2 & 4 & 2 \\ 0 & 0 & 1 & 2 \end{bmatrix} = \begin{bmatrix} a & b \\ c & d \end{bmatrix} \begin{bmatrix} 0 & 1 & 2 & 1 \\ 0 & 0 & 1 & 2 \end{bmatrix}$

$= \begin{bmatrix} a \cdot 0 + b \cdot 0 & a \cdot 1 + b \cdot 0 & a \cdot 2 + b \cdot 1 & a \cdot 1 + b \cdot 2 \\ c \cdot 0 + d \cdot 0 & c \cdot 1 + d \cdot 0 & c \cdot 2 + d \cdot 1 & c \cdot 1 + d \cdot 2 \end{bmatrix}$

$= \begin{bmatrix} 0 & a & 2a+b & a+2b \\ 0 & c & 2c+d & c+2d \end{bmatrix} \Rightarrow a = 2,\ c = 0$

$2a + b = 4 \Rightarrow 2(2) + b = 4 \Rightarrow b = 0$
$2c + d = 1 \Rightarrow 2(0) + d = 1 \Rightarrow d = 1$ $\Rightarrow A = \begin{bmatrix} 2 & 0 \\ 0 & 1 \end{bmatrix}$

65.

$AD = \begin{bmatrix} 0 & 1 & 4 & 5 \\ 0 & 0 & 1 & 2 \end{bmatrix} = \begin{bmatrix} a & b \\ c & d \end{bmatrix} \begin{bmatrix} 0 & 1 & 2 & 1 \\ 0 & 0 & 1 & 2 \end{bmatrix}$

$= \begin{bmatrix} a \cdot 0 + b \cdot 0 & a \cdot 1 + b \cdot 0 & a \cdot 2 + b \cdot 1 & a \cdot 1 + b \cdot 2 \\ c \cdot 0 + d \cdot 0 & c \cdot 1 + d \cdot 0 & c \cdot 2 + d \cdot 1 & c \cdot 1 + d \cdot 2 \end{bmatrix}$

$= \begin{bmatrix} 0 & a & 2a+b & a+2b \\ 0 & c & 2c+d & c+2d \end{bmatrix} \Rightarrow a = 1,\ c = 0$

$2a + b = 4 \Rightarrow 2(1) + b = 4 \Rightarrow b = 2$
$2c + d = 1 \Rightarrow 2(0) + d = 1 \Rightarrow d = 1$ $\Rightarrow A = \begin{bmatrix} 1 & 2 \\ 0 & 1 \end{bmatrix}$

6.2 Beyond the Basics

67. $AB = \begin{bmatrix} 0 & 3 \\ 0 & 0 \end{bmatrix}\begin{bmatrix} 2 & 1 \\ 3 & 0 \end{bmatrix} = \begin{bmatrix} 0(2)+3(3) & 0(1)+3(0) \\ 0(2)+0(3) & 0(1)+0(0) \end{bmatrix} = \begin{bmatrix} 9 & 0 \\ 0 & 0 \end{bmatrix}$

$AC = \begin{bmatrix} 0 & 3 \\ 0 & 0 \end{bmatrix}\begin{bmatrix} 5 & 4 \\ 3 & 0 \end{bmatrix} = \begin{bmatrix} 0(5)+3(3) & 0(4)+3(0) \\ 0(5)+0(3) & 0(4)+0(0) \end{bmatrix} = \begin{bmatrix} 9 & 0 \\ 0 & 0 \end{bmatrix}$

So, $AB = AC$ does not imply $B = C$.

69. Answers will vary. Let $A = \begin{bmatrix} 1 & 2 \\ 3 & 4 \end{bmatrix}$ and $B = \begin{bmatrix} -1 & 0 \\ 2 & -3 \end{bmatrix}$. Then $A + B = \begin{bmatrix} 1 & 2 \\ 3 & 4 \end{bmatrix} + \begin{bmatrix} -1 & 0 \\ 2 & -3 \end{bmatrix} = \begin{bmatrix} 0 & 2 \\ 5 & 1 \end{bmatrix}$

$(A+B)^2 = \begin{bmatrix} 0 & 2 \\ 5 & 1 \end{bmatrix}\begin{bmatrix} 0 & 2 \\ 5 & 1 \end{bmatrix} = \begin{bmatrix} 0(0)+2(5) & 0(2)+2(1) \\ 5(0)+1(5) & 5(2)+1(1) \end{bmatrix} = \begin{bmatrix} 10 & 2 \\ 5 & 11 \end{bmatrix}$

$A^2 = \begin{bmatrix} 1 & 2 \\ 3 & 4 \end{bmatrix}\begin{bmatrix} 1 & 2 \\ 3 & 4 \end{bmatrix} = \begin{bmatrix} 1(1)+2(3) & 1(2)+2(4) \\ 3(1)+4(3) & 3(2)+4(4) \end{bmatrix} = \begin{bmatrix} 7 & 10 \\ 15 & 22 \end{bmatrix}$

$2AB = \begin{bmatrix} 1 & 2 \\ 3 & 4 \end{bmatrix}\begin{bmatrix} -1 & 0 \\ 2 & -3 \end{bmatrix} = 2\begin{bmatrix} 1(-1)+2(2) & 1(0)+2(-3) \\ 3(-1)+4(2) & 3(0)+4(-3) \end{bmatrix} = 2\begin{bmatrix} 3 & -6 \\ 5 & -12 \end{bmatrix} = \begin{bmatrix} 6 & -12 \\ 10 & -24 \end{bmatrix}$

$B^2 = \begin{bmatrix} -1 & 0 \\ 2 & -3 \end{bmatrix}\begin{bmatrix} -1 & 0 \\ 2 & -3 \end{bmatrix} = \begin{bmatrix} -1(-1)+0(2) & -1(0)+0(-3) \\ 2(-1)-3(2) & 2(0)-3(-3) \end{bmatrix} = \begin{bmatrix} 1 & 0 \\ -8 & 9 \end{bmatrix}$

$A^2 + 2AB + B^2 = \begin{bmatrix} 7 & 10 \\ 15 & 22 \end{bmatrix} + \begin{bmatrix} 6 & -12 \\ 10 & -24 \end{bmatrix} + \begin{bmatrix} 1 & 0 \\ -8 & 9 \end{bmatrix} = \begin{bmatrix} 14 & -2 \\ 17 & 7 \end{bmatrix}$

So, $(A+B)^2 \neq A^2 + 2AB + B^2$.

71.
```
3[A]²
[[18 -27 21]
 [-9 21 12]
 [-3 12 24]]
```
```
2[A]
[[2 -4 6]
 [4 -8 2]
 [6 -10 4]]
```
```
[I]
[[1 0 0]
 [0 1 0]
 [0 0 1]]
```
```
3[A]²-2[A]+[I]
[[17 -23 15]
 [-13 30 10]
 [-9 22 21]]
```

73. Let $B = \begin{bmatrix} x & y \\ z & w \end{bmatrix}$

$\begin{bmatrix} 2 & 3 \\ 1 & 2 \end{bmatrix}\begin{bmatrix} x & y \\ z & w \end{bmatrix} = \begin{bmatrix} 1 & 0 \\ 0 & 1 \end{bmatrix} \Rightarrow \begin{bmatrix} 2x+3z & 2y+3w \\ x+2z & y+2w \end{bmatrix} = \begin{bmatrix} 1 & 0 \\ 0 & 1 \end{bmatrix} \Rightarrow \begin{cases} 2x+3z=1 \\ 2y+3w=0 \\ x+2z=0 \\ y+2w=1 \end{cases} \Rightarrow$

$\begin{bmatrix} 2 & 0 & 3 & 0 & | & 1 \\ 0 & 2 & 0 & 3 & | & 0 \\ 1 & 0 & 2 & 0 & | & 0 \\ 0 & 1 & 0 & 2 & | & 1 \end{bmatrix} \xrightarrow[-(R_2-2R_4)\to R_4]{-(R_1-2R_3)\to R_3} \begin{bmatrix} 2 & 0 & 3 & 0 & | & 1 \\ 0 & 2 & 0 & 3 & | & 0 \\ 0 & 0 & 1 & 0 & | & -1 \\ 0 & 0 & 0 & 1 & | & 2 \end{bmatrix} \xrightarrow[\frac{1}{2}(R_2-3R_4)\to R_2]{\frac{1}{2}(R_1-3R_3)\to R_1} \begin{bmatrix} 1 & 0 & 0 & 0 & | & 2 \\ 0 & 1 & 0 & 0 & | & -3 \\ 0 & 0 & 1 & 0 & | & -1 \\ 0 & 0 & 0 & 1 & | & 2 \end{bmatrix} \Rightarrow$

$x = 2, y = -3, z = -1, w = 2 \Rightarrow B = \begin{bmatrix} 2 & -3 \\ -1 & 2 \end{bmatrix}$

75. $\left(4\begin{bmatrix} 2 & -1 & 3 \\ 1 & 0 & 2 \end{bmatrix} - \begin{bmatrix} 1 & 2 & -1 \\ 2 & -3 & 4 \end{bmatrix}\right)\begin{bmatrix} 2 \\ -1 \\ 1 \end{bmatrix} = \begin{bmatrix} x \\ y \end{bmatrix} \Rightarrow \left(\begin{bmatrix} 8 & -4 & 12 \\ 4 & 0 & 8 \end{bmatrix} - \begin{bmatrix} 1 & 2 & -1 \\ 2 & -3 & 4 \end{bmatrix}\right)\begin{bmatrix} 2 \\ -1 \\ 1 \end{bmatrix} = \begin{bmatrix} x \\ y \end{bmatrix} \Rightarrow$

$\begin{bmatrix} 7 & -6 & 13 \\ 2 & 3 & 4 \end{bmatrix}\begin{bmatrix} 2 \\ -1 \\ 1 \end{bmatrix} = \begin{bmatrix} x \\ y \end{bmatrix} \Rightarrow \begin{bmatrix} 7(2)-6(-1)+13(1) \\ 2(2)+3(-1)+4(1) \end{bmatrix} = \begin{bmatrix} x \\ y \end{bmatrix} \Rightarrow \begin{bmatrix} 33 \\ 5 \end{bmatrix} = \begin{bmatrix} x \\ y \end{bmatrix} \Rightarrow x = 33, y = 5$

77. $AB = \begin{bmatrix} 0 & 1 \\ 0 & 0 \end{bmatrix}\begin{bmatrix} 1 & 0 \\ 0 & 0 \end{bmatrix} = \begin{bmatrix} 0(1)+1(0) & 0(0)+1(0) \\ 0(1)+0(0) & 0(0)+0(0) \end{bmatrix} = \begin{bmatrix} 0 & 0 \\ 0 & 0 \end{bmatrix} = 0$

$BA = \begin{bmatrix} 1 & 0 \\ 0 & 0 \end{bmatrix}\begin{bmatrix} 0 & 1 \\ 0 & 0 \end{bmatrix} = \begin{bmatrix} 1(0)+0(0) & 1(1)+0(0) \\ 0(0)+0(0) & 0(1)+0(0) \end{bmatrix} = \begin{bmatrix} 0 & 1 \\ 0 & 0 \end{bmatrix} \neq 0$

79. $2A - B = \begin{bmatrix} 6 & -6 & 0 \\ -4 & 2 & 1 \end{bmatrix}$ $A + B = \begin{bmatrix} 3 & 0 & 3 \\ -2 & 1 & -4 \end{bmatrix}$

$(2A - B) + (A + B) = 3A = \begin{bmatrix} 6 & -6 & 0 \\ -4 & 2 & 1 \end{bmatrix} + \begin{bmatrix} 3 & 0 & 3 \\ -2 & 1 & -4 \end{bmatrix} = \begin{bmatrix} 9 & -6 & 3 \\ -6 & 3 & -3 \end{bmatrix} \Rightarrow A = \begin{bmatrix} 3 & -2 & 1 \\ -2 & 1 & -1 \end{bmatrix}$

$A + B = \begin{bmatrix} 3 & -2 & 1 \\ -2 & 1 & -1 \end{bmatrix} + B = \begin{bmatrix} 3 & 0 & 3 \\ -2 & 1 & -4 \end{bmatrix} \Rightarrow B = \begin{bmatrix} 3 & 0 & 3 \\ -2 & 1 & -4 \end{bmatrix} - \begin{bmatrix} 3 & -2 & 1 \\ -2 & 1 & -1 \end{bmatrix} = \begin{bmatrix} 0 & 2 & 2 \\ 0 & 0 & -3 \end{bmatrix}$

81. $AB = \begin{bmatrix} 2 & -3 & -5 \\ -1 & 4 & 5 \\ 1 & -3 & -4 \end{bmatrix}\begin{bmatrix} 2 & -2 & -4 \\ -1 & 3 & 4 \\ 1 & -2 & -3 \end{bmatrix}$

$= \begin{bmatrix} 2(2)+(-3)(-1)+(-5)(1) & 2(-2)+(-3)(3)+(-5)(-2) & 2(-4)+(-3)(4)+(-5)(-3) \\ (-1)(2)+4(-1)+5(1) & (-1)(-2)+4(3)+5(-2) & (-1)(-4)+4(4)+5(-3) \\ 1(2)+(-3)(-1)+(-4)(1) & 1(-2)+(-3)(3)+(-4)(-2) & 1(-4)+(-3)(4)+(-4)(-3) \end{bmatrix} =$

$= \begin{bmatrix} 2 & -3 & -5 \\ -1 & 4 & 5 \\ 1 & -3 & -4 \end{bmatrix} = A$

$BA = \begin{bmatrix} 2 & -2 & -4 \\ -1 & 3 & 4 \\ 1 & -2 & -3 \end{bmatrix}\begin{bmatrix} 2 & -3 & -5 \\ -1 & 4 & 5 \\ 1 & -3 & -4 \end{bmatrix}$

$= \begin{bmatrix} 2(2)+(-2)(-1)+(-4)(1) & 2(-3)+(-2)(4)+(-4)(-3) & 2(-5)+(-2)(5)+(-4)(-4) \\ (-1)(2)+3(-1)+4(1) & (-1)(-3)+3(4)+4(-3) & (-1)(-5)+3(5)+4(-4) \\ 1(2)+(-2)(-1)+(-3)(1) & 1(-3)+(-2)(4)+(-3)(-3) & 1(-5)+(-2)(5)+(-3)(-4) \end{bmatrix}$

$= \begin{bmatrix} 2 & -2 & -4 \\ -1 & 3 & 4 \\ 1 & -2 & -3 \end{bmatrix} = B$

$A^2 = \begin{bmatrix} 2 & -3 & -5 \\ -1 & 4 & 5 \\ 1 & -3 & -4 \end{bmatrix}\begin{bmatrix} 2 & -3 & -5 \\ -1 & 4 & 5 \\ 1 & -3 & -4 \end{bmatrix}$

$= \begin{bmatrix} 2(2)+(-3)(-1)+(-5)(1) & 2(-3)+(-3)(4)+(-5)(-3) & 2(-5)+(-3)(5)+(-5)(-4) \\ (-1)(2)+4(-1)+5(1) & (-1)(-3)+4(4)+5(-3) & (-1)(-5)+4(5)+5(-4) \\ 1(2)+(-3)(-1)+(-4)(1) & 1(-3)+(-3)(4)+(-4)(-3) & 1(-5)+(-3)(5)+(-4)(-4) \end{bmatrix}$

$= \begin{bmatrix} 2 & -3 & -5 \\ -1 & 4 & 5 \\ 1 & -3 & -4 \end{bmatrix} = A$

83. $A^2 = \begin{bmatrix} 4 & -1 & -4 \\ 3 & 0 & -4 \\ 3 & -1 & -3 \end{bmatrix}\begin{bmatrix} 4 & -1 & -4 \\ 3 & 0 & -4 \\ 3 & -1 & -3 \end{bmatrix}$

$= \begin{bmatrix} 4(4)+(-1)(3)+(-4)(3) & 4(-1)+(-1)(0)+(-4)(-1) & 4(-4)+(-1)(-4)+(-4)(-3) \\ 3(4)+0(3)+(-4)(3) & 3(-1)+0(0)+(-4)(-1) & 3(-4)+0(-4)+(-4)(-3) \\ 3(4)+(-1)(3)+(-3)(3) & 3(-1)+(-1)(0)+(-3)(-1) & 3(-4)+(-1)(-4)+(-3)(-3) \end{bmatrix}$

$= \begin{bmatrix} 1 & 0 & 0 \\ 0 & 1 & 0 \\ 0 & 0 & 1 \end{bmatrix} = I$

6.2 Critical Thinking/Discussion/Writing

84. a. *AB* is defined when $n = 5$. The order of *AB* when this product is defined is $3 \times m$.

    b. *BA* is defined when $m = 3$. The order of *BA* when this product is defined is $5 \times n$.

85. $(CA)B$

86. a.
$$P^2 = \begin{bmatrix} 0.4 & 0.3 & 0.3 \\ 0.6 & 0.3 & 0.1 \\ 0.6 & 0.1 & 0.3 \end{bmatrix}\begin{bmatrix} 0.4 & 0.3 & 0.3 \\ 0.6 & 0.3 & 0.1 \\ 0.6 & 0.1 & 0.3 \end{bmatrix} = \begin{bmatrix} 0.52 & 0.24 & 0.24 \\ 0.48 & 0.28 & 0.24 \\ 0.48 & 0.24 & 0.28 \end{bmatrix} = \begin{bmatrix} \frac{13}{25} & \frac{6}{25} & \frac{6}{25} \\ \frac{12}{25} & \frac{7}{25} & \frac{6}{25} \\ \frac{12}{25} & \frac{6}{25} & \frac{7}{25} \end{bmatrix}$$

$$XP^2 = \begin{bmatrix} \frac{1}{3} & \frac{1}{3} & \frac{1}{3} \end{bmatrix}\begin{bmatrix} \frac{13}{25} & \frac{6}{25} & \frac{6}{25} \\ \frac{12}{25} & \frac{7}{25} & \frac{6}{25} \\ \frac{12}{25} & \frac{6}{25} & \frac{7}{25} \end{bmatrix} = \begin{bmatrix} \frac{37}{75} & \frac{19}{75} & \frac{19}{75} \end{bmatrix}$$

b.
$$P^3 = P^2 P = \begin{bmatrix} 0.4 & 0.3 & 0.3 \\ 0.6 & 0.3 & 0.1 \\ 0.6 & 0.1 & 0.3 \end{bmatrix}\begin{bmatrix} 0.52 & 0.24 & 0.24 \\ 0.48 & 0.28 & 0.24 \\ 0.48 & 0.24 & 0.28 \end{bmatrix} = \begin{bmatrix} 0.496 & 0.252 & 0.252 \\ 0.504 & 0.252 & 0.244 \\ 0.504 & 0.244 & 0.252 \end{bmatrix} = \begin{bmatrix} \frac{62}{125} & \frac{63}{250} & \frac{63}{250} \\ \frac{63}{125} & \frac{63}{250} & \frac{61}{250} \\ \frac{63}{125} & \frac{61}{250} & \frac{63}{250} \end{bmatrix}$$

$$XP^3 = \begin{bmatrix} \frac{1}{3} & \frac{1}{3} & \frac{1}{3} \end{bmatrix}\begin{bmatrix} \frac{62}{125} & \frac{63}{250} & \frac{63}{250} \\ \frac{63}{125} & \frac{63}{250} & \frac{61}{250} \\ \frac{63}{125} & \frac{61}{250} & \frac{63}{250} \end{bmatrix} = \begin{bmatrix} \frac{188}{375} & \frac{187}{750} & \frac{187}{750} \end{bmatrix}$$

$$P^4 = P^3 P = \begin{bmatrix} 0.496 & 0.252 & 0.252 \\ 0.504 & 0.252 & 0.244 \\ 0.504 & 0.244 & 0.252 \end{bmatrix}\begin{bmatrix} 0.4 & 0.3 & 0.3 \\ 0.6 & 0.3 & 0.1 \\ 0.6 & 0.1 & 0.3 \end{bmatrix} = \begin{bmatrix} 0.5008 & 0.2496 & 0.2496 \\ 0.4992 & 0.2512 & 0.2496 \\ 0.4992 & 0.2496 & 0.2512 \end{bmatrix} = \begin{bmatrix} \frac{313}{625} & \frac{156}{625} & \frac{156}{625} \\ \frac{312}{625} & \frac{157}{625} & \frac{156}{625} \\ \frac{312}{625} & \frac{156}{625} & \frac{157}{625} \end{bmatrix}$$

$$XP^4 = \begin{bmatrix} \frac{1}{3} & \frac{1}{3} & \frac{1}{3} \end{bmatrix}\begin{bmatrix} \frac{313}{625} & \frac{156}{625} & \frac{156}{625} \\ \frac{312}{625} & \frac{157}{625} & \frac{156}{625} \\ \frac{312}{625} & \frac{156}{625} & \frac{157}{625} \end{bmatrix} = \begin{bmatrix} \frac{937}{1875} & \frac{469}{1875} & \frac{469}{1875} \end{bmatrix}$$

(continued on next page)

(continued)

$$P^5 = P^4 P = \begin{bmatrix} 0.5008 & 0.2496 & 0.2496 \\ 0.4992 & 0.2512 & 0.2496 \\ 0.4992 & 0.2496 & 0.2512 \end{bmatrix} \begin{bmatrix} 0.4 & 0.3 & 0.3 \\ 0.6 & 0.3 & 0.1 \\ 0.6 & 0.1 & 0.3 \end{bmatrix} = \begin{bmatrix} 0.49984 & 0.25008 & 0.25008 \\ 0.50016 & 0.25008 & 0.24976 \\ 0.50016 & 0.24976 & 0.25008 \end{bmatrix}$$

$$= \begin{bmatrix} \dfrac{1562}{3125} & \dfrac{1563}{6250} & \dfrac{1563}{6250} \\ \dfrac{1563}{3125} & \dfrac{1563}{6250} & \dfrac{1561}{6250} \\ \dfrac{1563}{3125} & \dfrac{1561}{6250} & \dfrac{1563}{6250} \end{bmatrix}$$

$$XP^5 = \begin{bmatrix} \dfrac{1}{3} & \dfrac{1}{3} & \dfrac{1}{3} \end{bmatrix} \begin{bmatrix} \dfrac{1562}{3125} & \dfrac{1563}{6250} & \dfrac{1563}{6250} \\ \dfrac{1563}{3125} & \dfrac{1563}{6250} & \dfrac{1561}{6250} \\ \dfrac{1563}{3125} & \dfrac{1561}{6250} & \dfrac{1563}{6250} \end{bmatrix} = \begin{bmatrix} \dfrac{4688}{9375} & \dfrac{4687}{18,750} & \dfrac{4687}{18,750} \end{bmatrix} \Rightarrow$$

$$XP^n = \begin{bmatrix} \dfrac{1}{2} & \dfrac{1}{4} & \dfrac{1}{4} \end{bmatrix}$$

In the long term, the market shares for each company are: *A*, 50%; *B*, 25%; *C*, 25%.

6.2 Getting Ready for the Next Section

87. $\left(\dfrac{1}{2}\right)^{-1} = 2$

89. $x^{-1} = \dfrac{1}{8} \Rightarrow x = 8$

91. $\dfrac{1}{4}x - \dfrac{7}{12} = \dfrac{11}{12} - \dfrac{5}{4}x \Rightarrow \dfrac{6}{4}x = \dfrac{18}{12} \Rightarrow$
$x = \dfrac{18}{12} \cdot \dfrac{4}{6} = 1$

93. $\dfrac{2}{4x-1} = \dfrac{3}{4x+1} \Rightarrow 2(4x+1) = 3(4x-1) \Rightarrow$
$8x + 2 = 12x - 3 \Rightarrow -4x = -5 \Rightarrow x = \dfrac{5}{4}$

95. $\begin{cases} 3x + 5y = 2 & (1) \\ 6x + 10y = 4 & (2) \end{cases}$

Multiply equation (1) by −2, then add the two equations.
$$\begin{aligned} -6x - 10y &= -4 \\ 6x + 10y &= 4 \\ \hline 0 &= 0 \end{aligned}$$
The equations are dependent and the system has infinitely many solutions. Solve equation (1) for *y* in terms of *x* to find the general solution.

$3x + 5y = 2 \Rightarrow 5y = -3x + 2 \Rightarrow$
$y = -\dfrac{3}{5}x + \dfrac{2}{5} = \dfrac{2-3x}{5}$

Thus, the solution set can be written as
$$\left\{ \left(x, \dfrac{2-3x}{5}\right) \right\}.$$

97. $\begin{cases} 3x + 2y = 6 & (1) \\ 6x + 4y = -13 & (2) \end{cases}$

Multiply equation (1) by −2, then add the equations.
$$\begin{aligned} -6x - 4y &= -12 \\ 6x + 4y &= -13 \\ \hline 0 &= -25 \quad \text{False} \end{aligned}$$
The system is inconsistent and the solution set is $\varnothing$.

6.3 The Matrix Inverse

6.3 Practice Problems

1. We need to verify that $AB = BA = I$.

$$AB = \begin{bmatrix} 3 & 2 \\ 2 & 1 \end{bmatrix} \begin{bmatrix} -1 & 2 \\ 2 & -3 \end{bmatrix}$$
$$= \begin{bmatrix} 3(-1)+2(2) & 3(2)+2(-3) \\ 2(-1)+1(2) & 2(2)+1(-3) \end{bmatrix} = \begin{bmatrix} 1 & 0 \\ 0 & 1 \end{bmatrix}$$

(continued on next page)

(continued)

$$BA = \begin{bmatrix} -1 & 2 \\ 2 & -3 \end{bmatrix} \begin{bmatrix} 3 & 2 \\ 2 & 1 \end{bmatrix} = \begin{bmatrix} -1(3)+2(2) & -1(2)+2(1) \\ 2(3)-3(2) & 2(2)-3(1) \end{bmatrix} = \begin{bmatrix} 1 & 0 \\ 0 & 1 \end{bmatrix}$$

Thus B is the inverse of A.

2. Suppose A has an inverse B, where $B = \begin{bmatrix} x & y \\ z & w \end{bmatrix}$. Then

$$\begin{bmatrix} 3 & 1 \\ 3 & 1 \end{bmatrix} \begin{bmatrix} x & y \\ z & w \end{bmatrix} = \begin{bmatrix} 1 & 0 \\ 0 & 1 \end{bmatrix} \Rightarrow \begin{bmatrix} 3x+z & 3y+w \\ 3x+z & 3y+w \end{bmatrix} = \begin{bmatrix} 1 & 0 \\ 0 & 1 \end{bmatrix}.$$

Since the matrices are equal, the entries must be equal. So $3x+z = 1$ (in the 1, 1 position) and $3x+z = 0$ (in the 2, 1) position. This is a contradiction, so A does not have an inverse.

3. $[A \mid I] = \begin{bmatrix} 1 & 4 & -2 & | & 1 & 0 & 0 \\ -1 & 1 & 2 & | & 0 & 1 & 0 \\ 3 & 7 & -6 & | & 0 & 0 & 1 \end{bmatrix} \xrightarrow[-3R_1+R_3 \to R_3]{R_1+R_2 \to R_2} \begin{bmatrix} 1 & 4 & -2 & | & 1 & 0 & 0 \\ 0 & 5 & 0 & | & 1 & 1 & 0 \\ 0 & -5 & 0 & | & -3 & 0 & 1 \end{bmatrix}$

$\xrightarrow{R_2+R_3 \to R_3} \begin{bmatrix} 1 & 4 & -2 & | & 1 & 0 & 0 \\ 0 & 5 & 0 & | & 1 & 1 & 0 \\ 0 & 0 & 0 & | & -2 & 1 & 1 \end{bmatrix}$

The inverse of A does not exist.

4. $[A \mid I] = \begin{bmatrix} 1 & 2 & 3 & | & 1 & 0 & 0 \\ -2 & 3 & 1 & | & 0 & 1 & 0 \\ 4 & 5 & -2 & | & 0 & 0 & 1 \end{bmatrix} \xrightarrow[-4R_1+R_3 \to R_3]{2R_1+R_2 \to R_2} \begin{bmatrix} 1 & 2 & 3 & | & 1 & 0 & 0 \\ 0 & 7 & 7 & | & 2 & 1 & 0 \\ 0 & -3 & -14 & | & -4 & 0 & 1 \end{bmatrix}$

$\xrightarrow{\frac{1}{11}(2R_2+R_3) \to R_2} \begin{bmatrix} 1 & 2 & 3 & | & 1 & 0 & 0 \\ 0 & 1 & 0 & | & 0 & \frac{2}{11} & \frac{1}{11} \\ 0 & -3 & -14 & | & -4 & 0 & 1 \end{bmatrix} \xrightarrow[\left(-\frac{1}{14}\right)(3R_2+R_3) \to R_3]{R_1-2R_2 \to R_1} \begin{bmatrix} 1 & 0 & 3 & | & 1 & -\frac{4}{11} & -\frac{2}{11} \\ 0 & 1 & 0 & | & 0 & \frac{2}{11} & \frac{1}{11} \\ 0 & 0 & 1 & | & \frac{2}{7} & -\frac{3}{77} & -\frac{1}{11} \end{bmatrix}$

$\xrightarrow{R_1-3R_3 \to R_1} \begin{bmatrix} 1 & 0 & 0 & | & \frac{1}{7} & -\frac{19}{77} & \frac{1}{11} \\ 0 & 1 & 0 & | & 0 & \frac{2}{11} & \frac{1}{11} \\ 0 & 0 & 1 & | & \frac{2}{7} & -\frac{3}{77} & -\frac{1}{11} \end{bmatrix} \Rightarrow A^{-1} = \begin{bmatrix} \frac{1}{7} & -\frac{19}{77} & \frac{1}{11} \\ 0 & \frac{2}{11} & \frac{1}{11} \\ \frac{2}{7} & -\frac{3}{77} & -\frac{1}{11} \end{bmatrix}$

5. **a.** $ad - bc = (8)(1) - (2)(4) = 0$

The inverse of matrix A does not exist.

b. $ad - bc = (8)(1) - (-2)(3) = 14$

Thus, matrix B is invertible.

$$B^{-1} = \frac{1}{14} \begin{bmatrix} 1 & 2 \\ -3 & 8 \end{bmatrix} = \begin{bmatrix} \frac{1}{14} & \frac{1}{7} \\ -\frac{3}{14} & \frac{4}{7} \end{bmatrix}$$

6. $\begin{cases} 3x + 2y + 3z = 9 \\ 3x + y = 12 \\ x + z = 6 \end{cases} \Rightarrow AX = B \Rightarrow \begin{bmatrix} 3 & 2 & 3 \\ 3 & 1 & 0 \\ 1 & 0 & 1 \end{bmatrix}\begin{bmatrix} x \\ y \\ z \end{bmatrix} = \begin{bmatrix} 9 \\ 12 \\ 6 \end{bmatrix} \Rightarrow X = A^{-1}B$

Using a graphing calculator, we find that $A^{-1} = \begin{bmatrix} -\dfrac{1}{6} & \dfrac{1}{3} & \dfrac{1}{2} \\[6pt] \dfrac{1}{2} & 0 & -\dfrac{3}{2} \\[6pt] \dfrac{1}{6} & -\dfrac{1}{3} & \dfrac{1}{2} \end{bmatrix}$.

$X = A^{-1}B \Rightarrow \begin{bmatrix} x \\ y \\ z \end{bmatrix} = \begin{bmatrix} -\dfrac{1}{5} & \dfrac{2}{5} & \dfrac{2}{5} \\[6pt] \dfrac{3}{5} & -\dfrac{1}{5} & -\dfrac{6}{5} \\[6pt] \dfrac{1}{5} & -\dfrac{2}{5} & \dfrac{3}{5} \end{bmatrix}\begin{bmatrix} 9 \\ 12 \\ 6 \end{bmatrix} \Rightarrow X = \begin{bmatrix} \dfrac{11}{2} \\[6pt] -\dfrac{9}{2} \\[6pt] \dfrac{1}{2} \end{bmatrix}$

Solution set: $\left\{ \left(\dfrac{11}{2}, -\dfrac{9}{2}, \dfrac{1}{2} \right) \right\}$

7. Following the discussion given in Example 7, we have $X = (I - A)^{-1} D = \dfrac{1}{19}\begin{bmatrix} 36 & 24 \\ 16 & 36 \end{bmatrix}\begin{bmatrix} 800 \\ 3400 \end{bmatrix} = \begin{bmatrix} \dfrac{110,400}{19} \\[6pt] \dfrac{135,200}{19} \end{bmatrix}$

To meet the consumer demand for 800 units of energy and 3400 units of food, the energy produced must be $\dfrac{110,400}{19}$ units and food production must be $\dfrac{135,200}{19}$ units.

8. Associate each letter in the phrase with the number representing its position in the alphabet, and partition the numbers into groups of three (inserting two zeros at the right end), forming a 3×6 matrix:

J	A	C	K	I	S	N	O	W	S	A	F	E		
[10	1	3]	[11	0	9]	[19	0	14]	[15	23	0]	[19	1	6]	[5	0	0]

$M = \begin{bmatrix} 10 & 11 & 19 & 15 & 19 & 5 \\ 1 & 0 & 0 & 23 & 1 & 0 \\ 3 & 9 & 14 & 0 & 6 & 0 \end{bmatrix}$

Let $A = \begin{bmatrix} 1 & 2 & 3 \\ 1 & 3 & 3 \\ 1 & 2 & 4 \end{bmatrix}$. Then, $AM = \begin{bmatrix} 1 & 2 & 3 \\ 1 & 3 & 3 \\ 1 & 2 & 4 \end{bmatrix}\begin{bmatrix} 10 & 11 & 19 & 15 & 19 & 5 \\ 1 & 0 & 0 & 23 & 1 & 0 \\ 3 & 9 & 14 & 0 & 6 & 0 \end{bmatrix} = \begin{bmatrix} 21 & 38 & 61 & 61 & 39 & 5 \\ 22 & 38 & 61 & 84 & 40 & 5 \\ 24 & 47 & 75 & 61 & 45 & 5 \end{bmatrix}$

The cryptogram is 21 22 24 38 38 47 61 61 75 61 84 61 39 40 45 5 5 5.

6.3 Concepts and Vocabulary

1. For $n \times n$ matrices A and B, if $AB = I$, then B is called the <u>inverse</u> of A.

3. To find the inverse of an invertible matrix A, we transform $[A \,|\, I]$ by a sequence of row operations into $[I \,|\, B]$, where $B = \underline{A^{-1}}$.

5. False.   **7.** True

6.3 Building Skills

9. $AB = \begin{bmatrix} 1 & 2 \\ 1 & 3 \end{bmatrix}\begin{bmatrix} 3 & -2 \\ -1 & 1 \end{bmatrix} = \begin{bmatrix} 1 & 0 \\ 0 & 1 \end{bmatrix}$

$BA = \begin{bmatrix} 3 & -2 \\ -1 & 1 \end{bmatrix}\begin{bmatrix} 1 & 2 \\ 1 & 3 \end{bmatrix} = \begin{bmatrix} 1 & 0 \\ 0 & 1 \end{bmatrix} \Rightarrow$

B is the inverse of A.

11. $AB = \begin{bmatrix} 3 & 2 \\ 1 & 4 \end{bmatrix} \begin{bmatrix} \dfrac{2}{5} & -\dfrac{1}{5} \\ -\dfrac{1}{10} & \dfrac{3}{10} \end{bmatrix} = \begin{bmatrix} 1 & 0 \\ 0 & 1 \end{bmatrix}$

$BA = \begin{bmatrix} \dfrac{2}{5} & -\dfrac{1}{5} \\ -\dfrac{1}{10} & \dfrac{3}{10} \end{bmatrix} \begin{bmatrix} 3 & 2 \\ 1 & 4 \end{bmatrix} = \begin{bmatrix} 1 & 0 \\ 0 & 1 \end{bmatrix} \Rightarrow$

B is the inverse of A.

13. $AB = \begin{bmatrix} 1 & 0 & -1 \\ -1 & 1 & 1 \end{bmatrix} \begin{bmatrix} 2 & 1 \\ 1 & 1 \\ 1 & 1 \end{bmatrix} = \begin{bmatrix} 1 & 0 \\ 0 & 1 \end{bmatrix}$

$BA = \begin{bmatrix} 2 & 1 \\ 1 & 1 \\ 1 & 1 \end{bmatrix} \begin{bmatrix} 1 & 0 & -1 \\ -1 & 1 & 1 \end{bmatrix} = \begin{bmatrix} 1 & 1 & -1 \\ 0 & 1 & 0 \\ 0 & 1 & 0 \end{bmatrix} \Rightarrow$

B is not the inverse of A.

15. $AB = \begin{bmatrix} 1 & -2 & 1 \\ -8 & 6 & -2 \\ 5 & -3 & 1 \end{bmatrix} \left(\dfrac{1}{2} \begin{bmatrix} 0 & 1 & 2 \\ 1 & 2 & 3 \\ 3 & 1 & 1 \end{bmatrix} \right)$

$= \begin{bmatrix} \dfrac{1}{2} & -1 & -\dfrac{3}{2} \\ 0 & 1 & 0 \\ 0 & 0 & 1 \end{bmatrix} \Rightarrow B$ is not the inverse of A.

17. $AB = \begin{bmatrix} 1 & 1 & 1 \\ 1 & 2 & 3 \\ -1 & 1 & -1 \end{bmatrix} \left(\dfrac{1}{4} \begin{bmatrix} 5 & -2 & -1 \\ 2 & 0 & 2 \\ -3 & 2 & -1 \end{bmatrix} \right)$

$= \begin{bmatrix} 1 & 0 & 0 \\ 0 & 1 & 0 \\ 0 & 0 & 1 \end{bmatrix}$

$BA = \left(\dfrac{1}{4} \begin{bmatrix} 5 & -2 & -1 \\ 2 & 0 & 2 \\ -3 & 2 & -1 \end{bmatrix} \right) \begin{bmatrix} 1 & 1 & 1 \\ 1 & 2 & 3 \\ -1 & 1 & -1 \end{bmatrix}$

$= \begin{bmatrix} 1 & 0 & 0 \\ 0 & 1 & 0 \\ 0 & 0 & 1 \end{bmatrix} \Rightarrow B$ is the inverse of A.

In exercises 19–28, the steps to find the inverse may vary. Be sure to verify that $AA^{-1} = I$.

19. $[A \mid I] = \begin{bmatrix} 2 & 0 & | & 1 & 0 \\ 1 & 3 & | & 0 & 1 \end{bmatrix} \xrightarrow{\frac{1}{2}R_1} \begin{bmatrix} 1 & 0 & | & \dfrac{1}{2} & 0 \\ 1 & 3 & | & 0 & 1 \end{bmatrix}$

$\xrightarrow{R_1 - R_2 \to R_2} \begin{bmatrix} 1 & 0 & | & \dfrac{1}{2} & 0 \\ 0 & -3 & | & \dfrac{1}{2} & -1 \end{bmatrix}$

$\xrightarrow{-\frac{1}{3}R_2} \begin{bmatrix} 1 & 0 & | & \dfrac{1}{2} & 0 \\ 0 & 1 & | & -\dfrac{1}{6} & \dfrac{1}{3} \end{bmatrix} \Rightarrow$

$A^{-1} = \begin{bmatrix} \dfrac{1}{2} & 0 \\ -\dfrac{1}{6} & \dfrac{1}{3} \end{bmatrix}$

21. $[A \mid I] = \begin{bmatrix} 2 & 4 & | & 1 & 0 \\ 3 & 6 & | & 0 & 1 \end{bmatrix}$

$\xrightarrow{\frac{1}{2}R_1 \to R_1} \begin{bmatrix} 1 & 2 & | & \dfrac{1}{2} & 0 \\ 3 & 6 & | & 0 & 1 \end{bmatrix}$

$\xrightarrow{-3R_1 + R_2 \to R_2} \begin{bmatrix} 1 & 2 & | & \dfrac{1}{2} & 0 \\ 0 & 0 & | & -\dfrac{3}{2} & 1 \end{bmatrix} \Rightarrow$

The inverse of A does not exist.

23. $[A \mid I] = \begin{bmatrix} 1 & 6 & 4 & | & 1 & 0 & 0 \\ 0 & 2 & 3 & | & 0 & 1 & 0 \\ 0 & 1 & 2 & | & 0 & 0 & 1 \end{bmatrix}$

$\xrightarrow{2R_3 - R_2 \to R_3} \begin{bmatrix} 1 & 6 & 4 & | & 1 & 0 & 0 \\ 0 & 2 & 3 & | & 0 & 1 & 0 \\ 0 & 0 & 1 & | & 0 & -1 & 2 \end{bmatrix}$

$\xrightarrow{\frac{1}{2}(-3R_3 + R_2) \to R_2} \begin{bmatrix} 1 & 6 & 4 & | & 1 & 0 & 0 \\ 0 & 1 & 0 & | & 0 & 2 & -3 \\ 0 & 0 & 1 & | & 0 & -1 & 2 \end{bmatrix}$

$\xrightarrow[R_1 - 6R_2 \to R_1]{R_1 - 4R_3 \to R_1} \begin{bmatrix} 1 & 0 & 0 & | & 1 & -8 & 10 \\ 0 & 1 & 0 & | & 0 & 2 & -3 \\ 0 & 0 & 1 & | & 0 & -1 & 2 \end{bmatrix}$

$A^{-1} = \begin{bmatrix} 1 & -8 & 10 \\ 0 & 2 & -3 \\ 0 & -1 & 2 \end{bmatrix}$

25. $[A \mid I] = \begin{bmatrix} 2 & 3 & 1 & 1 & 0 & 0 \\ 2 & 4 & 1 & 0 & 1 & 0 \\ 3 & 7 & 2 & 0 & 0 & 1 \end{bmatrix}$

$\xrightarrow{R_2 - R_1 \to R_3} \begin{bmatrix} 2 & 3 & 1 & 1 & 0 & 0 \\ 0 & 1 & 0 & -1 & 1 & 0 \\ 3 & 7 & 2 & 0 & 0 & 1 \end{bmatrix}$

$\xrightarrow{2R_1 - R_3 \to R_1} \begin{bmatrix} 1 & -1 & 0 & 2 & 0 & -1 \\ 0 & 1 & 0 & -1 & 1 & 0 \\ 3 & 7 & 2 & 0 & 0 & 1 \end{bmatrix}$

$\xrightarrow[R_3 - 7R_2 \to R_3]{R_1 + R_2 \to R_1} \begin{bmatrix} 1 & 0 & 0 & 1 & 1 & -1 \\ 0 & 1 & 0 & -1 & 1 & 0 \\ 3 & 0 & 2 & 7 & -7 & 1 \end{bmatrix}$

$\xrightarrow{\frac{1}{2}(R_3 - 3R_1) \to R_3} \begin{bmatrix} 1 & 0 & 0 & 1 & 1 & -1 \\ 0 & 1 & 0 & -1 & 1 & 0 \\ 0 & 0 & 1 & 2 & -5 & 2 \end{bmatrix}$

$A^{-1} = \begin{bmatrix} 1 & 1 & -1 \\ -1 & 1 & 0 \\ 2 & -5 & 2 \end{bmatrix}$

27. $[A \mid I] = \begin{bmatrix} 3 & -3 & 4 & 1 & 0 & 0 \\ 2 & -3 & 4 & 0 & 1 & 0 \\ 0 & -1 & 1 & 0 & 0 & 1 \end{bmatrix}$

$\xrightarrow{R_1 - R_2 \to R_1} \begin{bmatrix} 1 & 0 & 0 & 1 & -1 & 0 \\ 2 & -3 & 4 & 0 & 1 & 0 \\ 0 & -1 & 1 & 0 & 0 & 1 \end{bmatrix}$

$\xrightarrow{2R_1 - R_2 \to R_2} \begin{bmatrix} 1 & 0 & 0 & 1 & -1 & 0 \\ 0 & 3 & -4 & 2 & -3 & 0 \\ 0 & -1 & 1 & 0 & 0 & 1 \end{bmatrix}$

$\xrightarrow{-(R_2 + 4R_3) \to R_2} \begin{bmatrix} 1 & 0 & 0 & 1 & -1 & 0 \\ 0 & 1 & 0 & -2 & 3 & -4 \\ 0 & -1 & 1 & 0 & 0 & 1 \end{bmatrix}$

$\xrightarrow{R_2 + R_3 \to R_3} \begin{bmatrix} 1 & 0 & 0 & 1 & -1 & 0 \\ 0 & 1 & 0 & -2 & 3 & -4 \\ 0 & 0 & 1 & -2 & 3 & -3 \end{bmatrix}$

$A^{-1} = \begin{bmatrix} 1 & -1 & 0 \\ -2 & 3 & -4 \\ -2 & 3 & -3 \end{bmatrix}$

29. $ad - bc = (1)(2) - (0)(3) = 2$

$A^{-1} = \frac{1}{2} \begin{bmatrix} 2 & 0 \\ -3 & 1 \end{bmatrix} = \begin{bmatrix} 1 & 0 \\ -\frac{3}{2} & \frac{1}{2} \end{bmatrix}$

31. $ad - bc = (2)(5) - (-3)(-3) = 1$

$A^{-1} = \begin{bmatrix} 5 & 3 \\ 3 & 2 \end{bmatrix}$

33. $ad - bc = (a)(-a) - (b)(-b) = -a^2 + b^2$

$A^{-1} = \frac{1}{-a^2 + b^2} \begin{bmatrix} -a & b \\ -b & a \end{bmatrix}$ where $a^2 \neq b^2$.

35. $\begin{cases} 2x + 3y = -9 \\ x - 3y = 13 \end{cases} \Rightarrow \begin{bmatrix} 2 & 3 \\ 1 & -3 \end{bmatrix} \begin{bmatrix} x \\ y \end{bmatrix} = \begin{bmatrix} -9 \\ 13 \end{bmatrix}$

37. $\begin{cases} 3x + 2y + z = 8 \\ 2x + y + 3z = 7 \\ x + 3y + 2z = 9 \end{cases} \Rightarrow \begin{bmatrix} 3 & 2 & 1 \\ 2 & 1 & 3 \\ 1 & 3 & 2 \end{bmatrix} \begin{bmatrix} x \\ y \\ z \end{bmatrix} = \begin{bmatrix} 8 \\ 7 \\ 9 \end{bmatrix}$

39. $\begin{bmatrix} 1 & -2 \\ 2 & 1 \end{bmatrix} \begin{bmatrix} x \\ y \end{bmatrix} = \begin{bmatrix} 0 \\ 5 \end{bmatrix} \Rightarrow \begin{cases} x - 2y = 0 \\ 2x + y = 5 \end{cases}$

41. $\begin{bmatrix} 2 & 3 & 1 \\ 5 & 7 & -1 \\ 4 & 3 & 0 \end{bmatrix} \begin{bmatrix} x_1 \\ x_2 \\ x_3 \end{bmatrix} = \begin{bmatrix} -1 \\ 5 \\ 5 \end{bmatrix} \Rightarrow \begin{cases} 2x_1 + 3x_2 + x_3 = -1 \\ 5x_1 + 7x_2 - x_3 = 5 \\ 4x_1 + 3x_2 = 5 \end{cases}$

43. $\begin{bmatrix} 1 & 2 & 5 \\ 2 & 3 & 8 \\ -1 & 1 & 2 \end{bmatrix} \begin{bmatrix} 2 & -1 & -1 \\ 12 & -7 & -2 \\ -5 & 3 & 1 \end{bmatrix} = \begin{bmatrix} 1 & 0 & 0 \\ 0 & 1 & 0 \\ 0 & 0 & 1 \end{bmatrix}$

45. $\begin{cases} x_1 + 2x_2 + 5x_3 = 1 \\ 2x_1 + 3x_2 + 8x_3 = 3 \\ -x_1 + x_2 + 2x_3 = -3 \end{cases} \Rightarrow AX = B \Rightarrow$

$\begin{bmatrix} 1 & 2 & 5 \\ 2 & 3 & 8 \\ -1 & 1 & 2 \end{bmatrix} \begin{bmatrix} x \\ y \\ z \end{bmatrix} = \begin{bmatrix} 1 \\ 3 \\ -3 \end{bmatrix} \Rightarrow X = A^{-1}B \Rightarrow$

$\begin{bmatrix} x \\ y \\ z \end{bmatrix} = \begin{bmatrix} 2 & -1 & -1 \\ 12 & -7 & -2 \\ -5 & 3 & 1 \end{bmatrix} \begin{bmatrix} 1 \\ 3 \\ -3 \end{bmatrix} \Rightarrow \begin{bmatrix} x \\ y \\ z \end{bmatrix} = \begin{bmatrix} 2 \\ -3 \\ 1 \end{bmatrix}$

The solution is $\{(2, -3, 1)\}$.

47. a. $[A \mid I] = \begin{bmatrix} 1 & 1 & 1 & 1 & 0 & 0 \\ 1 & 2 & 3 & 0 & 1 & 0 \\ 1 & 4 & 9 & 0 & 0 & 1 \end{bmatrix}$

$\xrightarrow[R_3 - R_1 \to R_3]{R_2 - R_1 \to R_2} \begin{bmatrix} 1 & 1 & 1 & 1 & 0 & 0 \\ 0 & 1 & 2 & -1 & 1 & 0 \\ 0 & 3 & 8 & -1 & 0 & 1 \end{bmatrix}$

$\xrightarrow[\frac{1}{2}(R_3 - 3R_2) \to R_3]{R_1 - R_2 \to R_1} \begin{bmatrix} 1 & 0 & -1 & 2 & -1 & 0 \\ 0 & 1 & 2 & -1 & 1 & 0 \\ 0 & 0 & 1 & 1 & -\frac{3}{2} & \frac{1}{2} \end{bmatrix}$

(*continued on next page*)

(*continued*)

$$\xrightarrow[\substack{R_1+R_3 \to R_1 \\ R_2-2R_3 \to R_2}]{} \left[\begin{array}{ccc|ccc} 1 & 0 & 0 & 3 & -\dfrac{5}{2} & \dfrac{1}{2} \\ 0 & 1 & 0 & -3 & 4 & -1 \\ 0 & 0 & 1 & 1 & -\dfrac{3}{2} & \dfrac{1}{2} \end{array}\right]$$

$$A^{-1} = \begin{bmatrix} 3 & -\dfrac{5}{2} & \dfrac{1}{2} \\ -3 & 4 & -1 \\ 1 & -\dfrac{3}{2} & \dfrac{1}{2} \end{bmatrix}$$

b. $\begin{cases} x+ y+ z = 6 \\ x+2y+3z = 14 \\ x+4y+9z = 36 \end{cases} \Rightarrow AX = B \Rightarrow$

$$\begin{bmatrix} 1 & 1 & 1 \\ 1 & 2 & 3 \\ 1 & 4 & 9 \end{bmatrix}\begin{bmatrix} x \\ y \\ z \end{bmatrix} = \begin{bmatrix} 6 \\ 14 \\ 36 \end{bmatrix} \Rightarrow X = A^{-1}B \Rightarrow$$

$$\begin{bmatrix} x \\ y \\ z \end{bmatrix} = \begin{bmatrix} 3 & -\dfrac{5}{2} & \dfrac{1}{2} \\ -3 & 4 & -1 \\ 1 & -\dfrac{3}{2} & \dfrac{1}{2} \end{bmatrix}\begin{bmatrix} 6 \\ 14 \\ 36 \end{bmatrix} \Rightarrow \begin{bmatrix} x \\ y \\ z \end{bmatrix} = \begin{bmatrix} 1 \\ 2 \\ 3 \end{bmatrix}$$

The solution is $\{(1,2,3)\}$.

6.3 Applying the Concepts

49. $\begin{cases} 3x+7y = 17 \\ -5x+4y = 13 \end{cases} \Rightarrow \begin{bmatrix} 3 & 7 \\ -5 & 4 \end{bmatrix}\begin{bmatrix} x \\ y \end{bmatrix} = \begin{bmatrix} 11 \\ 13 \end{bmatrix}$

$$[A \mid I] = \begin{bmatrix} 3 & 7 & 1 & 0 \\ -5 & 4 & 0 & 1 \end{bmatrix}$$

$$\xrightarrow{5R_1+3R_2 \to R_2} \begin{bmatrix} 3 & 7 & 1 & 0 \\ 0 & 47 & 5 & 3 \end{bmatrix}$$

$$\xrightarrow{\frac{1}{47}R_2 \to R_2} \left[\begin{array}{cc|cc} 3 & 7 & 1 & 0 \\ 0 & 1 & \dfrac{5}{47} & \dfrac{3}{47} \end{array}\right]$$

$$\xrightarrow{\frac{1}{3}(R_1-7R_2) \to R_1} \left[\begin{array}{cc|cc} 1 & 0 & \dfrac{4}{47} & -\dfrac{7}{47} \\ 0 & 1 & \dfrac{5}{47} & \dfrac{3}{47} \end{array}\right]$$

$$A^{-1} = \begin{bmatrix} \dfrac{4}{47} & -\dfrac{7}{47} \\ \dfrac{5}{47} & \dfrac{3}{47} \end{bmatrix}$$

$$X = A^{-1}B \Rightarrow \begin{bmatrix} x \\ y \end{bmatrix} = \begin{bmatrix} \dfrac{4}{47} & -\dfrac{7}{47} \\ \dfrac{5}{47} & \dfrac{3}{47} \end{bmatrix}\begin{bmatrix} 11 \\ 13 \end{bmatrix} = \begin{bmatrix} -1 \\ 2 \end{bmatrix}$$

The solution is $\{(-1,2)\}$.

51. $\begin{cases} x+ y+2z = 7 \\ x- y-3z = -6 \\ 2x+3y+ z = 4 \end{cases} \Rightarrow \begin{bmatrix} 1 & 1 & 2 \\ 1 & -1 & -3 \\ 2 & 3 & 1 \end{bmatrix}\begin{bmatrix} x \\ y \\ z \end{bmatrix} = \begin{bmatrix} 7 \\ -6 \\ 4 \end{bmatrix}$

$$[A \mid I] = \begin{bmatrix} 1 & 1 & 2 & 1 & 0 & 0 \\ 1 & -1 & -3 & 0 & 1 & 0 \\ 2 & 3 & 1 & 0 & 0 & 1 \end{bmatrix}$$

$$\xrightarrow[\substack{R_1-R_2 \to R_2 \\ R_3-2R_1 \to R_3}]{} \begin{bmatrix} 1 & 1 & 2 & 1 & 0 & 0 \\ 0 & 2 & 5 & 1 & -1 & 0 \\ 0 & 1 & -3 & -2 & 0 & 1 \end{bmatrix}$$

$$\xrightarrow[\substack{2R_1-R_2 \to R_1 \\ \frac{1}{11}(R_2-2R_3) \to R_3}]{} \left[\begin{array}{ccc|ccc} 2 & 0 & -1 & 1 & 1 & 0 \\ 0 & 2 & 5 & 1 & -1 & 0 \\ 0 & 0 & 1 & \dfrac{5}{11} & -\dfrac{1}{11} & -\dfrac{2}{11} \end{array}\right]$$

$$\xrightarrow[\substack{\frac{1}{2}(R_1+R_3) \to R_1 \\ \frac{1}{2}(R_2-5R_3) \to R_2}]{} \left[\begin{array}{ccc|ccc} 1 & 0 & 0 & \dfrac{8}{11} & \dfrac{5}{11} & -\dfrac{1}{11} \\ 0 & 1 & 0 & -\dfrac{7}{11} & -\dfrac{3}{11} & \dfrac{5}{11} \\ 0 & 0 & 1 & \dfrac{5}{11} & -\dfrac{1}{11} & -\dfrac{2}{11} \end{array}\right]$$

$$A^{-1} = \begin{bmatrix} \dfrac{8}{11} & \dfrac{5}{11} & -\dfrac{1}{11} \\ -\dfrac{7}{11} & -\dfrac{3}{11} & \dfrac{5}{11} \\ \dfrac{5}{11} & -\dfrac{1}{11} & -\dfrac{2}{11} \end{bmatrix}$$

$$X = A^{-1}B \Rightarrow$$

$$\begin{bmatrix} x \\ y \\ z \end{bmatrix} = \begin{bmatrix} \dfrac{8}{11} & \dfrac{5}{11} & -\dfrac{1}{11} \\ -\dfrac{7}{11} & -\dfrac{3}{11} & \dfrac{5}{11} \\ \dfrac{5}{11} & -\dfrac{1}{11} & -\dfrac{2}{11} \end{bmatrix}\begin{bmatrix} 7 \\ -6 \\ 4 \end{bmatrix} = \begin{bmatrix} 2 \\ -1 \\ 3 \end{bmatrix}$$

The solution is $\{(2,-1,3)\}$.

53. $\begin{cases} 2x+2y+3x=7 \\ 5x+3y+5z=3 \\ 3x+5y+z=-5 \end{cases} \Rightarrow \begin{bmatrix} 2 & 2 & 3 \\ 5 & 3 & 5 \\ 3 & 5 & 1 \end{bmatrix} \begin{bmatrix} x \\ y \\ z \end{bmatrix} = \begin{bmatrix} 7 \\ 3 \\ -5 \end{bmatrix}$

$$[A \mid I] = \begin{bmatrix} 2 & 2 & 3 & 1 & 0 & 0 \\ 5 & 3 & 5 & 0 & 1 & 0 \\ 3 & 5 & 1 & 0 & 0 & 1 \end{bmatrix} \xrightarrow[3R_1-2R_3 \to R_3]{5R_1-2R_2 \to R_2} \begin{bmatrix} 2 & 2 & 3 & 1 & 0 & 0 \\ 0 & 4 & 5 & 5 & -2 & 0 \\ 0 & -4 & 7 & 3 & 0 & -2 \end{bmatrix}$$

$$\xrightarrow[\frac{1}{12}(R_2+R_3) \to R_3]{R_2-2R_1 \to R_1} \begin{bmatrix} -4 & 0 & -1 & 3 & -2 & 0 \\ 0 & 4 & 5 & 5 & -2 & 0 \\ 0 & 0 & 1 & \frac{2}{3} & -\frac{1}{6} & -\frac{1}{6} \end{bmatrix} \xrightarrow[\frac{1}{4}(R_2-5R_3) \to R_2]{-\frac{1}{4}(R_1+R_3) \to R_1} \begin{bmatrix} 1 & 0 & 0 & -\frac{11}{12} & \frac{13}{24} & \frac{1}{24} \\ 0 & 1 & 0 & \frac{5}{12} & -\frac{7}{24} & \frac{5}{24} \\ 0 & 0 & 1 & \frac{2}{3} & -\frac{1}{6} & -\frac{1}{6} \end{bmatrix} \Rightarrow$$

$$A^{-1} = \begin{bmatrix} -\frac{11}{12} & \frac{13}{24} & \frac{1}{24} \\ \frac{5}{12} & -\frac{7}{24} & \frac{5}{24} \\ \frac{2}{3} & -\frac{1}{6} & -\frac{1}{6} \end{bmatrix}$$

$$X = A^{-1}B \Rightarrow \begin{bmatrix} x \\ y \\ z \end{bmatrix} = \begin{bmatrix} -\frac{11}{12} & \frac{13}{24} & \frac{1}{24} \\ \frac{5}{12} & -\frac{7}{24} & \frac{5}{24} \\ \frac{2}{3} & -\frac{1}{6} & -\frac{1}{6} \end{bmatrix} \begin{bmatrix} 7 \\ 3 \\ -5 \end{bmatrix} = \begin{bmatrix} -5 \\ 1 \\ 5 \end{bmatrix}$$

The solution is $\{(-5, 1, 5)\}$.

55. Let x = the amount invested in a treasury bill, y = the amount invested in bonds, and z = the amount invested in a mutual fund. Then we have

$$\begin{cases} x+y+z=90,000 \\ 0.03x+0.07y-0.11z=980 \\ 0.03x+0.07y+0.08z=5920 \end{cases} \Rightarrow \begin{bmatrix} 1 & 1 & 1 \\ 0.03 & 0.07 & -0.11 \\ 0.03 & 0.07 & 0.08 \end{bmatrix} \begin{bmatrix} x \\ y \\ z \end{bmatrix} = \begin{bmatrix} 90,000 \\ 980 \\ 5920 \end{bmatrix}$$

$$[A \mid I] = \begin{bmatrix} 1 & 1 & 1 & 1 & 0 & 0 \\ \frac{3}{100} & \frac{7}{100} & -\frac{11}{100} & 0 & 1 & 0 \\ \frac{3}{100} & \frac{7}{100} & \frac{8}{100} & 0 & 0 & 1 \end{bmatrix} \xrightarrow[-\frac{3}{100}R_3+R_3 \to]{-\frac{3}{100}R_1+R_2 \to R_2} \begin{bmatrix} 1 & 1 & 1 & 1 & 0 & 0 \\ 0 & \frac{1}{25} & -\frac{7}{50} & -\frac{3}{100} & 1 & 0 \\ 0 & \frac{1}{25} & \frac{1}{20} & -\frac{3}{100} & 0 & 1 \end{bmatrix}$$

$$\xrightarrow{-\frac{100}{19}(R_2-R_3) \to R_3} \begin{bmatrix} 1 & 1 & 1 & 1 & 0 & 0 \\ 0 & \frac{1}{25} & -\frac{7}{50} & -\frac{3}{100} & 1 & 0 \\ 0 & 0 & 1 & 0 & -\frac{100}{19} & \frac{100}{19} \end{bmatrix} \xrightarrow{25R_2 \to R_2} \begin{bmatrix} 1 & 1 & 1 & 1 & 0 & 0 \\ 0 & 1 & -\frac{7}{2} & -\frac{3}{4} & 25 & 0 \\ 0 & 0 & 1 & 0 & -\frac{100}{19} & \frac{100}{19} \end{bmatrix}$$

(continued on next page)

(continued)

$$\xrightarrow[\substack{R_2 + \frac{7}{2}R_3 \to R_2}]{R_1 - R_3 \to R_1} \begin{bmatrix} 1 & 1 & 0 & 1 & \frac{100}{19} & -\frac{100}{19} \\ 0 & 1 & 0 & -\frac{3}{4} & \frac{125}{19} & \frac{350}{19} \\ 0 & 0 & 1 & 0 & -\frac{100}{19} & \frac{100}{19} \end{bmatrix} \xrightarrow{R_1 - R_2 \to R_1} \begin{bmatrix} 1 & 1 & 0 & \frac{7}{4} & -\frac{25}{19} & -\frac{450}{19} \\ 0 & 1 & 0 & -\frac{3}{4} & \frac{125}{19} & \frac{350}{19} \\ 0 & 0 & 1 & 0 & -\frac{100}{19} & \frac{100}{19} \end{bmatrix}$$

$$A^{-1} = \begin{bmatrix} \frac{7}{4} & -\frac{25}{19} & -\frac{450}{19} \\ -\frac{3}{4} & \frac{125}{19} & \frac{350}{19} \\ 0 & -\frac{100}{19} & \frac{100}{19} \end{bmatrix} \Rightarrow X = A^{-1}B \Rightarrow \begin{bmatrix} x \\ y \\ z \end{bmatrix} = \begin{bmatrix} \frac{7}{4} & -\frac{25}{19} & -\frac{450}{19} \\ -\frac{3}{4} & \frac{125}{19} & \frac{350}{19} \\ 0 & -\frac{100}{19} & \frac{100}{19} \end{bmatrix} \begin{bmatrix} 90,000 \\ 980 \\ 5920 \end{bmatrix} = \begin{bmatrix} 16,000 \\ 48,000 \\ 26,000 \end{bmatrix}$$

Liz invested $16,000 in a treasury bill, $48,000 in bonds, and $26,000 in a mutual fund.

57. $\begin{cases} V - E + R = 2 \\ 2E = 3V \\ 2R = E - 1 \end{cases} \Rightarrow \begin{cases} V - E + R = 2 \\ -3V + 2E = 0 \\ -E + 2R = -1 \end{cases} \Rightarrow \begin{bmatrix} 1 & -1 & 1 \\ -3 & 2 & 0 \\ 0 & -1 & 2 \end{bmatrix} \begin{bmatrix} x \\ y \\ z \end{bmatrix} = \begin{bmatrix} 2 \\ 0 \\ 1 \end{bmatrix}$

$$[A \mid I] = \begin{bmatrix} 1 & -1 & 1 & 1 & 0 & 0 \\ -3 & 2 & 0 & 0 & 1 & 0 \\ 0 & -1 & 2 & 0 & 0 & 1 \end{bmatrix} \xrightarrow{3R_1 + R_2 \to R_2} \begin{bmatrix} 1 & -1 & 1 & 1 & 0 & 0 \\ 0 & -1 & 3 & 3 & 1 & 0 \\ 0 & -1 & 2 & 0 & 0 & 1 \end{bmatrix}$$

$$\xrightarrow{-R_2 \to R_2} \begin{bmatrix} 1 & -1 & 1 & 1 & 0 & 0 \\ 0 & 1 & -3 & -3 & -1 & 0 \\ 0 & -1 & 2 & 0 & 0 & 1 \end{bmatrix} \xrightarrow[\substack{-(R_3 + R_2) \to R_3}]{R_1 + R_2 \to R_1} \begin{bmatrix} 1 & 0 & -2 & -2 & -1 & 0 \\ 0 & 1 & -3 & -3 & -1 & 0 \\ 0 & 0 & 1 & 3 & 1 & -1 \end{bmatrix}$$

$$\xrightarrow[\substack{R_2 + 3R_3 \to R_2}]{R_1 + 2R_3 \to R_1} \begin{bmatrix} 1 & 0 & 0 & 4 & 1 & -2 \\ 0 & 1 & 0 & 6 & 2 & -3 \\ 0 & 0 & 1 & 3 & 1 & -1 \end{bmatrix}$$

$$A^{-1} = \begin{bmatrix} 4 & 1 & -2 \\ 6 & 2 & -3 \\ 3 & 1 & -1 \end{bmatrix} \Rightarrow X = A^{-1}B \Rightarrow \begin{bmatrix} x \\ y \\ z \end{bmatrix} = \begin{bmatrix} 4 & 1 & -2 \\ 6 & 2 & -3 \\ 3 & 1 & -1 \end{bmatrix} \begin{bmatrix} 2 \\ 0 \\ 1 \end{bmatrix} = \begin{bmatrix} 10 \\ 15 \\ 7 \end{bmatrix}$$

There are 10 vertices, 15 edges, and 7 regions.

59. $X = \left(\begin{bmatrix} 1 & 0 \\ 0 & 1 \end{bmatrix} - \begin{bmatrix} 0.1 & 0.4 \\ 0.5 & 0.2 \end{bmatrix} \right)^{-1} \begin{bmatrix} 50 \\ 30 \end{bmatrix}$; $(I-A)^{-1} = \begin{bmatrix} \dfrac{9}{10} & -\dfrac{2}{5} \\ -\dfrac{1}{2} & \dfrac{4}{5} \end{bmatrix}^{-1}$

$\left[\begin{array}{cc|cc} \dfrac{9}{10} & -\dfrac{2}{5} & 1 & 0 \\ -\dfrac{1}{2} & \dfrac{4}{5} & 0 & 1 \end{array} \right] \xrightarrow{\frac{1}{2}R_1 + \frac{9}{10}R_2 \to R_2} \left[\begin{array}{cc|cc} \dfrac{9}{10} & -\dfrac{2}{5} & 1 & 0 \\ 0 & \dfrac{13}{25} & \dfrac{1}{2} & \dfrac{9}{10} \end{array} \right] \xrightarrow{\frac{13}{25}R_1 + \frac{2}{5}R_2 \to R_1} \left[\begin{array}{cc|cc} \dfrac{117}{250} & 0 & \dfrac{18}{25} & \dfrac{9}{25} \\ 0 & \dfrac{13}{25} & \dfrac{1}{2} & \dfrac{9}{10} \end{array} \right]$

$\xrightarrow[\frac{25}{13}R_2 \to R_2]{\frac{250}{117}R_1 \to R_1} \left[\begin{array}{cc|cc} 1 & 0 & \dfrac{20}{13} & \dfrac{10}{13} \\ 0 & 1 & \dfrac{25}{26} & \dfrac{45}{26} \end{array} \right]$

$(I-A)^{-1} = \begin{bmatrix} \dfrac{20}{13} & \dfrac{10}{13} \\ \dfrac{25}{26} & \dfrac{45}{26} \end{bmatrix} \Rightarrow X = (I-A)^{-1}D = \begin{bmatrix} \dfrac{20}{13} & \dfrac{10}{13} \\ \dfrac{25}{26} & \dfrac{45}{26} \end{bmatrix} \begin{bmatrix} 50 \\ 30 \end{bmatrix} = \begin{bmatrix} 100 \\ 100 \end{bmatrix}$

61. $X = \left(\begin{bmatrix} 1 & 0 & 0 \\ 0 & 1 & 0 \\ 0 & 0 & 1 \end{bmatrix} - \begin{bmatrix} 0.2 & 0.2 & 0 \\ 0.1 & 0.1 & 0.3 \\ 0.1 & 0 & 0.2 \end{bmatrix} \right)^{-1} \begin{bmatrix} 400 \\ 600 \\ 800 \end{bmatrix}$

$(I-A)^{-1} = \begin{bmatrix} \dfrac{4}{5} & -\dfrac{1}{5} & 0 \\ -\dfrac{1}{10} & \dfrac{9}{10} & -\dfrac{3}{10} \\ -\dfrac{1}{10} & 0 & \dfrac{4}{5} \end{bmatrix}^{-1} \Rightarrow \left[\begin{array}{ccc|ccc} \dfrac{4}{5} & -\dfrac{1}{5} & 0 & 1 & 0 & 0 \\ -\dfrac{1}{10} & \dfrac{9}{10} & -\dfrac{3}{10} & 0 & 1 & 0 \\ -\dfrac{1}{10} & 0 & \dfrac{4}{5} & 0 & 0 & 1 \end{array} \right] \xrightarrow[\frac{1}{10}R_1 + \frac{4}{5}R_3 \to R_3]{\frac{1}{10}R_1 + \frac{4}{5}R_2 \to R_2} \left[\begin{array}{ccc|ccc} \dfrac{4}{5} & -\dfrac{1}{5} & 0 & 1 & 0 & 0 \\ 0 & \dfrac{7}{10} & -\dfrac{6}{25} & \dfrac{1}{10} & \dfrac{4}{5} & 0 \\ 0 & -\dfrac{1}{50} & \dfrac{16}{25} & \dfrac{1}{10} & 0 & \dfrac{4}{5} \end{array} \right]$

$\xrightarrow{\frac{625}{277}\left(\frac{1}{50}R_2 + \frac{7}{10}R_3 \right) \to R_3} \left[\begin{array}{ccc|ccc} \dfrac{4}{5} & -\dfrac{1}{5} & 0 & 1 & 0 & 0 \\ 0 & \dfrac{7}{10} & -\dfrac{6}{25} & \dfrac{1}{10} & \dfrac{4}{5} & 0 \\ 0 & 0 & 1 & \dfrac{45}{277} & \dfrac{10}{277} & \dfrac{350}{277} \end{array} \right]$

$\xrightarrow[\frac{10}{7}\left(\frac{6}{25}R_3 + R_2 \to R_2 \right)]{5R_1 \to R_1} \left[\begin{array}{ccc|ccc} 4 & -1 & 0 & 5 & 0 & 0 \\ 0 & 1 & 0 & \dfrac{55}{277} & \dfrac{320}{277} & \dfrac{120}{277} \\ 0 & 0 & 1 & \dfrac{45}{277} & \dfrac{10}{277} & \dfrac{350}{277} \end{array} \right] \xrightarrow{\frac{1}{4}(R_1 + R_2) \to R_1} \left[\begin{array}{ccc|ccc} 1 & 0 & 0 & \dfrac{360}{277} & \dfrac{80}{277} & \dfrac{30}{277} \\ 0 & 1 & 0 & \dfrac{55}{277} & \dfrac{320}{277} & \dfrac{120}{277} \\ 0 & 0 & 1 & \dfrac{45}{277} & \dfrac{10}{277} & \dfrac{350}{277} \end{array} \right] \Rightarrow$

$(I-A)^{-1} = \begin{bmatrix} \dfrac{360}{277} & \dfrac{80}{277} & \dfrac{30}{277} \\ \dfrac{55}{277} & \dfrac{320}{277} & \dfrac{120}{277} \\ \dfrac{45}{277} & \dfrac{10}{277} & \dfrac{350}{277} \end{bmatrix}$; $X = (I-A)^{-1}D = \begin{bmatrix} \dfrac{360}{277} & \dfrac{80}{277} & \dfrac{30}{277} \\ \dfrac{55}{277} & \dfrac{320}{277} & \dfrac{120}{277} \\ \dfrac{45}{277} & \dfrac{10}{277} & \dfrac{350}{277} \end{bmatrix} \begin{bmatrix} 400 \\ 600 \\ 800 \end{bmatrix} = \begin{bmatrix} \dfrac{216{,}000}{277} \\ \dfrac{310{,}000}{277} \\ \dfrac{304{,}000}{277} \end{bmatrix}$

Answers will vary in exercises 63−66.

63. a. Associate each letter in the phrase with the number representing its position in the alphabet, and partition the numbers into groups of two, forming a 2×9 matrix:

$$M = \begin{bmatrix} 3 & 14 & 15 & 6 & 14 & 3 & 12 & 13 & 15 \\ 1 & 14 & 20 & 9 & 4 & 15 & 15 & 2 & 0 \end{bmatrix} \text{ Let } A = \begin{bmatrix} 1 & 2 \\ 2 & 1 \end{bmatrix} \text{ Then,}$$

$$AM = \begin{bmatrix} 1 & 2 \\ 2 & 1 \end{bmatrix}\begin{bmatrix} 3 & 14 & 15 & 6 & 14 & 3 & 12 & 13 & 15 \\ 1 & 14 & 20 & 9 & 4 & 15 & 15 & 2 & 0 \end{bmatrix} = \begin{bmatrix} 5 & 42 & 55 & 24 & 22 & 33 & 42 & 17 & 15 \\ 7 & 42 & 50 & 21 & 32 & 21 & 39 & 28 & 30 \end{bmatrix}$$

The cryptogram is 5 7 42 42 55 50 24 21 22 32 33 21 42 39 17 28 15 30.

b. $A^{-1} = \begin{bmatrix} 1 & 2 & | & 1 & 0 \\ 2 & 1 & | & 0 & 1 \end{bmatrix} \xrightarrow{\frac{1}{3}(2R_1 - R_2) \to R_2} \begin{bmatrix} 1 & 2 & | & 1 & 0 \\ 0 & 1 & | & \frac{2}{3} & -\frac{1}{3} \end{bmatrix} \xrightarrow{R_1 - 2R_2 \to R_1} \begin{bmatrix} 1 & 0 & | & -\frac{1}{3} & \frac{2}{3} \\ 0 & 1 & | & \frac{2}{3} & -\frac{1}{3} \end{bmatrix} = \begin{bmatrix} -\frac{1}{3} & \frac{2}{3} \\ \frac{2}{3} & -\frac{1}{3} \end{bmatrix}$

$$M = A^{-1}(AM) = \begin{bmatrix} -\frac{1}{3} & \frac{2}{3} \\ \frac{2}{3} & -\frac{1}{3} \end{bmatrix}\begin{bmatrix} 5 & 42 & 55 & 24 & 22 & 33 & 42 & 17 & 15 \\ 7 & 42 & 50 & 21 & 32 & 21 & 39 & 28 & 30 \end{bmatrix}$$

$$= \begin{bmatrix} 3 & 14 & 15 & 6 & 14 & 3 & 12 & 13 & 15 \\ 1 & 14 & 20 & 9 & 4 & 15 & 15 & 2 & 0 \end{bmatrix}$$

65. a. Associate each letter in the phrase with the number representing its position in the alphabet, and partition the numbers into groups of three, forming a 3×6 matrix:

$$M = \begin{bmatrix} 3 & 14 & 6 & 4 & 12 & 2 \\ 1 & 15 & 9 & 3 & 15 & 15 \\ 14 & 20 & 14 & 15 & 13 & 0 \end{bmatrix} \text{ Let } A = \begin{bmatrix} 2 & 1 & 1 \\ 1 & 2 & 1 \\ 1 & 1 & 2 \end{bmatrix} \text{ Then,}$$

$$AM = \begin{bmatrix} 2 & 1 & 1 \\ 1 & 2 & 1 \\ 1 & 1 & 2 \end{bmatrix}\begin{bmatrix} 3 & 14 & 6 & 4 & 12 & 2 \\ 1 & 15 & 9 & 3 & 15 & 15 \\ 14 & 20 & 14 & 15 & 13 & 0 \end{bmatrix} = \begin{bmatrix} 21 & 63 & 35 & 26 & 52 & 19 \\ 19 & 64 & 38 & 25 & 55 & 32 \\ 32 & 69 & 43 & 37 & 53 & 17 \end{bmatrix}$$

The cryptogram is 21 39 32 63 64 69 35 38 43 26 25 37 52 55 53 19 32 17.

b. $A^{-1} = \begin{bmatrix} 2 & 1 & 1 & | & 1 & 0 & 0 \\ 1 & 2 & 1 & | & 0 & 1 & 0 \\ 1 & 1 & 2 & | & 0 & 0 & 1 \end{bmatrix} \xrightarrow[R_1 - 2R_3 \to R_3]{R_1 - 2R_2 \to R_2} \begin{bmatrix} 2 & 1 & 1 & | & 1 & 0 & 0 \\ 0 & -3 & -1 & | & 1 & -2 & 0 \\ 0 & -1 & -3 & | & 1 & 0 & -2 \end{bmatrix}$

$$\xrightarrow{\frac{1}{8}(R_2 - 3R_3) \to R_3} \begin{bmatrix} 2 & 1 & 1 & | & 1 & 0 & 0 \\ 0 & -3 & -1 & | & 1 & -2 & 0 \\ 0 & 0 & 1 & | & -\frac{1}{4} & -\frac{1}{4} & \frac{3}{4} \end{bmatrix} \xrightarrow[-\frac{1}{3}(R_3 + R_2) \to R_2]{R_1 - R_3 \to R_1} \begin{bmatrix} 2 & 1 & 0 & | & \frac{5}{4} & \frac{1}{4} & -\frac{3}{4} \\ 0 & 1 & 0 & | & -\frac{1}{4} & \frac{3}{4} & -\frac{1}{4} \\ 0 & 0 & 1 & | & -\frac{1}{4} & -\frac{1}{4} & \frac{3}{4} \end{bmatrix}$$

$$\xrightarrow{\frac{1}{2}(R_1 - R_2) \to R_1} \begin{bmatrix} 1 & 0 & 0 & | & \frac{3}{4} & -\frac{1}{4} & -\frac{1}{4} \\ 0 & 1 & 0 & | & -\frac{1}{4} & \frac{3}{4} & -\frac{1}{4} \\ 0 & 0 & 1 & | & -\frac{1}{4} & -\frac{1}{4} & \frac{3}{4} \end{bmatrix} = \begin{bmatrix} \frac{3}{4} & -\frac{1}{4} & -\frac{1}{4} \\ -\frac{1}{4} & \frac{3}{4} & -\frac{1}{4} \\ -\frac{1}{4} & -\frac{1}{4} & \frac{3}{4} \end{bmatrix}$$

$$M = A^{-1}(AM) = \begin{bmatrix} \frac{3}{4} & -\frac{1}{4} & -\frac{1}{4} \\ -\frac{1}{4} & \frac{3}{4} & -\frac{1}{4} \\ -\frac{1}{4} & -\frac{1}{4} & \frac{3}{4} \end{bmatrix} \begin{bmatrix} 21 & 63 & 35 & 26 & 52 & 19 \\ 19 & 64 & 38 & 25 & 55 & 32 \\ 32 & 69 & 43 & 37 & 53 & 17 \end{bmatrix} = \begin{bmatrix} 3 & 14 & 6 & 4 & 12 & 2 \\ 1 & 15 & 9 & 3 & 15 & 15 \\ 14 & 20 & 14 & 15 & 13 & 0 \end{bmatrix}$$

6.3 Beyond the Basics

67. a. Yes, by the definition of *inverse* if $AB = I$, then $BA = I$. So $B = A^{-1}$ and $A = B^{-1}$.

b. $\left(A^{-1}\right)^{-1} = A$

69. $I = A^2B = A(AB)$, so AB is the inverse of A.

71. $ABB^{-1}A^{-1} = AIA^{-1} = AA^{-1} = I$
$B^{-1}A^{-1}AB = B^{-1}IB = B^{-1}B = I \Rightarrow AB$ is
invertible and $(AB)^{-1} = B^{-1}A^{-1}$.

73. If A is invertible, then there exists a matrix D such that $AD = I$ and $DA = I$. Then
$AB = AC \Leftrightarrow DAB = DAC \Leftrightarrow IB = IC \Leftrightarrow B = C$.

75. If A is invertible, then $I = A^{-1}A = A^{-1}(AB) = \left(A^{-1}A\right)B = IB = B$.

77. a. $\begin{bmatrix} 3 & 4 \\ 2 & 3 \end{bmatrix}^2 - 6\begin{bmatrix} 3 & 4 \\ 2 & 3 \end{bmatrix} + \begin{bmatrix} 1 & 0 \\ 0 & 1 \end{bmatrix} = \begin{bmatrix} 17 & 24 \\ 12 & 17 \end{bmatrix} - \begin{bmatrix} 18 & 24 \\ 12 & 18 \end{bmatrix} + \begin{bmatrix} 1 & 0 \\ 0 & 1 \end{bmatrix} = \begin{bmatrix} 0 & 0 \\ 0 & 0 \end{bmatrix}$

b. $A^2 - 6A + I = 0 \Rightarrow I = 6A - A^2 \Rightarrow AA^{-1} = 6A - A^2 \Rightarrow A^{-1} = 6I - A$

c. $A^{-1} = 6I - A = \begin{bmatrix} 6 & 0 \\ 0 & 6 \end{bmatrix} - \begin{bmatrix} 3 & 4 \\ 2 & 3 \end{bmatrix} = \begin{bmatrix} 3 & -4 \\ -2 & 3 \end{bmatrix}$

79. $\begin{cases} x - y - z - w = -4 \\ x + y + z - w = 2 \\ 2x + y + z - w = 3 \\ x - y + z - w = -2 \end{cases} \Rightarrow \begin{bmatrix} 1 & -1 & -1 & -1 \\ 1 & 1 & 1 & -1 \\ 2 & 1 & 1 & -1 \\ 1 & -1 & 1 & -1 \end{bmatrix} \begin{bmatrix} x \\ y \\ z \\ w \end{bmatrix} = \begin{bmatrix} -4 \\ 2 \\ 3 \\ -2 \end{bmatrix} \Rightarrow \begin{bmatrix} 1 & -1 & -1 & -1 \\ 1 & 1 & 1 & -1 \\ 2 & 1 & 1 & -1 \\ 1 & -1 & 1 & -1 \end{bmatrix}^{-1} \begin{bmatrix} -4 \\ 2 \\ 3 \\ -2 \end{bmatrix} = \begin{bmatrix} x \\ y \\ z \\ w \end{bmatrix}$

Using a graphing calculator, we have

$$\begin{bmatrix} 1 & -1 & -1 & -1 \\ 1 & 1 & 1 & -1 \\ 2 & 1 & 1 & -1 \\ 1 & -1 & 1 & -1 \end{bmatrix}^{-1} = \begin{bmatrix} 0 & -1 & 1 & 0 \\ 0 & \frac{1}{2} & 0 & -\frac{1}{2} \\ -\frac{1}{2} & 0 & 0 & \frac{1}{2} \\ -\frac{1}{2} & -\frac{3}{2} & 1 & 0 \end{bmatrix} \quad \text{and then} \quad \begin{bmatrix} 0 & -1 & 1 & 0 \\ 0 & \frac{1}{2} & 0 & -\frac{1}{2} \\ -\frac{1}{2} & 0 & 0 & \frac{1}{2} \\ -\frac{1}{2} & -\frac{3}{2} & 1 & 0 \end{bmatrix} \begin{bmatrix} -4 \\ 2 \\ 3 \\ -2 \end{bmatrix} = \begin{bmatrix} 1 \\ 2 \\ 1 \\ 2 \end{bmatrix}$$

Solution set: $\{(1, 2, 1, 2)\}$

81. a. $ABB^{-1} = AI = A \Rightarrow A = \begin{bmatrix} -1 & 3 \\ 0 & 2 \end{bmatrix}\begin{bmatrix} 1 & 2 \\ 3 & 4 \end{bmatrix} = \begin{bmatrix} 8 & 10 \\ 6 & 8 \end{bmatrix}$

b. $B^{-1}BA = IA = A \Rightarrow A = \begin{bmatrix} 1 & 2 \\ 3 & 4 \end{bmatrix}\begin{bmatrix} -1 & 3 \\ 0 & 2 \end{bmatrix} = \begin{bmatrix} -1 & 7 \\ -3 & 17 \end{bmatrix}$

c. $B^{-1}BAB^{-1} = IAB^{-1} = AB^{-1};$ $\qquad AB^{-1} = \begin{bmatrix} 1 & 2 \\ 3 & 4 \end{bmatrix}\begin{bmatrix} 1 & 1 \\ 1 & 2 \end{bmatrix} = \begin{bmatrix} 3 & 5 \\ 7 & 11 \end{bmatrix}$

$\begin{bmatrix} a & b \\ c & d \end{bmatrix}\begin{bmatrix} 1 & 2 \\ 3 & 4 \end{bmatrix} = \begin{bmatrix} 3 & 5 \\ 7 & 11 \end{bmatrix} \Rightarrow \begin{bmatrix} a+3b & 2a+4b \\ c+3d & 2c+4d \end{bmatrix} = \begin{bmatrix} 3 & 5 \\ 7 & 11 \end{bmatrix} \Rightarrow \begin{cases} a+3b=3 \\ 2a+4b=5 \\ c+3d=7 \\ 2c+4d=11 \end{cases} \Rightarrow$

$a = \dfrac{3}{2}, b = \dfrac{1}{2}, c = \dfrac{5}{2}, d = \dfrac{3}{2} \Rightarrow A = \begin{bmatrix} \dfrac{3}{2} & \dfrac{1}{2} \\ \dfrac{5}{2} & \dfrac{3}{2} \end{bmatrix}$

d. $B^{-1}ABB^{-1} = B^{-1}AI = B^{-1}A \Rightarrow B^{-1}A = \begin{bmatrix} 1 & 1 \\ 1 & 2 \end{bmatrix}\begin{bmatrix} 1 & 2 \\ 3 & 4 \end{bmatrix} = \begin{bmatrix} 4 & 6 \\ 7 & 10 \end{bmatrix}$

$\begin{bmatrix} 1 & 2 \\ 3 & 4 \end{bmatrix}\begin{bmatrix} a & b \\ c & d \end{bmatrix} = \begin{bmatrix} 4 & 6 \\ 7 & 10 \end{bmatrix} \Rightarrow \begin{bmatrix} a+2c & b+2d \\ 3a+4c & 3b+4d \end{bmatrix} = \begin{bmatrix} 4 & 6 \\ 7 & 10 \end{bmatrix} \Rightarrow \begin{cases} a+2c=4 \\ b+2d=6 \\ 3a+4c=7 \\ 3b+4d=10 \end{cases} \Rightarrow$

$\begin{bmatrix} 1 & 0 & 2 & 0 \\ 0 & 1 & 0 & 2 \\ 3 & 0 & 4 & 0 \\ 0 & 3 & 0 & 4 \end{bmatrix}\begin{bmatrix} a \\ b \\ c \\ d \end{bmatrix} = \begin{bmatrix} 4 \\ 6 \\ 7 \\ 10 \end{bmatrix} \Rightarrow \begin{bmatrix} a \\ b \\ c \\ d \end{bmatrix} = \begin{bmatrix} 1 & 0 & 2 & 0 \\ 0 & 1 & 0 & 2 \\ 3 & 0 & 4 & 0 \\ 0 & 3 & 0 & 4 \end{bmatrix}^{-1}\begin{bmatrix} 4 \\ 6 \\ 7 \\ 10 \end{bmatrix} = \begin{bmatrix} -2 & 0 & 1 & 0 \\ 0 & -2 & 0 & 1 \\ \frac{3}{2} & 0 & -\frac{1}{2} & 0 \\ 0 & \frac{3}{2} & 0 & -\frac{1}{2} \end{bmatrix}\begin{bmatrix} 4 \\ 6 \\ 7 \\ 10 \end{bmatrix} = \begin{bmatrix} -1 \\ -2 \\ \frac{5}{2} \\ 4 \end{bmatrix} \Rightarrow$

$A = \begin{bmatrix} -1 & -2 \\ \dfrac{5}{2} & 4 \end{bmatrix}$

6.3 Critical Thinking/Discussion/Writing

83. a. True. $A^2B = AAB = ABA = BAA = BA^2$

b. False. For example, if $A = I$ and $B = -I$, then $A + B = \begin{bmatrix} 1 & 0 \\ 0 & 1 \end{bmatrix} + \begin{bmatrix} -1 & 0 \\ 0 & -1 \end{bmatrix} = \begin{bmatrix} 0 & 0 \\ 0 & 0 \end{bmatrix}$, which is not invertible.

84. Let $A = \begin{bmatrix} 0 & -1 \\ 0 & 0 \end{bmatrix}$. Then $A^2 = \begin{bmatrix} 0 & 0 \\ 0 & 0 \end{bmatrix}$.

$I + A = \begin{bmatrix} 1 & 0 \\ 0 & 1 \end{bmatrix} + \begin{bmatrix} 0 & -1 \\ 0 & 0 \end{bmatrix} = \begin{bmatrix} 1 & -1 \\ 0 & 1 \end{bmatrix}$

$(I + A)^{-1} = \begin{bmatrix} 1 & 1 \\ 0 & 1 \end{bmatrix}$

$I - A = \begin{bmatrix} 1 & 0 \\ 0 & 1 \end{bmatrix} - \begin{bmatrix} 0 & -1 \\ 0 & 0 \end{bmatrix} = \begin{bmatrix} 1 & 1 \\ 0 & 1 \end{bmatrix}$

85. True. A matrix has an inverse if and only if it is square.

6.3 Getting Ready for the Next Section

87. 1 **89.** -1

91. $\begin{bmatrix} -1 \\ -3 \end{bmatrix}\begin{bmatrix} 5 & 0 \end{bmatrix} = \begin{bmatrix} -1(5) & -1(0) \\ -3(5) & -3(0) \end{bmatrix} = \begin{bmatrix} -5 & 0 \\ -15 & 0 \end{bmatrix}$

93. $(-1)\begin{bmatrix} -6 & 7 \\ 4 & -1 \end{bmatrix} = \begin{bmatrix} 6 & -7 \\ -4 & 1 \end{bmatrix}$

In exercises 94–97, be sure to check the solution in the original equations.

95. $\begin{cases} 16x - 9y = -5 & (1) \\ 10x + 18y = -11 & (2) \end{cases}$

Multiply equation (1) by 2, then add the resulting equation and equation (2).

$32x - 18y = -10$
$\underline{10x + 18y = -11}$

$\qquad 42x = -21 \Rightarrow x = -\dfrac{1}{2}$

Substitute $x = -\dfrac{1}{2}$ in equation (2) and solve for y.

(continued on next page)

(*continued*)

$$10\left(-\frac{1}{2}\right) + 18y = -11 \Rightarrow -5 + 18y = -11 \Rightarrow$$

$$18y = -6 \Rightarrow y = -\frac{1}{3}$$

Solution set: $\left\{\left(-\frac{1}{2}, -\frac{1}{3}\right)\right\}$

97. $\begin{cases} 2x - y + 2z = 3 & (1) \\ 2x + 2y - z = 0 & (2) \\ -x + 2y + 2z = -12 & (3) \end{cases}$

Write the augmented matrix.

$$\begin{bmatrix} 2 & -1 & 2 & | & 3 \\ 2 & 2 & -1 & | & 0 \\ -1 & 2 & 2 & | & -12 \end{bmatrix}$$

Now use Gauss-Jordan elimination to solve the system.

$$\begin{bmatrix} 2 & -1 & 2 & | & 3 \\ 2 & 2 & -1 & | & 0 \\ -1 & 2 & 2 & | & -12 \end{bmatrix}$$

$$\xrightarrow[R_1 + 2R_3 \to R_3]{R_2 - R_1 \to R_2} \begin{bmatrix} 2 & -1 & 2 & | & 3 \\ 0 & 3 & -3 & | & -3 \\ 0 & 3 & 6 & | & -21 \end{bmatrix}$$

$$\xrightarrow{\frac{1}{9}(R_3 - R_2) \to R_3} \begin{bmatrix} 2 & -1 & 2 & | & 3 \\ 0 & 3 & -3 & | & -3 \\ 0 & 0 & 1 & | & -2 \end{bmatrix}$$

$$\xrightarrow[\frac{1}{3}(R_2 + 3R_3) \to R_2]{R_1 - 2R_3 \to R_1} \begin{bmatrix} 2 & -1 & 0 & | & 7 \\ 0 & 1 & 0 & | & -3 \\ 0 & 0 & 1 & | & -2 \end{bmatrix}$$

$$\xrightarrow{\frac{1}{2}(R_1 + R_2) \to R_1} \begin{bmatrix} 1 & 0 & 0 & | & 2 \\ 0 & 1 & 0 & | & -3 \\ 0 & 0 & 1 & | & -2 \end{bmatrix}$$

Solution set: $\{(2, -3, -2)\}$

6.4 Determinants and Cramer's Rule

6.4 Practice Problems

1. a. $\begin{vmatrix} 5 & 1 \\ 3 & -7 \end{vmatrix} = 5(-7) - 1(3) = -38$

   b. $\begin{vmatrix} 2 & -9 \\ -4 & 18 \end{vmatrix} = 2(18) - (-9)(-4) = 0$

2. a. $M_{11} = \begin{bmatrix} 3 & -1 & 2 \\ 4 & 5 & 6 \\ 7 & 1 & 2 \end{bmatrix} = \begin{vmatrix} 5 & 6 \\ 1 & 2 \end{vmatrix}$

   $= (5)(2) - 6(1) = 4$

$M_{23} = \begin{bmatrix} 3 & -1 & 2 \\ 4 & 5 & 6 \\ 7 & 1 & 2 \end{bmatrix} = \begin{vmatrix} 3 & -1 \\ 7 & 1 \end{vmatrix}$

$= 3(1) - (-1)(7) = 10$

$M_{32} = \begin{bmatrix} 3 & -1 & 2 \\ 4 & 5 & 6 \\ 7 & 1 & 2 \end{bmatrix} = \begin{vmatrix} 3 & 2 \\ 4 & 6 \end{vmatrix}$

$= 3(6) - 2(4) = 10$

b. $A_{11} = (-1)^{1+1} M_{11} = 4$

$A_{23} = (-1)^{2+3} M_{23} = -10$

$A_{32} = (-1)^{3+2} M_{32} = -10$

3. Expand by the third row:

$$\begin{vmatrix} 2 & -3 & 7 \\ -2 & -1 & 9 \\ 0 & 2 & -9 \end{vmatrix}$$

$$= a_{31}A_{31} + a_{32}A_{32} + a_{33}A_{33}$$

$$= 0 + 2(-1)^{3+2}\begin{vmatrix} 2 & 7 \\ -2 & 9 \end{vmatrix} - 9(-1)^{3+3}\begin{vmatrix} 2 & -3 \\ -2 & -1 \end{vmatrix}$$

$$= 0 - 2(32) - 9(-8) = 8$$

4. $D = \begin{vmatrix} 2 & 3 \\ 5 & 9 \end{vmatrix} = 3, \; D_x = \begin{vmatrix} 7 & 3 \\ 4 & 9 \end{vmatrix} = 51$

$D_y = \begin{vmatrix} 2 & 7 \\ 5 & 4 \end{vmatrix} = -27$

$x = \dfrac{D_x}{D} = \dfrac{51}{3} = 17; \; y = \dfrac{D_y}{D} = \dfrac{-27}{3} = -9$

The solution is $\{(17, -9)\}$.

5. $D = \begin{vmatrix} 3 & 2 & 1 \\ 4 & 3 & 1 \\ 5 & 1 & 1 \end{vmatrix} = -3, \; D_x = \begin{vmatrix} 4 & 2 & 1 \\ 5 & 3 & 1 \\ 9 & 1 & 1 \end{vmatrix} = -6$

$D_y = \begin{vmatrix} 3 & 4 & 1 \\ 4 & 5 & 1 \\ 5 & 9 & 1 \end{vmatrix} = 3, \; D_z = \begin{vmatrix} 3 & 2 & 4 \\ 4 & 3 & 5 \\ 5 & 1 & 9 \end{vmatrix} = 0$

$x = \dfrac{D_x}{D} = \dfrac{-6}{-3} = 2; \; y = \dfrac{D_y}{D} = \dfrac{3}{-3} = -1$

$z = \dfrac{D_z}{D} = \dfrac{0}{-3} = 0$

Solution set: $\{(2, -1, 0)\}$

6.
$$AD = \begin{bmatrix} -1 & 2 \\ 2 & -1 \end{bmatrix} \begin{matrix} P & Q & R & S \\ \begin{bmatrix} 0 & 4 & 2 & 1 \\ 0 & 0 & 2 & 2 \end{bmatrix} \end{matrix}$$
$$= \begin{matrix} P' & Q' & R' & S' \\ \begin{bmatrix} 0 & -4 & 2 & 3 \\ 0 & 6 & 2 & 0 \end{bmatrix} \end{matrix}$$

From Example 6, $A_{PQRS} = 5$ units^2.

$$\begin{aligned} A_{P'Q'R'S'} &= |\det A| \cdot A_{PQRS} \\ &= |-1(-1) - 2(2)| \cdot 5 \text{ units}^2 \\ &= |-3| \cdot 5 \text{ units}^2 \\ &= 15 \text{ units}^2 \end{aligned}$$

6.4 Concepts and Vocabulary

1. The determinant of $\begin{bmatrix} a & b \\ c & d \end{bmatrix}$ is $\underline{ad - bc}$.

3. To expand an $n \times n$ determinant, you multiply each element of some row (or column) by its <u>cofactor</u> and add the result.

5. False.

7. True

6.4 Building Skills

9. $\begin{vmatrix} 2 & 3 \\ 4 & 5 \end{vmatrix} = 2(5) - 4(3) = -2$

11. $\begin{vmatrix} 4 & -2 \\ 3 & -3 \end{vmatrix} = 4(-3) - (-2)(3) = -6$

13. $\begin{vmatrix} -1 & -3 \\ -4 & -5 \end{vmatrix} = (-1)(-5) - (-3)(-4) = -7$

15. $\begin{vmatrix} \frac{3}{8} & \frac{1}{2} \\ -\frac{1}{9} & 5 \end{vmatrix} = \left(\frac{3}{8}\right)(5) - \left(\frac{1}{2}\right)\left(-\frac{1}{9}\right) = \frac{139}{72}$

17. $\begin{vmatrix} \sqrt{a} & \sqrt{b} \\ \sqrt{b} & \sqrt{a} \end{vmatrix} = a - b$

19. a. $M_{21} = \begin{bmatrix} 2 & -3 & 4 \\ 1 & -1 & 2 \\ 0 & 1 & 2 \end{bmatrix} = \begin{vmatrix} -3 & 4 \\ 1 & 2 \end{vmatrix}$
$= (-3)(2) - 4(1) = -10$

b. $M_{23} = \begin{bmatrix} 2 & -3 & 4 \\ 1 & -1 & 2 \\ 0 & 1 & 2 \end{bmatrix} = \begin{vmatrix} 2 & -3 \\ 0 & 1 \end{vmatrix} = 2(1) - 0 = 2$

c. $M_{32} = \begin{bmatrix} 2 & -3 & 4 \\ 1 & -1 & 2 \\ 0 & 1 & 2 \end{bmatrix} = \begin{vmatrix} 2 & 4 \\ 1 & 2 \end{vmatrix} = 2(2) - 4(1) = 0$

21. a. $M_{11} = \begin{bmatrix} 2 & -3 & 4 \\ 1 & -1 & 2 \\ 0 & 1 & 2 \end{bmatrix} = \begin{vmatrix} -1 & 2 \\ 1 & 2 \end{vmatrix}$
$= (-1)(2) - 2(1) = -4$

b. $M_{22} = \begin{bmatrix} 2 & -3 & 4 \\ 1 & -1 & 2 \\ 0 & 1 & 2 \end{bmatrix} = \begin{vmatrix} 2 & 4 \\ 0 & 2 \end{vmatrix} = 4$

c. $M_{31} = \begin{bmatrix} 2 & -3 & 4 \\ 1 & -1 & 2 \\ 0 & 1 & 2 \end{bmatrix} = \begin{vmatrix} -3 & 4 \\ -1 & 2 \end{vmatrix}$
$= (-3)(2) - 4(-1) = -2$

23. Expand by the third row:
$$\begin{vmatrix} 1 & 0 & -1 \\ 0 & 2 & 2 \\ -1 & 0 & 0 \end{vmatrix} = a_{31}A_{31} + a_{32}A_{32} + a_{33}A_{33}$$
$$= -1(-1)^{3+1}\begin{vmatrix} 0 & -1 \\ 2 & 2 \end{vmatrix} + 0 + 0 = -2$$

25. Expand by the third row:
$$\begin{vmatrix} 1 & 2 & 3 \\ 0 & 3 & 4 \\ 0 & 0 & 4 \end{vmatrix} = a_{31}A_{31} + a_{32}A_{32} + a_{33}A_{33}$$
$$= 0 + 0 + 4(-1)^{3+3}\begin{vmatrix} 1 & 2 \\ 0 & 3 \end{vmatrix} = 12$$

27. Expand by the first row:

$$\begin{vmatrix} 1 & 0 & 0 \\ 2 & 0 & 0 \\ 3 & 4 & 5 \end{vmatrix} = a_{11}A_{11} + a_{12}A_{12} + a_{13}A_{13}$$

$$= 1(-1)^{1+1}\begin{vmatrix} 0 & 0 \\ 4 & 5 \end{vmatrix} + 0 + 0 = 0$$

29. Expand by the first row:

$$\begin{vmatrix} 1 & 6 & 0 \\ 2 & 5 & 3 \\ 3 & 4 & 0 \end{vmatrix} = a_{11}A_{11} + a_{12}A_{12} + a_{13}A_{13}$$

$$= 1(-1)^{1+1}\begin{vmatrix} 5 & 3 \\ 4 & 0 \end{vmatrix} + 6(-1)^{1+2}\begin{vmatrix} 2 & 3 \\ 3 & 0 \end{vmatrix} + 0$$

$$= -12 - 6(-9) = 42$$

31. Expand by the first row:

$$\begin{vmatrix} 3 & 4 & 1 \\ 1 & 4 & 3 \\ 4 & 3 & 1 \end{vmatrix} = a_{11}A_{11} + a_{12}A_{12} + a_{13}A_{13}$$

$$= 3(-1)^{1+1}\begin{vmatrix} 4 & 3 \\ 3 & 1 \end{vmatrix} + 4(-1)^{1+2}\begin{vmatrix} 1 & 3 \\ 4 & 1 \end{vmatrix}$$

$$+ 1(-1)^{1+3}\begin{vmatrix} 1 & 4 \\ 4 & 3 \end{vmatrix}$$

$$= 3(-5) - 4(-11) + 1(-13) = 16$$

33. Expand by the first row:

$$\begin{vmatrix} 0 & 1 & 6 \\ 1 & 0 & 4 \\ 8 & 3 & 1 \end{vmatrix} = a_{11}A_{11} + a_{12}A_{12} + a_{13}A_{13}$$

$$= 0 + 1(-1)^{1+2}\begin{vmatrix} 1 & 4 \\ 8 & 1 \end{vmatrix} + 6(-1)^{1+3}\begin{vmatrix} 1 & 0 \\ 8 & 3 \end{vmatrix}$$

$$= -1(-31) + 6(3) = 49$$

35. Expand by the first row:

$$\begin{vmatrix} a & b & 0 \\ 0 & a & b \\ b & 0 & a \end{vmatrix} = a_{11}A_{11} + a_{12}A_{12} + a_{13}A_{13}$$

$$= a(-1)^{1+1}\begin{vmatrix} a & b \\ 0 & a \end{vmatrix} + b(-1)^{1+2}\begin{vmatrix} 0 & b \\ b & a \end{vmatrix} + 0$$

$$= a(a^2) - b(-b^2) = a^3 + b^3$$

37. Expand by the first row:

$$\begin{vmatrix} a & b & c \\ c & a & b \\ b & c & a \end{vmatrix} = a_{11}A_{11} + a_{12}A_{12} + a_{13}A_{13}$$

$$= a(-1)^{1+1}\begin{vmatrix} a & b \\ c & a \end{vmatrix} + b(-1)^{1+2}\begin{vmatrix} c & b \\ b & a \end{vmatrix}$$

$$+ c(-1)^{1+3}\begin{vmatrix} c & a \\ b & c \end{vmatrix}$$

$$= a(a^2 - bc) - b(ac - b^2) + c(c^2 - ab)$$

$$= a^3 + b^3 + c^3 - 3abc$$

39. $D = \begin{vmatrix} 1 & 1 \\ 1 & -1 \end{vmatrix} = -2, D_x = \begin{vmatrix} 8 & 1 \\ -2 & -1 \end{vmatrix} = -6$

$$D_y = \begin{vmatrix} 1 & 8 \\ 1 & -2 \end{vmatrix} = -10$$

$$x = \frac{-6}{-2} = 3, y = \frac{-10}{-2} = 5$$

The solution is $\{(3, 5)\}$.

41. $D = \begin{vmatrix} 5 & 3 \\ 2 & 1 \end{vmatrix} = -1, D_x = \begin{vmatrix} 11 & 3 \\ 4 & 1 \end{vmatrix} = -1$

$$D_y = \begin{vmatrix} 5 & 11 \\ 2 & 4 \end{vmatrix} = -2$$

$$x = \frac{-1}{-1} = 1, y = \frac{-2}{-1} = 2$$

The solution is $\{(1, 2)\}$.

43. $D = \begin{vmatrix} 2 & 9 \\ 3 & -2 \end{vmatrix} = -31, D_x = \begin{vmatrix} 4 & 9 \\ 6 & -2 \end{vmatrix} = -62$

$$D_y = \begin{vmatrix} 2 & 4 \\ 3 & 6 \end{vmatrix} = 0$$

$$x = \frac{-62}{-31} = 2, y = \frac{0}{-31} = 0$$

The solution is $\{(2, 0)\}$.

45. $D = \begin{vmatrix} 2 & -3 \\ 4 & -6 \end{vmatrix} = 0 \Rightarrow$ there is not a unique

solution. $2x - 3y = 4 \Rightarrow x = \frac{3}{2}y + 2.$

The solution is $\left\{ \left(\frac{3}{2}y + 2, y \right) \right\}$.

47. Let $u = \dfrac{1}{x}$ and $v = \dfrac{1}{y}$. Then

$$\begin{cases} \dfrac{2}{x} + \dfrac{3}{y} = 2 \\ \dfrac{5}{x} + \dfrac{8}{y} = \dfrac{31}{6} \end{cases} \Rightarrow \begin{cases} 2u + 3v = 2 \\ 5u + 8v = \dfrac{31}{6} \end{cases} \Rightarrow$$

$$D = \begin{vmatrix} 2 & 3 \\ 5 & 8 \end{vmatrix} = 1, \; D_u = \begin{vmatrix} 2 & 3 \\ \dfrac{31}{6} & 8 \end{vmatrix} = \dfrac{1}{2},$$

$$D_v = \begin{vmatrix} 2 & 2 \\ 5 & \dfrac{31}{6} \end{vmatrix} = \dfrac{1}{3} \Rightarrow u = \dfrac{1}{2}, v = \dfrac{1}{3} \Rightarrow$$

$x = 2, y = 3$

The solution is $\{(2, 3)\}$.

49. $D = \begin{vmatrix} 1 & -2 & 1 \\ 3 & 1 & -1 \\ 0 & 1 & 1 \end{vmatrix} = 11, \; D_x = \begin{vmatrix} -1 & -2 & 1 \\ 4 & 1 & -1 \\ 1 & 1 & 1 \end{vmatrix} = 11,$

$D_y = \begin{vmatrix} 1 & -1 & 1 \\ 3 & 4 & -1 \\ 0 & 1 & 1 \end{vmatrix} = 11, \; D_z = \begin{vmatrix} 1 & -2 & -1 \\ 3 & 1 & 4 \\ 0 & 1 & 1 \end{vmatrix} = 0$

$x = \dfrac{11}{11} = 1, y = \dfrac{11}{11} = 1, z = \dfrac{0}{11} = 0$

The solution is $\{(1, 1, 0)\}$.

51. $D = \begin{vmatrix} 1 & 1 & -1 \\ 2 & 3 & 1 \\ 0 & 2 & 1 \end{vmatrix} = -5, \; D_x = \begin{vmatrix} -3 & 1 & -1 \\ 2 & 3 & 1 \\ 1 & 2 & 1 \end{vmatrix} = -5,$

$D_y = \begin{vmatrix} 1 & -3 & -1 \\ 2 & 2 & 1 \\ 0 & 1 & 1 \end{vmatrix} = 5, \; D_z = \begin{vmatrix} 1 & 1 & -3 \\ 2 & 3 & 2 \\ 0 & 2 & 1 \end{vmatrix} = -15$

$x = \dfrac{-5}{-5} = 1, y = \dfrac{5}{-5} = -1, z = \dfrac{-15}{-5} = 3$

The solution is $\{(1, -1, 3)\}$.

53. $D = \begin{vmatrix} 1 & 1 & 1 \\ 2 & -3 & 5 \\ 1 & 2 & -4 \end{vmatrix} = 22, \; D_x = \begin{vmatrix} 3 & 1 & 1 \\ 4 & -3 & 5 \\ -1 & 2 & -4 \end{vmatrix} = 22,$

$D_y = \begin{vmatrix} 1 & 3 & 1 \\ 2 & 4 & 5 \\ 1 & -1 & -4 \end{vmatrix} = 22, \; D_z = \begin{vmatrix} 1 & 1 & 3 \\ 2 & -3 & 4 \\ 1 & 2 & -1 \end{vmatrix} = 22$

$x = \dfrac{22}{22} = 1, y = \dfrac{22}{22} = 1, z = \dfrac{22}{22} = 1$

The solution is $\{(1, 1, 1)\}$.

55. $D = \begin{vmatrix} 2 & -3 & 5 \\ 3 & 5 & -2 \\ 1 & 2 & -3 \end{vmatrix} = -38,$

$D_x = \begin{vmatrix} 11 & -3 & 5 \\ 7 & 5 & -2 \\ -4 & 2 & -3 \end{vmatrix} = -38,$

$D_y = \begin{vmatrix} 2 & 11 & 5 \\ 3 & 7 & -2 \\ 1 & -4 & -3 \end{vmatrix} = -76,$

$D_z = \begin{vmatrix} 2 & -3 & 11 \\ 3 & 5 & 7 \\ 1 & 2 & -4 \end{vmatrix} = -114$

$x = \dfrac{-38}{-38} = 1, y = \dfrac{-76}{-38} = 2, z = \dfrac{-114}{-38} = 3$

The solution is $\{(1, 2, 3)\}$.

57. $D = \begin{vmatrix} 5 & 2 & 1 \\ 2 & 1 & 3 \\ 3 & 2 & 4 \end{vmatrix} = -7, \; D_x = \begin{vmatrix} 12 & 2 & 1 \\ 13 & 1 & 3 \\ 19 & 2 & 4 \end{vmatrix} = -7,$

$D_y = \begin{vmatrix} 5 & 12 & 1 \\ 2 & 13 & 3 \\ 3 & 19 & 4 \end{vmatrix} = -14, \; D_z = \begin{vmatrix} 5 & 2 & 12 \\ 2 & 1 & 13 \\ 3 & 2 & 19 \end{vmatrix} = -21$

$x = \dfrac{-7}{-7} = 1, y = \dfrac{-14}{-7} = 2, z = \dfrac{-21}{-7} = 3$

The solution is $\{(1, 2, 3)\}$.

6.4 Applying the Concepts

59. $A = |D| = \dfrac{1}{2} \begin{vmatrix} 1 & 2 & 1 \\ -3 & 4 & 1 \\ 4 & 6 & 1 \end{vmatrix} = 11$

61. $A = |D| = \dfrac{1}{2} \begin{vmatrix} -2 & 1 & 1 \\ -3 & -5 & 1 \\ 2 & 4 & 1 \end{vmatrix} = \dfrac{21}{2} = 10.5$

63. $\begin{vmatrix} 0 & 3 & 1 \\ -1 & 1 & 1 \\ 2 & 7 & 1 \end{vmatrix} = 0 \Rightarrow$ the points are collinear.

65. $\begin{vmatrix} 0 & -4 & 1 \\ 3 & -2 & 1 \\ 1 & -4 & 1 \end{vmatrix} = -2 \Rightarrow$ the points are not collinear.

67. $\begin{vmatrix} x & y & 1 \\ -1 & -1 & 1 \\ 1 & 3 & 1 \end{vmatrix} = 0 \Rightarrow x(-1)^{1+1} \begin{vmatrix} -1 & 1 \\ 3 & 1 \end{vmatrix} + y(-1)^{1+2} \begin{vmatrix} -1 & 1 \\ 1 & 1 \end{vmatrix} + 1(-1)^{1+3} \begin{vmatrix} -1 & -1 \\ 1 & 3 \end{vmatrix} = 0 \Rightarrow -4x + 2y - 2 = 0 \Rightarrow y = 2x + 1$

69. $\begin{vmatrix} x & y & 1 \\ 0 & \dfrac{1}{3} & 1 \\ 1 & 1 & 1 \end{vmatrix} = 0 \Rightarrow x(-1)^{1+1} \begin{vmatrix} \dfrac{1}{3} & 1 \\ 1 & 1 \end{vmatrix} + y(-1)^{1+2} \begin{vmatrix} 0 & 1 \\ 1 & 1 \end{vmatrix} + 1(-1)^{1+3} \begin{vmatrix} 0 & \dfrac{1}{3} \\ 1 & 1 \end{vmatrix} = 0 \Rightarrow -\dfrac{2}{3}x + y - \dfrac{1}{3} = 0 \Rightarrow y = \dfrac{2}{3}x + \dfrac{1}{3}$

For exercise 71–74, *PQRS* is a trapezoid. Its area is $\dfrac{1}{2}\left(\text{base}_1 + \text{base}_2\right) \cdot \text{altitude} = \dfrac{1}{2}(4 + 2) \cdot 2 = 6 \text{ units}^2$.

71. $A = \begin{bmatrix} 1 & -2 \\ 2 & 1 \end{bmatrix}$

$A_{P'Q'R'S'} = |\det A| \cdot A_{PQRS} = |1(1) - (-2)(2)| \cdot 6 \text{ units}^2 = |5| \cdot 6 \text{ units}^2 = 30 \text{ units}^2$

73. $A = \begin{bmatrix} -2 & 1 \\ 1 & 1 \end{bmatrix}$

$A_{P'Q'R'S'} = |\det A| \cdot A_{PQRS} = |-2(1) - 1(1)| \cdot 6 \text{ units}^2 = |-3| \cdot 6 \text{ units}^2 = 18 \text{ units}^2$

6.4 Beyond the Basics

75. a. $\begin{vmatrix} 0 & 0 \\ 2 & 5 \end{vmatrix} = 0(5) - 2(0) = 0$ **b.** $\begin{vmatrix} 1 & 0 \\ 3 & 0 \end{vmatrix} = 1(0) - 3(0) = 0$

c. Expand by the second row: $\begin{vmatrix} 1 & -2 & -3 \\ 0 & 0 & 0 \\ 4 & 4 & -7 \end{vmatrix} = 0(-1)^{2+1} \begin{vmatrix} -2 & -3 \\ 4 & -7 \end{vmatrix} + 0(-1)^{2+2} \begin{vmatrix} 1 & -3 \\ 4 & -7 \end{vmatrix} + 0(-1)^{2+3} \begin{vmatrix} 1 & -2 \\ 4 & 4 \end{vmatrix} = 0$

d. Expand by the third column: $\begin{vmatrix} 4 & 5 & 0 \\ 6 & -7 & 0 \\ 8 & 15 & 0 \end{vmatrix} = 0(-1)^{1+3} \begin{vmatrix} 6 & -7 \\ 8 & 15 \end{vmatrix} + 0(-1)^{2+3} \begin{vmatrix} 4 & 5 \\ 8 & 15 \end{vmatrix} + 0(-1)^{3+3} \begin{vmatrix} 4 & 5 \\ 6 & -7 \end{vmatrix} = 0$

77. a. Expand by the second row: $\begin{vmatrix} 2 & 3 & 5 \\ -1 & 4 & 8 \\ 2 & 3 & 5 \end{vmatrix} = -1(-1)^{2+1} \begin{vmatrix} 3 & 5 \\ 3 & 5 \end{vmatrix} + 4(-1)^{2+2} \begin{vmatrix} 2 & 5 \\ 2 & 5 \end{vmatrix} + 8(-1)^{2+3} \begin{vmatrix} 2 & 5 \\ 2 & 5 \end{vmatrix} = 0$

b. Expand by the second column: $\begin{vmatrix} 3 & 4 & 3 \\ 1 & 2 & 1 \\ -1 & 6 & -1 \end{vmatrix} = 4(-1)^{1+2} \begin{vmatrix} 1 & 1 \\ -1 & -1 \end{vmatrix} + 2(-1)^{2+2} \begin{vmatrix} 3 & 3 \\ -1 & -1 \end{vmatrix} + 6(-1)^{3+2} \begin{vmatrix} 3 & 3 \\ 1 & 1 \end{vmatrix} = 0$

79. $\begin{vmatrix} 2 & -3 & 4 \\ -4 & 7 & -8 \\ 5 & -1 & 3 \end{vmatrix} = -14, \begin{vmatrix} 2 & -3 & 4 \\ 0 & 1 & 0 \\ 5 & -1 & 3 \end{vmatrix} = -14$

81. $\begin{vmatrix} 2 & 0 & -3 & 4 \\ 0 & 1 & 0 & 5 \\ 5 & 0 & -9 & 8 \\ 1 & 2 & 0 & 7 \end{vmatrix} \xrightarrow[R_3 - 5R_4 \to R_3]{R_1 - 2R_4 \to R_1} \begin{vmatrix} 0 & -4 & -3 & -10 \\ 0 & 1 & 0 & 5 \\ 0 & -10 & -9 & -27 \\ 1 & 2 & 0 & 7 \end{vmatrix}$

Expand by column 1:

$\begin{vmatrix} 0 & -4 & -3 & -10 \\ 0 & 1 & 0 & 5 \\ 0 & -10 & -9 & -27 \\ 1 & 2 & 0 & 7 \end{vmatrix} = 0 + 0 + 0 + 1(-1)^{4+1} \begin{vmatrix} -4 & -3 & -10 \\ 1 & 0 & 5 \\ -10 & -9 & -27 \end{vmatrix} = 21$

83. $\begin{vmatrix} 5 & 7 & 1 & 2 \\ 6 & 8 & 9 & 3 \\ 24 & 22 & 6 & 10 \\ 21 & 17 & 7 & 10 \end{vmatrix} \xrightarrow[\substack{-6R_1 + R_3 \to R_3 \\ -7R_1 + R_4 \to R_4}]{-9R_1 + R_2 \to R_2} \begin{vmatrix} 5 & 7 & 1 & 2 \\ -39 & -55 & 0 & -15 \\ -6 & -20 & 0 & -2 \\ -14 & -32 & 0 & -4 \end{vmatrix}$

Expand by column 3: $\begin{vmatrix} 5 & 7 & 1 & 2 \\ -39 & -55 & 0 & -15 \\ -6 & -20 & 0 & -2 \\ -14 & -32 & 0 & -4 \end{vmatrix} = 1(-1)^{1+3} \begin{vmatrix} -39 & -55 & -15 \\ -6 & -20 & -2 \\ -14 & -32 & -4 \end{vmatrix} = 476$

85. $\begin{vmatrix} 3 & 2 \\ 6 & x \end{vmatrix} = 0 \Rightarrow 3x - 12 = 0 \Rightarrow x = 4$

87. $\begin{vmatrix} x & -2 \\ 1 & x-1 \end{vmatrix} = 0 \Rightarrow x^2 - x + 2 = 0 \Rightarrow x = \frac{1 \pm \sqrt{1-8}}{2} = \frac{1 \pm i\sqrt{7}}{2}$

89. Expand by the second row:

$\begin{vmatrix} 1 & -3 & 1 \\ 4 & 7 & x \\ 0 & 2 & 2 \end{vmatrix} = 0 \Rightarrow 4(-1)^{2+1} \begin{vmatrix} -3 & 1 \\ 2 & 2 \end{vmatrix} + 7(-1)^{2+2} \begin{vmatrix} 1 & 1 \\ 0 & 2 \end{vmatrix} + x(-1)^{2+3} \begin{vmatrix} 1 & -3 \\ 0 & 2 \end{vmatrix} = 0 \Rightarrow$

$-4(-8) + 7(2) - 2x = 46 - 2x = 0 \Rightarrow x = 23$

91. Expand by the second row:

$\begin{vmatrix} x & 0 & 1 \\ 0 & x & 0 \\ 1 & 0 & x \end{vmatrix} = 0 \Rightarrow 0 + x(-1)^{2+2} \begin{vmatrix} x & 1 \\ 1 & x \end{vmatrix} + 0 \Rightarrow x(x^2 - 1) = 0 \Rightarrow x(x-1)(x+1) = 0 \Rightarrow x = 0 \text{ or } x = 1 \text{ or } x = -1$

93. $\frac{1}{2} \begin{vmatrix} -4 & 2 & 1 \\ 0 & k & 1 \\ -2 & k & 1 \end{vmatrix} = 28 \Rightarrow 0 + k(-1)^{2+2} \begin{vmatrix} -4 & 1 \\ -2 & 1 \end{vmatrix} + 1(-1)^{2+3} \begin{vmatrix} -4 & 2 \\ -2 & k \end{vmatrix} = 56 \Rightarrow -2k - (-4k + 4) = 2k - 4 = 56 \Rightarrow k = 30$

6.4 Critical Thinking/Discussion/Writing

95. Let $u = \dfrac{1}{x-2}$ and $v = \dfrac{1}{y+1}$. Then $\begin{cases} \dfrac{1}{x-2} + \dfrac{3}{y+1} = 13 \\ \dfrac{4}{x-2} - \dfrac{5}{y+1} = 1 \end{cases} \Rightarrow \begin{cases} u + 3v = 13 \\ 4u - 5v = 1 \end{cases}$.

$D = \begin{vmatrix} 1 & 3 \\ 4 & -5 \end{vmatrix} = -17$, $D_u = \begin{vmatrix} 13 & 3 \\ 1 & -5 \end{vmatrix} = -68$, $D_v = \begin{vmatrix} 1 & 13 \\ 4 & 1 \end{vmatrix} = -51 \Rightarrow u = \dfrac{-68}{-14}$, $v = \dfrac{-51}{-17} = 3$

$4 = \dfrac{1}{x-2} \Rightarrow 4x - 8 = 1 \Rightarrow x = \dfrac{9}{4}$; $3 = \dfrac{1}{y+1} \Rightarrow 3y + 3 = 1 \Rightarrow y = -\dfrac{2}{3}$

The solution is $\left\{ \left(\dfrac{9}{4}, -\dfrac{2}{3} \right) \right\}$.

96. Let $u = \dfrac{1}{x+1}$ and $v = \dfrac{1}{y-1}$. Then $\begin{cases} \dfrac{6}{x+1} + \dfrac{4}{y-1} = 7 \\ \dfrac{8}{x+1} + \dfrac{5}{y-1} = 9 \end{cases} \Rightarrow \begin{cases} 6u + 4v = 7 \\ 8u + 5v = 9 \end{cases}$

$D = \begin{vmatrix} 6 & 4 \\ 8 & 5 \end{vmatrix} = -2$, $D_u = \begin{vmatrix} 7 & 4 \\ 9 & 5 \end{vmatrix} = -1$, $D_y = \begin{vmatrix} 6 & 7 \\ 8 & 9 \end{vmatrix} = -2$, $u = \dfrac{-1}{-2} = \dfrac{1}{2}$, $v = \dfrac{-2}{-2} = 1$

$\dfrac{1}{2} = \dfrac{1}{x+1} \Rightarrow x + 1 = 2 \Rightarrow x = 1$; $1 = \dfrac{1}{y-1} \Rightarrow y - 1 = 1 \Rightarrow y = 2$

The solution is $\{(1, 2)\}$.

97. Let $u = \dfrac{1}{3^x}$ and $v = \dfrac{1}{4^y}$. Then $\begin{cases} \dfrac{1}{3^x} + \dfrac{4}{4^y} = 25 \\ \dfrac{2}{3^x} - \dfrac{1}{4^y} = 14 \end{cases} \Rightarrow \begin{cases} u + 4v = 25 \\ 2u - v = 14 \end{cases}$

$D = \begin{vmatrix} 1 & 4 \\ 2 & -1 \end{vmatrix} = -9$, $D_u = \begin{vmatrix} 25 & 4 \\ 14 & -1 \end{vmatrix} = -81$, $D_y = \begin{vmatrix} 1 & 25 \\ 2 & 14 \end{vmatrix} = -36$, $u = \dfrac{-81}{-9} = 9$, $v = \dfrac{-36}{-9} = 4$

$9 = \dfrac{1}{3^x} \Rightarrow 3^x = \dfrac{1}{9} \Rightarrow x = -2$; $4 = \dfrac{1}{4^y} \Rightarrow 4^y = \dfrac{1}{4} \Rightarrow y = -1$

The solution is $\{(-2, -1)\}$.

98. We use the change of base formula along with the power rule of logarithms. See section 4.3.

a. Recall that $\log 8 = \log(2^3) = 3\log 2$.

$\begin{vmatrix} \log_2 3 & \log_8 3 \\ \log_3 4 & \log_3 4 \end{vmatrix} = \log_2 3 \cdot \log_3 4 - \log_8 3 \cdot \log_3 4 = \dfrac{\log 3 \cdot \log 4}{\log 2 \cdot \log 3} - \dfrac{\log 3 \cdot \log 4}{\log 8 \cdot \log 3} = \dfrac{\log 3 \cdot \log 4}{\log 2 \cdot \log 3} - \dfrac{\log 3 \cdot \log 4}{3\log 2 \cdot \log 3}$

$= \dfrac{\log 4}{\log 2} - \dfrac{\log 4}{3\log 2} = \dfrac{3\log 4 - \log 4}{3\log 2} = \dfrac{2\log 4}{3\log 2} = \dfrac{2\log(2^2)}{3\log 2} = \dfrac{4\log 2}{3\log 2} = \dfrac{4}{3}$

b. $\begin{vmatrix} \log_3 512 & \log_4 3 \\ \log_3 8 & \log_4 9 \end{vmatrix} = \log_3 512 \cdot \log_4 9 - \log_4 3 \cdot \log_3 8 = \dfrac{\log 512 \cdot \log 9}{\log 3 \cdot \log 4} - \dfrac{\log 3 \cdot \log 8}{\log 4 \cdot \log 3}$

$= \dfrac{\log(2 \cdot 16^2) \cdot 2\log 3}{\log 3 \cdot \log 4} - \dfrac{3\log 2}{\log 4} = \dfrac{(\log 2 + 2\log 16) \cdot 2}{2\log 2} - \dfrac{3\log 2}{2\log 2} = \dfrac{\log 2 + 2\log(2^4)}{\log 2} - \dfrac{3}{2}$

$= \dfrac{\log 2 + 8\log 2}{\log 2} - \dfrac{3}{2} = \dfrac{9\log 2}{\log 2} - \dfrac{3}{2} = 9 - \dfrac{3}{2} = \dfrac{15}{2}$

99. A system has no solution when $D = 0$.

$$D = \begin{vmatrix} k & 3 & -1 \\ 1 & 2 & 1 \\ -k & 1 & 2 \end{vmatrix}$$

$$= -1 \begin{vmatrix} 3 & -1 \\ 1 & 2 \end{vmatrix} + 2 \begin{vmatrix} k & -1 \\ -k & 2 \end{vmatrix} - 1 \begin{vmatrix} k & 3 \\ -k & 1 \end{vmatrix}$$

$$= -7 + 2(2k - k) - (k + 3k)$$

$$= -7 + 2k - 4k = -7 - 2k \Rightarrow -7 - 2k = 0 \Rightarrow$$

$$k = -\frac{7}{2}$$

The system has no solution for $k = -\frac{7}{2}$.

100. $D = \begin{vmatrix} 2 & a & 6 \\ 1 & 2 & b \\ 1 & 1 & 3 \end{vmatrix} = \begin{vmatrix} a & 6 \\ 2 & b \end{vmatrix} - \begin{vmatrix} 2 & 6 \\ 1 & b \end{vmatrix} + 3 \begin{vmatrix} 2 & a \\ 1 & 2 \end{vmatrix}$

$$= (ab - 12) - (2b - 6) + 3(4 - a)$$

$$= ab - 3a - 2b + 6$$

The system of equations has a unique solution when $D \neq 0$.

$$ab - 3a - 2b + 6 = 0 \Rightarrow ab - 3a = 2b - 6 \Rightarrow$$

$$a(b - 3) = 2(b - 3) \Rightarrow a = 2, b \neq 3$$

So, the system has a unique solution if $a \neq 2$ and $b \neq 3$.

If $a = 2$, then row 1 = 2 row 3, so there are infinitely many solutions.

If $a = b = 3$, then $D = 0$, but all rows are different, so there is no solution.

6.4 Getting Ready for the Next Section

You may want to refer to sections 1.2, 5.3, and 5.4.

101. $x^2 = 8y \Rightarrow \frac{1}{8} x^2 = y$

103. $(x - 2)^2 = 6(y - 3) \Rightarrow x^2 - 4x + 4 = 6y - 18 \Rightarrow$

$$x^2 - 4x + 22 = 6y \Rightarrow \frac{1}{6} x^2 - \frac{2}{3} x + \frac{11}{3} = y$$

105. $-2 = a(-2)^2 \Rightarrow -2 = 4a \Rightarrow -\frac{1}{2} = a$

The function is $y = -\frac{1}{2} x^2$.

107. The formula for the x-value of the vertex gives

$$x = -\frac{b}{2a} = -3 \Rightarrow b = 6a$$

Using $(0, 10)$ gives

$$10 = a(0^2) + b(0) + c \Rightarrow c = 10.$$

Substituting $(-3, -8)$, $b = 6a$, and $c = 10$ gives

$$-8 = a(-3)^2 + 6a(-3) + 10 \Rightarrow$$

$$-8 = 9a - 18a + 10 \Rightarrow -18 = -9a \Rightarrow 2 = a \Rightarrow$$

$$b = 12$$

The function is $y = 2x^2 + 12x + 10$.

109. $f(x) = 2x^2 + 4x - 6$

a. Opens up since $a > 0$.

b. $x = -\frac{b}{2a} = -\frac{4}{2(2)} = -1$

$$f(-1) = 2(-1)^2 + 4(-1) - 6 = -8$$

The vertex is $(-1, -8)$.

c. The axis of symmetry is $x = -1$.

d. To find the x-intercepts, let $y = 0$ and solve for x.

$$2x^2 + 4x - 6 = 0 \Rightarrow 2(x^2 + 2x - 3) = 0 \Rightarrow$$

$$(x + 3)(x - 1) = 0 \Rightarrow x = -3, \ x = 1$$

The x-intercepts are -3 and 1.

e. To find the y-intercept, let $x = 0$ and solve for y.

$$y = 2(0)^2 + 4(0) - 6 = -6$$

The y-intercept is -6.

Chapter 6 Review Exercises

Building Skills

1. 1×4 **3.** 3×2

5. $a_{12} = -1, a_{14} = -4, a_{23} = 3, a_{21} = 5$

7. $\begin{bmatrix} 2 & -3 & | & 7 \\ 3 & 1 & | & 6 \end{bmatrix}$

9. Answers may vary.

$$\begin{bmatrix} 0 & 1 & 2 & 1 \\ 2 & 0 & 3 & 4 \\ 1 & -2 & 1 & 7 \end{bmatrix} \xrightarrow{R_1 \leftrightarrow R_3} \begin{bmatrix} 1 & -2 & 1 & 7 \\ 2 & 0 & 3 & 4 \\ 0 & 1 & 2 & 1 \end{bmatrix}$$

$$\xrightarrow{2R_1 - R_2 \to R_2} \begin{bmatrix} 1 & -2 & 1 & 7 \\ 0 & -4 & -1 & 10 \\ 0 & 1 & 2 & 1 \end{bmatrix}$$

$$\xrightarrow{R_2 \leftrightarrow R_3} \begin{bmatrix} 1 & -2 & 1 & 7 \\ 0 & 1 & 2 & 1 \\ 0 & -4 & -1 & 10 \end{bmatrix}$$

$$\xrightarrow{\frac{1}{7}(4R_2 + R_3) \to R_3} \begin{bmatrix} 1 & -2 & 1 & 7 \\ 0 & 1 & 2 & 1 \\ 0 & 0 & 1 & 2 \end{bmatrix}$$

11. $\begin{bmatrix} 3 & 1 & 3 & 1 \\ 2 & 1 & 1 & 1 \\ 1 & -1 & -1 & 0 \end{bmatrix} \xrightarrow{R_1 \leftrightarrow R_3} \begin{bmatrix} 1 & -1 & -1 & 0 \\ 2 & 1 & 1 & 1 \\ 3 & 1 & 3 & 1 \end{bmatrix}$

$\xrightarrow[-3R_1+R_3 \to R_3]{-2R_1+R_2 \to R_2} \begin{bmatrix} 1 & -1 & -1 & 0 \\ 0 & 3 & 3 & 1 \\ 0 & 4 & 6 & 1 \end{bmatrix}$

$\xrightarrow{-\frac{1}{2}(-2R_2+R_3) \to R_2} \begin{bmatrix} 1 & -1 & -1 & 0 \\ 0 & 1 & 0 & \frac{1}{2} \\ 0 & 4 & 6 & 1 \end{bmatrix}$

$\xrightarrow[\frac{1}{6}(R_3-4R_2)\to R_3]{R_1+R_2\to R_1} \begin{bmatrix} 1 & 0 & -1 & \frac{1}{2} \\ 0 & 1 & 0 & \frac{1}{2} \\ 0 & 0 & 1 & -\frac{1}{6} \end{bmatrix}$

$\xrightarrow{R_1+R_3\to R_1} \begin{bmatrix} 1 & 0 & 0 & \frac{1}{3} \\ 0 & 1 & 0 & \frac{1}{2} \\ 0 & 0 & 1 & -\frac{1}{6} \end{bmatrix}$

13. $\begin{cases} x+y-z=0 \\ 2x+y-2z=3 \\ 3x-2y+3z=9 \end{cases} \Rightarrow \begin{bmatrix} 1 & 1 & -1 & 0 \\ 2 & 1 & -2 & 3 \\ 3 & -2 & 3 & 9 \end{bmatrix}$

$\xrightarrow[3R_1-R_3\to R_3]{2R_1-R_2\to R_2} \begin{bmatrix} 1 & 1 & -1 & 0 \\ 0 & 1 & 0 & -3 \\ 0 & 5 & -6 & -9 \end{bmatrix}$

$\xrightarrow[-\frac{1}{6}(R_3-5R_2)\to R_3]{R_1-R_2\to R_1} \begin{bmatrix} 1 & 0 & -1 & 3 \\ 0 & 1 & 0 & -3 \\ 0 & 0 & 1 & -1 \end{bmatrix}$

$\xrightarrow{R_1+R_3\to R_1} \begin{bmatrix} 1 & 0 & 0 & 2 \\ 0 & 1 & 0 & -3 \\ 0 & 0 & 1 & -1 \end{bmatrix}$

The solution is $\{(2,-3,-1)\}$.

15. $\begin{cases} x-2y+3z=-2 \\ 2x-3y+z=9 \\ 3x-y+2z=5 \end{cases} \Rightarrow \begin{bmatrix} 1 & -2 & 3 & -2 \\ 2 & -3 & 1 & 9 \\ 3 & -1 & 2 & 5 \end{bmatrix}$

$\xrightarrow[R_3-3R_1\to R_3]{R_2-2R_1\to R_2} \begin{bmatrix} 1 & -2 & 3 & -2 \\ 0 & 1 & -5 & 13 \\ 0 & 5 & -7 & 11 \end{bmatrix}$

$\xrightarrow{\frac{1}{18}(R_3-5R_2)\to R_3} \begin{bmatrix} 1 & -2 & 3 & -2 \\ 0 & 1 & -5 & 13 \\ 0 & 0 & 1 & -3 \end{bmatrix}$

$z=-3; y-5(-3)=13 \Rightarrow y=-2$
$x-2(-2)+3(-3)=-2 \Rightarrow x=3$
The solution is $\{(3,-2,-3)\}$.

17. $\begin{cases} x-2y-2z=11 \\ 3x+4y-z=-2 \\ 4x+5y+7z=7 \end{cases} \Rightarrow \begin{bmatrix} 1 & -2 & -2 & 1 \\ 3 & 4 & -1 & -2 \\ 4 & 5 & 7 & 7 \end{bmatrix}$

$\xrightarrow[R_3-4R_1\to R_3]{\frac{1}{5}(R_2-3R_1)\to R_2} \begin{bmatrix} 1 & -2 & -2 & 1 \\ 0 & 2 & 1 & -7 \\ 0 & 13 & 15 & -37 \end{bmatrix}$

$\xrightarrow[\frac{2}{17}(R_3-\frac{13}{2}R_2)\to R_3]{R_1+R_2\to R_1} \begin{bmatrix} 1 & 0 & -1 & 4 \\ 0 & 2 & 1 & -7 \\ 0 & 0 & 1 & 1 \end{bmatrix}$

$\xrightarrow[\frac{1}{2}(R_2-R_3)\to R_2]{R_1+R_3\to R_1} \begin{bmatrix} 1 & 0 & 0 & 5 \\ 0 & 1 & 0 & -4 \\ 0 & 0 & 1 & 1 \end{bmatrix}$

The solution is $\{(5,-4,1)\}$.

19. $\begin{cases} 2x-y+3z=4 \\ x+3y+3z=-2 \\ 3x+2y-6z=6 \end{cases} \Rightarrow \begin{bmatrix} 2 & -1 & 3 & 4 \\ 1 & 3 & 3 & -2 \\ 3 & 2 & -6 & 6 \end{bmatrix}$

$\xrightarrow{R_1\leftrightarrow R_2} \begin{bmatrix} 1 & 3 & 3 & -2 \\ 2 & -1 & 3 & 4 \\ 3 & 2 & -6 & 6 \end{bmatrix}$

$\xrightarrow[R_3-3R_1\to R_3]{2R_1-R_2\to R_2} \begin{bmatrix} 1 & 3 & 3 & -2 \\ 0 & 7 & 3 & -8 \\ 0 & -7 & -15 & 12 \end{bmatrix}$

$\xrightarrow{-\frac{1}{12}(R_2+R_3)\to R_3} \begin{bmatrix} 1 & 3 & 3 & -2 \\ 0 & 7 & 3 & -8 \\ 0 & 0 & 1 & -\frac{1}{3} \end{bmatrix}$

$\xrightarrow[\frac{1}{7}(R_2-3R_3)\to R_2]{R_1-3R_3\to R_1} \begin{bmatrix} 1 & 3 & 0 & -1 \\ 0 & 1 & 0 & -1 \\ 0 & 0 & 1 & -\frac{1}{3} \end{bmatrix}$

$\xrightarrow{R_1-3R_2\to R_1} \begin{bmatrix} 1 & 0 & 0 & 2 \\ 0 & 1 & 0 & -1 \\ 0 & 0 & 1 & -\frac{1}{3} \end{bmatrix}$

The solution is $\left\{\left(2,-1,-\frac{1}{3}\right)\right\}$.

21. $\begin{bmatrix} x-y & 0 \\ 1 & x+y \end{bmatrix} = \begin{bmatrix} 1 & 0 \\ 1 & 3 \end{bmatrix} \Rightarrow \begin{cases} x-y=1 \\ x+y=3 \end{cases} \Rightarrow$
$2x=4 \Rightarrow x=2, y=1$

23. a. $\begin{bmatrix} 1 & 2 \\ -3 & 4 \end{bmatrix} + \begin{bmatrix} 2 & -3 \\ -5 & 6 \end{bmatrix} = \begin{bmatrix} 3 & -1 \\ -8 & 10 \end{bmatrix}$

b. $\begin{bmatrix} 1 & 2 \\ -3 & 4 \end{bmatrix} - \begin{bmatrix} 2 & -3 \\ -5 & 6 \end{bmatrix} = \begin{bmatrix} -1 & 5 \\ 2 & -2 \end{bmatrix}$

c. $2\begin{bmatrix} 1 & 2 \\ -3 & 4 \end{bmatrix} = \begin{bmatrix} 2 & 4 \\ -6 & 8 \end{bmatrix}$

d. $-3\begin{bmatrix} 2 & -3 \\ -5 & 6 \end{bmatrix} = \begin{bmatrix} -6 & 9 \\ 15 & -18 \end{bmatrix}$

e. $2\begin{bmatrix} 1 & 2 \\ -3 & 4 \end{bmatrix} - 3\begin{bmatrix} 2 & -3 \\ -5 & 6 \end{bmatrix}$
$= \begin{bmatrix} 2 & 4 \\ -6 & 8 \end{bmatrix} - \begin{bmatrix} 6 & -9 \\ -15 & 18 \end{bmatrix} = \begin{bmatrix} -4 & 13 \\ 9 & -10 \end{bmatrix}$

25. $3A + 2B - 3X = 0 \Rightarrow$
$3\begin{bmatrix} 1 & 2 \\ -3 & 4 \end{bmatrix} + 2\begin{bmatrix} 2 & -3 \\ -5 & 6 \end{bmatrix} = 3X \Rightarrow$
$\begin{bmatrix} 3 & 6 \\ -9 & 12 \end{bmatrix} + \begin{bmatrix} 4 & -6 \\ -10 & 12 \end{bmatrix} = 3X \Rightarrow$
$\begin{bmatrix} 7 & 0 \\ -19 & 24 \end{bmatrix} = 3X \Rightarrow X = \begin{bmatrix} \frac{7}{3} & 0 \\ -\frac{19}{3} & 8 \end{bmatrix}$

27. a. $AB = \begin{bmatrix} 0 & 1 \\ 2 & 3 \end{bmatrix}\begin{bmatrix} -1 & -1 \\ -3 & 4 \end{bmatrix}$
$= \begin{bmatrix} 0(-1)+1(-3) & 0(-1)+1(4) \\ 2(-1)+3(-3) & 2(-1)+3(4) \end{bmatrix}$
$= \begin{bmatrix} -3 & 4 \\ -11 & 10 \end{bmatrix}$

b. $BA = \begin{bmatrix} -1 & -1 \\ -3 & 4 \end{bmatrix}\begin{bmatrix} 0 & 1 \\ 2 & 3 \end{bmatrix}$
$= \begin{bmatrix} -1(0)-1(2) & -1(1)-1(3) \\ -3(0)+4(2) & -3(1)+4(3) \end{bmatrix}$
$= \begin{bmatrix} -2 & -4 \\ 8 & 9 \end{bmatrix}$

29. a. $AB = \begin{bmatrix} 1 & 2 & -1 \end{bmatrix}\begin{bmatrix} 2 \\ 3 \\ 1 \end{bmatrix} = [1(2)+2(3)-1(1)]$
$= [7]$

b. $BA = \begin{bmatrix} 2 \\ 3 \\ 1 \end{bmatrix}\begin{bmatrix} 1 & 2 & -1 \end{bmatrix}$
$= \begin{bmatrix} 2(1) & 2(2) & 2(-1) \\ 3(1) & 3(2) & 3(-1) \\ 1(1) & 1(2) & 1(-1) \end{bmatrix} = \begin{bmatrix} 2 & 4 & -2 \\ 3 & 6 & -3 \\ 1 & 2 & -1 \end{bmatrix}$

31. $\begin{bmatrix} 1 & 2 \\ 3 & -1 \end{bmatrix}A = \begin{bmatrix} 5 & 6 \\ 1 & -3 \end{bmatrix} \Rightarrow A = \begin{bmatrix} 1 & 2 \\ 3 & -1 \end{bmatrix}^{-1}\begin{bmatrix} 5 & 6 \\ 1 & -3 \end{bmatrix}$
$\begin{bmatrix} 1 & 2 \\ 3 & -1 \end{bmatrix}^{-1} = \frac{1}{-1-6}\begin{bmatrix} -1 & -2 \\ -3 & 1 \end{bmatrix} = \begin{bmatrix} \frac{1}{7} & \frac{2}{7} \\ \frac{3}{7} & -\frac{1}{7} \end{bmatrix}$
$A = \begin{bmatrix} \frac{1}{7} & \frac{2}{7} \\ \frac{3}{7} & -\frac{1}{7} \end{bmatrix}\begin{bmatrix} 5 & 6 \\ 1 & -3 \end{bmatrix} = \begin{bmatrix} 1 & 0 \\ 2 & 3 \end{bmatrix}$

33. $\begin{bmatrix} 5 & 3 \\ 4 & 2 \end{bmatrix}A = \begin{bmatrix} 1 & 0 \\ 0 & 1 \end{bmatrix} \Rightarrow A = \begin{bmatrix} 5 & 3 \\ 4 & 2 \end{bmatrix}^{-1}\begin{bmatrix} 1 & 0 \\ 0 & 1 \end{bmatrix}$
$\begin{bmatrix} 5 & 3 \\ 4 & 2 \end{bmatrix}^{-1} = \frac{1}{10-12}\begin{bmatrix} 2 & -3 \\ -4 & 5 \end{bmatrix} = \begin{bmatrix} -1 & \frac{3}{2} \\ 2 & -\frac{5}{2} \end{bmatrix}$
$A = \begin{bmatrix} -1 & \frac{3}{2} \\ 2 & -\frac{5}{2} \end{bmatrix}\begin{bmatrix} 1 & 0 \\ 0 & 1 \end{bmatrix} = \begin{bmatrix} -1 & \frac{3}{2} \\ 2 & -\frac{5}{2} \end{bmatrix}$

35. a. $A^{-1} = \frac{1}{0-(-1)}\begin{bmatrix} 3 & -1 \\ 1 & 0 \end{bmatrix} = \begin{bmatrix} 3 & -1 \\ 1 & 0 \end{bmatrix}$

b. $(A^2)^{-1} = \left(\begin{bmatrix} 0 & 1 \\ -1 & 3 \end{bmatrix}^2\right)^{-1} = \begin{bmatrix} -1 & 3 \\ -3 & 8 \end{bmatrix}^{-1}$
$= \frac{1}{-8+9}\begin{bmatrix} 8 & -3 \\ 3 & -1 \end{bmatrix} = \begin{bmatrix} 8 & -3 \\ 3 & -1 \end{bmatrix}$

c. $(A^{-1})^2 = \begin{bmatrix} 3 & -1 \\ 1 & 0 \end{bmatrix}\begin{bmatrix} 3 & -1 \\ 1 & 0 \end{bmatrix} = \begin{bmatrix} 8 & -3 \\ 3 & -1 \end{bmatrix}$

d. $(A^2)(A^{-1})^2 = \begin{bmatrix} -1 & 3 \\ -3 & 8 \end{bmatrix}\begin{bmatrix} 8 & -3 \\ 3 & -1 \end{bmatrix} = \begin{bmatrix} 1 & 0 \\ 0 & 1 \end{bmatrix}$
So $(A^{-1})^2$ is the inverse of A^2.

For exercises 37–40, we show that $AB = I$. You should also show that $BA = I$ in order to show that B is the inverse of A.

37. $AB = \begin{bmatrix} 7 & 6 \\ 6 & 5 \end{bmatrix} \begin{bmatrix} -5 & 6 \\ 6 & -7 \end{bmatrix} = \begin{bmatrix} 1 & 0 \\ 0 & 1 \end{bmatrix}$

39. $AB = \begin{bmatrix} 2 & -1 & 0 \\ 1 & 0 & 4 \\ 1 & -1 & 1 \end{bmatrix} \left(\dfrac{1}{5}\right) \begin{bmatrix} 4 & 1 & -4 \\ 3 & 2 & -8 \\ -1 & 1 & 1 \end{bmatrix} = \left(\dfrac{1}{5}\right) \begin{bmatrix} 2 & -1 & 0 \\ 1 & 0 & 4 \\ 1 & -1 & 1 \end{bmatrix} \begin{bmatrix} 4 & 1 & -4 \\ 3 & 2 & -8 \\ -1 & 1 & 1 \end{bmatrix} = \left(\dfrac{1}{5}\right) \begin{bmatrix} 5 & 0 & 0 \\ 0 & 5 & 0 \\ 0 & 0 & 5 \end{bmatrix} = \begin{bmatrix} 1 & 0 & 0 \\ 0 & 1 & 0 \\ 0 & 0 & 1 \end{bmatrix}$

41. $\begin{bmatrix} 3 & 1 \\ 2 & 4 \end{bmatrix}^{-1} = \dfrac{1}{12-2} \begin{bmatrix} 4 & -1 \\ -2 & 3 \end{bmatrix} = \begin{bmatrix} \dfrac{2}{5} & -\dfrac{1}{10} \\ -\dfrac{1}{5} & \dfrac{3}{10} \end{bmatrix}$

43. $\begin{bmatrix} 1 & 2 & -2 & | & 1 & 0 & 0 \\ -1 & 3 & 0 & | & 0 & 1 & 0 \\ 0 & -2 & 1 & | & 0 & 0 & 1 \end{bmatrix} \xrightarrow{R_1 + R_2 \to R_2} \begin{bmatrix} 1 & 2 & -2 & | & 1 & 0 & 0 \\ 0 & 5 & -2 & | & 1 & 1 & 0 \\ 0 & -2 & 1 & | & 0 & 0 & 1 \end{bmatrix} \xrightarrow[R_2 + 2R_3 \to R_2]{R_1 + R_3 \to R_1} \begin{bmatrix} 1 & 0 & -1 & | & 1 & 0 & 1 \\ 0 & 1 & 0 & | & 1 & 1 & 2 \\ 0 & -2 & 1 & | & 0 & 0 & 1 \end{bmatrix}$

$\xrightarrow{2R_2 + R_3 \to R_3} \begin{bmatrix} 1 & 0 & -1 & | & 1 & 0 & 1 \\ 0 & 1 & 0 & | & 1 & 1 & 2 \\ 0 & 0 & 1 & | & 2 & 2 & 5 \end{bmatrix} \xrightarrow{R_1 + R_3 \to R_1} \begin{bmatrix} 1 & 0 & 0 & | & 3 & 2 & 6 \\ 0 & 1 & 0 & | & 1 & 1 & 2 \\ 0 & 0 & 1 & | & 2 & 2 & 5 \end{bmatrix}$

Thus, $\begin{bmatrix} 1 & 2 & -2 \\ -1 & 3 & 0 \\ 0 & -2 & 1 \end{bmatrix}^{-1} = \begin{bmatrix} 3 & 2 & 6 \\ 1 & 1 & 2 \\ 2 & 2 & 5 \end{bmatrix}$

45. $\begin{cases} x + 3y = 7 \\ 2x + 5y = 4 \end{cases} \Rightarrow \begin{bmatrix} 1 & 3 \\ 2 & 5 \end{bmatrix} \begin{bmatrix} x \\ y \end{bmatrix} = \begin{bmatrix} 7 \\ 4 \end{bmatrix} \Rightarrow \begin{bmatrix} x \\ y \end{bmatrix} = \begin{bmatrix} 1 & 3 \\ 2 & 5 \end{bmatrix}^{-1} \begin{bmatrix} 7 \\ 4 \end{bmatrix} = \begin{bmatrix} -5 & 3 \\ 2 & -1 \end{bmatrix} \begin{bmatrix} 7 \\ 4 \end{bmatrix} = \begin{bmatrix} -23 \\ 10 \end{bmatrix}$

The solution is $\{(-23, 10)\}$.

47. $\begin{cases} x + 3y + 3z = 3 \\ x + 4y + 3z = 5 \\ x + 3y + 4z = 6 \end{cases} \Rightarrow \begin{bmatrix} 1 & 3 & 3 \\ 1 & 4 & 3 \\ 1 & 3 & 4 \end{bmatrix} \begin{bmatrix} x \\ y \\ z \end{bmatrix} = \begin{bmatrix} 3 \\ 5 \\ 6 \end{bmatrix} \Rightarrow \begin{bmatrix} x \\ y \\ z \end{bmatrix} = \begin{bmatrix} 1 & 3 & 3 \\ 1 & 4 & 3 \\ 1 & 3 & 4 \end{bmatrix}^{-1} \begin{bmatrix} 3 \\ 5 \\ 6 \end{bmatrix} = \begin{bmatrix} 7 & -3 & -3 \\ -1 & 1 & 0 \\ -1 & 0 & 1 \end{bmatrix} \begin{bmatrix} 3 \\ 5 \\ 6 \end{bmatrix} = \begin{bmatrix} -12 \\ 2 \\ 3 \end{bmatrix}$

The solution is $\{(-12, 2, 3)\}$.

49. $\begin{cases} x - y + z = 3 \\ 4x + 2y = 5 \\ 7x - y - z = 6 \end{cases} \Rightarrow \begin{bmatrix} 1 & -1 & 1 \\ 4 & 2 & 0 \\ 7 & -1 & -1 \end{bmatrix} \begin{bmatrix} x \\ y \\ z \end{bmatrix} = \begin{bmatrix} 3 \\ 5 \\ 6 \end{bmatrix} \Rightarrow \begin{bmatrix} x \\ y \\ z \end{bmatrix} = \begin{bmatrix} 1 & -1 & 1 \\ 4 & 2 & 0 \\ 7 & -1 & -1 \end{bmatrix}^{-1} \begin{bmatrix} 3 \\ 5 \\ 6 \end{bmatrix} = \begin{bmatrix} \dfrac{1}{12} & \dfrac{1}{12} & \dfrac{1}{12} \\ -\dfrac{1}{6} & \dfrac{1}{3} & -\dfrac{1}{6} \\ \dfrac{3}{4} & \dfrac{1}{4} & -\dfrac{1}{4} \end{bmatrix} \begin{bmatrix} 3 \\ 5 \\ 6 \end{bmatrix} = \begin{bmatrix} \dfrac{7}{6} \\ \dfrac{1}{6} \\ 2 \end{bmatrix}$

The solution is $\left\{\left(\dfrac{7}{6}, \dfrac{1}{6}, 2\right)\right\}$.

51. $\begin{vmatrix} 2 & -5 \\ 3 & 4 \end{vmatrix} = 2(4) - (-5)(3) = 23$

53. $\begin{vmatrix} 12 & 21 \\ 4 & 7 \end{vmatrix} = 12(7) - 21(4) = 0$

55. a. $A = \begin{bmatrix} 4 & -1 & 2 \\ -2 & -3 & 5 \\ 0 & 2 & -4 \end{bmatrix}$

$M_{12} = \begin{vmatrix} -2 & 5 \\ 0 & -4 \end{vmatrix} = (-2)(-4) - 5(0) = 8 \qquad M_{23} = \begin{vmatrix} 4 & -1 \\ 0 & 2 \end{vmatrix} = 4(2) - (-1)(0) = 8$

$M_{22} = \begin{vmatrix} 4 & 2 \\ 0 & -4 \end{vmatrix} = 4(-4) - 2(0) = -16$

b. $A_{12} = (-1)^{1+2} M_{12} = -8 \qquad A_{23} = (-1)^{2+3} M_{23} = -8 \qquad A_{22} = (-1)^{2+2} M_{22} = -16$

57. a. $\begin{vmatrix} 1 & -2 & 3 \\ 4 & -1 & -2 \\ -2 & 1 & 5 \end{vmatrix} = a_{21}A_{21} + a_{22}A_{22} + a_{23}A_{23} = 4(-1)^{2+1}\begin{vmatrix} -2 & 3 \\ 1 & 5 \end{vmatrix} - 1(-1)^{2+2}\begin{vmatrix} 1 & 3 \\ -2 & 5 \end{vmatrix} - 2(-1)^{2+3}\begin{vmatrix} 1 & -2 \\ -2 & 1 \end{vmatrix}$

$= -4(-13) - 1(11) + 2(-3) = 35$

b. $\begin{vmatrix} 1 & -2 & 3 \\ 4 & -1 & -2 \\ -2 & 1 & 5 \end{vmatrix} = a_{13}A_{13} + a_{23}A_{23} + a_{33}A_{33} = 3(-1)^{1+3}\begin{vmatrix} 4 & -1 \\ -2 & 1 \end{vmatrix} - 2(-1)^{2+3}\begin{vmatrix} 1 & -2 \\ -2 & 1 \end{vmatrix} + 5(-1)^{3+3}\begin{vmatrix} 1 & -2 \\ 4 & -1 \end{vmatrix}$

$= 3(2) + 2(-3) + 5(7) = 35$

59. Expand by the first column: $\begin{vmatrix} 1 & 2 & 0 \\ 0 & 1 & 2 \\ 1 & 0 & 2 \end{vmatrix} = 1(-1)^{1+1}\begin{vmatrix} 1 & 2 \\ 0 & 2 \end{vmatrix} + 0 + 1(-1)^{3+1}\begin{vmatrix} 2 & 0 \\ 1 & 2 \end{vmatrix} = 2 + 4 = 6$

61. $D = \begin{vmatrix} 5 & 3 \\ 2 & 1 \end{vmatrix} = -1, \ D_x = \begin{vmatrix} 11 & 3 \\ 4 & 1 \end{vmatrix} = -1, \ D_y = \begin{vmatrix} 5 & 11 \\ 2 & 4 \end{vmatrix} = -2$

$x = \dfrac{-1}{-1} = 1, y = \dfrac{-2}{-1} = 2$

The solution is $\{(1, 2)\}$.

63. $D = \begin{vmatrix} 3 & 1 & -1 \\ 1 & 3 & -1 \\ 1 & 1 & -3 \end{vmatrix} = -20, \ D_x = \begin{vmatrix} 14 & 1 & -1 \\ 16 & 3 & -1 \\ -10 & 1 & -3 \end{vmatrix} = -100, \ D_y = \begin{vmatrix} 3 & 14 & -1 \\ 1 & 16 & -1 \\ 1 & -10 & -3 \end{vmatrix} = -120, \ D_z = \begin{vmatrix} 3 & 1 & 14 \\ 1 & 3 & 16 \\ 1 & 1 & -10 \end{vmatrix} = -140$

$x = \dfrac{-100}{-20} = 5, y = \dfrac{-120}{-20} = 6, z = \dfrac{-140}{-20} = 7$

The solution is $\{(5, 6, 7)\}$.

65. Expand by the third row:

$\begin{vmatrix} 1 & 2 & 4 \\ -3 & 5 & 7 \\ 1 & x & 4 \end{vmatrix} = 0 \Rightarrow 1(-1)^{3+1}\begin{vmatrix} 2 & 4 \\ 5 & 7 \end{vmatrix} + x(-1)^{3+2}\begin{vmatrix} 1 & 4 \\ -3 & 7 \end{vmatrix} + 4(-1)^{3+3}\begin{vmatrix} 1 & 2 \\ -3 & 5 \end{vmatrix} = 0 \Rightarrow -6 - 19x + 4(11) = 0 \Rightarrow x = 2$

67. $\begin{vmatrix} x & y & 1 \\ 2 & 3 & 1 \\ 1 & -4 & 1 \end{vmatrix} = 0 \Rightarrow x(-1)^{1+1}\begin{vmatrix} 3 & 1 \\ -4 & 1 \end{vmatrix} + y(-1)^{1+2}\begin{vmatrix} 2 & 1 \\ 1 & 1 \end{vmatrix} + 1(-1)^{1+3}\begin{vmatrix} 2 & 3 \\ 1 & -4 \end{vmatrix} = 0 \Rightarrow 7x - y - 11 = 0 \Rightarrow y = 7x - 11$

Using the given points to find the slope and y-intercept, we have $m = \dfrac{3 - (-4)}{2 - 1} = 7$ and

$3 = 7(2) + b \Rightarrow b = -11$.

Applying the Concepts

69.

$$\begin{array}{c} \text{Method 1} \\ \text{Method 2} \\ \text{Method 3} \end{array} \begin{array}{ccc} A & B & C \end{array} \\ \begin{bmatrix} 4 & 8 & 2 \\ 5 & 7 & 1 \\ 5 & 4 & 8 \end{bmatrix} \begin{bmatrix} 10 \\ 4 \\ 6 \end{bmatrix} = \begin{bmatrix} 84 \\ 84 \\ 114 \end{bmatrix} \Rightarrow \text{Method 3 is the most profitable.}$$

71. Let x = the speed of the plane. Let y = the velocity of the wind. Using Gauss-Jordan elimination, we have

$$\begin{cases} 3(x+y) = 1680 \\ 3.5(x-y) = 1680 \end{cases} \Rightarrow \begin{cases} 3x + 3y = 1680 \\ 3.5x - 3.5y = 1680 \end{cases} \Rightarrow$$

$$\begin{bmatrix} 3 & 3 & | & 1680 \\ 3.5 & -3.5 & | & 1680 \end{bmatrix} \xrightarrow[\frac{1}{3.5}R_2 \to R_2]{\frac{1}{3}R_1 \to R_1} \begin{bmatrix} 1 & 1 & | & 560 \\ 1 & -1 & | & 480 \end{bmatrix} \xrightarrow[\frac{1}{2}(R_1-R_2) \to R_2]{\frac{1}{2}(R_1+R_2) \to R_1} \begin{bmatrix} 1 & 0 & | & 520 \\ 0 & 1 & | & 40 \end{bmatrix} \Rightarrow x = 520, y = 40$$

The plane is traveling at 520 mph and the wind velocity is 40 mph.

73. Let x = Andrew's amount. Let y = Bonnie's amount. Let z = Chauncie's amount. Using Gaussian elimination, we have $\begin{cases} x+y+z = 320 \\ x = 2z \\ y+z = x-20 \end{cases} \Rightarrow \begin{cases} x+y+z = 320 \\ x \quad -2z = 0 \\ -x+y+z = -20 \end{cases} \Rightarrow$

$$\begin{vmatrix} 1 & 1 & 1 & | & 320 \\ 1 & 0 & -2 & | & 0 \\ -1 & 1 & 1 & | & -20 \end{vmatrix} \xrightarrow[\frac{1}{2}(R_1+R_3) \to R_3]{R_1-R_2 \to R_2} \begin{vmatrix} 1 & 1 & 1 & | & 320 \\ 0 & 1 & 3 & | & 320 \\ 0 & 1 & 1 & | & 150 \end{vmatrix} \xrightarrow{\frac{1}{2}(R_2-R_3) \to R_3} \begin{vmatrix} 1 & 1 & 1 & | & 320 \\ 0 & 1 & 3 & | & 320 \\ 0 & 0 & 1 & | & 85 \end{vmatrix} \Rightarrow$$

$z = 85, y + 3(85) = 320 \Rightarrow y = 65$
$x + 65 + 85 = 320 \Rightarrow x = 170$

Andrew has $170, Bonnie has $65, and Chauncie has $85.

75. Let x = the number of registered nurses. Let y = the number of licensed practical nurses. Let z = the number

of nurse's aids. Using Gaussian elimination, we have $\begin{cases} 75x + 20y + 30z = 4850 \\ 3x + 5y + 4z = 530 \\ x + y + z = 130 \end{cases} \Rightarrow$

$$\begin{bmatrix} 75 & 20 & 30 & | & 4850 \\ 3 & 5 & 4 & | & 530 \\ 1 & 1 & 1 & | & 130 \end{bmatrix} \xrightarrow{R_1 \leftrightarrow R_3} \begin{bmatrix} 1 & 1 & 1 & | & 130 \\ 3 & 5 & 4 & | & 530 \\ 75 & 20 & 30 & | & 4850 \end{bmatrix} \xrightarrow[75R_1 - R_3 \to R_3]{\frac{1}{2}(R_2-3R_1) \to R_2} \begin{bmatrix} 1 & 1 & 1 & | & 130 \\ 0 & 1 & \frac{1}{2} & | & 70 \\ 0 & 55 & 45 & | & 4900 \end{bmatrix}$$

$$\xrightarrow{\frac{2}{35}(R_3 - 55R_2) \to R_3} \begin{bmatrix} 1 & 1 & 1 & | & 130 \\ 0 & 1 & \frac{1}{2} & | & 70 \\ 0 & 0 & 1 & | & 60 \end{bmatrix} \Rightarrow z = 60, y + \frac{1}{2}(60) = 70 \Rightarrow y = 40, \ x + 60 + 40 = 130 \Rightarrow x = 30$$

There are 30 registered nurses, 40 licensed practical nurses, and 60 nurses aides.

Chapter 6 Practice Test A

1. 5×4, $a_{43} = 9$

2. $\begin{bmatrix} 7 & -3 & 9 & | & 5 \\ -2 & 4 & 3 & | & -12 \\ 8 & -5 & 1 & | & -9 \end{bmatrix}$

3. $\begin{cases} 4x \quad - z = -3 \\ x + 3y \quad = 9 \\ 2x + 7y + 5z = 8 \end{cases}$

4. $x - 5 = 5 \Rightarrow x = 10; y + 3 = 13 \Rightarrow y = 10; z = 0$
The solution is $\{(10, 10, 0)\}$.

5. $\begin{cases} x+2y+z=6 \\ x+y-z=7 \\ 2x-y+2z=-3 \end{cases} \Rightarrow \begin{bmatrix} 1 & 2 & 1 & | & 6 \\ 1 & 1 & -1 & | & 7 \\ 2 & -1 & 2 & | & -3 \end{bmatrix}$

$\xrightarrow[\frac{1}{5}(2R_1-R_3)\rightarrow R_3]{R_1-R_2\rightarrow R_2} \begin{bmatrix} 1 & 2 & 1 & | & 6 \\ 0 & 1 & 2 & | & -1 \\ 0 & 1 & 0 & | & 3 \end{bmatrix}$

$\xrightarrow{\frac{1}{2}(R_2-R_3)\rightarrow R_3} \begin{bmatrix} 1 & 2 & 1 & | & 6 \\ 0 & 1 & 2 & | & -1 \\ 0 & 0 & 1 & | & -2 \end{bmatrix}$

$\xrightarrow[R_2-2R_3\rightarrow R_2]{R_1-R_3\rightarrow R_1} \begin{bmatrix} 1 & 2 & 0 & | & 8 \\ 0 & 1 & 0 & | & 3 \\ 0 & 0 & 1 & | & -2 \end{bmatrix}$

$\xrightarrow{R_1-2R_2\rightarrow R_1} \begin{bmatrix} 1 & 0 & 0 & | & 2 \\ 0 & 1 & 0 & | & 3 \\ 0 & 0 & 1 & | & -2 \end{bmatrix}$

The solution is $\{(2,3,-2)\}$.

6. $\begin{cases} 2x+y-4z=6 \\ -x+3y-z=-2 \\ 2x-6y+2z=4 \end{cases} \Rightarrow \begin{bmatrix} 2 & 1 & -4 & | & 6 \\ -1 & 3 & -1 & | & -2 \\ 2 & -6 & 2 & | & 4 \end{bmatrix}$

$\xrightarrow[R_1-R_3\rightarrow R_3]{R_1+2R_2\rightarrow R_2} \begin{bmatrix} 2 & 1 & -4 & | & 6 \\ 0 & 7 & -6 & | & 2 \\ 0 & 7 & -6 & | & 2 \end{bmatrix}$

$\xrightarrow[R_2-R_3\rightarrow R_3]{\frac{1}{14}(7R_1-R_2)\rightarrow R_1} \begin{bmatrix} 1 & 0 & -\frac{11}{7} & | & \frac{20}{7} \\ 0 & 7 & -6 & | & 2 \\ 0 & 0 & 0 & | & 0 \end{bmatrix}$

$\xrightarrow{\frac{1}{7}R_2\rightarrow R_2} \begin{bmatrix} 1 & 0 & -\frac{11}{7} & | & \frac{20}{7} \\ 0 & 1 & -\frac{6}{7} & | & \frac{2}{7} \\ 0 & 0 & 0 & | & 0 \end{bmatrix} \Rightarrow$

$x-\frac{11}{7}z=\frac{20}{7} \Rightarrow x=\frac{11}{7}z+\frac{20}{7}$

$y-\frac{6}{7}z=\frac{2}{7} \Rightarrow y=\frac{6}{7}z+\frac{2}{7}$

The solution is $\left\{\left(\frac{11}{7}z+\frac{20}{7},\frac{6}{7}z+\frac{2}{7},z\right)\right\}$.

7. $A-B=\begin{bmatrix} 5 & -2 \\ 4 & 0 \\ 7 & 6 \end{bmatrix}-\begin{bmatrix} -3 & -1 \\ 0 & -8 \\ -4 & 6 \end{bmatrix}=\begin{bmatrix} 8 & -1 \\ 4 & 8 \\ 11 & 0 \end{bmatrix}$

8. $AB=\begin{bmatrix} 3 & -7 & 2 \end{bmatrix}\begin{bmatrix} 0 \\ 1 \\ 4 \end{bmatrix}=[3(0)-7(1)+2(4)]=[1]$

9. The product AB is not defined.

10. $2A=2\begin{bmatrix} -3 & 1 & 0 \\ 5 & 7 & 2 \end{bmatrix}=\begin{bmatrix} -6 & 2 & 0 \\ 10 & 14 & 4 \end{bmatrix}$

11. $A+BA=\begin{bmatrix} -3 & 1 & 0 \\ 5 & 7 & 2 \end{bmatrix}+\begin{bmatrix} -1 & 4 \\ 8 & 2 \end{bmatrix}\begin{bmatrix} -3 & 1 & 0 \\ 5 & 7 & 2 \end{bmatrix}$

$=\begin{bmatrix} -3 & 1 & 0 \\ 5 & 7 & 2 \end{bmatrix}+\begin{bmatrix} 23 & 27 & 8 \\ -14 & 22 & 4 \end{bmatrix}$

$=\begin{bmatrix} 20 & 28 & 8 \\ -9 & 29 & 6 \end{bmatrix}$

12. $C^2=\begin{bmatrix} 1 & 5 \\ 0 & 4 \end{bmatrix}\begin{bmatrix} 1 & 5 \\ 0 & 4 \end{bmatrix}=\begin{bmatrix} 1 & 25 \\ 0 & 16 \end{bmatrix}$

13. $C^{-1}=\begin{bmatrix} 1 & 5 \\ 0 & 4 \end{bmatrix}=\frac{1}{4}\begin{bmatrix} 4 & -5 \\ 0 & 1 \end{bmatrix}=\begin{bmatrix} 1 & -\frac{5}{4} \\ 0 & \frac{1}{4} \end{bmatrix}$

14. $\begin{bmatrix} 2 & 1 & 3 \\ 1 & 2 & -1 \\ 3 & 1 & 5 \end{bmatrix}^{-1}=\begin{bmatrix} 2 & 1 & 3 & | & 1 & 0 & 0 \\ 1 & 2 & -1 & | & 0 & 1 & 0 \\ 3 & 1 & 5 & | & 0 & 0 & 1 \end{bmatrix}$

$\xrightarrow{R_1\leftrightarrow R_2} \begin{bmatrix} 1 & 2 & -1 & | & 0 & 1 & 0 \\ 2 & 1 & 3 & | & 1 & 0 & 0 \\ 3 & 1 & 5 & | & 0 & 0 & 1 \end{bmatrix}$

$\xrightarrow[3R_1-R_3\rightarrow R_3]{2R_1-R_2\rightarrow R_2} \begin{bmatrix} 1 & 2 & -1 & | & 0 & 1 & 0 \\ 0 & 3 & -5 & | & -1 & 2 & 0 \\ 0 & 5 & -8 & | & 0 & 3 & -1 \end{bmatrix}$

$\xrightarrow[3R_3-5R_2\rightarrow R_3]{5R_1-R_2\rightarrow R_1} \begin{bmatrix} 5 & 7 & 0 & | & 1 & 3 & 0 \\ 0 & 3 & -5 & | & -1 & 2 & 0 \\ 0 & 0 & 1 & | & 5 & -1 & -3 \end{bmatrix}$

$\xrightarrow{\frac{1}{3}(R_2+5R_3)\rightarrow R_2} \begin{bmatrix} 5 & 7 & 0 & | & 1 & 3 & 0 \\ 0 & 1 & 0 & | & 8 & -1 & -5 \\ 0 & 0 & 1 & | & 5 & -1 & -3 \end{bmatrix}$

$\xrightarrow{\frac{1}{5}(R_1-7R_2)\rightarrow R_1} \begin{bmatrix} 1 & 0 & 0 & | & -11 & 2 & 7 \\ 0 & 1 & 0 & | & 8 & -1 & -5 \\ 0 & 0 & 1 & | & 5 & -1 & -3 \end{bmatrix}$

$\begin{bmatrix} 2 & 1 & 3 \\ 1 & 2 & -1 \\ 3 & 1 & 5 \end{bmatrix}^{-1}=\begin{bmatrix} -11 & 2 & 7 \\ 8 & -1 & -5 \\ 5 & -1 & -3 \end{bmatrix}$

15. $\begin{bmatrix} 5 & 2 \\ 3 & 1 \end{bmatrix}\begin{bmatrix} x \\ y \end{bmatrix} = \begin{bmatrix} 32 \\ 18 \end{bmatrix}$

16. $\begin{cases} 12x - 3y = 5 \\ -2x + 7y = -9 \end{cases}$

17. $\begin{bmatrix} 1 & 5 & -2 \\ 4 & -2 & 7 \end{bmatrix} - 5X = 2\begin{bmatrix} 2 & 5 & -11 \\ 18 & 8 & 11 \end{bmatrix} \Rightarrow$

$\begin{bmatrix} 1 & 5 & -2 \\ 4 & -2 & 7 \end{bmatrix} - 5X = \begin{bmatrix} 4 & 10 & -22 \\ 36 & 16 & 22 \end{bmatrix} \Rightarrow$

$\begin{bmatrix} 1 & 5 & -2 \\ 4 & -2 & 7 \end{bmatrix} - \begin{bmatrix} 4 & 10 & -22 \\ 36 & 16 & 22 \end{bmatrix} = 5X \Rightarrow$

$\begin{bmatrix} -3 & -5 & 20 \\ -32 & -18 & -15 \end{bmatrix} = 5X \Rightarrow$

$\begin{bmatrix} -\dfrac{3}{5} & -1 & 4 \\ -\dfrac{32}{5} & -\dfrac{18}{5} & -3 \end{bmatrix} = X$

18. $\begin{vmatrix} \dfrac{1}{2} & -\dfrac{1}{4} \\ \dfrac{1}{2} & \dfrac{3}{4} \end{vmatrix} = \left(\dfrac{1}{2}\right)\left(\dfrac{3}{4}\right) - \left(-\dfrac{1}{4}\right)\left(\dfrac{1}{2}\right) = \dfrac{1}{2}$

19. Expand by the second row:

$\begin{vmatrix} 1 & 3 & 5 \\ 2 & 0 & 10 \\ -3 & 1 & -15 \end{vmatrix}$

$= 2(-1)^{2+1}\begin{vmatrix} 3 & 5 \\ 1 & -15 \end{vmatrix} + 0 + 10(-1)^{2+3}\begin{vmatrix} 1 & 3 \\ -3 & 1 \end{vmatrix}$

$= -2(-50) - 10(10) = 0$

20. $\begin{cases} 2x - y + z = 3 \\ x + y + z = 6 \\ 4x + 3y - 2z = 4 \end{cases} \Rightarrow D = \begin{vmatrix} 2 & -1 & 1 \\ 1 & 1 & 1 \\ 4 & 3 & -2 \end{vmatrix}$

$D_x = \begin{vmatrix} 3 & -1 & 1 \\ 6 & 1 & 1 \\ 4 & 3 & -2 \end{vmatrix}, D_y = \begin{vmatrix} 2 & 3 & 1 \\ 1 & 6 & 1 \\ 4 & 4 & -2 \end{vmatrix},$

$D_z = \begin{vmatrix} 2 & -1 & 3 \\ 1 & 1 & 6 \\ 4 & 3 & 4 \end{vmatrix}$

$x = \dfrac{\begin{vmatrix} 3 & -1 & 1 \\ 6 & 1 & 1 \\ 4 & 3 & -2 \end{vmatrix}}{\begin{vmatrix} 2 & -1 & 1 \\ 1 & 1 & 1 \\ 4 & 3 & -2 \end{vmatrix}}, y = \dfrac{\begin{vmatrix} 2 & 3 & 1 \\ 1 & 6 & 1 \\ 4 & 4 & -2 \end{vmatrix}}{\begin{vmatrix} 2 & -1 & 1 \\ 1 & 1 & 1 \\ 4 & 3 & -2 \end{vmatrix}},$

$z = \dfrac{\begin{vmatrix} 2 & -1 & 3 \\ 1 & 1 & 6 \\ 4 & 3 & 4 \end{vmatrix}}{\begin{vmatrix} 2 & -1 & 1 \\ 1 & 1 & 1 \\ 4 & 3 & -2 \end{vmatrix}}$

Chapter 6 Practice Test B

1. D   **2.** D   **3.** D

4. $x + 3 = 9 \Rightarrow x = 6; y + 4 = 2 \Rightarrow y = -2; z = 3$
The answer is C.

5. $\begin{cases} 2x + y = 15 \\ 2y + z = 25 \\ 2z + x = 26 \end{cases} \Rightarrow \begin{bmatrix} 2 & 1 & 0 & | & 15 \\ 0 & 2 & 1 & | & 25 \\ 1 & 0 & 2 & | & 26 \end{bmatrix}$

$\xrightarrow{R_1 - 2R_3 \to R_3} \begin{bmatrix} 2 & 1 & 0 & | & 15 \\ 0 & 2 & 1 & | & 25 \\ 0 & 1 & -4 & | & -37 \end{bmatrix}$

$\xrightarrow{\frac{1}{9}(R_2 - 2R_3) \to R_3} \begin{bmatrix} 2 & 1 & 0 & | & 15 \\ 0 & 2 & 1 & | & 25 \\ 0 & 0 & 1 & | & 11 \end{bmatrix} \Rightarrow z = 11$

The answer is C.

6. $\begin{cases} 2x + y = 17 \\ y + 2z = 15 \\ x + z = 9 \end{cases} \Rightarrow \begin{bmatrix} 2 & 1 & 0 & | & 17 \\ 0 & 1 & 2 & | & 15 \\ 1 & 0 & 1 & | & 9 \end{bmatrix}$

$\xrightarrow{2R_3 - R_1 \to R_3} \begin{bmatrix} 2 & 1 & 0 & | & 17 \\ 0 & 1 & 2 & | & 15 \\ 0 & -1 & 2 & | & 1 \end{bmatrix}$

$\xrightarrow[\frac{1}{4}(R_2 + R_3) \to R_3]{\frac{1}{2}(R_1 + R_3) \to R_1} \begin{bmatrix} 1 & 0 & 1 & | & 9 \\ 0 & 1 & 2 & | & 15 \\ 0 & 0 & 1 & | & 4 \end{bmatrix}$

$\xrightarrow[R_2 - 2R_3 \to R_2]{R_1 - R_3 \to R_1} \begin{bmatrix} 1 & 0 & 0 & | & 5 \\ 0 & 1 & 0 & | & 7 \\ 0 & 0 & 1 & | & 4 \end{bmatrix} \Rightarrow$

$x = 5, y = 7, z = 4 \Rightarrow 4(5) + 3(7) + 4 = 45$
The answer is D.

7. $\begin{bmatrix} -1 & 4 \\ 0 & 4 \\ 8 & -4 \end{bmatrix} - \begin{bmatrix} 7 & 2 \\ 17 & 4 \\ 2 & 2 \end{bmatrix} = \begin{bmatrix} -8 & 2 \\ -17 & 0 \\ 6 & -6 \end{bmatrix}$

The answer is C.

8. $AB = \begin{bmatrix} -8 & 2 & 9 \end{bmatrix} \begin{bmatrix} 3 \\ 0 \\ -3 \end{bmatrix} = \begin{bmatrix} -8(3) + 2(0) + 9(-3) \end{bmatrix}$

$= \begin{bmatrix} -51 \end{bmatrix}$. The answer is B.

9. AB is not defined. The answer is A.

10. $2A = 2\begin{bmatrix} 2 & 1 & -3 \\ -5 & 2 & 1 \end{bmatrix} = \begin{bmatrix} 4 & 2 & -6 \\ -10 & 4 & 2 \end{bmatrix}$

The answer is B.

11. $A + BA = \begin{bmatrix} 2 & 1 & -3 \\ -5 & 2 & 1 \end{bmatrix} + \begin{bmatrix} -3 & 7 \\ 2 & 4 \end{bmatrix}\begin{bmatrix} 2 & 1 & -3 \\ -5 & 2 & 1 \end{bmatrix}$

$= \begin{bmatrix} 2 & 1 & -3 \\ -5 & 2 & 1 \end{bmatrix} + \begin{bmatrix} -41 & 11 & 16 \\ -16 & 10 & -2 \end{bmatrix}$

$= \begin{bmatrix} -39 & 12 & 13 \\ -21 & 12 & -1 \end{bmatrix}$

The answer is C.

12. $C^2 = \begin{bmatrix} 5 & 4 \\ 1 & 0 \end{bmatrix}\begin{bmatrix} 5 & 4 \\ 1 & 0 \end{bmatrix} = \begin{bmatrix} 29 & 20 \\ 5 & 4 \end{bmatrix}$

The answer is C.

13. $C^{-1} = \begin{bmatrix} 5 & 4 \\ 1 & 0 \end{bmatrix}^{-1} = -\frac{1}{4}\begin{bmatrix} 0 & -4 \\ -1 & 5 \end{bmatrix} = \begin{bmatrix} 0 & 1 \\ \frac{1}{4} & -\frac{5}{4} \end{bmatrix}$

The answer is B.

14. $\begin{bmatrix} 1 & 0 & 0 \\ 2 & 1 & 0 \\ 3 & -4 & 1 \end{bmatrix}^{-1} = \begin{bmatrix} 1 & 0 & 0 & | & 1 & 0 & 0 \\ 2 & 1 & 0 & | & 0 & 1 & 0 \\ 3 & -4 & 1 & | & 0 & 0 & 1 \end{bmatrix}$

$\xrightarrow[3R_1 - R_3 \to R_3]{R_2 - 2R_1 \to R_2} \begin{bmatrix} 1 & 0 & 0 & | & 1 & 0 & 0 \\ 0 & 1 & 0 & | & -2 & 1 & 0 \\ 0 & 4 & -1 & | & 3 & 0 & -1 \end{bmatrix}$

$\xrightarrow{4R_2 - R_3 \to R_3} \begin{bmatrix} 1 & 0 & 0 & | & 1 & 0 & 0 \\ 0 & 1 & 0 & | & -2 & 1 & 0 \\ 0 & 0 & 1 & | & -11 & 4 & 1 \end{bmatrix} \Rightarrow$

$\begin{bmatrix} 1 & 0 & 0 \\ 2 & 1 & 0 \\ 3 & -4 & 1 \end{bmatrix}^{-1} = \begin{bmatrix} 1 & 0 & 0 \\ -2 & 1 & 0 \\ -11 & 4 & 1 \end{bmatrix}$

The answer is D.

15. A   **16.** A

17. $\begin{bmatrix} -2 & -3 & 1 \\ -5 & 3 & -2 \end{bmatrix} - 3X = -5\begin{bmatrix} -1 & -2 & -1 \\ 1 & 0 & 1 \end{bmatrix} \Rightarrow$

$\begin{bmatrix} -2 & -3 & 1 \\ -5 & 3 & -2 \end{bmatrix} - 3X = \begin{bmatrix} 5 & 10 & 5 \\ -5 & 0 & -5 \end{bmatrix} \Rightarrow$

$\begin{bmatrix} -2 & -3 & 1 \\ -5 & 3 & -2 \end{bmatrix} - \begin{bmatrix} 5 & 10 & 5 \\ -5 & 0 & -5 \end{bmatrix} = 3X \Rightarrow$

$3X = \begin{bmatrix} -7 & -13 & -4 \\ 0 & 3 & 3 \end{bmatrix} \Rightarrow$

$X = \begin{bmatrix} -\dfrac{7}{3} & -\dfrac{13}{3} & -\dfrac{4}{3} \\ 0 & 1 & 1 \end{bmatrix}$

The answer is B.

18. $\begin{vmatrix} -8 & 5 \\ -4 & -1 \end{vmatrix} = (-8)(-1) - 5(-4) = 28$

The answer is D.

19. $\begin{vmatrix} 2 & 3 & -2 \\ 3 & 0 & -3 \\ -3 & 0 & -5 \end{vmatrix} = 3(-1)^{1+2}\begin{vmatrix} 3 & -3 \\ -3 & -5 \end{vmatrix} = -3(-24) = 72$

The answer is C.

20. $\begin{cases} x + y + z = -6 \\ x - y + 3z = -22 \\ 2x + y + z = -10 \end{cases} \Rightarrow D = \begin{vmatrix} 1 & 1 & 1 \\ 1 & -1 & 3 \\ 2 & 1 & 1 \end{vmatrix} = 4$

$D_x = \begin{vmatrix} -6 & 1 & 1 \\ -22 & -1 & 3 \\ -10 & 1 & 1 \end{vmatrix} = -16, D_y = \begin{vmatrix} 1 & -6 & 1 \\ 1 & -22 & 3 \\ 2 & -10 & 1 \end{vmatrix} = 12$

$D_z = \begin{vmatrix} 1 & 1 & -6 \\ 1 & -1 & -22 \\ 2 & 1 & -10 \end{vmatrix} = -20$

$x = \dfrac{-16}{4} = -4, y = \dfrac{12}{4} = 3, z = \dfrac{-20}{4} = -5$

The answer is B.

Cumulative Review Exercises (Chapters P–6)

1. $\sqrt{(-1-2)^2 + (y - (-3))^2} = 5 \Rightarrow$
$9 + (y + 3)^2 = 25 \Rightarrow y^2 + 6y - 7 = 0 \Rightarrow$
$(y + 7)(y - 1) = 0 \Rightarrow y = -7$ or $y = 1$

2. $\dfrac{4}{x-1} - \dfrac{3}{x+2} = \dfrac{18}{(x+2)(x-1)} \Rightarrow$
$4(x + 2) - 3(x - 1) = 18 \Rightarrow x + 11 = 18 \Rightarrow x = 7$

3. $|2x - 5| = 3 \Rightarrow 2x - 5 = 3$ or $2x - 5 = -3 \Rightarrow$
$x = 4$ or $x = 1$

4. $4x^2 = 8x - 13 \Rightarrow 4x^2 - 8x + 13 = 0 \Rightarrow$

$x = \dfrac{8 \pm \sqrt{64 - 4(4)(13)}}{2(4)} = \dfrac{8 \pm \sqrt{-144}}{8} = 1 \pm \dfrac{3}{2}i$

5. $\left(\dfrac{3x-1}{x+5}\right)^2 - 3\left(\dfrac{3x-1}{x+5}\right) - 28 = 0$

Let $u = \dfrac{3x-1}{x+5}$. Then we have $u^2 - 3u - 28 = 0 \Rightarrow$

$(u+4)(u-7) = 0 \Rightarrow u = -4$ or $u = 7$

$\dfrac{3x-1}{x+5} = -4 \Rightarrow 3x - 1 = -4x - 20 \Rightarrow x = -\dfrac{19}{7}$

$\dfrac{3x-1}{x+5} = 7 \Rightarrow 3x - 1 = 7x + 35 \Rightarrow x = -9$

The solution is $\left\{-\dfrac{19}{7}, -9\right\}$.

6. $\log_2 |x| + \log_2 |x+6| = 4 \Rightarrow$

$\log_2 (|x| \cdot |x+6|) = 4 \Rightarrow 2^4 = |x^2 + 6x| \Rightarrow$

$x^2 + 6x = 16$ or $x^2 + 6x = -16$

$x^2 + 6x - 16 = 0 \Rightarrow (x+8)(x-2) = 0 \Rightarrow$

$x = -8$ or $x = 2$

$x^2 + 6x + 16 = 0 \Rightarrow$

$x = \dfrac{-6 \pm \sqrt{36 - 64}}{2} = -3 \pm i\sqrt{7}$ (reject this)

The solution is $\{-8, 2\}$.

7. Solve $x + 2 = 0 \Rightarrow x = -2$ and

$2x - 1 = 0 \Rightarrow x = \dfrac{1}{2}$. The intervals to be tested

are $(-\infty, -2), \left(-2, \dfrac{1}{2}\right)$, and $\left(\dfrac{1}{2}, \infty\right)$.

Interval	Test point	Value of $\dfrac{x+2}{2x-1}$	Result
$(-\infty, -2)$	-3	$1/7$	$+$
$(-2, 1/2)$	0	-2	$-$
$(1/2, \infty)$	1	3	$+$

The solution set is $(-\infty, -2) \cup \left(\dfrac{1}{2}, \infty\right)$.

8. Solve the associated equation:

$x^2 - 7x + 6 = 0 \Rightarrow (x-6)(x-1) = 0 \Rightarrow x = 6$

or $x = 1$. The intervals are $(-\infty, 1], [1, 6]$, and $[6, \infty)$.

Interval	Test point	Value of $x^2 - 7x + 6$	Result
$(-\infty, 1]$	0	6	$+$
$[1, 6]$	2	-4	$-$
$(6, \infty)$	7	6	$+$

The solution set is $[1, 6]$.

9. The factors of the constant term are $\{\pm 1, \pm 3\}$.
The factors of the leading coefficient are
$\{\pm 1, \pm 2, \pm 4\}$. The possible rational zeros are
$\left\{\pm 1, \pm\dfrac{1}{2}, \pm\dfrac{1}{4}, \pm 3, \pm\dfrac{3}{2}, \pm\dfrac{3}{4}\right\}$.

10. Using synthetic division, we have

$$\begin{array}{r|rrrr}
\frac{1}{2} & 4 & 8 & -11 & 3 \\
 & & 2 & 5 & -3 \\
\hline
 & 4 & 10 & -6 & 0
\end{array}$$

$4x^3 + 8x^2 - 11x + 3 = \left(x - \dfrac{1}{2}\right)\left(4x^2 + 10x - 6\right)$

The zeros of the depressed function
$4x^2 + 10x - 6$ are also zeros of the original
function.

$4x^2 + 10x - 6 = 0 \Rightarrow 2(x+3)(2x-1) = 0 \Rightarrow$

$x = -3$ or $x = \dfrac{1}{2}$.

So $\dfrac{1}{2}$ is a zero of multiplicity 2.

11. $h = kr^3 \Rightarrow 10{,}125 = k(15^3) \Rightarrow k = 3$

$h = 3(20^3) = 24{,}000$ horsepower

12. Let $x =$ the number of acres to be annexed.
Then
$0.12(400 + x) = 0.02(400) + 0.2x \Rightarrow$
$48 + 0.12x = 8 + 0.2x \Rightarrow x = 500$ square miles

13. Using substitution, we have

$\begin{cases} 3x + y = 2 \\ 4x + 5y = -1 \end{cases} \Rightarrow y = -3x + 2$

$4x + 5(-3x + 2) = -1 \Rightarrow -11x + 10 = -1 \Rightarrow$
$x = 1; y = -3(1) + 2 = -1$
The solution is $\{(1, -1)\}$.

14. Using elimination, we have

$$\begin{cases} 2x + y = 5 \\ y^2 - 2y = -3x + 5 \end{cases} \Rightarrow \begin{cases} 2x + y = 5 \\ 3x + y^2 - 2y = 5 \end{cases} \Rightarrow$$

$$\begin{cases} -6x - 3y = -15 \\ 6x + 2y^2 - 4y = 10 \end{cases} \Rightarrow 2y^2 - 7y = -5 \Rightarrow$$

$$2y^2 - 7y + 5 = 0 \Rightarrow (y-1)(2y-5) = 0 \Rightarrow$$

$$y = 1 \text{ or } y = \frac{5}{2}$$

$$2x + 1 = 5 \Rightarrow x = 2$$

$$2x + \frac{5}{2} = 5 \Rightarrow x = \frac{5}{4}$$

The solution is $\left\{ (2,1), \left(\frac{5}{4}, \frac{5}{2} \right) \right\}$.

15. Shift the graph of $f(x) = |x|$ one unit left, stretch by a factor of 3, then shift the resulting graph two units up.

16.

$$\begin{bmatrix} 1 & 2 & -2 \\ -1 & 3 & 0 \\ 0 & -2 & 1 \end{bmatrix}^{-1} = \begin{bmatrix} 1 & 2 & -2 & | & 1 & 0 & 0 \\ -1 & 3 & 0 & | & 0 & 1 & 0 \\ 0 & -2 & 1 & | & 0 & 0 & 1 \end{bmatrix}$$

$$\xrightarrow{R_1 + R_2 \to R_2} \begin{bmatrix} 1 & 2 & -2 & | & 1 & 0 & 0 \\ 0 & 5 & -2 & | & 1 & 1 & 0 \\ 0 & -2 & 1 & | & 0 & 0 & 1 \end{bmatrix}$$

$$\xrightarrow[2R_2 + 5R_3 \to R_3]{R_1 + R_3 \to R_1} \begin{bmatrix} 1 & 0 & -1 & | & 1 & 0 & 1 \\ 0 & 5 & -2 & | & 1 & 1 & 0 \\ 0 & 0 & 1 & | & 2 & 2 & 5 \end{bmatrix}$$

$$\xrightarrow[\frac{1}{5}(R_2 + 2R_3) \to R_2]{R_1 + R_3 \to R_1} \begin{bmatrix} 1 & 0 & 0 & | & 3 & 2 & 6 \\ 0 & 1 & 0 & | & 1 & 1 & 2 \\ 0 & 0 & 1 & | & 2 & 2 & 5 \end{bmatrix} \Rightarrow$$

$$\begin{bmatrix} 1 & 2 & -2 \\ -1 & 3 & 0 \\ 0 & -2 & 1 \end{bmatrix}^{-1} = \begin{bmatrix} 3 & 2 & 6 \\ 1 & 1 & 2 \\ 2 & 2 & 5 \end{bmatrix}$$

17.

$$\begin{bmatrix} 1 & 2 & -2 \\ -1 & 3 & 0 \\ 0 & -2 & 1 \end{bmatrix} \begin{bmatrix} x \\ y \\ z \end{bmatrix} = \begin{bmatrix} 5 \\ 2 \\ -3 \end{bmatrix} \Rightarrow$$

$$\begin{bmatrix} x \\ y \\ z \end{bmatrix} = \begin{bmatrix} 1 & 2 & -2 \\ -1 & 3 & 0 \\ 0 & -2 & 1 \end{bmatrix}^{-1} \begin{bmatrix} 5 \\ 2 \\ -3 \end{bmatrix}$$

$$= \begin{bmatrix} 3 & 2 & 6 \\ 1 & 1 & 2 \\ 2 & 2 & 5 \end{bmatrix} \begin{bmatrix} 5 \\ 2 \\ -3 \end{bmatrix} = \begin{bmatrix} 1 \\ 1 \\ -1 \end{bmatrix}$$

The solution is $\{(1, 1, -1)\}$.

18. a. $F(x) = (x+2)^2 + 3(x+2) - 1 = x^2 + 7x + 9$

    b. $F(4) = 4^2 + 7(4) + 9 = 53$

19. $y = \dfrac{x}{x+4}$. Switch the variables, and then

solve for y to find $f^{-1}(x)$:

$$x = \frac{y}{y+4} \Rightarrow$$

$$xy + 4x = y \Rightarrow xy - y = -4x \Rightarrow$$

$$y(x-1) = -4x \Rightarrow y = -\frac{4x}{x-1} \Rightarrow$$

$$f^{-1}(x) = -\frac{4x}{x-1}$$

20. Domain: $(-\infty, -4) \cup (-4, \infty)$

Range: $(-\infty, 1) \cup (1, \infty)$

Chapter 7 Conic Sections

7.2 The Parabola

7.2 Practice Problems

1. a.

Vertex: (0, 0)
$x^2 = 12y = 4py \Rightarrow p = 3$, so the focus is (0, 3).
The directrix is $y = -3$.
The axis is the y-axis.

b.

Vertex: (0, 0)
$y^2 = -6x = 4px \Rightarrow p = -\dfrac{3}{2}$, so the focus is $\left(-\dfrac{3}{2}, 0\right)$.
The directrix is $x = \dfrac{3}{2}$.
The axis is the x-axis.

2. $y^2 = -8x = 4px \Rightarrow p = -2$
Vertex: (0, 0)
Focus: (−2, 0)
Directrix: $x = 2$
Focal diameter: $-4p = -4(-2) = 8$

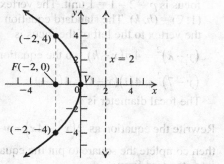

3. a. The vertex is (0, 0) and the focus is (0, 2), so the graph opens up. The general form is $x^2 = 4py$.

$p = 2 \Rightarrow x^2 = 4(2)y \Rightarrow x^2 = 8y$.

b. Since the vertex is (0, 0), the axis of the parabola is the x-axis, and the parabola passes through (1, 2) which is to the right of the vertex, the parabola opens to the right and the general form is $y^2 = 4px$. We need to solve for p:

$2^2 = 4p(1) \Rightarrow 4 = 4p \Rightarrow 1 = p$.

The equation is $y^2 = 4x$.

4. a. Focus: (0, 2), directrix: $y = -10$
The distance between the focus (0, 2) and the directrix $y = -10$ is $2 - (-10) = 12$ units.

Then the vertex is $\dfrac{1}{2}(12) = 6$ units below the focus, so $p = 6$, and the vertex is $(0, 2 - 6) = (0, -4) = (h, k)$. The standard equation when the directrix is below the vertex is $(x - h)^2 = 4p(y - k)$, so the equation is

$x^2 = 4(6)(y - (-4)) = 24(y + 4)$.

The focal diameter is 24.

b. Vertex: (1, 2), directrix: $x = 5$
The distance between the vertex (1, 2) and the directrix $x = 5$ is $p = 5 - 1 = 4$ units.
The vertex $(1, 2) = (h, k)$. The standard equation when the directrix is to the right of the vertex is $(y - k)^2 = -4p(x - h)$, so the equation is

$(y - 2)^2 = -4(4)(x - 1) = -16(x - 1)$.

The focal diameter is 16.

c. Vertex: $(1, 7)$, focus: $(2, 7)$

The distance between the vertex and the focus is $p = 2 - 1 = 1$ unit. The vertex $(1, 7) = (h, k)$. The standard equation when the vertex to the left of the focus is

$(y - k)^2 = 4p(x - h)$, so the equation is

$(y - 7)^2 = 4(1)(x - 1) = 4(x - 1)$.

The focal diameter is 4.

5. Rewrite the equation as $2(x^2 - 4x) = y - 7$, then complete the square to put the equation into standard form:

$$2x^2 - 8x - y + 7 = 0$$
$$2x^2 - 8x = y - 7$$
$$2(x^2 - 4x) = y - 7$$
$$2(x^2 - 4x + 4) = y - 7 + 2(4)$$
$$2(x - 2)^2 = y + 1$$
$$(x - 2)^2 = \frac{1}{2}(y + 1)$$

$4p = \frac{1}{2} \Rightarrow p = \frac{1}{8}$; vertex $= (h, k) = (2, -1)$

The parabola opens up, so the focus is at

$(h, k + p) = \left(2, -1 + \frac{1}{8}\right) = \left(2, -\frac{7}{8}\right)$. The

directrix is located at

$y = k - p = -1 - \frac{1}{8} = -\frac{9}{8}$.

6. a. The vertex $(h, k) = (0, 0)$, so $h = 0$ and $k = 0$. The distance between the vertex $(0, 0)$ and the focus $(0, -2)$ is 2 units, so $p = 2$. The equation for a parabola that opens down is $(x - h)^2 = -4p(y - k)$, so the equation of the graph is

$x^2 = -4(2)y = -8y$.

b. The vertex $(h, k) = (-2, -1)$, so $h = -2$ and $k = -1$. The distance between the vertex and the focus is the same as the distance between the vertex and the directrix. So p is the distance between the vertex $(-2, -1)$ and the point $(3, -1)$ on the directrix $x = 3$. So, $p = 3 - (-2) = 5$ units. The equation for a parabola that opens left is

$(y - k)^2 = -4p(x - h)$, so the equation of the graph is

$(y - (-1))^2 = -4(5)(x - (-2))$, or

$(y + 1)^2 = -20(x + 2)$.

7. The equation is $x^2 = 4py \Rightarrow x^2 = 4(7.3)y \Rightarrow$

$x^2 = 29.2y$. To find the thickness y of the mirror at the edge, substitute $x = 1.5$ (half the diameter) in the equation and solve for y.

$1.5^2 = 29.2y \Rightarrow y \approx 0.0771$

The mirror is about 0.0771 in. thick at the edge.

7.2 Concepts and Vocabulary

1. A parabola is the set of all points P in the plane that are equidistant from a fixed line called the <u>directrix</u> and a fixed point not on the line called the <u>focus</u>.

3. The point at which the axis intersects the parabola is called the <u>vertex</u> of the parabola.

5. True

7. False

7.2 Building Skills

9. $x^2 = 2y = 4py \Rightarrow p = \frac{1}{2}$

Focus: $\left(0, \frac{1}{2}\right)$, directrix: $y = -\frac{1}{2}$, graph (e)

11. $16x^2 = -9y \Rightarrow x^2 = -\frac{9}{16}y = 4py \Rightarrow p = -\frac{9}{64}$

Focus: $\left(0, -\frac{9}{64}\right)$, directrix: $y = \frac{9}{64}$, graph (d)

13. $y^2 = 2x = 4px \Rightarrow p = \frac{1}{2}$

Focus: $\left(\frac{1}{2}, 0\right)$, directrix: $x = -\frac{1}{2}$, graph (g)

15. $9y^2 = -16x \Rightarrow y^2 = -\dfrac{16}{9}x = 4px \Rightarrow p = -\dfrac{4}{9}$

Focus: $\left(-\dfrac{4}{9}, 0\right)$, directrix: $x = \dfrac{4}{9}$, graph (f)

For exercises 17–28, the vertex is (0, 0).

17. The vertex and the focus (2, 0) lie on the
x-axis, so the parabola opens to the right. The
equation is of the form $y^2 = 4px$ with $p = 2$.
The equation of the parabola is
$y^2 = 4(2)x = 8x$.

19. The vertex and the focus (0, –3) lie on the
y-axis, so the parabola opens down. The
equation is of the form $x^2 = -4py$ with $p = 3$.
The equation of the parabola is
$x^2 = -4(3)y = -12y$.

21. The vertex is (0, 0), the axis of the parabola is
the y-axis, and the parabola passes through
(–6, –3) which is below the x-axis, so the
parabola opens down and the standard form is
$x^2 = 4py$. We need to solve for p:
$(-6)^2 = -4p(-3) \Rightarrow 36 = 12p \Rightarrow 3 = p$.
The equation is $x^2 = -4(3)y = -12y$.

23. The vertex is (0, 0), the axis of the parabola is
the x-axis, and the parabola passes through
(3, 6) which is to the right of the y-axis, so the
parabola opens to the right and the standard
form is $y^2 = 4px$. We need to solve for p:
$(6)^2 = 4p(3) \Rightarrow 36 = 12p \Rightarrow 3 = p$.
The equation is $y^2 = 4(3)x = 12x$.

25. The endpoints of the latus rectum are (4, 8)
and (4, –8), so the focus is (4, 0), the axis of
the parabola is the x-axis, $p = 4$, and the
parabola opens to the right.
The equation is $y^2 = 4(4)y = 16x$.

27. The endpoints of the latus rectum are (4, 2)
and (–4, 2), so the focus is (0, 2), the axis of
the parabola is the y-axis, $p = 2$, and the
parabola opens up.
The equation is $x^2 = 4(2)y = 8y$.

29. The focus is (0, 2) and the directrix is $y = 4$, so
the vertex is (0, 3), and the graph opens down.
The general form is $(x - h)^2 = -4p(y - k)$.
$p = 3 - 2 = 1 \Rightarrow x^2 = -4(y - 3)$. The focal
diameter $= 4(1) = 4$.

31. The focus is (–2, 0) and the directrix is $x = 3$,
so the vertex is $(1/2, 0)$, and the graph opens
to the left. The general form is
$(y - k)^2 = -4a(x - h)$. $a = \dfrac{1}{2} - (-2) = \dfrac{5}{2} \Rightarrow$
$y^2 = -4\left(\dfrac{5}{2}\right)\left(x - \dfrac{1}{2}\right) \Rightarrow y^2 = -10\left(x - \dfrac{1}{2}\right)$.
The focal diameter $= 4\left(\dfrac{5}{2}\right) = 10$.

33. The vertex is (1, 1) and the directrix is $x = 3$,
so the graph opens to the left. The general
form is $(y - k)^2 = -4p(x - h)$.
$p = 3 - 1 = 2 \Rightarrow$
$(y - 1)^2 = -4(2)(x - 1) \Rightarrow (y - 1)^2 = -8(x - 1)$.
The focal diameter $= 4(2) = 8$.

35. The vertex is (1, 1) and the directrix is $y = -3$,
so the graph opens up. The general form is
$(x - h)^2 = 4p(y - k)$. $p = 1 - (-3) = 4 \Rightarrow$
$(x - 1)^2 = 4(4)(y - 1) \Rightarrow (x - 1)^2 = 16(y - 1)$.
The focal diameter $= 4(4) = 16$.

37. The vertex is (1, 0) and the focus is (3, 0), so
the graph opens to the right. The general form
is $(y - k)^2 = 4p(x - h)$. $p = 3 - 1 = 2 \Rightarrow$
$(y - 0)^2 = 4(2)(x - 1) \Rightarrow y^2 = 8(x - 1)$.
The focal diameter $= 4(2) = 8$.

39. The vertex is (0, 1) and the focus is (0, –2), so
the graph opens down. The general form is
$(x - h)^2 = -4p(y - k)$. $p = 1 - (-2) = 3 \Rightarrow$
$(x - 0)^2 = -4(3)(y - 1) \Rightarrow x^2 = -12(y - 1)$.
The focal diameter $= 4(3) = 12$.

41. The vertex is (2, 3) and the directrix is $x = 4$,
so the graph opens to the left. The general
form is $(y - k)^2 = -4p(x - h)$.
$p = 4 - 2 = 2 \Rightarrow$
$(y - 3)^2 = -4(2)(x - 2) \Rightarrow (y - 3)^2 = -8(x - 2)$
. The focal diameter $= 4(2) = 8$.

43. The vertex is (2, 3) and the directrix is $y = 1$ so the graph opens up. The general form is $(x - h)^2 = 4p(y - k)$. $p = 3 - 1 = 2 \Rightarrow$ $(x - 2)^2 = 4(2)(y - 3) \Rightarrow (x - 2)^2 = 8(y - 3)$. The focal diameter $= 4(2) = 8$.

45. $4p = 2 \Rightarrow p = 1/2$. The graph opens to the right, so the focus is $(h + p, k)$, and the directrix is $x = h - p$. Vertex: $(-1, 1)$, focus $(-1/2, 1)$, directrix: $x = -3/2$.

47. $4p = 3 \Rightarrow p = 3/4$. The graph opens up, so the focus is $(h, k + p)$, and the directrix is $y = k - p$. Vertex: $(-2, 2)$, focus $(-2, 11/4)$, directrix: $y = 5/4$.

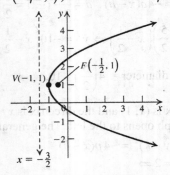

49. $4p = 6 \Rightarrow p = 3/2$. The graph opens to the left, so the focus is $(h - p, k)$, and the directrix is $x = h + p$. Vertex: $(2, -1)$, focus $(1/2, -1)$, directrix: $x = 7/2$.

51. $4p = 10 \Rightarrow p = \dfrac{5}{2}$. The graph opens down, so the focus is $(h, k - p)$, and the directrix is $y = k + p$. Vertex: $(1, 3)$, focus $\left(1, \dfrac{1}{2}\right)$, directrix: $y = \dfrac{11}{2}$.

53. Rewrite the equation as $y - 2 = x^2 + 2x$, then complete the square to put the equation into standard form: $y - 2 + 1 = x^2 + 2x + 1 \Rightarrow$ $y - 1 = (x + 1)^2 \Rightarrow 4a = 1 \Rightarrow a = 1/4$. The graph opens up, so the focus is $(h, k + a)$, and the directrix is $y = k - a$. Vertex: $(-1, 1)$, focus $(-1, 5/4)$, directrix: $y = 3/4$.

55. Rewrite the equation as $2(y^2 + 2y) = 2x - 1$, then complete the square to put the equation into standard form:
$$2(y^2 + 2y + 1) = 2x - 1 + 2 \Rightarrow$$
$$2(y + 1)^2 = 2\left(x + \frac{1}{2}\right) \Rightarrow (y + 1)^2 = x + \frac{1}{2} \Rightarrow$$
$$4p = 1 \Rightarrow p = \frac{1}{4}.$$

The graph opens to the right, so the focus is $(h + p, k)$, and the directrix is $x = h - p$.

(continued on next page)

(*continued*)

Vertex: $\left(-\dfrac{1}{2},-1\right)$, focus $\left(-\dfrac{1}{4},-1\right)$,

directrix: $x=-\dfrac{3}{4}$.

57. Complete the square to put the equation into standard form:

$x+2-48=-3\left(y^2-8y+16\right)\Rightarrow$

$x-46=-3(y-4)^2\Rightarrow-\dfrac{1}{3}(x-46)=(y-4)^2\Rightarrow$

$4p=\dfrac{1}{3}\Rightarrow p=\dfrac{1}{12}$. The graph opens to the left, so the focus is $(h-p,k)$, and the directrix is $x=h+p$. Vertex: $(46,4)$, focus $\left(\dfrac{551}{12},4\right)$,

directrix: $x=\dfrac{553}{12}$.

59. The vertex is $(0, 0) = (h, k)$, the focus is $(0, 3)$, and the graph opens up. The general form is $(x-h)^2=4p(y-k)$. $p=3-0=3$, so the equation of the graph is

$(x-0)^2=4(3)(y-0)$, or $x^2=12y$.

61. The vertex $(h, k) = (-1, 2)$, the directrix is $x = 2$, and the graph opens to the left. The general form for a parabola that opens left is $(y-k)^2=-4p(x-h)$. $p=2-(-1)=3$, so the equation of the graph is

$(y-2)^2=-4(3)(x-(-1))$, or

$(y-2)^2=-12(x+1)$.

63. The vertex is $(1, -1) = (h, k)$, the focal diameter $= 4p = 4$, and the graph opens down. The general form is $(x-h)^2=-4p(y-k)$, so the equation of the graph is

$(x-1)^2=-4(y-(-1))\Rightarrow(x-1)^2=-4(y+1)$.

65. If the axis of symmetry is the y-axis, point P is above the vertex, and the parabola opens up. Substitute the coordinates of the vertex and P into the standard equation to find a:

$(x-h)^2=4p(y-k)\Rightarrow(1-0)^2=4p(2-0)\Rightarrow$

$p=\dfrac{1}{8}\Rightarrow$ the equation is $x^2=\dfrac{1}{2}y$. If the axis of symmetry is the x-axis, point P is to the right of the vertex, and the parabola opens to the right. Substitute the coordinates of the vertex and P into the standard equation to find p:

$(y-k)^2=4a(x-h)\Rightarrow(2-0)^2=4a(1-0)\Rightarrow$

$a=1\Rightarrow y^2=4x$.

67. If the axis of symmetry is parallel to the y-axis, point P is above the vertex, and the parabola opens up. Substitute the coordinates of the vertex and P into the standard equation to find p: $(x-h)^2=4p(y-k)\Rightarrow$

$(2-0)^2=4p(3-1)\Rightarrow p=\dfrac{1}{2}\Rightarrow$ the equation

is $x^2=2(y-1)$. If the axis of symmetry is parallel to the x-axis, point P is to the right of the vertex, and the parabola opens to the right. Substitute the coordinates of the vertex and P into the standard equation to find p:

$(y-k)^2=4p(x-h)\Rightarrow$

$(3-1)^2=4p(2-0)\Rightarrow p=\dfrac{1}{2}\Rightarrow$ the equation

is $(y-1)^2=2x$.

7.2 Applying the Concepts

69. If we sketch the parabola so that the focus is on the y-axis, and the vertex is at (0, 0), then the points (20, 20) and (−20, 20) must lie on the parabola.

$x^2 = 4py \Rightarrow 20^2 = 4p(20) \Rightarrow p = 5 \Rightarrow$ the receptor should be placed at the focus 5 inches from the vertex.

71. If we sketch the parabola so that the focus is on the y-axis, and the vertex is at (0, 0), then the points (9, 6) and (−9, 6) must lie on the parabola.

$x^2 = 4py \Rightarrow 9^2 = 4p(6) \Rightarrow p = \dfrac{81}{24} = \dfrac{27}{8} \Rightarrow$

the heating element should be placed at the focus $3\dfrac{3}{8}$ feet from the vertex.

73. $x = 4y^2 \Rightarrow \dfrac{x}{4} = y^2 \Rightarrow 4p = \dfrac{1}{4} \Rightarrow p = \dfrac{1}{16} \Rightarrow$ the bulb should be placed at the focus, $\left(\dfrac{1}{16}, 0\right)$.

75. $y = 4x^2 \Rightarrow \dfrac{1}{4}y = x^2 \Rightarrow 4p = \dfrac{1}{4} \Rightarrow p = \dfrac{1}{16} \Rightarrow$ the microphone should be placed at the focus, $\left(0, \dfrac{1}{16}\right)$.

77. If we sketch the parabola so that the focus is on the y-axis, the roadbed is the x-axis, and the vertex is at (0, 0), then the points (400, 120) and (−400, 120) must lie on the parabola.

$x^2 = 4py \Rightarrow 400^2 = 4p(120) \Rightarrow p = \dfrac{1000}{3} \Rightarrow$

the equation of the parabola is

$x^2 = \dfrac{4000}{3}y \Rightarrow y = \dfrac{3}{4000}x^2$. The point on the cable 250 feet from the tower has coordinates $(150, f(150))$. The height of the cable is $f(150) = \dfrac{135}{8} = 16.875$ feet.

79. The vertex of the parabola occurs at (35, 30). The parabola also passes through the origin.

The standard form of the equation is $(x-35)^2 = -4p(y-30)$. Substitute (0, 0) for (x, y) and solve for a:

$(0-35)^2 = -4p(0-30) \Rightarrow p = \dfrac{245}{24}$. The equation of the parabola is

$(x-35)^2 = -\dfrac{245}{6}(y-30) \Rightarrow$

$y = -\dfrac{6}{245}(x-35)^2 + 30$.

Now find $f(65) = \dfrac{390}{49} \approx 7.96$ yards.

81. Rewrite the equation as $10y - 500 = x^2 - 30x$, then complete the square to put the equation in standard form:

$10y - 500 + 225 = x^2 - 30x + 225 \Rightarrow$

$10(y - 27.5) = (x - 15)^2$

So, the vertex is (15, 27.5).
The output is 15 tons at a cost of $27.50.

7.2 Beyond the Basics

83. Solve the system $\begin{cases} 2x - 3y = -16 \\ y^2 = 16x \end{cases}$ using

substitution: $2x - 3y = -16 \Rightarrow x = \dfrac{3}{2}y - 8 \Rightarrow$

$y^2 = 16\left(\dfrac{3}{2}y - 8\right) \Rightarrow y^2 = 24y - 128 \Rightarrow$

$y^2 - 24y + 128 = 0 \Rightarrow (y - 8)(y - 16) = 0 \Rightarrow$
$y = 8$ or $y = 16$.
$2x - 3(8) + 16 = 0 \Rightarrow x = 4$
$2x - 3(16) + 16 = 0 \Rightarrow x = 16$. The line and the parabola intersect at (16, 16) and (4, 8). Verify using a graphing calculator.

85. The parabola opens up or down, so the equation is of the form $(x - h)^2 = \pm 4p(y - k)$. Substitute the coordinates of the given points into the equation $ax^2 + bx + c = y$, and solve the system
$\begin{cases} a(0^2) + b(0) + c = 5 \\ a(1^2) + b(1) + c = 4 \\ a(2^2) + b(2) + c = 7 \end{cases} \Rightarrow a = 2, b = -3, c = 5$.

Now rewrite the equation $2x^2 - 3x + 5 = y$ as

$2\left(x^2 - \dfrac{3}{2}x\right) = y - 5$ and complete the square:

$2\left(x^2 - \dfrac{3}{2}x + \dfrac{9}{16}\right) = y - 5 + \dfrac{9}{8} \Rightarrow$

$\left(x - \dfrac{3}{4}\right)^2 = \dfrac{1}{2}\left(y - \dfrac{31}{8}\right)$.

87. Rewrite the equation as $x^2 - 8x = -2y - 4$, and then complete the square to put the equation in standard form:

$x^2 - 8x + 16 = -2y - 4 + 16 \Rightarrow$
$(x - 4)^2 = -2(y - 6)$. The vertex is (4, 6).

$4a = -2 \Rightarrow p = -\dfrac{1}{2}$, so the focus is $\left(4, \dfrac{11}{2}\right)$.

The directrix is $y = \dfrac{13}{2}$. The axis of the

parabola is $x = 4$.

89. Rewrite the equation as $y^2 - 6y = -3x - 15$, and then complete the square to put the equation in standard form:

$y^2 - 6y + 9 = -3x - 15 + 9 \Rightarrow$
$(y - 3)^2 = -3(x + 2)$. The vertex is (−2, 3).

$4p = -3 \Rightarrow p = -\dfrac{3}{4}$, so the focus is

$\left(-\dfrac{11}{4}, 3\right)$. The directrix is $x = -\dfrac{5}{4}$. The axis

is $y = 3$.

91. Step 1: Let m = the slope of the tangent line. Then the equation of the tangent line is
$y - 3 = m(x - 1) \Rightarrow y = mx - m + 3$.

Step 2: $y = 3x^2 \Rightarrow mx - m + 3 = 3x^2$.

Step 3:
$mx - m + 3 = 3x^2 \Rightarrow 3x^2 - mx + m - 3 = 0 \Rightarrow$
$a = 3, b = -m, c = m - 3$
$b^2 - 4ac = (-m)^2 - 4(3)(m - 3)$, so
$b^2 - 4ac = 0 \Rightarrow (-m)^2 - 4(3)(m - 3) = 0 \Rightarrow$
$m^2 - 12m + 36 = 0 \Rightarrow (m - 6)^2 = 0 \Rightarrow m = 6$

Step 4: The equation of the tangent line is
$y - 3 = 6(x - 1) \Rightarrow y = 6x - 3$.

93. Step 1: Let m = the slope of the tangent line. Then the equation of the tangent line is

$y - y_1 = m(x - x_1) \Rightarrow \dfrac{y - y_1}{m} + x_1 = x$.

Step 2:

$x = 4py^2 \Rightarrow \dfrac{y - y_1}{m} + x_1 = 4py^2$.

$x_1 = 4py_1^2 \Rightarrow \dfrac{y - y_1}{m} + x_1 = 4py^2 \Rightarrow$

$\dfrac{y - y_1}{m} + 4py_1^2 = 4py^2$

Step 3:

$\dfrac{y - y_1}{m} + 4py_1^2 = 4py^2 \Rightarrow$

$4py^2 - \dfrac{y}{m} + \left(\dfrac{y_1}{m} - 4py_1^2\right) = 0 \Rightarrow$

$4py^2 - \dfrac{y}{m} + \left(\dfrac{y_1 - 4mpy_1^2}{m}\right) = 0$

(continued on next page)

(*continued*)

$$a = 4p, \quad b = -\frac{1}{m}, \quad c = \frac{y_1}{m} - 4py_1^2$$

$$b^2 - 4ac = \left(-\frac{1}{m}\right)^2 - 4(4p)\left(\frac{y_1}{m} - 4py_1^2\right), \text{ so}$$

$$b^2 - 4ac = 0 \Rightarrow \frac{1}{m^2} - \frac{16py_1}{m} + 64p^2y_1^2 = 0 \Rightarrow$$

$$\left(\frac{1}{m} - 8py_1\right)^2 = 0 \Rightarrow \frac{1}{m} = 8py_1 \Rightarrow m = \frac{1}{8py_1}$$

Step 4: The equation of the tangent line is

$$y - y_1 = \frac{1}{8py_1}(x - x_1) \Rightarrow$$

$$y - y_1 = \frac{1}{8py_1}x - \frac{1}{8py_1}x_1 \Rightarrow$$

$$y - y_1 = \frac{1}{8py_1}x - \frac{1}{8py_1}\left(4py_1^2\right) \Rightarrow$$

$$y - y_1 = \frac{1}{8py_1}x - \frac{1}{2}y_1 \Rightarrow y = \frac{1}{8py_1}x + \frac{1}{2}y_1$$

95. Rewrite the equation in standard form to find the vertex and focus.

$$x^2 - 2x - 12y + 25 = 0$$
$$x^2 - 2x = 12y - 25$$
$$x^2 - 2x + 1 = 12y - 24$$
$$(x - 1)^2 = 12(y - 2) = 4(3)(y - 2)$$

The vertex is $(1, 2)$, and $p = 3$, so the focus is $(1, 2 + 3) = (1, 5)$. Find the endpoints of the latus rectum by substituting $y = 5$ into the equation and solving for x.

$$x^2 - 2x - 12(5) + 25 = 0$$
$$x^2 - 2x - 35 = 0$$
$$(x + 5)(x - 7) = 0 \Rightarrow x = -5, \ x = 7$$

The endpoints of the latus rectum are $A(-5, 5)$ and $B(7, 5)$.

$$AB = 12, \ FV = h = 3$$

$$A_{\triangle ABV} = \frac{1}{2}(12)(3) = 18 \text{ units}^2$$

7.2 Critical Thinking/Discussion/Writing

97. a. No. A parabola that opens to the right or left is not the graph of a function because it does not pass the vertical line test.

b. A parabola is always a function if the directrix is parallel to the x-axis, so the slope of the directrix is 0.

98.

The parabola widens as $|k|$ increases.

99. For a point $P(x_1, y_1)$ on a parabola, the distance from P to the focus is the same as the distance from P to the directrix. Using the given formula, the distance from P to the directrix is $\dfrac{\left|3x_1 + 4y_1 - 7\right|}{\sqrt{3^2 + 4^2}} = \dfrac{\left|3x_1 + 4y_1 - 7\right|}{5}$.

The distance from P to the focus is $\sqrt{(x_1 - 4)^2 + (y_1 - 5)^2}$. Find the equation of the parabola by setting

$$\frac{\left|3x_1 + 4y_1 - 7\right|}{5} = \sqrt{(x_1 - 4)^2 + (y_1 - 5)^2} \text{ and}$$

simplifying:

$$16x^2 + 9y^2 - 24xy - 158x - 194y + 976 = 0$$

The axis is perpendicular to the directrix and passes through the focus. The slope of the directrix is $-\dfrac{3}{4}$, so the slope of the axis is $\dfrac{4}{3}$.

The equation of the line through $(4, 5)$ with slope $\dfrac{4}{3}$ is $y - 5 = \dfrac{4}{3}(x - 4) \Rightarrow y = \dfrac{4}{3}x - \dfrac{1}{3}$.

100. Using the formula given in exercise 88, the distance from the vertex to the directrix is

$$\frac{\left|3(6) - 5(-3) + 1\right|}{\sqrt{3^2 + (-5)^2}} = \frac{34}{\sqrt{34}} = \sqrt{34}. \text{ The distance}$$

from the focus (x, y) to the vertex is

(1) $\sqrt{(x - 6)^2 + (y + 3)^2} = \sqrt{34}$.

The distance from the focus (x, y) to the directrix is twice the distance from the focus to the vertex:

(2) $\dfrac{\left|3x - 5y + 1\right|}{\sqrt{3^2 + (-5)^2}} = 2\sqrt{34} \Rightarrow \left|3x - 5y + 1\right| = 68$.

(*continued on next page*)

(continued)

Solve the system consisting of equations (1) and (2) to find the coordinates of the focus:

$$\begin{cases} (x-6)^2 + (y+3)^2 = 34 \\ 3x - 5y = 67 \end{cases} \Rightarrow$$

$$\begin{cases} x^2 + y^2 - 12x + 6y + 11 = 0 \\ x = \dfrac{5}{3}y + \dfrac{67}{3} \end{cases} \Rightarrow$$

$$y^2 + 16y + 64 = 0 \Rightarrow y = -8$$
$$3x - 5(-8) = 67 \Rightarrow x = 9$$

The focus is at $(9, -8)$.

7.2 Getting Ready for the Next Section

101. $d = \sqrt{(x_2 - x_1)^2 + (y_2 - y_1)^2}$
$= \sqrt{(3 - (-1))^2 + (-2 - 4)^2} = \sqrt{52} = 2\sqrt{13}$

103. $M = \left(\dfrac{x_1 + x_2}{2}, \dfrac{y_1 + y_2}{2} \right) = \left(\dfrac{3 + (-7)}{2}, \dfrac{5 + 5}{2} \right)$
$= (-2, 5)$

105. $\dfrac{x^2}{9} + \dfrac{y^2}{25} = 1$

x-intercepts: $\dfrac{x^2}{9} + \dfrac{0^2}{25} = 1 \Rightarrow x^2 = 9 \Rightarrow x = \pm 3$

y-intercepts:

$\dfrac{0^2}{9} + \dfrac{y^2}{25} = 1 \Rightarrow y^2 = 25 \Rightarrow y = \pm 5$

107. $x^2 + 6x + 1 = 0 \Rightarrow x^2 + 6x = -1 \Rightarrow$
$x^2 + 6x + 9 = -1 + 9 \Rightarrow (x+3)^2 = 8$

109. $y - y_1 = m(x - x_1)$
$y - (-4) = -2(x - 3)$
$y + 4 = -2x + 6$
$y = -2x + 2$

7.3 The Ellipse

7.3 Practice Problems

1. Since the foci are $(0, -8)$ and $(0, 8)$, the major axis is on the y-axis, and $c = 8$. One vertex is $(0, 10)$, so the other vertex is $(0, -10)$, and $a = 10$.

$b^2 = a^2 - c^2 \Rightarrow b^2 = 10^2 - 8^2 \Rightarrow b = 6.$

Thus, the equation is $\dfrac{x^2}{36} + \dfrac{y^2}{100} = 1$.

2. $4x^2 + y^2 = 16 \Rightarrow \dfrac{x^2}{4} + \dfrac{y^2}{16} = 1 \Rightarrow$
$a^2 = 16 \Rightarrow a = 4; \quad b^2 = 4 \Rightarrow b = 2$

The length of the major axis $= 2a = 8$.
The length of the minor axis $= 2b = 4$.

3. a. Since the foci $(2, -3)$ and $(2, 5)$ lie on the vertical line $x = 2$, the ellipse is a vertical ellipse. The center of the ellipse is at $\left(2, \dfrac{-3 + 5}{2} \right) = (2, 1) = (h, k)$. Since the major axis has length 10, the vertices are 5 units from the center, so $a = 5$. The foci are 4 units from the center, so $c = 4$.

$b^2 = a^2 - c^2 \Rightarrow b^2 = 5^2 - 4^2 \Rightarrow b^2 = 9.$
Thus, the equation is

$\dfrac{(x-2)^2}{9} + \dfrac{(y-1)^2}{25} = 1.$

b. Since the vertices $(0, 0)$ and $(0, 10)$ lie on the y-axis, the ellipse is a vertical axis. The center is $(0, 5)$. The distance a from the center to a vertex is $a = 5$. The distance c from the center to a focus $(0, 8)$ is $c = 3$.

$c^2 = a^2 - b^2 \Rightarrow 3^2 = 5^2 - b^2 \Rightarrow$
$9 = 25 - b^2 \Rightarrow b^2 = 16$
Therefore, the equation is

$\dfrac{x^2}{16} + \dfrac{(y-5)^2}{25} = 1.$

4. Rewrite the equation as $\left(x^2 - 6x\right) + 4\left(y^2 + 2y\right) = 29$, then complete both squares and write the equation in standard form:

$$\left(x^2 - 6x\right) + 4\left(y^2 + 2y\right) = 29$$
$$\left(x^2 - 6x + 9\right) + 4\left(y^2 + 2y + 1\right) = 29 + 9 + 4$$
$$(x - 3)^2 + 4(y + 1)^2 = 42$$
$$\frac{(x-3)^2}{42} + \frac{2(y+1)^2}{21} = 1$$
$$\frac{(x-3)^2}{42} + \frac{(y+1)^2}{21/2} = 1$$

The center is $(3, -1)$.

$a^2 = 42 \Rightarrow a = \sqrt{42}$, so the vertices are $\left(3 - \sqrt{42}, -1\right)$ and $\left(3 + \sqrt{42}, -1\right)$.

$$b^2 = a^2 - c^2 \Rightarrow \frac{21}{2} = 42 - c^2 \Rightarrow c^2 = \frac{63}{2} \Rightarrow$$
$$c = \sqrt{\frac{63}{2}} = \frac{\sqrt{126}}{2} = \frac{3\sqrt{14}}{2}, \text{ the foci are}$$
$$\left(3 - \frac{3\sqrt{14}}{2}, -1\right) \text{ and } \left(3 + \frac{3\sqrt{14}}{2}, -1\right).$$

5. **a.** The center is $(1, -2)$, so $h = 1$ and $k = -2$. The distance a between the center $(1, -2)$ and the vertex $(1, 1)$ is $a = 1 - (-2) = 3$. The distance b between the center and $(-1, -2)$ is $b = 1 - (-1) = 2$. The equation for a vertical ellipse is $\frac{(x-h)^2}{b^2} + \frac{(y-k)^2}{a^2} = 1$, so the equation of the ellipse is

$$\frac{(x-1)^2}{4} + \frac{(y+2)^2}{9} = 1.$$

   b. The center is $(-1, 1)$ so $h = -1$ and $k = 1$. The distance c between the center and the focus is $c = \left(-1 + \sqrt{3}\right) - (-1) = \sqrt{3}$. The focus and the center lie on the horizontal line $y = 1$, so the ellipse is horizontal, and $(-1, 0)$ is an endpoint of the minor axis. The distance b between the center and $(-1, 0)$ is $b = 1 - 0 = 1$.

$$c^2 = a^2 - b^2 \Rightarrow \left(\sqrt{3}\right)^2 = a^2 - 1^2 \Rightarrow$$
$$a^2 = 3 + 1 = 4$$

The equation for a horizontal ellipse is $\frac{(x-h)^2}{a^2} + \frac{(y-k)^2}{b^2} = 1$, so the equation of the ellipse is

$$\frac{(x+1)^2}{4} + \frac{(y-1)^2}{1} = \frac{(x+1)^2}{4} + (y-1)^2 = 1.$$

6. Since the length of the major axis of the ellipse is 8 feet, $a = 4$. Since the length of the minor axis is 4 feet, $b = 2$.

$2^2 = 4^2 - c^2 \Rightarrow c^2 = 12 \Rightarrow c = 2\sqrt{3}$. If we position the center of the ellipse at $(0, 0)$ and the major axis along the x-axis, the foci of the ellipse are $\left(-2\sqrt{3}, 0\right)$ and $\left(2\sqrt{3}, 0\right)$. The distance between the two foci is $4\sqrt{3} \approx 6.9282$ feet. Thus the stone should be positioned 6.9282 feet from the source.

7.3 Concepts and Vocabulary

1. An ellipse is the set of all points in the plane, the <u>sum</u> of whose distances from two fixed points is a constant.

3. The standard equation of an ellipse with center $(0, 0)$, vertices $(\pm a, 0)$, foci $(\pm c, 0)$ is $\frac{x^2}{a^2} + \frac{y^2}{b^2} = 1$, where $\underline{b^2 = a^2 - c^2}$.

5. True

7. True

7.3 Building Skills

9. The major axis is on the x-axis.

$$b^2 = a^2 - c^2 \Rightarrow b^2 = 3^2 - 1^2 = 8.$$

The equation is $\frac{x^2}{9} + \frac{y^2}{8} = 1$.

11. The major axis is on the y-axis.

$b^2 = a^2 - c^2 \Rightarrow b^2 = 4^2 - 2^2 = 12$.

The equation is $\dfrac{x^2}{12} + \dfrac{y^2}{16} = 1$.

13. The major axis is on the x-axis.

$b^2 = a^2 - c^2 \Rightarrow 3^2 = a^2 - 4^2 \Rightarrow a^2 = 25$.

The equation is $\dfrac{x^2}{25} + \dfrac{y^2}{9} = 1$.

15. The major axis is on the y-axis.

$b^2 = a^2 - c^2 \Rightarrow 4^2 = a^2 - 2^2 \Rightarrow a^2 = 20$.

The equation is $\dfrac{x^2}{16} + \dfrac{y^2}{20} = 1$.

17. The major axis is on the y-axis.
Major axis length $= 10 \Rightarrow a = 5$ and minor axis length $= 6 \Rightarrow b = 3$. The equation is

$\dfrac{x^2}{9} + \dfrac{y^2}{25} = 1$.

19. The major axis is on the x-axis. $a = 6$.

$b^2 = a^2 - c^2 \Rightarrow b^2 = 6^2 - 3^2 = 27$.

The equation is $\dfrac{x^2}{36} + \dfrac{y^2}{27} = 1$.

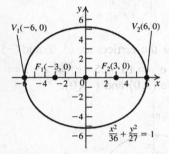

21. The major axis is on the y-axis. $c = 2$.

$b^2 = a^2 - c^2 \Rightarrow 3^2 = a^2 - 2^2 \Rightarrow a^2 = 13$.

The equation is $\dfrac{x^2}{9} + \dfrac{y^2}{13} = 1$.

23. $a^2 = 16 \Rightarrow$ the vertices are $(4, 0)$ and $(-4, 0)$.
$b^2 = a^2 - c^2 \Rightarrow 4 = 16 - c^2 \Rightarrow c = 2\sqrt{3} \Rightarrow$
the foci are $\left(2\sqrt{3}, 0\right)$ and $\left(-2\sqrt{3}, 0\right)$.

25. $a^2 = 9 \Rightarrow$ the vertices are $(3, 0)$ and $(-3, 0)$.
$b^2 = a^2 - c^2 \Rightarrow 1 = 9 - c^2 \Rightarrow c = 2\sqrt{2} \Rightarrow$
the foci are $\left(2\sqrt{2}, 0\right)$ and $\left(-2\sqrt{2}, 0\right)$.

27. $a^2 = 25 \Rightarrow$ the vertices are $(5, 0)$ and $(-5, 0)$.
$b^2 = a^2 - c^2 \Rightarrow 16 = 25 - c^2 \Rightarrow c = 3 \Rightarrow$
the foci are $(3, 0)$ and $(-3, 0)$.

29. $a^2 = 36 \Rightarrow$ the vertices are $(0, 6)$ and $(0, -6)$.
$b^2 = a^2 - c^2 \Rightarrow 16 = 36 - c^2 \Rightarrow c = 2\sqrt{5} \Rightarrow$
the foci are $\left(0, 2\sqrt{5}\right)$ and $\left(0, -2\sqrt{5}\right)$.

31. A circle with radius 2, centered at the origin.

33. $x^2 + 4y^2 = 4 \Rightarrow \dfrac{x^2}{4} + y^2 = 1.$ $a^2 = 4 \Rightarrow$ the
vertices are $(2, 0)$ and $(-2, 0)$.
$b^2 = a^2 - c^2 \Rightarrow 1 = 4 - c^2 \Rightarrow c = \sqrt{3} \Rightarrow$
the foci are $\left(\sqrt{3}, 0\right)$ and $\left(-\sqrt{3}, 0\right)$.

35. $9x^2 + 4y^2 = 36 \Rightarrow \dfrac{x^2}{4} + \dfrac{y^2}{9} = 1.$ $a^2 = 9 \Rightarrow$
the vertices are $(0, 3)$ and $(0, -3)$.
$b^2 = a^2 - c^2 \Rightarrow 4 = 9 - c^2 \Rightarrow c = \pm\sqrt{5} \Rightarrow$
the foci are $\left(0, \sqrt{5}\right)$ and $\left(0, -\sqrt{5}\right)$.

37. $3x^2 + 4y^2 = 12 \Rightarrow \dfrac{x^2}{4} + \dfrac{y^2}{3} = 1.$ $a^2 = 4 \Rightarrow$ the vertices are $(2, 0)$ and $(-2, 0)$.
$b^2 = a^2 - c^2 \Rightarrow 3 = 4 - c^2 \Rightarrow c = 1 \Rightarrow$ the foci are $(1, 0)$ and $(-1, 0)$.

39. $5x^2 - 10 = -2y^2 \Rightarrow 5x^2 + 2y^2 = 10 \Rightarrow$
$\dfrac{x^2}{2} + \dfrac{y^2}{5} = 1.$ $a^2 = 5 \Rightarrow$ the vertices are
$\left(0, \sqrt{5}\right)$ and $\left(0, -\sqrt{5}\right)$.
$b^2 = a^2 - c^2 \Rightarrow 2 = 5 - c^2 \Rightarrow c = \sqrt{3} \Rightarrow$ the foci are $\left(0, \sqrt{3}\right)$ and $\left(0, -\sqrt{3}\right)$.

41. $2x^2 + 3y^2 = 7 \Rightarrow \dfrac{x^2}{7/2} + \dfrac{y^2}{7/3} = 1 \Rightarrow a^2 = \dfrac{7}{2} \Rightarrow$

the vertices are $\left(\dfrac{\sqrt{14}}{2}, 0\right)$ and $\left(-\dfrac{\sqrt{14}}{2}, 0\right)$.

$b^2 = a^2 - c^2 \Rightarrow \dfrac{7}{3} = \dfrac{7}{2} - c^2 \Rightarrow c = \dfrac{\sqrt{42}}{6} \Rightarrow$

the foci are $\left(\dfrac{\sqrt{42}}{6}, 0\right)$ and $\left(-\dfrac{\sqrt{42}}{6}, 0\right)$.

43. The major axis has length 10, so $a = 5$. The minor axis has length 6, so $b = 3$. The ellipse is oriented vertically with center $(-2, 3)$, so the vertices are $(-2, 3 - 5) = (-2, -2)$ and $(-2, 3 + 5) = (-2, 8)$.
$b^2 = a^2 - c^2 \Rightarrow 9 = 25 - c^2 \Rightarrow c^2 = 16 \Rightarrow c = 4$
Thus, the foci are located at $(-2, 3 - 4) = (-2, -1)$ and $(-2, 3 + 4) = (-2, 7)$.

45. The vertices are located at $(1, 0)$ and $(9, 0)$, so the major axis has length 8 and $a = 4$. The

center of the ellipse is $\left(\dfrac{1+9}{2}, 0\right) = (5, 0)$.

Since a focus is located at $(7, 0)$, $c = 7 - 5 = 2$.
$b^2 = a^2 - c^2 \Rightarrow b^2 = 16 - 4 \Rightarrow b^2 = 12$
Thus, the equation of the ellipse is

$\dfrac{(x-5)^2}{16} + \dfrac{y^2}{12} = 1.$

47. The vertices are at $(2, -5)$ and $(2, 3)$, so the ellipse is positioned vertically and the length of the major axis is 8. Therefore, $a = 4$. The minor axis has length 2, so $b = 1$. The center

of the ellipse is $\left(\dfrac{2+2}{2}, \dfrac{-5+3}{2}\right) = (2, -1)$.

Thus, the equation of the ellipse is

$(x - 2)^2 + \dfrac{(y+1)^2}{16} = 1.$

49. The foci are located at $(-2, 3)$ and $(4, 3)$ and a vertex is located at $(-5, 3)$. Therefore, the ellipse is positioned horizontally. The center of the ellipse is $\left(\dfrac{-2+4}{2}, \dfrac{3+3}{2} \right) = (1, 3)$.

The distance from the given vertex $(-5, 3)$ to the center is 6, so $a = 6$. The distance from the focus $(-2, 3)$ to the center is 3, so $c = 3$.

$b^2 = a^2 - c^2 \Rightarrow b^2 = 36 - 9 \Rightarrow b^2 = 27$

Thus, the equation of the ellipse is

$$\dfrac{(x-1)^2}{36} + \dfrac{(y-3)^2}{27} = 1.$$

51. Center: $(1, 1)$; $a = 3 \Rightarrow$ vertices: $(1, 4)$ and $(1, -2)$.

$b^2 = a^2 - c^2 \Rightarrow 4 = 9 - c^2 \Rightarrow c = \sqrt{5}$. The foci are $\left(1, 1+\sqrt{5}\right)$ and $\left(1, 1-\sqrt{5}\right)$.

53. $4(x+3)^2 + 5(y-1)^2 = 20 \Rightarrow$

$\dfrac{(x+3)^2}{5} + \dfrac{(y-1)^2}{4} = 1 \Rightarrow$ center: $(-3, 1)$.

$a = \sqrt{5} \Rightarrow$ vertices: $\left(-3+\sqrt{5}, 1\right)$ and $\left(-3-\sqrt{5}, 1\right)$. $b^2 = a^2 - c^2 \Rightarrow 4 = 5 - c^2 \Rightarrow$

$c = \sqrt{5}$. The foci are $(-2, 1)$ and $(-4, 1)$.

55. Rewrite the equation as

$5(x^2 + 2x) + 9(y^2 - 4y) = 4$, then complete both squares:

$5(x^2 + 2x + 1) + 9(y^2 - 4y + 4) = 4 + 5 + 36 \Rightarrow$

$5(x+1)^2 + 9(y-2)^2 = 45 \Rightarrow$

$\dfrac{(x+1)^2}{9} + \dfrac{(y-2)^2}{5} = 1 \Rightarrow$ center: $(-1, 2)$.

$a = 3 \Rightarrow$ vertices: $(2, 2)$ and $(-4, 2)$.

$b^2 = a^2 - c^2 \Rightarrow 5 = 9 - c^2 \Rightarrow c = 2$.

The foci are $(1, 2)$ and $(-3, 2)$.

57. Rewrite the equation as

$9(x^2 + 4x) + 5(y^2 - 8y) = -71$, then complete both squares: $9(x^2 + 4x + 4) + 5(y^2 - 8y + 16)$

$= -71 + 36 + 80 \Rightarrow$

$9(x+2)^2 + 5(y-4)^2 = 45 \Rightarrow$

$\dfrac{(x+2)^2}{5} + \dfrac{(y-4)^2}{9} = 1 \Rightarrow$ center: $(-2, 4)$.

$a = 3 \Rightarrow$ vertices: $(-2, 7)$ and $(-2, 1)$.

$b^2 = a^2 - c^2 \Rightarrow 5 = 9 - c^2 \Rightarrow c = 2$.

The foci are $(-2, 2)$ and $(-2, 6)$.

59. Rewrite the equation as
$(x^2 - 2x) + 2(y^2 + 2y) = -1$, then complete both squares:
$(x^2 - 2x + 1) + 2(y^2 + 2y + 1) = -1 + 1 + 2 \Rightarrow$
$(x-1)^2 + 2(y+1)^2 = 2 \Rightarrow$
$\dfrac{(x-1)^2}{2} + (y+1)^2 = 1 \Rightarrow$ center: $(1, -1)$.
$a = \sqrt{2} \Rightarrow$ vertices: $\left(1+\sqrt{2}, -1\right)$ and $\left(1-\sqrt{2}, -1\right)$.
$b^2 = a^2 - c^2 \Rightarrow 1 = 2 - c^2 \Rightarrow c = 1$. The foci are $(2, -1)$ and $(0, -1)$.

61. Rewrite the equation as
$2(x^2 - 2x) + 9(y^2 + 2y) = -12$, then complete both squares:
$2(x^2 - 2x + 1) + 9(y^2 + 2y + 1) = -12 + 2 + 9 \Rightarrow$
$2(x-1)^2 + 9(y+1)^2 = -1 \Rightarrow$ there is no graph.

63. The major axis is on the y-axis. Major axis length $= 6 \Rightarrow a = 3$ and minor axis length $= 2$
$\Rightarrow b = 1$. The equation is $x^2 + \dfrac{y^2}{9} = 1$.

65. The major axis is on the x-axis. $c = 4$, $b = 2$.
$b^2 = a^2 - c^2 \Rightarrow 2^2 = a^2 - 4^2 \Rightarrow a^2 = 20$.
The equation is $\dfrac{x^2}{20} + \dfrac{y^2}{4} = 1$.

67. The center is $(0, 1) = (h, k)$. The distance a between the center and the vertex $(3, 1)$ is $3 - 0 = 3$. The distance b between the center and the vertex of the minor axis $(0, 3)$ is $3 - 1 = 2$.
The equation is $\dfrac{x^2}{9} + \dfrac{(y-1)^2}{4} = 1$.

7.3 Applying the Concepts

69. Sketch the ellipse so that its center is at the origin and the major axis lies on the x-axis.

Then $a = 25$ and $b = 20$. The equation of the ellipse is $\dfrac{x^2}{25^2} + \dfrac{y^2}{20^2} = 1$. Let $x = 10$, then solve for y:
$\dfrac{10^2}{25^2} + \dfrac{y^2}{20^2} = 1 \Rightarrow y = 4\sqrt{21} \approx 18.3$ ft.

71. Sketch the ellipse so that its center is at the origin and the major axis lies on the x-axis.

Then $a = 75$ and $b = 45$. The equation of the ellipse is $\dfrac{x^2}{75^2} + \dfrac{y^2}{45^2} = 1$. Let $x = 25$, then solve for y:
$\dfrac{25^2}{75^2} + \dfrac{y^2}{45^2} = 1 \Rightarrow y = 30\sqrt{2} \approx 42.4$ m.

73. $a = 48$ and $b = 23$, so $b^2 = a^2 - c^2 \Rightarrow$
$23^2 = 48^2 - c^2 \Rightarrow c = 5\sqrt{71} \approx 42.13$ feet from the center or $48 - 42.13 \approx 5.9$ feet from the wall on the opposite side, along the major axis.

75. Using the hints given, we have
$a = \dfrac{128.49 + 154.83}{2} = 141.66 \Rightarrow$
$a^2 = 20{,}067.5556$ and
$c = 141.66 - 128.49 = 13.17$.
$b^2 = 141.66^2 - 13.17^2 = 19{,}894.1067$. The equation is $\dfrac{x^2}{20{,}067.5556} + \dfrac{y^2}{19{,}894.1067} = 1$.

77. Using the hints given in exercise 75, we have

$a = \dfrac{28.56 + 43.88}{2} = 36.22 \Rightarrow a^2 = 1311.8884$

and $c = 36.22 - 28.56 = 7.66$.

$b = 36.22^2 - 7.66^2 = 1253.2128$. The

equation is $\dfrac{x^2}{1311.8884} + \dfrac{y^2}{1253.2128} = 1$.

79. $a = \dfrac{768,806}{2} = 384,403$ and

$b = \dfrac{767,746}{2} = 383,873$. So $b^2 = a^2 - c^2 \Rightarrow$

$383,873^2 = 384,403^2 - c^2 \Rightarrow c \approx 20,179$.

The perihelion $= a - c \approx 364,224$ km.

The aphelion $= a + c \approx 404,582$ km.

81. The perihelion of the satellite is 4070 miles and the aphelion is 4124 miles.

Not drawn to scale

$a = \dfrac{4070 + 4124}{2} = 4097 \Rightarrow a^2 = 16,785,409$

and $c = 4097 - 4070 = 27$. Then

$b^2 = 16,784,680$. The equation is

$\dfrac{x^2}{16,785,409} + \dfrac{y^2}{16,784,680} = 1$.

7.3 Beyond the Basics

83. The center is (0, 0) and $a = 3$, so the equation

is either $\dfrac{x^2}{3^2} + \dfrac{y^2}{b^2} = 1$ or $\dfrac{x^2}{b^2} + \dfrac{y^2}{3^2} = 1$.

Substitute the coordinates of the point for x and y, then solve for b:

$\dfrac{(1)^2}{5^2} + \dfrac{\left(3\sqrt{5}/2\right)^2}{b^2} = 1 \Rightarrow \dfrac{1}{25} + \dfrac{45}{4b^2} = 1 \Rightarrow$

$b^2 = \dfrac{1125}{96} \Rightarrow b \approx 3.4$. Because the exercise

states that the length of the major axis is 6 there is no solution.

85. The general form of the equation is

$\dfrac{(x-2)^2}{a^2} + \dfrac{(y-1)^2}{b^2} = 1$. Substitute the

coordinates of the points into the equation of the ellipse and then solve the system:

$\begin{cases} \dfrac{(10-2)^2}{a^2} + \dfrac{(1-1)^2}{b^2} = 1 \\ \dfrac{(6-2)^2}{a^2} + \dfrac{(2-1)^2}{b^2} = 1 \end{cases} \Rightarrow \begin{cases} \dfrac{8^2}{a^2} + \dfrac{0}{b^2} = 1 \\ \dfrac{16}{a^2} + \dfrac{1}{b^2} = 1 \end{cases} \Rightarrow$

$a^2 = 64, b^2 = 4/3$. Then the equation is

$\dfrac{(x-2)^2}{64} + \dfrac{(y-1)^2}{4/3} = 1 \Rightarrow$

$(x-2)^2 + 48(y-1)^2 = 64$

87. $a = 4, b^2 = a^2 - c^2 \Rightarrow 9 = 16 - c^2 \Rightarrow c = \sqrt{7}$

$e = \sqrt{7}/4$

89. $x^2 + 4y^2 = 1 \Rightarrow x^2 + \dfrac{y^2}{1/4} = 1$

$a = 1, b^2 = a^2 - c^2 \Rightarrow \dfrac{1}{4} = 1 - c^2 \Rightarrow c = \dfrac{\sqrt{3}}{2}$

$e = \sqrt{3}/2$

91. Rewrite the equation as

$(x^2 - 2x) + 2(y^2 + 2y) = -1$, then complete

both squares:

$(x^2 - 2x + 1) + 2(y^2 + 2y + 1) = -1 + 1 + 2 \Rightarrow$

$(x+1)^2 + 2(y+1)^2 = 2 \Rightarrow \dfrac{(x+1)^2}{2} + (y+1)^2 = 1$

$a = \sqrt{2}, b^2 = a^2 - c^2 \Rightarrow 1 = 2 - c^2 \Rightarrow c = 1$

$e = \dfrac{1}{\sqrt{2}} = \dfrac{\sqrt{2}}{2}$

93. $a = 3; e = \dfrac{c}{a} \Rightarrow \dfrac{1}{3} = \dfrac{c}{3} \Rightarrow c = 1$

$b^2 = a^2 - c^2 \Rightarrow b^2 = 9 - 1 \Rightarrow b^2 = 8$

The equation is $\dfrac{x^2}{9} + \dfrac{y^2}{8} = 1$.

95. Using substitution, we have

$$\begin{cases} x + 3y = -2 \\ 4x^2 + 3y^2 = 7 \end{cases} \Rightarrow \begin{cases} x = -3y - 2 \\ 4x^2 + 3y^2 = 7 \end{cases} \Rightarrow 4(-3y - 2)^2 + 3y^2 = 7 \Rightarrow 39y^2 + 48y + 9 = 0 \Rightarrow$$

$$y = -1 \text{ or } y = -\frac{3}{13}$$

$$x + 3(-1) = -2 \Rightarrow x = 1$$

$$x + 3\left(-\frac{3}{13}\right) = -2 \Rightarrow x = -\frac{17}{13}$$

The points of intersection are $\left(-\dfrac{17}{13}, -\dfrac{3}{13}\right)$ and $(1, -1)$.

97. Using elimination, we have $\begin{cases} x^2 + y^2 = 20 \\ 9x^2 + y^2 = 36 \end{cases} \Rightarrow 8x^2 = 16 \Rightarrow x = \pm\sqrt{2}$

$$2 + y^2 = 20 \Rightarrow y = \pm 3\sqrt{2}$$

The points of intersection are $\left(-\sqrt{2}, -3\sqrt{2}\right), \left(-\sqrt{2}, 3\sqrt{2}\right), \left(\sqrt{2}, -3\sqrt{2}\right)$, and $\left(\sqrt{2}, 3\sqrt{2}\right)$.

99. $\dfrac{x^2}{a^2} + \dfrac{y^2}{b^2} = 1 \Rightarrow \dfrac{x^2}{(5/2)^2} + \dfrac{y^2}{5^2} = 1 \Rightarrow \dfrac{4x^2}{25} + \dfrac{y^2}{25} = 1 \Rightarrow 4x^2 + y^2 = 25 \Rightarrow y^2 = 25 - 4x^2$

The tangent line equation is $y = m(x - 2) + 3$. Solve $25 - 4x^2 = [m(x - 2) + 3]^2$ to find m.

$$25 - 4x^2 = m^2(x - 2)^2 + 6m(x - 2) + 9 \Rightarrow$$

$$(4 + m^2)x^2 + (6m - 4m^2)x + (4m^2 - 12m - 16) = 0 \Rightarrow (6m - 4m^2)^2 - 4(4 + m^2)(4m^2 - 12m - 16) = 0 \Rightarrow$$

$$36m^2 + 192m + 256 = 0 \Rightarrow 9m^2 + 48m + 64 = 0 \Rightarrow (3m + 8)^2 = 0 \Rightarrow m = -\frac{8}{3}$$

For a unique solution, the discriminant must be zero. The tangent line as equation

$$y = -\frac{8}{3}(x - 2) + 3 = -\frac{8}{3}x + \frac{25}{3}.$$

101. $9(x+1)^2 + 16(y-2)^2 = 144 \Rightarrow$

$$\frac{(x+1)^2}{16} + \frac{(y-2)^2}{9} = 1 \Rightarrow a = 4, b = 3$$

$$A = \pi ab = 12\pi$$

103. $16(x-1)^2 + 9(y+2)^2 = 144 \Rightarrow$

$$\frac{(x-1)^2}{9} + \frac{(y+2)^2}{16} = 1$$

This ellipse has center $(1, -2)$ and is oriented vertically. $a = 3, b = 4$, so $A = 12\pi$.

$$16(x-1)^2 + 4(y+2)^2 = 64 \Rightarrow$$

$$\frac{(x-1)^2}{4} + \frac{(y+2)^2}{16} = 1$$

This ellipse has center $(1, -2)$ and is oriented vertically. $a = 2, b = 4$, so $A = 8\pi$.
The area between the two ellipses is $12\pi - 8\pi = 4\pi$.

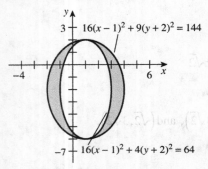

105. Let the length of $P_1F_1 = s$. Then, the coordinates of P_1 are (c, s). But, $ae = c$, so the coordinates of P_1 are (ae, s). Since P_1 lies on the ellipse

$$\frac{x^2}{a^2} + \frac{y^2}{b^2} = 1 \Rightarrow \frac{c^2}{a^2} + \frac{s^2}{b^2} = 1 \Rightarrow$$

$$\frac{a^2 - b^2}{a^2} + \frac{s^2}{b^2} = 1 \Rightarrow$$

$$b^2(a^2 - b^2) + a^2 s^2 = a^2 b^2 \Rightarrow$$

$$s^2 = \frac{a^2 b^2 - b^2(a^2 - b^2)}{a^2}$$

$$= \frac{b^2(a^2 - a^2 + b^2)}{a^2} \Rightarrow s = \frac{b^2}{a}$$

The length of the ellipse is symmetric with respect to the x-axis, so the length of the latus rectum, P_1P_2, is $\frac{2b^2}{a}$.

107. $\frac{x^2}{16} + \frac{y^2}{8} = 1 \Rightarrow a^2 = 16 \Rightarrow a = 4, b^2 = 8$

$$\text{L.R.} = \frac{2b^2}{a} = \frac{2(8)}{4} = 4$$

109. $5x^2 + 9y^2 = 45 \Rightarrow \frac{x^2}{9} + \frac{y^2}{5} = 1 \Rightarrow$

$$a^2 = 9 \Rightarrow a = 3, b^2 = 5$$

$$\text{L.R.} = \frac{2b^2}{a} = \frac{2(5)}{3} = \frac{10}{3}$$

111. The length of the major axis
$2a = 3(2b) \Rightarrow a = 3b$.

$$\text{L.R.} = 2 = \frac{2b^2}{a} = \frac{2b^2}{3b} \Rightarrow 2 = \frac{2b}{3} \Rightarrow b = 3 \text{ and }$$

$a = 9$.
The center of the ellipse is at $(0, 0)$, so an equation of the ellipse is

$$\frac{x^2}{9^2} + \frac{y^2}{3^2} = 1 \Rightarrow \frac{x^2}{81} + \frac{y^2}{9} = 1.$$

Another ellipse with major axis $3b$, latus rectum $= 2$, and center $(0, 0)$ is

$$\frac{x^2}{3^2} + \frac{y^2}{9^2} = 1 \Rightarrow \frac{x^2}{9} + \frac{y^2}{81} = 1.$$

7.3 Critical Thinking/Discussion/Writing

113. There are four circles, one in each quadrant:
$(x-3)^2 + (y-3)^2 = 9, (x+3)^2 + (y-3)^2 = 9,$
$(x-3)^2 + (y+3)^2 = 9,$ and
$(x+3)^2 + (y+3)^2 = 9.$

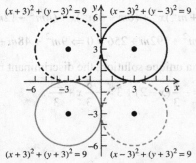

114. There are eight ellipses, two in each quadrant:

$$\frac{(x-3)^2}{9} + \frac{(y-2)^2}{4} = 1, \frac{(x-2)^2}{4} + \frac{(y-3)^2}{9} = 1,$$

$$\frac{(x+3)^2}{9} + \frac{(y-2)^2}{4} = 1, \frac{(x+2)^2}{4} + \frac{(y-3)^2}{9} = 1,$$

$$\frac{(x+3)^2}{9} + \frac{(y+2)^2}{4} = 1, \frac{(x+2)^2}{4} + \frac{(y+3)^2}{9} = 1,$$

$$\frac{(x-3)^2}{9} + \frac{(y+2)^2}{4} = 1, \frac{(x-2)^2}{4} + \frac{(y+3)^2}{9} = 1$$

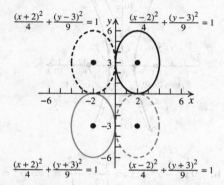

7.3 Getting Ready for the Next Section

115. $y - y_1 = m(x - x_1) \Rightarrow y - 5 = -(x + 3) \Rightarrow$
$y = -x - 3 + 5 = -x + 2$

117. $\begin{cases} y = 3x + 6 \\ y = -3x \end{cases} \Rightarrow 2y = 6 \Rightarrow y = 3$
$y = -3x \Rightarrow 3 = -3x \Rightarrow x = -1$
The point of intersection is (–1, 3).

119. $4y^2 - x^2 = 1$
x-intercepts:
$4(0)^2 - x^2 = 1 \Rightarrow x^2 = -1 \Rightarrow$ there are no x-intercepts.
y-intercepts:
$4y^2 - (0)^2 = 1 \Rightarrow y^2 = \frac{1}{4} \Rightarrow y = \pm\frac{1}{2}$

121. $\left(\sqrt{29}\right)^2 = a^2 + 4^2 \Rightarrow a^2 = 29 - 16 = 13 \Rightarrow$
$a = \pm\sqrt{13}$

123. $f(x) = \frac{x^2 + 1}{x - 1}$
$x - 1 = 0 \Rightarrow x = 1$ is the vertical asymptote.

$$\begin{array}{r} x+1 \\ x-1\overline{)x^2 + 0x + 1} \\ \underline{x^2 - x} \\ x + 1 \\ \underline{x - 1} \\ 2 \end{array}$$

$\frac{x^2 + 1}{x - 1} = x + 1 + \frac{2}{x - 1} \Rightarrow y = x + 1$ is the slant asymptote.

7.4 The Hyperbola

7.4 Practice Problems

1. The transverse axis is on the y-axis.

2. Rewrite the equation in standard form to determine a and b. $x^2 - 4y^2 = 8 \Rightarrow$
$\frac{x^2}{8} - \frac{y^2}{2} = 1 \Rightarrow a^2 = 8 \Rightarrow a = 2\sqrt{2}$ and
$b^2 = 2 \Rightarrow b = \sqrt{2}$.
The transverse axis of the hyperbola is along the x-axis, so the vertices are $\left(-2\sqrt{2}, 0\right)$ and $\left(2\sqrt{2}, 0\right)$.
To find the foci, we need c:
$c^2 = a^2 + b^2 \Rightarrow c^2 = 8 + 2 \Rightarrow c = \sqrt{10}$.
The foci are $\left(-\sqrt{10}, 0\right)$ and $\left(\sqrt{10}, 0\right)$.

3. **a.** Since the foci of the hyperbola, (0, –6) and (0, 6) lie on the y-axis, the transverse axis also lies on the y-axis, and $c = 6$. The center of the hyperbola is (0, 0), so the standard form of this hyperbola is
$\frac{y^2}{a^2} - \frac{x^2}{b^2} = 1$.
The vertices are (0, –3) and (0, 3), so $a = 3$.
$c^2 = a^2 + b^2 \Rightarrow 6^2 = 3^2 + b^2 \Rightarrow b^2 = 27$.
The equation is $\frac{y^2}{9} - \frac{x^2}{27} = 1$.

b. The center $(0, 0)$ and the vertex $(5, 0)$ are on the x-axis, so the transverse axis is on the x-axis. The distance between the center and the vertex is 5, so $a = 5$. The distance between the center and the focus $(7, 0)$ is 7, so $c = 7$.

$c^2 = a^2 + b^2 \Rightarrow 7^2 = 5^2 + b^2 \Rightarrow$
$b^2 = 49 - 25 = 24$

The equation is $\dfrac{x^2}{25} - \dfrac{y^2}{24} = 1$.

c. The foci are on the y-axis, so the tranverse axis is also on the y-axis. The length of the transverse axis is 10, so $2a = 10 \Rightarrow a = 5$. The center is $(0, 0)$ and the distance between the center and the focus $(0, 6)$ is $c = 6$.

$c^2 = a^2 + b^2 \Rightarrow 6^2 = 5^2 + b^2 \Rightarrow$
$b^2 = 36 - 25 = 11$

The equation is $\dfrac{y^2}{25} - \dfrac{x^2}{11} = 1$.

4. The hyperbola $\dfrac{y^2}{4} - \dfrac{x^2}{9} = 1$ is of the form

$\dfrac{y^2}{a^2} - \dfrac{x^2}{b^2} = 1$, so $b = 3$ and $a = 2$. The

asymptotes are of the form $y = \dfrac{a}{b}x$ and

$y = -\dfrac{a}{b}x$. They are $y = \dfrac{2}{3}x$ and $y = -\dfrac{2}{3}x$.

5. a. The foci are on the x-axis, so the transverse axis is on the x-axis and the asymptotes are

$y = \pm\dfrac{b}{a}x$. The asymptotes for this hyperbola

are $y = \pm x$, so $\dfrac{b}{a} = 1 \Rightarrow a = b$. For this

hyperbola, $c = 2\sqrt{2}$.

$c^2 = a^2 + b^2 \Rightarrow \left(2\sqrt{2}\right)^2 = a^2 + b^2 \Rightarrow$
$8 = a^2 + a^2 \Rightarrow 8 = 2a^2 \Rightarrow a^2 = 4$
$a = b \Rightarrow a^2 = b^2 = 4$

The equation is $\dfrac{x^2}{4} - \dfrac{y^2}{4} = 1$.

b. The vertices are on the y-axis, so the so the transverse axis is on the y-axis and the

asymptotes are $y = \pm\dfrac{a}{b}x$.

The asymptotes for this hyperbola are

$y = \pm\dfrac{1}{3}x$, so $\dfrac{a}{b} = \dfrac{1}{3} \Rightarrow a = \dfrac{1}{3}b$. For this

hyperbola, $a = 1$, so $b = 3$. The equation is

$y^2 - \dfrac{x^2}{9} = 1$.

6. a. $25x^2 - 4y^2 = 100 \Rightarrow \dfrac{x^2}{4} - \dfrac{y^2}{25} = 1 \Rightarrow a = 2$,
$b = 5$.
$c^2 = a^2 + b^2 \Rightarrow c^2 = 4 + 25 \Rightarrow c = \sqrt{29}$.
The vertices are $(2, 0)$ and $(-2, 0)$. The endpoints of the conjugate axis are $(0, 5)$ and $(0, -5)$, and the foci are $\left(-\sqrt{29}, 0\right)$ and $\left(\sqrt{29}, 0\right)$. The asymptotes are

$y = \dfrac{b}{a}x = \dfrac{5}{2}x$ and $y = -\dfrac{b}{a}x = -\dfrac{5}{2}x$.

b. $9y^2 - x^2 = 1 \Rightarrow \dfrac{y^2}{1/9} - \dfrac{x^2}{1} = 1 \Rightarrow a^2 = \dfrac{1}{9} \Rightarrow$
$a = \dfrac{1}{3}$ and $b = 1$.
$c^2 = a^2 + b^2 \Rightarrow c^2 = \dfrac{1}{9} + 1 \Rightarrow c = \dfrac{\sqrt{10}}{3}$.

The vertices are $\left(0, \dfrac{1}{3}\right)$ and $\left(0, -\dfrac{1}{3}\right)$.

The endpoints of the conjugate axis are $(1, 0)$ and $(-1, 0)$, and the foci are

$\left(0, \dfrac{\sqrt{10}}{3}\right)$ and $\left(0, -\dfrac{\sqrt{10}}{3}\right)$.

The asymptotes are $y = \dfrac{a}{b}x = \dfrac{1/3}{1}x = \dfrac{1}{3}x$

and $y = -\dfrac{a}{b}x = -\dfrac{1/3}{1}x = -\dfrac{1}{3}x$.

(continued on next page)

(*continued*)

7. Complete the square to put the equation in standard form:

$$\left(x^2 - 2x\right) - 4\left(y^2 - 4y\right) = 20$$

$$\left(x^2 - 2x + 1\right) - 4\left(y^2 - 4y + 4\right) = 20 + 1 - 16$$

$$(x-1)^2 - 4(y-2)^2 = 5$$

$$\frac{(x-1)^2}{5} - \frac{4(y-2)^2}{5} = 1$$

$$\frac{(x-1)^2}{5} - \frac{(y-2)^2}{5/4} = 1$$

The center is $(1, 2)$, $a = \sqrt{5}$ and $b = \dfrac{\sqrt{5}}{2}$.

The vertices are $\left(1 + \sqrt{5}, 2\right)$ and $\left(1 - \sqrt{5}, 2\right)$.

The endpoints of the conjugate axis are

$$\left(1, 2 + \frac{\sqrt{5}}{2}\right) \text{ and } \left(1, 2 - \frac{\sqrt{5}}{2}\right).$$

The asymptotes are $y - k = \pm \dfrac{b}{a}(x - h) \Rightarrow$

$$y - 2 = \pm \frac{\sqrt{5}/2}{\sqrt{5}}(x - 1) \Rightarrow y - 2 = \pm \frac{1}{2}(x - 1).$$

$c^2 = a^2 + b^2 \Rightarrow c^2 = 5 + \dfrac{5}{4} = \dfrac{25}{4} \Rightarrow c = \dfrac{5}{2}$. The

foci are $\left(1 + \dfrac{5}{2}, 2\right) = \left(\dfrac{7}{2}, 2\right)$ and

$$\left(1 - \frac{5}{2}, 2\right) = \left(-\frac{3}{2}, 2\right).$$

8. **a.** The center is $(0, -2)$, so $h = 0$ and $k = -2$. The distance from the center to the vertex is $a = 1$. The distance between the center and the focus is $c = 3$. The transverse axis lies on the horizontal line $y = -2$.

$$c^2 = a^2 + b^2 \Rightarrow 3^2 = 1^2 + b^2 \Rightarrow b^2 = 8.$$

The equation is

$$\frac{x^2}{1} - \frac{(y+2)^2}{8} = x^2 - \frac{(y+2)^2}{8} = 1.$$

b. The center is $(-1, 0)$, so $h = -1$ and $k = 0$. The distance between the center and the vertex $(-1, 1)$ is $a = 1$. The slope of the asymptote with positive slope is $\dfrac{1}{2}$ and the transverse axis is on the vertical line

$x = -1$, so $\dfrac{a}{b} = \dfrac{1}{2} \Rightarrow b = 2a = 2(1) = 2$.

The equation is

$$\frac{y^2}{1} - \frac{(x+1)^2}{4} = y^2 - \frac{(x+1)^2}{4} = 1.$$

9. The hyperbola has foci $(-150, 0)$ and $(150, 0)$, so $c = 150$. The distance from A to B is 260 miles, so $a = 130$. $c^2 = a^2 + b^2 \Rightarrow$

$150^2 = 130^2 + b^2 \Rightarrow b^2 = 5600$.

The equation of the hyperbola is

$$\frac{x^2}{130^2} - \frac{y^2}{5600} = 1 \Rightarrow \frac{x^2}{16,900} - \frac{y^2}{5600} = 1.$$

7.4 Concepts and Vocabulary

1. A hyperbola is a set of all points in the plane, the <u>difference</u> of whose distances form two fixed points is constant.

3. The standard equation of the hyperbola with center $(0, 0)$, vertices $(\pm a, 0)$, and foci

$(\pm c, 0)$ is $\dfrac{x^2}{a^2} - \dfrac{y^2}{b^2} = 1$, where

$b^2 = \underline{c^2 - a^2}$.

5. True

7. False. If one asymptote of a hyperbola has equation $y = -2x$, then the other has the equation $y = 2x$.

7.4 Building Skills

9. G   11. H

13. D   15. B

13. D   **15.** B

17. $a^2 = 1 \Rightarrow$ the vertices are $(1, 0)$ and $(-1, 0)$.
$c^2 = a^2 + b^2 \Rightarrow c^2 = 1 + 4 \Rightarrow c = \sqrt{5} \Rightarrow$
the foci are $\left(\sqrt{5}, 0\right)$ and $\left(-\sqrt{5}, 0\right)$.

Transverse axis: x-axis. Hyperbola opens left and right. Vertices of the fundamental rectangle: $(1, 2), (-1, 2), (-1, -2), (1, -2)$. Asymptotes: $y = \pm 2x$.

19. $x^2 - y^2 = -1 \Rightarrow y^2 - x^2 = 1 \Rightarrow a^2 = 1 \Rightarrow$ the vertices are $(0, 1)$ and $(0, -1)$.
$c^2 = a^2 + b^2 \Rightarrow c^2 = 1 + 1 \Rightarrow c = \sqrt{2} \Rightarrow$
the foci are $\left(0, \sqrt{2}\right)$ and $\left(0, -\sqrt{2}\right)$.

Transverse axis: y-axis. Hyperbola opens up and down. Vertices of the fundamental rectangle: $(1, 1), (-1, 1), (-1, -1), (1, -1)$. Asymptotes: $y = \pm x$.

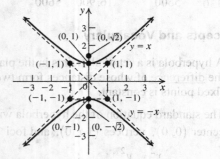

21. $9y^2 - x^2 = 36 \Rightarrow \dfrac{y^2}{4} - \dfrac{x^2}{36} = 1 \Rightarrow a^2 = 4 \Rightarrow$ the vertices are $(0, 2)$ and $(0, -2)$.
$c^2 = a^2 + b^2 \Rightarrow c^2 = 4 + 36 \Rightarrow c = 2\sqrt{10} \Rightarrow$
the foci are $\left(0, 2\sqrt{10}\right)$ and $\left(0, -2\sqrt{10}\right)$.

Transverse axis: y-axis. Hyperbola opens up and down. Vertices of the fundamental rectangle: $(6, 2), (-6, 2), (-6, -2), (6, -2)$. Asymptotes: $y = \pm \dfrac{a}{b} x = \pm \dfrac{1}{3} x$.

23. $4x^2 - 9y^2 - 36 = 0 \Rightarrow \dfrac{x^2}{9} - \dfrac{y^2}{4} = 1 \Rightarrow$
$a^2 = 9 \Rightarrow$ the vertices are $(3, 0)$ and $(-3, 0)$.
$c^2 = a^2 + b^2 \Rightarrow c^2 = 9 + 4 \Rightarrow c = \sqrt{13} \Rightarrow$
the foci are $\left(\sqrt{13}, 0\right)$ and $\left(-\sqrt{13}, 0\right)$.

Transverse axis: x-axis. Hyperbola opens left and right. Vertices of the fundamental rectangle: $(3, 2), (-3, 2), (-3, -2), (3, -2)$.

Asymptotes: $y = \pm \dfrac{b}{a} x = \pm \dfrac{2}{3} x$.

25. $y = \pm\sqrt{4x^2 + 1} \Rightarrow y^2 - \dfrac{x^2}{1/4} = 1 \Rightarrow a^2 = 1 \Rightarrow$
the vertices are $(0, 1)$ and $(0, -1)$.

$c^2 = a^2 + b^2 \Rightarrow c^2 = 1 + \dfrac{1}{4} \Rightarrow c = \dfrac{\sqrt{5}}{2} \Rightarrow$ the

foci are $\left(0, \dfrac{\sqrt{5}}{2}\right)$ and $\left(0, -\dfrac{\sqrt{5}}{2}\right)$. Transverse

axis: y-axis. Hyperbola opens up and down. Vertices of the fundamental rectangle:
$\left(\dfrac{1}{2}, 1\right), \left(-\dfrac{1}{2}, 1\right), \left(-\dfrac{1}{2}, -1\right),$ and $\left(\dfrac{1}{2}, -1\right)$.

Asymptotes: $y = \pm \dfrac{a}{b} x = \pm 2x$.

(continued on next page)

(*continued*)

27. $y = \pm\sqrt{9x^2 - 1} \Rightarrow 9x^2 - y^2 = 1 \Rightarrow$

$\dfrac{x^2}{1/9} - y^2 = 1 \Rightarrow a^2 = \dfrac{1}{9} \Rightarrow$ the vertices are

$\left(\dfrac{1}{3}, 0\right)$ and $\left(-\dfrac{1}{3}, 0\right)$. $c^2 = a^2 + b^2 \Rightarrow$

$c^2 = \dfrac{1}{9} + 1 \Rightarrow c = \dfrac{\sqrt{10}}{3} \Rightarrow$ the foci are $\left(\dfrac{\sqrt{10}}{3}, 0\right)$

and $\left(-\dfrac{\sqrt{10}}{3}, 0\right)$. Transverse axis:

x-axis. Hyperbola opens left and right. Vertices of the fundamental rectangle:

$\left(\dfrac{1}{3}, 1\right), \left(-\dfrac{1}{3}, 1\right), \left(-\dfrac{1}{3}, -1\right)$, and $\left(\dfrac{1}{3}, -1\right)$.

Asymptotes: $y = \pm\dfrac{b}{a}x \Rightarrow y = \pm 3x$.

29. The transverse axis is the x-axis and the center is $(0, 0)$, so the equation is of the form

$\dfrac{x^2}{a^2} - \dfrac{y^2}{b^2} = 1$. $a = 2, c = 3 \Rightarrow 9 = 4 + b^2 \Rightarrow$

$b^2 = 5$. The equation is $\dfrac{x^2}{4} - \dfrac{y^2}{5} = 1$.

31. The transverse axis is the y-axis and the center is $(0, 0)$, so the equation is of the form

$\dfrac{y^2}{a^2} - \dfrac{x^2}{b^2} = 1$. $a = 4, c = 6 \Rightarrow 36 = 16 + b^2 \Rightarrow$

$b^2 = 20$. The equation is $\dfrac{y^2}{16} - \dfrac{x^2}{20} = 1$.

33. The transverse axis is the y-axis and the center is $(0, 0)$, so the equation is of the form

$\dfrac{y^2}{a^2} - \dfrac{x^2}{b^2} = 1$. $a = 2, c = 5 \Rightarrow 25 = 4 + b^2 \Rightarrow$

$b^2 = 21$. The equation is $\dfrac{y^2}{4} - \dfrac{x^2}{21} = 1$.

35. The transverse axis is the x-axis and the center is $(0, 0)$, so the equation is of the form

$$\frac{x^2}{a^2} - \frac{y^2}{b^2} = 1 . \ a = 1, c = 5 \Rightarrow 25 = 1 + b^2 \Rightarrow$$

$b^2 = 24$. The equation is $x^2 - \frac{y^2}{24} = 1$.

37. The transverse axis is the y-axis and the center is $(0, 0)$, so the equation is of the form

$$\frac{y^2}{a^2} - \frac{x^2}{b^2} = 1. \text{ Length of the transverse axis} = 6,$$

so, $a = 3. c = 5 \Rightarrow 25 = 9 + b^2 \Rightarrow b^2 = 16$.

The equation is $\frac{y^2}{9} - \frac{x^2}{16} = 1$.

39. The transverse axis is the x-axis and the center is $(0, 0)$, so the equation is of the form

$$\frac{x^2}{a^2} - \frac{y^2}{b^2} = 1. \ y = 2x = \frac{b}{a} x \Rightarrow b = 2a .$$

$$c^2 = a^2 + b^2 \Rightarrow 5 = a^2 + (2a)^2 \Rightarrow a = 1, b = 2.$$

The equation is $x^2 - \frac{y^2}{4} = 1$.

41. The transverse axis is the y-axis and the center is $(0, 0)$, so the equation is of the form

$$\frac{y^2}{a^2} - \frac{x^2}{b^2} = 1. \ y = x = \frac{a}{b} x \Rightarrow a = b .$$

$a = 4 \Rightarrow b = 4$. The equation is $\frac{y^2}{16} - \frac{x^2}{16} = 1$.

43. Center: $(1, -1)$; vertices: $(4, -1)$, $(-2, -1)$; transverse axis: $y = -1$; asymptotes:

$$y - k = \pm \frac{b}{a}(x - h) \Rightarrow y + 1 = \pm \frac{4}{3}(x - 1)$$

45. Center: $(-2, 0)$; vertices: $(3, 0)$, $(-7, 0)$; transverse axis: x-axis ($y = 0$); asymptotes:

$$y - k = \pm\frac{b}{a}(x - h) \Rightarrow y = \pm\frac{7}{5}(x + 2)$$

47. Center: $(-4, -3)$; vertices: $(1, -3)$, $(-9, -3)$; transverse axis: $y = -3$; asymptotes:

$$y - k = \pm\frac{b}{a}(x - h) \Rightarrow y + 3 = \pm\frac{7}{5}(x + 4).$$

49. $4x^2 - (y + 1)^2 = 25 \Rightarrow \dfrac{x^2}{25/4} - \dfrac{(y+1)^2}{25} = 1 \Rightarrow$

Center: $(0, -1)$; vertices: $\left(\dfrac{5}{2}, -1\right), \left(-\dfrac{5}{2}, -1\right)$;

transverse axis: $y = -1$; asymptotes:

$$y - k = \pm\frac{b}{a}(x - h) \Rightarrow y + 1 = \pm 2x.$$

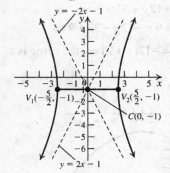

51. $(y + 1)^2 - 9(x - 2)^2 = 25 \Rightarrow$

$$\frac{(y+1)^2}{25} - \frac{(x-2)^2}{25/9} = 1 \Rightarrow \text{ center: } (2, -1);$$

vertices: $(2, 4)$, $(2, -6)$;
transverse axis: $x = 2$;

asymptotes: $y - k = \pm\dfrac{a}{b}(x - h) \Rightarrow$

$$y + 1 = \pm 3(x - 2).$$

53. Complete the square to put the equation in standard form: $x^2 - y^2 + 6x = 36 \Rightarrow$

$$x^2 + 6x + 9 - y^2 = 36 + 9 \Rightarrow$$

$$\frac{(x+3)^2}{45} - \frac{y^2}{45} = 1 \Rightarrow \text{ center: } (-3, 0);$$

vertices: $\left(-3 + 3\sqrt{5}, 0\right), \left(-3 - 3\sqrt{5}, 0\right)$;

transverse axis:
x-axis ($y = 0$); asymptotes:

$$y - k = \pm\frac{b}{a}(x - h) \Rightarrow y = \pm(x + 3).$$

55. Complete the square to put the equation in standard form: $x^2 + 4x - 4y^2 = 12 \Rightarrow$

$x^2 + 4x + 4 - 4y^2 = 12 + 4 \Rightarrow$

$\dfrac{(x+2)^2}{16} - \dfrac{y^2}{4} = 1 \Rightarrow$ center: $(-2, 0)$; vertices:

$(2, 0), (-6, 0)$; transverse axis: x-axis $(y = 0)$;

asymptotes: $y - k = \pm \dfrac{b}{a}(x - h) \Rightarrow$

$y = \pm \dfrac{1}{2}(x + 2)$.

57. Complete the square to put the equation in standard form: $2x^2 - y^2 + 12x - 8y + 3 = 0 \Rightarrow$

$2(x^2 + 6x + 9) - (y^2 + 8y + 16) = -3 + 18 - 16 \Rightarrow$

$(y+4)^2 - \dfrac{(x+3)^2}{1/2} = 1 \Rightarrow$ center: $(-3, -4)$;

vertices: $(-3, -3), (-3, -5)$; transverse axis:

$x = -3$; asymptotes: $y - k = \pm \dfrac{a}{b}(x - h) \Rightarrow$

$y + 4 = \pm\sqrt{2}(x + 3)$.

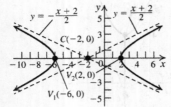

59. Complete the square to put the equation in standard form: $3x^2 - 18x - 2y^2 - 8y + 1 = 0 \Rightarrow$

$3(x^2 - 6x + 9) - 2(y^2 + 4y + 4) = -1 + 27 - 8 \Rightarrow$

$\dfrac{(x-3)^2}{6} - \dfrac{(y+2)^2}{9} = 1 \Rightarrow$ center: $(3, -2)$;

vertices: $\left(3 + \sqrt{6}, -2\right), \left(3 - \sqrt{6}, -2\right)$; transverse

axis: $y = -2$; asymptotes:

$y - k = \pm \dfrac{b}{a}(x - h) \Rightarrow y + 2 = \pm \dfrac{\sqrt{6}}{2}(x - 3)$

61. Complete the square to put the equation in standard form: $y^2 + 2\sqrt{2} - x^2 + 2\sqrt{2}x = 1 \Rightarrow$

$y^2 - \left(x^2 - 2\sqrt{2}x + 2\right) = 1 - 2\sqrt{2} - 2 \Rightarrow$

$\dfrac{(x-\sqrt{2})^2}{1+2\sqrt{2}} - \dfrac{y^2}{1+2\sqrt{2}} = 1 \Rightarrow$ center: $\left(\sqrt{2}, 0\right)$;

vertices: $\left(\sqrt{2} + \sqrt{1+2\sqrt{2}}, 0\right)$,

$\left(\sqrt{2} - \sqrt{1+2\sqrt{2}}, 0\right)$; transverse axis: x-axis

$(y = 0)$; asymptotes: $y - k = \pm \dfrac{b}{a}(x - h) \Rightarrow$

$y = \pm\left(x - \sqrt{2}\right)$

63. $x^2 - 6x + 12y + 33 = 0 \Rightarrow$

$x^2 - 6x + 9 = -12y - 33 + 9 \Rightarrow$

$(x - 3)^2 = -12(y + 2) \Rightarrow$ the conic is a parabola.

65. $y^2 - 9x^2 = -1 \Rightarrow \dfrac{x^2}{1/9} - y^2 = 1 \Rightarrow$ the conic is a hyperbola.

67. $x^2 + y^2 - 4x + 8y = 16 \Rightarrow$
$x^2 - 4x + 4 + y^2 + 8y + 16 = 16 + 4 + 16 \Rightarrow$
$(x-2)^2 + (y+4)^2 = 36 \Rightarrow$ the conic is a circle.

69. $2x^2 - 4x + 3y + 8 = 0 \Rightarrow$
$2(x^2 - 2x + 1) = -3y - 8 + 2 \Rightarrow$
$2(x-1)^2 = -3(y+2) \Rightarrow$ the conic is a parabola.

71. $4x^2 + 9y^2 + 8x - 54y + 49 = 0 \Rightarrow$
$4(x^2 + 2x + 1) + 9(y^2 - 6y + 9) = -49 + 4 + 81 \Rightarrow$
$\dfrac{(x+1)^2}{9} + \dfrac{(y-3)^2}{4} = 1 \Rightarrow$ the conic is an ellipse.

73. The distance from the center to the vertex is $a = 3$. The distance from the center to the focus is $c = \sqrt{17}$.
$c^2 = a^2 + b^2 \Rightarrow 17 = 9 + b^2 \Rightarrow b^2 = 8$.
The transverse axis is on the *y*-axis, so the equation is $\dfrac{y^2}{9} - \dfrac{x^2}{8} = 1$.

75. The distance from the center to the vertex is $a = 3$. The distance from the center to the focus is $c = 4$.
$c^2 = a^2 + b^2 \Rightarrow 16 = 9 + b^2 \Rightarrow b^2 = 7$.
The transverse axis is on the horizontal line $y = 1$, so the equation
is $\dfrac{(x+2)^2}{9} - \dfrac{(y-1)^2}{7} = 1$.

77. The vertex and center are on the *x*-axis, so the transverse axis is on the *x*-axis and the asymptotes are $y = \pm\dfrac{b}{a}x$. The asymptote with positive slope has slope $\dfrac{5}{4}$, so
$\dfrac{b}{a} = \dfrac{5}{4} \Rightarrow b = \dfrac{5}{4}a$. For this hyperbola, $a = 2$, so
$b = \dfrac{5}{4}(2) = \dfrac{5}{2}$.
The equation is
$\dfrac{(x+1)^2}{4} - \dfrac{y^2}{(5/2)^2} = 1 \Rightarrow \dfrac{(x+1)^2}{4} - \dfrac{4y^2}{25} = 1$.

7.4 Applying the Concepts

79. A hyperbola is the set of all points in the plane whose distances from two fixed points have a constant difference. The difference in the distances of the location of the explosion from point A to point B is 600 meters, so A and B are the foci of the hyperbola. Let the coordinates of A be $(-500, 0)$ and those of B be $(500, 0)$, so $c = 500$. The distance between V_1 and V_2 is 600, so $a = 300$.
$$c^2 = a^2 + b^2 \Rightarrow$$
$$500^2 = 300^2 + b^2 \Rightarrow b^2 = 160,000$$

The equation is $\dfrac{x^2}{90,000} - \dfrac{y^2}{160,000} = 1$.

81. Using the same reasoning as in exercises 79 and 80, Nicole and Juan are located at the foci of the hyperbola, $(-4000, 0)$ and $(4000, 0)$. The distance between the vertices is 3300, so $a = 1650$.
$$c^2 = a^2 + b^2 \Rightarrow 4000^2 = 1650^2 + b^2 \Rightarrow$$
$$b^2 = 13,277,500 . \text{ The equation is}$$
$$\dfrac{x^2}{2,722,500} - \dfrac{y^2}{13,277,500} = 1.$$
Because Nicole hears the thunder first, the graph consists only of the part of the hyperbola closest to Nicole.

83. Let the coordinates of A and B be $(-150, 0)$ and $(150, 0)$, respectively. At 300,000 km per sec, the difference in the distances the signals travel is 150 km, so $a = 75$. $c^2 = a^2 + b^2 \Rightarrow$ $150^2 = 75^2 + b^2 \Rightarrow b^2 = 16,875$. The equation is $\dfrac{x^2}{5625} - \dfrac{y^2}{16,875} = 1$.

85. Let the coordinates of A be $(-125, 0)$ and those of B be $(125, 0)$. The difference in the distances of the location of the plane from point A to point B is 500 microseconds × 980 feet per microsecond = 490,000 feet = 92.8 miles. So $a = 245,000$ feet ≈ 46.4015 miles and $c = 125$ miles.
$$c^2 = a^2 + b^2 \Rightarrow 125^2 = 46.4015^2 + b^2 \Rightarrow$$
$$b^2 = 13,471.9008 . \text{ The equation is}$$
$$\dfrac{x^2}{2153.0992} - \dfrac{y^2}{13,471.9008} = 1.$$
The plane is flying along the line $y = 50$, so
$$\dfrac{x^2}{2153.0992} - \dfrac{50^2}{13,471.9008} = 1 \Rightarrow$$
$x = \pm 50.5238$ miles
Since the signal from point B arrives sooner than the signal from point A, the plane is located closer to B. It is at approximately $(50.5238, 50)$.

87. The fishing boat is located at the intersection of two hyperbolas, one with foci at A and B, and the other with foci at A and C. For the first hyperbola, $c = 200$ and $a = 150$.

$c^2 = a^2 + b^2 \Rightarrow 200^2 = 150^2 + b^2 \Rightarrow$

$b^2 = 17,500$. The equation of the first hyperbola

is $\dfrac{x^2}{22,500} - \dfrac{y^2}{17,500} = 1$. For the second

hyperbola, $c = 500$ and $a = 200$.

$500^2 = 200^2 + b^2 \Rightarrow b^2 = 210,000$. The center of the second hyperbola is (200, 500), so the equation of the second hyperbola is

$\dfrac{(y-500)^2}{40,000} - \dfrac{(x-200)^2}{210,000} = 1$. Solve the system

$$\begin{cases} \dfrac{x^2}{22,500} - \dfrac{y^2}{17,500} = 1 \\ \dfrac{(y-500)^2}{40,000} - \dfrac{(x-200)^2}{210,000} = 1 \end{cases}$$

or use a graphing calculator to find the intersections (the possible locations of the boat).

7.4 Beyond the Basics

89.

91.

93. Assume that the transverse axis is the x-axis. Then the equation of the hyperbola is

$\dfrac{x^2}{a^2} - \dfrac{y^2}{b^2} = 1$. The hyperbola is equilateral,

so $a = b$, and the equations of the asymptotes are $y = \pm x$. These lines are perpendicular to each other. Similarly, it can be shown that the asymptotes are perpendicular to each other if the transverse axis is the y-axis.

95. Since $b > 0$ and $c^2 = a^2 + b^2$, it follows that $c^2 > a^2$ and $c > a$. So, $e > 1$. When $e = 1$, the hyperbola becomes the union of two rays.

97. $36x^2 - 25y^2 = 900 \Rightarrow \dfrac{x^2}{25} - \dfrac{y^2}{36} = 1$

$a = 5, b = 6 \Rightarrow c^2 = 25 + 36 \Rightarrow c = \sqrt{61}$

$e = \dfrac{c}{a} = \dfrac{\sqrt{61}}{5}$.

Length of latus rectum $= \dfrac{2b^2}{a} = \dfrac{72}{5}$.

99. $8(x-1)^2 - (y+2)^2 = 2 \Rightarrow$

$\dfrac{(x-1)^2}{1/4} - \dfrac{(y+2)^2}{2} = 1$

$a = \dfrac{1}{2}, b = \sqrt{2} \Rightarrow c^2 = \dfrac{1}{4} + 2 \Rightarrow c = \dfrac{3}{2}$

$e = \dfrac{c}{a} = 3$.

Length of latus rectum $= \dfrac{2b^2}{a} = 8$.

101. $e = \dfrac{c}{a} \Rightarrow e^2 = \dfrac{c^2}{a^2} \Rightarrow$

$e^2 = \dfrac{a^2 + b^2}{a^2} = 1 + \dfrac{b^2}{a^2} \Rightarrow$

$e^2 - 1 = \dfrac{b^2}{a^2} \Rightarrow a^2(e^2 - 1) = b^2$

103. The foci are $(\pm 6,\ 0) \Rightarrow c = 6$ and the center is $(0,\ 0)$. The transverse axis lies along the x-axis.

Length of the latus rectum =

$$\frac{2b^2}{a} = 10 \Rightarrow b^2 = 5a$$

$$c^2 = a^2 + b^2 \Rightarrow 36 = a^2 + 5a \Rightarrow$$

$$a^2 + 5a - 36 = 0 \Rightarrow (a-4)(a+9) = 0 \Rightarrow$$

$$a = 4,\ a = -9 \text{ (disregard)}$$

$$b^2 = 5(4) = 20,\ \text{so an equation of the hyperbola}$$

is $\dfrac{x^2}{4^2} - \dfrac{y^2}{20} = 1 \Rightarrow \dfrac{x^2}{16} - \dfrac{y^2}{20} = 1.$

105. The foci are $(0,\ \pm 12) \Rightarrow c = 12$ and the center is $(0,\ 0)$. The transverse axis lies along the y-axis.

Length of the latus rectum =

$$\frac{2b^2}{a} = 36 \Rightarrow b^2 = 18a$$

$$c^2 = a^2 + b^2 \Rightarrow 144 = a^2 + 18a \Rightarrow$$

$$a^2 + 18a - 144 = 0 \Rightarrow (a-6)(a+24) = 0 \Rightarrow$$

$$a = 6,\ a = -24 \text{ (disregard)}$$

$$b^2 = 18(6) = 108,\ \text{so an equation of the}$$

hyperbola is $\dfrac{y^2}{6^2} - \dfrac{x^2}{108} = 1 \Rightarrow \dfrac{y^2}{36} - \dfrac{x^2}{108} = 1.$

107. The foci are $\left(0,\ \pm\sqrt{10}\right) \Rightarrow c = \sqrt{10} \Rightarrow c^2 = 10.$

The transverse axis lies along the y-axis.

$$a^2 + b^2 = c^2 \Rightarrow a^2 + b^2 = 10 \Rightarrow$$

$$b^2 = 10 - a^2$$

The graph passes through $(2, 3)$ so

$$\frac{y^2}{a^2} - \frac{x^2}{b^2} = 1 \Rightarrow \frac{3^2}{a^2} - \frac{2^2}{10-a^2} = 1 \Rightarrow$$

$$9\left(10-a^2\right) - 4a^2 = a^2\left(10-a^2\right)$$

$$90 - 9a^2 - 4a^2 = 10a^2 - a^4$$

$$a^4 - 23a^2 + 90 = 0 \Rightarrow$$

$$\left(a^2 - 18\right)\left(a^2 - 5\right) = 0 \Rightarrow a^2 = 18,\ a^2 = 5$$

If $a^2 = 18$, then $a^2 + b^2 = 10 \Rightarrow$

$18 + b^2 = 10 \Rightarrow b^2 = -8$, which is impossible.

If $a^2 = 5$, then $a^2 + b^2 = 10 \Rightarrow$

$5 + b^2 = 10 \Rightarrow b^2 = 5.$

The equation is $\dfrac{y^2}{5} - \dfrac{x^2}{5} = 1.$

109. Solve using substitution:

$$\begin{cases} y - 2x - 20 = 0 \\ y^2 - 4x^2 = 36 \end{cases} \Rightarrow \begin{cases} y = 2x + 20 \\ y^2 - 4x^2 = 36 \end{cases} \Rightarrow$$

$$(2x+20)^2 - 4x^2 = 36 \Rightarrow$$

$$80x + 400 = 36 \Rightarrow x = -\frac{91}{20}$$

$$y = 2\left(-\frac{91}{20}\right) + 20 = \frac{109}{10}$$

The only point of intersection is $\left(-\dfrac{91}{20}, \dfrac{109}{10}\right).$

111. Solve using elimination:

$$\begin{cases} x^2 + y^2 = 15 \\ x^2 - y^2 = 1 \end{cases} \Rightarrow 2x^2 = 16 \Rightarrow x = \pm 2\sqrt{2}$$

$$\left(\pm 2\sqrt{2}\right)^2 + y^2 = 15 \Rightarrow y^2 = 7 \Rightarrow y = \pm\sqrt{7}$$

The points of intersection are

$$\left(2\sqrt{2}, \sqrt{7}\right),\ \left(2\sqrt{2}, -\sqrt{7}\right),\ \left(-2\sqrt{2}, -\sqrt{7}\right),\ \text{and}$$

$$\left(-2\sqrt{2}, \sqrt{7}\right).$$

113. The equation of the hyperbola is

$$\frac{x^2}{5/3} - \frac{y^2}{5/7} = 1 \Rightarrow \frac{3x^2}{5} - \frac{7y^2}{5} = 1 \Rightarrow 3x^2 - 7y^2 = 5 \Rightarrow 7y^2 = 3x^2 - 5$$

Step 1: Let m = the slope of the tangent line. Then the equation of the tangent line is

$$y - 1 = m(x - 2) \Rightarrow y = m(x - 2) + 1$$

Step 2: $7y^2 = 7\left[m(x - 2) + 1\right]^2 = 7\left[m^2(x - 2)^2 + 2m(x - 2) + 1\right] = 7m^2(x - 2)^2 + 14m(x - 2) + 7$

Step 3:

$$3x^2 - 5 = 7m^2(x - 2)^2 + 14m(x - 2) + 7$$
$$3x^2 - 5 = 7m^2(x^2 - 4x + 4) + 14mx - 28m + 7$$
$$3x^2 - 5 = 7m^2x^2 - 28m^2x + 28m^2 + 14mx - 28m + 7$$
$$3x^2 - 7m^2x^2 + 28m^2x - 14mx - 28m^2 + 28m - 12 = 0$$
$$\left(3 - 7m^2\right)x^2 + \left(28m^2 - 14m\right)x - \left(28m^2 - 28m + 12\right) = 0$$

Now set the discriminant equal to 0 and solve for x:

$$\left(28m^2 - 14m\right)^2 - 4\left(3 - 7m^2\right)\left(28m^2 - 28m + 12\right) = 0$$
$$784m^4 - 784m^3 + 196m^2 - 784m^4 + 784m^3 - 336m + 144 = 0$$
$$196m^2 - 336m + 144 = 0$$
$$49m^2 - 84m + 36 = 0$$
$$\left(7m - 6\right)^2 = 0$$
$$m = \frac{6}{7}$$

Step 4: The equation of the tangent line is $y - 1 = \frac{6}{7}(x - 2) \Rightarrow y = \frac{6}{7}x - \frac{5}{7}$.

7.4 Critical Thinking/Writing/Discussion

115. a. $4x^2 - 9y^2 = 0 \Rightarrow (2x + 3y)(2x - 3y) = 0$.

The graph consists of two lines, $y = \pm\frac{2}{3}x$.

b. All the variable terms on the left side of the equation are always nonnegative, so there are no x or y values that would satisfy the equation. There is no graph.

c. Both x^2 and y^2 are nonnegative, so the only solution of the equation is $(0, 0)$.

d. Complete the square:
$$x^2 + y^2 - 4x + 8y = -20 \Rightarrow$$
$$(x^2 - 4x + 4) + (y^2 + 8y + 16) = -20 + 4 + 16 \Rightarrow$$
$$(x - 2)^2 + (y + 4)^2 = 0 \Rightarrow x = 2 \text{ and } y = -4.$$

(continued on next page)

(*continued*)

e. $y^2 - 2x^2 = 0 \Rightarrow y^2 = 2x^2 \Rightarrow y = \pm x\sqrt{2}$.
The graph consists of two lines.

116. $Ax^2 + Cy^2 + Dx + Ey + F = 0, \ AC \neq 0$
Rewriting the equation, we have

$$A\left(x^2 + \frac{D}{A}x + \frac{D^2}{4A^2}\right) + C\left(y^2 + \frac{E}{C}y + \frac{E^2}{4C^2}\right)$$

$$= \frac{D^2}{4A} + \frac{E^2}{4C} - F \Rightarrow$$

$$A\left(x + \frac{D}{2A}\right)^2 + C\left(y + \frac{E}{2C}\right)^2 = \frac{D^2}{4A} + \frac{E^2}{4C} - F.$$

If the term on the right side is not zero, then the
equation can be rewritten as

$$\frac{A\left(x + \frac{D}{2A}\right)^2}{\frac{D^2}{4A} + \frac{E^2}{4C} - F} + \frac{C\left(y + \frac{E}{2C}\right)^2}{\frac{D^2}{4A} + \frac{E^2}{4C} - F} = 1 \Rightarrow$$

$$\frac{\left(x + \frac{D}{2A}\right)^2}{\frac{D^2}{4A^2} + \frac{E^2}{4AC} - \frac{F}{A}} + \frac{\left(y + \frac{E}{2C}\right)^2}{\frac{D^2}{4AC} + \frac{E^2}{4C^2} - \frac{F}{C}} = 1$$

(i) If $A > 0$ and $C > 0$, and if $\frac{D^2}{4A} + \frac{E^2}{4C} - F = 0$,
the graph consists of one point,
$\left(-\frac{D}{2A}, -\frac{E}{2C}\right)$. If $\frac{D^2}{4A} + \frac{E^2}{4C} - F > 0$, the
graph is an ellipse. If $\frac{D^2}{4A} + \frac{E^2}{4C} - F < 0$, then
there is no graph.

(ii) If $A > 0$ and $C < 0$, and if $\frac{D^2}{4A} + \frac{E^2}{4C} - F = 0$,
the graph consists of two lines,
$\sqrt{A}\left(x + \frac{D}{2A}\right) = \pm\sqrt{-C}\left(y + \frac{E}{2C}\right)$. If
$\frac{D^2}{4A} + \frac{E^2}{4C} - F > 0$, the graph is a hyperbola
with a horizontal transverse axis. If
$\frac{D^2}{4A} + \frac{E^2}{4C} - F < 0$, then the graph is a
hyperbola with a vertical transverse axis.

(iii) If $A < 0$ and $C > 0$, and if $\frac{D^2}{4A} + \frac{E^2}{4C} - F = 0$,
the graph consists of two lines,
$\sqrt{-A}\left(x + \frac{D}{2A}\right) = \pm\sqrt{C}\left(y + \frac{E}{2C}\right)$.
If $\frac{D^2}{4A} + \frac{E^2}{4C} - F > 0$, the graph is a hyperbola
with a vertical transverse axis. If
$\frac{D^2}{4A} + \frac{E^2}{4C} - F < 0$, then the graph is a
hyperbola with a horizontal transverse axis.

(iv) If $A < 0$ and $C < 0$, and if $\frac{D^2}{4A} + \frac{E^2}{4C} - F = 0$,
the graph consists of one point,
$\left(-\frac{D}{2A}, -\frac{E}{2C}\right)$. If $\frac{D^2}{4A} + \frac{E^2}{4C} - F > 0$, there is
no graph. If $\frac{D^2}{4A} + \frac{E^2}{4C} - F < 0$, then the
graph is an ellipse.

7.4 Getting Ready for the Next Section

117. $f(3) = \frac{1}{1+3} = \frac{1}{4}$

119. $h(5) = (-1)^3 3^{5-1} = -3^4 = -81$

121. $(-1)^{17} = -1$

123. $(-1)^5 2^5 = 32$

125. $\dfrac{3 \cdot \pi \cdot e^2 \cdot 7 \cdot 11 \cdot 19 \cdot a}{a \cdot 3 \cdot \pi \cdot e^2 \cdot 19} = 7 \cdot 11 = 77$

127. $c(1 - 5a + 3b) = c - 5ac + 3bc$
True

Chapter 7 Review Exercises

1. $y^2 = -6x \Rightarrow -6 = 4a \Rightarrow a = -\dfrac{3}{2}$

Vertex: (0, 0), focus: $\left(-\dfrac{3}{2}, 0\right)$, axis: x-axis,

directrix: $x = \dfrac{3}{2}$.

3. $x^2 = 7y \Rightarrow 7 = 4a \Rightarrow \dfrac{7}{4} = a$.

Vertex: (0, 0), focus: $\left(0, \dfrac{7}{4}\right)$, axis: y-axis,

directrix: $y = -\dfrac{7}{4}$.

5. $(x - 2)^2 = -(y + 3) \Rightarrow -1 = 4a \Rightarrow -\dfrac{1}{4} = a$.

Vertex: (2, −3), focus: $\left(2, -\dfrac{13}{4}\right)$, axis: $x = 2$,

directrix: $y = -\dfrac{11}{4}$.

7. Rearrange the equation and complete the square to put the equation in standard form:
$$y^2 = -4y + 2x + 1 \Rightarrow y^2 + 4y + 4 = 2x + 1 + 4 \Rightarrow$$
$$(y + 2)^2 = 2\left(x + \dfrac{5}{2}\right) \Rightarrow 2 = 4a \Rightarrow a = \dfrac{1}{2}.$$

Vertex: $\left(-\dfrac{5}{2}, -2\right)$, focus: (−2, −2),
axis: $y = -2$, directrix: $x = -3$.

9. The vertex is (0, 0) and the focus is (−3, 0), so the graph opens to the left. The general form is $(y - k)^2 = -4a(x - h)$.
$a = -3 \Rightarrow y^2 = -12x$.

11. The focus is (0, 4) and the directrix is $y = -4$, so the vertex is (0, 0), and the graph opens up. The general form is $(x - h)^2 = 4a(y - k)$.
$a = 4 \Rightarrow x^2 = 16y$.

13. $a^2 = 25 \Rightarrow$ the vertices are (5, 0) and (−5, 0).
$b^2 = a^2 - c^2 \Rightarrow 4 = 25 - c^2 \Rightarrow c = \pm\sqrt{21} \Rightarrow$ the foci are $\left(\sqrt{21}, 0\right)$ and $\left(-\sqrt{21}, 0\right)$.
Endpoints of the minor axis: (0, −2) and (0, 2).

(*continued on next page*)

(*continued*)

15. $4x^2 + y^2 = 4 \Rightarrow x^2 + \dfrac{y^2}{4} = 1$.

$a^2 = 4 \Rightarrow$ the vertices are $(0, 2)$ and $(0, -2)$.
$b^2 = a^2 - c^2 \Rightarrow 1 = 4 - c^2 \Rightarrow c = \pm\sqrt{3} \Rightarrow$
the foci are $\left(0, \sqrt{3}\right)$ and $\left(0, -\sqrt{3}\right)$.
Endpoints of the minor axis: $(1, 0)$ and $(-1, 0)$.

17. $16(x+1)^2 + 9(y+4)^2 = 144 \Rightarrow$

$\dfrac{(x+1)^2}{9} + \dfrac{(y+4)^2}{16} = 1$.

The center is $(-1, -4)$ and $a^2 = 16 \Rightarrow$ the vertices are $(-1, -8)$ and $(-1, 0)$.
$b^2 = a^2 - c^2 \Rightarrow 9 = 16 - c^2 \Rightarrow c = \pm\sqrt{7}$.
The foci are $\left(-1, -4 - \sqrt{7}\right)$ and $\left(-1, -4 + \sqrt{7}\right)$.
Endpoints of the minor axis: $(-4, -4)$ and $(2, -4)$.

19. Rearrange the equation and complete the square to put the equation in standard form:
$x^2 + 9y^2 + 2x - 18y + 1 = 0 \Rightarrow$
$x^2 + 2x + 1 + 9(y^2 - 2y + 1) = -1 + 1 + 9 \Rightarrow$

$(x+1)^2 + 9(y-1)^2 = 9 \Rightarrow \dfrac{(x+1)^2}{9} + (y-1)^2 = 1$

The center is $(-1, 1)$ and $a^2 = 9 \Rightarrow$ the vertices are $(-4, 1)$ and $(2, 1)$.
$b^2 = a^2 - c^2 \Rightarrow 1 = 9 - c^2 \Rightarrow c = \pm 2\sqrt{2}$
The foci are $\left(-1 - 2\sqrt{2}, 1\right)$ and $\left(-1 + 2\sqrt{2}, 1\right)$.
Endpoints of the minor axis: $(-1, 2)$ and $(-1, 0)$.

21. $a = 4$, $b = 2$, major axis: x-axis. $\dfrac{x^2}{16} + \dfrac{y^2}{4} = 1$

23. $a = 10$, center $(0, 0)$, major axis: x-axis, $c = 5$, so $b^2 = 100 - 25 = 75$. $\dfrac{x^2}{100} + \dfrac{y^2}{75} = 1$

25. $a^2 = 16 \Rightarrow$ the vertices are $(0, 4)$ and $(0, -4)$.
$c^2 = a^2 + b^2 \Rightarrow c^2 = 16 + 4 \Rightarrow c = \pm 2\sqrt{5} \Rightarrow$
the foci are $\left(0, 2\sqrt{5}\right)$ and $\left(0, -2\sqrt{5}\right)$.

Asymptotes: $y = \pm\dfrac{a}{b}x = \pm 2x$.

27. $8x^2 - y^2 = 8 \Rightarrow x^2 - \dfrac{y^2}{8} = 1 \Rightarrow a^2 = 1 \Rightarrow$ the vertices are $(-1, 0)$ and $(1, 0)$.

$c^2 = a^2 + b^2 \Rightarrow c^2 = 1 + 8 \Rightarrow c = \pm 3 \Rightarrow$ the foci are $(3, 0)$ and $(-3, 0)$. Asymptotes:

$y = \pm\dfrac{b}{a}x = \pm 2\sqrt{2}x$.

29. The center is $(-2, 3)$. $a^2 = 9 \Rightarrow$ the vertices are $(1, 3)$ and $(-5, 3)$.

$c^2 = a^2 + b^2 \Rightarrow c^2 = 9 + 4 \Rightarrow c = \pm\sqrt{13} \Rightarrow$ the foci are $\left(-2+\sqrt{13}, 3\right)$ and $\left(-2-\sqrt{13}, 3\right)$.

Asymptotes:

$y - k = \pm\dfrac{b}{a}(x-h) \Rightarrow y - 3 = \pm\dfrac{2}{3}(x+2)$.

31. Rearrange the equation and complete the square to put the equation in standard form:

$4y^2 - x^2 + 40y - 4x + 60 = 0 \Rightarrow$

$4(y^2 + 10y + 25) - (x^2 + 4x + 4)$
$\qquad\qquad = -60 + 100 - 4$

$4(y+5)^2 - (x+2)^2 = 36 \Rightarrow$

$\dfrac{(y+5)^2}{9} - \dfrac{(x+2)^2}{36} = 1$.

The center is $(-2, -5)$. $a^2 = 9 \Rightarrow$ the vertices are $(-2, -5-3) = (-2, -8)$ and $(-2, -5+3) = (-2, -2)$.

$c^2 = a^2 + b^2 \Rightarrow c^2 = 9 + 36 \Rightarrow c = \pm 3\sqrt{5} \Rightarrow$ the foci are $\left(-2, -5+3\sqrt{5}\right)$ and $\left(-2, -5-3\sqrt{5}\right)$. Asymptotes:

$y - k = \pm\dfrac{a}{b}(x-h) \Rightarrow y + 5 = \pm\dfrac{1}{2}(x+2)$.

33. The vertices are $(\pm 1, 0)$ and the foci are $(\pm 2, 0)$, so the center is $(0, 0)$, $a = 1$, $c = 2$, and the transverse axis is the x-axis.

$c^2 = a^2 + b^2 \Rightarrow 4 = 1 + b^2 \Rightarrow b^2 = 3$. The equation is $x^2 - \dfrac{y^2}{3} = 1$.

35. The vertices are $(\pm 2, 0)$ so the center is $(0, 0)$, $a = 2$, and the transverse axis is the x-axis.

The asymptotes are $y = \pm 3x \Rightarrow \pm\dfrac{b}{a} = \pm 3 \Rightarrow \dfrac{b}{2} = 3 \Rightarrow b = 6$. The equation is $\dfrac{x^2}{4} - \dfrac{y^2}{36} = 1$.

37. Hyperbola. In standard form, the equation is $\dfrac{x^2}{4} - \dfrac{y^2}{5} = 1$.

39. Ellipse. In standard form, the equation is
$$\frac{(x-2)^2}{22/3}+\frac{(y+1)^2}{11/2}=1.$$

41. Circle. In standard form, the equation is
$$(x+1)^2+y^2=4.$$

43. Hyperbola. In standard form, the equation is
$$y^2-x^2=3.$$

45. Circle. In standard form, the equation is
$$(x-1)^2+(y+2)^2=\frac{10}{3}$$

47. Ellipse. In standard form, the equation is
$$\frac{x^2}{4}+\frac{y^2}{9/2}=1.$$

49. $4x^2+9y^2=36\Rightarrow\dfrac{x^2}{9}+\dfrac{y^2}{4}=1\Rightarrow$ the vertices of the ellipse are $(3,0)$ and $(-3,0)$. $b^2=a^2-c^2\Rightarrow 4=9-c^2\Rightarrow c=\pm\sqrt5\Rightarrow$ the foci are $\left(\sqrt5,0\right)$ and $\left(-\sqrt5,0\right)$. For the hyperbola, the transverse axis is the x-axis, $a=\sqrt5$, and $c=3$, so $c^2=a^2+b^2\Rightarrow$ $9=5+b^2\Rightarrow b^2=4$. The equation of the hyperbola is $\dfrac{x^2}{5}-\dfrac{y^2}{4}=1$.

51. Solve using substitution:
$$\begin{cases}x^2-4y^2=36\\x-2y-20=0\end{cases}\Rightarrow\begin{cases}x^2-4y^2=36\\x=2y+20\end{cases}\Rightarrow$$
$$(2y+20)^2-4y^2=36\Rightarrow 80y+400=36\Rightarrow$$
$$y=-\frac{91}{20};\ x=2\left(-\frac{91}{20}\right)+20=\frac{109}{10}$$

The only point of intersection is $\left(\dfrac{109}{10},-\dfrac{91}{20}\right)$.

53. Solve using elimination:

$$\begin{cases} 3x^2 - 7y^2 = 5 \\ 9y^2 - 2x^2 = 1 \end{cases} \Rightarrow \begin{cases} 6x^2 - 14y^2 = 10 \\ -6x^2 + 27y^2 = 3 \end{cases} \Rightarrow$$

$$13y^2 = 13 \Rightarrow y = \pm 1$$

$$3x^2 - 7 = 5 \Rightarrow x^2 = 4 \Rightarrow x = \pm 2$$

The points of intersection are $(-2, -1)$,
$(-2, 1)$, $(2, -1)$, and $(2, 1)$.

55. Let the end of the pipe, and therefore the vertex of the parabola, be at $(0, 0)$. Then the ground is at $y = -20$. The water goes through the point $(8, -6)$.

The equation is of the form $y = -4ax^2$.
Substitute $(8, -6)$ into the equation and solve for

a: $-6 = -4a(64) \Rightarrow a = \dfrac{3}{128}$ and the equation of

the parabola is $y = -\dfrac{3}{32}x^2$.

Let $y = -20$ and solve for x:

$$-20 = -\frac{3}{32}x^2 \Rightarrow x^2 = \frac{640}{3} \Rightarrow x = \pm 14.61.$$

We are looking for the positive solution. The water hits the ground 14.6 feet from the end of the pipe.

57. The equation of the ellipse is $\dfrac{x^2}{36} + \dfrac{y^2}{49/4} = 1$.

At $x = 4$, $y = \dfrac{7\sqrt{5}}{6} \approx 2.61$. (Reject the negative

solution.) This is the radius of the cross section.
The circumference $= 2\pi(2.61) \approx 16.4$ inches.

Chapter 7 Practice Test A

1. The focus is at $(0, 12)$ and the directrix is $x = -12$, so the vertex is at $(-6, 12)$ and the parabola opens to the right. The general form is $(y - k)^2 = 4a(x - h)$. $a = 0 - (-6) = 6 \Rightarrow$

$$(y - 12)^2 = 4(6)(x + 6) = 24(x + 6).$$

2. $y^2 - 2y + 8x + 25 = 0 \Rightarrow$
$y^2 - 2y + 1 = -8x - 25 + 1 \Rightarrow$
$(y - 1)^2 = -8(x + 3)$

3. $x^2 = -9y = 4ay \Rightarrow a = -\dfrac{9}{4}$

Focus: $\left(0, -\dfrac{9}{4}\right)$, directrix: $y = \dfrac{9}{4}$.

4.

5. $(x + 2)^2 = -8(y - 1) = 4a(y - 1) \Rightarrow a = -2$.
Vertex: $(-2, 1)$, focus: $(-2, -1)$,
directrix: $y = 3$.

6. Let the vertex be at $(0, 0)$. Then the parabola goes through the point $(3, 2)$. The general

form of the equation is $(x - h)^2 = 4a(y - k)$.
Substitute the values for x, y, h, and k, then

solve for a: $3^2 = 4a(2) \Rightarrow a = \dfrac{9}{8}$. The

equation of the parabola is $x^2 = \dfrac{9}{2}y$. Now

find the value of y for $x = 1$:

$$1 = \frac{9}{2}y \Rightarrow y = \frac{2}{9} \approx 0.22 \text{ feet.}$$

7. Vertex $(3, -1)$ and directrix $x = -3 \Rightarrow a = 6$. The parabola opens to the right, so the general form of the equation is $(y-k)^2 = 4a(x-h)$. The equation is $(y+1)^2 = 24(x-3)$.

8. Foci $(0, -2)$ and $(0, 2) \Rightarrow c = 2$.
    Vertices $(0, -4)$ and $(0, 4) \Rightarrow a = 4$.
    $b^2 = a^2 - c^2 \Rightarrow b^2 = 4^2 - 2^2 = 12$.
    The equation is $\dfrac{x^2}{12} + \dfrac{y^2}{16} = 1$.

9. Foci $(-2, 0)$ and $(2, 0) \Rightarrow c = 2$.
    y-intercepts -5 and $5 \Rightarrow b = 5$.
    $b^2 = a^2 - c^2 \Rightarrow 25 = a^2 - 2^2 \Rightarrow a^2 = 29$.
    The equation is $\dfrac{x^2}{29} + \dfrac{y^2}{25} = 1$.

10. $a = 9$, $b = 2$. The equation is $\dfrac{x^2}{81} + \dfrac{y^2}{4} = 1$.

11.

12. $49y^2 - x^2 = 49 \Rightarrow y^2 - \dfrac{x^2}{49} = 1 \Rightarrow$
    $a^2 = 1, b^2 = 49$. The vertices are at $(0, 1)$ and $(0, -1)$.
    $c^2 = a^2 + b^2 \Rightarrow c^2 = 1 + 49 \Rightarrow c = \pm 5\sqrt{2}$. The foci are at $\left(0, 5\sqrt{2}\right)$ and $\left(0, -5\sqrt{2}\right)$.

13.

14. The center is $(0, 0)$ and the transverse axis is the y-axis, so the equation is of the form
    $\dfrac{y^2}{a^2} - \dfrac{x^2}{b^2} = 1$.
    $a = 6, c = \sqrt{45} \Rightarrow 45 = 36 + b^2 \Rightarrow b^2 = 9$.
    The equation is $\dfrac{y^2}{36} - \dfrac{x^2}{9} = 1$.

15. $y^2 - x^2 + 2x = 2 \Rightarrow$
    $y^2 - (x^2 - 2x + 1) = 2 - 1 \Rightarrow y^2 - (x-1)^2 = 1$

16. $x^2 - 2x - 4y^2 - 16y = 19 \Rightarrow$
    $(x^2 - 2x + 1) - 4(y^2 + 4y + 4) = 19 + 1 - 16 \Rightarrow$
    $(x-1)^2 - 4(y+2)^2 = 4 \Rightarrow$
    $\dfrac{(x-1)^2}{4} - (y+2)^2 = 1$

17. Hyperbola

18. Parabola

19. Circle

20. Ellipse

Chapter 7 Practice Test B

1. The focus is at $(-10, 0)$ and the directrix is $x = 10$, so the vertex is at $(0, 0)$ and the parabola opens to the left. The general form is

   $(y - k)^2 = -4a(x - h)$.

   $a = 0 - (-10) = 10 \Rightarrow y^2 = -4(10)x = -40x$.
   The answer is B.

2. $y^2 - 4y - 5x + 24 = 0 \Rightarrow$

   $y^2 - 4y + 4 = 5x - 24 + 4 \Rightarrow (y - 2)^2 = 5(x - 4)$
   The answer is B.

3. $x = 7y^2 \Rightarrow \dfrac{x}{7} = y^2 \Rightarrow \dfrac{1}{7} = 4a \Rightarrow \dfrac{1}{28} = a \Rightarrow$ the

   focus at $\left(\dfrac{1}{28}, 0\right)$ and the directrix is $x = -\dfrac{1}{28}$.
   The answer is A.

4. The answer is A.

5. $(x - 3)^2 = 12(y - 1) \Rightarrow$ the vertex is $(3, 1)$.

   $12 = 4a \Rightarrow a = 3 \Rightarrow$ the focus is at $(3, 4)$, and the directrix is $y = -2$. The answer is A.

6. Let the vertex be located at $(0, 25)$ and one end of the base is at $(90, 0)$. The general form of the equation is $(x - h)^2 = -4a(y - k)$. Substitute the values for x, y, h, and k, then solve for a:

   $(90 - 0)^2 = -4a(0 - 25) \Rightarrow a = 81$
   The equation of the parabola is

   $x^2 = -4(81)(y - 25) \Rightarrow x^2 = -324(y - 25)$.
   Find the value of y for

   $x = 45$: $45^2 = -324(y - 25) \Rightarrow y = 18.75$.
   The answer is D.

7. Vertex $(-2, 1)$ and directrix $x = 2 \Rightarrow a = 4$.
   The parabola opens to the left, so the general form of the equation is

   $(y - k)^2 = -4a(x - h)$.

   The equation is $(y - 1)^2 = -16(x + 2)$.
   The answer is A.

8. Foci $(-3, 0)$ and $(3, 0) \Rightarrow c = 3$.
   Vertices $(-5, 0)$ and $(5, 0) \Rightarrow a = 5$.

   $b^2 = a^2 - c^2 \Rightarrow b^2 = 5^2 - 3^2 = 16$

   The equation is $\dfrac{x^2}{25} + \dfrac{y^2}{16} = 1$.
   The answer is B.

9. Foci $(0, -3)$ and $(0, 3) \Rightarrow c = 3$. y-intercepts

   -7 and $7 \Rightarrow a = 7$. $b^2 = a^2 - c^2 \Rightarrow$

   $49 = 3^2 - b^2 \Rightarrow b^2 = 40$. The equation is

   $\dfrac{x^2}{40} + \dfrac{y^2}{49} = 1$. The answer is D.

10. $a = 8$, $b = 4$. The equation is $\dfrac{x^2}{16} + \dfrac{y^2}{64} = 1$.
    The answer is A.

11. $9(x - 1)^2 + 4(y - 2)^2 = 36 \Rightarrow$

    $\dfrac{(x - 1)^2}{4} + \dfrac{(y - 2)^2}{9} = 1$. The answer is A.

12. $a = 11$, $b = 2$. The vertices are at $(11, 0)$ and

    $(-11, 0)$. $c^2 = a^2 + b^2 \Rightarrow c^2 = 121 + 4 \Rightarrow$

    $c = 5\sqrt{5}$. The foci are at $\left(-5\sqrt{5}, 0\right)$ and

    $\left(5\sqrt{5}, 0\right)$. The answer is C.

13. The answer is B.

14. The center is $(0, 0)$ and the transverse axis is the y-axis, so the equation is of the form

    $\dfrac{y^2}{a^2} - \dfrac{x^2}{b^2} = 1$.

    $a = 5, c = 10 \Rightarrow 100 = 25 + b^2 \Rightarrow b^2 = 75$.

    The equation is $\dfrac{y^2}{25} - \dfrac{x^2}{75} = 1$.
    The answer is B.

15. $y^2 - 4x^2 - 2y - 16x - 19 = 0 \Rightarrow$
$y^2 - 2y + 1 - 4(x^2 + 4x + 4) = 19 + 1 - 16 \Rightarrow$
$(y-1)^2 - 4(x+2)^2 = 4 \Rightarrow$
$\dfrac{(y-1)^2}{4} - (x+2)^2 = 1$. The answer is C.

16. $4y^2 - 9x^2 - 16y - 36x - 56 = 0 \Rightarrow$
$4(y^2 - 4y + 4) - 9(x^2 + 4x + 4)$
$\qquad = 56 + 16 - 36 \Rightarrow$
$4(y-2)^2 - 9(x+2)^2 = 36 \Rightarrow$
$\dfrac{(y-2)^2}{9} - \dfrac{(x+2)^2}{4} = 1$. The answer is C.

17. C   18. A   19. D   20. D

Cumulative Review Exercises (Chapters P–7)

1. $\dfrac{\left((x+h)^2 - 3(x+h) + 2\right) - \left(x^2 - 3x + 2\right)}{h}$
$= \dfrac{x^2 + 2xh + h^2 - 3x - 3h + 2 - x^2 + 3x - 2}{h}$
$= \dfrac{h^2 + 2xh - 3h}{h} = h + 2x - 3$

2.

3. Switch the variables, then solve for y:
$y = 2x - 3 \Rightarrow x = 2y - 3 \Rightarrow \dfrac{x+3}{2} = y = f^{-1}(x)$
$f(f^{-1}(x)) = 2\left(\dfrac{x+3}{2}\right) - 3 = x$

4.

5. $\log_5(x-1) + \log_5(x-2) = 3\log_5 \sqrt[3]{6} \Rightarrow$
$\log_5\left((x-1)(x-2)\right) = \log_5\left(\sqrt[3]{6}\right)^3 \Rightarrow$
$x^2 - 3x + 2 = 6 \Rightarrow x^2 - 3x - 4 = 0 \Rightarrow$
$(x-4)(x+1) = 0 \Rightarrow x = 4$ or $x = -1$. (Reject the negative solution.) The solution is $\{4\}$.

6. $\log_a \sqrt[3]{x\sqrt{yz}} = \log_a \left(x\sqrt{yz}\right)^{1/3}$
$= \dfrac{1}{3}\log_a x + \dfrac{1}{3}\log_a \sqrt{yz}$
$= \dfrac{1}{3}\log_a x + \dfrac{1}{3}\log_a (yz)^{1/2}$
$= \dfrac{1}{3}\log_a x + \dfrac{1}{3}\cdot\dfrac{1}{2}\log_a (yz)$
$= \dfrac{1}{3}\log_a x + \dfrac{1}{6}\log_a y + \dfrac{1}{6}\log_a z$

7. $\dfrac{x}{x-2} \geq 1 \Rightarrow \dfrac{x}{x-2} - 1 \geq 0 \Rightarrow \dfrac{2}{x-2} \geq 0$.
$x - 2 > 0 \Rightarrow x > 2$. The solution is $(2, \infty)$.

8. $I = \dfrac{V}{R}(1 - e^{-0.3t}) \Rightarrow 1 - \dfrac{IR}{V} = e^{-0.3t} \Rightarrow$
$\ln\left(1 - \dfrac{IR}{V}\right) = -0.3t \Rightarrow t = -\dfrac{\ln\left(1 - \dfrac{IR}{V}\right)}{0.3}$

9. Solve using elimination:
$\begin{cases} 1.4x - 0.5y = 1.3 \\ 0.4x + 1.1y = 4.1 \end{cases} \Rightarrow \begin{cases} 0.56x - 0.2y = 0.52 \\ -0.56x - 1.54y = -5.74 \end{cases} \Rightarrow$
$-1.74y = -5.22 \Rightarrow y = 3$
$0.4x + 1.1(3) = 4.1 \Rightarrow x = 2$
The solution is $\{(2, 3)\}$.

10. Switch the first and second equations:
$\begin{cases} 2x + y - 4z = 3 \\ x - 2y + 3z = 4 \\ -3x + 4y - z = -2 \end{cases} \Rightarrow \begin{cases} x - 2y + 3z = 4 \\ 2x + y - 4z = 3 \\ -3x + 4y - z = -2 \end{cases}$.

Multiply the first equation by -2, add the result to the second equation, and replace the second equation with the new equation:
$\begin{cases} x - 2y + 3z = 4 \\ 2x + y - 4z = 3 \\ -3x + 4y - z = -2 \end{cases} \Rightarrow \begin{cases} x - 2y + 3z = 4 \\ 5y - 10z = -5 \\ -3x + 4y - z = -2 \end{cases}$

Divide the second equation by 5 and replace the equation with the result. Then multiply the first equation by 3, add the result to the third equation, and replace the third equation with the new equation.
$\begin{cases} x - 2y + 3z = 4 \\ 5y - 10z = -5 \\ -3x + 4y - z = -2 \end{cases} \Rightarrow \begin{cases} x - 2y + 3z = 4 \\ y - 2z = -1 \\ -2y + 8z = 10 \end{cases}$.

(*continued on next page*)

(*continued*)

Divide the third equation by -2 and replace the equation with the result:

$$\begin{cases} x - 2y + 3z = 4 \\ \quad\; y - 2z = -1 \\ -2y + 8z = 10 \end{cases} \Rightarrow \begin{cases} x - 2y + 3z = 4 \\ \quad\; y - 2z = -1 \\ \quad\; y - 4z = -5 \end{cases}.$$

Subtract the third equation from the second equation and solve for z:

$$\begin{cases} x - 2y + 3z = 4 \\ \quad\; y - 2z = -1 \\ \quad\; y - 4z = -5 \end{cases} \Rightarrow \begin{cases} x - 2y + 3z = 4 \\ \quad\; y - 2z = -1 \\ \qquad\quad z = 2 \end{cases}$$

Substitute $z = 2$ into the second equation and solve for y: $y - 2(2) = -1 \Rightarrow y = 3$. Substitute the values for y and z into the first equation and solve for x: $x - 2(3) + 3(2) = 4 \Rightarrow x = 4$. The solution is $\{(4, 3, 2)\}$.

11. Solve using substitution:

$$\begin{cases} y = 2 - \log x \\ y - \log(x + 3) = 1 \end{cases} \Rightarrow$$

$2 - \log x - \log(x + 3) = 1 \Rightarrow$

$\log x + \log(x + 3) = 1 \Rightarrow \log\left(x(x+3)\right) = 1 \Rightarrow$

$x^2 + 3x = 10 \Rightarrow x^2 + 3x - 10 = 0 \Rightarrow$

$(x + 5)(x - 2) = 0 \Rightarrow x = -5$ or $x = 2$

(Reject the negative solution.)

The solution is $\{(2, 2 - \log 2)\}$.

12. Solve using substitution:

$$\begin{cases} y = x^2 - 1 \\ 3x^2 + 8y^2 = 8 \end{cases} \Rightarrow \begin{cases} y + 1 = x^2 \\ 3x^2 + 8y^2 = 8 \end{cases} \Rightarrow$$

$3(y + 1) + 8y^2 = 8 \Rightarrow 8y^2 + 3y - 5 = 0 \Rightarrow$

$(y + 1)(8y - 5) = 0 \Rightarrow y = -1$ or $y = \dfrac{5}{8}$

$-1 = x^2 - 1 \Rightarrow x^2 = 0 \Rightarrow x = 0$

$\dfrac{5}{8} = x^2 - 1 \Rightarrow \dfrac{13}{8} = x^2 \Rightarrow x = \pm\dfrac{\sqrt{26}}{4}$

The solutions are $\left\{ (0, -1), \left(\dfrac{\sqrt{26}}{4}, \dfrac{5}{8} \right), \right.$

$\left. \left(-\dfrac{\sqrt{26}}{4}, \dfrac{5}{8} \right) \right\}$.

13. Expand by the first column:

$$\begin{vmatrix} 1 & 4 & 7 \\ 2 & 5 & 8 \\ 3 & 6 & 9 \end{vmatrix} = a_{11}A_{11} + a_{21}A_{21} + a_{31}A_{31}$$

$$= 1(-1)^{1+1}\begin{vmatrix} 5 & 8 \\ 6 & 9 \end{vmatrix} + 2(-1)^{2+1}\begin{vmatrix} 4 & 7 \\ 6 & 9 \end{vmatrix}$$

$$+ 3(-1)^{3+1}\begin{vmatrix} 4 & 7 \\ 5 & 8 \end{vmatrix}$$

$$= -3 - 2(-6) + 3(-3) = 0$$

14. $\begin{cases} 2x - 3y = -4 \\ 5x + 7y = 1 \end{cases} \Rightarrow D = \begin{vmatrix} 2 & -3 \\ 5 & 7 \end{vmatrix} = 29$

$$D_x = \begin{vmatrix} -4 & -3 \\ 1 & 7 \end{vmatrix} = -25, D_y = \begin{vmatrix} 2 & -4 \\ 5 & 1 \end{vmatrix} = 22$$

$$x = -\frac{25}{29}, y = \frac{22}{29}$$

15. $A = \begin{vmatrix} 3 & -2 \\ -5 & 4 \end{vmatrix} \Rightarrow A^{-1} = \dfrac{1}{12-10}\begin{vmatrix} 4 & 2 \\ 5 & 3 \end{vmatrix} = \begin{vmatrix} 2 & 1 \\ \dfrac{5}{2} & \dfrac{3}{2} \end{vmatrix}$

16. To find the point of intersection, solve the system

$$\begin{cases} x + 2y - 3 = 0 \\ 3x + 4y - 5 = 0 \end{cases} \Rightarrow \begin{cases} -2x - 4y + 6 = 0 \\ 3x + 4y - 5 = 0 \end{cases} \Rightarrow$$

$x = -1; -1 + 2y - 3 = 0 \Rightarrow y = 2$

The point of intersection is $(-1, 2)$. The line

$x - 3y + 5 = 0 \Rightarrow y = \dfrac{1}{3}x + \dfrac{5}{3} \Rightarrow$ its slope is

$1/3$. The slope of the perpendicular is -3. The equation of the line through $(-1, 2)$ with slope -3 is $y - 2 = -3(x + 1) \Rightarrow y = -3x - 1$.

17. Let $u = x^2$. Then the equation becomes

$2u^2 - 5u + 3 = 0 \Rightarrow (2u - 3)(u - 1) = 0 \Rightarrow$

$u = \dfrac{3}{2}$ or $u = 1$. Now solve for x:

$x^2 = \dfrac{3}{2} \Rightarrow x = \pm\dfrac{\sqrt{6}}{2}; x^2 = 1 \Rightarrow x = \pm 1$.

The solutions are $\left\{ \pm\dfrac{\sqrt{6}}{2}, \pm 1 \right\}$.

18. Hyperbola

19. Circle

20.

Chapter 8 Further Topics in Algebra

8.1 Sequences and Series

8.1 Practice Problems

1. $a_n = -2^n \Rightarrow a_1 = -2^1 = -2;\ a_2 = -2^2 = -4;$
$a_3 = -2^3 = -8;\ a_4 = -2^4 = -16$

2. $a_n = 1 + \left(\dfrac{1}{2}\right)^n$

$a_1 = 1 + \left(\dfrac{1}{2}\right)^1 = \dfrac{3}{2}$

$a_2 = 1 + \left(\dfrac{1}{2}\right)^2 = \dfrac{5}{4}$

$a_3 = 1 + \left(\dfrac{1}{2}\right)^3 = \dfrac{9}{8}$

$a_4 = 1 + \left(\dfrac{1}{2}\right)^4 = \dfrac{17}{16}$

$a_5 = 1 + \left(\dfrac{1}{2}\right)^5 = \dfrac{33}{32}$

3.

n:	1	2	3	4	5	...	n
term:	0	$-\dfrac{1}{2}$	$\dfrac{2}{3}$	$-\dfrac{3}{4}$	$\dfrac{4}{5}$	...	a_n

The terms alternate, and the negative terms are the even-numbered terms, so use the factor $(-1)^{n+1}$ to alternate the signs. The absolute value of each term is $\dfrac{n-1}{n}$, so the general term is $(-1)^{n+1}\left(\dfrac{n-1}{n}\right)$ or $(-1)^{n+1}\left(1 - \dfrac{1}{n}\right)$.

4. $a_1 = -3,\ a_{n+1} = 2a_n + 5$
$a_2 = 2a_1 + 5 = 2(-3) + 5 = -1$
$a_3 = 2a_2 + 5 = 2(-1) + 5 = 3$
$a_4 = 2a_3 + 5 = 2(3) + 5 = 11$
$a_5 = 2a_4 + 5 = 2(11) + 5 = 27$

5. The female bee is a_0. Since female bees have two parents, $a_1 = 2$. The female parent has two parents and the male parent has one, so $a_2 = 3$. There are two females and one male in this generation, so the next generation has three females and two males. We can use a tree to represent this:

$a_k = a_{k-2} + a_{k-1}$ for $k \geq 3$.

6. a. $\dfrac{13!}{12!} = \dfrac{13 \cdot 12!}{12!} = 13$

b. $\dfrac{n!}{(n-3)!} = \dfrac{n(n-1)(n-2)(n-3)!}{(n-3)!}$
$= n(n-1)(n-2)$

7. $a_n = \dfrac{(-1)^n 2^n}{n!} \Rightarrow a_1 = \dfrac{(-1)^1 2^1}{1!} = -2$

$a_2 = \dfrac{(-1)^2 2^2}{2!} = \dfrac{4}{2 \cdot 1} = 2$

$a_3 = \dfrac{(-1)^3 2^3}{3!} = \dfrac{-8}{3 \cdot 2 \cdot 1} = -\dfrac{8}{6} = -\dfrac{4}{3}$

$a_4 = \dfrac{(-1)^4 2^4}{4!} = \dfrac{16}{4 \cdot 3 \cdot 2 \cdot 1} = \dfrac{16}{24} = \dfrac{2}{3}$

$a_5 = \dfrac{(-1)^5 2^5}{5!} = \dfrac{-32}{5 \cdot 4 \cdot 3 \cdot 2 \cdot 1} = -\dfrac{32}{120} = -\dfrac{4}{15}$

8. $\displaystyle\sum_{k=0}^{3} (-1)^k k!$
$= (-1)^0(0!) + (-1)^1(1!) + (-1)^2(2!) + (-1)^3(3!)$
$= 1 + (-1) + (2 \cdot 1) + (-1)(3 \cdot 2 \cdot 1) = -4$

9. $2 - 4 + 6 - 8 + 10 - 12 + 14$ alternates addition and subtraction of consecutive even integers from 2 to 14. The even integers can be represented as $2k$ with k ranging from 1 to 7. We alternate the signs using the factor $(-1)^{k+1}$. Thus, the expression is

$\displaystyle\sum_{k=1}^{7} (-1)^{k+1}(2k).$

8.1 Concepts and Vocabulary

1. An infinite sequence is a function whose domain is the set of <u>positive integers</u>.

3. By definition, $0! = \underline{1}$.

5. False. $a_1 = 3$, $a_2 = a_1^2 = 3^2 = 9$,
$$a_3 = a_2^2 = 9^2 = 81$$

7. True

8.1 Building Skills

9. $a_1 = 3(1) - 2 = 1, a_2 = 3(2) - 2 = 4,$
$a_3 = 3(3) - 2 = 7, a_4 = 3(4) - 2 = 10$

11. $a_1 = 1 - \dfrac{1}{1} = 0, a_2 = 1 - \dfrac{1}{2} = \dfrac{1}{2},$
$a_3 = 1 - \dfrac{1}{3} = \dfrac{2}{3}, a_4 = 1 - \dfrac{1}{4} = \dfrac{3}{4}$

13. $a_1 = -1^2 = -1, a_2 = -2^2 = -4,$
$a_3 = -3^2 = -9, a_4 = -4^2 = -16$

15. $a_1 = \dfrac{2(1)}{1+1} = 1, a_2 = \dfrac{2(2)}{2+1} = \dfrac{4}{3}, a_3 = \dfrac{2(3)}{3+1} = \dfrac{3}{2},$
$a_4 = \dfrac{2(4)}{4+1} = \dfrac{8}{5}$

17. $a_1 = (-1)^{1+1} = 1, a_2 = (-1)^{2+1} = -1,$
$a_3 = (-1)^{3+1} = 1, a_4 = (-1)^{4+1} = -1$

19. $a_1 = 3 - \dfrac{1}{2^1} = \dfrac{5}{2}, a_2 = 3 - \dfrac{1}{2^2} = \dfrac{11}{4},$
$a_3 = 3 - \dfrac{1}{2^3} = \dfrac{23}{8}, a_4 = 3 - \dfrac{1}{2^4} = \dfrac{47}{16}$

21. $a_1 = a_2 = a_3 = a_4 = 0.6$

23. $a_1 = \dfrac{(-1)^1}{1!} = -1, a_2 = \dfrac{(-1)^2}{2!} = \dfrac{1}{2},$
$a_3 = \dfrac{(-1)^3}{3!} = -\dfrac{1}{6}, a_4 = \dfrac{(-1)^4}{4!} = \dfrac{1}{24}$

25. $a_1 = (-1)^1 3^{-1} = -\dfrac{1}{3}, a_2 = (-1)^2 3^{-2} = \dfrac{1}{9},$
$a_3 = (-1)^3 3^{-3} = -\dfrac{1}{27}, a_4 = (-1)^4 3^{-4} = \dfrac{1}{81}$

27. $a_1 = \dfrac{e^1}{2(1)} = \dfrac{e}{2}, a_2 = \dfrac{e^2}{2(2)} = \dfrac{e^2}{4}, a_3 = \dfrac{e^3}{2(3)} = \dfrac{e^3}{6},$
$a_4 = \dfrac{e^4}{2(4)} = \dfrac{e^4}{8}$

29. $a_n = 2n - 3$
$a_1 = 2(1) - 3 = -1 \quad a_2 = 2(2) - 3 = 1$
$a_3 = 2(3) - 3 = 3 \quad a_4 = 2(4) - 3 = 5$
The points to be plotted are $(1, -1)$, $(2, 1)$, $(3, 3)$, and $(4, 5)$.

31. $a_n = \dfrac{1}{2}n^2 - n$
$a_1 = \dfrac{1}{2}(1)^2 - 1 = -\dfrac{1}{2} \quad a_2 = \dfrac{1}{2}(2)^2 - 2 = 0$
$a_3 = \dfrac{1}{2}(3)^2 - 3 = \dfrac{3}{2} \quad a_4 = \dfrac{1}{2}(4)^2 - 4 = 4$
The points to be plotted are $\left(1, -\dfrac{1}{2}\right)$, $(2, 0)$, $\left(3, \dfrac{3}{2}\right)$, $(4, 4)$.

33. $a_n = (n-2)^2 - 1$
$a_1 = (1-2)^2 - 1 = 0 \quad a_2 = (2-2)^2 - 1 = -1$
$a_3 = (3-2)^2 - 1 = 0 \quad a_4 = (4-2)^2 - 1 = 3$
The points to be plotted are $(1, 0)$, $(2, -1)$, $(3, 0)$, $(4, 3)$.

35. $a_n = 3 - \dfrac{1}{n}$

$a_1 = 3 - \dfrac{1}{1} = 2 \qquad a_2 = 3 - \dfrac{1}{2} = \dfrac{5}{2}$

$a_3 = 3 - \dfrac{1}{3} = \dfrac{8}{3} \qquad a_4 = 3 - \dfrac{1}{4} = \dfrac{11}{4}$

The points to be plotted are $\left(1,\,2\right)$, $\left(2,\,\dfrac{5}{2}\right)$,

$\left(3,\,\dfrac{8}{3}\right)$, $\left(4,\,\dfrac{11}{4}\right)$.

37. $a_n = 2^{n-1} - 3$

$a_1 = 2^{1-1} - 3 = -2 \qquad a_2 = 2^{2-1} - 3 = -1$

$a_3 = 2^{3-1} - 3 = 1 \qquad a_4 = 2^{4-1} - 3 = 5$

The points to be plotted are $(1, -2)$, $(2, -1)$, $(3, 1)$, and $(4, 5)$.

39. $a_n = 3n - 2$ **41.** $a_n = \dfrac{1}{n+1}$

43. $a_n = (-1)^{n-1}(2)$ **45.** $a_n = \dfrac{3^{n-1}}{2^n}$

47. $a_n = n(n+1)$ **49.** $a_n = 2 - \dfrac{(-1)^n}{n+1}$

51. $a_n = \dfrac{3^{n+1}}{n+1}$

53. $a_1 = 2, a_2 = 2 + 3 = 5, a_3 = 5 + 3 = 8,$
$a_4 = 8 + 3 = 11, a_5 = 11 + 3 = 14$

55. $a_1 = 3, a_2 = 2(3) = 6, a_3 = 2(6) = 12,$
$a_4 = 2(12) = 24, a_5 = 2(24) = 48$

57. $a_1 = 7, a_2 = -2(7) + 3 = -11,$
$a_3 = -2(-11) + 3 = 25, a_4 = -2(25) + 3 = -47,$
$a_5 = -2(-47) + 3 = 97$

59. $a_1 = 2, a_2 = \dfrac{1}{2}, a_3 = \dfrac{1}{1/2} = 2, a_4 = \dfrac{1}{2},$

$a_5 = \dfrac{1}{1/2} = 2$

61. $a_1 = 25, a_2 = \dfrac{(-1)^1}{5(25)} = -\dfrac{1}{125},$

$a_3 = \dfrac{(-1)^2}{5(-1/125)} = -25, a_4 = \dfrac{(-1)^3}{5(-25)} = \dfrac{1}{125},$

$a_5 = \dfrac{(-1)^4}{5(1/125)} = 25$

In exercises 63–68, we show only the first screen to determine the terms of the sequence before we list the ten terms.

63. a.

2, 11, 26, 47, 74, 107, 146, 191, 242, 299

b.

65. a.

0, 1, 2, 3, 4, 5, 6, 7, 8, 9

b.

67. a.

0.5, −1, 2, 0.5, −1, 2, 0.5, −1, 2, 0.5

b.

69. $\dfrac{3!}{5!} = \dfrac{3!}{5 \cdot 4 \cdot 3!} = \dfrac{1}{20}$

71. $\dfrac{12!}{11!} = \dfrac{12 \cdot 11!}{11!} = 12$

73. $\dfrac{n!}{(n+1)!} = \dfrac{n!}{(n+1) \cdot n!} = \dfrac{1}{n+1}$

75. $\dfrac{(2n+1)!}{(2n)!} = \dfrac{(2n+1) \cdot (2n)!}{(2n)!} = 2n+1$

77. $\displaystyle\sum_{k=1}^{7} 5 = 5+5+5+5+5+5+5 = 35$

79. $\displaystyle\sum_{j=0}^{5} j^2 = 0^2 + 1^2 + 2^2 + 3^2 + 4^2 + 5^2 = 55$

81. $\displaystyle\sum_{i=1}^{5} (2i-1) = (2(1)-1) + (2(2)-1) + (2(3)-1)$
$$+ (2(4)-1) + (2(5)-1) = 25$$

83. $\displaystyle\sum_{j=3}^{7} \dfrac{j+1}{j} = \dfrac{4}{3} + \dfrac{5}{4} + \dfrac{6}{5} + \dfrac{7}{6} + \dfrac{8}{7} = \dfrac{853}{140}$

85. $\displaystyle\sum_{i=2}^{6} (-1)^i 3^{i-1}$
$$= (-1)^2(3^1) + (-1)^3(3^2) + (-1)^4(3^3)$$
$$+ (-1)^5(3^4) + (-1)^6(3^5)$$
$$= 183$$

87. $\displaystyle\sum_{k=4}^{7} (2-k^2)$
$$= \left(2-4^2\right) + \left(2-5^2\right) + \left(2-6^2\right) + \left(2-7^2\right)$$
$$= -118$$

89. $\displaystyle\sum_{k=1}^{51} (2k-1)$ **91.** $\displaystyle\sum_{k=1}^{11} \dfrac{1}{5k}$

93. $\displaystyle\sum_{k=1}^{50} \dfrac{(-1)^{k+1}}{k}$ **95.** $\displaystyle\sum_{k=1}^{10} \dfrac{k}{k+1}$

97.
```
sum(seq(12n²,n,1
,10))
            4620
```

99.
```
sum(seq(7/(1-n²)
,n,5,30))
     -1.345430108
```

101.
```
sum(seq((-1)^n(n
),n,1,100))
              50
```

8.1 Applying the Concepts

103. a. Use a table to determine the pattern. It appears that each second, the body falls 32 feet per second faster than it did in the second before.

Time	Distance each second
1	16
2	48
3	80
4	112
5	144
6	176
7	208

b. $16 + 32(n-1) = 32n - 16$ feet

105. The fine is $50 plus $25 for each additional day the work is not done. So on the ninth day, she will be fined $50 + 8($25) = $250.

107. In three years, there are 6 six-month periods. For the first six-month period, cell phone usage was 600 minutes per month. For the next 5 six-month periods, the usage doubled during each period. At the end of three years, cell phone use was $600 \cdot 2^5 = 19,200$ minutes per month.

109. a. $A_1 = 10,000\left(1 + \dfrac{0.06}{2}\right)^1 = \$10,300$

$A_2 = 10,000\left(1 + \dfrac{0.06}{2}\right)^2 = \$10,609$

$A_3 = 10,000\left(1 + \dfrac{0.06}{2}\right)^3 = \$10,927.27$

$A_4 = 10,000\left(1 + \dfrac{0.06}{2}\right)^4 = \$11,255.09$

$A_5 = 10,000\left(1 + \dfrac{0.06}{2}\right)^5 = \$11,592.74$

$A_6 = 10,000\left(1 + \dfrac{0.06}{2}\right)^6 = \$11,940.52$

b. The interest is compounded semiannually, so in eight years, there are 16 compounding periods.

$A_{16} = 100\left(1 + \dfrac{0.08}{4}\right)^{16} = \$16,047.10$

111. $A_1 = (100,000)(1.05) = \$105,000$
$A_2 = (105,000)(1.05) = \$110,250$
$A_3 = (110,250)(1.05) = \$115,762.50$
$A_4 = (115,762.50)(1.05) = \$121,550.63$
$A_5 = (121,550.63)(1.05) = \$127,628.16$
$A_6 = (127,628.16)(1.05) = \$134,009.56$
$A_7 = (134,009.56)(1.05) = \$140,710.04$

The formula is $100,000\left(1.05^n\right)$.

8.1 Beyond the Basics

113. $a_1 = 2^{1/2}, a_2 = \left(2 \cdot 2^{1/2}\right)^{1/2} = 2^{3/4},$

$a_3 = \left(2 \cdot 2^{3/4}\right)^{1/2} = 2^{7/8},$

$a_4 = \left(2 \cdot 2^{7/8}\right)^{1/2} = 2^{15/16},$

$a_5 = \left(2 \cdot 2^{15/16}\right)^{1/2} = 2^{31/32},$

$a_n = 2^{(2^n - 1)/2^n} = 2^{1 - 1/2^n}$

115. a. $a_1 = 1, a_2 = \dfrac{1}{2}, a_3 = \dfrac{1}{4}, a_4 = \dfrac{1}{8}, a_5 = \dfrac{1}{16}$

b. $a_n = \dfrac{1}{2^{n-1}}$

117. $a_1 = a_2 = 1,$
$a_3 = a_{a_2} + a_{3 - a_2} = a_1 + a_2 = 1 + 1 = 2,$
$a_4 = a_{a_3} + a_{4 - a_3} = a_2 + a_2 = 1 + 1 = 2,$
$a_5 = a_{a_4} + a_{5 - a_4} = a_2 + a_3 = 1 + 2 = 3,$
$a_6 = a_{a_5} + a_{6 - a_5} = a_3 + a_3 = 2 + 2 = 4,$
$a_7 = a_{a_6} + a_{7 - a_6} = a_4 + a_3 = 2 + 2 = 4,$
$a_8 = a_{a_7} + a_{8 - a_7} = a_4 + a_4 = 2 + 2 = 4,$
$a_9 = a_{a_8} + a_{9 - a_8} = a_4 + a_5 = 2 + 3 = 5,$
$a_{10} = a_{a_9} + a_{10 - a_9} = a_5 + a_5 = 3 + 3 = 6$

119. $\displaystyle\sum_{n=1}^{20} n^2 = \sum_{m=0}^{19} a_m$

Write the first five terms to determine the pattern: $\displaystyle\sum_{n=1}^{1} n^2 = 1 = \sum_{m=0}^{0} a_m = a_0 \Rightarrow a_0 = 1,$

$\displaystyle\sum_{n=1}^{2} n^2 = 1 + 2^2 = 5 = \sum_{m=0}^{1} a_m = a_0 + a_1 \Rightarrow$
$5 = 1 + a_1 \Rightarrow a_1 = 4 = 2^2.$

$\displaystyle\sum_{n=1}^{3} n^2 = 1 + 2^2 + 3^2 = 14$

$\displaystyle = \sum_{m=0}^{2} a_m = a_0 + a_1 + a_2 \Rightarrow$
$14 = 1 + 4 + a_2 \Rightarrow a_2 = 9 = 3^2$

$\displaystyle\sum_{n=1}^{4} n^2 = 1 + 2^2 + 3^2 + 4^2 = 30$

$\displaystyle = \sum_{m=0}^{3} a_m = a_0 + a_1 + a_2 + a_3 \Rightarrow$
$30 = 1 + 4 + 9 + a_3 \Rightarrow a_3 = 16 = 4^2$

$\displaystyle\sum_{n=1}^{5} n^2 = 1 + 2^2 + 3^2 + 4^2 + 5^2 = 55$

$\displaystyle = \sum_{m=0}^{4} a_m = a_0 + a_1 + a_2 + a_3 + a_4 \Rightarrow$
$55 = 1 + 4 + 9 + 16 + a_4 \Rightarrow a_4 = 25 = 5^2$

Following the pattern, we see that
$a_m = (m + 1)^2.$

121. $\displaystyle\sum_{n=0}^{10} n^3 = \sum_{m=p}^{q} (m-2)^3 \Rightarrow n^3 = (m-2)^3 \Rightarrow$
$n = m - 2 \Rightarrow m = n + 2$. So, $p = 2$ and $q = 12$.

123. $7! - 2(5!) = (7 \cdot 6)(5!) - 2(5!)$
$= (42 - 2)(5!) = 40(5!)$

125. $\dfrac{1}{5!}+\dfrac{1}{6!}=\dfrac{x}{7!}\Rightarrow\dfrac{1}{5!}+\dfrac{1}{6\cdot5!}=\dfrac{x}{7\cdot6\cdot5!}$

Multiply both sides of the equation by $7!=7\cdot6\cdot5!$ and simplify. Then solve for *x*.

$\dfrac{7\cdot6\cdot5!}{5!}+\dfrac{7\cdot6\cdot5!}{6\cdot5!}=\dfrac{7\cdot6\cdot5!x}{7\cdot6\cdot5!}\Rightarrow 42+7=x\Rightarrow x=49$

8.1 Critical Thinking/Discussion/Writing

127.

The general term formula $a_n=1$ better represents the fact that the given sequence is constant.

128.

$a_n=1+\dfrac{(-1)^n}{n}$

$a_1=1+\dfrac{(-1)^1}{1}=0 \quad a_3=1+\dfrac{(-1)^3}{3}=\dfrac{2}{3} \quad a_5=1+\dfrac{(-1)^5}{5}=\dfrac{4}{5} \quad a_7=1+\dfrac{(-1)^7}{7}=\dfrac{6}{7} \quad a_9=1+\dfrac{(-1)^9}{9}=\dfrac{8}{9}$

$a_2=1+\dfrac{(-1)^2}{2}=\dfrac{3}{2} \quad a_4=1+\dfrac{(-1)^4}{4}=\dfrac{5}{4} \quad a_6=1+\dfrac{(-1)^6}{6}=\dfrac{7}{6} \quad a_8=1+\dfrac{(-1)^8}{8}=\dfrac{9}{8} \quad a_{10}=1+\dfrac{(-1)^{10}}{10}=\dfrac{11}{10}$

The points representing a_n alternate between $f(x)=1-\dfrac{1}{x}$ and $g(x)=1+\dfrac{1}{x}$. As $n\to\infty$, the values of a_n approach 1.

129. $n^2+n=(n-1)(n-2)(n-3)(n-4)(n-5)+n^2+n\Rightarrow 0=(n-1)(n-2)(n-3)(n-4)(n-5)\Rightarrow n=1,2,3,4,5$, so $k=5$ is the upper limit.

130. If $a_0=13$, then $a_1=3(13)+1=40$, $a_2=a_1/2=40/2=20$, $a_3=a_2/2=20/2=10$, $a_4=a_3/2=10/2=5$, $a_5=3a_4+1=3(5)+1=16$, $a_6=a_5/2=16/2=8$, $a_7=a_6/2=8/2=4$, $a_8=a_7/2=4/2=2$, . $a_9=a_8/2=2/2=1$

So the conjecture is true for $a_0=13$.

131. $\dfrac{(2n)!}{n!}=\dfrac{(2n)(2n-1)(2n-2)(2n-3)(2n-4)(2n-5)\cdots}{n(n-1)(n-2)\cdots}=\dfrac{2n(2n-1)2(n-1)(2n-3)2(n-2)(2n-5)\cdots}{n(n-1)(n-2)\cdots}$

$=2^n(2n-1)(2n-3)(2n-5)\cdots$

Note that the odd numbers are represented by $2n-1$, $2n-3$, $2n-5$, etc. Therefore,

$\dfrac{(2n)!}{n!}=2^n(2n-1)(2n-3)(2n-5)\cdots=2^n\left[1\cdot3\cdot5\cdot(2n-1)\right].$

132. *n*! is divisible by all integers between 2 and *n* (from the definition of *n*!). So, if we divide $(n!+1)$ by any integer between 2 and *n*, we will end up with remainder 1. Hence, $(n!+1)$ is not divisible by any integer between 2 and *n*.

8.1 Getting Ready for the Next Section

133.
$$a_2 - a_1 = 5 - 1 = 4$$
$$a_3 - a_2 = 9 - 5 = 4$$
$$a_4 - a_3 = 13 - 9 = 4$$
$$a_5 - a_4 = 17 - 13 = 4$$

135.
$$a_2 - a_1 = 1 - 6 = -5$$
$$a_3 - a_2 = -4 - 1 = -5$$
$$a_4 - a_3 = -9 - (-4) = -5$$
$$a_5 - a_4 = -14 - (-9) = -5$$

137.
$$a_2 = a_1 + d = 3 + 4 = 7$$
$$a_3 = a_2 + d = 7 + 4 = 11$$
$$a_4 = a_3 + d = 11 + 4 = 15$$
$$a_5 = a_4 + d = 15 + 4 = 19$$

139.
$$a_2 = a_1 + d = 7 + (-3) = 4$$
$$a_3 = a_2 + d = 4 + (-3) = 1$$
$$a_4 = a_3 + d = 1 + (-3) = -2$$
$$a_5 = a_4 + d = -2 + (-3) = -5$$

141.
$$a_n = 3n + 5$$
$$a_1 = 3(1) + 5 = 8$$
$$a_2 = 3(2) + 5 = 11$$
$$a_3 = 3(3) + 5 = 14$$
$$a_4 = 3(4) + 5 = 17$$
$$a_5 = 3(5) + 5 = 20$$
$$a_2 - a_1 = 11 - 8 = 3$$
$$a_3 - a_2 = 14 - 11 = 3$$
$$a_4 - a_3 = 17 - 14 = 3$$
$$a_5 - a_4 = 20 - 17 = 3$$
Note that
$$a_n - a_{n-1} = (3n + 5) - (3(n-1) + 5)$$
$$= (3n + 5) - (3n - 3 + 5)$$
$$= (3n + 5) - (3n + 2) = 3$$

143.
$$a_n = -2n - 6$$
$$a_1 = -2(1) - 6 = -8$$
$$a_2 = -2(2) - 6 = -10$$
$$a_3 = -2(3) - 6 = -12$$
$$a_4 = -2(4) - 6 = -14$$
$$a_5 = -2(5) - 6 = -16$$
$$a_2 - a_1 = -10 - (-8) = -2$$
$$a_3 - a_2 = -12 - (-10) = -2$$
$$a_4 - a_3 = -14 - (-12) = -2$$
$$a_5 - a_4 = -16 - (-14) = -2$$

Note that
$$a_n - a_{n-1} = (-2n - 6) - (-2(n-1) - 6)$$
$$= (-2n - 6) - (-2n + 2 - 6)$$
$$= (-2n - 6) - (-2n - 4) = -2$$

145.
$$a_n = 2n - 5$$
$$a_{n+1} = 2(n+1) - 5 = 2n + 2 - 5 = 2n - 3$$
$$a_{n-1} = 2(n-1) - 5 = 2n - 2 - 5 = 2n - 7$$

8.2 Arithmetic Sequences; Partial Sums

8.2 Practice Problems

1. $(-2) - 3 = -5$, $-7 - (-2) = -5$, etc. The common difference is -5.

2. $a_1 = -3$ and $d = 1 - (-3) = 4$, so the expression is
$$a_n = -3 + 4(n-1) = -3 + 4n - 4 = 4n - 7.$$

3. $a_n = 7, 4, 1, -2, -5, \ldots$

We will use the points $(1, 7)$ and $(2, 4)$ to find the linear function that produces the sequence.
$$m = \frac{7 - 4}{1 - 2} = -3$$
$$y - 7 = -3(x - 1) \Rightarrow y = -3x + 10$$
The general term for this sequence is
$$a_n = f(n) = -3n + 10.$$

4.
$$a_4 = a_1 + d(4-1) \Rightarrow 41 = a_1 + 3d$$
$$a_{15} = a_1 + d(15-1) \Rightarrow 8 = a_1 + 14d$$
$$\begin{cases} a_1 + 3d = 41 \\ a_1 + 14d = 8 \end{cases} \Rightarrow -11d = 33 \Rightarrow d = -3, a_1 = 50$$
$$a_n = 50 - 3(n-1) = -3n + 53$$

5. $d = \dfrac{5}{6} - \dfrac{2}{3} = \dfrac{1}{6}; \ n = 10; \ a_1 = \dfrac{2}{3}; \ a_{10} = \dfrac{13}{6}$

$$S_{10} = 10 \left(\dfrac{\frac{2}{3} + \frac{13}{6}}{2} \right) = \dfrac{85}{6}$$

6. $d = 32; \ n = 5; \ a_{11} = 32(11) - 16 = 336;$
$a_{15} = 32(15) - 16 = 464$

$$S = 5 \left(\dfrac{336 + 464}{2} \right) = 2000$$

The object fell 2000 feet during the 11th through 15th seconds.

8.2 Concepts and Vocabulary

1. If 5 is the common difference of an arithmetic sequence with general term a_n, then
$a_{17} - a_{16} = \underline{5}$.

3. If 14 is the term immediately following the sequence term 17 is an arithmetic sequence, then the common difference is $\underline{-3}$.

5. False. The common difference of an arithmetic sequence may be positive or negative.

7. False. $a_n = a_1 + (n - 1)d$

8.2 Building Skills

9. The sequence is arithmetic. $a_1 = 1, d = 1$

11. The sequence is arithmetic. $a_1 = 2, d = 3$

13. The sequence is not arithmetic.

15. The sequence is not arithmetic.

17. The sequence is arithmetic. $a_1 = 0.6, d = -0.4$

19. Write the first four terms of the sequence:
$a_1 = 2(1) + 6 = 8, a_2 = 2(2) + 6 = 10,$
$a_3 = 2(3) + 6 = 12, a_4 = 2(4) + 6 = 14$.

It appears that the sequence is arithmetic, with a common difference of 2. Verify as follows:
$d = a_{n+1} - a_n = (2(n+1) + 6) - (2n + 6) = 2$.

21. Write the first four terms of the sequence:
$a_1 = 1 - 1^2 = 0, a_2 = 1 - 2^2 = -3,$
$a_3 = 1 - 3^2 = -8, a_4 = 1 - 4^2 = -15$.

There is no common difference, so the sequence is not arithmetic.

23. $d = 3, a_1 = 5 \Rightarrow a_n = 5 + 3(n-1) = 3n + 2$

25. $d = -5, a_1 = 11 \Rightarrow a_n = 11 - 5(n-1) = 16 - 5n$

27. $d = -\dfrac{1}{4}, a_1 = \dfrac{1}{2} \Rightarrow a_n = \dfrac{1}{2} - \dfrac{1}{4}(n-1) = \dfrac{3-n}{4}$

29. $d = -\dfrac{2}{5}, a_1 = -\dfrac{3}{5} \Rightarrow$
$a_n = -\dfrac{3}{5} - \dfrac{2}{5}(n-1) = \dfrac{-1-2n}{5} = -\dfrac{2n+1}{5}$

31. $d = 3, a_1 = e \Rightarrow a_n = e + 3(n-1) = e + 3n - 3$

33. $a_n = -1, -3, -5, -7, -9, \ldots$

We will use the points $(1, -1)$ and $(2, -3)$ to find the linear function that produces the sequence.
$$m = \dfrac{-3 - (-1)}{2 - 1} = -2$$
$$y - (-1) = -2(x - 1) \Rightarrow y = -2x + 1$$
The general term for this sequence is
$a_n = f(n) = -2n + 1$.

35. $a_n = \dfrac{5}{2}, 3, \dfrac{7}{2}, 4, \dfrac{9}{2}, \ldots$

We will use the points $(2, 3)$ and $(4, 4)$ to find the linear function that produces the sequence.
$$m = \dfrac{4 - 3}{4 - 2} = \dfrac{1}{2}$$
$$y - 3 = \dfrac{1}{2}(x - 2) \Rightarrow y = \dfrac{1}{2}x + 2$$
The general term for this sequence is
$a_n = f(n) = \dfrac{1}{2}n + 2$.

37. $a_n = -\dfrac{5}{4}, -\dfrac{1}{2}, \dfrac{1}{4}, 1, \dfrac{7}{4}, \ldots$

We will use the points $\left(3, \dfrac{1}{4}\right)$ and $\left(5, \dfrac{7}{4}\right)$ to find the linear function that produces the sequence.

$$m = \dfrac{\dfrac{7}{4} - \dfrac{1}{4}}{5 - 3} = \dfrac{\dfrac{3}{2}}{2} = \dfrac{3}{4}$$

$$y - \dfrac{1}{4} = \dfrac{3}{4}(x - 3) \Rightarrow y = \dfrac{3}{4}x - 2$$

The general term for this sequence is

$$a_n = f(n) = \dfrac{3}{4}n - 2.$$

39. $a_4 = a_1 + d(4 - 1) \Rightarrow 21 = a_1 + 3d$
$a_{10} = a_1 + d(10 - 1) \Rightarrow 60 = a_1 + 9d$
$\begin{cases} a_1 + 3d = 21 \\ a_1 + 9d = 60 \end{cases} \Rightarrow -6d = -39 \Rightarrow d = \dfrac{13}{2}, a_1 = \dfrac{3}{2}$
$a_n = \dfrac{3}{2} + \dfrac{13}{2}(n - 1) = \dfrac{13}{2}n - 5$

41. $a_7 = a_1 + d(7 - 1) \Rightarrow 8 = a_1 + 6d$
$a_{15} = a_1 + d(15 - 1) \Rightarrow -8 = a_1 + 14d$
$\begin{cases} a_1 + 6d = 8 \\ a_1 + 14d = -8 \end{cases} \Rightarrow 8d = -16 \Rightarrow d = -2, a_1 = 20$
$a_n = 20 - 2(n - 1) = 22 - 2n$

43. $a_3 = a_1 + d(3 - 1) \Rightarrow 7 = a_1 + 2d$
$a_{23} = a_1 + d(23 - 1) \Rightarrow 17 = a_1 + 22d$
$\begin{cases} a_1 + 2d = 7 \\ a_1 + 22d = 17 \end{cases} \Rightarrow 20d = 10 \Rightarrow d = \dfrac{1}{2}, a_1 = 6$
$a_n = 6 + \dfrac{1}{2}(n - 1) = \dfrac{n + 11}{2}$

45. $n = 50; S = 50\left(\dfrac{1 + 50}{2}\right) = 1275$

47. $d = 2, a_1 = 1, a_n = 99 \Rightarrow 99 = 1 + 2(n - 1) \Rightarrow$
$n = 50; S = 50\left(\dfrac{1 + 99}{2}\right) = 2500$

49. $d = 3, a_1 = 3, a_n = 300 \Rightarrow 300 = 3 + 3(n - 1) \Rightarrow$
$n = 100; S = 100\left(\dfrac{3 + 300}{2}\right) = 15,150$

51. $d = -3, a_1 = 2, a_n = -34 \Rightarrow$
$-34 = 2 - 3(n - 1) \Rightarrow n = 13$
$S = 13\left(\dfrac{2 - 34}{2}\right) = -208$

53. $d = \dfrac{2}{3}, a_1 = \dfrac{1}{3}, a_n = 7 \Rightarrow$
$7 = \dfrac{1}{3} + \dfrac{2}{3}(n - 1) \Rightarrow n = 11$
$S = 11\left(\dfrac{1/3 + 7}{2}\right) = \dfrac{121}{3}$

55. $d = 5, n = 50, a_1 = 2 \Rightarrow a_n = 2 + 5(50 - 1) = 247$
$S = 50\left(\dfrac{2 + 247}{2}\right) = 6225$

57. $d = 4, n = 20, a_1 = -15 \Rightarrow$
$a_n = -15 + 4(20 - 1) = 61$
$S = 20\left(\dfrac{-15 + 61}{2}\right) = 460$

59. $d = 0.2, n = 100, a_1 = 3.5 \Rightarrow$
$a_n = 3.5 + 0.2(100 - 1) = 23.3$
$S = 100\left(\dfrac{3.5 + 23.3}{2}\right) = 1340$

In exercises 61–66, use the formula for finding the terms in an arithmetic sequence, $a_n = a_1 + d(n - 1)$.

61. $a_n = 75, a_1 = 1, d = 2$
$75 = 1 + 2(n - 1) \Rightarrow 37 = n - 1 \Rightarrow n = 38$

63. $a_n = 95, a_1 = -1, d = 4$
$95 = -1 + 4(n - 1) \Rightarrow 24 = n - 1 \Rightarrow n = 25$

65. $a_n = 50\sqrt{3}, a_1 = 2\sqrt{3}, d = 2\sqrt{3}$
$50\sqrt{3} = 2\sqrt{3} + 2\sqrt{3}(n - 1) \Rightarrow$
$48\sqrt{3} = 2\sqrt{3}(n - 1) \Rightarrow 24 = n - 1 \Rightarrow n = 25$

8.2 Applying the Concepts

67. $d = 2, n = 30, a_1 = 10 \Rightarrow a_n = 10 + 2(30 - 1) = 68$
$S = 30\left(\dfrac{10 + 68}{2}\right) = 1170$

69. $d = 25, n = 30, a_1 = 50 \Rightarrow$
$a_{30} = 50 + 25(30 - 1) = 775$
$S = 30\left(\dfrac{50 + 775}{2}\right) = \$12,375$

71. Antonio has 17 different hourly wages over the four years (his original wage plus 16 raises), so $n = 17$. $a_1 = \$12.75, d = 0.25 \Rightarrow$
$a_{13} = 12.75 + 0.25(17 - 1) = \$16.75$.

73. $n = 25, a_1 = 20, d = 2 \Rightarrow$
$a_{25} = 20 + 2(25 - 1) = 68$
$S = 25\left(\dfrac{20 + 68}{2}\right) = 1100$.

There are 1100 seats in the theater.

75. $n = 17, a_1 = 40, a_{28} = 8 \Rightarrow$
$S = 17\left(\dfrac{40 + 8}{2}\right) = 408$

There are 408 boxes in the stack.

77. Using the formula for the sum of an arithmetic sequence with $a_1 = 3$, we have
$192 = n\left(\dfrac{3 + a_n}{2}\right) \Rightarrow 384 = 3n + na_n$. The last

term in the sequence, given $d = 6$, is
$a_n = 3 + 6(n - 1) = 6n - 3$. Solve the system:
$\begin{cases} 3n + na_n = 384 \\ a_n = 6n - 3 \end{cases} \Rightarrow 3n + n(6n - 3) = 384 \Rightarrow$

$6n^2 = 384 \Rightarrow n^2 = 64 \Rightarrow n = \pm 8$ (Reject the negative solution.) The flea market was open for 8 days.

79. 1, 3, 5, …
The common difference is 2.
$a_{20} = 1 + 2(20 - 1) = 39$
At the 20th stage, there will be 39 dots.

81. 13, 16, 19, …
The common difference is 3.
$a_{20} = 13 + 3(20 - 1) = 70$
At the 20th stage, there will be 70 dots.

8.2 Beyond the Basics

83. $a_n = \log\left(\dfrac{a^n}{b^{n-1}}\right)$, $a_{n+1} = \log\left(\dfrac{a^{n+1}}{b^n}\right)$

$a_{n+1} - a_n = \log\left(\dfrac{a^{n+1}}{b^n}\right) - \log\left(\dfrac{a^n}{b^{n-1}}\right)$

$= \log\left(\dfrac{\dfrac{a^{n+1}}{b^n}}{\dfrac{a^n}{b^{n-1}}}\right) = \log\left(\dfrac{a^{n+1}}{b^n} \cdot \dfrac{b^{n-1}}{a^n}\right)$

$= \log\dfrac{a}{b}$

85. $S_n = 3n^2 + 4n$
$S_{n-1} = 3(n-1)^2 + 4(n-1)$
$= 3(n^2 - 2n + 1) + 4n - 4$
$= 3n^2 - 6n + 3 + 4n - 4$
$= 3n^2 - 2n - 1$
$a_n = S_n - S_{n-1} = (3n^2 + 4n) - (3n^2 - 2n - 1)$
$= 6n + 1$
$d = a_n - a_{n-1} = (6n + 1) - (6(n-1) + 1)$
$= (6n + 1) - (6n - 5) = 6$
Since $d = 6$ is a constant, the sequence a_n is an arithmetic sequence.

87. 45 is divisible by 3 and the largest number less than 100 that is divisible by 3 is 99, so $a_1 = 45$,
$d = 3, a_n = 99 \Rightarrow 99 = 45 + 3(n-1) \Rightarrow n = 19$.
$S = 19\left(\dfrac{45 + 99}{2}\right) = 1368$.

89. The sequence of the reciprocals is
$2, \dfrac{5}{3}, \dfrac{4}{3}, 1, \ldots$. The common difference is $-\dfrac{1}{3}$,
so the sequence is arithmetic, and the original sequence is harmonic.

91. $m = a + d$ and $b = a + 2d$. So
$\dfrac{a + b}{2} = \dfrac{a + a + 2d}{2} = \dfrac{2a + 2d}{2} = a + d = m$

8.2 Critical Thinking/Discussion/Writing

93. $a_1 = 10, a_{21} = a_1 + d(21 - 1) \Rightarrow 0 = 10 + 20d \Rightarrow$
$d = -\dfrac{1}{2} \Rightarrow a_n = 10 - \dfrac{1}{2}(n-1) = \dfrac{21 - n}{2}$

94. First term $= -a_1$; difference $= -d$.

95. There are $100 - 22 + 1 = 79$ terms in the series.

96. In the set of counting numbers, $a_n = n$.
$12,403 = n\left(\dfrac{1 + n}{2}\right) \Rightarrow 24,806 = n^2 + n \Rightarrow$
$n^2 + n - 24,806 = 0 \Rightarrow (n - 157)(n + 158) = 0 \Rightarrow$
$n = -158$ (reject this) or $n = 157$.
The sum of the first 157 counting number is 12,403.

8.2 Getting Ready for the Next Section

97. $\dfrac{a_2}{a_1} = \dfrac{6}{3} = 2$

$\dfrac{a_3}{a_2} = \dfrac{12}{6} = 2$

$\dfrac{a_4}{a_3} = \dfrac{24}{12} = 2$

$\dfrac{a_5}{a_4} = \dfrac{48}{24} = 2$

99. $\dfrac{a_2}{a_1} = \dfrac{-12}{4} = -3$

$\dfrac{a_3}{a_2} = \dfrac{36}{-12} = -3$

$\dfrac{a_4}{a_3} = \dfrac{-108}{36} = -3$

$\dfrac{a_5}{a_4} = \dfrac{324}{-108} = -3$

101. $a_2 = a_1 r = 2 \cdot 3 = 6$

$a_3 = a_2 r = 6 \cdot 3 = 18$

$a_4 = a_3 r = 18 \cdot 3 = 54$

$a_5 = a_4 r = 54 \cdot 3 = 162$

103. $a_2 = a_1 r = 1(-2) = -2$

$a_3 = a_2 r = -2(-2) = 4$

$a_4 = a_3 r = 4(-2) = -8$

$a_5 = a_4 r = -8(-2) = 16$

105. $a_n = 5 \cdot 2^n$

$a_1 = 5 \cdot 2^1 = 10$ $\dfrac{a_2}{a_1} = \dfrac{20}{10} = 2$

$a_2 = 5 \cdot 2^2 = 20$ $\dfrac{a_3}{a_2} = \dfrac{40}{20} = 2$

$a_3 = 5 \cdot 2^3 = 40$ $\dfrac{a_4}{a_3} = \dfrac{80}{40} = 2$

$a_4 = 5 \cdot 2^4 = 80$ $\dfrac{a_5}{a_4} = \dfrac{160}{80} = 2$

$a_5 = 5 \cdot 2^5 = 160$

Note that

$\dfrac{a_{n+1}}{a_n} = \dfrac{5 \cdot 2^{n+1}}{5 \cdot 2^n} = \dfrac{5 \cdot 2^n \cdot 2}{5 \cdot 2^n} = 2$

107. $a_n = 2(-3)^n$

$a_1 = 2(-3)^1 = -6$ $\dfrac{a_2}{a_1} = \dfrac{18}{-6} = -3$

$a_2 = 2(-3)^2 = 18$ $\dfrac{a_3}{a_2} = \dfrac{-54}{18} = -3$

$a_3 = 2(-3)^3 = -54$ $\dfrac{a_4}{a_3} = \dfrac{162}{-54} = -3$

$a_4 = 2(-3)^4 = 162$ $\dfrac{a_5}{a_4} = \dfrac{-486}{162} = -3$

$a_5 = 2(-3)^5 = -486$

Note that

$\dfrac{a_{n+1}}{a_n} = \dfrac{2(-3)^{n+1}}{2(-3)^n} = \dfrac{2(-3)^n(-3)}{2(-3)^n} = -3$

109. $a_n = -3 \cdot 2^n$

$a_{n+1} = -3 \cdot 2^{n+1}$

$a_{n-1} = -3 \cdot 2^{n-1}$

8.3 Geometric Sequences and Series

8.3 Practice Problems

1. $r = \dfrac{18}{6} = 3$

2. The first four terms of the sequence are

$\dfrac{3}{2}, \left(\dfrac{3}{2}\right)^2 = \dfrac{9}{4}, \left(\dfrac{3}{2}\right)^3 = \dfrac{27}{8}, \left(\dfrac{3}{2}\right)^4 = \dfrac{81}{16}$.

Since the common ratio is $\dfrac{3}{2}$, the sequence is geometric.

3. a. $a_1 = 2$ **b.** $r = \dfrac{6/5}{2} = \dfrac{3}{5}$

c. $a_n = 2 \cdot \left(\dfrac{3}{5}\right)^{n-1}$

4. $a_n = \dfrac{1}{3}, \dfrac{2}{3}, \dfrac{4}{3}, \dfrac{8}{3}, \dfrac{16}{3}, \ldots$

We will use the points $\left(1, \dfrac{1}{3}\right)$ and $\left(2, \dfrac{2}{3}\right)$ to find the exponential function that produces the sequence. First, find the common ratio B.

$B = \dfrac{\frac{2}{3}}{\frac{1}{3}} = 2$

Use one of the points to find the constant A.

$y = A \cdot 2^x \Rightarrow \dfrac{1}{3} = A \cdot 2^1 \Rightarrow \dfrac{1}{6} = A$

The function that produces this sequence is

$y = \dfrac{1}{6} \cdot 2^x$ and the general term is $a_n = \dfrac{1}{6} \cdot 2^n$.

5. $a_{18} = a_1 r^{18-1} = 7(1.5^{17}) \approx 6896.8288$

6. $a_1 = \dfrac{1}{9}$; $r = \dfrac{1/3}{1/9} = 3$

$a_n = a_1 r^{n-1} \Rightarrow \dfrac{1}{9} \cdot 3^{n-1} = 243 \Rightarrow$

$3^{n-1} = 2187 \Rightarrow 3^{n-1} = 3^7 \Rightarrow$

$n - 1 = 7 \Rightarrow n = 8$

There are 8 terms in the sequence.

7. $a_1 = 3(0.4)^1 = 1.2$; $r = 0.4$

$S_{17} = \displaystyle\sum_{i=1}^{17} 3(0.4)^i = 1.2\left(\dfrac{1 - 0.4^{17}}{1 - 0.4}\right) \approx 2$

8. $A = 1500\left[\dfrac{\left(1 + \dfrac{0.045}{1}\right)^{(1)(30)} - 1}{\dfrac{0.045}{1}}\right]$

$= \$91,510.60$

9. $a_1 = 3$, $r = \dfrac{6/3}{3} = \dfrac{2}{3}$

Since $|r| < 1$, we can use the formula for the sum of an infinite geometric series.

$S = \dfrac{a_1}{1 - r} = \dfrac{3}{1 - \dfrac{2}{3}} = 9$

10. $\displaystyle\sum_{i=1}^{\infty} (10,000,000)(0.85)^{i-1} = \dfrac{a_1}{1 - r}$

$= \dfrac{10,000,000}{1 - 0.85}$

$\approx \$66,666,666.67$

8.3 Basic Concepts and Skills

1. If 5 is the common ratio of a geometric sequence with general term a_n, then

$\dfrac{a_{63}}{a_{62}} = \underline{5}.$

3. If 24 is the term immediately following the sequence term 8 is a geometric sequence, then the common ratio is $\underline{3}$.

5. True

7. False. $a_n = a_1 r^{n-1}$

8.3 Building Skills

9. The sequence is geometric. $a_1 = 3, r = 2$

11. The sequence is not geometric.

13. The sequence is geometric. $a_1 = 1, r = -3$

15. The sequence is geometric. $a_1 = 7, r = -1$

17. The sequence is geometric. $a_1 = 9, r = \dfrac{1}{3}$

19. The sequence is geometric. $a_1 = -\dfrac{1}{2}, r = -\dfrac{1}{2}$

21. The sequence is geometric. $a_1 = 1, r = 2$

23. The sequence is not geometric.

25. The sequence is geometric. $a_1 = \dfrac{1}{3}, r = \dfrac{1}{3}$

27. The sequence is geometric. $a_1 = \sqrt{5}, r = \sqrt{5}$

39. $a_1 = 2, r = 5, a_n = 2 \cdot 5^{n-1}$

31. $a_1 = 5, r = \dfrac{2}{3}, a_n = 5\left(\dfrac{2}{3}\right)^{n-1}$

33. $a_1 = 0.2, r = -3, a_n = 0.2(-3)^{n-1}$

35. $a_1 = \pi^4, r = \pi^2, a_n = \pi^4(\pi^2)^{n-1} = \pi^{2n+2}$

37. $a_n = 6, 12, 24, 48, 96, \ldots$

We will use the points $(1, 6)$ and $(2, 12)$ to find the exponential function that produces the sequence. First, find the common ratio B.

$B = \dfrac{12}{6} = 2$

Use one of the points to find the constant A.

$y = A \cdot 2^x \Rightarrow 6 = A \cdot 2^1 \Rightarrow 3 = A$

The function that produces this sequence is $y = 3 \cdot 2^x$ and the general term is $a_n = 3 \cdot 2^n$.

39. $a_n = 36, 12, 4, \dfrac{4}{3}, \dfrac{4}{9}, \ldots$

We will use the points $(1, 36)$ and $(2, 12)$ to find the exponential function that produces the sequence. First, find the common ratio B.

$$B = \dfrac{12}{36} = \dfrac{1}{3}$$

Use one of the points to find the constant A.

$$y = A \cdot \left(\dfrac{1}{3}\right)^x \Rightarrow 36 = A \cdot \left(\dfrac{1}{3}\right)^1 \Rightarrow 108$$

The function that produces this sequence is

$$y = 108 \cdot \left(\dfrac{1}{3}\right)^x \quad \text{and the general term is}$$

$$a_n = 108 \cdot \left(\dfrac{1}{3}\right)^n.$$

41. $a_n = -\dfrac{4}{5}, -\dfrac{16}{5}, -\dfrac{64}{5}, -\dfrac{256}{5}, -\dfrac{1024}{5}, \ldots$

We will use the points $\left(1, -\dfrac{4}{5}\right)$ and $\left(2, -\dfrac{16}{5}\right)$ to find the exponential function that produces the sequence. First, find the common ratio B.

$$B = \dfrac{-\frac{16}{5}}{-\frac{4}{5}} = 4$$

Use one of the points to find the constant A.

$$y = A \cdot 4^x \Rightarrow -\dfrac{4}{5} = A \cdot 4^1 \Rightarrow -\dfrac{1}{5} = A$$

The function that produces this sequence is

$$y = -\dfrac{1}{5} \cdot 4^x \quad \text{and the general term is}$$

$$a_n = -\dfrac{1}{5} \cdot 4^n.$$

$$y = -\dfrac{1}{5} \cdot 4^x$$

43. $a_7 = a_1 r^{7-1} = 5(2^6) = 320$

45. $a_{10} = a_1 r^{10-1} = 3(-2)^9 = -1536$

47. $a_6 = a_1 r^{6-1} = \dfrac{1}{16}(3)^5 = \dfrac{243}{16}$

49. $a_9 = a_1 r^{9-1} = -1\left(\dfrac{5}{2}\right)^8 = -\dfrac{390,625}{256}$

51. $a_{20} = a_1 r^{20-1} = 500\left(-\dfrac{1}{2}\right)^{19} = -\dfrac{125}{131,072}$

53. $a_1 = 5; \; r = \dfrac{10}{5} = 2$

$$a_n = a_1 r^{n-1} \Rightarrow 5 \cdot 2^{n-1} = 5120 \Rightarrow$$
$$2^{n-1} = 1024 \Rightarrow 2^{n-1} = 2^{10} \Rightarrow$$
$$n - 1 = 10 \Rightarrow n = 11$$

There are 11 terms in the sequence.

55. $a_1 = 16; \; r = \dfrac{8}{16} = \dfrac{1}{2}$

$$a_n = a_1 r^{n-1} \Rightarrow 16 \cdot \left(\dfrac{1}{2}\right)^{n-1} = \dfrac{1}{32} \Rightarrow$$
$$\left(\dfrac{1}{2}\right)^{n-1} = \dfrac{1}{16 \cdot 32} \Rightarrow \left(\dfrac{1}{2}\right)^{n-1} = \dfrac{1}{2^4 \cdot 2^5} = \dfrac{1}{2^9} \Rightarrow$$
$$n - 1 = 9 \Rightarrow n = 10$$

There are 10 terms in the sequence.

57. $a_1 = 18; r = \dfrac{-12}{18} = -\dfrac{2}{3}$

$a_n = a_1 r^{n-1} \Rightarrow 18 \cdot \left(-\dfrac{2}{3}\right)^{n-1} = \dfrac{512}{729} \Rightarrow$

$\left(-\dfrac{2}{3}\right)^{n-1} = \dfrac{512}{729 \cdot 18} \Rightarrow$

$\left(-\dfrac{2}{3}\right)^{n-1} = \dfrac{2^9}{3^6 \cdot 3^2 \cdot 2} = \dfrac{2^8}{3^8} = \left(\dfrac{2}{3}\right)^8 = \left(-\dfrac{2}{3}\right)^8 \Rightarrow$

$n - 1 = 8 \Rightarrow n = 9$

There are 9 terms in the sequence.

59. $r = 5, S_{10} = \dfrac{(1/10)(1 - 5^{10})}{1 - 5} = \dfrac{1,220,703}{5}$

61. $r = -5, S_{12} = \dfrac{(1/25)(1 - 5^{12})}{1 - (-5)} = -\dfrac{40,690,104}{25}$

63. $r = \dfrac{1}{4}, S_8 = \dfrac{5(1 - (1/4)^8)}{1 - 1/4} = \dfrac{109,225}{16,384}$

65. $\displaystyle\sum_{i=1}^{5} \left(\dfrac{1}{2}\right)^{i-1} = \dfrac{(1)(1 - (1/2)^5)}{1 - 1/2} = \dfrac{31}{16}$

67. $a_1 = 3, r = \dfrac{2}{3}$

$\displaystyle\sum_{i=1}^{8} 3\left(\dfrac{2}{3}\right)^{i-1} = \dfrac{3(1 - (2/3)^8)}{1 - (2/3)} = \dfrac{6305}{729}$

69. $a_1 = \dfrac{1}{4}, r = 2$

$\displaystyle\sum_{i=3}^{10} \dfrac{2^{i-1}}{4} = \sum_{i=1}^{10} \dfrac{2^{i-1}}{4} - \sum_{i=1}^{2} \dfrac{2^{i-1}}{4}$

$= \dfrac{(1/4)(1 - 2^{10})}{1 - 2} - \dfrac{(1/4)(1 - 2^2)}{1 - 2}$

$= \dfrac{1023}{4} - \dfrac{3}{4} = 255$

71. $a_1 = -\dfrac{3}{5}, r = -\dfrac{5}{2}$

$\displaystyle\sum_{i=1}^{20} \left(-\dfrac{3}{5}\right)\left(-\dfrac{5}{2}\right)^{i-1} = \dfrac{(-3/5)(1 - (-5/2)^{20})}{1 - (-5/2)}$

$= \dfrac{3(5^{20} - 2^{20})}{35(2^{19})}$

73. $a_1 = \dfrac{1}{5}; r = \dfrac{1}{2}$

$\dfrac{1}{320} = \dfrac{1}{5}\left(\dfrac{1}{2}\right)^{n-1} \Rightarrow \dfrac{1}{64} = \left(\dfrac{1}{2}\right)^{n-1} \Rightarrow$

$\left(\dfrac{1}{2}\right)^6 = \left(\dfrac{1}{2}\right)^{n-1} \Rightarrow 6 = n - 1 \Rightarrow n = 7$

Thus, the sum is $\displaystyle\sum_{n=1}^{7} \dfrac{1}{5}\left(\dfrac{1}{2}\right)^{n-1}$.

```
sum(seq((1/5)(1/
2)^(n-1),n,1,7)
          .396875
Ans▶Frac
          127/320
■
```

75.
```
sum(seq(3^(n-2),
n,1,10)
        9841.333333
```

77.
```
sum(seq((-1)^(n-
1)*3^(2-n),n,1,1
0)
        2.249961896
```

79. $a_1 = \dfrac{1}{3}, r = \dfrac{1}{3}$

$\dfrac{1}{3} + \dfrac{1}{9} + \dfrac{1}{27} + \dfrac{1}{81} + \cdots = \dfrac{1/3}{1 - (1/3)} = \dfrac{1}{2}$

81. $a_1 = -\dfrac{1}{2}, r = -\dfrac{1}{2}$

$-\dfrac{1}{2} + \dfrac{1}{4} - \dfrac{1}{8} + \dfrac{1}{16} + \cdots = \dfrac{-1/2}{1 - (-1/2)} = -\dfrac{1}{3}$

83. $a_1 = 8, r = -\dfrac{1}{4}$

$8 - 2 + \dfrac{1}{2} - \dfrac{1}{8} + \cdots = \dfrac{8}{1 - (-1/4)} = \dfrac{32}{5}$

85. $a_1 = 5, r = \dfrac{1}{3}; \displaystyle\sum_{n=0}^{\infty} 5\left(\dfrac{1}{3}\right)^n = \dfrac{5}{1 - (1/3)} = \dfrac{15}{2}$

87. $a_1 = 1, r = -\dfrac{1}{4}; \displaystyle\sum_{n=0}^{\infty} \left(-\dfrac{1}{4}\right)^n = \dfrac{1}{1 - (-1/4)} = \dfrac{4}{5}$

8.3 Applying the Concepts

89. $a_5 = 20,000(1.03^5) = 23,185$

91. $a_{36} = 100\left(\dfrac{\left(1 + \dfrac{0.06}{12}\right)^{36} - 1}{\dfrac{0.06}{12}}\right) = \$3933.61$

93. $a_{10} = 2^{10} = 1024$.

A person has 1024 ancestors in the tenth generation back.

95. $a_1 = \$36,000, r = 1.05$

$$S_{20} = \frac{36,000(1-1.05^{20})}{1-1.05} = \$1,190,374.35$$

97. $a_1 = 56.25, r = 0.8$

$$S_5 = \frac{56.25(1-0.8^5)}{1-0.8} = 189.09 \text{ cm}$$

99. $D = 5 + 2(5)\left(\frac{3}{5}\right) + 2(5)\left(\frac{3}{5}\right)^2 + 2(5)\left(\frac{3}{5}\right)^3 + \ldots$

$$= 5 + 10\left(\frac{3}{5}\right) + 10\left(\frac{3}{5}\right)^2 + 10\left(\frac{3}{5}\right)^3 + \ldots$$

$$= 5 + \frac{6}{1-(3/5)} = 20 \text{ m}$$

101. Number of squares: 1, 2, 4, …

Common ratio $r = \frac{2}{1} = 2$

$a_{10} = a_1 \cdot r^9 = 1 \cdot 2^9 = 512$

At the 10th stage, there will be 512 squares.

103. The ratio of the sides of the triangles is $\frac{1}{2}$, so the ratio of the areas of the triangles is

$\left(\frac{1}{2}\right)^2 = \frac{1}{4}$. The areas of the shaded triangles

are $\frac{1}{4}, \frac{1}{16}, \frac{1}{64}, \frac{1}{256}, \ldots$

So, $a_n = \frac{1}{4} \cdot \left(\frac{1}{4}\right)^{n-1} = \left(\frac{1}{4}\right)^n$. Because

$|r| = \frac{1}{4} < 1$, we can use the formula for the sum of an infinite geometric series.

$$S = \frac{a_1}{1-r} = \frac{\frac{1}{4}}{1-\frac{1}{4}} = \frac{\frac{1}{4}}{\frac{3}{4}} = \frac{1}{3}$$

This sum of the areas of the shaded triangles is $\frac{1}{3}$.

8.3 Beyond the Basics

105. The area of each square is 1/2 the area of the next larger square. However, we are including the shaded regions only, so $r = -1/2$. (This will eliminate the white squares.)

$$\sum_{i=1}^{\infty}\left(-\frac{1}{2}\right)^{i-1} = \frac{1}{1-(-1/2)} = \frac{2}{3}.$$

107. Since $a_1, a_2, a_3, \ldots$ is a geometric sequence, $a_n = a_{n-1}r$ for every $n \geq 1$, where r is the common ratio. Taking the logarithm of both sides, we have $\ln a_n = \ln\left(a_{n-1}r\right)$

$= \ln a_{n-1} + \ln r$. So the sequence $\ln a_1, \ln a_2, \ln a_3, \ldots$ is arithmetic with common difference $\ln r$.

109. Since $a_1, a_2, a_3, \ldots$ is a geometric sequence, $a_n = a_{n-1}r$ for every $n \geq 1$, where r is the common ratio. Taking the reciprocal of both sides, we have $\frac{1}{a_n} = \frac{1}{a_{n-1}r} = \frac{1}{a_{n-1}} \cdot \frac{1}{r}$. So the

sequence $\frac{1}{a_1}, \frac{1}{a_2}, \frac{1}{a_3}, \cdots$ is geometric with common ratio $1/r$.

111. Since $a_n = \frac{2a_{n-1}}{x}$ for every $n \geq 1$, the sequence is geometric with common ratio $2/x$.

113. Let the general term be represented by b_n.

Since $b_n = \frac{b_{n-1}}{y}$, the sequence is geometric with common ratio $1/y$.

8.3 Critical Thinking/Discussion/Writing

115. $a_n = 2^{4-n} = \frac{2^4}{2^n} = 16 \cdot \left(\frac{1}{2}\right)^n$, which is a geometric sequence.

$$b_n = \frac{4}{2^{n-2}} = \frac{2^2}{\frac{2^n}{2^2}} = \frac{2^4}{2^n} = 16 \cdot \left(\frac{1}{2}\right)^n$$

The sequences are the same.

116. $4, -2, 1, -\dfrac{1}{2}, \dfrac{1}{4}, \ldots$

$$r = \dfrac{-2}{4} = -\dfrac{1}{2}$$

$$y = A \cdot \left(-\dfrac{1}{2}\right)^n \Rightarrow 4 = A \cdot \left(-\dfrac{1}{2}\right)^1 \Rightarrow A = -8$$

So, $y = -8 \cdot \left(-\dfrac{1}{2}\right)^x$ is a function that

produces this sequence and $a_n = -8 \cdot \left(-\dfrac{1}{2}\right)^n$

is the general term for the sequence.

Another way of looking at this is as follows:

Examine just the positive values: $4, 1, \dfrac{1}{4}, \ldots$

These values are generated by

$$f(x) = 8 \cdot \left(\dfrac{1}{2}\right)^x.$$

Now, look at just the negative values:

$$-2, -\dfrac{1}{2}, -\dfrac{1}{8}, \ldots$$

These values are generated by

$$g(x) = -8 \cdot \left(\dfrac{1}{2}\right)^x.$$

The points representing a_n alternate between $f(x)$ and $g(x)$. As $x \to \infty$, $f(x) \to 0$, and $g(x) \to 0$.

117. $a_{1001} = 10 + 1000(2.7) = 2710$.

$b_{1001} = 10(2^{1000})$, so b_{1001} is larger.

118. $\dfrac{a_1 \left(1 - \left(\dfrac{1}{2}\right)^{10}\right)}{1 - \dfrac{1}{2}} = 1023 \Rightarrow a_1 = 512$

119. $5 + 5(2) + 5(2^2) + \cdots + 5(2^{15}) = \displaystyle\sum_{i=0}^{15} 5(2^i)$ or

$$\displaystyle\sum_{i=1}^{16} 5(2^{i-1}).$$

120. Let $\dfrac{a}{r}, a,$ and ar be the three terms of the geometric sequence. Then we have

$$\begin{cases} \dfrac{a}{r} + a + ar = 35 \\ \left(\dfrac{a}{r}\right) a(ar) = 1000 \end{cases} \Rightarrow \begin{cases} a + ar + ar^2 = 35r \\ a^3 = 1000 \end{cases} \Rightarrow$$

$a = 10$

$10 + 10r + 10r^2 = 35r \Rightarrow 2r^2 - 5r + 2 = 0 \Rightarrow$

$(2r - 1)(r - 2) \Rightarrow r = \dfrac{1}{2}$ or $r = 2$.

If $r = \dfrac{1}{2}$, then the three numbers are 20, 10, and 5. If $r = 2$, then the numbers are also 5, 10, and 20.

121. Let $a, a + d,$ and $a + 2d$ be the three terms of the arithmetic sequence. Then

$a + (a + d) + (a + 2d) = 3a + 3d = 15 \Rightarrow$

$a + d = 5$.

The geometric sequence is $a + 1, a + d + 4,$ and $a + 2d + 19$, and

$$\dfrac{a + d + 4}{a + 1} = \dfrac{a + 2d + 19}{a + d + 4}.$$

Use substitution to solve the system:

$$\begin{cases} a + d = 5 \\ \dfrac{a + d + 4}{a + 1} = \dfrac{a + 2d + 19}{a + d + 4} \end{cases} \Rightarrow$$

$$\begin{cases} d = 5 - a \\ \dfrac{a + d + 4}{a + 1} = \dfrac{a + 2d + 19}{a + d + 4} \end{cases} \Rightarrow$$

$$\dfrac{a + (5 - a) + 4}{a + 1} = \dfrac{a + 2(5 - a) + 19}{a + (5 - a) + 4} \Rightarrow$$

$$\dfrac{9}{a + 1} = \dfrac{29 - a}{9} \Rightarrow 29 + 28a - a^2 = 81 \Rightarrow$$

$-(a^2 - 28a + 52) = 0 \Rightarrow (a - 2)(a - 26) = 0 \Rightarrow$

$a = 2$ or $a = 26$.

If $a = 2$, then $d = 3$, and the arithmetic sequence is 2, 5, 8. If $a = 26$, then $d = -21$, and the sequence is 26, 5, -16.

122. Let $a, ar,$ and ar^2 be the three terms of the geometric sequence. So,

$a \cdot ar \cdot ar^2 = a^3 r^3 = 1000 \Rightarrow ar = 10$.

The arithmetic sequence is $a, ar + 6,$ and $ar^2 + 7$. Then,

$$(ar + 6) - a = \left(ar^2 + 7\right) - (ar + 6).$$

(continued on next page)

(continued)

Substituting $ar = 10$ gives

$(10 + 6) - a = (10r + 7) - (10 + 6) \Rightarrow$

$16 - a = 10r - 9 \Rightarrow 25 = 10r + a$

$ar = 10 \Rightarrow r = \dfrac{10}{a}$

Substituting, we have

$25 = 10\left(\dfrac{10}{a}\right) + a \Rightarrow 25a = 100 + a^2 \Rightarrow$

$a^2 - 25a + 100 = 0 \Rightarrow (a - 5)(a - 20) = 0 \Rightarrow$

$a = 5,\ a = 20$

If $a = 5$, then we have $5r = 10 \Rightarrow r = 2$, and
the geometric sequence is 5, 10, 20.

If $a = 20$, then we have $20r = 10 \Rightarrow r = \dfrac{1}{2}$, and

the geometric sequence is 20, 10, 5.

8.3 Getting Ready for the Next Section

123. $P_n : n(n + 1)$ is even.
  $P_3 : 3(3 + 1)$ is even.   True
  $P_4 : 4(4 + 1)$ is even.   True

125. $P_n : 1 + 2 + 3 + \cdots + n = \dfrac{n(n + 1)}{2}$

  $P_3 : 1 + 2 + 3 = \dfrac{3(3 + 1)}{2}$ True

  $P_4 : 1 + 2 + 3 + 4 = \dfrac{4(4 + 1)}{2}$ True

127. $P_n : n^2 + n$ is divisible by 2.
  $P_{n+1} : (n + 1)^3 + (n + 1)$ is divisible by 2.

129. $P_n : 3^n > 5n$
  $P_{n+1} : 3^{n+1} > 5(n + 1)$

131. $\dfrac{k(k + 1)}{2} + (k + 1) = \dfrac{k(k + 1)}{2} + \dfrac{2(k + 1)}{2}$

$= \dfrac{k^2 + k + 2k + 2}{2}$

$= \dfrac{k^2 + 3k + 2}{2}$

$= \dfrac{(k + 1)(k + 2)}{2}$

8.4 Mathematical Induction

8.4 Practice Problems

1. $P_{k+1} : \left[(k + 1) + 3\right]^2 > (k + 1)^2 + 9 \Rightarrow$
  $(k + 4)^2 > (k + 1)^2 + 9$

2. For $n = 1 : 1 = \dfrac{1(1 + 1)}{2}$ is true. Assume that it is

true for $n = k : 1 + 2 + 3 + \ldots + k = \dfrac{k(k + 1)}{2}$.

Then for $n = k + 1 : 1 + 2 + 3 + \ldots + k + (k + 1)$

$= \dfrac{k(k + 1)}{2} + (k + 1) = \dfrac{k^2 + k + 2k + 2}{2}$

$= \dfrac{(k + 1)(k + 2)}{2}$, which is exactly the

statement for $n = k + 1$. Therefore the formula
is true for all natural numbers.

3. For $n = 1 : 3^1 > 1$ is true. Assume that the
inequality is true for $n = k : 3^k > k$. Then, we
must use this fact to prove that for
$n = k + 1 : 3^{k+1} > k + 1$.
$3^k > k \Rightarrow 3^k \cdot 3 > 3 \cdot k \Rightarrow 3^{k+1} > 3k > k + 1$,
which is exactly the statement for $n = k + 1$.
Therefore the formula is true for all natural
numbers.

8.4 Concepts and Vocabulary

1. Mathematical induction can only be used to
prove statements about the <u>natural numbers</u>.

3. The second step in a mathematical induction
proof is to assume that P_k is true for a natural
number k, and then show that $\underline{P_{k+1}}$ is true.

5. True

7. False. The number e is a constant. The
statement $e^n \geq n$ can be proven by
mathematical induction for all natural
numbers n.

8.4 Building Skills

9. $P_{k+1} : ((k + 1) + 1)^2 - 2(k + 1)$
        $= k^2 + 4k + 4 - 2k - 2$
        $= k^2 + 2k + 2 = (k + 1)^2 + 1$

11. $P_{k+1} : 2^{k+1} > 5(k + 1)$

13. For $n=1:2=2$ is true. Assume that it is true for $n=k$:
$$2+4+6+\cdots+2k=k^2+k=k(k+1)$$
Then for $n=k+1$:
$$2+4+6+...+2k+2(k+1)=k(k+1)+2(k+1)=k^2+3k+2=(k+1)(k+2),$$
which is exactly the statement for $n=k+1$. Therefore the formula is true for all natural numbers.

15. For $n=1:4=2(1)(1+1)$ is true. Assume that it is true for $n=k$:
$$4+8+12+\cdots+4k=2k(k+1)$$
Then for $n=k+1$:
$$4+8+12+\cdots+4k+(4k+4)=2k(k+1)+(4k+4)=2k^2+2k+4k+4$$
$$=2(k^2+3k+2)=2(k+1)(k+2),$$
which is exactly the statement for $n=k+1$. Therefore the formula is true for all natural numbers.

17. For $n=1:1=1(2-1)$ is true. Assume that it is true for $n=k$:
$$1+5+9+\cdots+(4k-3)=k(2k-1)$$
Then for $n=k+1$:
$$1+5+9+\cdots+(4k-3)+(4k+1)=k(2k-1)+(4k+1)=2k^2-k+4k+1$$
$$=2k^2+3k+1=(k+1)(2k+1)=(k+1)(2(k+1)-1),$$
which is exactly the statement for $n=k+1$. Therefore the formula is true for all natural numbers.

19. For $n=1:3=\dfrac{3(3-1)}{2}$ is true. Assume that it is true for $n=k$:
$$3+9+27+\cdots+3^k=\frac{3(3^k-1)}{2}$$
Then for $n=k+1$:
$$3+9+27+\cdots+3^k+3^{k+1}=\frac{3(3^k-1)}{2}+3^{k+1}=\frac{3^{k+1}-3+2(3^{k+1})}{2}=\frac{3(3^{k+1})-3}{2}=\frac{3(3^{k+1}-1)}{2},$$
which is exactly the statement for $n=k+1$. Therefore the formula is true for all natural numbers.

21. For $n=1:1\cdot2=\dfrac{1}{3}\cdot(1+1)(4\cdot1-1)$ is true. Assume that the statement is true for $n=k$:
$$1\cdot2+3\cdot4+5\cdot6+\cdots+(2k-1)(2k)=\frac{1}{3}k(k+1)(4k-1)$$
Then, for $n=k+1$:
$$\frac{1}{3}k(k+1)(4k-1)+(2(k+1)-1)(2(k+1))=\frac{1}{3}k(k+1)(4k-1)+(2k+1)(2k+2)$$
$$=\frac{4}{3}k^3+k^2-\frac{1}{3}k+4k^2+6k+2$$
$$=\frac{4}{3}k^3+5k^2+\frac{17}{3}k+2$$
$$=\frac{4k^3+15k^2+17k+6}{3}$$
$$=\frac{(k+1)(k+2)(4k+3)}{3}$$
$$=\frac{1}{3}(k+1)(k+2)(4(k+1)-1)$$
This is exactly the statement for $n=k+1$. Therefore, the formula is true for all natural numbers.

23. For $n = 1: \dfrac{1}{(1)(2)} = \dfrac{1}{1+1}$ is true. Assume that it is true for $n = k$:

$$\dfrac{1}{(1)(2)} + \dfrac{1}{(2)(3)} + \dfrac{1}{(3)(4)} + \cdots + \dfrac{1}{k(k+1)} = \dfrac{k}{k+1}. \text{ Then for } n = k+1:$$

$$\dfrac{1}{(1)(2)} + \dfrac{1}{(2)(3)} + \dfrac{1}{(3)(4)} + \cdots + \dfrac{1}{k(k+1)} + \dfrac{1}{(k+1)(k+2)} = \dfrac{k}{k+1} + \dfrac{1}{(k+1)(k+2)} = \dfrac{k^2 + 2k + 1}{(k+1)(k+2)}$$

$$= \dfrac{(k+1)^2}{(k+1)(k+2)} = \dfrac{k+1}{k+2}$$

which is exactly the statement for $n = k+1$. Therefore the formula is true for all natural numbers.

25. For $n = 1: 2 \le 2$ is true. Assume that it is true for $n = k$: $2 \le 2^k$. Then for $n = k+1$:

$2 \le 2^k \Rightarrow 2 \cdot 2 \le 2(2^k) \Rightarrow 4 \le 2^{k+1} \Rightarrow 2 \le 2^{k+1}$ Therefore the formula is true for all natural numbers.

27. For $n = 1: 1(1+2) < (1+1)^2$ is true. Assume that it is true for $n = k$: $k(k+2) < (k+1)^2$. Then for $n = k+1$:

$(k+1)(k+1+2) = (k+1)(k+3) < (k+1+1)^2 \Rightarrow k^2 + 4k + 3 < k^2 + 4k + 4 \Rightarrow$

$k^2 + 2k + 2k + 3 < k^2 + 2k + 1 + 2k + 3 \Rightarrow k(k+2) + (2k+3) < (k+1)^2 + (2k+3) \Rightarrow k(k+2) < (k+1)^2$,

which is exactly the statement for $n = k+1$. Therefore the formula is true for all natural numbers.

29. For $n = 1: \dfrac{1!}{1} = (1-1)!$ is true. Assume that it is true for $n = k$: $\dfrac{k!}{k} = (k-1)!$. Then for $n = k+1$:

$\dfrac{(k+1)!}{k+1} = \dfrac{(k+1)k!}{k+1} = k!$, which is exactly the statement for $n = k+1$. Therefore the formula is true for all natural numbers.

31. For $n = 1: 1^2 = \dfrac{1(1+1)(2+1)}{6}$ is true. Assume that it is true for $n = k$:

$$1^2 + 2^2 + 3^2 + \cdots + k^2 = \dfrac{k(k+1)(2k+1)}{6}.$$

Then for $n = k+1$:

$$1^2 + 2^2 + 3^2 + \cdots + k^2 + (k+1)^2 = \dfrac{k(k+1)(2k+1)}{6} + (k+1)^2 = \dfrac{k(k+1)(2k+1) + 6(k+1)^2}{6}$$

$$= \dfrac{(k+1)\big(k(2k+1) + 6(k+1)\big)}{6} = \dfrac{(k+1)(2k^2 + 7k + 6)}{6}$$

$$= \dfrac{(k+1)(k+2)(2k+3)}{6}$$

which is exactly the statement for $n = k+1$. Therefore the formula is true for all natural numbers.

33. For $n = 1$: 2 is a factor of $1^2 + 1 = 2$.

For $n = k$: assume that 2 is a factor of $n^2 + n \Rightarrow k^2 + k = 2p$ for some integer p. Then for $n = k+1$:

$(k+1)^2 + (k+1) = k^2 + 3k + 2 = (k^2 + k) + (2k + 2) = 2p + 2(k+1) = 2(p + k + 1)$

$(p + k + 1)$ is an integer, so $2(p + k + 1)$ is divisible by 2. Therefore the formula is true for all natural numbers.

35. For $n=1: 6$ is a factor of $1(1+1)(1+2)=6$. For $n=k$: Assume that 6 is a factor of
$n(n+1)(n+2) \Rightarrow k(k+1)(k+2)=6p$ for some integer p.
Then for $n=k+1$:
$(k+1)(k+2)(k+3)=k(k+1)(k+2)+3(k+1)(k+2)=6p+3(k+1)(k+2)$
Either $k+1$ or $k+2$ is even, so $3(k+1)(k+2)$ is divisible by 6 and, thus, $6p+3(k+1)(k+2)$ is divisible by 6. Therefore the formula is true for all natural numbers.

37. For $n=1:(1+a)^1 \geq 1 \cdot a, a>1$ is true. Assume that the statement is true for $n=k$,:

$(1+a)^k \geq ka, a>1$. Then for $n=k+1$,

$(1+a)^k \geq ka \Rightarrow (1+a)(1+a)^k \geq ka(1+a) \Rightarrow (1+a)^{k+1} \geq ka+ka^2 \geq (k+1)a, a>1$
Therefore the formula is true for all natural numbers $a>1$.

39. For $n=1: \sum_{i=1}^{1} 3 \cdot 4^1 = 4(4^1-1)=12$ is true. Assume that the statement is true for $n=k$: $\sum_{i=1}^{k} 3 \cdot 4^i = 4(4^k-1)$.
Then, for $n=k+1$,
$\sum_{i=1}^{k+1} 3 \cdot 4^i = 4(4^k-1)+3 \cdot 4^{k+1} = 4(4^k-1)+4(4^k \cdot 3) = 4(4^k-1+4^k \cdot 3)$
$= 4((4^k+4^k \cdot 3)-1) = 4(4^k(1+3)-1) = 4(4 \cdot 4^k-1) = 4(4^{k+1}-1),$
which is exactly the statement for $n=k+1$. Therefore the formula is true for all natural numbers.

41. For $n=1: 1+\frac{1}{1}=1+1$ is true. Assume that the statement is true for $n=k$:

$\left(1+\frac{1}{1}\right)\left(1+\frac{1}{2}\right)\left(1+\frac{1}{3}\right)+\cdots+\left(1+\frac{1}{k}\right)=k+1.$
Then, for $n=k+1$,
$(k+1)\left(1+\frac{1}{k+1}\right)=(k+1)\left(\frac{k+2}{k+1}\right)=k+2=(k+1)+1$
This is exactly the statement for $n=k+1$. Therefore the formula is true for all natural numbers.

43. For $n=1:(ab)^1=a^1b^1$ is true. Assume that it is true for $n=k:(ab)^k=a^kb^k$.

Then for $n=k+1: (ab)^{k+1}=(ab)^k(ab)=a^kb^kab=a^{k+1}b^{k+1}$, which is exactly the statement for $n=k+1$. Therefore the formula is true for all natural numbers.

8.4 Applying the Concepts

45. For $n=2:\frac{2^2-2}{2}=1$, which is the correct number of hugs for two people. Assume that the number of hugs

for k people is $\frac{k^2-k}{2}$. If there are $k+1$ people, then we can separate one of them. The remaining k have

$\frac{k^2-k}{2}$ hugs by the hypothesis. The $k+1$ st person has to hug each of the other k people. So the number of

hugs is $\frac{k^2-k}{2}+k = \frac{k^2-k+2k}{2} = \frac{k^2+2k+1-(k+1)}{2} = \frac{(k+1)^2-(k+1)}{2}$, which is exactly the statement

for $n=k+1$. Therefore the formula is true for all natural numbers $n \geq 2$.

47. a. The number of sides of the *n*th figure is $3\left(4^{n-1}\right)$. For $n=1:3\left(4^0\right)=3$, which is the number of sides of the triangle. Assume the number of sides of the *k*th figure is $3\left(4^{k-1}\right)$. When making the $k+1$st figure, each new small triangle splits the original side into two sides, thus doubling the number of sides. In addition, each new triangle adds on two new sides per original side. So, the number of sides of the $k+1$st figure is four times the number of sides of the *k*th figure, i.e., $4(3)\left(4^{k-1}\right)=3\left(4^k\right)$, which is exactly the formula for $n=k+1$. Therefore the formula is true for all natural numbers.

b. The perimeter of the *n*th figure is $3\left(\dfrac{4}{3}\right)^{n-1}$.

For $n=1:3\left(\dfrac{4}{3}\right)^0=3$, which is the perimeter of the equilateral triangle with side length 1. Assume that the perimeter of the *k*th figure is $3\left(\dfrac{4}{3}\right)^{k-1}$. When making the $k+1$st figure, each new small triangle adds on two new sides and deletes one small side, each with length 1/3 of the original side. So, each new triangle increases the perimeter by 1/3 of an original side. Therefore the perimeter of the $k+1$st figure is 4/3 times that of the *k*th figure, i.e., $\dfrac{4}{3}(3)\left(\dfrac{4}{3}\right)^{k-1}=3\left(\dfrac{4}{3}\right)^k$, which is exactly the formula for $n=k+1$. Therefore the formula is true for all natural numbers.

49. The smallest number of moves to accomplish the transfer is 2^n-1. Prove as follows: for $n=1:2^1-1=1$ is the number of necessary moves for one ring. Assume that the smallest number of moves for *k* rings is 2^k-1. When we move the biggest peg from the bottom to another peg, all the other rings must be stacked on the third peg in decreasing order from bottom to top. This requires 2^k-1 moves by the hypothesis. It takes one move to put the biggest ring onto the other peg.

Then we have to move all the other rings on top of it, with again requires 2^k-1 moves. Altogether, then, there are

$$(2^k-1)+1+(2^k-1)=2(2^k)-1=2^{k+1}-1$$

moves, which is exactly the formula for $n=k+1$. Therefore the formula is true for all natural numbers.

8.4 Beyond the Basics

51. For $n=1$, 5 is a factor of $8^1-3^1=5$. For $n=k$, assume that 5 is a factor of $8^n-3^n \Rightarrow 8^k-3^k=5p$ for some integer *p*. Then for $n=k+1$: $8^{k+1}-3^{k+1}=8(8^k)-3(3^k)$
$=8(8^k)-8(3^k)+5(3^k)=8(8^k-3^k)+5(3^k)$
$8(5p)+5(3^k)=5(8p+3^k)$. Since $8p+3^k$ is an integer, $5(8p+3^k)$ is divisible by 5. Therefore the formula is true for all natural numbers.

53. For $n=1$, 64 is a factor of $3^{2(1)+2}-8(1)-9=64$. For $n=k$, assume that 64 is a factor of $3^{2n+2}-8n-9 \Rightarrow 3^{2k+2}-8k-9=64p$ for some integer *p*. Then for $n=k+1$:
$3^{2(k+1)+2}-8(k+1)-9$
$=3^{2k+4}-8(k+1)-9$
$=3^2(3^{2k+2})-8k-8-9$
$=9(3^{2k+2})-9(8k)-9(9)+8(8k)+64$
$=9(3^{2k+2}-8k-9)+64(k+1)$
$=9(64p)+64(k+1)=64(9p+k+1)$
Since $9p+k+1$ is an integer, $64(9p+k+1)$ is divisible by 64. Therefore the formula is true for all natural numbers.

55. For $n=1$, 3 is a factor of $2^{2(1)+1}+1=9$. For $n=k$, assume that 3 is a factor of $2^{2n+1}+1 \Rightarrow 2^{2k+1}+1=3p$ for some integer *p*. Then for $n=k+1$:
$2^{2(k+1)+1}+1=2^{2k+3}+1=2^2 2^{2k+1}+1$
$=2^2 2^{2k+1}+4-3$
$=4(2^{2k+1}+1)-3$
$=4(3p)-3=3(4p-1)$
Since $4p-1$ is an integer, $3(4p-1)$ is divisible by 3. Therefore the formula is true for all natural numbers.

57. For $n = 1$, $a - b$ is a factor of $a^1 - b^1 = a - b$. For $n = k$, assume that $a - b$ is a factor of

$$a^n - b^n \Rightarrow a^k - b^k = (a - b)P(a, b)$$ where $P(a, b)$ is a polynomial of a and b. Then for $n = k + 1$:

$$a^{k+1} - b^{k+1} = a\left(a^k - b^k\right) + b^k(a - b) = a(a - b)P(a, b) + b^k(a - b) = (a - b)\left(aP(a, b) + b^k\right)$$

Since $aP(a, b) + b^k$ is a polynomial of a and b, $a - b$ is a factor of $(a - b)\left(aP(a, b) + b^k\right)$. Therefore the statement is true for all natural numbers.

59. For $n = 1$: $\displaystyle\sum_{k=1}^{2} \frac{1}{k+1} = \frac{1}{2} + \frac{1}{3} \le \frac{5}{6}$ is true. Assume that it is true for $n = m$: $\displaystyle\sum_{k=1}^{m+1} \frac{1}{k+m} \le \frac{5}{6}$.

Then for $n = m + 1$: $\displaystyle\sum_{k=1}^{m+2} \frac{1}{k+m+1} = \sum_{k=2}^{m+3} \frac{1}{k+m} = \left(\sum_{k=1}^{m+1} \frac{1}{k+m}\right) + \frac{1}{2m+2} + \frac{1}{2m+3} - \frac{1}{m+1}$.

In the last expression, $\dfrac{1}{2m+2} + \dfrac{1}{2m+3} - \dfrac{1}{m+1} = \dfrac{(2m+3) + (2m+2) - 2(2m+3)}{2(m+1)(2m+3)} = -\dfrac{1}{2(m+1)(2m+3)} < 0$

So, $\displaystyle\sum_{k=1}^{m+2} \frac{1}{k+m+1} = \left(\sum_{k=1}^{m+1} \frac{1}{k+m}\right) + \frac{1}{2m+2} + \frac{1}{2m+3} - \frac{1}{m+1} \quad \left(\sum_{k=1}^{m+1} \frac{1}{k+m}\right) \le \frac{5}{6}$

This is exactly the formula for $n = k + 1$. Therefore the formula is true for all natural numbers.

8.4 Critical Thinking/Discussion/Writing

60. The proof is not valid because the first step of the mathematical induction, which is to show that the statement is true for P_1, is missing from this proof.

61. For $n = m = 4$: $2^4 \ge 4^2 \Rightarrow 16 \ge 16$ is true. Now assume that the statement is true for $n = k$: $2^k \ge k^2$. Now we must use P_k to prove that P_{k+1} is true. That is, $P_{k+1} : 2^{k+1} \ge (k+1)^2$.

$2 \cdot 2^k \ge 2k^2 \Rightarrow 2^{k+1} \ge 2k^2 \Rightarrow 2^{k+1} \ge k^2 + k^2$

Since $k \ge 4$, we have $2k \ge 8 \Rightarrow 2k > 1$. Also, since $k \ge 4$, we have $k^2 \ge 4k$.

$2^{k+1} > k^2 + k^2 > k^2 + 4k \Rightarrow 2^{k+1} > k^2 + 4k \Rightarrow 2^{k+1} > k^2 + 2k + 2k \Rightarrow$

$2^{k+1} > k^2 + 2k + 2k > k^2 + 2k + 1 \Rightarrow 2^{k+1} > k^2 + 2k + 1 \Rightarrow 2^{k+1} > (k+1)^2$

Thus, $P_k \Rightarrow P_{k+1}$, and the statement is true for all natural numbers.

62. For $n = m = 6$, $6! > 6^3 \Rightarrow 720 > 216$ is true. Now assume that the statement is true for $n = k$: $k! > k^3$.

Now we must use P_k to prove that P_{k+1} is true. That is, $P_{k+1} : (k+1)! \ge (k+1)^3$.

$k! > k^3 \Rightarrow (k+1) \cdot k! \ge (k+1) \cdot k^3 \Rightarrow (k+1)! \ge (k+1) \cdot k^3$.

Since $k + 1 \ge 7$, we have $(k+1)! \ge (k+1) \cdot k^3 \Rightarrow (k+1)! \ge 7k^3$.

We need to show that $7k^3 \ge (k+1)^3$ for $k \ge 6$:

$(k+1)^3 = k^3 + 3k^2 + 3k + 1 \Rightarrow 7k^3 \ge k^3 + 3k^2 + 3k + 1 \Rightarrow 6k^3 \ge 3k^2 + 3k + 1 \Rightarrow$

$k^3 \ge \frac{1}{2}k^2 + \frac{1}{2}k + \frac{1}{6} \Rightarrow k^3 \ge \frac{1}{2}\left(k^2 + k + \frac{1}{3}\right) \Rightarrow 7k^3 \ge \frac{7}{2}\left(k^2 + k + \frac{1}{3}\right)$

$k \ge 6 \Rightarrow 7k^2 \cdot k \ge 7k^2 \cdot 6 \Rightarrow 7k^3 \ge 42k^2$

(continued on next page)

(continued)

Now compare $\frac{7}{2}\left(k^2+k+\frac{1}{3}\right)$ with $42k^2$:

$$12\left[\frac{7}{2}\left(k^2+k+\frac{1}{3}\right)\right]=42k^2+42k+14\Rightarrow 42k^2\geq\frac{7}{2}\left(k^2+k+\frac{1}{3}\right),\text{ so}$$

$$7k^3\geq\frac{7}{2}\left(k^2+k+\frac{1}{3}\right)\Rightarrow(k+1)!\geq\frac{7}{2}\left(k^2+k+\frac{1}{3}\right)\Rightarrow(k+1)!\geq(k+1)^3.$$

8.4 Getting Ready for the Next Section

63. $(x+y)^2=(x+y)(x+y)=x^2+2xy+y^2$

65. $(x+y)^4=(x+y)(x+y)^3=(x+y)\left(x^3+3x^2y+3xy^2+y^3\right)$

$$=x^4+3x^3y+3x^2y^2+xy^3+yx^3+3x^2y^2+3xy^3+y^4=x^4+4x^3y+6x^2y^2+4xy^3+y^4$$

67. The exponents of x decrease from n to 1 when expanding $(x+y)^n$.

69. In exercise 63, we see that the sum of the exponents in each term is 2. In exercise 64, the sum of the exponents in each term is 3. In exercise 65, the sum of the exponents in each term is 4, and in exercise 66, the sum is 5. Thus, we can say that the sum of the exponents on x and y in each term in the expansion of $(x+y)^n$ is n.

8.5   The Binomial Theorem

8.5 Practice Problems

1. $(3y-x)^6=1(3y)^6+6(3y)^5(-x)+15(3y)^4(-x)^2+20(3y)^3(-x)^3+15(3y)^2(-x)^4+6(3y)(-x)^5+1(-x)^6$

$=729y^6-1458y^5x+1215y^4x^2-540y^3x^3+135y^2x^4-18yx^5+x^6$

2. a. $\binom{6}{2}=\frac{6!}{2!(6-2)!}=\frac{6\cdot5\cdot4!}{2!4!}=15$

b. $\binom{12}{9}=\frac{12!}{9!(12-9)!}=\frac{12\cdot11\cdot10\cdot9!}{9!3!}=\frac{12\cdot11\cdot10}{3\cdot2\cdot1}=220$

3. $(3x-y)^4=\binom{4}{0}(3x)^4+\binom{4}{1}(3x)^3(-y)+\binom{4}{2}(3x)^2(-y)^2+\binom{4}{3}(3x)(-y)^3+\binom{4}{4}(-y)^4$

$=\frac{4!}{0!4!}\left(81x^4\right)-\frac{4!}{1!3!}\left(27x^3y\right)+\frac{4!}{2!2!}\left(9x^2y^2\right)-\frac{4!}{3!1!}\left(3xy^3\right)+\frac{4!}{4!0!}y^4$

$=81x^4-108x^3y+54x^2y^2-12xy^3+y^4$

4. The term x^3y^9 is the tenth term in the expansion of $(x+y)^{12}$. So its coefficient is $\binom{12}{9}$.

$$\binom{12}{9}=\frac{12!}{9!(12-9)!}=\frac{12\cdot11\cdot10\cdot9!}{9!3!}=\frac{12\cdot11\cdot10}{3\cdot2\cdot1}=220$$

5. $\binom{n}{n-r}x^r(2a)^{n-r}=\binom{15}{15-3}x^3(2a)^{15-3}=\binom{15}{12}x^3(2a)^{12}=\frac{15!}{12!(15-12)!}x^3(2a)^{12}=\frac{15!}{12!3!}x^3(2a)^{12}$

$=\frac{15\cdot14\cdot13\cdot12!}{12!3!}x^3(2a)^{12}=\frac{15\cdot14\cdot13}{3\cdot2\cdot1}x^3a^{12}=455x^3(2a)^{12}=1,863,680x^3a^{12}$

6. The fourth term in the expansion of $(x-3)^{12}$ is

$$\binom{12}{3}x^{12-4+1}(-3)^{4-1} = \frac{12!}{3!(12-3)!}x^9(-3)^3$$
$$= 220(-3)^3 x^9$$
$$= -5940x^9$$

8.5 Concepts and Vocabulary

1. The expansion of $(x+y)^n$ has $\underline{n+1}$ terms.

3. Expanding a difference, such as $(2x-y)^{10}$, results in <u>alternating signs</u> between terms.

5. False. If n is even, then there are an odd number of terms in the expansion, so one coefficient appears just once.

7. True

8.5 Building Skills

9. $\dfrac{6!}{3!} = \dfrac{6\cdot5\cdot4\cdot3!}{3!} = 120$

11. $\dfrac{12!}{11!} = \dfrac{12\cdot11!}{11!} = 12$

13. $\binom{6}{4} = \dfrac{6!}{4!(6-4)!} = \dfrac{6\cdot5\cdot4!}{4!2!} = 15$

15. $\binom{9}{0} = \dfrac{9!}{0!(9-0)!} = \dfrac{9!}{0!9!} = 1$

17. $\binom{7}{1} = \dfrac{7!}{1!(7-1)!} = \dfrac{7\cdot6!}{1!6!} = 7$

19. $\binom{45}{45} = \dfrac{45!}{45!(45-45)!} = 1$

21. $\binom{100}{98} = \dfrac{100!}{98!(100-98)!} = \dfrac{100\cdot99\cdot98!}{98!2!} = 4950$

23. $\binom{21}{19} = \dfrac{21!}{19!(21-19)!} = \dfrac{21\cdot20\cdot19!}{19!2!} = 210$

25. $(x+2)^4 = x^4 + 4(2x^3) + 6(2^2x^2) + 4(2^3x) + 2^4 = x^4 + 8x^3 + 24x^2 + 32x + 16$

27. $(x-2)^5 = x^5 + 5(-2)x^4 + 10(-2)^2x^3 + 10(-2)^3x^2 + 5(-2)^4x + (-2)^5 = x^5 - 10x^4 + 40x^3 - 80x^2 + 80x - 32$

29. $(2-3x)^3 = 2^3 + 3(2^2)(-3x) + 3(2)(-3x)^2 + (-3x)^3 = 8 - 36x + 54x^2 - 27x^3$

31. $(2x+3y)^4 = (2x)^4 + 4(2x)^3(3y) + 6(2x)^2(3y)^2 + 4(2x)(3y)^3 + (3y)^4$
$= 16x^4 + 96x^3y + 216x^2y^2 + 216xy^3 + 81y^4$

33. $(x+1)^4 = x^4 + 4(1x^3) + 6(1^2x^2) + 4(1^3x) + 1^4 = x^4 + 4x^3 + 6x^2 + 4x + 1$

35. $(x-1)^5 = x^5 + 5(-1)x^4 + 10(-1)^2x^3 + 10(-1)^3x^2 + 5(-1)^4x + (-1)^5 = x^5 - 5x^4 + 10x^3 - 10x^2 + 5x - 1$

37. $(y-3)^3 = y^3 + 3(-3)y^2 + 3(-3)^2y + (-3)^3 = y^3 - 9y^2 + 27y - 27$

39. $(x+y)^6$
$$= \binom{6}{0}x^6 + \binom{6}{1}x^5y + \binom{6}{2}x^4y^2 + \binom{6}{3}x^3y^3 + \binom{6}{4}x^2y^4 + \binom{6}{5}xy^5 + \binom{6}{6}y^6$$
$$= \frac{6!}{0!(6-0)!}x^6 + \frac{6!}{1!(6-1)!}x^5y + \frac{6!}{2!(6-2)!}x^4y^2 + \frac{6!}{3!(6-3)!}x^3y^3 + \frac{6!}{4!(6-4)!}x^2y^4$$
$$+ \frac{6!}{5!(6-5)!}xy^5 + \frac{6!}{6!(6-6)!}y^6$$
$$= x^6 + 6x^5y + 15x^4y^2 + 20x^3y^3 + 15x^2y^4 + 6xy^5 + y^6$$

41. $(1+3y)^5 = 1^5 + 5(1^4)(3y) + (10)(1^3)(3y)^2 + (10)(1^2)(3y)^3 + (5)(1)(3y)^4 + (3y)^5$
$= 1 + 15y + 90y^2 + 270y^3 + 405y^4 + 243y^5$

43. $(2x+1)^4 = (2x)^4 + 4(2x)^3(1) + 6(2x)^2(1)^2 + 4(2x)(1)^3 + 1^4 = 16x^4 + 32x^3 + 24x^2 + 8x + 1$

45. $(x-2y)^3 = x^3 + 3(x^2)(-2y) + 3(x)(-2y)^2 + (-2y)^3 = x^3 - 6x^2y + 12xy^2 - 8y^3$

47. $(2x+y)^4 = (2x)^4 + 4(2x)^3y + 6(2x)^2y^2 + 4(2x)y^3 + y^4 = 16x^4 + 32x^3y + 24x^2y^2 + 8xy^3 + y^4$

49. $\left(\dfrac{x}{2}+2\right)^7 = \binom{7}{0}\left(\dfrac{x}{2}\right)^7 + \binom{7}{1}\left(\dfrac{x}{2}\right)^6(2) + \binom{7}{2}\left(\dfrac{x}{2}\right)^5(2^2) + \binom{7}{3}\left(\dfrac{x}{2}\right)^4(2^3) + \binom{7}{4}\left(\dfrac{x}{2}\right)^3(2^4) + \binom{7}{5}\left(\dfrac{x}{2}\right)^2(2^5)$
$$+ \binom{7}{6}\left(\dfrac{x}{2}\right)(2^6) + \binom{7}{7}(2^7)$$
$$= \dfrac{7!}{0!(7-0)!}\left(\dfrac{x}{2}\right)^7 + \dfrac{7!}{1!(7-1)!}\left(\dfrac{x}{2}\right)^6(2) + \dfrac{7!}{2!(7-2)!}\left(\dfrac{x}{2}\right)^5(2^2) + \dfrac{7!}{3!(7-3)!}\left(\dfrac{x}{2}\right)^4(2^3)$$
$$+ \dfrac{7!}{4!(7-4)!}\left(\dfrac{x}{2}\right)^3(2^4) + \dfrac{7!}{5!(7-5)!}\left(\dfrac{x}{2}\right)^2(2^5) + \dfrac{7!}{6!(7-6)!}\left(\dfrac{x}{2}\right)(2^6) + \dfrac{7!}{7!(7-7)!}(2^7)$$
$$= \dfrac{1}{128}x^7 + \dfrac{7}{32}x^6 + \dfrac{21}{8}x^5 + \dfrac{35}{2}x^4 + 70x^3 + 168x^2 + 224x + 128$$

51. $\left(a^2 - \dfrac{1}{3}\right)^4 = (a^2)^4 + 4(a^2)^3\left(-\dfrac{1}{3}\right) + 6(a^2)^2\left(-\dfrac{1}{3}\right)^2 + 4(a^2)\left(-\dfrac{1}{3}\right)^3 + \left(-\dfrac{1}{3}\right)^4 = a^8 - \dfrac{4}{3}a^6 + \dfrac{2}{3}a^4 - \dfrac{4}{27}a^2 + \dfrac{1}{81}$

53. $\left(\dfrac{1}{x}+y\right)^3 = \left(\dfrac{1}{x}\right)^3 + 3\left(\dfrac{1}{x}\right)^2 y + 3\left(\dfrac{1}{x}\right)y^2 + y^3 = \dfrac{1}{x^3} + \dfrac{3y}{x^2} + \dfrac{3y^2}{x} + y^3$

55. The coefficient of the x^2y^7 term in $(x+y)^9$ is $\binom{9}{9-2} = \dfrac{9!}{2!7!} = \dfrac{9\cdot8\cdot7!}{2!7!} = 36$

57. The x^5 in $(x+3)^8$ has a factor of $3^3 = 27$ in its coefficient. The coefficient is
$$\binom{8}{8-5}3^3 = \left(\dfrac{8!}{5!3!}\right)3^3 = \left(\dfrac{8\cdot7\cdot6\cdot5!}{5!3!}\right)3^3 = 1512.$$

59. The x^2y^7 term in $(2x+3y)^9$ has a factor of $2^2\cdot3^7 = 8748$ in its coefficient. The coefficient is
$$\binom{9}{9-2}(8748) = \dfrac{9!}{7!2!}(8748) = \dfrac{9\cdot8\cdot7!}{7!2!}(8748) = 314,928.$$

61. The $x^{10}y^{12}$ term in $(x^2+y^3)^9$ is the $(x^2)^5(y^3)^4$ term in the expansion. The coefficient is
$$\binom{9}{9-5} = \dfrac{9!}{4!5!} = \dfrac{9\cdot8\cdot7\cdot6\cdot5!}{4!5!} = 126.$$

63. The term containing x^7 is the fourth term in the expansion: $\binom{10}{7}x^7y^3 = \dfrac{10!}{7!(10-7)!}x^7y^3 = 120x^7y^3$

65. The term containing x^3 is the tenth term in the expansion:
$$\binom{12}{3}x^3(-2)^9 = \dfrac{12!}{3!(12-3)!}(-512)x^3 = -112,640x^3.$$

67. The term containing x^6 is the third term in the expansion:
$$\binom{8}{2}(2x)^6(3y)^2 = \dfrac{8!}{2!(8-2)!}(64x^6)(9y^2) = 16,128x^6y^2.$$

69. The term containing y^9 is the tenth term in the expansion:
$$\binom{11}{9}(5x)^2(-2y)^9 = \frac{11!}{2!(11-2)!}\left(25x^2\right)\left(-512y^9\right) = -704,000x^2y^9.$$

71. The seventh term in the expansion of $(x-1)^9$ is $\binom{9}{7-1}x^{9-7+1}(-1)^{7-1} = \frac{9!}{6!3!}x^3(-1)^6 = 84x^3.$

73. The fifth term in the expansion of $(2x-y)^7$ is $\binom{7}{5-1}(2x)^{7-5+1}(-y)^{5-1} = \frac{7!}{4!3!}(2x)^3(-y)^4 = 280x^3y^4.$

75. The third term in the expansion of $(x+3y)^{10}$ is $\binom{10}{3-1}x^{10-3+1}(3y)^{3-1} = \frac{10!}{2!8!}x^8(3y)^2 = 405x^8y^2$

77. $(1.2)^5 = (1+0.2)^5 = 1^5 + 5(1)^4(0.2) + 10(1)^3(0.2)^2 + 10(1)^2(0.2)^3 + 5(1)(0.2)^4 + (0.2)^5 = 2.48832$

8.5 Beyond the Basics

79. The middle term is the sixth term in the expansion:
$$\binom{10}{5}(\sqrt{x})^5\left(-\frac{2}{x^2}\right)^5 = \frac{10!}{5!(10-5)!}x^2\sqrt{x}\left(-\frac{32}{x^{10}}\right) = -\frac{8064\sqrt{x}}{x^8}$$

81. The middle term is the seventh term in the expansion: $\binom{12}{6}(1^6)(-x^2y^{-3})^6 = \frac{12!}{6!(12-6)!}\left(\frac{x^{12}}{y^{18}}\right) = \frac{924x^{12}}{y^{18}}$

83. By the Binomial Theorem, we have
$$2^n = (1+1)^n = \binom{n}{0}(1)^n(1)^0 + \binom{n}{1}(1)^{n-1}(1)^1 + \binom{n}{2}(1)^{n-2}(1)^2 + \cdots + \binom{n}{n}(1)^0(1)^n$$
$$= \binom{n}{0} + \binom{n}{1} + \binom{n}{2} + \ldots + \binom{n}{n}$$

85. $\binom{k}{j} + \binom{k}{j-1} = \frac{k!}{j!(k-j)!} + \frac{k!}{(j-1)!(k-j+1)!} = \frac{k!(j-1)!(k-j+1)! + k!j!(k-j)!}{j!(k-j)!(j-1)!(k-j+1)!}$
$$= \frac{[k!(j-1)!(k-j)!][j+(k-j+1)]}{j!(k-j)!(j-1)!(k-j+1)!} = \frac{k!(j+(k-j+1))}{j!(k+1-j)!} = \frac{k!(k+1)}{j!(k+1-j)!} = \frac{(k+1)!}{j!(k+1-j)!} = \binom{k+1}{j}$$

87. $(2x-1)^4 + 4(2x-1)^3(3-2x) + 6(2x-1)^2(3-2x)^2 + 4(2x-1)(3-2x)^3 + (3-2x)^4$
$$= ((2x-1)+(3-2x))^4 = 2^4 = 16$$

89. $(3x-1)^5 + 5(3x-1)^4(1-2x) + 10(3x-1)^3(1-2x)^2 + 10(3x-1)^2(1-2x)^3 + 5(3x-1)(1-2x)^4 + (1-2x)^5$
$$= ((3x-1)+(1-2x))^5 = x^5$$

91. The constant term is the third term in the expansion:
$$240 = \binom{6}{2}(kx)^4\left(-\frac{1}{x^2}\right)^2 = \frac{6!}{2!(6-2)!}k^4x^4\left(\frac{1}{x^4}\right) = 15k^4 \Rightarrow k^4 = 16 \Rightarrow k = \pm 2$$

93. The general term in the expansion has the form $\binom{11}{k}(2x^2)^k\left(-\frac{1}{4x}\right)^{11-k} = \binom{11}{k}(2^k)\left(-\frac{1}{4}\right)^{11-k}\frac{x^{2k}}{x^{11-k}}.$

In order to get a constant term, $2k$ must equal $11 - k$. However, the solution of this equation is not an integer, so there is no constant term in the expansion.

8.5 Critical Thinking/Discussion/Writing

94. $2^6 = (1+1)^6$

$$= \binom{6}{0}(1)^6(1)^0 + \binom{6}{1}(1)^5(1)^1 + \binom{6}{2}(1)^4(1)^2 + \binom{6}{3}(1)^3(1)^3 + \binom{6}{4}(1)^2(1)^4 + \binom{6}{5}(1)^1(1)^5 + \binom{6}{6}(1)^0(1)^6$$

$$= \binom{6}{0} + \binom{6}{1} + \binom{6}{2} + \binom{6}{3} + \binom{6}{4} + \binom{6}{5} + \binom{6}{6}$$

95. $0 = (1-1)^{10}$

$$= \binom{10}{0}(1)^{10}(1)^0 - \binom{10}{1}(1)^9(1)^1 + \binom{10}{2}(1)^8(1)^2 - \binom{10}{3}(1)^7(1)^3 + \binom{10}{4}(1)^6(1)^4 - \binom{10}{5}(1)^5(1)^5 + \binom{10}{6}(1)^4(1)^6$$

$$- \binom{10}{7}(1)^3(1)^7 + \binom{10}{8}(1)^2(1)^8 - \binom{10}{9}(1)^1(1)^9 + \binom{10}{10}(1)^9(1)^{10}$$

$$= \binom{10}{0} - \binom{10}{1} + \binom{10}{2} - \binom{10}{3} + \binom{10}{4} - \binom{10}{5} + \binom{10}{6} - \binom{10}{7} + \binom{10}{8} - \binom{10}{9} + \binom{10}{10}$$

96. $(x+y)^2 = x^2 + 2xy + y^2$. If $x > 0$ and $y > 0$, then $2xy > 0$ and $x^2 + 2xy + y^2 > x^2 + y^2$. Therefore, $(x+y)^2 > x^2 + y^2$.

97. $(x+y)^n = x^n + \binom{n}{1}x^{n-1}y + \binom{n}{2}x^{n-2}y^2 + \cdots + \binom{n}{n-1}xy^{n-1} + y^n$. If $x > 0$ and $y > 0$, then all the intermediate terms in the expansion are positive. Therefore $(x+y)^n > x^n + y^n$ if $n > 1$.

98. $(1+x)^n = 1 + \binom{n}{1}x + \binom{n}{2}x^2 + \ldots + \binom{n}{n}x^n = 1 + nx + \binom{n}{1}x + \binom{n}{2}x^2 + \ldots + \binom{n}{n}x^n$. If $x > 0$, then all the terms in the sum are positive, and $1 + nx + \binom{n}{1}x + \binom{n}{2}x^2 + \ldots + \binom{n}{n}x^n > 1 + nx$. Therefore, $(1+x)^n > 1 + nx$ if $n > 1$.

8.5 Getting Ready for the Next Section

99. $5 \cdot 6 \cdot 7 \cdot 8 = \dfrac{8!}{4!}$

101. $2 \cdot 4 \cdot 6 \cdot 8 \cdot 10 \cdot 12$
$= 2 \cdot (2 \cdot 2) \cdot (2 \cdot 3) \cdot (2 \cdot 4) \cdot (2 \cdot 5) \cdot (2 \cdot 6)$
$= 2^6 \cdot 6!$

103. $\dfrac{12!}{10!} = 12 \cdot 11 = 132$

105. $\dfrac{8!}{5!3!} = \dfrac{8 \cdot 7 \cdot 6}{3 \cdot 2 \cdot 1} = 56$

8.6 Counting Principles

8.6 Practice Problems

1.

There are 9 ways to choose a course.

2. There are 7 restaurants and 5 movies, so there are $7 \cdot 5 = 35$ different choices.

3. There are $10^6 = 1,000,000$ possibilities.

4. There are $7! = 7 \cdot 6 \cdot 5 \cdot 4 \cdot 3 \cdot 2 \cdot 1 = 5040$ possible arrangements.

5. There are $9 \cdot 8 \cdot 7 \cdot 6 = 3024$ possibilities.

6. a. $P(9,2) = \dfrac{9!}{(9-2)!} = \dfrac{9 \cdot 8 \cdot 7!}{7!} = 72$

   b. $P(n,0) = \dfrac{n!}{(n-0)!} = \dfrac{n!}{n!} = 1$

7. {bears, bulls, lions}, {bears, bulls, tigers}, {bears, lions, tigers}, {bulls, lions, tigers}
$C(4, 3) = 4$

8. a. $C(9,2) = \dfrac{9!}{(9-2)!2!} = \dfrac{9 \cdot 8 \cdot 7!}{7!2!} = 36$

   b. $C(n,0) = \dfrac{n!}{(n-0)!0!} = \dfrac{n!}{n!} = 1$

   c. $C(n,n) = \dfrac{n!}{(n-n)!n!} = \dfrac{n!}{n!} = 1$

9. $C(12,3) = \dfrac{12!}{(12-3)!3!} = \dfrac{12 \cdot 11 \cdot 10 \cdot 9!}{9!3!} = \dfrac{12 \cdot 11 \cdot 10}{3 \cdot 2 \cdot 1} = 220$
There are 220 ways that three of the twelve chocolates can be chosen.

10. $\dfrac{6!}{2!2!2!} = 90$
There are 90 ways to send six counselors in pairs to three different locations.

8.6 Concepts and Vocabulary

1. Any arrangement of *n* distinct objects in a fixed order in which no object is used more than once is called a <u>permutation</u>.

3. When *r* objects are chosen from *n* distinct objects, the set of *r* objects is called a <u>combination</u> of *n* objects taken *r* at a time.

5. True

7. False. Three Coke, two Sprite, and three Pepsi cans can be arranged in $\dfrac{8!}{3!2!3!} = 560$ ways.

8.6 Building Skills

9. $P(6,1) = \dfrac{6!}{(6-1)!} = 6$

11. $P(8,2) = \dfrac{8!}{(8-2)!} = 56$

13. $P(9,9) = \dfrac{9!}{(9-9)!} = 362,880$

15. $P(7,0) = \dfrac{7!}{(7-0)!} = 1$

17. $C(8,3) = \dfrac{8!}{(8-3)!3!} = 56$

19. $C(9,4) = \dfrac{9!}{(9-4)!(4!)} = 126$

21. $C(5,5) = \dfrac{5!}{(5-5)!(5!)} = 1$

23. $C(3,0) = \dfrac{3!}{(3-0)!0!} = 1$

25. a. The first letter can be chosen in 26 different ways. The second letter can also be chosen in 26 different ways, so the number of different 2-letter codes is $26 \cdot 26 = 676$.

   b. The first letter can be chosen in 26 different ways. The second letter can be chosen in 25 different ways, so the number of different 2-letter codes is $26 \cdot 25 = 650$.

27. The president can be chosen in 50 ways, while the vice-president can be chosen in 49 ways. So the number of possibilities is $50 \cdot 49 = 2450$.

29. There are four people to be seated in a row of four chairs, so there are
$P(4,4) = \dfrac{4!}{(4-4)!} = 24$ different ways to arrange the people.

31. There are five letters in the word ANGLE. Using three letters at a time, there are
$P(5,3) = \dfrac{5!}{(5-3)!} = 5 \cdot 4 \cdot 3 = 60$ ways to arrange the letters.

33. There are five different toppings. So there are
$C(5,2) = \dfrac{5!}{(5-2)!2!} = 10$ different pizzas with two additional toppings.

35. There are eleven problems, so it is possible to choose nine of them in

$$C(11,9) = \frac{11!}{(11-9)!(9!)} = 55 \text{ different ways.}$$

37. Order is not important, so find the number of combinations. There are $C(11,4) = 330$ different combinations.

39. Order is important, so find the number of permutations. There are 8 seats and six people will be seated, so there are

$$P(8,6) = \frac{8!}{(8-6)!} = 20,160 \text{ different ways to}$$

seat 6 people.

41. Order is important, so find the number of permutations or use the Fundamental Counting Principle. There are

$$P(5,5) = \frac{5!}{(5-5)!} = 120 \text{ ways to visit the five}$$

colleges.

8.6 Applying the Concepts

43. a. The first digit cannot be a zero or a one, so there are eight possibilities for the first digit. There are two possibilities for the second digit and ten possibilities for the third digit. So there are $8 \cdot 2 \cdot 10 = 160$ possible three-digit area codes.

b. There are eight possibilities for the first digit, ten possibilities for the second digit, and ten possibilities for the third digit. So there are $8 \cdot 10 \cdot 10 = 800$ possible three-digit area codes.

45. a. If no repetitions are allowed, there are $24 \cdot 23 \cdot 22 = 12,144$ three-letter fraternity names.

b. If repetitions are allowed, there are $24^3 = 13,824$ three-letter fraternity names.

47. To reach a majority decision, five, six, seven, eight or all nine of the nine justices must agree. Since order is not important, find the sum of the combinations
$$C(9,5) + C(9,6) + C(9,7) + C(9,8) + C(9,9)$$
$$= \frac{9!}{(9-5)!(5!)} + \frac{9!}{(9-6)!(6!)} + \frac{9!}{(9-7)!(7!)}$$
$$+ \frac{9!}{(9-8)!(8!)} + \frac{9!}{(9-9)!(9!)}$$
$$= 126 + 84 + 36 + 9 + 1 = 256$$

There are 256 ways to form a majority of five from the nine justices.

49. Al can choose from $8 \cdot 8 \cdot 8 = 512$ different outfits. Since there are 365 days in a year, he can wear a different outfit every day.

51. a. There are 7! = 5040 different ways to arrange the houses since any arrangement of the seven houses could be made to correspond to a permutation of the seven designs by agreeing that the first four go on one side of the street and the remaining three go on the other side of the street.

b. There are 7! = 5040 different ways to arrange the houses since any arrangement of the seven houses could be made to correspond to a permutation of the seven designs by agreeing that the first five go on one side of the street and the remaining two go on the other side of the street.

53. a. There are seven letters in the word CHARITY. If two letters in each four-letter group must be an A and an R, then there are five letters that can be chosen in groups of two: $C(5,2) = \frac{5!}{(5-2)!(2)!} = 10$ possibilities.

b. If one letter in each four-letter group must be an A and none of the letters in the group can be an R, then there are five letters that can be chosen in groups of three: $C(5,3) = \frac{5!}{(5-3)!(3)!} = 10$ possibilities.

c. If none of the letters in each four-letter group can be an A or an R, there there are five letters that can be chosen in groups of four: $C(5,4) = \frac{5!}{(5-4)!(4!)} = 5$ possibilities.

55. The store will need to stock $5 \cdot 7 \cdot 3 = 105$ shirts to have one of each type.

57. The maximum number of folders is necessary if all of the questionnaires have different rankings. There are 10! = 3,628,800 possible rankings, so they would need that many folders.

59. In each group of four, if one child is "constant", then the other three children can be chosen in $C(6,3) = \dfrac{6!}{(6-3)!3!} = 20$ ways. So the maximum number of times any one child will go to the circus is 20.

There are $C(7,4) = \dfrac{7!}{(7-4)!4!} = 35$ combinations of four children, so the father will go to the circus 35 times.

61. Because order is not important, we use combinations. Under the "B", there are $C(15,5) = \dfrac{15!}{(15-5)!5!} = 3003$ combinations of numbers. There are the same number under the "I", "G", and "O". Under the "N", there are $C(15,4) = \dfrac{15!}{(15-4)!4!} = 1365$. So there are

$3003^4 \cdot 1365 = 111,007,923,832,370,565$ different bingo cards.

63. There are 10 people and four are needed for the committee. Once the first person is chosen, that person's spouse is eliminated, so there are 8 ways to choose the next person. That person's spouse cannot be chosen next, so there are 6 ways to choose the third person, and 4 ways to choose the fourth person. That gives 1920 possible permutations. However, the order that people are chosen for the committee is not important, so the number of combinations is $\dfrac{1920}{4!} = 80$.

65. If Cory wants to give his sister one coin, there are 4 different combinations of one coin. If he wants to give his sister two coins, there are $C(4,2) = 6$ different combinations. If he wants to give his sister three coins, there are 4 combinations, and if he wants to give his sister four coins, there is one combination. So there are $4 + 6 + 4 + 1 = 15$ different combinations of coins.

67. Out of six letters, there are two "A"s and the rest are singles. So, there are $\dfrac{6!}{2!1!1!1!1!} = 360$ distinguishable ways to arrange the letters.

69. Out of seven letters, there are three "S"s, two "C"s, and the rest are singles. So, there are $\dfrac{7!}{3!2!1!1!!} = 420$ distinguishable ways to arrange the letters.

8.6 Beyond the Basics

71. The company needs n workers in combinations of three. So
$$C(n,3) = \dfrac{n!}{(n-3)!3!} \geq 20 \Rightarrow$$
$$\dfrac{n(n-1)(n-2)(n-3)!}{(n-3)!3!} \geq 20 \Rightarrow$$
$$n^3 - 3n^2 + 2n \geq 120 \Rightarrow$$
$$n^3 - 3n^2 + 2n - 120 \geq 0$$
The possible rational zeros are
$\pm 1, \pm 2, \pm 3, \pm 4, \pm 5, \pm 6, \pm 8, \pm 10, \pm 12,$
$\pm 15, \pm 20, \pm 24, \pm 30, \pm 60, \pm 120$.
Using synthetic division, we find that 6 is a zero:

$$\begin{array}{r|rrrr} 6 & 1 & -3 & 2 & -120 \\ & & 6 & 18 & 120 \\ \hline & 1 & 3 & 20 & 0 \end{array}$$

and $(n-6)(n^2+3n+20) = 0$.

The two zeros of $n^2 + 3n + 20 = 0$ are complex, so we reject them. Therefore, six different workers are necessary.

73. There are $C(30,2) = \dfrac{30!}{(30-2)!2!} = 435$ lines.

75. $\dbinom{m+1}{m-1} = 3! \Rightarrow$
$$\dfrac{(m+1)!}{(m-1)!((m+1)-(m-1))!} = 6 \Rightarrow$$
$$\dfrac{(m+1)!}{(m-1)!2!} = 6 \Rightarrow \dfrac{(m+1)m(m-1)!}{(m-1)!} = 12 \Rightarrow$$
$$m^2 + m - 12 = 0 \Rightarrow (m+4)(m-3) = 0 \Rightarrow$$
$m = 3$ or $m = -4$
Reject the negative solution because factorials are not defined for negative numbers. The solution is $\{3\}$.

77. $2\dbinom{n-1}{2} = \dbinom{n}{3} \Rightarrow \dfrac{2(n-1)!}{2!(n-3)!} = \dfrac{n!}{3!(n-3)!} \Rightarrow$
$$\dfrac{(n-1)(n-2)(n-3)!}{(n-3)!} = \dfrac{n(n-1)(n-2)(n-3)!}{6(n-3)!} \Rightarrow$$
$$6(n-1)(n-2) = n(n-1)(n-2) \Rightarrow n = 6$$

79. Using the Binomial Theorem,

$$\binom{n}{0} + \binom{n}{1} + \binom{n}{2} + \cdots + \binom{n}{n} = 64 \text{ is the sum of}$$

the coefficients in the nth row of the

expansion of $(x+y)^n$. If $x = 1$ and $y = 1$,

then

$$(1+1)^n = \binom{n}{0} + \binom{n}{1} + \binom{n}{2} + \cdots + \binom{n}{n} = 64 \Rightarrow$$

$$2^n = 64 \Rightarrow n = 6.$$

81. $\binom{m}{4} = \binom{m}{5} \Rightarrow \dfrac{m!}{(m-4)!4!} = \dfrac{m!}{(m-5)!5!} \Rightarrow$

$$\dfrac{5!}{4!} = \dfrac{(m-4)!}{(m-5)!} \Rightarrow 5 = \dfrac{(m-4)(m-5)!}{(m-5)!} \Rightarrow$$

$$5 = m - 4 \Rightarrow m = 9$$

8.6 Critical Thinking/Discussion/Writing

83. There are $5 \cdot 5 = 25$ terms of the form $x^n y^m$ if n and m are any integers such that $1 \le n \le 5$ and $1 \le m \le 5$.

84. There are two terms in the first factor and three terms in the second factor, so there are $2 \cdot 3 = 6$ terms.

8.6 Preparing for the Next Section

85. $E = \{2, 5, 7\} \Rightarrow n(E) = 3$

87. $E = \{0\} \Rightarrow n(E) = 1$

For exercises 89–92, $A = \{1, 2, 3, 5\}$ and $B = \{3, 4, 5, 6, 7\}$.

89. $n(A) = 4$, $n(B) = 5$

91. $A \cap B = \{3, 5\}$
$n(A \cap B) = 2$

For exercises 93–96, $A = \{a, b, c\}$ and $B = \{2, 4, 6, 8\}$.

93. $n(A) = 3$, $n(B) = 4$

95. $A \cap B = \varnothing$
$n(A \cap B) = 0$

8.7 Probability

8.7 Practice Problems

1. a. $E_1 = \{2, 4, 6\}$, $P(E_1) = \dfrac{3}{6} = \dfrac{1}{2}$

 b. $E_2 = \{5, 6\}$, $P(E_2) = \dfrac{2}{6} = \dfrac{1}{3}$

2. $S = \{(H, H), (H, T), (T, H), (T, T)\}$
$E = \{(H, T), (T, H)\}$
$P(E) = \dfrac{2}{4} = \dfrac{1}{2}$

3. $E = \{(6, 1), (5, 2), (4, 3), (3, 4), (2, 5), (6, 1)\}$
There are 36 possible outcomes, so

$$P(E) = \dfrac{6}{36} = \dfrac{1}{6}$$

4. Since the order in which the numbers are selected is not important, the sample space S consists of all sets of 6 numbers that can be selected from 50 numbers. So,

$$n(S) = C(50, 6) = \dfrac{50!}{(50-6)!6!}$$

$$= \dfrac{50 \cdot 49 \cdot 48 \cdot 47 \cdot 46 \cdot 45}{6 \cdot 5 \cdot 4 \cdot 3 \cdot 2 \cdot 1} = 15{,}890{,}700$$

$$P(\text{winning the lottery}) = \dfrac{1}{15{,}890{,}700}$$

5. $P(\text{jack or king}) = P(\text{jack}) + P(\text{king})$

$$= \dfrac{4}{52} + \dfrac{4}{52} = \dfrac{8}{52} = \dfrac{2}{13}$$

6. $P(\text{Brandy is not selected}) = \dfrac{C(9,3)}{C(10,3)}$

$$= \dfrac{9!}{3!6!} \cdot \dfrac{3!7!}{10!} = \dfrac{7}{10}$$

$P(\text{Brandy is selected})$

$$= 1 - P(\text{Brandy is not selected}) = 1 - \dfrac{7}{10} = \dfrac{3}{10}$$

7. The probability that a student selected at random is a male is $\dfrac{45,143}{82,074} \approx 0.55$. The probability that a student selected at random attends Sacramento College is $\dfrac{20,878}{82,074} \approx 0.25$.

The probability that a student selected at random is a male attending Sacramento College is $\dfrac{9040}{82,074} \approx 0.11$. So, the probability that a student selected at random is a male or attends Sacramento College is approximately $0.55 + 0.25 - 0.11 = 0.69$.

8.7 Concepts and Vocabulary

1. If no outcome of an experiment results more often than any other outcome, then the outcomes are said to be <u>equally likely</u>.

3. If it is impossible for two events to occur simultaneously, then the events are said to be <u>mutually exclusive</u>.

5. True (if "between 0 and 1" includes 0 and 1.)

7. False. The probability will be $\frac{1}{10}$ only if each outcome is equally likely.

8.7 Building Skills

9. $S = \{$(Wendy's, McDonald's, Burger King), (Wendy's, Burger King, McDonald's), (McDonald's, Wendy's, Burger King), (McDonald's, Burger King, Wendy's), (Burger King, McDonald's, Wendy's), (Burger King, Wendy's, McDonald's)$\}$

11. $S = \{\{$*Thriller, The Wall*$\}\}$, $\{$*Thriller, Eagles: Their Greatest Hits*$\}$, $\{$*Thriller, Led Zeppelin IV*$\}$, $\{$*The Wall, Eagles: Their Greatest Hits*$\}$, $\{$*The Wall, Led Zeppelin IV*$\}$, $\{$*Eagles: Their Greatest Hits, Led Zeppelin IV*$\}\}$

13. $S = \{$(white, male), (white, female), (African-American, male), (African-American, female), (Native American, male), (Native American, female), (Asian, male), (Asian, female), (other, male), (other, female), (multiracial, male), (multiracial, female)$\}$

15. $P = 0$.

17. $P = 1$

19. experimental

21. theoretical

23. experimental

25. theoretical

27. experimental

29. An event that is very likely to happen has probability 0.999. An event that will surely happen has probability 1. An event that is a rare event has probability 0.001. An event that is equally likely to happen or not happen has probability 0.5. An event that will never happen has probability 0.

31. $E = \{1, 6\}$, $P(E) = \dfrac{1}{3}$

33. $E = \{5, 6\}$, $P(E) = \dfrac{1}{3}$

35. $E = \{2, 4, 6\}$, $P(E) = \dfrac{1}{2}$

37. $E = \{$club queen, spade queen, heart queen, diamond queen$\}$, $P(E) = \dfrac{1}{13}$

39. $E = \{$heart ace, heart 2, heart 3, heart 4, heart 5, heart 6, heart 7, heart 8, heart 9, heart 10, heart jack, heart queen, heart king$\}$,

$P(E) = \dfrac{1}{4}$

41. $E = \{$heart jack, heart queen, heart king, diamond jack, diamond queen, diamond king, club jack, club queen, club king, spade jack, spade queen, spade king$\}$, $P(E) = \dfrac{3}{13}$

43. $E = \{8 + 3\}$, $P(E) = \dfrac{1}{10}$

45. $E = \{0 + 3\}$, $P(E) = \dfrac{1}{10}$

47. $E = \{1 + 3, 3 + 3, 5 + 3, 7 + 3, 9 + 3\}$, $P(E) = \dfrac{1}{2}$

8.7 Applying the Concepts

49. The probability the Tony will not like his blind date is $1 - 0.3 = 0.7$

51. $P = \dfrac{20}{100} = \dfrac{1}{5}$

53. 6% of the Marines were women, so 94% were men. The probability that a Marine chosen at random is a man is $\dfrac{94}{100} = \dfrac{47}{50}$.

55. If you have 10 tickets and there is only one prize, then the probability of winning is $\frac{10}{125} = \frac{2}{25}$. If there are two prizes, then there are $C(125, 2) = 7750$ possible winning combinations, and you have $C(10, 2) = 45$ possible winning tickets. So the probability of winning is $\frac{45}{7750} = \frac{9}{1550}$.

57. a. There are $2^2 = 4$ possible combinations (bb, bg, gb, gg). The probability of having two boys is 1/4.

   b. The probability of having two girls is 1/4.

   c. The probability of having one boy and one girl is 1/2.

59. a. Out of the 4440 families, 2370 families own two cars, so the probability of a family owning two cars is $\frac{2370}{4440} = \frac{79}{148}$.

   b. Out of the 4440 families, 37 own no cars, 1256 own one car, and 2370 own two cars. So the probability of a family owning at most two cars is $\frac{3663}{4440} = \frac{33}{40}$.

   c. Out of the 4440 families, none own six cars, so the probability of a family owning six cars is 0.

   d. All of the families own fewer than six cars, so the probability of a family owning at most six cars is 1.

   e. Of the 4440 families, all but 37 own at least one car, so the probability of a family owning at least one car is $\frac{4403}{4340} = \frac{119}{120}$.

61. The probability that a non-Hispanic white American has no lactose intolerance is $1 - 0.2 = 0.8 = \frac{4}{5}$. The probability that an African-American, Asian, or Native American has no lactose intolerance is $1 - 0.75 = 0.25 = 1/4$.

63. a. There are $C(10, 2) = 10$ different combinations of people for the committee. There is only one combination of the two oldest members, so the probability of the committee consisting of the two oldest members is 1/10.

   b. There is only one combination of the oldest and youngest members, so the probability is 1/10.

65.

Number of people who are	Depressed according to the test	Normal according to the test
Actually depressed	90	10
Actually normal	135	765

If 10% of the population suffers from depression, then we expect that 10% of the 1000 people tested = 100 people will suffer from depression and 900 will not. Of the 100 people, the test will work for 90% of them, or 90 people, and the remaining 10 people are normal according to the test. Of the 900 people who don't suffer from depression, the test works for 85% of them, or 765 people. The remaining 135 are depressed according to the test.

67. The possible combinations are {bbbb, bbbg, bbgg, bggg, gggg}. (Birth order is not important.) So, there are two ways to have three children of one gender and one of the other gender, while there is only one way to have two children of each gender. Therefore, it is more likely to have three children of one gender and one of the other.

69. a. Out of 180 people, 50 people received the placebo, a total of 56 people had no pain relief, while 34 people received the placebo and had no pain relief. So the probability that a person either received the placebo or had no pain relief is $\frac{50}{180} + \frac{56}{180} - \frac{34}{180} = \frac{2}{5}$.

   b. Out of 180 people, 70 people received the new medicine, a total of 69 people had complete pain relief, while 40 people who received the new medicine also had complete pain relief. So, the probability that a person either received the new medicine or had complete pain relief is $\frac{70}{180} + \frac{69}{180} - \frac{40}{180} = \frac{99}{180}$.

8.7 Beyond the Basics

71. To find the probability that a number between 1 and 1,000,000 does not contain the digit 3, note that each place value through the hundred-thousand's place can contain one of nine digits. So the probability that a number between 1 and 1,000,000 will not contain the digit 3 is $\dfrac{9^6}{1,000,000} = \dfrac{531,441}{1,000,000}$. So the probability that a number between 1 and 1,000,000 will contain the digit 3 is $1 - \dfrac{531,441}{1,000,000} = \dfrac{468,559}{1,000,000}$.

73. Because one of the digits is repeated, there are $\dfrac{4!}{2!1!1!} = 12$ different ways to arrange the digits. The probability that Maggie will dial the correct number is $1/12$.

8.7 Critical Thinking/Discussion/Writing

74. The probability that a person chosen at random in the survey would report his or her highest level of education as high school graduation is $\dfrac{59,840,000}{187,000,000} = \dfrac{8}{25}$.

75. If there are x dark caramels in the box, then there are $2x$ light caramels, and a total of $3x$ caramels in the box. So the probability of choosing a dark caramel is $\dfrac{x}{3x} = \dfrac{1}{3}$.

76. If a single die is rolled two times, then there are $6 \cdot 6 = 36$ possible outcomes. If the first number rolled is n, then there are $6 - n$ possible rolls in which the second roll will be greater than the first roll. So there are $\displaystyle\sum_{n=1}^{5} 6 - n = 15$ possible rolls. The probability that the second roll will result in a number larger than the first roll is $\dfrac{15}{36} = \dfrac{5}{12}$.

Chapter 8 Review Exercises

1. $a_1 = 2(1) - 3 = -1, a_2 = 2(2) - 3 = 1,$
$a_3 = 2(3) - 3 = 3, a_4 = 2(4) - 3 = 5,$
$a_5 = 2(5) - 3 = 7$

3. $a_1 = \dfrac{1}{2(1)+1} = \dfrac{1}{3}, a_2 = \dfrac{2}{2(2)+1} = \dfrac{2}{5},$
$a_3 = \dfrac{3}{2(3)+1} = \dfrac{3}{7}, a_4 = \dfrac{4}{2(4)+1} = \dfrac{4}{9},$
$a_5 = \dfrac{5}{2(5)+1} = \dfrac{5}{11}$

5. This is an arithmetic sequence with a common difference of -2. $a_n = 32 - 2n$ for $n \geq 1$.

7. $\dfrac{9!}{8!} = 9$

9. $\dfrac{(n+1)!}{n!} = n+1$

11. $\displaystyle\sum_{k=1}^{4} k^3 = 1^3 + 2^3 + 3^3 + 4^3 = 100$

13. $\displaystyle\sum_{k=1}^{7} \dfrac{k+1}{k}$
$= \dfrac{1+1}{1} + \dfrac{2+1}{2} + \dfrac{3+1}{3} + \dfrac{4+1}{4} + \dfrac{5+1}{5}$
$+ \dfrac{6+1}{6} + \dfrac{7+1}{7} = \dfrac{1343}{140}$

15. $\displaystyle\sum_{k=1}^{50} \dfrac{1}{k}$

17. The sequence is arithmetic. $a_1 = 11, d = -5$

19. The sequence is not arithmetic.

21. $a_n = 3n$

23. $a_n = x + n - 1$

25. $a_3 = a_1 + d(3-1) \Rightarrow 7 = a_1 + 2d$
$a_8 = a_1 + d(8-1) \Rightarrow 17 = a_1 + 7d$
$\begin{cases} a_1 + 2d = 7 \\ a_1 + 7d = 17 \end{cases} \Rightarrow -5d = -10 \Rightarrow d = 2, a_1 = 3$
$a_n = 3 + 2(n-1) = 2n + 1$

27. $d = 2, a_1 = 7, a_n = 37 \Rightarrow$
$37 = 7 + 2(n-1) \Rightarrow n = 16$
$S = 16\left(\dfrac{7+37}{2}\right) = 352$

29. $d = 5, a_1 = 3, n = 40 \Rightarrow$
$a_n = 3 + 5(40-1) \Rightarrow a_n = 198$
$S = 40\left(\dfrac{3+198}{2}\right) = 4020$

31. The sequence is geometric. $a_1 = 4, r = -2$

33. The sequence is not geometric.

35. $a_1 = 16, r = -\dfrac{1}{4}, a_n = 16 \cdot \left(-\dfrac{1}{4}\right)^{n-1} = \dfrac{(-1)^{n-1}}{4^{n-3}}$

37. $a_{10} = a_1 r^{10-1} = 2 \cdot 3^9 = 39,366$

39. $r = \dfrac{1/5}{1/10} = 2; S_{12} = \dfrac{(1/10)\left(1-2^{12}\right)}{1-2} = \dfrac{819}{2}$

41. $r = \dfrac{1/6}{1/2} = \dfrac{1}{3}; S = \dfrac{1/2}{1-(1/3)} = \dfrac{3}{4}$

43. $a_1 = \dfrac{3}{5}, r = \dfrac{3}{5}; \sum\limits_{i=1}^{\infty}\left(\dfrac{3}{5}\right)^i = \dfrac{3/5}{1-3/5} = \dfrac{3}{2}$

45. For $n=1, \sum\limits_{k=1}^{1} 2^k = 2 = 2^{1+1} - 2$ is true.

Assume that it is true for

$n = m : \sum\limits_{k=1}^{m} 2^k = 2^{m+1} - 2$.

Then for $n = m+1$,

$\sum\limits_{k=1}^{m+1} 2^k = \left(\sum\limits_{k=1}^{m} 2^k\right) + 2^{m+1} = 2^{m+1} - 2 + 2^{m+1}$

$= 2(2^{m+1}) - 2 = 2^{m+2} - 2$

which is exactly the statement for $n = m+1$. Therefore the formula is true for all natural numbers.

47. $\dbinom{12}{7} = \dfrac{12!}{7!(12-7)!} = \dfrac{12\cdot11\cdot10\cdot9\cdot8\cdot7!}{7!5!}$

$= \dfrac{12\cdot11\cdot10\cdot9\cdot8}{5\cdot4\cdot3\cdot2\cdot1} = 792$

49. $(x-3)^4$
$= x^4 + 4(-3)x^3 + 6(-3)^2 x^2 + 4(-3)^3 x + (-3)^4$
$= x^4 - 12x^3 + 54x^2 - 108x + 81$

51. The term containing x^5 is the eighth term.

$\dbinom{12}{5}x^5(2^7) = \dfrac{12!}{5!(12-5)!}x^5(2^7) = 101,376x^5$

53. If no repetitions are allowed, using the Fundamental Counting Principle, there are $4\cdot3\cdot2\cdot1 = 24$ different numbers that can be written.

55. Since order is important, find the number of permutations of 7 taken 7 at a time:

$P(7,7) = \dfrac{7!}{(7-7)!} = 5040$.

There are 5040 different ways that seven people can line up.

57. The five movies can be listed in $5! = 120$ different ways.

59. Order is not important, so find the number of combinations of three candies from a group of ten. There $C(10,3) = \dfrac{10!}{3!(10-3)!} = 120$ ways to choose three candies from a box of ten candies.

61. There are $C(12,2) = 66$ ways to choose two shirts from 12, and $C(8,3) = 56$ ways to choose three pairs of pants from eight. So, there are $66\cdot56 = 3696$ ways to choose two shirts and three pairs of pants.

63. There are nine letters, so $n = 9$. There are two R's, three E's, and two T's. So there are $\dfrac{9!}{2!3!2!1!1!} = 15,120$ distinguishable ways to arrange the letters.

65. Pair the two coins that appear together as one so that there are three positions to fill. Then, (with the paired coins counted as one) there are $3! = 6$ ways to fill these positions. Since the paired coins can be inserted into any position in $2! = 2$ ways, there are $3!2! = 6\times2 = 12$ allowable ways to arrange the coins. The total number of ways to arrange the coins is $4! = 24$, so the probability of arranging the coins so that the two most recent dates will be next to each other is $\dfrac{12}{24} = \dfrac{1}{2}$.

67. There are six letters, so there are $C(6,2) = 15$ combinations of two letters. There are four consonants, so there are $C(4,2) = 6$ combinations of two consonants. So the probability of choosing two consonants is $\dfrac{6}{15} = \dfrac{2}{5}$.

69. There are $C(8, 2) = 28$ ways to choose two people from the group of either. There are five ways to choose one man and three ways to choose one woman, so there are $5 \cdot 3 = 15$ ways to choose one man and one woman. So the probability of choosing one man and one woman is $\frac{15}{28}$.

71. There are 13 clubs, so there are $C(13, 2) = 78$ ways to choose two clubs. There are $C(52, 2) = 1326$ ways to choose two cards from the entire deck, so the probability of choosing two clubs is $\frac{78}{1326} = \frac{1}{17}$.

73. a. $\frac{1}{9}$ **b.** $\frac{4}{9}$ **c.** $\frac{5}{9}$

Chapter 8 Practice Test A

1. $a_1 = 3(5 - 4(1)) = 3, a_2 = 3(5 - 4(2)) = -9,$
$a_3 = 3(5 - 4(3)) = -21, a_4 = 3(5 - 4(4)) = -33,$
$a_5 = 3(5 - 4(5)) = -45$.
The sequence is arithmetic.

2. $a_1 = -3(2^1) = -6, a_2 = -3(2^2) = -12,$
$a_3 = -3(2^3) = -24, a_4 = -3(2^4) = -48,$
$a_5 = -3(2^5) = -96$. The sequence is geometric.

3. $a_1 = -2, a_2 = 3a_1 + 5 = 3(-2) + 5 = -1,$
$a_3 = 3a_2 + 5 = 3(-1) + 5 = 2,$
$a_4 = 3a_3 + 5 = 3(2) + 5 = 11,$
$a_5 = 3a_4 + 5 = 3(11) + 5 = 38$

4. $\frac{(n-1)!}{n!} = \frac{(n-1)!}{n(n-1)!} = \frac{1}{n}$

5. $a_7 = -3 + 6(4) = 21$

6. $a_8 = 13\left(-\frac{1}{2}\right)^7 = -\frac{13}{128}$

7. $a_1 = 1, a_{20} = 58$
$\sum_{k=1}^{20} (3k - 2) = 20\left(\frac{1 + 58}{2}\right) = 590$

8. $a_1 = \frac{3}{8}, a_5 = \frac{3}{128}$
$\sum_{k=1}^{5} \left(\frac{3}{4}\right)(2^{-k}) = \frac{\frac{3}{8}\left(1 - \left(\frac{1}{2}\right)^5\right)}{1 - \frac{1}{2}} = \frac{93}{128}$

9. $a_1 = \frac{9}{50}; \sum_{k=1}^{\infty} 18\left(\frac{1}{100}\right)^k = \frac{9/50}{1 - 1/100} = \frac{2}{11}$

10. $\binom{13}{0} = \frac{13!}{0!(13 - 0)!} = 1$

11. $(1 - 2x)^4$
$= 1^4 + 4(1^3)(-2x) + 6(1^2)(-2x)^2$
$\qquad + 4(1)(-2x)^3 + (-2x)^4$
$= 16x^4 - 32x^3 + 24x^2 - 8x + 1$

12. The term containing x^1 is the fourth term.
$\binom{4}{1}(2x)^1(1)^3 = \frac{4!}{1!(4 - 1)!}(2x)^1(1)^3 = 8x$

13. There are $2^{10} = 1024$ different ways to answer every question on a ten-question true-false test.

14. $P(9, 2) = \frac{9!}{(9 - 2)!} = 72$

15. $C(7, 5) = \frac{7!}{(7 - 5)!5!} = 21$

16. There are $6 \cdot 10 = 60$ ways to fill the positions.

17. There are $26 \cdot 26 \cdot 10^4 = 6,760,000$ possible license plate numbers.

18. $\frac{5}{12}$ **19.** $\frac{5}{6}$

20. There are 18 coins and $C(18, 2) = 153$ ways to choose two coins. There are $C(7, 2) = 21$ ways to choose two quarters. So the probability of choosing two quarters is $\frac{21}{153} = \frac{7}{51}$.

Chapter 8 Practice Test B

1. $a_1 = 4(2(1) - 3) = -4, a_2 = 4(2(2) - 3) = 4,$
$a_3 = 4(2(3) - 3) = 12, a_4 = 4(2(4) - 3) = 20,$
$a_5 = 4(2(5) - 3) = 28$. The sequence is arithmetic. The answer is C.

2. $a_1 = 2(4^1) = 8, a_2 = 2(4^2) = 32,$
$a_3 = 2(4^3) = 128, a_4 = 2(4^4) = 512,$
$a_5 = 2(4^5) = 2048$. The sequence is geometric. The answer is D.

3. $a_1 = 5, a_2 = 2a_1 + 4 = 2(5) + 4 = 14,$
$a_3 = 2a_2 + 4 = 2(14) + 4 = 32,$
$a_4 = 2a_3 + 4 = 2(32) + 4 = 68,$
$a_5 = 2a_4 + 4 = 2(68) + 4 = 140$.
The answer is A.

4. $\dfrac{(n+2)!}{(n+2)} = \dfrac{(n+2)(n+1)!}{(n+2)} = (n+1)!$.
The answer is C.

5. $a_8 = a_1 + 7d = -6 + 7(3) = 15$
The answer is A.

6. $a_{10} = 247\left(\dfrac{1}{3}\right)^9 = \dfrac{247}{19,683}$
The answer is B.

7. $a_1 = -3, a_{45} = 4(45) - 7 = 173$
$\displaystyle\sum_{k=1}^{45}(4k-7) = 45\left(\dfrac{-3+173}{2}\right) = 3825$
The answer is D.

8. $a_1 = \dfrac{8}{3}, r = 2; \displaystyle\sum_{k=1}^{5}\left(\dfrac{4}{3}\right)(2^k) = \dfrac{\frac{8}{3}\left(1-2^5\right)}{1-2} = \dfrac{248}{3}$.
The answer is A.

9. $a_1 = 8; \displaystyle\sum_{k=1}^{\infty} 8(-0.3)^{k-1} = \dfrac{8}{1-(-0.3)} \approx 6.15$.
The answer is B.

10. $\dbinom{12}{11} = \dfrac{12!}{11!(12-11)!} = 12$
The answer is B.

11. $(3x-1)^4$
$= (3x)^4 + 4(3x)^3(-1) + 6(3x)^2(-1)^2$
$\qquad + 4(3x)(-1)^3 + (-1)^4$
$= 81x^4 - 108x^3 + 54x^2 - 12x + 1$
The answer is A.

12. The term containing x is the third term.
$\dbinom{3}{1}(2x)^1(3)^2 = \dfrac{3!}{1!(3-1)!}(2x)^1(9) = 54x$
The answer is D.

13. There are $4 \cdot 10 \cdot 3 = 120$ ways to choose a necklace, a pair of earrings, and a bracelet. The answer is C.

14. $P(7,3) = \dfrac{7!}{(7-3)!} = 210$
The answer is A.

15. $C(10,7) = \dfrac{10!}{7!(10-7)!} = 120$
The answer is D.

16. The first digit can be chosen in nine ways (note that the first digit cannot be zero), and the second digit can be chosen in nine ways. So there are $9 \cdot 9 = 81$ two-digit numbers that can be formed without using a digit more than once. The answer is B.

17. Since one CD is a "must buy", four CDs must be chosen from the remaining seven. There are $C(7,4) = \dfrac{7!}{4!(7-4)!} = 35$ ways to choose the CDs. The answer is D.

18. The answer is C.

19. There are $6 \cdot 6 = 36$ possible outcomes. The outcomes in which the sum is greater than 9 are {(4, 6), (6, 4), (5, 5), (5, 6), (6, 5) (6, 6)}, so the probability of getting a sum greater than 9 when two die are rolled is $\dfrac{6}{36} = \dfrac{1}{6}$.
The answer is A.

20. There are $52 - 13 = 39$ non-spades in the deck, so the probability of choosing a card that is not a spade is $\dfrac{39}{52} = \dfrac{3}{4}$.
The answer is D.

Cumulative Review Exercises (Chapters P–8)

1. $|3x-8| = |x| \Rightarrow 3x - 8 = x$ or $3x - 8 = -x \Rightarrow$
$x = 4$ or $x = 2$. The solution is {2, 4}.

2. $x^2(x^2 - 5) = -4 \Rightarrow x^4 - 5x^2 + 4 = 0 \Rightarrow$
$(x^2 - 4)(x^2 - 1) = 0 \Rightarrow$
$(x-2)(x+2)(x-1)(x+1) = 0 \Rightarrow$
$x = 2$ or $x = -2$ or $x = 1$ or $x = -1$
The solution is {−2, −1, 1, 2}.

3. $\log_2(3x-5)+\log_2 x = 1 \Rightarrow$
$\log_2 x(3x-5) = 1 \Rightarrow 3x^2 - 5x = 2 \Rightarrow$
$3x^2 - 5x - 2 = 0 \Rightarrow (3x+1)(x-2) = 0 \Rightarrow$
$x = -\dfrac{1}{3}$ or $x = 2$
Reject the negative answer. The solution is $\{2\}$.

4. Let $u = e^x$. Then $e^{2x} - e^x - 2 = 0 \Rightarrow$
$u^2 - u - 2 = 0 \Rightarrow (u-2)(u+1) = 0 \Rightarrow$
$u = 2$ or $u = -1$. If $u = -1$, then
$-1 = e^x \Rightarrow \ln(-1) = x$, which is impossible,
so reject -1. If $u = 2$, then $2 = e^x \Rightarrow x = \ln 2$.
The solution is $\{\ln 2\}$.

5. $\dfrac{6}{x+2} = \dfrac{4}{x} \Rightarrow 6x = 4x + 8 \Rightarrow x = 4$

6. $2(x-8)^{-1} = (x-2)^{-1} \Rightarrow \dfrac{2}{x-8} = \dfrac{1}{x-2} \Rightarrow$
$2x - 4 = x - 8 \Rightarrow x = -4$

7. $\sqrt{x-3} = \sqrt{x} + 1 \Rightarrow \left(\sqrt{x-3}\right)^2 = \left(\sqrt{x}-1\right)^2 \Rightarrow$
$x - 3 = x - 2\sqrt{x} + 1 \Rightarrow 2 = \sqrt{x} \Rightarrow x = 4$

8. $x^2 + 4x \ge 0 \Rightarrow x(x+4) \ge 0$
Solving the associated equation, we have $x = 0$
or $x = -4$. The intervals to be tested are
$(-\infty, -4], [-4, 0], [0, \infty)$.

Interval	Test point	Value of $x(x+4)$	Result
$(-\infty, -4]$	-5	5	$+$
$[-4, 0]$	-1	-3	$-$
$[0, \infty)$	1	5	$+$

The solution is $(-\infty, -4] \cup [0, \infty)$.

9. $18 \le x^2 + 6x \Rightarrow x^2 + 6x - 18 \ge 0$
Solving the associated equation, we have
$$x = \dfrac{-6 \pm \sqrt{36 - 4(-18)}}{2} = \dfrac{-6 \pm \sqrt{108}}{2}$$
$$= \dfrac{-6 \pm 6\sqrt{3}}{2} = -3 \pm 3\sqrt{3}$$
The intervals to be tested are $\left(-\infty, -3 - 3\sqrt{3}\right]$,
$\left[-3 - 3\sqrt{3}, -3 + 3\sqrt{3}\right], \left[-3 + 3\sqrt{3}, \infty\right)$.

Interval	Test point	Value of $x^2 + 6x - 18$	Result
$\left(-\infty, -3 - 3\sqrt{3}\right]$	-10	22	$+$
$\left[-3 - 3\sqrt{3}, -3 + 3\sqrt{3}\right]$	0	-18	$-$
$\left[-3 + 3\sqrt{3}, \infty\right)$	3	9	$+$

The solution is
$\left(-\infty, -3 - 3\sqrt{3}\right] \cup \left[-3 + 3\sqrt{3}, \infty\right)$.

10. $\begin{cases} 5x + 4y = 6 \\ 4x - 3y = 11 \end{cases} \Rightarrow \begin{cases} 15x + 12y = 18 \\ 16x - 12y = 44 \end{cases} \Rightarrow$
$31x = 62 \Rightarrow x = 2$
$5(2) + 4y = 6 \Rightarrow y = -1$

11. $\begin{cases} x - 3y + 6z = -8 \\ 5x - 6y - 2z = 7 \\ 3x - 2y - 10z = 11 \end{cases} \Rightarrow \begin{bmatrix} 1 & -3 & 6 & -8 \\ 5 & -6 & -2 & 7 \\ 3 & -2 & -10 & 11 \end{bmatrix}$

$\xrightarrow[\substack{R_2 - 5R_1 \to R_2 \\ R_3 - 3R_1 \to R_3}]{} \begin{bmatrix} 1 & -3 & 6 & -8 \\ 0 & 9 & -32 & 47 \\ 0 & 7 & -28 & 35 \end{bmatrix}$

$\xrightarrow[-\frac{1}{28}(9R_3 - 7R_2) \to R_3]{} \begin{bmatrix} 1 & -3 & 6 & -8 \\ 0 & 9 & -32 & 47 \\ 0 & 0 & 1 & \frac{1}{2} \end{bmatrix}$

$\xrightarrow[\frac{1}{9}(R_2 + 32R_3) \to R_2]{} \begin{bmatrix} 1 & -3 & 6 & -8 \\ 0 & 1 & 0 & 7 \\ 0 & 0 & 1 & \frac{1}{2} \end{bmatrix} \Rightarrow$

$y = 7, z = \dfrac{1}{2}; x - 3(7) + 6\left(\dfrac{1}{2}\right) = -8 \Rightarrow x = 10$

The solution is $\left\{\left(10, 7, \dfrac{1}{2}\right)\right\}$.

12.

13.
$$\sin\left(-\frac{7\pi}{12}\right) = -\sin\left(\frac{7\pi}{12}\right) = -\sin\left(\frac{\pi}{4}+\frac{\pi}{3}\right)$$
$$= -\left(\sin\frac{\pi}{4}\cos\frac{\pi}{3}+\cos\frac{\pi}{4}\sin\frac{\pi}{3}\right)$$
$$= -\left(\frac{\sqrt{2}}{2}\cdot\frac{1}{2}+\frac{\sqrt{2}}{2}\cdot\frac{\sqrt{3}}{2}\right)$$
$$= -\left(\frac{\sqrt{2}+\sqrt{6}}{4}\right) = \frac{-\sqrt{2}-\sqrt{6}}{4}$$

14.
$$r = \sqrt{2^2+\left(2\sqrt{3}\right)^2} = 4$$
$$\tan\theta = \sqrt{3} \Rightarrow \theta = \frac{\pi}{3} \text{ or } \theta = \frac{4\pi}{3}$$

$2+2\sqrt{3}i$ lies in Quadrant I, so $\theta = \frac{\pi}{3}$

The polar coordinates for $\left(2+2\sqrt{3}i\right)$ are

$$4\left(\cos\frac{\pi}{3}+i\sin\frac{\pi}{3}\right).$$

15. Shift the graph of $y = \sqrt{x}$ three units to the right.

16. The domain is $(-\infty,-1)\cup(-1,0)\cup(0,\infty)$.

17. $7\ln x - \ln(x-5) = \ln\left(\dfrac{x^7}{x-5}\right)$

18.
$$[A\,|\,I] = \begin{bmatrix} 1 & 0 & -2 & 1 & 0 & 0 \\ 4 & 1 & 0 & 0 & 1 & 0 \\ 1 & 1 & 7 & 0 & 0 & 1 \end{bmatrix}$$
$$\xrightarrow[R_3-R_1\to R_3]{-4R_1+R_2\to R_2} \begin{bmatrix} 1 & 0 & -2 & 1 & 0 & 0 \\ 0 & 1 & 8 & -4 & 1 & 0 \\ 0 & 1 & 9 & -1 & 0 & 1 \end{bmatrix}$$
$$\xrightarrow{R_3-R_2\to R_3} \begin{bmatrix} 1 & 0 & -2 & 1 & 0 & 0 \\ 0 & 1 & 8 & -4 & 1 & 0 \\ 0 & 0 & 1 & 3 & -1 & 1 \end{bmatrix}$$
$$\xrightarrow[R_2-8R_3\to R_2]{R_1+2R_3\to R_1} \begin{bmatrix} 1 & 0 & 0 & 7 & -2 & 2 \\ 0 & 1 & 0 & -28 & 9 & -8 \\ 0 & 0 & 1 & 3 & -1 & 1 \end{bmatrix} \Rightarrow$$
$$A^{-1} = \begin{bmatrix} 7 & -2 & 2 \\ -28 & 9 & -8 \\ 3 & -1 & 1 \end{bmatrix}$$

19. Since order is important, find the number of permutations of four people taken two at a time: $P(4,2) = \dfrac{4!}{(4-2)!} = 12$.

There are 12 arrangements.

20. There are 18 people in total. Three people who are 25 years or older have received speeding tickets, and six people whose ages are between 17 and 24 have received a speeding ticket. The probability that a person chosen at random will be 25 years old or older is $\dfrac{6}{18}$. The probability that a person chosen at random will have received a speeding ticket is $\dfrac{9}{18}$. The probability that a person 25 years old or older will have received a speeding ticket is $\dfrac{3}{18}$. So, the probability that a person chosen at random from the room will be a person who is 25 years or older or has gotten a speeding ticket is $\dfrac{6}{18}+\dfrac{9}{18}-\dfrac{3}{18}=\dfrac{12}{18}=\dfrac{2}{3}$.

